The Physiology of Training for High Performance

To Betty Jo, for her patience, invaluable technological assistance and editorial input over the years that this book was in progress.

Duncan

To Valerie. Thank you for your enduring patience, support, and encouragement during the course of writing this book.

Digby

Acknowledgments

The authors would like to acknowledge the graduate students with whom we have worked over the years, many of whom have gone on to highly successful positions in academia and medicine. Your insightful questions, suggestions, and assistance in our laboratory were an important and memorable part of our research careers. We also want to acknowledge and thank the professional colleagues whose assistance and advice has contributed so much to our work and our pleasure in it for over four decades. Although the list is too long for us to name you individually, you know who you are.

Thanks too to our publishing editor Dewi Jackson, production editor Joanna Hardern and free-lance copy editor Nik Prowse for their patience and perceptive suggestions. It was a pleasure working with you.

The Physiology of Training for High Performance

Duncan MacDougall

Professor Emeritus at McMaster University, Canada

Digby Sale

Professor Emeritus at McMaster University, Canada

Great Clarendon Street, Oxford, OX2 6DP,
United Kingdom

Oxford University Press is a department of the University of Oxford.
It furthers the University's objective of excellence in research, scholarship,
and education by publishing worldwide. Oxford is a registered trade mark of
Oxford University Press in the UK and in certain other countries

Impression: 1

Published in the United States of America by Oxford University Press
198 Madison Avenue, New York, NY 10016, United States of America

British Library Cataloguing in Publication Data
Data available

Library of Congress Control Number: 2014934428

ISBN 978–0–19–965064–4

Printed in Great Britain by
Ashford Colour Press Ltd, Gosport, Hampshire

Brief Contents

Detailed Contents

PART I
Physiological Bases for Athletic Training

1

Introduction to Training for High Performance

On any given occasion, the quality of an athlete's performance is the result of a complex blend of many factors, as summarized in Figure 1.1. In most sports, the major factor that determines the athlete's potential to excel is his or her genetic endowment. Factors such as body size and proportionality are genetically predetermined. To a large extent, this is also true of body composition, many cardiovascular traits (5), proportions of muscle fiber types, and gross motor coordination (14). In addition, there is evidence that an athlete's ability to improve through training may be inherited (3,4,11). Next to heredity, the most important factor affecting athletic performance is usually the amount and suitability of the training that precedes the competition.

What is Training?

Training can be defined as the stimulation of biological adaptations that result in an improvement in performance in a given task. For example, it is known that, when a muscle is required to contract more forcefully than normally (resistance training), over a period of time it will adapt by becoming larger and stronger. The processes that cause this increase in

Figure 1.1 A summary of the interacting factors affecting sport performance.

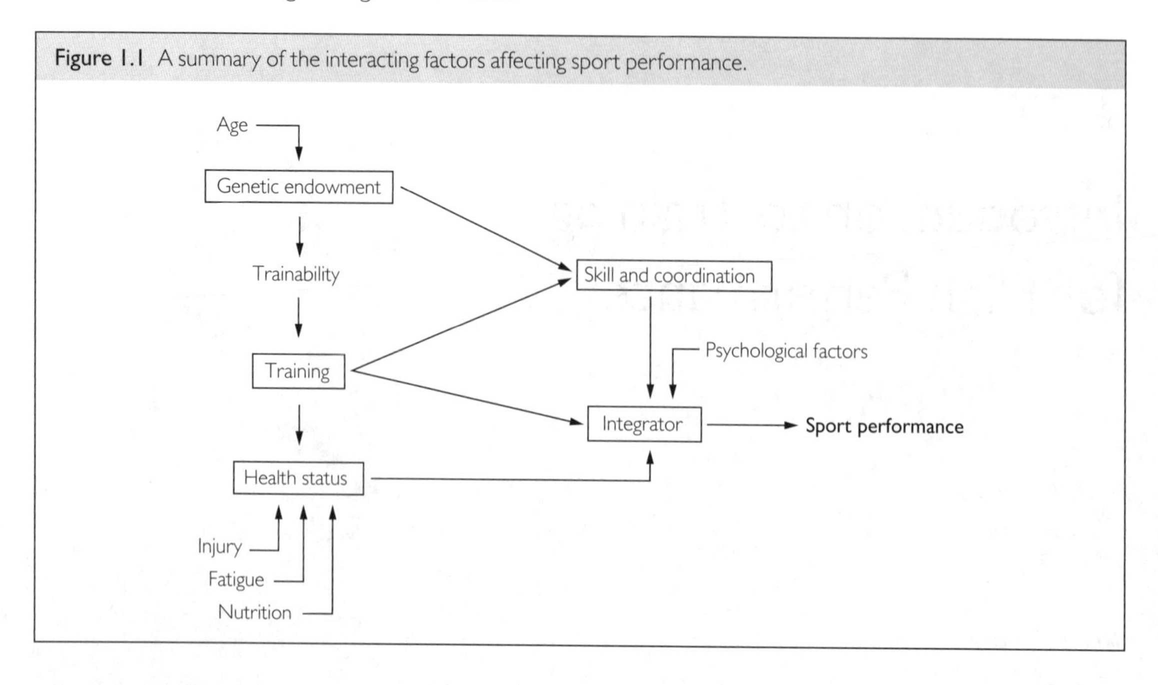

muscle size are complex and not yet fully understood but, as documented later in this book, probably represent the body's attempt to protect itself from the contractile protein damage that is caused by such training. From a purely biological perspective, the benefits of this adaptation are that the muscle becomes more resistant to future damage; from the athlete's perspective, however, the benefit is that it is now capable of developing more force and thus improving sport performance. Over the years (and often through trial and error), athletes and coaches have begun to learn how to exploit the body's ability to adapt to potentially harmful stimuli in order to effect biological changes that will improve performance. The challenge lies in determining the most effective stimulus that will result in the desired adaptation. This book will focus on the information necessary for making these decisions.

Adaptation

The human body is a marvelous machine, but unlike a machine it also has the capacity to alter its structure and function to meet the demands that are placed on it. To varying extents, almost all the tissues of the body are capable of adapting to conditions or events that challenge their normal balance or homeostasis. Adaptations to exercise and training may be acute or chronic.

Acute vs Chronic Adaptations

Acute adaptations are functional changes that occur immediately in response to exercise, but then revert back to baseline shortly after the exercise is terminated. Chronic adaptations are longer-lasting and involve progressive alterations in the cell's gene expression and protein synthesis in response to repeated exercise bouts. For example, the increase in the stroke volume of the heart above its normal resting value, as an athlete goes from rest to the exercise state, is an acute adaptation. In contrast, the increase in ventricular chamber size that occurs over several months of repeated exercise is considered a chronic adaptation.

The Adaptive Process

One of our most rapidly adapting tissues is skin. Its adaptability can be illustrated by the following two examples, with which we are probably all familiar.

- The weekend gardener or occasional manual laborer is well aware that the unaccustomed use of hand tools can result in irritated skin and painful blisters. Gradual use of such tools over a few days will, however, cause the epidermal layer of the skin to adapt by forming calluses. The friction and pressure caused by the task stimulates a localized hyperplasia that increases the

Figure 1.2 Suntan: an example of adaptation to help protect our skin from damage caused by excessive exposure to ultraviolet-wavelength light.

thickness of the stratum corneum and the amount of the protein keratin, forming hard, protective pads at the base of the fingers and in the palms that protect the hands from further injury.

- When we overexpose ourselves to direct sunlight, the ultraviolet-wavelength light causes damage to the dermis of our skin. We know this as **sunburn**. The same ultraviolet waves also stimulate our melanocytes, which are found between the dermal and epidermal layers, to increase their synthesis and secretion of melanin, causing greater pigmentation of our skin. We know this as a **suntan** (Figure 1.2). The darkened skin absorbs much of the ultraviolet light and reduces (but does not eliminate) further damage to the dermis.

In this second example, the **stimulus** for adaptation is the exposure to ultraviolet light and the **effect** is the increase in pigmentation. Up to a point, the greater the stimulus (the intensity of the sunlight and/or the duration of the exposure), the greater the effect and the darker our tan. In order for the adaptive process to continue, the stimulus must be constantly adjusted upwards. This progression must, however, be gradual so that there is adequate time for adaptation to occur. If the stimulus is increased too rapidly, serious damage (sunburn) will occur. When we remove the stimulus (stop going out in the sun), the effect disappears and we lose our tan.

Adaptation in Muscle

For most athletes, it is the adaptations that can occur in skeletal and cardiac muscle that will have the greatest effects on performance. Muscle is an extremely dynamic tissue with a remarkable capacity for adapting its structure and function to a wide range of increased demand. For example, an increase in demand would occur if the muscle were required to contract more forcefully than it does normally. A different form of increased demand would occur if the muscle were required to contract more frequently than normal. Both of these changes in demand would cause adaptation in the muscle, but the nature of the response would be different in each case.

Over the last decade, molecular biologists have begun to unravel the transcriptional and translational processes involved in gene expression and the signaling pathways that result in adaptation in muscle. While these processes are of vital interest to the scientist, it is knowing the most effective stimulus to trigger these processes that is most important from the purely practical perspective of the athlete and coach. We have therefore chosen to include only brief coverage of these molecular events and to concentrate on the stimuli that cause them to occur and their time course.

Evolution

It is no accident that muscle is by far the largest tissue component in the body, making up 50% or more of our total mass. This is because our evolution and survival have depended upon our ability for locomotion and mobility.

Although as a species we are extremely new, the first hominids (*Australopithecus*) probably appeared as our upright-walking ancestors sometime between 3.5 and 4 million years ago. Modern humans (*Homo sapiens sapiens*) evolved approximately 50 000 years ago and, like their forebears, depended upon scavenging, hunting, and gathering for food, often in competition with larger and stronger predators. As a result, our skeletal system and muscles evolved to give us the endurance capacity

Figure 1.3 The evolution of modern man (*Homo sapiens sapiens*) over approximately 4 million years. Note that, until about only 10 000 years ago, we survived by hunting and gathering.

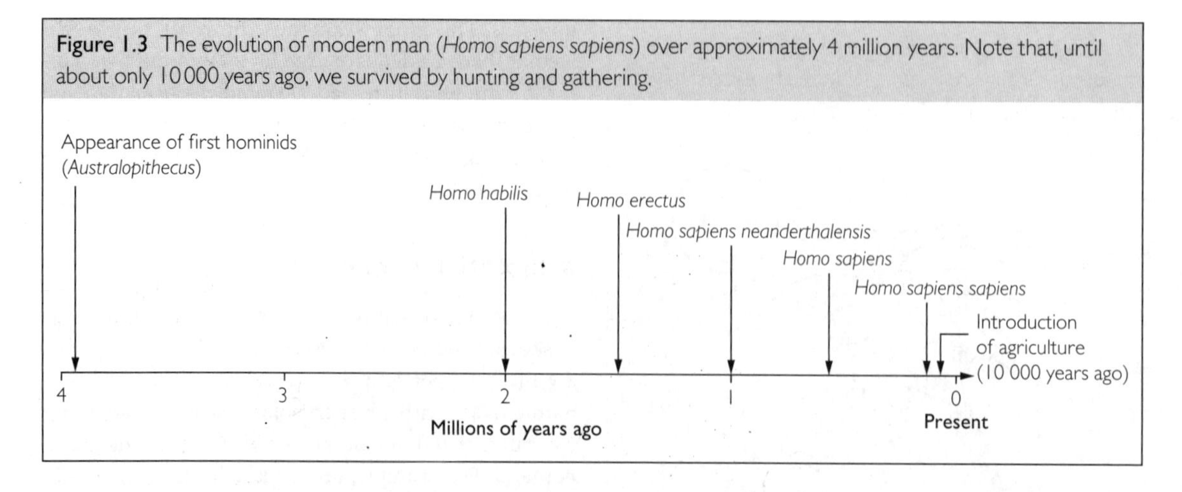

required to follow our game-herd food sources as they migrated vast distances on the savanna, as well as the capacity for the rapid powerful movements needed to kill prey or escape from other predators. It was probably only 10 000 years ago that humans began to exchange the hunting and gathering existence for agriculture and farming. Thus, for more than 99.75% of human evolution, our survival has been directly linked to our muscular prowess. Figure 1.3 provides an evolutionary timescale to illustrate this point.

Although our lifestyle today does not include following game herds or sprinting to escape predators, these demands are mimicked by the wide variety of sports in which we compete, and our muscles have retained their ability to adapt.

Training Terminology

Overload

The stimulus for most training adaptations is provided by increasing the demands on the system, creating what is known as a functional **overload**. As we will see, there is an almost infinite number of ways to overload our muscles and the supporting biological systems. A functional overload occurs when there is an increase in exercise intensity, duration, or frequency.

Intensity

The intensity of the training stimulus is usually expressed relative to the maximum performance capacity of the muscle or muscle groups. For example, for many forms of training, intensity might be expressed as a percentage of the maximum aerobic power of the muscle or muscle group (% $\dot{V}O_{2max}$), or, perhaps, percentage of maximum effort, maximum heart rate, or best performance time. For resistance training it could be expressed as a percentage of the maximum amount of weight that the athlete can lift with a single attempt (percentage of one-repetition maximum or % 1RM), or the maximum voluntary force-generating capacity of the muscle (percentage of maximum voluntary contraction or % MVC).

Duration

Duration refers to the total time that the stimulus is provided over a single training session and is usually quantified in conjunction with intensity. For example, on a given day, a middle-distance athlete might train for 40 minutes at 70% $\dot{V}O_{2max}$ and for 10 minutes at 85% $\dot{V}O_{2max}$.

Frequency

Frequency refers to how often an athlete trains in a given time: for example, two times per day, six times per week, etc. Most coaches and athletes tend to express frequency as the number of training sessions per week.

Volume

The product of intensity, duration, and frequency is commonly referred to as the volume of training. Volume is often quantified in units of distance covered

Figure 1.4 The relationship between the magnitude of the adaptation (the training effect) and the magnitude of the training stimulus.

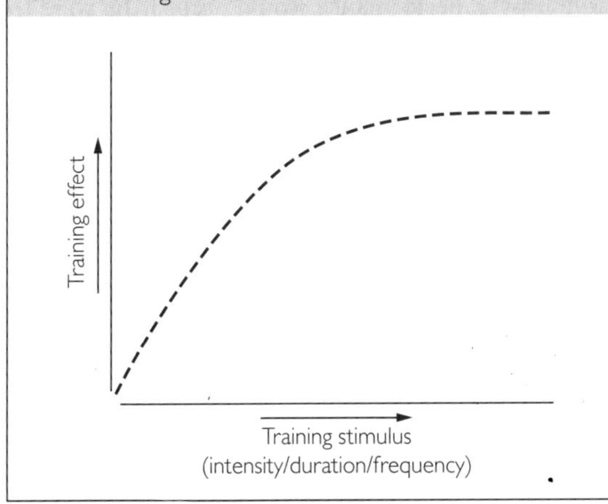

per unit time (e.g., kilometers per week) or work performed (e.g., kilograms lifted per week). Note that an increase in intensity that is accompanied by a decrease in frequency or duration may result in a *decrease* in volume. Similarly, an increase in volume can occur despite a decrease in training intensity.

Progression

Since adaptation occurs to protect the body and to prepare it for the demands that are placed upon it, the magnitude of the training stimulus must be progressively increased if adaptation is to continue. Otherwise, performance will improve only to a certain level and then cease to change. For many training adaptations there also appears to be a critical threshold above which the magnitude of the stimulus must be adjusted. For example, little or no adaptation would occur in an athlete who trains by lifting 10 kg weights when he is capable of lifting 50 kg weights. Similarly, little or no adaptation would occur if the athlete trained by lifting a 50 kg weight, but did so only once per week.

The rate at which most training adaptations occur is depicted in Figure 1.4. Notice that the shape of the stimulus / response curve is such that the athlete can expect relatively rapid gains when a program is first begun, but will require considerably more training stimulus to cause even a slight improvement as he / she becomes more highly trained. Athletes who have been training for several or more years fall along the upper or flatter portion of this curve. However, since competitions at this level are often decided by very narrow differences in performance, even minimal changes are critical. As the response curve becomes flatter and flatter, the athlete is approaching his or her maximum capacity for further adaptation and theoretically will reach a point beyond which no adaptation will occur, no matter how great the training stimulus. Whether or not athletes ever achieve this potential ceiling effect before their performance begins to be affected by other factors such as injury or aging is not really known. What we do know is that the level of this ceiling is at least partially determined by heredity (3), and that some athletes are capable of much greater adaptive response than others (see Box 1.1).

BOX 1.1 BOB SWIFT, MIDDLE-DISTANCE RUNNER

Note: although the names are fictitious and some of the details embellished for illustrative purposes, the examples given throughout this book are based on actual case studies from the authors' laboratory.

At age 17, Bob was one of the premier 800 and 1500 m runners in the province. Although he had been running competitively with the local track club for only 3 years, he had broken his high-school records in these two events and won the provincial high school championship in the 1500 m. In a laboratory treadmill test his $\dot{V}O_{2max}$ was measured at 71 ml·kg^{-1}·min^{-1} (this value will mean more to you by the end of this chapter), and he and his coach had aspirations that he would make the Olympic team in 3 years' time. He was widely recruited by a number of universities and accepted a track scholarship at a northeastern American university. The local sports reporters, his friends, and his family followed his career closely, with every expectation that it would be a spectacular one. Alas, it was not to be. He trained hard but, while he had a few good races, he was never able to achieve the goals that he and his coaches established for him. After his junior year he retired from running. His $\dot{V}O_{2max}$ was 73 ml·kg^{-1}·min^{-1} and his personal best for the 1500 m was only 1.5 seconds faster than his best high-school time.

In hindsight, it is probable that, by the time that he had graduated from high school, Bob had achieved almost his maximum capacity for adaptation. Although genetics had been very kind to him in giving him the physical attributes that would make him a top runner early in his career, they were also cruel in that he was dealt a very narrow window for improvement.

Detraining

When the athlete stops training, the training stimulus is removed and, over time, the body reverts back to its pre-training status. This process is sometimes called **detraining**. Upon complete cessation of training, the rate of loss for most adaptations mimics the rate at which they were achieved. This pattern is depicted in Figure 1.5. In this example, a sedentary individual begins a running program 5 days per week. The program is progressive so that the training load (in terms of both intensity and duration) is increased slightly each week and, each week, the subject visits the laboratory for an assessment of $\dot{V}O_{2max}$. Notice that the gains are most rapid over the first month or two and then become more gradual over the next 3 or 4 months. If, at the end of 6 months, the subject suddenly ceases all training, loss in adaptation (decrease in $\dot{V}O_{2max}$) would occur in almost mirror-image opposite fashion. The gains would be preserved for perhaps 6–10 days but then decline rapidly over the first 2 months and more gradually over the next 4 months. Six months after the subject stopped training, values would have returned to their normal pre-training level.

Studies show, however, that if, at the end of the training period, instead of stopping training altogether, the subject had simply reduced the volume of training by up to 40%, the gains in $\dot{V}O_{2max}$ that had occurred would have been maintained. Putting this into perspective, if, for example, the subject was running 48 km a week after 6 months of training, little or no loss in performance would occur if he were to reduce his training to only 3 days per week, or to reduce his weekly distance to 29 km (assuming that he did not also reduce his training intensity). Thus, as will be emphasized frequently throughout this book, *once a training adaptation has been gained, considerably less training stimulus is required to maintain it than to produce it in the first place.*

Rest and Recovery

Athletes and coaches sometimes lose sight of the fact that, whereas the training stimulus is applied during each training session, the adaptation occurs *after* the session; that is, during the recovery period before the next training session. For some forms of training stimulus, two or more days are necessary for the protein synthesis that has been stimulated to occur, and more frequent training could actually be counterproductive. Thus, adequate rest and recovery are a necessary component of any effective training program.

Interval Training

It is well known that, when short rest periods are inserted between bouts of heavy exercise, the total

Figure 1.5 Typical changes in $\dot{V}O_{2max}$ when an untrained subject undergoes 6 months of endurance training followed by 6 months of no training, or by a 40% reduction in training volume. See text for details. Compiled from (6,7,12,13,15).

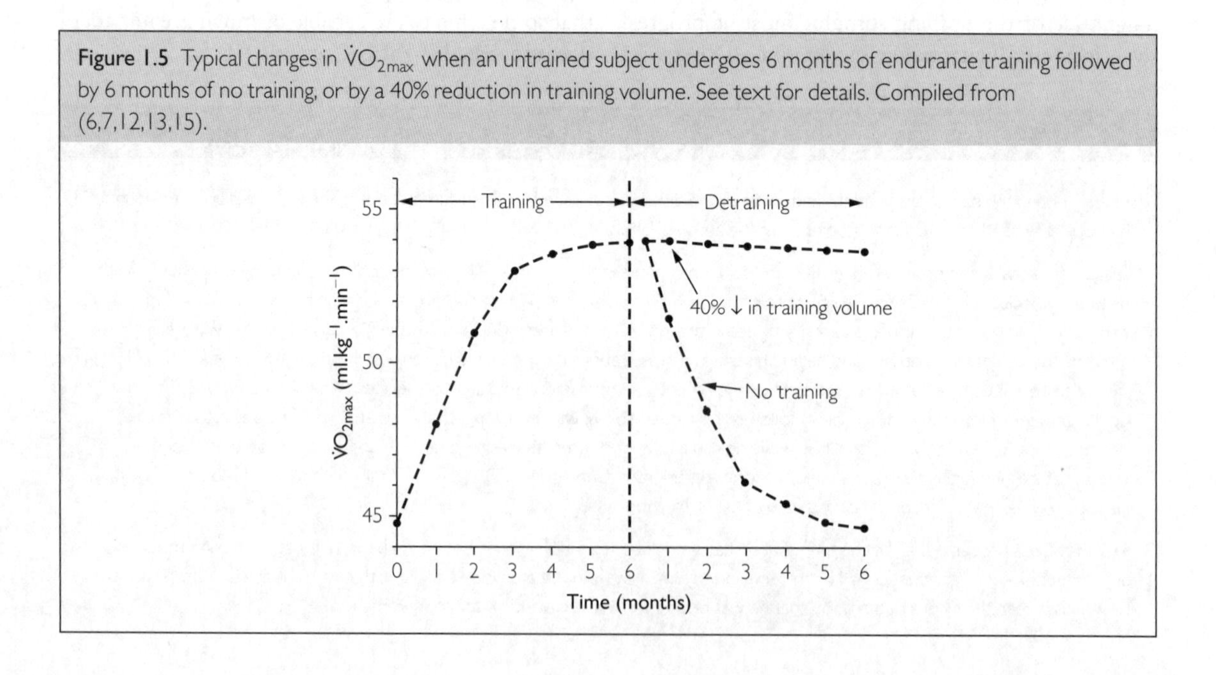

accumulated exercise time over these intervals can be increased far beyond what is possible in a single continuous bout at the same intensity to the point of exhaustion. For example, a highly motivated athlete, in a single maximal effort, could probably train continuously at an intensity requiring 100% of $\dot{V}O_{2max}$ for approximately 6–8 minutes before becoming exhausted. On the other hand, if the athlete exercised at this same intensity for 2–3 minutes, followed by 2–3 minutes of recovery, he/she could complete 10 or more intervals before experiencing the same degree of fatigue. In the continuous-exercise situation, the athlete would have received 6–8 minutes of training stimulus, whereas in the interval-training situation the total training stimulus would have been increased to 30 minutes or more, or over four times as long! Shorter work and rest intervals (e.g., 30 seconds at the same intensity, followed by 30 seconds of recovery) would be even less fatiguing, but would not stress the oxidative capacity of the muscle to the same extent (1,9). As we shall see, there is an almost infinite number of methods of varying the intensity, duration, and exercise:rest ratio of the intervals, depending upon the desired adaptation outcome. When the goal is a high-intensity exercise stimulus, the most effective program will invariably be some form of interval training.

Peaking and Tapering

The ultimate goal of any athlete is to achieve the maximum possible adaptation prior to what she or he considers to be the most important competition or competitions. However, since considerably greater training stimulus is necessary to promote further adaptation as athletes approach their biological limits, performance gains at this point are achieved only at the price of increased risk of illness and overuse injury. Although there are large inter-individual differences, athletes have learned that they cannot simply continue to increase the training stimulus indefinitely day after day without encountering chronic fatigue, sickness, or injury. The solution lies in determining the optimal balance between training volume, intensity, and duration that will provide maximum adaptation for a given athlete without exceeding the individual's capacity for adaptation. Many sports also require adaptation in a number of different physiological systems and, since adaptation often occurs at different rates in different systems, this variation must also be factored into the planning of the athlete's program. This sequencing of different forms of training and their timing so that maximum adaptation coincides with the competition date is called **peaking**.

It is also well known that an athlete's performance on any given occasion can be enhanced if it is preceded by a brief period in which the normal training load is sharply reduced. This process is known as **tapering**. Here too timing is crucial, so that the maximum adaptation is preserved over the competition period before the reduced training stimulus results in detraining. Tapering will be discussed in more detail in Chapter 9.

Overtraining

When the magnitude of the training stimulus chronically exceeds the body's capacity to adapt to it, a condition known as **overtraining** results. The symptoms of overtraining vary widely among individual athletes, but it is almost always reflected by a sudden decrease in performance and, often, general fatigue or illness. Milder incidents of overtraining, such as that brought on by sudden inappropriate increases in training stimulus without adequate rest, are sometimes known as **overreaching**. With overreaching, the more serious consequences can often be avoided by modifications of the athlete's program.

What Determines Performance?

Muscles contract when stored chemical energy is converted to mechanical energy. This chemical energy is in the form of the compound **adenosine triphosphate** or **ATP**, which is formed in the muscle by the bonding of inorganic phosphate (P_i) to adenosine diphosphate (ADP). When this bond is broken by the activation of an enzyme, the potential energy that it contains is released, causing muscle filaments to slide past each other, the muscle to shorten, and mechanical work to be performed. In resting muscle, this energy reserve of ATP molecules is stored throughout the fiber, with concentrations found in the vicinity of the cross-bridge contractile sites. The problem is that the total amount of stored ATP is very small and sufficient only for several forceful contractions. Thus, if mechanical work is to continue for more than a few seconds, processes must be activated to resynthesize ATP at the rate at which it is being used. These processes are known as **metabolic pathways**, and include four main possibilities:

- the synthesis of ATP from the splitting of a second high-energy phosphate compound, phosphocreatine (PCr; this pathway is also known as the phosphagen system),
- the synthesis of ATP from the anaerobic breakdown of muscle glycogen or glucose to lactic acid (anaerobic glycogenolysis or glycolysis, sometimes called the lactic acid system),
- the synthesis of ATP from the oxidation of glycogen or glucose,
- the synthesis of ATP from the oxidation of fatty acids.

Since the first two of these pathways do not require oxygen to function, they are known as **anaerobic** pathways, whereas the latter two are considered oxidative or **aerobic** pathways. Pathways also exist for the generation of ATP from the oxidation of proteins, but their involvement is minimal for most sports. The aerobic pathways occur in the thousands of mitochondria that are found in each muscle fiber, whereas the anaerobic pathways occur outside the mitochondria, in the cytosol of the fiber (Figure 1.6).

Major differences exist in the maximal rates at which these four pathways can deliver ATP (Figure 1.7). The most rapid is the phosphagen system, followed by anaerobic glycogenolysis, the oxidation of glycogen, and finally the oxidation of fats. The enzymes that activate and control these pathways attempt to maintain a constant level of ATP in the fiber (sometimes called the "energy charge" of the cell). This means that, in situations involving a rapid rate of ATP usage (as in sprinting), rapid replacement pathways, such as the phosphagen system and anaerobic glycogenolysis, must be recruited to preserve ATP. On the other hand, for activities in which the rate of ATP utilization is relatively slow (as in distance running), it can be restored through the two oxidative pathways.

Figure 1.7 The four energy-producing pathways are preferentially recruited so that the rate of ATP production matches the rate at which it is used.

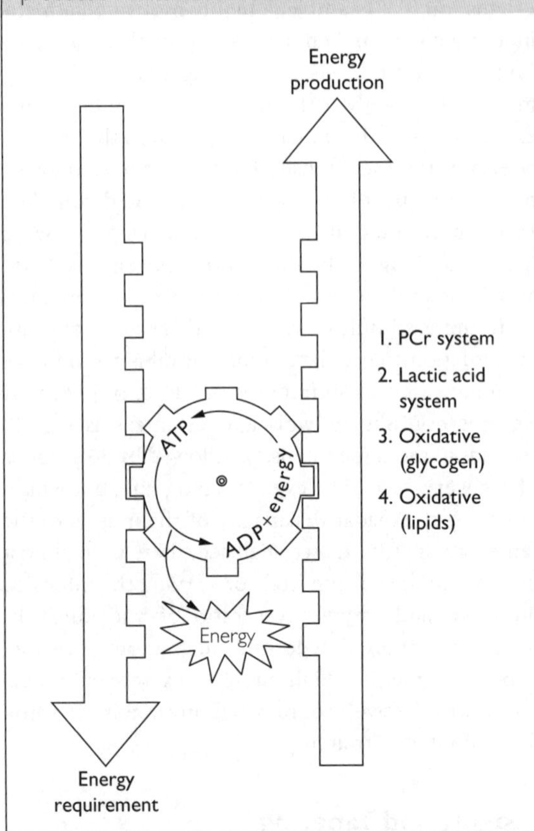

Figure 1.6 The three systems for generating ATP in skeletal muscle. The aerobic system can oxidize either carbohydrate or fats. The other two systems do not require oxygen and are therefore referred to as anaerobic systems.

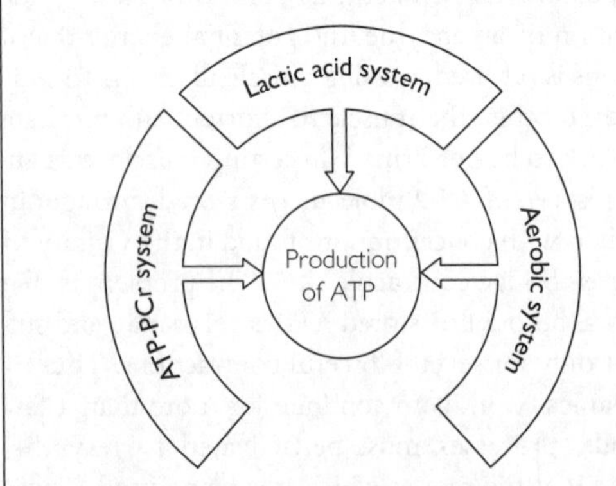

Power vs Capacity

The maximal rate at which a pathway can synthesize ATP is termed its **power**. The maximum amount of ATP that can be derived from the pathway in a single maximal effort is termed its **capacity**. An inverse relationship exists between the power of the four energy-delivery pathways

Table 1.1 Estimated maximal power and capacity of the four metabolic pathways in skeletal muscle of a trained 80 kg athlete. From (2,10).

Pathway	Maximal power (kJ·min^{-1})	Duration	Maximal capacity (total kJ)
Phosphagen system	400	10–15 s	≈100
Anaerobic glycogenolysis	200	45–60 s	≈200
Oxidation of glycogen and glucose	125	≈2 h	15 000
Oxidation of fats	110	8+ h	52 800+

and their capacities: those with the greatest power have the least capacity, and vice versa. For example, since PCr exists in muscle only in low concentrations, it is rapidly depleted over 10–15 seconds of maximal effort. In contrast, fat reserves in the body are infinitely more abundant and thus, although the pathway for lipid oxidation is the least powerful, it has by far the greatest capacity. Estimates of the power and capacity of the four pathways are provided in Table 1.1.

At any point in time, the total ATP available for exercise is the sum of the amounts that can be synthesized via each of the individual pathways. However, their different capacities dictate their relative importance over different durations of effort. This is illustrated in Figure 1.8.

It can be seen that, in an event such as throwing or jumping that requires a maximum explosive effort lasting less than 1–2 seconds, almost all of the ATP would be derived from the phosphagen system. In an event such as a 100 m sprint, almost all of the energy would be derived equally from the two anaerobic pathways. A 400 m sprint (e.g., 50 seconds) would involve the oxidation of glycogen in addition to the phosphagen and anaerobic glycogenolytic pathways, but the predominant pathway would be anaerobic glycogenolysis. For the 1500 m runner (e.g., 3 minutes 50 seconds), approximately 60% of the ATP would be supplied by the oxidation of glycogen and 40% by phosphagen splitting and anaerobic glycogenolysis. The 10 000 m runner (e.g., 30 minutes) would derive more than 95% of the necessary energy from the oxidative system (mostly glycogen), while the marathon runner (e.g., 2 hour 20 minutes) would derive more than 99% from the oxidative pathways, with an almost equal balance between carbohydrate and fat oxidation.

Figure 1.8 The relative contribution of the four energy-delivery systems according to the duration of the event (time of maximal effort). See text for details.

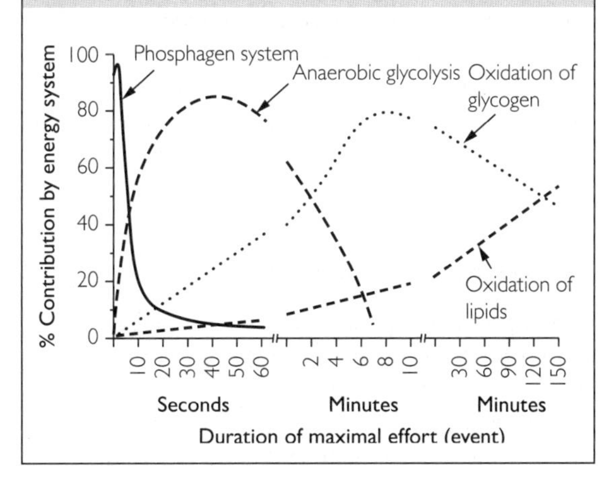

The Cardiorespiratory Systems

Since the two oxidative pathways are dependent upon a constant supply of oxygen, the maximal rate at which they can generate ATP will be dependent upon the maximal rate at which oxygen can be delivered to the mitochondria. This rate is limited by:

- the rate at which oxygen can be delivered to the muscle, and
- the rate at which it can be taken up or extracted by the muscle.

The first of these factors is determined by the capacity of the lungs to load oxygen into the blood and the capacity of the heart to pump the blood to the muscles. Together these two systems are often referred to as **central** factors. As we shall see in a later chapter, the lungs are normally able to saturate blood fully with oxygen during maximal exercise and are not considered to be the limiting central factor. We can thus conclude that the main limiting central factor is the maximal capacity of the athlete's heart for pumping blood (also known as maximal cardiac output). This capacity can, however, be greatly enhanced with training.

The capacity of the muscle to extract and utilize the oxygen delivered to it depends upon the distribution of blood flow throughout the muscle, the maximal rate at which oxygen can diffuse from blood to muscle fibers, and the capacity of the enzymes in the mitochondria to utilize this oxygen in the oxidation of carbohydrates and fats. Collectively, these are often referred to as **peripheral** factors; they too are profoundly affected by training.

In addition to delivering oxygen to the muscle, these central and peripheral factors play a vital role during exercise by removing carbon dioxide and other metabolites, such as lactate and H^+, which can have negative effects on performance.

Maximal Aerobic Power

Maximal aerobic power is the maximum rate at which ATP can be synthesized aerobically in the muscle. It can be indirectly quantified by measuring the rate at which oxygen is taken up by the body while performing maximal exercise, on the assumption that the rate at which oxygen is consumed equals the rate of aerobic ATP production. Thus, in the world of the athlete, the term **maximal aerobic power** is conventionally used synonymously with the term **maximal oxygen consumption**, or $\dot{V}O_{2max}$. Units for measuring $\dot{V}O_{2max}$ are normally absolute units (liters per minute) or units relative to the athlete's body weight (milliliters per kilogram per minute). In most sports (e.g., running, skating, Nordic skiing), the athlete must support or project the body weight against gravity and thus $\dot{V}O_{2max}$ is most meaningful when expressed in relative units (milliliters per kilogram per minute). On the other hand, in sports where body weight is fully or partially supported, such as rowing, cycling, or swimming, the athlete's absolute $\dot{V}O_{2max}$ (liters per minute) will be more important. A more detailed discussion of $\dot{V}O_{2max}$ is provided in Chapter 3.

Maximal Aerobic Capacity

Maximal aerobic capacity refers to the total amount of ATP that can be generated by the oxidative processes. This is indirectly quantified as the maximal amount of work that can be performed aerobically in a given time or at a given percentage of $\dot{V}O_{2max}$ (e.g., 80% of $\dot{V}O_{2max}$) before fatigue. Thus, *for most sports, an athlete's maximal aerobic capacity is a more important determinant of performance than maximal aerobic power.* Consider, for example, two 10 000 m runners. Athlete A has a $\dot{V}O_{2max}$ of 75 $ml{\cdot}kg^{-1}{\cdot}min^{-1}$ and can manage to run his race at a pace that averages 70% of this value (52.5 $ml{\cdot}kg^{-1}{\cdot}min^{-1}$). Athlete B has a $\dot{V}O_{2max}$ of 70 $ml{\cdot}kg^{-1}{\cdot}min^{-1}$, but can average a pace equivalent to 80% of this value (56 $ml{\cdot}kg^{-1}{\cdot}min^{-1}$). Although A has the greater maximal aerobic power, B has the greater maximal aerobic capacity and will invariably win the race. Proportionately, maximal aerobic capacity can be improved to a greater extent than can maximal aerobic power. The factors that determine an athlete's maximal aerobic capacity are discussed in more detail in Chapter 3.

Maximal Anaerobic Power and Capacity

The preceding concepts of maximal power and capacity can also be applied to the anaerobic energy-delivery pathways, but here the distinction between the two terms is somewhat less apparent. Theoretically, **maximal anaerobic power** would be the highest rate at which ATP can be generated (e.g., moles of ATP per second) by the splitting of PCr and by anaerobic glycogenolysis. The *total* amount of ATP that could be generated from these two pathways over a short-duration (e.g., 30 s or 45 s) maximum effort would be the athlete's **maximal anaerobic capacity**. Maximal anaerobic power output can be sustained for only 2 or 3 seconds and is largely dependent on the maximal enzyme kinetics rate of the two pathways. Maximal anaerobic capacity, in contrast, is limited primarily by the muscle's concentration of PCr and its ability to resist the fatigue caused by the increase in lactic acid and other metabolic byproducts. Specific strategies for training to improve anaerobic capacity are presented in Chapter 7.

Muscle Strength and Power

Among athletes, performance is not measured in units of ATP generated or oxygen consumed, but in units of distance, force, or time. Thus, in addition to biochemical and cardiovascular components, performance is also dependent upon the strength and power characteristics of the muscles involved. In general, the shorter the duration of the effort, the greater the relative importance of strength and power to performance. **Strength** is defined as the peak force that can be developed during a single maximal voluntary contraction (MVC). **Power** is defined as the rate at which mechanical work is performed. Since power reflects the rate at which force can be exerted, it is normally of greater importance to sport performance than is strength. In explosive events, where a segment of the body, the whole body, or an external object must be accelerated, as in kicking, jumping, sprinting, or throwing, the duration of the contraction is usually too brief to allow the muscle group to develop its maximal strength. Thus, success will be determined by the muscle group's maximal rate of force development (power).

Conversely, in relatively slow, high-force-dependent movements, such as weight lifting and some wrestling or gymnastics movements, strength may be of greater importance than power. While there is usually a high correlation between a muscle's strength and its power, this is not always the case for some athletes (Figure 1.9) and different methods may be required for the most effective training of different individuals. A number of factors affect maximal strength and power, but the most important are muscle size (cross-sectional area) and the capacity of the athlete's central nervous system to activate or drive the muscle. Further details are presented in Chapters 4 and 5.

Figure 1.9 The force/time curve for two athletes performing a maximal isometric leg press with the intent to generate force as rapidly as possible. Athlete A attains a greater peak force, but B achieves a greater rate of force development (RFD), particularly over the first 200 ms of contraction. Greater RFD would translate into greater power for dynamic actions such as jumping or throwing.

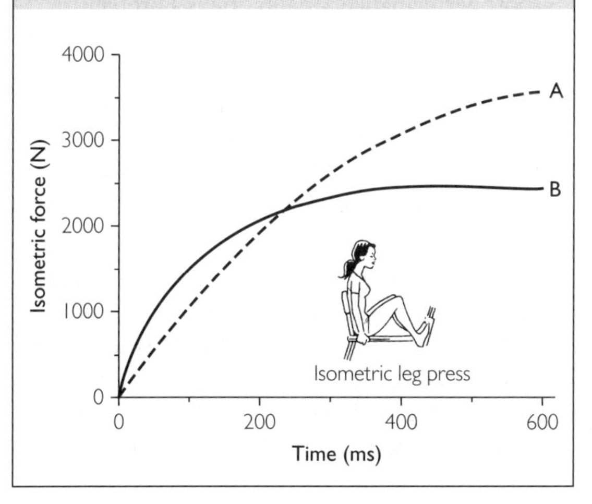

Skill and Coordination

The neural processes that are involved in the execution of even the simplest of motor tasks are extremely complex and as yet only partially understood. A discussion of these processes lies beyond the scope of this book, but we must accept that, in many sports, such factors as motor coordination, reaction time, balance, timing, and motor decision-making are the prime determinants of performance. Even standard locomotive tasks, such as running, have a skill component that affects the athlete's mechanical efficiency or economy of movement. We will not address the training principles for acquiring sports skills and refer interested readers to the psychology and motor learning literature for such information.

Fatigue

An inevitable consequence of sustained exercise is fatigue. Fatigue can be defined as an exercise-induced decrease in force or power-generating capacity (8). This is illustrated in Figure 1.10, which shows data from a subject who performed 50 consecutive maximal concentric contractions on an isokinetic dynamometer. The dynamometer was set to a moderately fast velocity (3.14 rad·s^{-1} or 180°·s^{-1}), and each contraction was done at the set velocity. The vertical bars indicate the force (torque) obtained in each of the 50 contractions. Note the approximately 50% decline in force over the 50 contractions. There was a corresponding decline in power, because power = force × velocity. In this example, velocity was held constant, but in many sports fatigue is associated with a decline in both force and velocity, and therefore a very large decrease in power. Much of our knowledge of the physiology of training for high performance has come from studies of the possible sites and causes of fatigue during exercise. The consequence of fatigue is inability or failure to maintain a given exercise intensity. Depending upon the intensity of the exercise, fatigue can be traced to several possible sites (both within and outside the muscle) and may involve a number of different mechanisms.

Figure 1.10 The changes in force and power output (vertical bars) that occur as a subject performs 50 consecutive maximal contractions on an isokinetic dynamometer set at a velocity of $180° \cdot s^{-1}$. See text for further explanation. From the authors' laboratory.

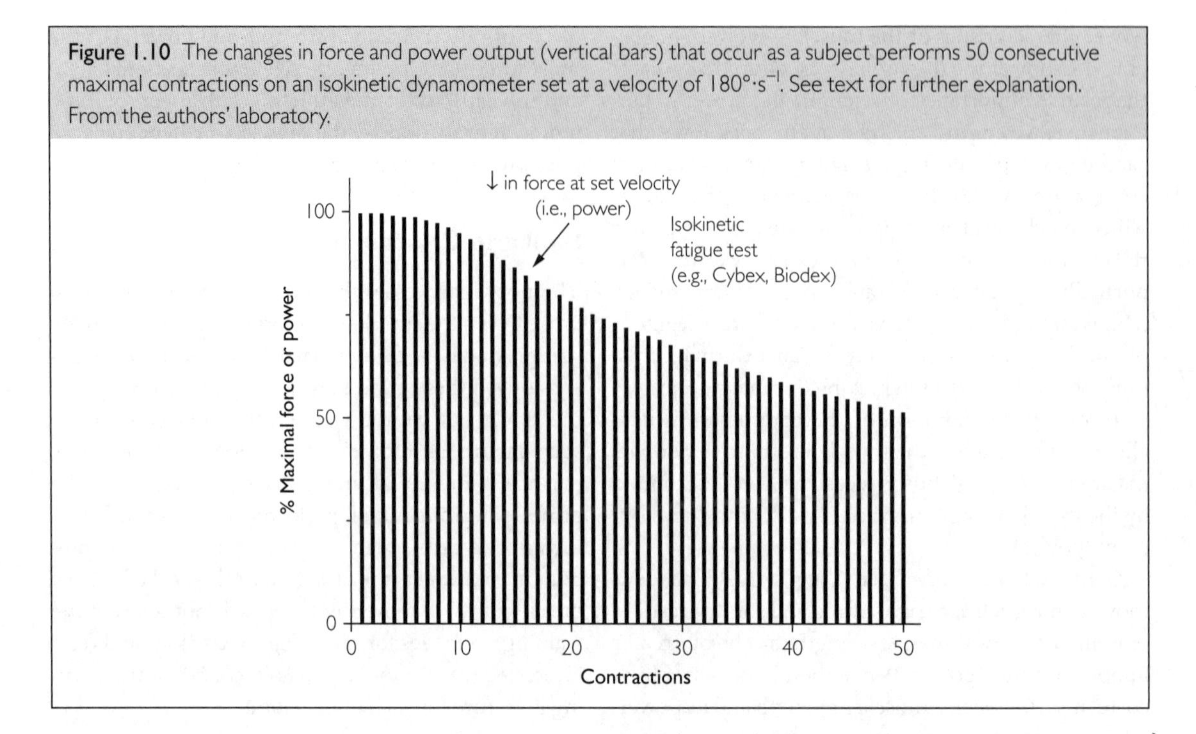

Figure 1.11 summarizes the causes of fatigue for running events of different intensities and durations. Not surprisingly, the higher the intensity of the event, the more rapid the onset of fatigue, and the shorter the event in distance and time. As the figure illustrates, some causes of fatigue, such as failure of neural activation of muscle or impaired excitation–contraction coupling, may be involved in all events, whereas others may be involved mainly in a restricted range of events, such as depletion of PCr and accumulation of H^+. Knowledge of fatigue mechanisms can assist in the design of training programs and in the planning of pre-event and within-event strategies. Fatigue and recovery from fatigue will be covered in more detail in Chapters 3 and 5.

Determining the Weak Links to Performance

A common theme throughout the following chapters will be the search for the weak links that limit performance in a given sport. Based on the scientific literature and experimental evidence, we will attempt to identify those factors that exert the greatest influence on components such as aerobic and anaerobic power and capacity, strength or power for each different sport. Then, because a chain is only as strong as its weakest link, our task will be to develop a training program that addresses these weak points and stimulates specific adaptations to strengthen them.

Summary of Key Points

- Training is the stimulation of biological adaptations that result in an improvement in sport performance. Adaptation is a protective mechanism that has evolved in humans; in order for it to occur, the body must be presented with stimuli that upset its normal balance or homeostasis. For the coach and athlete, the challenge lies in determining the specific stimulus that will bring about the desired adaptation.
- For the adaptive process to continue, the stimulus must be constantly adjusted upwards, so that the athlete is functionally overloaded. Muscles and the supporting biological systems can be overloaded by increasing the intensity, duration, or frequency of the training stimulus, but progression must be

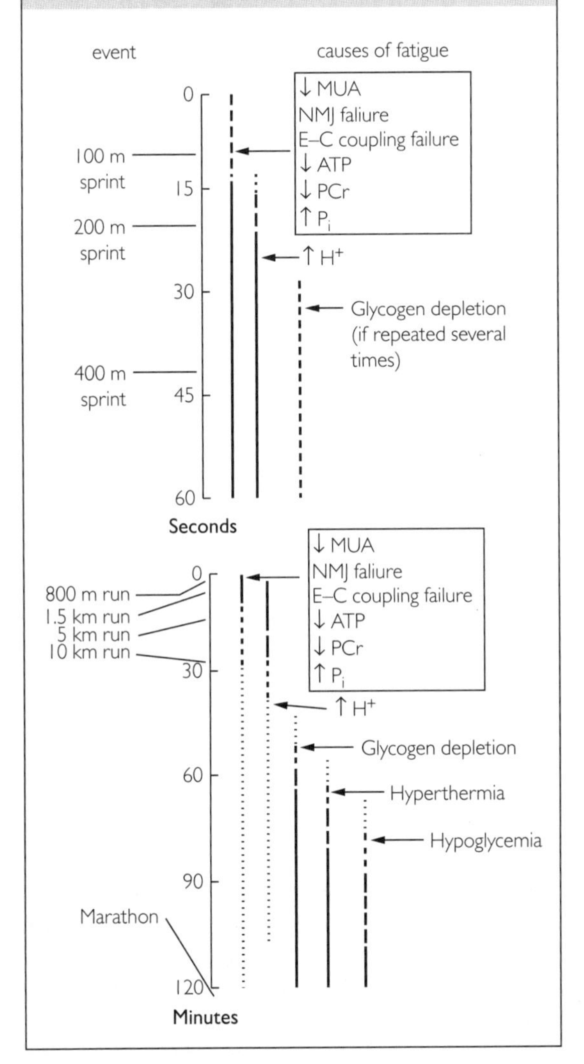

Figure 1.11 A summary of the causes of fatigue for running events of different distances (100 m to the marathon). A solid line indicates that factors are more important, whereas dashed lines indicate less importance. ATP, adenosine triphosphate; E–C, excitation–contraction; H^+, hydrogen ions; MUA, motor unit activation; NMJ, neuromuscular junction; PCr, creatine phosphate; P_i, inorganic phosphate.

gradual and there must be adequate recovery to allow time for adaptation to occur. Otherwise, overtraining results and there is a decrease in performance and often illness.

- When an athlete begins a training program, the rate of adaptation is most rapid over the first few months and then much more gradual. For veteran athletes, considerably more training stimulus is required in order to achieve even slight improvements in performance. Once a training adaptation has been gained, less training stimulus is required to maintain it than to produce it in the first place. However, excessive reductions in training will result in a loss of adaptation.
- An athlete's performance is determined by the maximal rate at which his/her muscles can generate energy in the form of ATP, the maximal rate at which the cardiovascular system can deliver oxygen to the muscles and remove metabolites, and the strength and power of the active muscles. There are four different pathways in human muscle for the generation of ATP. Two of these, the phosphagen system and glycolysis, do not require oxygen and are therefore described as anaerobic. The other two, the oxidation of glycogen and the oxidation of fats, are aerobic. These pathways differ considerably as to the maximal rate at which they can generate ATP and the total amount that can be generated.
- The relative importance of each of the factors that determine athletic performance depends upon the event. In general, the shorter the duration of maximal effort, the greater the importance of strength and power and the enzyme kinetics of the two anaerobic pathways. The longer the duration of the event, the greater the importance of aerobic metabolism and a well-trained cardiovascular system.

2

Biochemical Bases for Performance

Adenosine Triphosphate

The function of a metabolic pathway is to generate **free energy**. In our muscles, **adenosine triphosphate (ATP)** is the donor of the free energy that is used for muscle contraction, for the supporting ionic pumps that regulate muscle electrical activity, and for biosynthesis. A molecule of ATP is made up of a molecule of **adenosine**, consisting of a purine base (**adenine**) and a five-carbon sugar (**ribose**), and a chain of three **phosphate** groups (Figure 2.1). The phosphate groups are attached by what are known as high-energy bonds, which means that considerable energy is required (in this case, 30.6 kJ·mol^{-1}) to produce the attaching reaction. Since energy can be neither created nor destroyed (the first law of thermodynamics), the energy required to form the bond is released as free energy (i.e., 30.6 kJ·mol^{-1}) when that bond is broken. This step is triggered by the activation of the enzyme **myosin ATPase**. The formation of the high-energy bond is known as **phosphorylation**, while the breaking of the bond is known as **hydrolysis**.

When a molecule of ATP is hydrolyzed and loses one of its phosphate groups, it becomes **adenosine diphosphate (ADP)**. A new molecule of ATP can be generated by the re-phosphorylation of this molecule of ADP (Figure 2.2), with the energy required

Figure 2.1 The structure of ATP, which consists of a molecule of adenosine, attached to three phosphates by high-energy bonds.

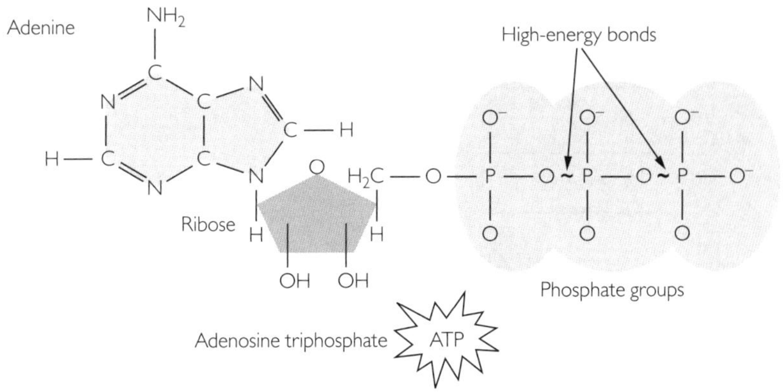

Figure 2.2 (A) A simplified illustration of a molecule of ATP. (B) A schematic illustration of the hydrolysis of ATP to ADP and the release of energy for muscle contraction. The reaction is triggered by the enzyme myosin ATPase. (C) The re-phosphorylation of ADP to ATP with the energy for re-forming the bond coming from one of the four metabolic pathways.

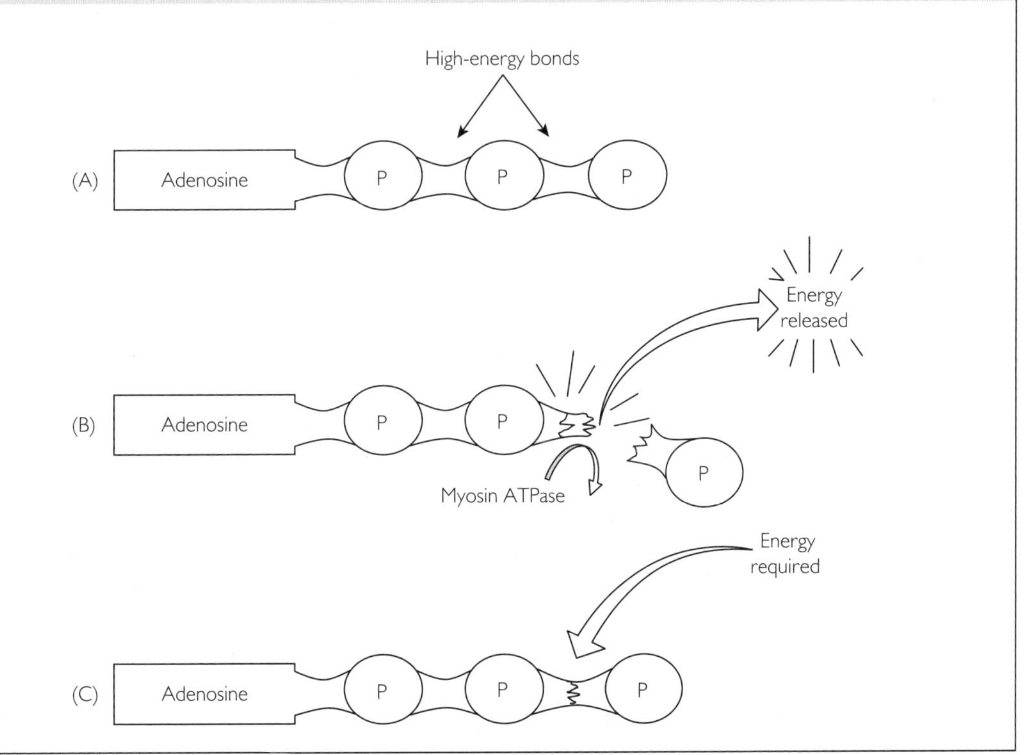

Figure 2.3 (A) A simplified illustration of a molecule of PCr. (B) The hydrolysis of PCr to creatine and phosphate (P_i) and the release of energy. The reaction is triggered by the enzyme creatine kinase. (C) The re-phosphorylation of ADP to ATP using the energy released in the hydrolysis of PCr (above).

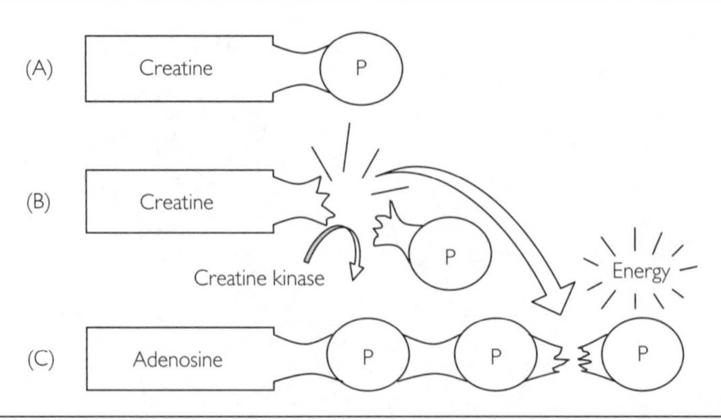

for the phosphorylation reaction coming from one of the four pathways listed in Chapter 1:

- the hydrolysis of phosphocreatine (PCr),
- anaerobic glycolysis,
- the oxidation of glycogen or glucose, or
- the oxidation of fatty acids.

Each of these pathways is characterized by specific **enzymes**, specialized proteins that trigger the individual chemical reactions in a sequential manner, so that the **product** of each preceding reaction becomes the **substrate** for the next. Enzyme activity can also be accelerated or slowed, based on the presence of certain **stimulators** or **inhibitors**. Some reactions occur more readily than others and give off free energy that can be used to drive a second reaction, while others require the addition of free energy in order to proceed.

The concentration of ATP in resting muscle is very low (< 20 $mM{\cdot}kg^{-1}$ dry weight[1]) and would become exhausted after only 1–2 seconds of high-intensity exercise if it were not replaced by the re-phosphorylation of ADP. At first glance, this might seem like a very inefficient arrangement, compared to storing a much larger portion of energy in the form of ATP. However, this pay-as-you-go system is in fact an energy conserver, since energy is produced only on demand, and enables us to function with a much lower muscle mass. Contrast this system with that of a fossil-fuel engine such as the Space Shuttle or a trans-Atlantic airliner, where a large proportion of the total mass consists of fuel.

[1] Concentrations of various substances in muscle are conventionally expressed in units relative to muscle weight. Since approximately 75% of a muscle's weight is water, a distinction must be made as to whether or not the measurements were made after the water had been removed from the sample. When the water has been removed, the term "dry weight" is used; if not, "wet weight" is used.

The High-Energy Phosphate System

In addition to its small store of ATP, skeletal muscle also contains a somewhat larger reserve of the high-energy phosphate compound **phosphocreatine (PCr)**. A molecule of PCr consists of a molecule of creatine and a single phosphate group, attached by a high-energy bond. When a molecule of PCr is hydrolyzed to creatine and phosphate by the action of the enzyme **creatine kinase**, the free energy that is released is used to phosphorylate a molecule of ADP back to ATP (Figure 2.3). Thus, PCr provides an important energy buffer that can maintain muscle ATP concentration during intense exercise until it reaches a critical depletion level. This occurs after approximately 15 seconds of sprint exercise and results in a decline in ATP and the onset of muscle fatigue (Figure 2.4).

Resynthesis of PCr

Once PCr becomes depleted in muscle, it is resynthesized by the reversal of the creatine kinase reaction described in Figure 2.3, with the phosphorylation energy being derived from oxidative pathways (21). The

Figure 2.4 The utilization of PCr hydrolysis to resynthesize muscle ATP over 16 seconds of maximal exercise. Notice the progressive decline in muscle concentration of PCr, but the maintenance of muscle ATP over the exercise period. Concentrations are expressed per kilogram of dry muscle weight (dw). From (44), a study in which subjects performed maximal isometric contractions of the quadriceps and PCr and ATP were measured in needle biopsy samples taken every 2 seconds. Copyright © 1995, The American College of Sports Medicine.

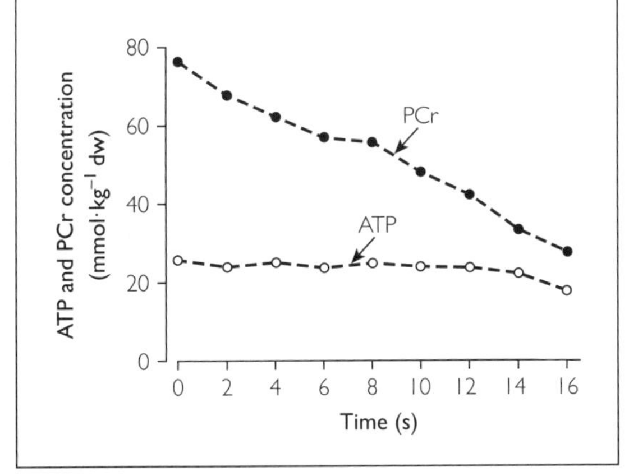

Figure 2.5 The time course for the resynthesis of PCr following its depletion during a single exhaustive sprint interval and after several such intervals. From (14,21,44).

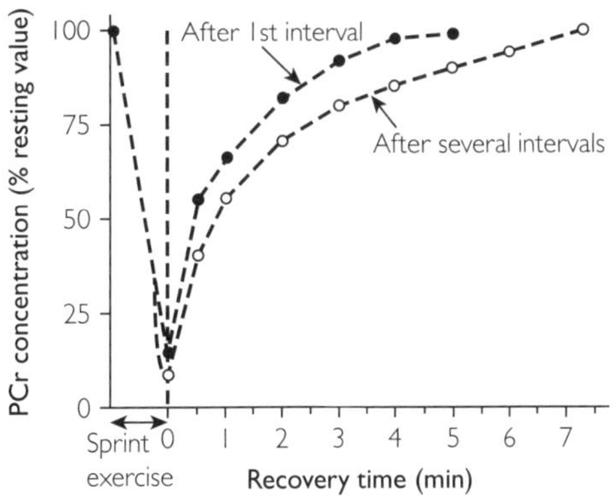

energy buffer, which briefly supplies ATP during transitional phases of exercise or during rapid powerful movements of short duration, such as those that occur in field events, sprinting, and most team sports. Since oxygen is not required for the hydrolysis of PCr, it is known as an **anaerobic** pathway.

time course for this process is illustrated in Figure 2.5. The time required for complete restoration of PCr following a single exhaustive sprint interval and a resting recovery would be approximately 3–5 minutes (with a half-time of approximately 30 seconds). If submaximal exercise is performed during the recovery or several sprint intervals are repeated, we can expect the recovery curve to be flatter, and full resynthesis would thus take longer to occur. This means, for example, that the athlete who is performing sprint interval training or the bodybuilder performing multiple sets should extend the recovery time to perhaps 4–5 minutes as the number of intervals or sets increases. Similarly, consider the example of the cross-country runner who encounters a steep hill and depletes muscle PCr early in the race. Since the running pace for such an athlete might average as much as 85–90% of $\dot{V}O_{2max}$, it is probable that only a small portion of the PCr muscle reserve will be restored in time for a kick at the finish line.

To summarize, as outlined in Chapter 1, hydrolysis of PCr is the most rapid of all the pathways that can supply ATP, but the total amount available is very limited. This limitation relegates this pathway to the role of an

Anaerobic Glycolysis

A second anaerobic system for the rapid resynthesis of ATP also exists within the muscle. The glycolytic pathway, or **glycolysis**, results in the stepwise breakdown of either muscle glycogen or free glucose to lactic acid (or lactate). The process requires 11 chemical reactions, each of which is catalyzed by a specific enzyme, as summarized in Figure 2.6. Although one of these reactions actually consumes energy (ATP), two release enough free energy for the phosphorylation of ADP to ATP, so that there is a net gain of ATP (3 molecules of ATP for each molecule of glycogen broken down to lactic acid). This pathway rapidly consumes carbohydrate stores, but the kinetics of the glycolytic enzymes are such that the rate of ATP production is quite rapid as well (although only about 50% as rapid as PCr hydrolysis).

Carbohydrate Storage

Approximately 510 g of carbohydrate are stored in a 75 kg athlete (≈440 g in the muscles, ≈65 g in the liver, and ≈5 g circulating in the blood as glucose). When

Figure 2.6 A summary of the glycolytic pathway, which includes 11 chemical reactions that break down a molecule of glycogen or free glucose to two molecules of glyceraldehyde 3-phosphate and two molecules of lactic acid. Note that some reactions consume ATP while others produce ATP, so that the net gain is three molecules of ATP for the breakdown of glycogen, or two molecules of ATP for the breakdown of glucose.

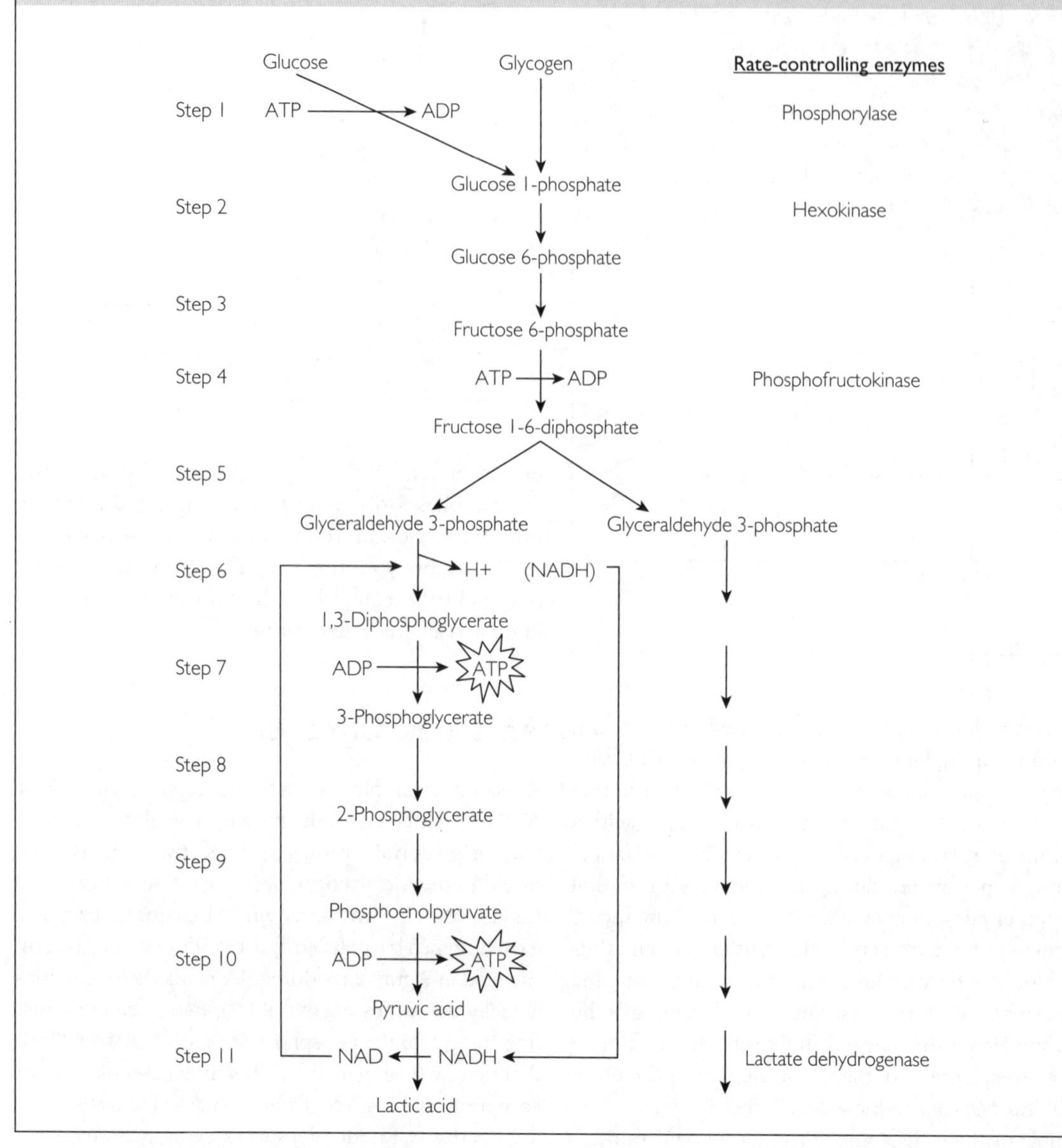

carbohydrate is ingested, it is converted to the simple six-carbon sugar **glucose**, which is stored in liver and muscle as branched chains of glucose molecules, known as **glycogen**. The concentration of glycogen in muscle is ≈14 $g \cdot kg^{-1}$ wet weight and may be as high as 25 $g \cdot kg^{-1}$ in a trained athlete. The concentration in liver is ≈44 $g \cdot kg^{-1}$ and can also be increased with training or with a high-carbohydrate diet. Liver glycogen provides an indispensable reservoir that can be released as glucose to maintain a constant blood glucose concentration over extended periods between meals or during long-term exercise. Release of glucose from the liver and its uptake by the liver are regulated by the pancreatic hormones **glucagon** and **insulin** respectively. Unlike the liver, muscle lacks the enzyme necessary for converting glycogen back to free glucose

(**glucose-6-phosphatase**) and thus cannot release its carbohydrate reserves back into the circulation for direct maintenance of blood glucose.

Activation and Rate-Controlling Steps

Glycolysis, like all the metabolic pathways, is exquisitely controlled, so that the concentration of ATP, or energy charge, of the muscle cells is maintained over a wide range of exercise intensities. Stated another way, when exercise intensity increases and ATP is used more rapidly, the rate of ATP production increases proportionately; when exercise intensity decreases, the rate of ATP resynthesis via glycolysis decreases as well. A large proportion of this control is exerted at the first, or mobilization, step (Step 1, Figure 2.6), in which glycogen is phosphorylated to glucose 1-phosphate. This reaction is catalyzed by the enzyme **glycogen phosphorylase** (normally referred to simply as phosphorylase), which is activated and upregulated by the appearance of ADP, adenosine monophosphate (AMP), inorganic phosphate (P_i), and calcium ions (Ca^{2+}) (37,42). Since the rate at which each of these substances appears is proportional to exercise intensity, the rate of glycolytic energy supply is tightly coupled to energy demand.

A second important rate-controlling enzyme in glycolysis acts to catalyze the phosphorylation of fructose 6-phosphate to fructose 1,6-diphosphate (Step 4, Figure 2.6). This enzyme is **phosphofructokinase** (**PFK**) and, like phosphorylase, its activity can be upregulated or downregulated by the appearance rates of metabolites such as ADP, AMP, and P_i. Enzymes such as PFK and phosphorylase, whose reaction rates can be stimulated or inhibited by the concentration of certain compounds or metabolic byproducts, are known as **rate-limiting** or **allosteric** enzymes. It is important to note here that the activity of PFK can also be inhibited or downregulated if intense sprint-type exercise is prolonged. This downregulation is largely due to the inhibitory effects of the increased concentration of hydrogen ions (decreased pH)[2] generated by the product lactic acid (51,55).

[2] The concentration of hydrogen ions, or $[H^+]$ (throughout this book, square brackets [ ] are used to denote the concentration of a substance), is expressed in pH units. The term pH refers to the negative logarithm of $[H^+]$, or $1/\log [H^+]$. Thus, as $[H^+]$ increases, pH decreases and vice versa. The pH of arterial blood and most cells is approximately 7.4, and the body can tolerate only minor deviations from this level. A decrease in pH from 7.4 is termed **acidosis**, whereas an increase from 7.4 is termed **alkalosis**.

The Glycolytic Pathway

During exercise, glycolysis can be viewed as beginning with the phosphorylation of glycogen to glucose 1-phosphate. This step is catalyzed by the activation of glycogen phosphorylase, which sequentially removes glucose units from the outer branches of glycogen and adds a phosphate group to the first carbon of this six-carbon sugar. Glucose 1-phosphate is then converted to glucose 6-phosphate (i.e., the phosphate is transferred to the sixth carbon) by the enzyme **phosphoglucomutase**. Once it is in this form the pathway is self-propagating and reactions will continue sequentially (Steps 3–10, Figure 2.6) until two molecules of **pyruvic acid** (or pyruvate) have been produced. As an alternative to using glycogen, the muscle cell can also derive energy from free glucose by phosphorylating it directly to glucose 6-phosphate by means of the enzyme **hexokinase**. However, since this step requires the addition of free energy and since the concentration of glucose in the cell is very low compared to that of glycogen, during heavy exercise the vast majority of energy is derived from the breakdown of muscle glycogen.

Glucose 6-phosphate is then converted to a similar six-carbon sugar: fructose 6-phosphate (Step 3), which is phosphorylated to fructose 1,6-diphosphate (i.e., an additional phosphate is added to the first carbon) by the enzyme PFK. This compound is then split into two molecules of glyceraldehyde 3-phosphate (a three-carbon sugar with a phosphate on the third carbon). Each of these molecules undergoes five reactions (Steps 5–10), eventually becoming pyruvic acid. Two of these reactions (Steps 7 and 10) are energy-releasing reactions that generate ATP while, in another (Step 6), hydrogen ions are removed from the compound and can be used for the aerobic generation of ATP in the mitochondria. Because each molecule of glycogen generates *two* three-carbon sugars (Step 5), the energy yield from this point onwards is effectively doubled. The breakdown of a molecule of glycogen to pyruvic acid thus results in a net gain of three ATP molecules, while the breakdown of glucose to pyruvic acid results in a net gain of two ATP.

In the resting state or during mild- to moderate-intensity exercise, most of the pyruvic acid generated through glycolysis will be oxidized in the mitochondria to provide a rich yield of ATP. However, as exercise intensity and the rate of ATP utilization increase, more and more of the pyruvic acid will be

Figure 2.7 The transformation of pyruvic acid to lactic acid, with the two hydrogens being supplied by the carrier NAD (shown here as 2 NADH). The reaction is triggered by the enzyme lactate dehydrogenase.

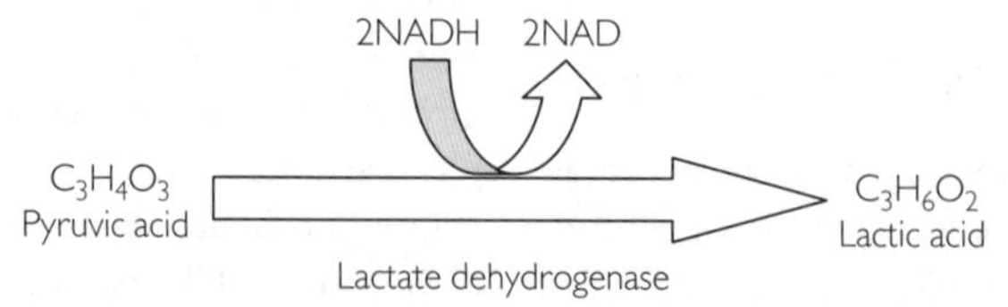

transformed to **lactic acid**. The enzyme for this reaction is **lactate dehydrogenase** (**LDH**). In this transformation, the hydrogen ions that were removed in Step 6 are added to pyruvic acid to produce lactic acid (Figure 2.7). The significance of this step is that it frees the hydrogen carrier **nicotinamide adenine dinucleotide** (**NAD**), increasing its capacity to accept more hydrogens from Step 6, and thus allows glycolysis to proceed more rapidly. From the athlete's perspective, the advantage is that the ATP-production rate can be greatly accelerated; the disadvantage is that it does so at the expense of an accelerated rate of lactic acid production. Lactic acid is known as a **strong acid**, which means that it readily gives up (dissociates) its H^+ into the intracellular fluid of the muscle or into the blood (lactic acid → $lactate^-$ and H^+). There is thus a direct relationship between the production rate of lactic acid and $[H^+]$.

Aerobic Metabolism

The metabolic pathways involved in the aerobic generation of ATP occur within the thousands of mitochondria that are found in a muscle fiber. During exercise, most ATP resynthesis is derived from the oxidation of the pyruvic acid generated by glycolysis or from the oxidation of fatty acids. **Oxidation** is the term used to describe the *removal of hydrogen atoms* from various compounds. It is the potential energy contained in the bonding of the hydrogen proton ($^+$) and its electron ($^-$) that is released through a series of steps to re-phosphorylate ADP to ATP. These steps are known as the **electron transport chain** (ETC) and the total process is known as **oxidative phosphorylation**. In the final step, the electrons and protons that have been transported within the mitochondrial inner membrane are reunited and accepted by oxygen to form water. Thus, this pathway can occur only in the presence of oxygen and the maximal rate at which it can generate ATP will be dependent upon the maximal rate at which oxygen can be made available to the mitochondria.

The Krebs Cycle

There are a variety of ways by which hydrogens can be removed from carbohydrates, fats, or proteins, but the most common is to convert each to the compound **acetyl coenzyme A** (**acetyl-CoA**). The energy (ATP) required for this conversion differs for these three substrates, with glycogen having the lowest net ATP requirement per molecule of acetyl-CoA produced. Thus, the oxidation of pyruvic acid provides the greatest net energy yield per unit of oxygen consumed.

The pyruvic acid that has been generated by glycolysis is transported into the mitochondria by the carrier protein **coenzyme A**. This reaction is catalyzed by the enzyme **pyruvate dehydrogenase** (or **PDH**) and results in the transformation of pyruvic acid to acetyl-CoA and the removal of a pair of hydrogen atoms. Once inside the mitochondrial membrane, the acetyl-CoA gives up its acetyl group by combining with the four-carbon compound **oxaloacetic acid** to form a six-carbon compound, **citric acid**, and becomes free to pick up and transport another pyruvic acid. The citric acid now undergoes a series of enzyme-driven decarboxylations (removal of carbon dioxide) and dehydrogenations (removal of hydrogens) until oxaloacetic acid is again regenerated and the cycle repeats itself (Figure 2.8).

In four of the eight reactions of the Krebs cycle (also known as the citric acid cycle and the tricarboxylic acid, or TCA, cycle), a pair of hydrogen atoms is removed and transported to the ETC for the phosphorylation of ADP to ATP. In addition, the cleavage of the energy-rich bond in **succinyl-CoA** to produce **succinic acid** (reaction 5) releases sufficient energy for

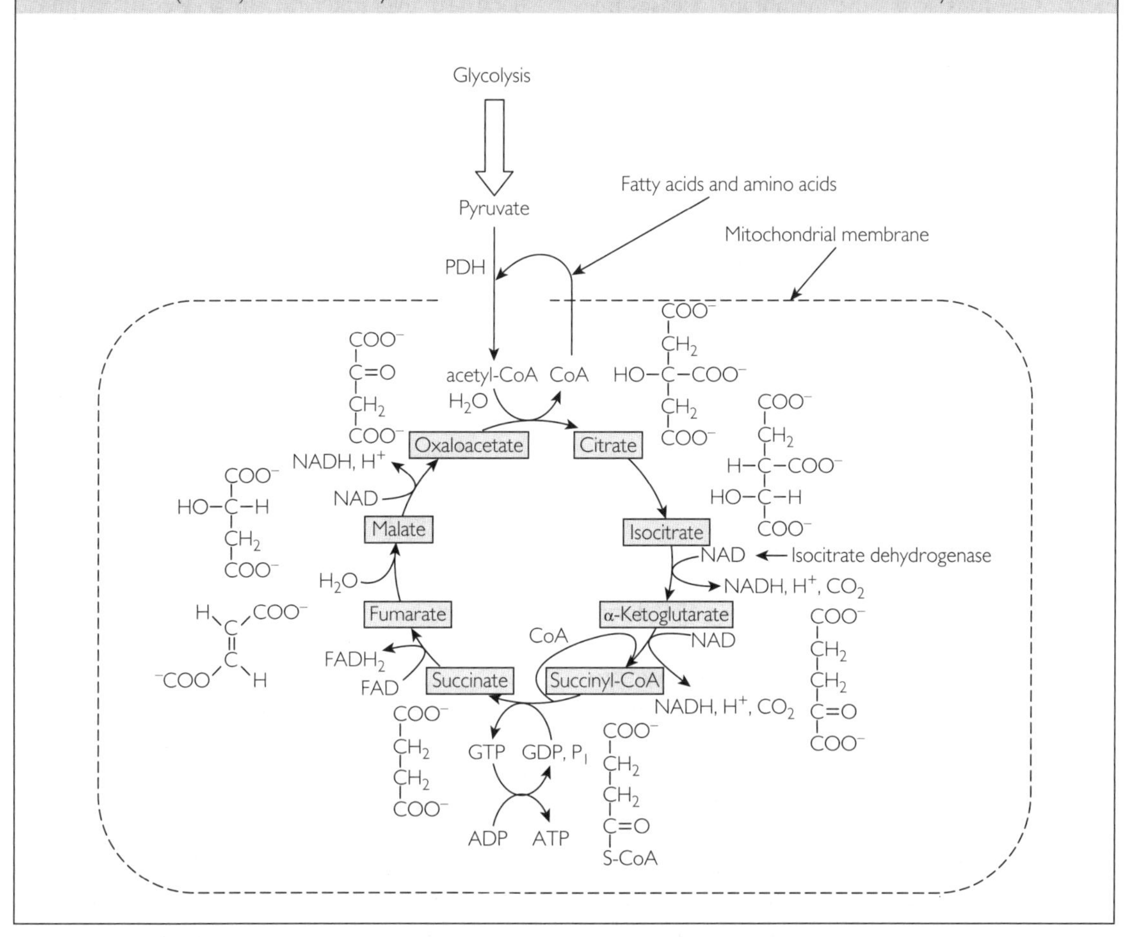

Figure 2.8 A summary of the reactions of the Krebs cycle. The cycle begins with a molecule of pyruvic acid (generated through glycolysis) being converted to acetyl-CoA and entering the mitochondrion to combine with oxaloacetate to form citric acid (citrate). Note that fatty acids and amino acids can also be used as substrates for the cycle.

the direct generation of a molecule of ATP. In two of the reactions, carbon dioxide (CO_2) is also removed and diffuses into the blood, from which it is eventually eliminated in expired gas.

Since two molecules of acetyl-CoA are produced from each molecule of glucose (or glycogen), the Krebs cycle must rotate twice per molecule of glucose broken down, giving rise to the following products: four molecules of CO_2, two molecules of directly produced ATP, and eight pairs of hydrogen atoms. To these can be added the two ATP directly generated through glycolysis, the two pairs of hydrogen atoms from glycolysis, and the two pairs of hydrogen atoms removed in the transformation of pyruvic acid to acetyl-CoA (for a total of 12 pairs of hydrogen atoms).

Electron Transport Chain

The hydrogen atoms that have been removed up to this point are transported to the inner membrane of the mitochondrion by the reduced carrier coenzymes **NAD** and **flavin adenine dinucleotide** (**FAD**). When these carrier proteins combine with hydrogen (e.g., $NADH_2$), they are said to be **reduced**. When the hydrogens are removed (e.g., NAD), they are said to be **oxidized**. In a sequence of reactions, the electrons are separated from the hydrogen atoms and passed down a series of iron-containing carriers, known as **cytochromes**, while the protons are ejected into the fluid-filled space between the two mitochondrial membranes. The carriers are alternately reduced and oxidized as the electrons are transferred from one carrier to the next; as a result, the reactions are described as **redox** (*red*uction/*ox*idation) reactions. Three of these carriers are large protein complexes that span the membrane and act as proton pumps. The movement of protons across the membrane releases free energy for phosphorylating ADP to ATP. In the final step, the electrons and protons are reunited to form hydrogen atoms and combine with oxygen to form water. Each molecule of $NADH_2$ that enters the ETC releases enough energy to produce three molecules of ATP. $FADH_2$ contains less potential energy and releases only enough energy to generate two molecules of ATP. However, since 10 of the 12 pairs of hydrogens that have been removed are transported by NAD (with each pair generating three ATP molecules), the result is a very rich energy yield. In fact, if we want to keep score, the total energy production from the complete aerobic breakdown of one molecule of glucose is 36 molecules of ATP!

Fat as an Exercise Fuel Source

Although the fuel source for generating ATP during heavy exercise is predominantly glycogen, pathways also exist within the muscle for using fat and even protein. During low- to moderate-intensity exercise, as much as 50% of the energy may be derived from the oxidation of fats. Fat is thus an important fuel source for prolonged exercise (or for recovery from high-intensity exercise), helping to preserve body carbohydrate stores and thus maintain blood glucose.

Fat is abundantly stored in the muscle and in the fat cells (adipocytes) throughout the body as **triglyceride**. A molecule of triglyceride consists of a molecule of **glycerol** binding three molecules of **free fatty acids** (**FFAs**). A number of different fatty acids occur naturally in the body, or are ingested in the diet, but they all have a similar molecular structure: an even-numbered chain of carbons (16–24), each of which is bound to two hydrogen atoms. Because of their hydrogen-rich structure, fatty acids are the most energy-dense fuel (i.e., when oxidized they generate more ATP per gram than carbohydrate or protein). The processes by which fats are metabolized are summarized in Figure 2.9.

Before fatty acids can be oxidized, they must first be **mobilized** from their triglyceride storage sites. This occurs as a result of the activation of the enzyme **hormone-sensitive lipase**, which is found in fat cells. The reaction, known as **lipolysis**, converts the triglyceride back to glycerol and three FFAs, which are released into the blood and transported to the muscle cells, bound to albumin. FFAs that are mobilized directly in the muscle are spared this transportation step.

Because of their solubility, fatty acids cross the muscle cell membrane relatively easily. They are then activated by combining with coenzyme A to form **fatty acyl-CoA**, at the expense of utilizing an ATP for each molecule formed. The fatty acyl-CoA must next be transported across the mitochondrial membrane before it can be oxidized. This is accomplished by the carrier protein **carnitine** and the enzyme **carnitine transferase**, through a process known as the **carnitine shuttle**.

Once inside the mitochondrion, the fatty acyl-CoA undergoes **β-oxidation**. In one turn of the β-oxidation cycle, two carbon atoms are cleaved off to form a molecule of acetyl-CoA and two hydrogen atoms are removed. The acetyl-CoA then enters the Krebs cycle (see Figure 2.8), while the hydrogens are transported by NAD and FAD to the ETC for a rich yield of ATP. The cycle continues until all the carbons in the fatty acyl-CoA have been utilized in the production of acetyl-CoA. The energy yield for a single fatty acid molecule will depend upon the length of the hydrocarbon chain, but may be as great as 160 ATPs!

Efficiency of Fat as a Fuel

The efficiency of fatty acids as a fuel for exercise can be compared to that of glycogen in two different ways:

- by comparing the ATP produced by the oxidation of a given amount (e.g., 1 g) of each substance, or
- by comparing the amount of oxygen required to produce a given amount of ATP (e.g., one molecule) for each fuel.

Using the former comparison, fat is much more efficient. When oxidized in the muscle, 1 g of fat will produce approximately 9.5 kcal of energy, whereas 1 g of glycogen would produce only 4.2 kcal. However, the picture changes considerably when the comparison is made on the basis of oxygen cost. Because of its oxygen-poor structure, oxidation of fat requires the addition of considerably more oxygen than does glycogen. In fact, *per molecule of ATP produced, the oxidation of fat requires approximately 12% more oxygen than does the oxidation of glycogen*. In other words, when exercising at 100% $\dot{V}O_{2max}$, the maximal rate of ATP production would be approximately 12% greater when oxidizing glycogen than when oxidizing lipid.

Figure 2.9 A summary of the steps by which fats are oxidized in the muscle cell. The process begins with the mobilization of FFAs, which are transported by the blood to the cell. After diffusing across the cell membrane, they are transported across the mitochondrial membrane as fatty acyl-carnitine to form Acyl CoA (the carnitine shuttle). The acyl-CoA then enters a series of reactions known as the β-oxidation cycle, eventually forming acetyl-CoA. With each turn of the cycle, two H^+ ions and two carbons are removed. The H^+ ions enter the electron transport system and the acetyl-CoA enters the Krebs cycle to liberate more H^+.

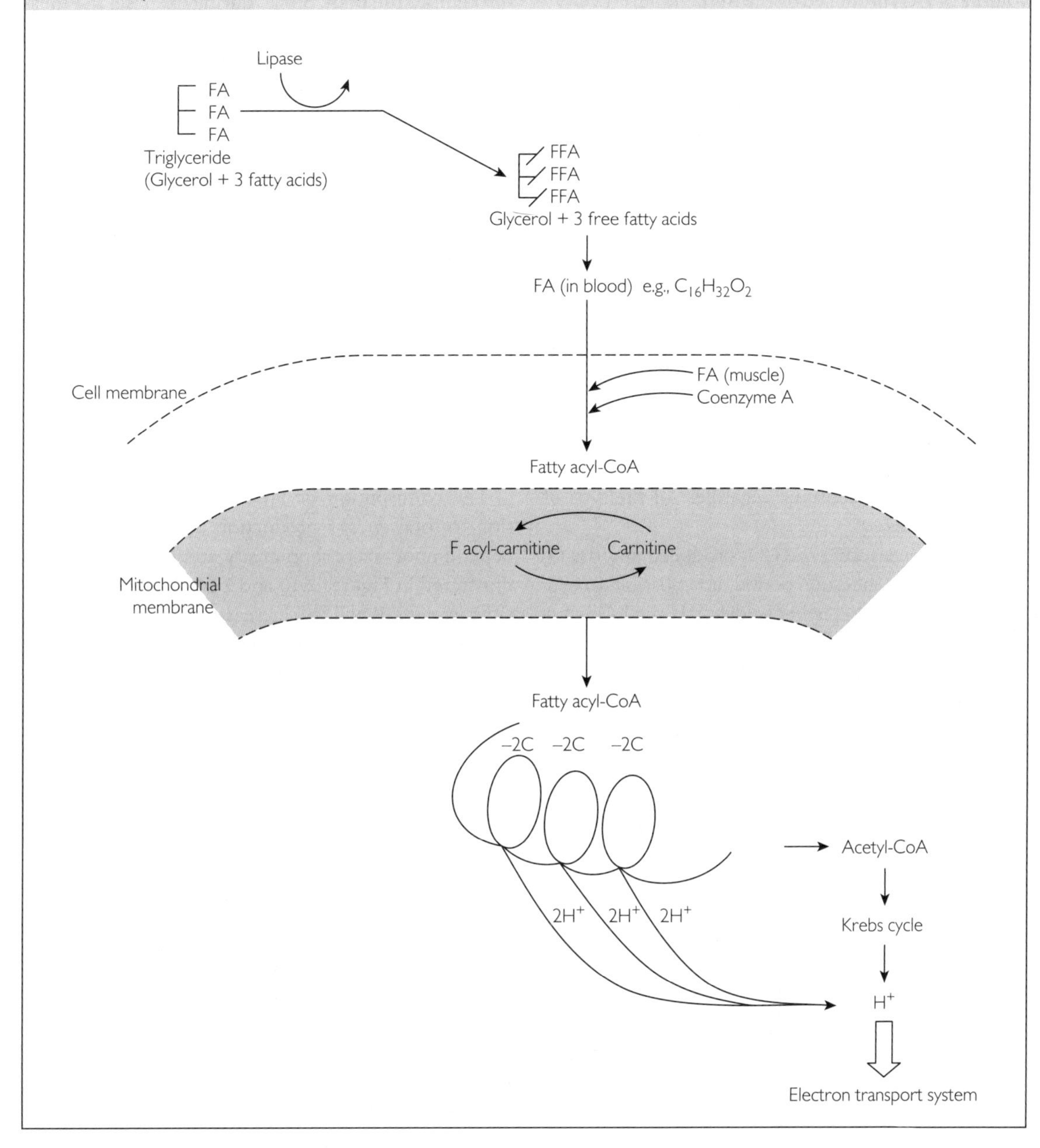

Respiratory Quotient

A simple yet effective method for estimating the proportional contribution of carbohydrate and fat oxidation as fuel sources is to examine the **respiratory quotient** (**RQ**). The RQ is defined as the ratio of the volume of CO_2 produced by the mitochondria ($\dot{V}CO_2$) to the volume of O_2 consumed ($\dot{V}O_2$), over the same time:

$$RQ = \dot{V}CO_2 / \dot{V}O_2$$

Because of carbohydrate's structure ($C_6H_{12}O_6$), when one molecule of glycogen is completely oxidized, it produces six molecules of CO_2 and consumes six molecules of O_2. The RQ is thus 6/6 or 1.0. On the other hand, the structure of a fatty acid (e.g., palmitic acid, $C_{16}H_{32}O_2$) is such that approximately 44% more O_2 is consumed than CO_2 produced. For example, in the complete oxidation of a molecule of palmitic acid, 23 molecules of O_2 would be consumed and 16 molecules of CO_2 produced. The RQ would thus be 16/23, or 0.7. Because the proportion of hydrogens to carbons is the same for all fatty acids, this ratio remains constant, regardless of the type of fatty acid oxidized.

Thus, by measuring the CO_2 produced and the O_2 consumed over the same period, it is possible to estimate the type of fuel being oxidized. If the RQ is 1.0, we can assume that 100% of the energy production is from carbohydrate. If the RQ is 0.7, we can assume that 100% of the energy is from fat. An RQ of 0.85 would indicate a 50/50 mixture of the two, and so on.

Although RQ should theoretically be measured in a single cell or muscle, this is technically very difficult to do. It is therefore most common to measure $\dot{V}CO_2$ and $\dot{V}O_2$ within the *whole* body, by analyzing respiratory gases. This gives us the **respiratory exchange ratio** (**R**), which is assumed to reflect the average RQ throughout the body. Since, during exercise, the muscles are by far the greatest source of oxygen consumption and carbon dioxide production, measurement of R provides a reasonable estimate of fuel proportions. As exercise intensity increases and muscle metabolism contributes more and more to the total that is measured, this estimate becomes more and more valid. The validity of the process is also based on the assumptions that little or no protein is being oxidized and that, at the time of measurement, the body is in steady state, with the rate of CO_2 production as measured at the lungs being equal to its rate of production in the muscles. For this reason, it would be invalid to use R to estimate fuel utilization during maximal or near-maximal exercise. At these intensities, the buffering of lactic acid (by bicarbonate) generates additional CO_2, which would result in R values in excess of 1.0.

The contribution of fat and carbohydrate oxidation to total energy production during exercise will depend upon exercise intensity and duration. This is illustrated in Figures 2.10 and 2.11. As exercise intensity increases, R will also increase from an average resting value of approximately 0.82 to values greater than 1.0 at 100% $\dot{V}O_{2max}$ (Figure 2.10). From this observation, we can conclude that, over this time, there has been a progressive increase in carbohydrate (primarily

Figure 2.10 The relative contribution of carbohydrate and fats during exercise at increasing intensities and the effect on the respiratory exchange ratio (R). Note that, at 100% $\dot{V}O_{2max}$, R is greater than 1.

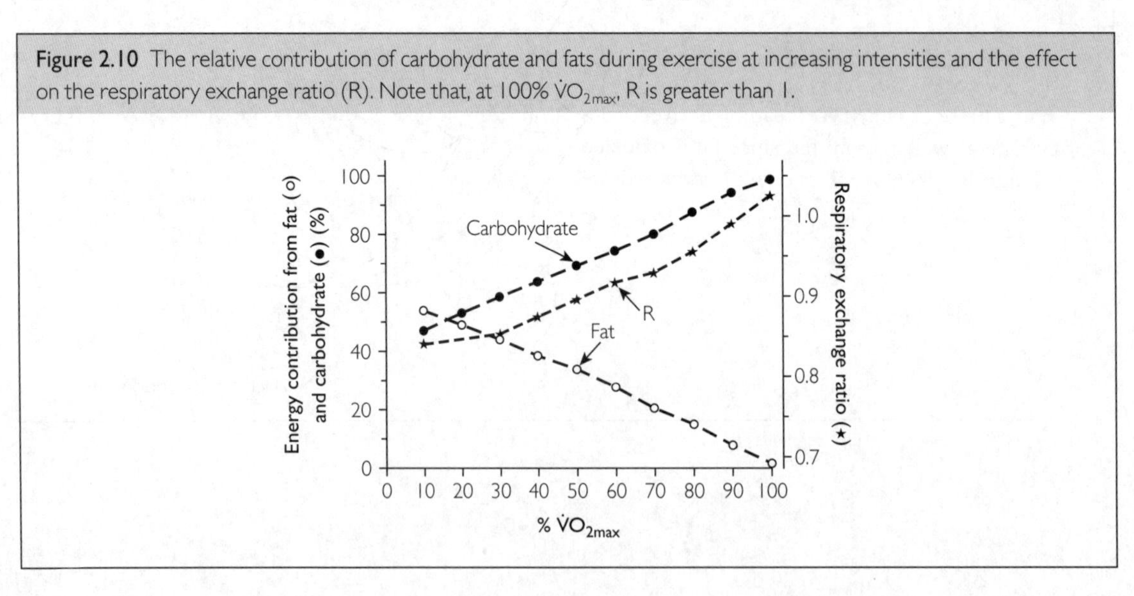

Figure 2.11 The relative contribution of carbohydrate and fats, and changes in the respiratory exchange ratio (R), during prolonged exercise (as in a marathon race).

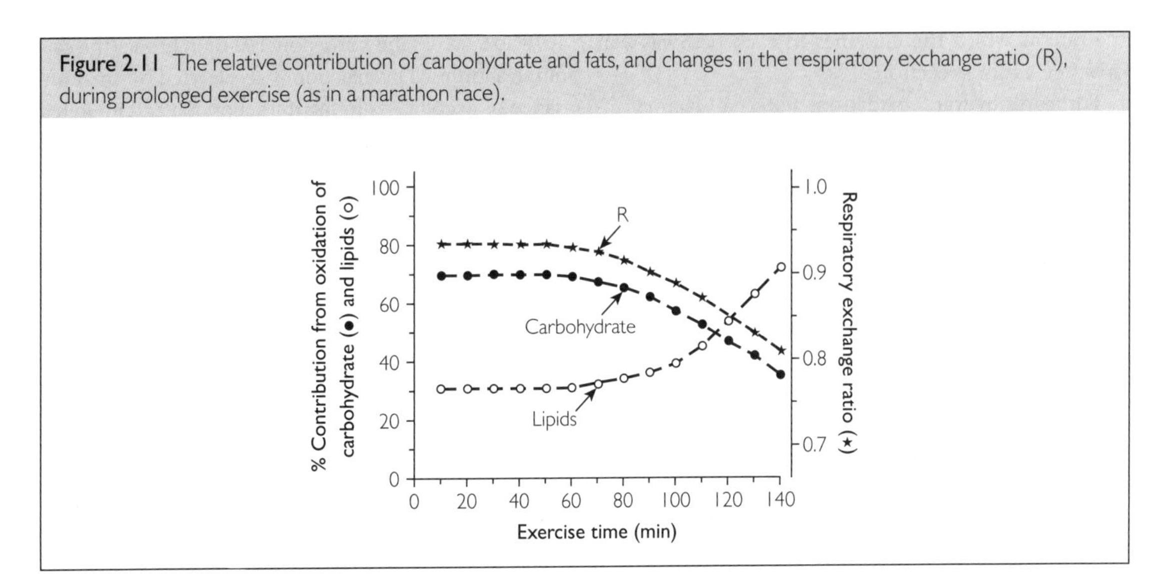

muscle glycogen) oxidation and a progressive decline in the relative contribution of fat oxidation. The advantage of this shift towards glycogen use is that more ATP can be produced per unit of oxygen consumed. Thus, as the athlete approaches his or her maximal capacity for supplying oxygen to the mitochondria, the maximal rate at which ATP can be resynthesized is greater than could be achieved by metabolizing a mixture of fuels.

In contrast, when submaximal exercise is continued for prolonged periods, as in marathon running, R values tend to decrease after approximately 40–60 minutes, indicating a decrease in the proportion of energy derived from carbohydrates and an increase in that from fat (Figure 2.11). This is largely due to the depletion that occurs in muscle and liver glycogen stores. The summed result is a progressive decline in the overall rate at which ATP can be derived from carbohydrate as well as from fat, since fat is oxidized most effectively when combined with carbohydrate (52). The athlete thus finds it more and more difficult to maintain the earlier pace.

Protein as a Fuel Source

Pathways exist in muscle for removing the nitrogen component from several amino acids and funneling their carbon skeletons into the Krebs cycle as acetyl-CoA or as other intermediates. In addition, many amino acids in muscle can be converted to the common amino acid **alanine**, which can then be converted in the liver to glucose through a pathway known as the **glucose-alanine cycle**, thus helping to maintain blood glucose during prolonged exercise. Although these pathways are important for sustaining life during starvation, they are used to only a minimal extent during exercise in the fed state. During prolonged exercise, as carbohydrate stores become depleted, protein oxidation increases but still remains a minor contributor to the total. For example, during 1 hour of heavy exercise, amino acid oxidation may account for 2–5% of the total energy, whereas over 2–3 hours of heavy exercise its contribution may approach 10% (18,59). In the nutritionally conscious athlete it is thus unlikely that protein availability would ever become a limiting factor for performance; however, athletes who train intensively for 2–3 hours each day clearly have higher dietary requirements than the normal population. The significance of this fact is discussed in Chapter 11.

Rate-Controlling Steps for Aerobic Metabolism

A number of factors, both outside and within the muscle cell, work together to determine substrate availability and the rate at which oxidative phosphorylation proceeds. External factors include oxygen delivery to the muscle, changes in blood hormone concentrations and the delivery rate of glucose and FFAs from storage sites outside the cell. Within the cell, the maximal rate at which each fuel source can be oxidized will also be determined by the availability of the substrate and by the rate at which the enzymes in each pathway can drive the various reactions.

For carbohydrate oxidation, external factors include the concentration of the hormone **epinephrine** (also known as adrenaline), which accelerates glycogen mobilization (at the phosphorylase step) and, if exercise is prolonged, the availability of blood glucose as muscle glycogen becomes depleted. The availability of blood glucose is dependent upon the ability of the liver to release glucose into the blood and the rate at which it can be transported across the muscle cell membrane. As liver stores of glycogen become reduced, the supply of blood glucose to the muscle will diminish, thus reducing the muscle's overall ability to utilize carbohydrate as an energy source. Within the muscle cell, control is exerted by the enzyme pyruvate dehydrogenase, which must increase its activity to transform the pyruvic acid that has been generated through glycolysis to acetyl-CoA for subsequent oxidation in the mitochondria. The more pyruvic acid that can be accommodated at this step, the less that will be converted to lactic acid. Within the Krebs cycle, the key regulatory enzymes are **citrate synthase**, which is responsible for the first reaction in the cycle, and **isocitrate dehydrogenase**, which catalyzes the third reaction and removes the first pair of hydrogens. The activity of both of these enzymes is stimulated by the appearance of ADP and a decrease in the NAD/NADH ratio, causing the Krebs cycle to turn more rapidly when more ATP is required.

External factors regulating lipid oxidation include those involved in mobilizing FFAs from their triglyceride storage sites throughout the body (lipolysis). As previously noted, the controlling enzyme for this reaction is hormone-sensitive lipase. A similar enzyme, **lipoprotein lipase**, performs the same function on triglycerides circulating in the blood, bound to albumin. The substances known to accelerate this reaction, for both enzymes, include epinephrine, glucagon, growth hormone, and caffeine. Inhibitors include insulin and a decrease in pH. Although adequate delivery of FFAs to the muscle is necessary during prolonged exercise, it is unlikely that the mobilization step ever becomes a major limiting factor. This is because the amount of FFAs delivered to the muscle is always many times greater than the amount used, and further increases in the concentration of blood FFAs do not increase their rate of utilization (53). Within the cell, it is generally believed that the rate-limiting enzyme is **carnitine acyltransferase**, which is involved in the transport of fatty acyl-CoA across the mitochondrial membrane (the carnitine shuttle). This reaction is accelerated by an abundance of acetyl-CoA in the mitochondria and inhibited as acetyl-CoA decreases. The significance of this control is that, when energy is abundant (as in low-intensity exercise), oxidation of fats is encouraged, whereas in high-intensity exercise oxidation of glycogen is encouraged.

Although of minor importance, the rate of protein oxidation is accelerated by a decrease in glycogen and blood glucose and an increase in the blood concentration of the hormone **cortisol**. Cortisol is known as a catabolic hormone, in that it stimulates the degradation of body protein stores. It will be discussed further in Chapter 9, which deals with overtraining.

The key regulatory enzyme for the ETC is **cytochrome oxidase**. Since it catalyzes the final reaction for all oxidative pathways, its activity can be considered to be the ultimate rate-controlling step. Its activity increases in proportion to the appearance rate of ADP, provided there is adequate oxygen available. Its activity decreases as oxygen availability decreases.

Possible Sites for Metabolic Limitations to Performance

Explosive Events (less than 2 seconds)

In those sports that call for maximal efforts over 2 seconds or less, energy is supplied almost exclusively from the resting store of ATP and from the hydrolysis of PCr. Such activities would include field events such as throwing and jumping, weight lifting, and the brief explosive movements that occur in team sports and in events like gymnastics and figure skating.

Theoretically, a metabolic limitation could occur at the creatine kinase step (the reaction that hydrolyzes PCr to rapidly regenerate ATP). In reality, however, the activity of creatine kinase is so high that the ratio of the concentration of PCr to Cr remains almost in equilibrium with that of ATP to ADP, thus ensuring almost instantaneous regeneration of ATP (39).

$$\mathrm{PCr} \rightarrow \mathrm{Cr} + \mathrm{P_i} \quad \text{such that} \frac{[\mathrm{PCr}]}{[\mathrm{Cr}]} \approx \frac{[\mathrm{ATP}]}{[\mathrm{ADP}]}$$

In addition, a metabolic limitation could theoretically occur if the PCr reserve in muscle were depleted prior to the beginning of the exercise bout. In most sports, however, the activity pattern is so brief that this would be a rarity. The exception would be situations where explosive bursts occur in rapid succession or in combination with sustained high-intensity activity (e.g., wrestling, or an unusually long period of uninterrupted high-tempo ice hockey, soccer, or basketball). In such situations, there might not be adequate recovery time between bursts for resynthesis of PCr and/or the activity of creatine kinase might be slowed because of the increase in $[H^+]$ generated by anaerobic glycolysis (39).

The question occasionally arises as to whether or not the creatine kinase reaction could be accelerated by elevating the muscle's store of PCr above its normal concentration (a popular claim made by some in the nutritional supplement business who market creatine monohydrate). The answer is no, since the K_m for this reaction is so low that, at normal resting PCr concentrations, the activity of creatine kinase is virtually at V_{max} (see Box 2.1).

We must therefore conclude that performance in very brief explosive events is influenced primarily by those factors that determine the strength and power of the muscle group (see Chapters 4 and 5) and only rarely limited by metabolic factors.

Figure 2.12 The proportion of energy derived from PCr hydrolysis and from anaerobic glycogenolysis during a maximal isometric contraction over 14 seconds. Note that, after the first 2 seconds, the energy contribution from the two pathways is almost equal. From (44).

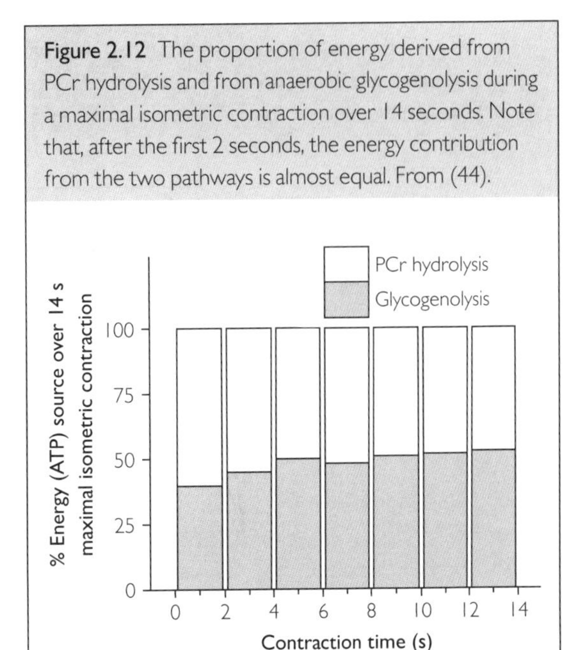

Maximal Efforts (over 12–15 seconds or less)

These include activities such as sprinting, hurdling, most team sports, and racquet sports. When the effort extends beyond 1–2 seconds, energy is derived in almost equal proportions from the high-energy phosphate system and from anaerobic glycolysis (see Figure 2.12). A potential metabolic limitation is thus the maximal rate at which the glycolytic enzymes can break down glycogen to lactic acid and generate ATP. As previously noted, the key regulatory enzymes here are phosphorylase, PFK, and LDH.

Enzyme kinetics are affected by a number of factors, including the concentration of the substrate for the enzyme to act upon and the concentration of the enzyme itself. Simply put, the greater the substrate concentration, or the greater the amount of enzyme, the greater the rate of the reaction (see Figure 2.13). With training, it is possible to increase the concentration of certain muscle enzymes and thus to accelerate their reaction rates. With training, it is also possible to increase the muscle concentration of certain exercise substrates. It is well known, for example, that resting muscle glycogen concentration can be increased by training or by a high-carbohydrate diet; consequently, it is attractive to hypothesize that this increase might result in an increase in the maximal rate of the phosphorylase reaction. Although in theory this is true, in actual fact the K_m of this reaction is so low that any manipulation of muscle glycogen concentration within the physiological range falls along the flat portion of the velocity curve (approaching V_{max}), so that further increases have no significant effect (20,64). Indeed, muscle glycogen concentration must be reduced to extremely low levels before it can be shown to exert an inhibitory effect on sprint performance.

If the effort is confined to a single sprint (e.g., high hurdles or 100 m), the duration is brief enough that the amount of lactic acid produced would not be sufficient to inhibit enzyme reaction rates. However, in repeated efforts, as in team sports, this becomes a major limitation, since recovery time between bursts is often not adequate for sufficient removal of the lactic acid

BOX 2.1

The K_m for an enzyme is the substrate concentration at which the reaction will proceed at 50% of its maximal possible rate (V_{max}). This concept is explained in more detail later in this chapter.

Figure 2.13 As the substrate concentration for an enzyme or the concentration of the enzyme itself increases, the reaction proceeds more rapidly. V_{max} refers to the theoretical maximal reaction rate that the enzyme can generate and K_m is the substrate concentration that results in 50% of V_{max}.

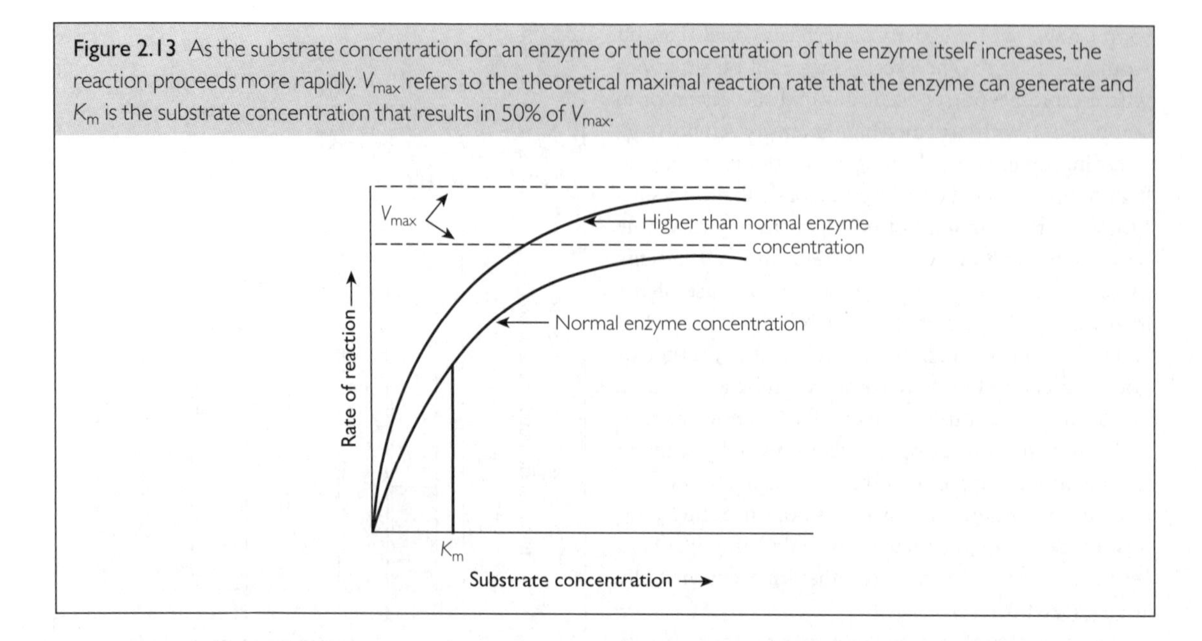

that has been generated, and each subsequent burst begins with an already elevated concentration of H^+. In such instances, the accumulation of H^+ in the muscle will have an inhibitory effect on regulatory enzymes such as PFK and slow the rate of anaerobic energy production.

To summarize, in maximal efforts over 12–15 seconds or less, an important metabolic limitation is the maximal rate at which the glycolytic enzymes can drive their various reactions. In repeated exercise bursts, the accumulation of H^+ also becomes a major limiting factor. In such activities, a training program that stimulates an increase in muscle glycolytic enzyme concentrations and/or an increased capacity to remove or buffer H^+ will have a beneficial effect on performance.

Sustained Sprinting (maximal efforts over 15–60 seconds)

These activities include longer sprints, such as 200 and 400 m in track, 50 and 100 m swimming events, and those team sports (e.g., ice hockey and soccer) where intense activity may be sustained for up to 60 seconds without a stoppage in play or a slowing in tempo. In such events, anaerobic glycolysis is the dominant pathway and performance will be determined by the maximal rate of glycolytic enzyme kinetics (anaerobic power) and the muscle's ability to tolerate the H^+ generated by the production of lactic acid (anaerobic capacity). In addition, one could theorize that an increase in resting muscle PCr concentration might delay depletion of this energy buffer and extend ATP production from this source. In support of this hypothesis, several studies have demonstrated greater short-term power outputs following creatine supplementation (4).

As the duration of the effort increases, the contribution of oxidative metabolism also increases, so that, in a maximal effort over 60 seconds, it can be estimated that approximately 30% of the total ATP is generated by the oxidation of muscle glycogen. Thus, factors that affect maximal aerobic power and the kinetics of oxygen delivery and utilization could also be expected to have a small but significant effect on total power output. Such factors also determine recovery rate during team sports, since lactic acid removal and the rate of PCr resynthesis are influenced by an athlete's maximal aerobic power.

Middle-Distance Events (up to 6 minutes)

These are events where maximal efforts are sustained for 1 to approximately 6 minutes. They include the 800 and 1500 m in track, 200 m swimming events, speed skating, rowing, and paddling. In these sports, maximum energy production is derived from both anaerobic and oxidative pathways. The crossover point occurs at approximately 2 minutes, which means that in a 2 minute effort (e.g., 800 m in track), approximately 50% of the total ATP would be derived from the anaerobic pathways and 50% from the oxidation of glycogen. Over a 4 minute effort (e.g., 1500 m in track), approximately 60% of the total energy would be derived aerobically and 40% anaerobically. In a 6 minute effort, such as heavyweight rowing, approximately 75% would be derived aerobically and 25% anaerobically, and so on. In such activities, maximal muscle concentrations of lactic acid are achieved and maximal oxygen uptake ($\dot{V}O_{2max}$) is attained after approximately 3 minutes.

In sports that fall into this category, the metabolic limitations to performance are those that limit anaerobic power (glycolytic enzyme kinetics) and capacity (H^+ tolerance) in addition to those that limit maximal aerobic power (oxygen availability and oxidative enzyme kinetics). An athlete's performance in these sports will also be affected by the rate at which $\dot{V}O_{2max}$ is achieved (sometimes referred to as $\dot{V}O_2$ kinetics). The more rapid this response, the greater the average power output over the duration of the event.

Endurance Events (up to 40 minutes)

These include distance running (5000 and 10 000 m), cross-country skiing, cycling, and each stage of the Olympic distance triathlon. In such events, almost all of the ATP is generated from oxidative pathways and exercise intensity would range from 100% $\dot{V}O_{2max}$ (or above) in events lasting approximately 10 minutes to approximately 75–85% $\dot{V}O_{2max}$ in a 40 minute event. Although muscle glycogen is the primary fuel for such activities, exercise is usually terminated before exhaustion of the glycogen stores.

Performance is largely determined by the athlete's $\dot{V}O_{2max}$ and, more importantly, by his/her **maximal aerobic capacity** (the ability to maintain exercise at a high percentage of $\dot{V}O_{2max}$). The major metabolic limitation to maximal aerobic capacity is the level of the athlete's **lactate threshold** (see Box 2.2). If exercise intensity exceeds the lactate threshold, time to fatigue is accelerated because of the accumulation of H^+ in the muscle and blood. In muscle, increased $[H^+]$ will inhibit the activity of certain regulatory enzymes and in blood it can inhibit the mobilization and uptake of fatty acids and stimulate an increase in pulmonary ventilation (Chapter 3). The lactate threshold can be increased through training, probably as a result of an increase in mitochondrial enzyme activity. Since there is metabolic generation of heat in such events, performance is also affected by the ambient temperature and humidity and the athlete's heat acclimatization status.

BOX 2.2

The **lactate threshold** is conventionally defined as the exercise intensity above which there is a non-linear accumulation of lactic acid in the blood. It is measured by sampling blood frequently while the athlete undergoes progressive increases in exercise intensity (see Chapter 11). Since lactic acid is removed by tissues such as liver and other muscle fibers, it is only when its production rate exceeds its removal rate that it begins to accumulate significantly in the blood. The level of the lactate threshold, whether expressed as a percentage of $\dot{V}O_{2max}$ or as an absolute exercise intensity, is known to increase with training. Lactate thresholds may range from 55% $\dot{V}O_{2max}$ in untrained individuals to as high as 85% $\dot{V}O_{2max}$ in well-trained endurance athletes. The lactate threshold is probably caused by several mechanisms. These include: (1) an inability on the part of the activity of the mitochondrial oxidative enzymes to keep pace with the activity of those of glycolysis, (2) an increase in the recruitment of type II muscle fibers (see Chapter 5), (3) a decrease in hepatic blood flow and lactic acid removal, and (4) inadequate oxygen supply to the muscle.

Long-Distance Events (several hours or more)

These include events such as the marathon, road cycling, Ironman triathlon, 50 km cross-country skiing, and ultra-distance running, swimming, or skiing events. In these sports, more than 99% of the energy is generated aerobically through the oxidation of both carbohydrates and fatty acids. As the duration of the event increases, the contribution of fat oxidation also increases. For example, over the course of a 2 hour 20 minute marathon, approximately 50% of the energy would be generated by the oxidation of muscle glycogen and plasma glucose and 50% by the oxidation of fatty acids. Over a 5 hour cross-country ski race, more than 70% of the total energy would be derived from fat oxidation. Average exercise intensities in these two events would range from approximately 80% of $\dot{V}O_{2max}$ for the well-trained marathoner to approximately 60% of $\dot{V}O_{2max}$ in the 5 hour skiing event.

In such sports, the major metabolic limitation is the exhaustion of body carbohydrate stores, both in muscle and in liver. Over the first hour of the above marathon, muscle glycogen would be depleted by approximately 55%; by the end of the second hour by approximately 72%; and at the finish line by more than 80% (11). In the ski race, the pattern of glycogen depletion would be similar, but extended over 5 hours because of the lower exercise intensity.

Over the first 30 or 40 minutes of these types of events, when muscle glycogen stores are still relatively high, uptake and oxidation of blood glucose provides a small portion of the total carbohydrate utilized. However, as muscle glycogen diminishes, blood glucose will make a progressively larger contribution. As glucose is taken up by the muscles, the liver must increase its release of glucose in order to maintain a constant blood concentration. This mobilization of glucose from the liver is stimulated by the hormones **glucagon** (released from the pancreas) and epinephrine (released from the adrenal glands), which increase in concentration in the blood as exercise progresses. In the well-trained and well-fed athlete, blood glucose can be maintained by this process for approximately 70–80 minutes. Thereafter, without oral glucose supplementation, it will begin to decrease as liver stores of glycogen become depleted. This decrease in blood glucose poses a double problem as the athlete attempts to continue exercise at the same pace: it accelerates muscle glycogen depletion and, more importantly, it deprives the brain and nervous system of its primary fuel source. A drop in blood glucose to abnormal levels is known as **hypoglycemia**. Thought processes and motor coordination are impaired and the condition, if unchecked, can result in coma. The importance of maintaining blood glucose and sparing muscle glycogen by ingesting carbohydrates during such events is discussed in Chapter 11.

Success in the long-distance events is thus profoundly affected by the athlete's initial muscle and liver stores of glycogen, as well as by the capacity to spare these stores by metabolizing a greater amount of fat early in the event. This latter ability is largely determined by the availability of fatty acids, the kinetics of the carnitine acyltransferase step, and the activity of the β-oxidation enzymes.

Metabolic Adaptations to Training

Endurance Training

The majority of investigations appearing in the scientific literature have focused on endurance training. In those studies that have used human subjects, the training stimulus is usually either **continuous** exercise at 65–80% of $\dot{V}O_{2max}$ for 30–60 minutes, or higher-intensity exercise (90–100% of $\dot{V}O_{2max}$) performed in **intervals** of 1–4 minutes. Training frequency is typically three to five times per week and the duration of the training period varies from several weeks to 5 or 6 months. The most common mode of training in a laboratory setting is cycle ergometry, although a number of studies have used running, swimming, or rowing as the exercise stimulus. The standard practice in such investigations is to obtain measurements of exercise performance, $\dot{V}O_{2max}$, and various muscle (by needle biopsy) and blood parameters, before and after the training period. These studies are often described as **longitudinal** studies and are considered the most effective design for examining the adaptations that occur in response to the training intervention. A second type of study involves examining the same parameters in a group of athletes who are already well trained, and then comparing them with those in a matched group of sedentary individuals. A study of this kind is known as a **cross-sectional** study and lacks the validity of the longitudinal design, since it does not control for inherited traits.

Substrate (fuel) supply A common adaptation to endurance training is an increase in the amount of glycogen stored in the trained muscles. Depending upon the intensity and duration of the training, glycogen concentration in humans may increase by as much as 100% or more (16,19), but increases in the 40–60% range are more common. Endurance training also causes an elevation in the concentration of liver glycogen. Because of the technical difficulty of obtaining direct measurements of liver glycogen concentration, there have been very few studies with humans (43,46), but this change has been well documented in rodents (3,40,67). The mechanisms for such adaptations probably relate to the repeated cycles of depletion (during each training session) and repletion (following each training session), since adaptation occurs only in those muscles that have performed the exercise. An increase in glucose transport protein (41) and perhaps an up-regulation of the enzymes for glycogen synthesis are thought to be the cause of this increase in glycogen storage.

Endurance training may also result in an increase in intramuscular fat content. This possibility has been suggested by a cross-sectional study of well-trained athletes and untrained subjects (26) but has not yet been demonstrated in a longitudinal investigation with humans.

Number and size of mitochondria It is well known that endurance training causes an increase in the number and size of the mitochondria in the trained muscles. In untrained subjects, the average mitochondrial volume density (the percentage of the fiber that is comprised of mitochondria) of the muscles of the thigh is approximately 5%, ranging from about 3.5% in type 2 or fast-twitch fibers to about 6% in type 1 or slow-twitch fibers (2). Using electron microscopy to identify the mitochondria, investigators have demonstrated in longitudinal training studies that mitochondrial density can be expected to increase by 50–60% (25,27,58) and, in one cross-sectional study, mitochondrial density in well-trained endurance athletes was more than double that of sedentary control subjects (26).

Oxidative enzymes Since endurance training causes an increase in the number of mitochondria in each muscle fiber, it is not surprising that there is also a parallel increase in the total amount of the various mitochondrial enzymes. This increase in enzyme concentration means that there will now be a greater oxidative reaction rate for a given concentration of pyruvic acid, or FFA (Figure 2.13). Stated another way, for the same oxygen consumption, the addition of more mitochondria ensures that each mitochondrion now functions at a lower percentage of its maximal capacity during exercise and thus has a greater reserve.

A number of investigators have documented an increase in maximal activity of the enzymes of the Krebs cycle in human muscle following training (16,23,38,47,49). These changes generally parallel the increases in mitochondrial volume density and are accompanied by an increase in enzyme activity for the ETC as well (23,65). As with mitochondrial density, cross-sectional studies have shown that oxidative enzyme activities of elite endurance athletes are more than double the levels found in untrained subjects (13).

The increase in mitochondrial density also results in an increase in the activity of the enzymes responsible for β-oxidation (17), as well as carnitine transferase, the enzyme responsible for the transport of long-chain FFAs into the mitochondria (6,10). More mitochondria also result in a greater total mitochondrial surface membrane area for FFA transport.

As a result of these adaptations, glycogen can be oxidized more rapidly during maximal exercise, while at lower exercise intensities there will be an increase in the proportion of FFA oxidation, thus sparing muscle glycogen and blood glucose during endurance events.

Glycolytic enzymes Moderate-intensity endurance training (>75% $\dot{V}O_{2max}$) appears to have a minimal effect on the enzymes of the glycolytic pathway (7,15,22,24); however, following endurance training that included high-intensity intervals, significant increases in the maximal activity rate of PFK have been noted (38).

Muscle fiber types Although, in certain animal models, chronic low-frequency stimulation of skeletal muscle by implanted electrodes is known to result in a transformation of fast-twitch fibers to slow (48), it is generally accepted that endurance training in humans has little or no effect on the proportion of slow-twitch or type 1 fibers. It can, however, affect the proportions of the type 2 subtypes, resulting in an increase in the proportion of fibers that are classified as 2A at the expense of those classified as 2X. Theoretically, such an adaptation should result in a lower production of lactic acid for a given recruitment of type 2 motor units and may thus be a mechanism for the elevation of the lactate threshold following training.

The endocrine system Endurance training has a marked effect on the blood concentration of a number of hormones that indirectly control muscle metabolism. This is most apparent in the stress-response hormones: epinephrine and **norepinephrine** (these are also often called **catecholamines**; note that norepinephrine is also known as noradrenaline); cortisol; and the glucose-regulatory hormone glucagon. When exercise exceeds approximately 60% of $\dot{V}O_{2max}$, the concentrations of these hormones begin to increase exponentially with exercise intensity (Figure 2.14). Short-duration (e.g., 20 minutes), moderate-intensity (e.g., 55% $\dot{V}O_{2max}$) exercise has little or no effect on these hormones, but if that exercise is continued for a prolonged period (e.g., 45 minutes or more) they then begin to increase with exercise duration. Following training, the same absolute exercise intensity elicits a reduced response in these stress hormones (Figure 2.15). This blunting effect downregulates glycogen mobilization in both muscle and liver, thus sparing

Figure 2.14 The changes in blood concentration of the catecholamines epinephrine and norepinephrine as a subject performs progressive exercise to exhaustion. Although not illustrated, the changes in cortisol and glucagon would show a very similar pattern, but of much smaller magnitude.

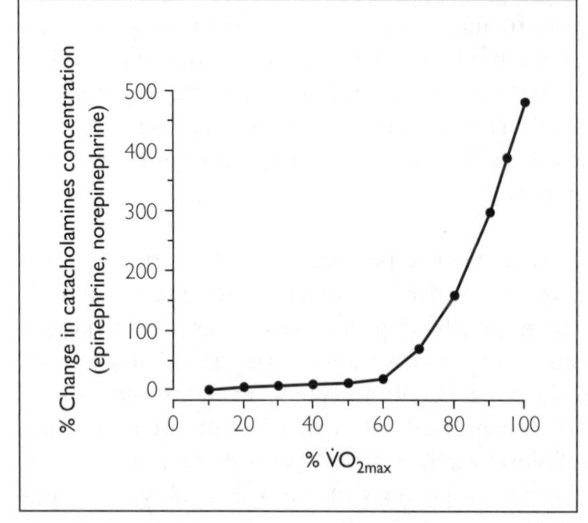

glycogen and protein stores, while at the same time facilitating more fatty acid utilization.

Following training, when an athlete exercises at the same relative intensity (i.e., the same percentage of the new $\dot{V}O_{2max}$), hormonal concentrations will be the same as, or higher than, before training (19). At maximal exercise, trained subjects thus exhibit higher concentrations of the stress-response hormones than before training.

Sprint Training

"Sprint training" is a general term that is used to describe repeated intervals at maximal or near-maximal effort. Since the power outputs that are generated during such intervals considerably exceed those that can be achieved at 100% $\dot{V}O_{2max}$ (where, theoretically, all of the ATP is supplied oxidatively), this is sometimes known as supra-maximal training. The duration of the exercise intervals may vary considerably between one program and another and may be as short as 5 seconds or as long as 45 seconds. Average power output (rate of ATP production) would also vary depending upon the duration of the interval. For example, over a 5 second interval, maximal power output would be approximately three times greater than can be achieved aerobically, whereas, over a 45 second interval, power output would be approximately two times greater. These averages would decline as the athlete fatigues while performing subsequent intervals. Recovery times between intervals and the number of intervals completed in a given training session vary considerably as well. In contrast to the literature on adaptations to endurance training, there are relatively few studies of the effects of sprint training in humans.

Figure 2.15 The changes in blood catecholamines concentration as a subject performs prolonged exercise at the same absolute intensity (75% of the pre-training $\dot{V}O_{2max}$) before and after several months of endurance training.

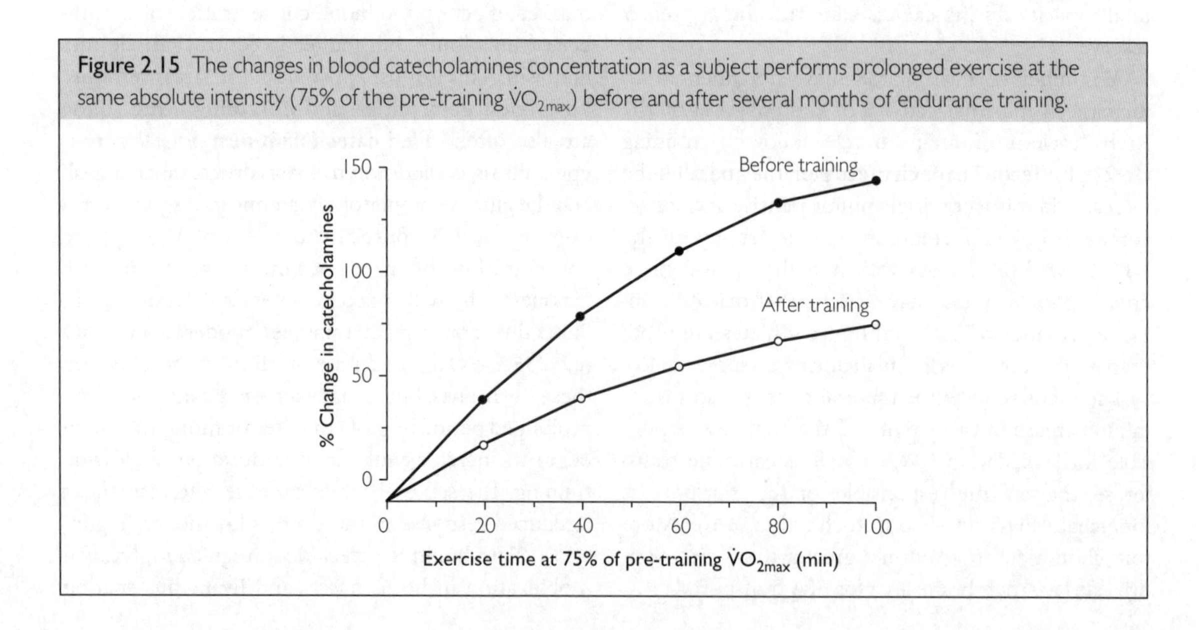

Substrate (fuel) supply As with endurance training, sprint training results in an increase in muscle glycogen storage (1,9,29). Although it has not yet been examined in single fibers, it is probable that this increase would be most evident in the type 2 motor units, since these fibers would be expected to deplete their glycogen more rapidly than type 1 fibers during such training. Sprint training appears to have minimal effect on the PCr concentration of resting muscle (14).

Muscle enzymes Sprint training is known to result in an increase in the activity of the allosteric glycolytic enzymes PFK, hexokinase (9,33,50), and phosphorylase (9). In addition, several authors have documented an increase in LDH activity (30,50), but this is not a consistent finding (9,33). The significance of the increased activity of these regulatory enzymes is that, during maximal efforts, the athlete can achieve greater rates of ATP production from anaerobic glycolysis and thus higher peak and average short-term power outputs (33).

In addition to its effect on glycolytic enzymes, sprint training can also have a profound effect on certain *oxidative* enzymes. Following training programs involving 20–40 second intervals, increased activity has been noted for a number of Krebs cycle enzymes, including citrate synthase, **succinate dehydrogenase**, and **malate dehydrogenase** (8,12,33,50). These findings are somewhat surprising given the relatively short total duration of exercise (usually less than 5 minutes in a typical training session) and the relatively minor contribution of oxidative metabolism, and suggest that increased pyruvic acid flux may be the stimulus for upregulating these mitochondrial enzymes (33).

Muscle buffering capacity One of the most effective adaptations to sprint training may be enhanced muscle buffering capacity (45). A **buffer** is a substance that functions to minimize a change in pH. As pH tends to decrease, buffers remove H^+ ions and, as pH tends to increase, they respond by releasing H^+. They accomplish this either by converting strong acids into weak acids (as in the bicarbonate buffer system) or by acting as proton acceptors that neutralize the H^+ (as in protein buffers). Within skeletal muscle, substances that act as buffers include bicarbonate, phosphate compounds such as ATP and PCr, amino acids such as histidine, and muscle proteins such as carnosine and myoglobin (28,45). An increase in the muscle concentration of any of these substances would mean that, for a given production of lactic acid, there would be a lower concentration of free H^+ ions. Thus, following sprint training, the athlete would be able to tolerate a greater production of lactic acid before fatigue.

By examining the ratio of the muscle's lactic acid concentration to its H^+ concentration following intense exercise, several investigators have shown that sprint training improves muscle buffering capacity and performance (5,54). The mechanism causing this adaptation is not known but, since buffering capacity is not enhanced following endurance training (54), it may be related to the repeated bouts of muscle acidosis that are specific to sprint training.

Resistance Training

Resistance training refers to repeated contractions of certain muscles or muscle groups, where force development is 60–100% of maximum (see Chapter 8). The resistance may be provided by conventional free weights, devices that utilize weight stacks or hydraulic resistance, elastic cables, or the athlete's own body weight. The exercises are usually performed in multiple **sets** of three to 15 **repetitions**. For example, for a given exercise, a typical training session might include four sets of eight to 10 repetitions, with 3–4 minutes between sets, performed three times per week.

As detailed in Chapter 8, the most obvious adaptation to a program of resistance training is the increase in muscle size (hypertrophy) that occurs. This increase in size is due to an increased cross-sectional area of the muscle fibers in response to greater synthesis of contractile protein. Hypertrophy occurs in all fiber types, but there is a greater relative increase in the size of the type 2 fibers. In addition to these structural changes, there are also a number of biochemical adaptations to resistance training.

Substrate (fuel) supply Resistance training results in an increase in resting muscle glycogen and PCr concentration (36). Since anaerobic glycolysis is the primary metabolic pathway for resistance exercise (34), as with endurance and sprint training, the repeated cycles of glycogen usage and restoration may be the stimulus for increased glycogen storage. It is not yet known whether the increase in total muscle PCr concentration is due to increased concentrations in each fiber, or to the greater relative hypertrophy of the type 2 units (which have higher PCr concentrations than type 1 fibers).

Muscle enzymes Resistance training appears to have little or no effect on the activity of the glycolytic enzymes (60,63) and may even cause a reduction in the activity of oxidative enzymes per gram of muscle tissue (61,62). Although recent studies suggest that there is some increase in mitochondrial protein synthesis when subjects first begin a resistance training program, as training progresses this response is attenuated, while the synthesis of myofibrillar protein continues (66). As a result, there is probably a dilution in mitochondrial volume density with increased contractile protein (10,31,35). The functional significance of this change is that, when resistance training has resulted in significant muscle hypertrophy, there would be an actual *reduction* in oxidative capacity relative to muscle mass and such training could therefore prove counterproductive for the endurance athlete.

Muscle fiber types Resistance training does not alter the proportion of type 1 to type 2 fibers (32), but normally results in an increase in the proportion of type 2A fibers at the expense of the type 2X population (56,57).

Summary of Key Points

- The metabolic pathways that provide energy for muscle contraction differ as to their maximal rates of ATP generation. The recruitment of these pathways is tightly controlled by specific enzymes, so that the rate of ATP production during exercise matches its rate of utilization.
- In brief explosive activities, such as throwing or jumping, energy can be supplied almost exclusively from the resting stores of ATP and from the hydrolysis of PCr. Performance in such activities is limited primarily by the strength and power of the muscle group, rather than by metabolic characteristics.
- In sprint-type events and team sports where the effort extends beyond 1–2 seconds and may last as long as 12–15 seconds, energy is derived in approximately equal portions from PCr hydrolysis and anaerobic glycogenolysis. Performance in these activities will be limited by the initial concentration of PCr in the muscle and by the maximal rate at which the glycolytic enzymes can break down muscle glycogen to lactic acid. If such bursts are repeated in rapid succession, as in many team sports, the accumulation of H^+ ions and other metabolic byproducts in the muscles will also be a major limiting factor to performance.
- In longer-duration sprints, or in those team sports where maximal effort may be sustained for 15–60 seconds, anaerobic glycogenolysis is the dominant pathway. Performance will be determined primarily by the maximal activity of the glycolytic enzymes and the muscle's ability to tolerate the increase in H^+ generated by the production of lactic acid. Performance will also be affected by the initial PCr concentration in the muscle (in shorter efforts) and by the kinetics of oxygen delivery and utilization (in efforts of longer duration).
- Middle-distance events that require a peak effort for 1 to approximately 6 minutes demand maximal energy delivery from both anaerobic and oxidative pathways. Consequently, muscle glycogen is the major fuel source and limitations to performance include maximal glycolytic enzyme kinetics, tolerance to H^+, and those factors that determine maximal aerobic power (oxygen availability and oxidative enzyme kinetics).
- In endurance events that are typically completed in 40 minutes or less, almost all of the energy is derived from the oxidation of muscle glycogen, with minor contributions from the oxidation of blood glucose and fat. Performance is thus determined largely by the athlete's maximal aerobic power ($\dot{V}O_{2max}$) and, more importantly, by his/her ability to maintain exercise at a high percentage of $\dot{V}O_{2max}$ (maximal aerobic capacity). The latter is dependent upon the level of the athlete's lactate threshold, which may be affected by a number of mechanisms, including the inability of the kinetics of the oxidative enzymes to keep pace with those of glycolysis and a decrease in lactic acid removal by muscle and liver.
- In longer-distance events that last several hours or more, the energy is generated by the oxidation of both carbohydrate (muscle glycogen and blood glucose) and fat. The major metabolic limitation

is the exhaustion of carbohydrate stores in muscle and liver. Because fat oxidation requires more oxygen per unit of ATP produced and because the brain and nervous system rely on blood glucose as a substrate, performance is profoundly affected by the athlete's initial muscle and liver glycogen stores, as well as by the capacity to spare these stores by metabolizing a higher proportion of fat early in the event. This latter ability is limited by the activity of the enzymes responsible for FFA mobilization, uptake and oxidation.

- Most of the preceding limitations to performance are affected by appropriate training. For example, endurance training is known to increase muscle glycogen and lipid stores, the number and size of the mitochondria, and the maximal activities of the enzymes involved in oxidation of glycogen and FFAs. Sprint training also elevates muscle glycogen stores, increases the maximal activity of both glycolytic and oxidative enzymes, and can improve muscle buffering capacity and the athlete's ability to tolerate H^+. Resistance training causes an increase in muscle size and contractile protein and an increase in glycogen and PCr concentration, but can result in a decrease in oxidative enzyme activity relative to muscle mass.

3

Cardiorespiratory Bases for Performance

The cardiovascular and respiratory systems help to maintain body homeostasis during exercise in a number of different ways. From the perspective of the athlete, however, their most important function is to deliver oxygen to the exercising muscles and to remove carbon dioxide, while at the same time maintaining blood flow to vital organs such as the brain and the heart. The cardiovascular system does this by matching the blood flow to skeletal muscle to its metabolic rate, by increasing cardiac output and mean blood pressure, and by regional adjustments in peripheral resistance.

In simplest terms, the cardiovascular system consists of a pump (the heart), a series of distributing and collecting tubes (the arteries and veins), and a system of thin-walled vessels (the capillaries) that permit rapid exchange of gases and other substances between the blood and the tissues of the body. The system involves two circuits that are arranged both in parallel and in series: the **pulmonary circuit**, which conducts blood from the right side of the heart to the lungs and back to the left side of the heart; and the **systemic circuit**, which conducts blood from the left side of the heart to all the tissues in the body and back to the right side of the heart (Figure 3.1). The arteries carry blood away from the heart and the veins carry blood toward the heart. The walls of the arteries and veins contain smooth

Figure 3.1 A schematic illustration of the cardiovascular system and the heart. The system consists of two separate circuits: the pulmonary circuit, which carries blood from the right ventricle through the lungs and back to the left atrium, and the systemic circuit, which carries blood from the left ventricle to all tissues in the body and back to the right atrium.

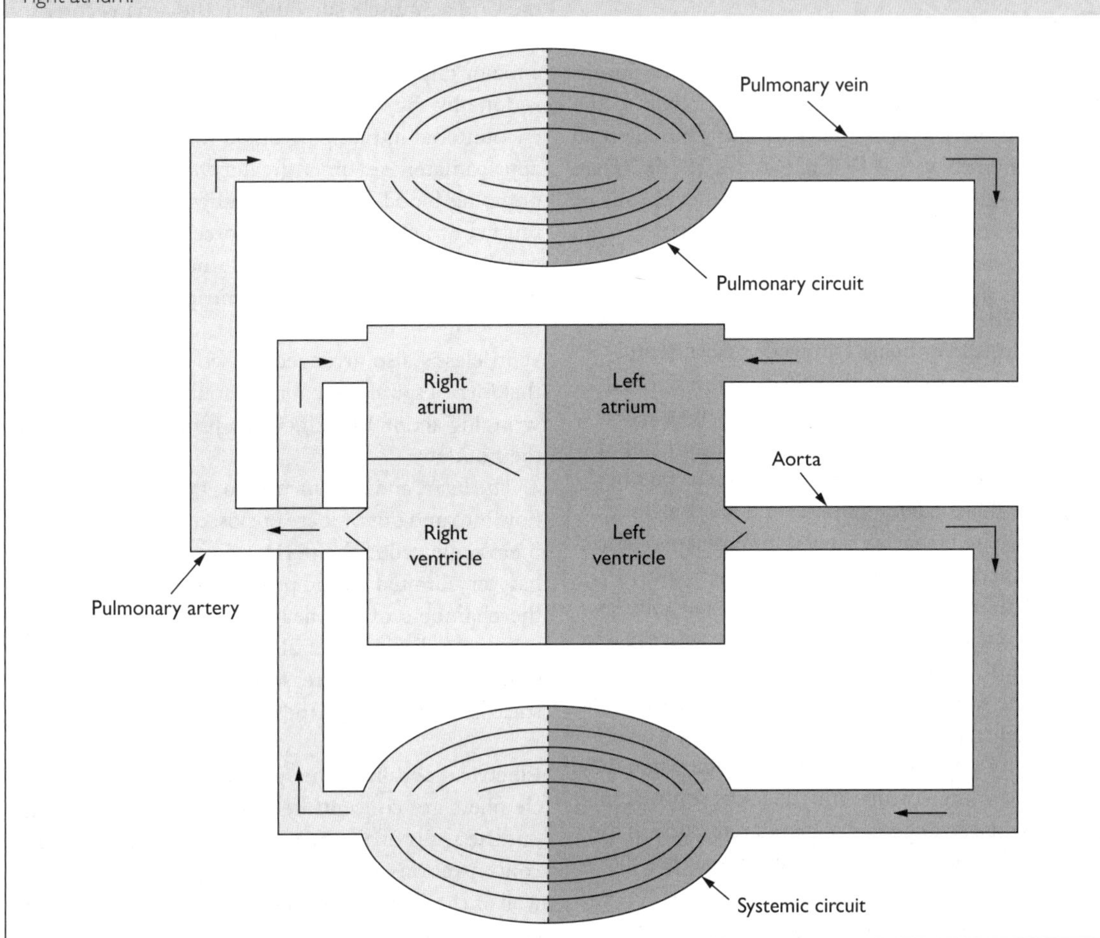

muscle that is innervated by the autonomic nervous system. Contraction or relaxation of this smooth muscle changes the diameter of the blood vessel and thus the amount and velocity of the blood that flows through it.

Complex interactions between the autonomic nerves that regulate the rate at which the heart pumps blood and those that control the diameter of the blood vessels allow muscle blood flow to increase during exercise to as much as 30 times its resting value. In some sports and activities, however, even this tremendous reserve for increasing blood flow to muscle is still inadequate to meet the muscle's oxygen requirements. The maximum rate at which oxygen can be delivered to the exercising muscle is determined essentially by three factors:

1. the amount of blood that can be pumped by the heart in a given period (the **cardiac output**),
2. the proportion of this cardiac output that can be directed to the muscle, and
3. the amount of oxygen carried by the arterial blood.

Each of these factors will be examined in detail in this chapter.

Factors That Determine Cardiac Output

The Heart

Functionally, the heart consists of two separate pumps in series: the right heart, which pumps blood to the lungs, and the left heart, which pumps blood to all other tissues (Figure 3.1). Cardiac muscle fibers are similar to skeletal muscle fibers in that they consist of myofibrils containing contractile filaments of myosin and actin, arranged in sarcomeres (see Chapter 4). They differ, however, in a number of ways.

- First, instead of being arranged in parallel and joining bone to bone, they form an interlocking lattice arrangement, where one fiber is attached end-to-end to another by means of **intercalated discs**. These discs are formed by the fusion of two or more cell membranes and allow electrical impulses to be passed on with minimal resistance. Because of its lattice-type arrangement (or **syncytium**), the heart follows the all-or-none law: when one fiber depolarizes and contracts, the electrical impulse is passed on to all the fibers, so that the whole heart contracts as if it were a single large fiber. Unlike skeletal muscle fibers, which cause movement to occur about a joint when they contract, when cardiac muscle fibers contract their syncythial arrangement causes a reduction in the size of the cavity that they enclose and thus the ejection of blood.
- Secondly, in addition to having contractile properties, some fibers have electrical generating and conductance properties. Consequently, the heart does not require external nerve impulses in order to contract, but generates its own beating.
- Thirdly, cardiac muscle fibers are richly endowed for oxidative metabolism, with mitochondrial volume densities and capillary densities up to four and five times those of skeletal muscle fibers.

The human heart has four cavities: the right and left **atria** and the right and left **ventricles**. The atria act as conduit chambers, through which blood passes on its way to the corresponding ventricles. Compared to the ventricles, the forces generated by atrial contraction are quite low; in fact, they contribute only about 10% to the total filling of the ventricles, with most of the blood simply being sucked through the atria as a result of the negative pressures created by the increase in ventricular volume during relaxation. For this reason, the cardiac muscle surrounding the atria is considerably thinner than that of the ventricles. There is also a major difference between the amount of work that is done by the two ventricles. Because the resistance encountered in pumping blood to the whole body is approximately six times greater than that involved in pumping blood through the lungs, the left ventricle must contract six times more forcefully than the right ventricle to eject the same volume of blood. Consequently, the wall of the left ventricle is much thicker than that of the right ventricle. The muscle of the ventricles is also arranged in two overlapping layers that form a spiral, so that their contraction produces a wringing action that efficiently propels blood toward their outlets.

The heart and the junctions of the large vessels that flow into and out of it are enclosed in a thin but tough membrane called the **pericardium**. The pericardium has very limited elastic properties and thus prevents the chambers of the heart from overstretching and becoming distended. It also acts as a pressure buffer so that any sudden increase in the pressure of one ventricle is transmitted to the other.

Electrical activity The specialized neural-type muscle fibers are concentrated in certain areas throughout the walls of the heart (Figure 3.2). One of these concentrations occurs in the upper portion of the wall of the right atrium and is known as the **sinoatrial node**, or **SA node**. A second concentration is found in the lower portion of the right atrium close to the septum that separates it from the right ventricle. This is termed the **atrioventricular node**, or **AV node**. Extending from the AV node is an elongated bundle of fibers that runs through the septum separating the two ventricles and is known as the **bundle of His**. This conducting system divides into the right and left bundle branches, terminating in further branches known as **Purkinje fibers** that extend to all parts of the ventricles.

Each of these specialized fibers is capable of generating its own action potential, but those in the SA node are more excitable than those in other parts of the heart and fire at an intrinsic rate of approximately 70 times per minute. (The fibers comprising the AV node fire approximately 60 times per minute and those in

Figure 3.2 The neural conduction system of the heart. Under normal conditions, conduction of the action potential begins at the SA node, is passed on to the AV node, and then to the bundle of His and to the Purkinje fibers throughout the ventricles. © McGraw-Hill Education.

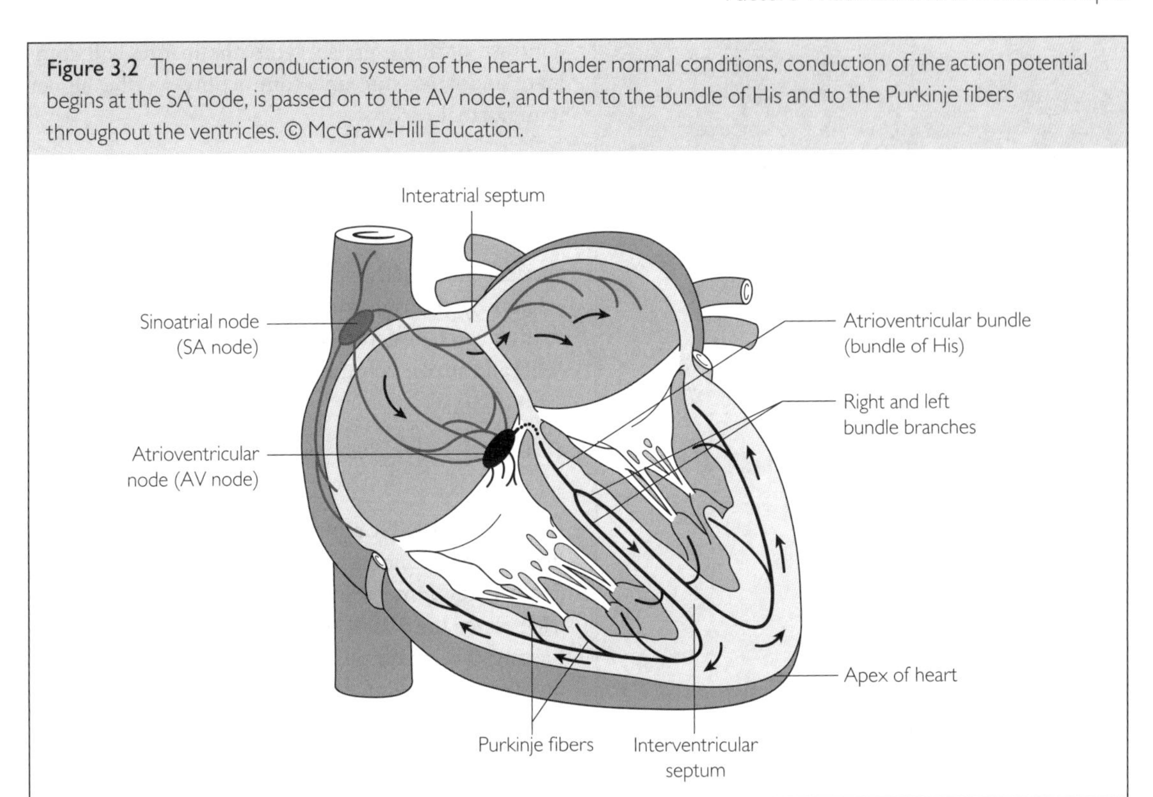

the ventricles approximately 40 times per minute.) Because of its more rapid firing rate, the action potentials generated by the SA node upstage and override those that would have been generated further down the heart and act as the pacemaker that sets the timing of the heartbeat.

Nerve supply Although the heart generates its own contraction, it also receives nerve impulses from cardiovascular control centers in the brain by way of the autonomic nervous system. The cardiovascular control centers are located in the medulla and consist of two different types of neuron: some that exert an inhibitory effect on the heart's normal intrinsic rate (the cardiac inhibitory center, CIC) and others that stimulate it (the cardiac acceleratory center, CAC). These give rise to the two separate components of the autonomic system: the sympathetic branch and the parasympathetic branch, respectively (Figure 3.3). The parasympathetic fibers originate in the cardiac inhibitory center and travel by way of the vagus nerve (tenth cranial nerve) to focus primarily on the SA and AV nodes, although some fibers innervate the ventricles as well. In contrast, the sympathetic fibers originate in the cardiac acceleratory center, descend the spinal cord and branch off to form synapses with additional sympathetic fibers in the cervical and upper thoracic portion of the sympathetic ganglion chain. These postganglionic fibers then richly innervate all parts of the heart, including the SA and AV nodes, with the greatest concentration of fibers going to the ventricles.

When activated, the sympathetic fibers release **norepinephrine** at their nerve endings, whereas the parasympathetic fibers release **acetylcholine** (**ACh**). These neural transmitters then combine with receptors on the myocardium to affect heart rate and the force of contraction. The release of ACh at the SA and AV nodes increases the permeability of these cells' membranes for K^+. This results in a more negative resting potential, or **hyperpolarization**, which delays the rate at which depolarization occurs, thus slowing heart rate. ACh released at the nerve endings in the ventricles also results in a slight reduction in the force of the cardiac contraction.

In contrast, the release of norepinephrine accelerates the rate at which the SA node depolarizes, increasing heart rate and increasing the force of ventricular

Figure 3.3 The heart is connected to the cardiovascular control centers in the medulla by the autonomic nervous system. It receives impulses from the cardiac inhibitory center (CIC) by way of parasympathetic nerve fibers and from the cardiac acceleratory center (CAC) by way of sympathetic fibers that have synapsed in the sympathetic ganglia. When activated, the sympathetic fibers release norepinephrine at their nerve endings and the parasympathetic fibers release acetylcholine. Notice also that the CIC and CAC interact in a negative feedback loop, so that when one center is activated the other is inhibited.

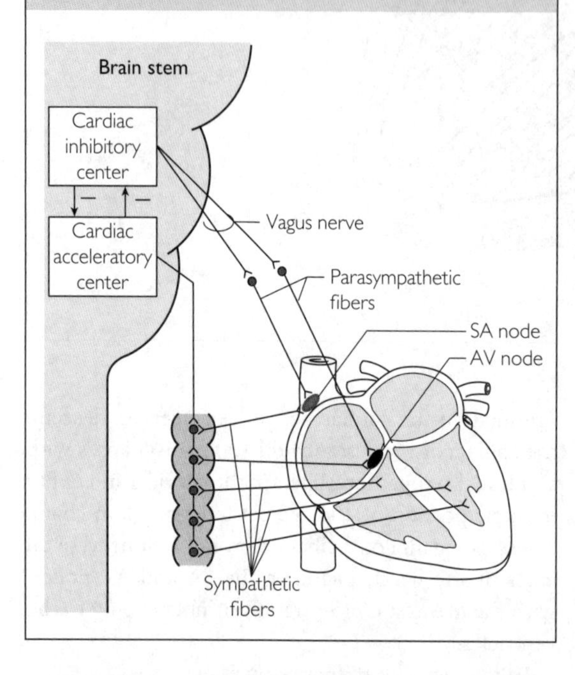

contraction. These effects are thought to be due to enhanced membrane permeability to Na^+ and an increased Ca^{2+} release by the sarcoplasmic reticulum of the cardiac muscle fibers, respectively. Thus, the two branches of the autonomic system exert opposing effects on heart rate and contractility, with sympathetic activity causing heart rate and contractile force to increase above their normal intrinsic levels and parasympathetic activity causing them to decrease. At any given point in time, there is always some activity, or tone, present in both systems, and cardiac function is determined by their *net* effect. For example, in the resting state, parasympathetic tone predominates over sympathetic tone, and heart rate and force of contraction will be lower than their normal intrinsic values. In contrast, during exercise, sympathetic tone will dominate and heart rate and contractility will be higher than normal.

Cardiac Output

Cardiac output (Q_c)[1] is the amount of blood that is ejected by the right or left ventricle per minute. As such, it can be calculated as the product of heart rate (HR) and stroke volume (SV):

$$Q_c\,(\text{L}\cdot\text{min}^{-1}) = \text{HR (beats per minute)} \times \text{average SV (mL)}$$

Although cardiac output is normally thought of as the volume of blood pumped by the left ventricle, in fact it is the same volume pumped by the right ventricle, since the SVs of the two ventricles are identical. Resting cardiac output in a normal-sized adult is approximately 5 $\text{L}\cdot\text{min}^{-1}$ and may increase to more than 35 $\text{L}\cdot\text{min}^{-1}$ in a well-trained athlete during maximal exercise.

Heart Rate

The time from the beginning of one ventricular contraction to the beginning of the next is called the **cardiac cycle**. The period during which the ventricles contract is called **systole** and the period during which they are relaxed is called **diastole**. In the resting state, the heart is in diastole for approximately 65% of the cardiac cycle. As heart rate increases, the cardiac cycle shortens as a result of a decrease in both systole and diastole, with a proportionally much greater decrease in diastole: at a heart rate of 180 beats per minute (bpm) diastole would account for only about 30% of the cardiac cycle. Thus, increases in heart rate occur largely at the expense of a decrease in diastole.

Resting heart rate for a typical untrained individual is approximately 70–80 bpm and is determined by the

[1] Q_c, the symbol for cardiac output, refers to the minute blood flow through the pulmonary capillaries of the lungs, which can be measured relatively easily using non-invasive techniques. Since all of the cardiac output of the right ventricle flows through the lungs and is equivalent to that of the left ventricle, Q_c equals the cardiac output of the left ventricle.

net balance between parasympathetic and sympathetic tone. Resting heart rate for a highly trained endurance athlete might be as low as 35–40 bpm because of a more dominant parasympathetic (vagal) tone. As an individual goes from the resting state to the exercise state there is an immediate increase in sympathetic tone and reduction in parasympathetic tone. These changes are due to the excitation of the cardiovascular control centers in the brain by the nerve impulses traveling from the motor cortex to the skeletal muscles. This has been described as a form of feed-forward **central command** and is initiated instantaneously with the commencement of exercise. The sudden reduction in vagal tone, together with the increase in sympathetic tone, causes an immediate increase in heart rate that is proportional to the neural traffic descending from the motor cortex and thus the exercise intensity. As exercise continues, there is a further increase in sympathetic activity that is stimulated by feedback from mechanical and chemical receptors in the muscles themselves, resulting in a further upward adjustment of the heart rate response. This has been referred to as feedback **peripheral command** (30).

The time course for the increase in heart rate is dependent upon the type and intensity of exercise. If, for example, the exercise being performed is at a constant, submaximal intensity (e.g., 60% of $\dot{V}O_{2max}$), most of the increase in heart rate (approximately 80%) would be apparent within 20 seconds of beginning exercise, with the final fine-tuning occurring over the next 90 seconds, so that the heart rate plateaus in approximately 2 minutes (Figure 3.4). If exercise were to continue at this intensity for more than 10 minutes, there would be an additional gradual increase in heart rate in response to the norepinephrine released into the blood by the adrenal glands. The circulating norepinephrine affects the receptors in the SA and AV nodes in the same way as that released at the sympathetic nerve endings, leading to a further increase in heart rate and contractility.

If, in the above example, the athlete were to continue the exercise for a prolonged period (e.g., from 40 minutes to 1–2 hours), there would be a further increase in heart rate over time. This phenomenon is known as **cardiovascular drift** and is reflexively related to the progressive decline in SV that occurs, so that cardiac output can be maintained (26). The decrease in SV is thought to be caused by reduced venous return to the heart, due to a reduction in plasma volume lost by sweating (31). The magnitude of the drift is thus affected by ambient temperature and humidity and can be attenuated by adequate fluid intake during exercise.

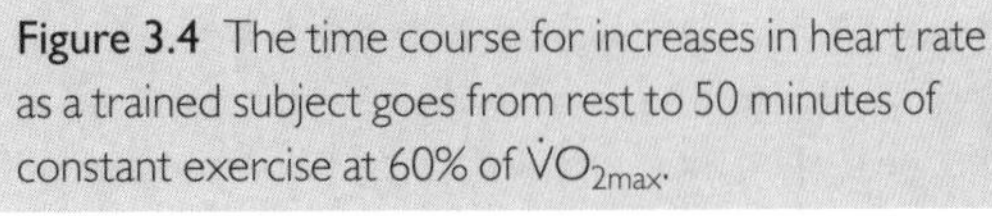

Figure 3.4 The time course for increases in heart rate as a trained subject goes from rest to 50 minutes of constant exercise at 60% of $\dot{V}O_{2max}$.

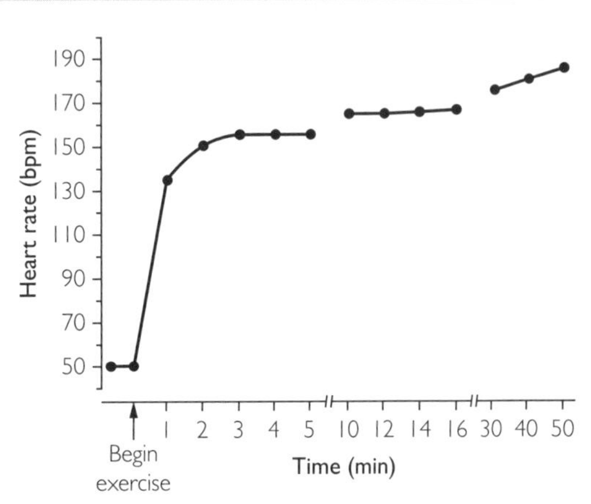

In competitive situations (for example, prior to a 100 m sprint or the opening bell in a boxing match), heart rates often exceed 130 bpm before exercise even begins. This anticipatory response is due to heightened sympathetic activity originating from higher centers in the brain and accelerates the rate at which the heart rate response occurs upon beginning exercise.

Maximal heart rate Once cardiac muscle cells have depolarized, there is a brief period during which the cell cannot be depolarized again. This period is known as the **absolute refractory period** and is similar in duration to the actual contraction period, thus protecting the cell from random electrical activity, serial contractions, and tetany. This suppressive effect is particularly evident in the cells of the AV node and sets an upper limit on the rate at which the heart is able to contract. Without such a mechanism, heart rate would become too rapid during intense exercise to allow adequate time for ventricular filling between contractions. This upper limit or **maximal heart rate** varies considerably among individual athletes and declines almost linearly with age. The reason for this decrease as individuals age is not fully understood, but probably reflects a change in the refractory period of the cells at the AV node. An approximate estimate of maximal heart rate can be gained by subtracting the athlete's age from 220 bpm (see Figure 3.5). For

Figure 3.5 The decline in maximal heart rate that occurs with age. The solid line indicates the mean value and the dashed lines the standard deviation about the mean (approximately ±10 bpm).

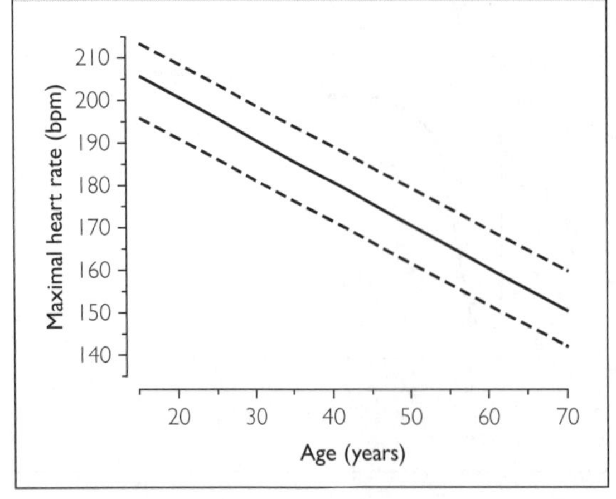

Figure 3.6 The heart rate recovery response for a trained and untrained subject following constant submaximal exercise at the same relative intensity. Note that the trained subject recovers more quickly than the untrained subject.

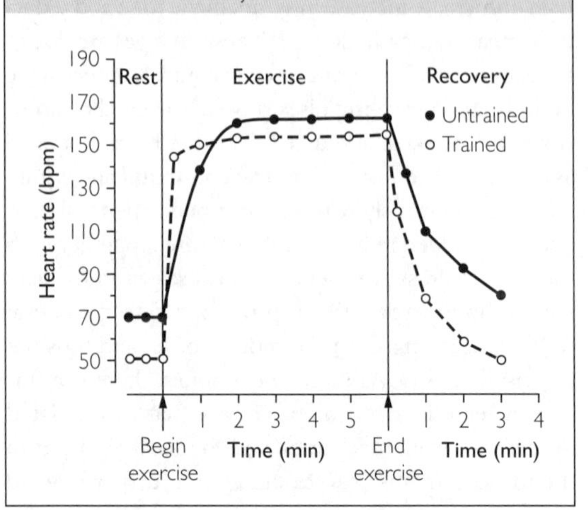

example, maximal heart rate would be approximately 205 bpm for a typical 15-year-old athlete, 195 bpm for a 25-year-old, and 180 bpm for a 40-year-old. The standard deviation around this method, however, exceeds ±10 bpm (4). Thus an athlete's true maximal heart rate can be determined only by measuring it during maximal exercise to exhaustion.

Recovery heart rate Following exercise, the mechanisms that caused heart rate to increase are rapidly switched off, so that the recovery response is similar in pattern to the increase at the beginning of the exercise. As a result, the heart rate recovery curve is almost the mirror-image opposite of its increase curve (Figure 3.6). The time course to regain normal resting heart rate will, however, be dependent upon the intensity and/or duration of the exercise. For example, the higher the exercise intensity, the longer the recovery time. Similarly, if moderate-intensity exercise is continued for a long duration, so that there has been a significant increase in body temperature, recovery time will be delayed. It is also well known that recovery heart rate is affected by the athlete's training status and that well-trained endurance athletes will recover more quickly than lesser-trained individuals (Figure 3.6).

Stroke Volume

As the ventricles relax, their volumes increase, causing the pressure in the chambers to drop to zero. Because this pressure is lower than the blood pressure in the veins returning blood to the heart, blood flows into them during diastole, causing them to fill. In the resting state, at the end of diastole, the ventricles of a normal-sized athlete will each contain approximately 150 mL of blood. When they contract, the pressures in the ventricles increase above the levels in the large arteries leading away from them,[2] causing blood to be pumped into the arteries.

In the resting state, normally about two-thirds of ventricular volume, or about 100 mL, is emptied with each contraction. This is the **stroke volume**, or SV, of the heart. The SV, expressed as a percentage of the ventricular volume at the end of diastole, is known as the **ejection fraction**. With exercise, several events occur that result in an increase in SV. These include both an increase in ejection fraction and an increase in end-diastolic ventricular volume.

Factors Affecting Stroke Volume

During exercise, SV is determined by two factors: (1) the amount of ventricular filling that occurs during

[2] Blood flows into the right heart from the superior and inferior venae cavae and into the left heart from the pulmonary vein. The right ventricle empties into the pulmonary artery and the left ventricle into the aorta (Figure 3.1).

diastole and (2) the proportion of this volume that can be ejected in a single contraction (ejection fraction).

Ventricular filling Ventricular filling depends upon the amount of blood returned to the heart through the venae cavae (often called **venous return**) and the time available for filling (diastole). During dynamic exercise there is a marked increase in venous return attributable to three main mechanisms. One of these is the pumping effect provided by the contracting muscles themselves, or what is known as the **muscle pump**. With each contraction and relaxation, the veins in the muscles are rhythmically squeezed shut and then opened. Veins contain one-way valves that open only towards the heart, thus preventing backwards flow and creating a milking action that forces blood back to the heart with each contraction and increases venous return pressure.

A second important mechanism that aids venous return and ventricular filling during exercise is the act of breathing, or the **respiratory pump**. When the athlete inhales, the pressure inside the thorax drops below atmospheric pressure, increasing venous return to the right heart, right ventricular filling pressure and the amount of blood in the pulmonary circuit. When he or she exhales, there is a sudden increase in thoracic pressure that acts to squeeze blood out of the lungs and to enhance left ventricular filling. As exercise intensity increases, the force and/or frequency of the muscle contractions increase, as do the depth and frequency of breathing, causing both the muscle and respiratory pumps to become more effective.

A third mechanism that enhances venous return to the heart during exercise is the vasoconstriction that occurs in all veins in response to sympathetic stimulation. Although the veins and venules are relatively thin-walled compared to the arteries and arterioles, they too are composed of smooth muscle and capable of considerable contraction. In the resting state, approximately four times more blood is present in the venous system than in the arteries. Vasoconstriction of the veins reduces their diameter, increases venous pressure, and accelerates the rate at which blood is returned to the heart. Since the pressures on the venous side of the circulation are much lower than those on the arterial side, the back-pressure created by such vasoconstriction is minimal and is offset by the increased pressures that occur in the arteries.

As heart rate increases, the duration of diastole decreases, shortening the time available for diastolic filling. Even so, the increase in venous filling pressure caused by these factors is such that, in trained athletes, ventricular filling continues to increase until maximal, or near-maximal, heart rate is achieved (15).

Ejection fraction The extent to which the ventricles empty with each contraction depends upon the force of the contraction and the resistance to blood flow that is encountered in the arteries (i.e., the pressures in the aorta and pulmonary artery). During exercise, there is an increase both in the peak force of contraction and in the rate at which this force develops, causing more blood to be ejected in a shorter period of time. The increase in contractility of the ventricles is the result of the increase in sympathetic stimulation, described above, as well as a phenomenon known as the **Starling effect** (see Box 3.1). The Starling effect refers to the fact that the force or tension generated by the contracting cardiac fibers increases as the fibers are stretched by an increase in venous return. Since this response also occurs in denervated hearts, it is a separate mechanism from the increase in contractility caused by sympathetic stimulation. The increase in venous return that occurs during exercise is often referred to as **preload**.

SV also depends upon the difference in pressure between that generated in the ventricle and in the aorta. Stated another way, for a given force of contraction, SV will be lower if the mean pressure in the aorta is high, but greater if this pressure is low. This resistance

BOX 3.1

E.H. Starling was an English physiologist who in the early 1900s described the intrinsic response of the heart to changes in atrial filling pressures. In following up previous observations by another physiologist, **Otto Frank**, Starling used an isolated heart and lung preparation in dogs that allowed him to vary the amount of blood being returned to the right atrium while at the same time measuring the output of the right ventricle. He noted that an increase in atrial filling pressure resulted in a stretching of muscle fibers of the ventricle and that this somehow facilitated contraction and SV, so that the cardiac output matched the augmented venous return. This adaptation is known as **Starling's law of the heart** or the **Frank–Starling mechanism**, and is thought to be attributable to the fact that, as the fibers are stretched, more cross-bridges become available between the actin and myosin filaments (see force–length relationship in Chapter 4).

to flow that is encountered on the arterial side is described as **afterload**.

During exercise, the extent to which afterload changes depends upon the type of exercise. In dynamic aerobic-type exercise, the change is minimal because of the vasodilation that occurs in the arteries supplying the exercising muscles. In contrast, during resistance exercise or static contractions, afterload can be extremely high.

Cardiac Output During Exercise

Cardiac output increases during exercise as a result of increases in both heart rate and SV. The way in which these two components combine in order to match cardiac output to the oxygen cost of the exercise is illustrated in Figure 3.7. In this example, a trained athlete ($\dot{V}O_{2max}$ = 5.0 L·min^{-1}) undergoes progressive dynamic exercise on a treadmill until exhaustion. It will be noted that, while heart rate increases almost linearly with oxygen consumption, SV increases dramatically until about 40% $\dot{V}O_{2max}$, after which point the increase is much more gradual. The result is that cardiac output, the product of the two factors, increases linearly with $\dot{V}O_2$ throughout light to moderate exercise, but tends to flatten out as the athlete approaches maximum. In untrained individuals, the response is somewhat different: the increase in SV plateaus at approximately 40% $\dot{V}O_{2max}$, with little or no change thereafter (5,15), resulting in a more pronounced flattening of cardiac output as maximum is approached.

Figure 3.7 The increases in heart rate, SV, and cardiac output that occur as a trained subject undergoes progressive-intensity exercise until exhaustion.

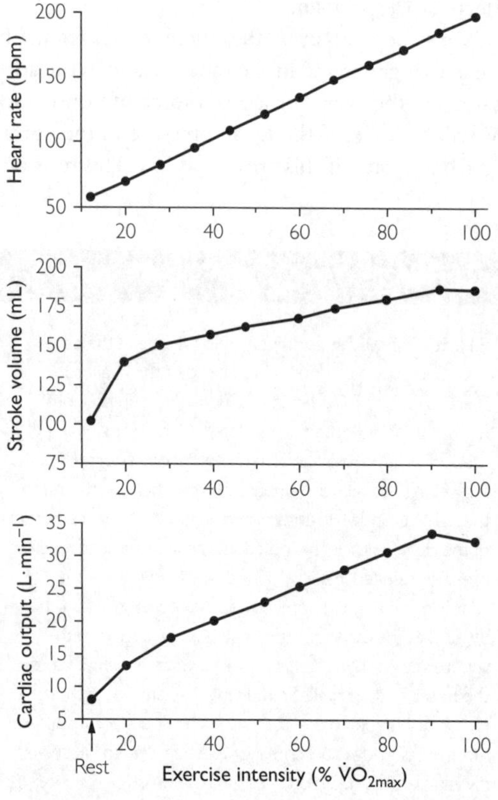

The major difference in cardiac function between a well-trained athlete and an untrained individual is thus the magnitude of SV that their hearts are capable of generating during exercise. This is demonstrated in Table 3.1, where typical values for heart rate, SV, and Q_c are presented for a sedentary individual ($\dot{V}O_{2max}$ = 3.0 L·min^{-1}) and a trained endurance athlete ($\dot{V}O_{2max}$ = 5.0 L·min^{-1}) of the same body size. Notice that, at rest, Q_c is the same for both, whereas heart rate is considerably lower (and SV considerably higher) in the athlete. During maximal exercise, there is little difference between the heart rate achieved by the two individuals, but SV almost doubles in the athlete while the untrained subject shows a considerably more modest increase. The higher SV in the athlete is the result of both a larger ventricular chamber and the ability to fill more rapidly during diastole. In large well-trained endurance athletes such as heavyweight rowers, whose $\dot{V}O_{2max}$ values typically exceed 6.0 L·min^{-1}, maximal cardiac outputs approach 40 L·min^{-1}. This means that their SVs during maximal exercise are 200 mL or more!

Factors That Determine Blood Flow

Fluid flows through a tube when pressure is applied to it, moving from areas of high pressure to areas of low pressure. The rate at which this occurs depends directly on the difference in pressure between the two ends of the tube and inversely on the resistance that the tube offers to flow. The major cause of this resistance is the diameter of the tube: the smaller the diameter, the greater the resistance. If the circulatory system consisted of rigid pipes, like the plumbing in our homes, it would be relatively simple to calculate blood flow and the velocity of flow to any given

Table 3.1 Heart rate, stroke volume, and cardiac output for an untrained subject and a trained subject performing light and maximal exercise on a cycle ergometer. Exercise intensities are expressed in watts (W).

	Untrained subject			Trained subject		
	Heart rate (bpm)	SV (mL)	Cardiac output ($L \cdot min^{-1}$)	Heart rate (bpm)	SV (mL)	Cardiac output ($L \cdot min^{-1}$)
Resting	80	70	5.6	56	100	5.6
Light exercise, e.g., 100 W	135	120	16.2	112	145	16.2
Maximal exercise, e.g., 250 W (untrained), 400 W (trained)	200	120	24.0	200	170	34.0

region by applying two basic principles of fluid mechanics. The first of these is **Poiseuille's law**,[3] which states that flow through a rigid tube varies directly according to the pressure and radius of the tube to the fourth power, and inversely according to the length of the tube and the viscosity of the fluid:

$$Q = \frac{(\text{Pressure} \times \text{radius}^4)}{(\text{Length} \times \text{viscosity})}$$

The second principle is that the velocity of flow is inversely related to the cross-sectional area of the tube:

$$Velocity = \frac{Q}{A}$$

Of course, the circulatory system is not composed of rigid tubes, but rather of multi-branched blood vessels of varying sizes that are constantly changing in diameter. Nonetheless, blood flow is still governed by these basic principles, although direct application is overly simplistic.

Of the variables that determine flow (Poiseuille's law) during exercise, length remains quite constant and viscosity changes only slightly, making pressure and blood vessel diameter (cross-sectional area) the most important controlling factors. Similarly, of these two, since flow varies as to the radius to the fourth power, the cross-sectional area of the vessel is by far the most important variable in determining blood flow through a tissue. For example, a change in vessel radius of only 10% would result in more than a two-fold change in flow! During exercise, large increases in muscle blood flow are achieved by a combination of vasodilation in the arterioles leading to the muscles and vasoconstriction of those leading to areas such as the kidney and the gut.

[3] In the late 1840s Jean Léonard Marie Poiseuille, a French physician, studied the factors that affect the flow of fluid through cylindrical tubes. He examined non-pulsatile flow using configurations of fixed glass tubing for his experiments, and although his observations were not directly applicable to the human circulatory system, they greatly advanced the knowledge of cardiovascular physiology.

Vasomotor Control

All arteries and veins are lined with endothelial cells and their walls consist of smooth muscle and a combination of both structural and elastic tissue. The endothelial cells comprising the innermost layer are smooth and flat, offering minimal resistance to flow. The structural tissue (**collagen**) gives a semi-rigid shape to the vessel, while the elastic component (**elastin**) allows it to expand and recoil. The smooth muscle cells are spindle-shaped, have a single nucleus and are arranged in helical or circular layers around the blood vessel. When the smooth muscle layer contracts, the diameter of the vessel decreases; when it relaxes, the diameter increases. The proportions of smooth muscle, elastin and collagen vary considerably, however, depending upon the type and size of the vessel. The aorta and the large arteries have a much higher proportion of elastin in their walls than other vessels. This makes them capable of more stretch and recoil than smaller arteries and allows them to store pressure energy temporarily as the heart contracts and to release it during diastole, propelling blood into the smaller arteries and capillaries. In contrast, the small arteries and arterioles have less elastic tissue and proportionately much more smooth muscle, allowing them to constrict to a greater degree than the large arteries. The veins have thinner walls than the arteries and proportionately less elastin and smooth muscle, making them quite distensible but still capable of vasoconstriction.

The capillaries are the smallest blood vessels and consist of a single layer of endothelial cells and a thin basement membrane. They contain no smooth muscle and do not change in diameter. Between adjacent endothelial cells there are extremely small slits or pores that allow plasma and small molecules to pass through but prevent the passage of larger molecules such as red blood cells and most plasma proteins. Lipid-soluble gases such as oxygen (O_2) and carbon dioxide (CO_2) diffuse easily through the lipoprotein membranes of the endothelial cells, while water-soluble substances, such as glucose, must diffuse or be filtered through the pores. Since diffusion and filtration occur only in the capillaries, they are the most important functional units of the cardiovascular system. At the ends of the terminal arterioles, just before they branch off into capillaries, there is a thin band of smooth muscle known as the **precapillary sphincter**. Constriction or dilation of the precapillary sphincters is thus the main determinant of capillary flow.

The smooth muscle in the walls of the arteries and veins is controlled by input from two sources: regional changes in tissue chemistry (**local control**) and changes in sympathetic nerve activity, originating in the brain (**central control**). The effects of circulating hormones, such as norepinephrine, are also considered a form of central control, since they originate from a source remote from the capillary bed.

Local control When a tissue increases its metabolic rate, it uses more oxygen and produces more carbon dioxide. This is immediately reflected in the partial pressure of oxygen (P_{O_2}) and carbon dioxide (P_{CO_2}) in the interstitial fluid of that tissue. If the tissue happens to be muscle, in addition to a decrease in P_{O_2} and an increase in P_{CO_2} there is also an increase in the concentration of K^+, H^+, and adenosine, as well as an increase in temperature. Each of these byproducts of metabolism and contraction, when studied in isolation, has been shown to result in a local dilation of the arterioles that supply that tissue. The mechanisms by which these local vasodilators are able to exert an effect upstream from the capillary bed are not yet fully understood, but appear to involve the ascending transmission of electrical signals along the endothelial cells of the capillaries, backwards to those of the terminal arterioles, to those of the small arteries, and even extending to those of the larger arteries (39). These events then trigger the release of several chemical agents from the endothelial cells, which cause dilation of vascular beds. These agents are collectively referred to as **endothelium-derived relaxing factors**, and include **nitric oxide** (**NO**). It also appears that, once blood flow to a region has increased, the increased pressure on the walls of the vessels (sometimes referred to as shear stress) will, in itself, promote further release of endothelial-derived relaxing factors (40). To a certain extent, then, flow begets more flow.

Local control of blood flow is thus an autoregulatory response that attempts to match flow with the metabolic rate of the tissue. Further increases in metabolism lead to further drops in P_{O_2} and increases in metabolic byproducts, resulting in further vasodilation and so on. When metabolic activity decreases, the tissue concentration of these vasodilators declines, signaling removal of the vasodilator response, an increase in precapillary resistance and a decrease in blood flow.

Central control The smooth muscle in the walls of all arteries, arterioles, and veins is innervated by sympathetic nerve fibers that originate in cardiovascular control centers in the medulla (Figure 3.8). The pathway consists of myelinated fibers that descend the spinal cord and synapse with other myelinated fibers that branch off the cord and lead to the sympathetic ganglia. They then synapse with the unmyelinated postganglionic fibers that innervate the blood vessels. There is always some degree of sympathetic activity present, creating what is known as sympathetic tone and causing a release of norepinephrine at the nerve endings in the vascular walls. In most of the body's blood vessels, the presence of norepinephrine causes vasoconstriction, by activating what are known as **alpha receptors** in the smooth muscle. The exceptions are found in the vessels of skeletal muscle and the heart, which are dominated by a different type of receptor, called **beta receptors**. The appearance of norepinephrine in the presence of beta receptors has the opposite effect and results in vasodilation.

Although the majority of sympathetic nerve fibers release norepinephrine when activated (**adrenergic fibers**), a small proportion release acetylcholine. These are known as sympathetic **cholinergic fibers** and are found in blood vessels of the skin and, to a lesser

Figure 3.8 The smooth muscles in the walls of all blood vessels receive input from the sympathetic nervous system, which arises from the vasoconstrictor center in the brain stem. Release of norepinephrine at the nerve endings causes vasoconstriction in vessels with alpha receptors and vasodilation in those with beta receptors. Input to the control centers from the baroreceptors regulates blood pressure by increasing or decreasing heart rate and by constricting or dilating blood vessels. See Figure 3.9 for an example.

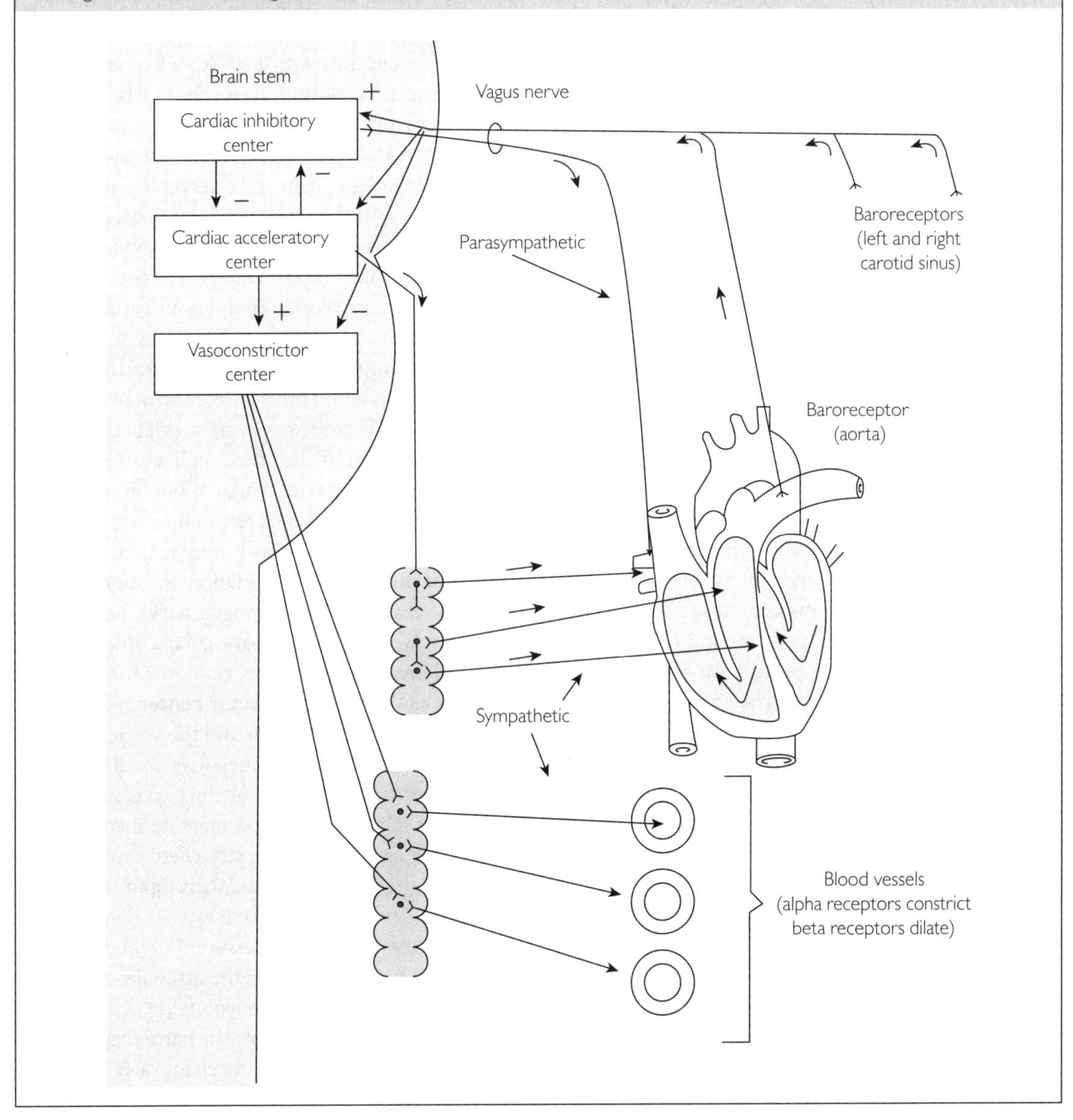

extent, in those of skeletal muscle. In contrast to the passive dilation that occurs in response to a reduction in vasoconstrictor tone, their activation is thought to promote **active vasodilation** and to play an important role in temperature regulation during exercise.

When an athlete begins exercise, in addition to the immediate locally controlled vasomotor changes, the increase in sympathetic nerve activity results in two important responses:

1. beta-adrenergic vasodilation in the blood vessels of the skeletal muscles and the heart, and
2. alpha-adrenergic vasoconstriction of blood vessels in most other parts of the body.

This double-barreled effect of dilating muscle and heart vessels while at the same time constricting those elsewhere serves to re-route blood flow, resulting in dramatic increases in muscle and coronary flow. The adrenal glands are also innervated by the sympathetic nervous system and release norepinephrine and epinephrine in response to increased sympathetic activity. As a result, if the exercise is intense or prolonged, the concentration of these circulating hormones begins to rise, further accentuating this response. When these vasomotor changes are combined with the large increases in cardiac output that occur with exercise, blood flow to the active muscles can increase by as much as 30 times!

Control of blood pressure Since blood flow requires a difference in pressure from one end of the vessel to the other in order to push the blood along, all of these changes in vessel size would be ineffective unless pressure was maintained. The body has a number of mechanisms for controlling blood pressure. Some of these are rapidly responding mechanisms that are activated by changes in body position and physical activity, while others (such as those originating from the kidneys; see Box 3.2) are more chronic in nature and provide long-term control. Because pressure must be maintained to provide flow to vital organs, most of the sensors involved in control are on the arterial side of the circulation. Thus, when we refer to blood pressure, it is usually the *arterial* blood pressure that is being considered.

In the simplest analysis, arterial pressure at any given point in time is determined by the amount of blood in the arterial tree (cardiac output) and the total peripheral resistance (the summed effect of the average radii of the arterioles in each vascular bed). Blood pressure can thus be controlled either by increasing or decreasing cardiac output, or by constricting or dilating the small arteries and arterioles. This is, however, somewhat of an oversimplification, since the two variables are not independent and interact in a number of different ways. In the resting state, it is usually resistance that is adjusted to control pressure, with minimal changes in cardiac output. During exercise, the situation becomes more complex and both variables must be adjusted.

The most important method of regulating short-term changes in blood pressure is control by the **baroreceptors**. The baroreceptors are specialized neurons located in the wall of the aortic arch and the internal carotid arteries (the vessels that supply blood to the brain). Their nerve endings respond to stretch of the endothelial lining and smooth muscle of these arteries, and transmit this information through connections that travel along the vagus nerve back to the inhibitory area of the cardiovascular control center, to a third area that controls vasomotor function and is known as the **vasoconstrictor center** (Figure 3.8). The cardiac inhibitory center and the vasoconstrictor center work together in a negative feedback loop. In other words, when either of the two is stimulated, the other is inhibited. If blood pressure increases, the walls of the arteries become stretched, causing them to increase their firing rate and their input to the control center. The change in firing rate is related almost directly to the change in pressure. If blood pressure drops, there is less stretch on the arterial wall and firing frequency decreases proportionally.

A simple illustration of how the baroreceptors control blood pressure during acute changes in body position or posture is provided in Figure 3.9. In situation A, an individual who has been standing suddenly lies down. Because of the reduced effects of gravity on the column of blood between the heart and the brain, there is a sudden rise of blood pressure in the vessels leading to the brain. This stretches the arteries and stimulates the carotid baroreceptors to increase their neural traffic back to the cardiovascular control centers. This then stimulates the cardiac inhibitor center

BOX 3.2

Students are sometimes surprised to learn that the kidney is a major controller of blood pressure but, in fact, it has a vested interest in maintaining it. The kidney's main function is to filter water and waste products from the blood and to eliminate them as urine. Since the blood must be under pressure for filtration to occur, it is vital that renal blood pressure does not decline. The kidney has several mechanisms for preventing this. One elevates blood volume by retaining water and sodium (activated by antidiuretic hormone and aldosterone) and another by releasing angiotensin into the blood. Angiotensin is a powerful vasoconstrictor of vascular smooth muscle throughout the body, which results in an increase in arterial blood pressure. The release of angiotensin is controlled by the renal hormone renin. Recent advances in the understanding of the **renin-angiotensin system** have fostered the development of angiotensin-converting enzyme inhibitors (ACE inhibitors) as an effective means of controlling hypertension in many individuals.

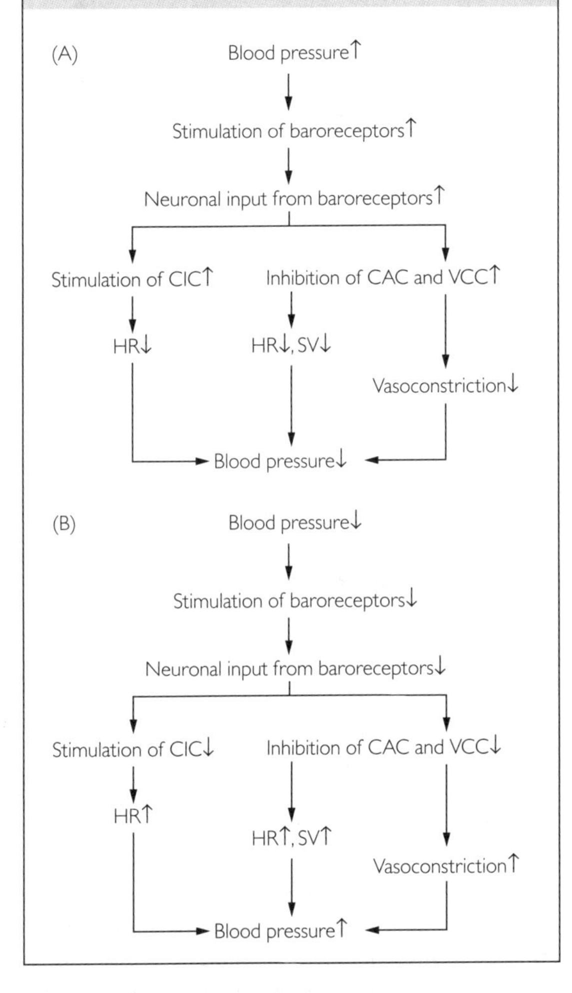

Figure 3.9 An illustration of how blood pressure is controlled by the baroreceptors, located in the carotid sinus and the aortic arch. Figure A shows how normal blood pressure is restored following a sudden increase in pressure, while Figure B shows how it is restored following a sudden decrease in pressure. See text for further explanation. HR, heart rate; VCC, vasoconstrictor center.

and inhibits the vasoconstrictor center. The result is increased parasympathetic output to the heart, leading to a reduction in heart rate and SV (reducing blood pressure), as well as reduced sympathetic vasoconstrictor tone, which decreases peripheral resistance (also reducing blood pressure).

In situation B, an individual who has been lying down suddenly stands up. The sudden drop in pressure at the brain reduces blood flow to an extent that would lead the individual to faint, were it not for the following adjustments. The decrease in pressure is detected by the aortic and carotid baroreceptors, causing them to reduce their input to the cardiovascular control centers, thus decreasing parasympathetic output to the heart (heart rate and SV increase) and removing some of the inhibition of the vasoconstrictor center, resulting in an increase in sympathetic vascular tone (an increase in blood pressure).

During exercise, the baroreceptors are faced with the complex challenge of maintaining blood pressure so that flow to vital organs is not affected (the rapid locally controlled vasodilation in the exercising muscles would cause a drop in pressure), while at the same time preventing extreme increases in blood pressure that could damage the vascular walls. The task is made even more difficult with different forms of exercise. For example, during prolonged heavy exercise or exercise in a hot or humid environment, vasodilation of skin vessels for temperature regulation and loss of blood volume due to sweating tend to reduce blood pressure. In contrast, the massive increases in intramuscular pressure and resistance to flow that occur during weight lifting sharply increase blood pressure.

The typical blood pressure response to three different types of exercise is illustrated in Figure 3.10. During progressive dynamic exercise such as running or cycling (A), systolic pressure increases almost linearly with exercise intensity to values of 200 mmHg or more at maximal exercise. During this time, the four- to six-fold increase in cardiac output would have resulted in much more extreme increases in blood pressure were it not for the counteractive effect of the progressive vasodilation of the arterioles leading to the muscles. This decreases peripheral resistance to the extent that there is little or no change in diastolic pressure and even a slight decrease at higher workloads. As a result, mean blood pressure (the average pressure on the arterial wall over time) during this type of exercise seldom exceeds 120 mmHg.

During a static or isometric contraction (Figure 3.10B) at increasing percentages of maximal voluntary contraction (MVC) force, there is a proportional increase in intramuscular compression that is mechanically caused by the contraction itself. This tends to occlude muscle blood flow and negates the pressure-buffering effect of the local muscle vasodilation that occurs with dynamic exercise. As a result, diastolic pressure increases in proportion to the increase in systolic pressure, resulting in large increases in mean blood pressure. Resistive exercise such as

Figure 3.10 Typical blood pressure response to three different types of exercise. (A) The changes that occur during progressive dynamic exercise such as running or cycling to exhaustion. (B) The response to progressive intensities of isometric contraction (MVC denotes maximal voluntary contraction). (C) An actual tracing of intra-arterial blood pressure (upper panel) and esophageal pressure (lower panel) as a subject performs double leg-press exercise to failure at 95% of his maximal strength. The inset indicates pressure changes at the lifting, lockout, and lowering phases of the lift. From (24). © John Wiley & Sons.

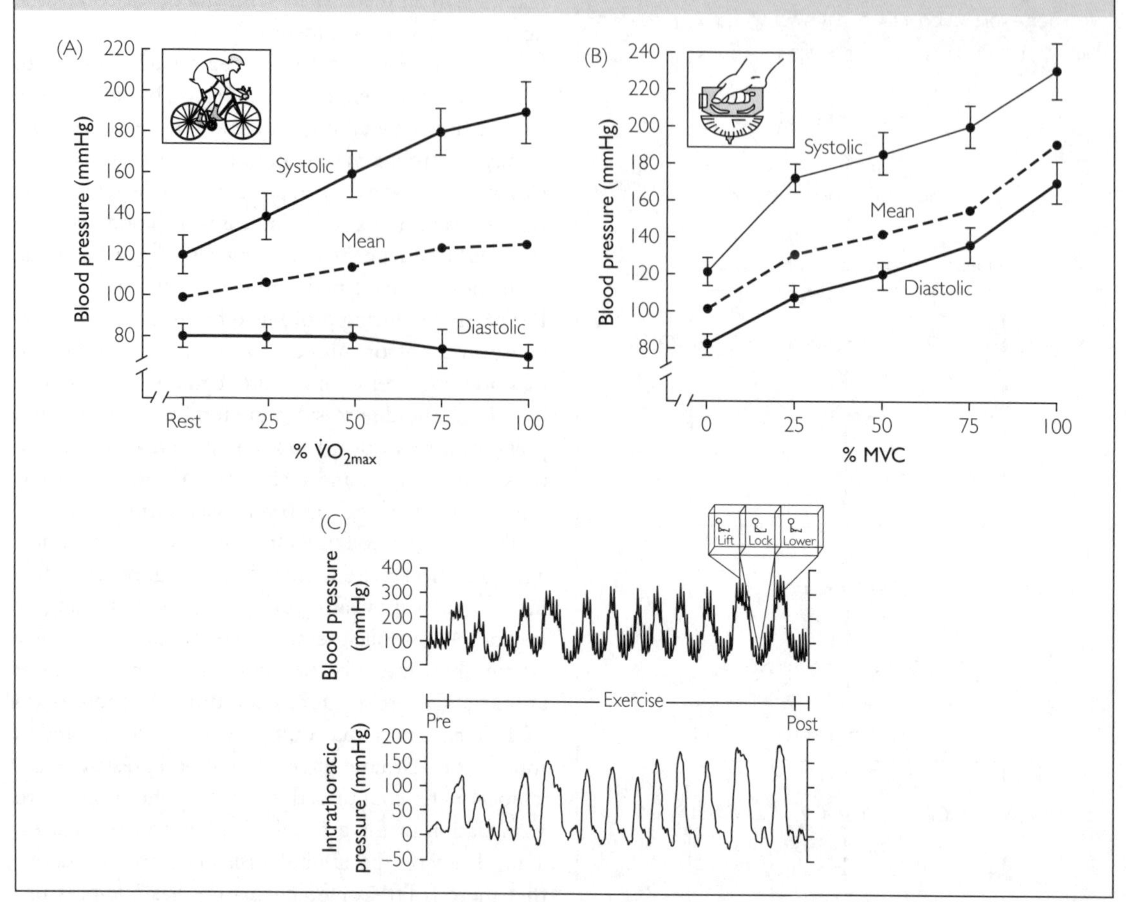

weight lifting can be considered a combination of static and dynamic exercise, in that each movement begins with a static contraction (until force production exceeds the weight of the object to be lifted), followed by a forceful concentric contraction (while the weight is being raised), an eccentric contraction (while the weight is being lowered), and finally a brief relaxation before the beginning of the next repetition. During weight lifting (Figure 3.10C), extreme increases in both systolic and diastolic pressures occur with the exertional phase of each lift. In maximal efforts that involve a large muscle mass, pressures in excess of 350/250 mmHg are not uncommon in healthy young individuals (27). These extreme increases in pressure reflect the summed effects of a potent pressor response (increased cardiac output and vasoconstriction in non-exercising vasculature), mechanical compression of the blood vessels in the contracting muscles, and performance of the Valsalva maneuver (see Box 3.3).

Muscle Blood Flow

The artery that supplies blood to a muscle and the motor nerve that innervates it enter the muscle at its approximate midpoint. Normally there is one main artery and one main motor nerve (containing hundreds of motor neurons) per muscle, but in longer muscles

BOX 3.3

In 1707, Antonio Valsalva, an Italian anatomist, first described the maneuver whereby an individual performs a forced expiration against a closed or partially closed glottis. When this occurs, there is an abrupt increase in intrathoracic pressure that is transmitted to the arterial tree, causing an immediate increase in systolic and diastolic blood pressure. During weight lifting, the Valsalva maneuver provides a mechanical advantage by helping to stabilize the trunk, and is thus an integral component of the lifting process in maximal or near-maximal efforts. Although a major portion of the increase in blood pressure during very heavy lifting is caused by the Valsalva maneuver, the increased intrathoracic pressure is also transmitted to the cerebrospinal fluid. This serves to *reduce* the transmural pressures in the cerebral vessels and the risk of vascular damage under the extremes of pressure that are generated by heavy weight lifting. Similarly, the compressive effect of the elevated intrathoracic pressure on the heart reduces myocardial transmural pressure and helps to maintain SV against peak peripheral resistance. Thus, performing a **brief** Valsalva maneuver (the natural response to lifting) may constitute an important protective function during very heavy lifting, and should not be discouraged in healthy young individuals (16,24).

there may be more. The artery then branches into smaller arteries that form a network along the whole length of the muscle. Each of these then branches again into still smaller arteries that penetrate the interior of the muscle and branch yet again to form the terminal arterioles. With each branching, the diameters of the vessels decrease, but the exponential increase in their numbers is such that there is a large increase in their *total* cross-sectional area.

The arterioles and the million or so capillaries that arise from them are arranged so that they generally run parallel to each muscle fiber. Although the exact architecture of the muscle capillary bed and their numbers are not yet known, because of their parallel arrangement it is possible to estimate their density by microscopic examination of small samples of muscle that have been sectioned at right angles to the direction of the fibers. When this is done, it appears that human skeletal muscle has a capillary density of 300–500 capillaries·mm^{-2}, depending upon the athlete's state of training (1,9). Further examination indicates that these are distributed so that there is an average of two to four capillaries surrounding each fiber, with a tendency for more in close proximity to type 1 fibers (three to five capillaries) than to type 2 fibers (one to three capillaries) (21). It has been estimated that, if all the existing capillaries in muscle were open during maximal whole-body exercise, the surface area of the capillary bed could be as much as 200 m^2 (4)! Although it is unlikely that this peak capacity for receiving blood is ever fully achieved during maximal exercise with a large muscle mass (36), blood flows of 200–250 mL·100 g^{-1}·min^{-1} have been recorded in the human quadriceps during maximal single-leg exercise (35).

In the resting state, muscle receives very little blood flow and, in fact, next to bone, is the least perfused tissue in the body. The vasoconstrictor tone of the precapillary sphincters is high and is only overridden by locally induced dilation when P_{O_2} in a given region drops to a critical level. This results in a condition where alternate capillary beds within the same muscle are continuously opening and closing. When the athlete begins exercise, there is an immediate drop in P_{O_2} and an increase in metabolic byproducts in the active fibers that cause local dilation. Simultaneously, there is an increase in sympathetic output from the cardiovascular control centers, resulting in centrally controlled dilation in the muscles as well as constriction on the venous side and in the vessels leading to non-active tissues. The latter two adjustments serve to augment cardiac output by facilitating venous return, while at the same time re-routing blood flow from tissues such as kidney and liver to the exercising muscles. As the intensity of the exercise increases, the magnitude of these vasomotor changes also increases, as does cardiac output.

During dynamic exercise, such as running or cycling, the muscle contractions are rhythmic so that muscles are active during a portion of the movement cycle and relaxed during the remainder. Surges of blood flow occur during each relaxation phase, while the compression-aided muscle pump increases venous return during each contraction phase. As the speed of the movement increases, the duration of each phase decreases, but the proportion of time spent in contraction or relaxation remains quite constant. As a result, during maximal dynamic exercise, as much as 85% of the total cardiac output is directed to the exercising muscles (Figure 3.11). During static or isometric exercise, there is no cycling or muscle-pumping effect between an active and relaxed state and blood flow will occur only when arterial pressure is greater than intramuscular pressure. The intensity at which blood flow becomes completely occluded during a sustained

Figure 3.11 The changes in the distribution of blood flow as a trained subject goes from rest to maximal dynamic exercise. Note that blood flow to the active muscles can increase to more than 30 times its resting level and represent up to 85% of total cardiac output.

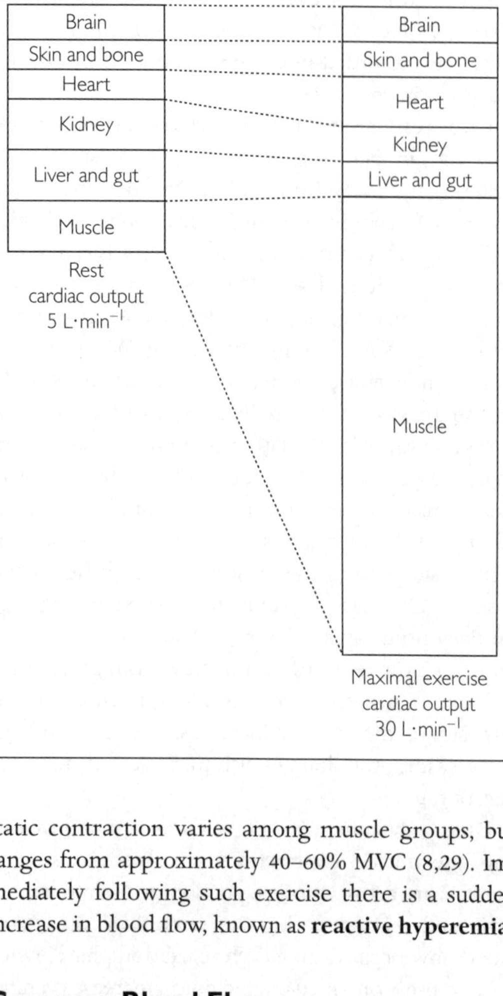

static contraction varies among muscle groups, but ranges from approximately 40–60% MVC (8,29). Immediately following such exercise there is a sudden increase in blood flow, known as **reactive hyperemia**.

Coronary Blood Flow

Although 8000–10 000 L of blood pass through the chambers of the heart each day, the heart cannot use this blood directly to support its pumping and must rely on the coronary arteries to redirect some of this blood back to cardiac muscle after it has left the ventricles. The left and right coronary arteries are the first vessels to branch off the aorta, about 1 cm from its attachment to the left ventricle (Figure 3.12). The left coronary artery then branches into the circumflex and anterior descending arteries. These three main arteries pass across the surface of the heart, giving rise to smaller arteries that branch off at right angles to penetrate deep into the heart and end in a rich network of arterioles and capillaries. Because the heart is its own pump, a unique problem arises in supplying itself with blood. As in the case of skeletal muscle, when contraction occurs, shortening of cardiac muscle fibers causes a mechanical compression of the smaller blood vessels in the interior. Thus, unlike all other tissues in the body, the heart can receive only minimal flow during systole and must function on the flow that occurs during diastole. Since aortic blood pressure falls during diastole, this drop would normally be expected to reduce the perfusion pressure and greatly decrease blood flow. The heart solves this problem by taking advantage of the elastic energy that is temporarily stored in the wall of the aorta following each ventricular contraction. This situation exists because, with each contraction, the bolus of blood that is ejected causes a bulging of the elastic walls of the aorta, which then snap back to their normal shape, creating a pressure head for driving blood into the coronary arteries (Figure 3.12).

During the resting state (the term "resting" is a misnomer here, since the heart is always beating), coronary blood flow is approximately 250 mL·min^{-1} and accounts for about 5% of total cardiac output. Because of its constant activity, high capillary density, and abundant mitochondria, the heart extracts more than 75% of the oxygen from the blood flowing through it. In contrast, the oxygen extraction by resting skeletal muscle would be less than 25% of that delivered in the blood. As will be described later in this chapter, skeletal muscle is able to increase its extraction of oxygen significantly during exercise but, since extraction is already close to maximal in cardiac muscle, the heart has very little reserve to draw on. The oxygen costs of increasing heart rate and cardiac contraction force during exercise must therefore be borne by large increases in coronary flow.

A second problem that must be overcome during exercise is the fact that, as heart rate increases, the amount of time that the heart is in diastole decreases. Since the majority of coronary flow occurs during diastole, this would compromise blood flow to the heart, were it not for other compensating adjustments. These include a pronounced dilation of the coronary arteries and arterioles, as well as an increased driving pressure that stems from the greater aortic recoil caused by an increase in SV. The vasodilatory response is primarily a locally controlled phenomenon that is

Figure 3.12 The heart receives its blood flow from a series of arteries that originate with the right and left coronary arteries. With each ventricular contraction, the ejection of blood causes the elastic walls of the aorta to bulge out and then snap back into place. This aortic recoil provides a pressure head for driving blood into the coronary arteries.

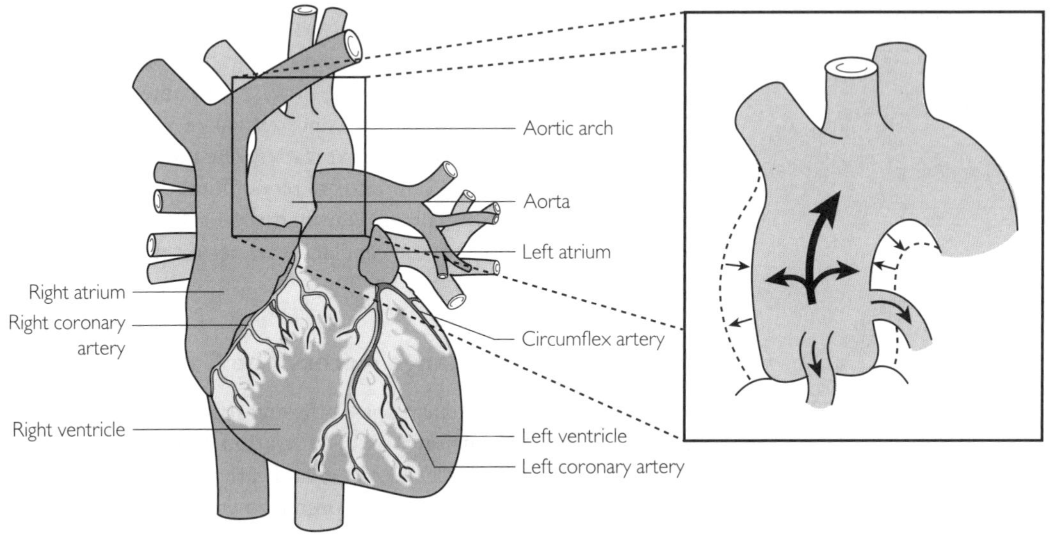

triggered by the increased metabolic rate in the cardiac muscle. Although metabolic byproducts such as adenosine and increased P_{CO_2} are also known to cause a reflex dilation, the coronary arteries are particularly sensitive to hypoxia and even a slight drop in P_{O_2} results in an exaggerated response. During intense or prolonged exercise, circulating catecholamines (primarily epinephrine) will also act to dilate coronary arteries directly. These adjustments permit coronary blood flow to increase up to 1.5 $L{\cdot}min^{-1}$ during maximal aerobic exercise.

Although such increases in blood flow are impressive, as exercise becomes more and more intense, a point will eventually be reached where coronary flow will be inadequate. Unlike skeletal muscle, cardiac muscle is severely limited in its capacity for anaerobic metabolism and its force of contraction will diminish as soon as flow fails to match the oxygen cost. This begins a cycle of events whereby ventricular SV decreases, leading to less aortic recoil, leading to a further decline in coronary flow, and so on. Since heart rate is maximal at this point, the decrease in SV results in decreased cardiac output and blood flow to the exercising muscles, forcing the athlete to terminate exercise or to modify its intensity. Thus, while the final cause for fatigue may have been inadequate muscle blood flow, the precipitating events can be traced back to inadequate coronary blood flow.

Velocity of Flow

Extrapolating from Poiseuille's law, it is apparent that the average velocity of blood flow through a given region in the circulatory system will vary inversely as the total cross-sectional area of the blood vessels in that region:

$$\text{Velocity} = \frac{\text{Flow}}{\text{Cross-sectional area}}$$

The circulatory system is arranged so that its total cross-sectional area (i.e., the sum of the areas of all the blood vessels) increases as the distance from the heart increases. This is because of the profuse exponential branching that occurs as the vessels become smaller in size. Thus, although vessels such as the aorta and venae cavae have the largest diameter, they have the smallest *total* cross-sectional area, whereas the capillaries have the largest total cross-sectional area. The relationship between area and velocity of flow is depicted in Figure 3.13. As can be seen, blood travels most rapidly as it leaves the heart through

Figure 3.13 A summary of the pressure changes (A), the distribution of blood volume (B), and the cross-sectional area to velocity of flow relationship (C) that exist throughout the systemic circuit. Note that there is an inverse relationship between pressure and volume in different parts of the system (A and B) and between total cross-sectional area and velocity of flow (C). Lt. vent., left ventricle; Lg. art., large arteries; Sm. art., small arteries; Caps., capillaries; Rt. ven., right ventricle; Pul. art., pulmonary artery. From (42). © Wolters Kluwer Health.

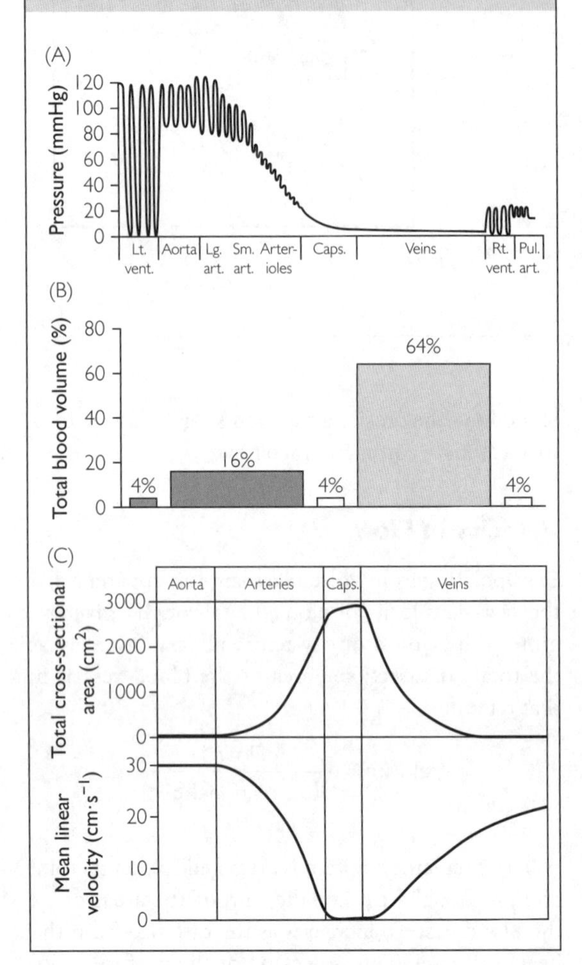

the aorta. (Its velocity would also be fluctuating at this point because of the pulsating oscillations in pressure, being most rapid during systole and slowest during diastole, with an average flow rate of approximately 30 cm·s^{-1}.) As the aorta branches into smaller and smaller arteries, their total area increases and there is a rapid decrease in velocity. (There is also a progressive drop in pressure and, by the time that blood has reached the arterioles, the fluctuations in pressure will have disappeared.) As blood reaches the capillary bed, its velocity of flow becomes extremely slow, but accelerates again on the venous side on its way back to the heart. From a functional standpoint, this pattern of rapid flow in the conducting arteries and veins and the slowest flow rate in the capillaries is important, since exchange with tissues occurs only across the capillaries. The time during which exchange can occur as blood flows through a capillary bed is known as **transit time**. The prolonged transit times at the capillary level permit maximal exchange to occur.

Exchange Across The Capillary Wall

Substances pass through the capillary walls by **diffusion** or by **filtration**. Lipid-soluble materials such as oxygen and carbon dioxide diffuse easily through the lipoprotein membrane of the capillary wall. Since diffusion occurs from high to low concentration, oxygen normally diffuses out of the capillary while carbon dioxide diffuses into the capillary. The exception to this is at the lung, where the directions are reversed.

Blood is approximately 55% water. This aqueous component, along with the substances dissolved in it, is known as **plasma**. Because of its small molecular size, water also diffuses rapidly through the endothelial cells of the capillary wall, as well as through the pores between the endothelial cells. Small molecules that are dissolved in the plasma, such as glucose, NaCl, urea, and hormones, move easily with the water through the capillary pores and continue to diffuse until their concentration reaches equilibrium with that outside the capillary. Large non-lipid-soluble molecules, such as most plasma proteins, as well as blood cells, cannot pass through the pores and thus remain in the blood.

In addition to diffusion, plasma is also forced out of, or into, the capillary by differences in pressure. This process is known as filtration. Since filtration occurs in both directions, filtration back into the capillary is often called **re-absorption** in order to distinguish the direction of movement. The amount of plasma that is filtered and the direction in which it moves are determined essentially by the algebraic sum of two different pressures: the pressure forcing it to move outwards and the pressure drawing it into the capillary.

Figure 3.14 A simplified illustration of the pressures that cause plasma to be exchanged across the capillary wall. All pressures are expressed in millimeters of mercury (mmHg). While the ingoing osmotic pressure remains constant (25 mmHg) throughout the length of the capillary, the outgoing hydrostatic pressure declines from approximately 35 mmHg at the arterial end to 15 mmHg at the venous end of the capillary. As a result, there is a net movement of plasma out of the capillary at the arterial end and a net movement into the capillary at the venous end.

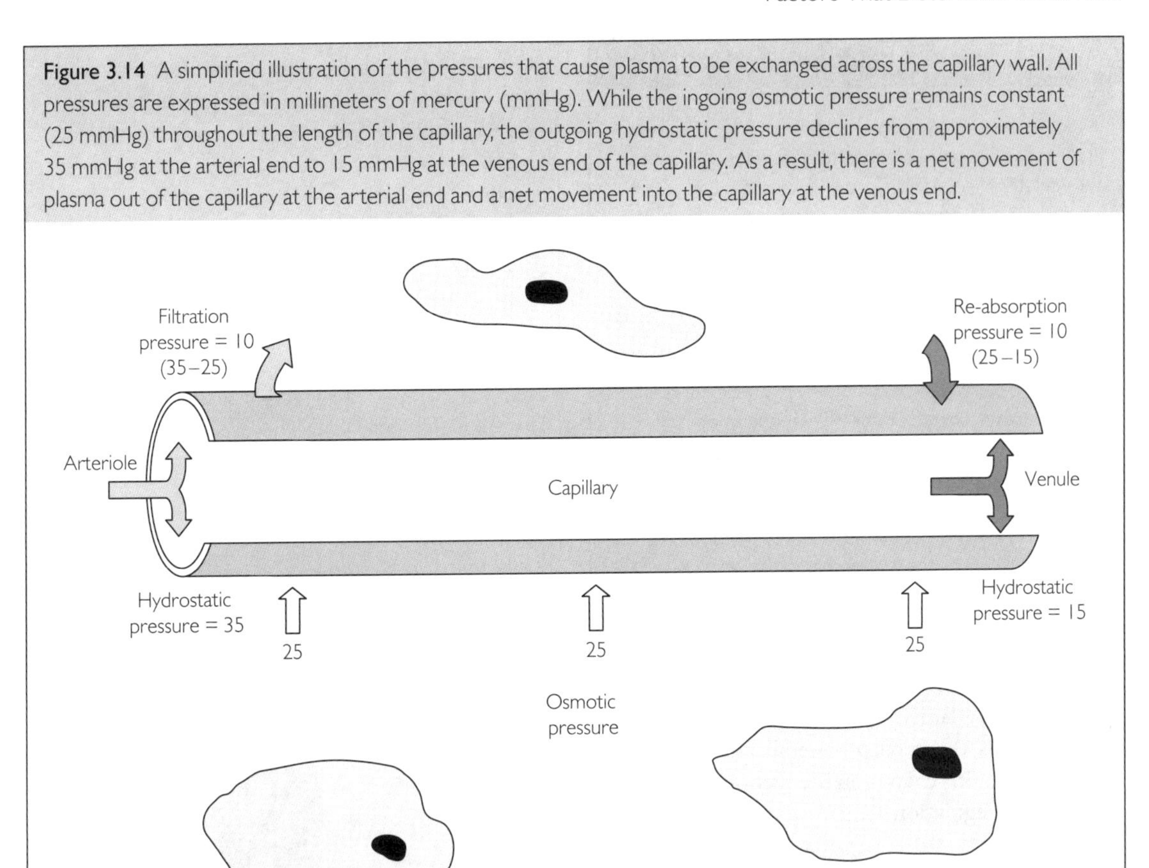

The main component determining outgoing pressure is that provided by the heart; that is, blood pressure. This is known as the capillary **hydrostatic pressure** and gradually declines as blood flows from one end of the capillary to the other. In most tissues, in the resting state, hydrostatic pressure is typically about 35 mmHg at the arterial end of the capillary and about 15 mmHg at the venous end. An opposite or ingoing pressure is caused by the plasma proteins trapped in the blood. Their presence creates an **osmotic pressure** that attempts to draw plasma from the more dilute solution (interstitial fluid) to the more concentrated solution in the capillary. The osmotic pressure is determined primarily by the concentration of the plasma protein albumin and is normally about 25 mmHg. The dynamics as to how these two pressures affect filtration and re-absorption of plasma along the capillary are illustrated in Figure 3.14.

At the arterial end of the capillary, where the hydrostatic pressure is 35 mmHg and the osmotic pressure 25 mmHg, there is a net excess of hydrostatic over osmotic pressure (10 mmHg), triggering filtration (movement of plasma out of the capillary into the interstitial space). In contrast, at the venous end, where the hydrostatic pressure is only 15 mmHg and the osmotic pressure 25 mmHg, there is a net excess of osmotic over hydrostatic pressure (10 mmHg), causing plasma to be re-absorbed. Thus, plasma moves out until approximately the midpoint of the capillary; thereafter it moves in (see Box 3.4). During exercise, this balance can be altered by any changes in arterial blood pressure. For example, the slight increase in mean blood pressure that accompanies endurance exercise elevates the capillary hydrostatic pressure, resulting in a slight plasma loss from the capillaries into the interstitial fluid compartment. In contrast, the extreme increases in blood pressure that occur with heavy-resistance exercise result in major shifts of fluid from the capillaries to the interstitial space. This causes a temporary localized **edema** or swelling in the exercised muscles. Bodybuilders refer to this swelling as "the pump" and strive to achieve this effect

BOX 3.4

Although the capillary hydrostatic and osmotic pressures are the major determinants of filtration and re-absorption, the interstitial fluid also has its own hydrostatic and osmotic pressures. These pressures are relatively low and variable in different tissues but, on average, their net effect is a negative pressure of approximately 4–6 mmHg (6, p.502), tending to assist filtration by drawing plasma out of the capillary along its entire length. Thus, while the above example (Figure 3.14) appears to indicate that the amount of plasma that is filtered is completely re-absorbed, in actual fact the amount filtered slightly exceeds the amount re-absorbed. It is estimated that, over 24 hours, a total of approximately 20 L of plasma are filtered, of which 16–18 L are returned by re-absorption. The extra 2–4 L are removed by the lymphatic system, cleansed, and returned to the circulation on the venous side.

immediately before a contest, because it makes their muscles appear larger. Other factors affecting fluid balance across the capillary membrane will be discussed later in this chapter in the section dealing with temperature regulation.

Factors That Determine the Oxygen Content of Blood and Muscle

Blood

Blood consists of two main components: specialized cells and plasma, the water medium in which they are suspended. There are three types of cells: **red blood cells (erythrocytes)**, **white blood cells (leukocytes)**, and **platelets**. Together, the cells make up about 45% of the total blood volume. When a tube of whole blood is centrifuged for a few minutes, the denser cells will separate out from the plasma and be clearly visible in the bottom of the tube (Figure 3.15). The volume of these packed cells relative to the total volume is called the **hematocrit**. Red blood cells make up more than 99% of the hematocrit, while white blood cells and platelets together account for less than 1%. The latter cells are less dense than the red cells and appear as a thin whitish band on top of the column of red cells. A single microliter of blood contains approximately 4–6 million red cells, 5000–10 000 white cells, and 150 000–350 000 platelets. The main function of the red cells is transporting oxygen and carbon dioxide, while the white cells combat bacterial infection and provide immunity. The platelets are crucial for blood clotting.

Figure 3.15 After a sample of whole blood has been centrifuged, it separates into plasma and hematocrit. More than 99% of the hematocrit consists of red blood cells, with white blood cells and platelets accounting for less than 1%.

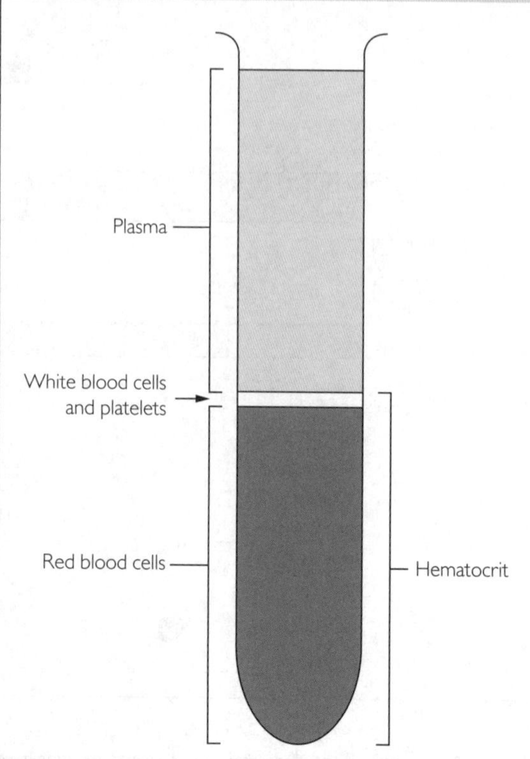

If a tube of whole blood is allowed to stand, it will form a large clot after approximately 6–7 minutes. After another 20–25 minutes the clot will shrink and settle to the bottom of the tube, leaving the water component on top as if it had been centrifuged. This component, known as **serum**, is the same as the plasma in Figure 3.15, except that it does not contain the protein fibrinogen (now part of the clot). It is generally estimated that blood comprises approximately 8% of the average individual's total body weight. Thus, a 75 kg athlete would have approximately 6.0 kg of blood. Since blood has virtually the same specific gravity as water, this athlete's total blood volume would be approximately 6.0 L.

Red Blood Cells

Mature red blood cells are unique in that they do not have nuclei. They are enclosed in a very durable yet flexible membrane that allows them to bend and deform in order to pass through small capillaries. In addition, they are flattened or disc-shaped, with biconcave depressions that give them a very large surface area for their relatively small mass. This characteristic shape makes it easy for gas to diffuse in and out of the cell.

Next to water, the main component of the red blood cell is the protein **hemoglobin**, which makes up approximately one third of its weight. Although its structure is very complex, hemoglobin is a small protein that would be rapidly filtered out in the kidneys and excreted in urine were it not protectively confined by the red cell membrane. Each molecule of hemoglobin consists of four iron-containing structures known as **hemes** attached to four polypeptide chains that, grouped together, are called **globin**. The iron in each heme binds readily with oxygen and it is this iron that gives the red color to the cell. The amount of oxygen that becomes bound to hemoglobin depends upon its partial pressure (P_{O_2}). In regions where the local P_{O_2} is high, hemoglobin binds or takes up oxygen; in regions where the local P_{O_2} is low, it releases oxygen. It is important to note that this is not a direct, linear relationship, a point that will be discussed later in this chapter. At a supernormal P_{O_2} (for example, if an individual inhales 100% oxygen), each heme would bind one molecule of oxygen, for a total of four molecules of oxygen per molecule of hemoglobin. In such a situation, where all available hemes are bound to oxygen, the blood is considered to be 100% saturated. If half the hemes were bound to oxygen, it would be 50% saturated, and so on. To put this into perspective, under normal conditions the P_{O_2} in the lungs is such that the hemoglobin becomes approximately 97% saturated as blood flows through the lungs.

A typical adult red blood cell contains approximately 200–300 million molecules of hemoglobin, each capable of binding and transporting four molecules of oxygen. If we consider that 1 μL (less than a drop) of whole blood contains approximately 5 million red cells, we can appreciate the impressive oxygen-carrying capacity of blood. Expressed in units of mass and volume, at normal saturation (97%), 1 g of hemoglobin will bind with approximately 1.3 mL of oxygen. Blood hemoglobin concentration is usually expressed as grams per 100 mL ($g{\cdot}100\ mL^{-1}$) of whole blood. In individuals with a normal red blood cell concentration (e.g., a hematocrit of ≈45; see Box 3.5), hemoglobin concentration would be ≈15 $g{\cdot}100\ mL^{-1}$. The normal range for hematocrit is considered to be 39–55 for adult males and 36–48 for adult females. The normal range for hemoglobin concentration would be 14–16 $g{\cdot}100\ mL^{-1}$ in men and 13–15 $g{\cdot}100\ mL^{-1}$ in women. Higher-than-normal values are known as **polycythemia**, and lower-than-normal values as **anemia**.

> **BOX 3.5**
>
> Hematocrit can easily be calculated by drawing a small sample (1 mL) of blood into a narrow, uniform capillary tube and centrifuging it for several minutes. The height of the packed red cell column, expressed as a percentage of the total column of blood, equals the hematocrit. Hemoglobin concentration can be accurately measured only by chemical analyses, but it can be estimated if we know the hematocrit. Since red cells are normally about one third hemoglobin by mass and the specific gravity of blood is almost the same as that of water, hemoglobin concentration (in $g{\cdot}100mL^{-1}$) is approximately equal to hematocrit divided by 3. For example, in an individual whose hematocrit is 45, hemoglobin concentration would be approximately 15 $g{\cdot}100\ mL^{-1}$ of blood; in an individual whose hematocrit is 40, it would be approximately 13.3 $g{\cdot}100\ mL^{-1}$, and so on.

Oxygen-carrying capacity of blood Oxygen is only moderately soluble in water. In fact, at body temperature, 1 L of plasma would contain less than 5 mL of dissolved oxygen. In contrast, the hemoglobin present in 1 L of blood (150 g) can combine with and transport approximately 200 mL of oxygen, almost 50 times more. This means that, if blood did not contain hemoglobin, cardiac output would have to be 50 times greater in order to deliver the same amount of oxygen!

The volume of a gas contained in a liquid is known as its **content**. The oxygen content of blood is the sum of the oxygen dissolved in the plasma plus the oxygen combined with hemoglobin. For example, the arterial oxygen content of an individual with a hemoglobin concentration of 15 $g{\cdot}100\ mL^{-1}$ would be $15\ (g) \times 1.3\ (mL{\cdot}g^{-1}) = 19.5\ mL{\cdot}100\ mL^{-1}$ of blood, plus the oxygen dissolved in the plasma (≈0.5 mL), for a total of 20 mL of oxygen$\cdot$100 mL^{-1} of blood after it has passed through the lungs. This means that the arterial blood contains 20% of its own volume as oxygen or, as it is more commonly expressed, 20 **volumes %** (vol %).

Since gases cannot diffuse through the walls of the arteries and arterioles, this content is maintained until the blood reaches the capillaries. At the capillaries, however, the hemoglobin encounters a local P_{O_2} that is considerably lower than in the lungs, forcing it to release some of its bound oxygen. For example, the P_{O_2} in resting muscle and most tissues of the body is only about 40% of that in the lungs. At this P_{O_2}, the hemoglobin can bind only 75% as much oxygen as it did in the arteries, or 15 vol %. This means that 5 vol % (5 mL of oxygen for every 100 mL of blood) will have been released as the blood flows through the capillaries. Once the blood reaches the veins, it maintains this content of 15 vol % until it again passes through the lungs. The amount of oxygen that is released in the capillaries (i.e., the difference between the arterial content and the venous content) is known as the oxygen **extraction**. During exercise, the P_{O_2} in the muscles drops to extremely low values and extraction is greatly enhanced. The kinetics of oxygen transport will be presented in considerably more detail later in this chapter.

Life cycle Because they lack a nucleus, red blood cells cannot be produced by mitosis and must be produced by specialized cells in the bone marrow. This lack of a nucleus also makes them incapable of synthesizing proteins to replace enzymes or of repairing damage to their membranes. As a result, red cells exist for only about 120 days. This is still an amazing feat, however, when we consider that over this time they will travel more than 100 miles throughout the body. Since almost 1% of our total number of red blood cells die per day, we can calculate that, in the time that it takes you to read this paragraph, approximately 75 million cells have been removed from your circulation. You will not miss them, however, because over the same time approximately 100 million red blood cells have been added.

In the adult, blood cells develop from stem cells in the marrow of the bones of the upper body. Those destined to become red blood cells are differentiated by a process that involves the hormone **erythropoietin** (also known as EPO). EPO is produced and stored in the kidneys and its release is stimulated by a decrease in tissue oxygen content, as would occur with anemia or ascent to altitude. The development of red blood cells is known as **erythropoiesis**, a process that requires approximately 6 days. For the first 3–4 days, the infant cells (or erythroblasts) retain their nuclei and rapidly synthesize hemoglobin. They then lose their nuclei, but retain RNA so that hemoglobin synthesis continues. At this stage the still-immature cells, known as **reticulocytes**, are released from the bone marrow into the circulation. After about 2 days they attain the size of an adult erythrocyte and are no longer capable of any protein synthesis. After approximately 120 days their cell membranes rupture and the hemoglobin that they contained is transported by the protein **haptoglobin** to the bone marrow for synthesis of new cells, or its iron content is stored in the blood as **ferritin**. The ratio of reticulocytes to mature erythrocytes provides an indirect index as to how quickly red blood cells are being replaced. For example, during periods of intensive training, especially in runners, a higher than normal number of reticulocytes indicates a shorter lifespan for the red blood cells. An understanding of the time course for erythropoiesis is important in planning strategies for training or competing at altitude.

White Blood Cells

There are five different types of white cells or leukocytes. In order of their normal frequency of occurrence, they are the neutrophils, lymphocytes, monocytes, eosinophils, and basophils. Like red cells, they develop from the bone marrow but, unlike red cells, they can pass through the capillary membrane, so that at any point in time only a small proportion of their total number is actively in circulation, while the remainder are stored in the spleen or in other extravascular compartments. The lifespan for most white cells is less than 10 days, but some lymphocytes (T-cells) can live and provide immunity for up to 10 years or longer. Neutrophils, monocytes, eosinophils, and basophils have phagocytic properties that allow them to ingest and destroy damaged tissue or invading organisms such as bacteria. The manner in which each type of cell performs this function differs slightly, but does not require a memory of each specific type of bacterium. As a result, this process is known as **non-specific immunity**.

The lymphocytes can be subclassified into three types: T-cells, B-cells, and null cells. The T- and B-cells have the capacity to become programmed with a memory that enables them to recognize specific foreign proteins (antigens) and to transform themselves into antibodies that will destroy the invaders when reinfection occurs. As a result, this form of immunity is known as **specific immunity**. Null cells do not require specific recognition of an antigen and are able to destroy different types of tumor cells and viruses. The effects that exercise and training have on the immune system will be discussed in Chapter 9, which deals with overtraining.

Plasma

Plasma is the most important transporting medium in the body. Because it diffuses through the capillary wall, it can deliver materials to all cells. Dissolved substances that are transported in plasma include glucose, amino acids, lipids, enzymes, hormones, vitamins, oxygen, and carbon dioxide. Also dissolved in plasma are the positive ions (cations) such as sodium, potassium, and hydrogen, and the negative ions (anions) such as chloride, bicarbonate, and lactate. The balance between the cations and anions, especially hydrogen and bicarbonate, determines the pH of the blood. Plasma volume in an adult, since it comprises approximately 55% of total blood volume, is approximately 3 L.

Oxygen Loading

Oxygen is loaded into arterial blood as it passes through the lungs, transported by way of progressively smaller arteries and arterioles to the tissues, then unloaded as the blood passes through the tissues. In the resting state, or during low-intensity exercise, considerably more oxygen is delivered to the muscles than is extracted and utilized. However, during maximal efforts in endurance sports, a point is reached at which the rate of oxygen delivery is insufficient to meet the oxygen needs of the muscles. Recall that the maximal amount of oxygen that can be delivered to a muscle is determined by its arterial content and the maximal rate of blood flow. Arterial oxygen content is a function of hemoglobin concentration and the lung's ability to increase ventilation and to match it with blood flow. In the majority of situations the lung is capable of maintaining the same arterial content as at rest, even during maximal exercise. This is an impressive feat when we consider that the oxygen cost of this exercise may be up to 25 times more than that at rest.

The Lung

In simplest terms, the primary function of the lung is to move oxygen into the blood and carbon dioxide out of it. In accomplishing these tasks, the lung also performs other important functions, such as acting as a reservoir for blood and helping to control the acid/base balance of the body, but these are secondary to its role as a gas exchanger. In the resting state, the respiratory system moves about 6–8 L (see Box 3.6) of air in and out of the lungs each minute and extracts about 250 mL of oxygen from it. During maximal exercise, it is capable of moving more than 200 L of air per minute and extracting 6 L or more of oxygen. As will be discussed later in this section, the respiratory system has an extremely large reserve capacity that is normally not exceeded in healthy young athletes performing exhaustive exercise at sea level. As a result, training programs designed specifically to improve lung function are generally ineffective and unnecessary. For this reason, lung and respiratory physiology will not be presented in the same detail as the cardiovascular system.

> **BOX 3.6**
>
> Since gas volumes are affected by the temperature, barometric pressure, and water content at which they are collected, inspired volumes are normally expressed as what they would be at a standard temperature (0°C), pressure at sea level (760 mmHg), and without water vapor (dry) (STPD). Expired volumes are normally expressed as what they are at body temperature (37°C) and pressure, and saturated with water vapor (BTPS).

Anatomy The right and left lungs are housed within the rib cage and connected to the ambient air by the **bronchi** and **trachea**. The trachea leads from the larynx, in the middle of the throat, and branches into the right and left bronchi (Figure 3.16). The walls of the trachea and bronchi consist of rings of semi-rigid cartilage that allow them flexibility but prevent them from collapsing. Each bronchus divides and subdivides approximately 25 times into progressively smaller bronchioles, terminating in approximately 3 million **terminal bronchioles** in each lung. Each terminal bronchus leads into approximately 50 spherically shaped **alveoli**, which look like clusters of grapes at the end of a stalk. Although each alveolus is less than 0.3 mm in diameter, their shape provides a maximum of surface area per volume of lung. In a young adult there are approximately 300–400 million alveoli with a combined total surface area of more than 80 m^2, almost the size of a singles tennis court.

Figure 3.16 A simplified illustration of the anatomy of the lungs and connecting airways. The right and left bronchi divide and subdivide approximately 25 times into progressively smaller bronchioles, ending in approximately 3 million terminal bronchioles. Each terminal bronchus leads into approximately 50 alveoli, for a total of 300–400 million alveoli in a typical young adult.

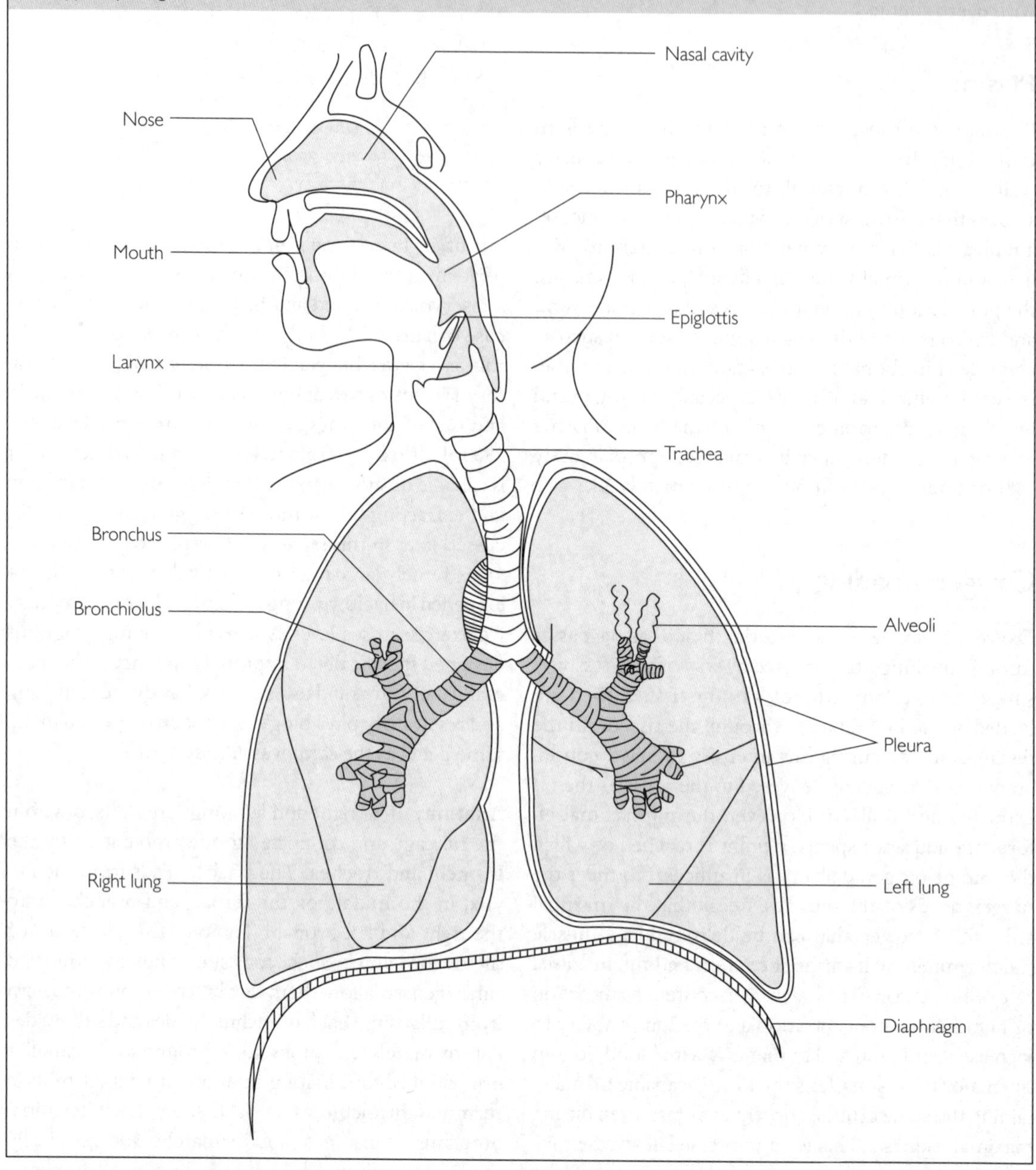

Like the alveolar membrane, which consists of a single layer of epithelial cells, the rich network of capillaries embedded in each alveolus is only one layer thick, creating extremely short diffusion distances for gas exchange. To protect the thin-walled alveoli from collapsing during forced expirations, the inner epithelial layer is coated with a specialized liquid film called **surfactant**. Surfactant is a detergent-like substance secreted by certain epithelial cells that reduces the surface tension in the alveolar wall to aid inflation, while keeping them stable and resistant to collapse. The entire cardiac output flows through the lung, with approximately 450 mL of blood in the pulmonary capillaries at rest at any one point in time. To appreciate the functional significance of this, imagine 450 mL of blood spread so thinly that it covers our tennis court: a truly optimal situation for gas exchange!

The lungs are enclosed by two membranes, known as **pleura**. The inner membrane adheres directly to the lungs while the outer membrane adheres to the chest wall and **diaphragm**. The two pleural layers are separated by a thin layer of fluid that has both cohesive and adhesive properties, so that they slide easily past each other but the seal between them remains unbroken. As a result, any movement of the chest wall or diaphragm is transmitted directly to the lungs. The dome-shaped diaphragm muscle is structured in such a way that, when it contracts, it flattens and descends into the abdominal cavity. The insertions of the external intercostal muscles are designed to raise the ribs and chest wall when they contract. Both of these actions cause an increase in the volume of the intrathoracic cavity. When the internal intercostal muscles contract, they decrease the volume of the chest.

Mechanics of ventilation An increase in intrathoracic volume creates a negative pressure in the lungs relative to atmospheric pressure, causing air to be drawn into them. Conversely, a decrease in intrathoracic volume increases the pressure in the lungs to a level above that in the atmosphere, forcing the lungs to deflate as air is expelled. During quiet breathing, the movement of air into and out of the lungs is accomplished almost exclusively by contraction and relaxation of the diaphragm. The diaphragm is innervated by branches of the phrenic nerve and receives bursts of neural traffic from respiratory control centers in the brain, causing it to contract approximately 10–15 times per minute. After each contraction, the diaphragm then relaxes and returns to its normal position, moving air out of the lungs. Expiration is thus largely a passive process that is also assisted by the elastic properties of lung. Normally, only a limited number of the total diaphragm fibers are activated during resting ventilation and the floor of the diaphragm descends only about 1 cm for each breath. However, during maximal ventilation or forced breathing, it may descend as much as 10 cm (45). With deeper inspirations or during exercise, the external intercostals, together with other chest muscles, are recruited to assist the diaphragm and cause more forceful contractions. In such situations, expiration becomes active as well, as the internal intercostals and abdominal muscles are activated.

The volume of gas that is moved into and out of the lungs in a single breath is called the **tidal volume** (V_T). Since the conventional unit for expressing pulmonary ventilation is liters per minute, it can be calculated as the product of frequency of breathing (f) per minute and the average tidal volume over that time:

$$V_E = f \times V_T$$

where V_E is expired ventilatory volume. In the resting state, the average tidal volume for a 70 kg adult would be approximately 600 mL and breathing frequency 12–15 times per minute, resulting in a ventilation of 7–8 $L \cdot min^{-1}$. During maximal exercise, tidal volumes could approach 3500 mL and breathing frequency 40–50 times per minute, resulting in a ventilation of 130–170 $L \cdot min^{-1}$. In large well-trained athletes, such as heavyweight rowers, it is not uncommon to record ventilations in excess of 200 $L \cdot min^{-1}$ during maximum exercise.

Pulmonary blood flow The pulmonary arteries and veins are much shorter and have larger diameters and thinner, more compliant walls than those in the systemic circuit. As a result they offer only about one sixth as much total resistance to flow as systemic vessels. At rest, mean pressure in the pulmonary artery is approximately 15 mmHg, compared to approximately 95 mmHg in the aorta. A consequence of this relatively low pressure is that, when an individual is in the upright position, flow is affected by the hydrostatic pressure effect of gravity and is not evenly distributed throughout the lung. Those branches of the pulmonary artery that supply blood to the upper portions of the lung (above the heart) will have a lower pressure and receive less flow than those at the bottom of the lung. Because the alveolar and capillary membranes are so thin and pliant and in contact with each other, the pressure in the alveoli is normally very similar to that in the surrounding capillaries. The exceptions are at the top portions of the lung, where alveolar pressures will exceed capillary pressures, and the bottom portions of the lung, where capillary pressures will exceed alveolar pressures. Thus, some alveoli at the top of the lung will be underperfused and overventilated, while some at the bottom will be overperfused and underventilated. As a consequence, some of the red blood cells passing through the bottom of the lung do not become saturated with oxygen. These imbalances in the **ventilation/perfusion ratio** have a minimal effect on total gas exchange in the healthy individual, but can pose problems in certain pathological conditions. During exercise, the increase in cardiac output elevates pulmonary arterial blood pressure, increasing the proportion of flow to the top of the lung and greatly alleviating any inequalities in ventilation/perfusion.

In the resting condition, red blood cells move through the pulmonary capillaries in about 0.75 seconds. As will be discussed in the subsection on oxygen, this transit time is more than adequate for complete gas exchange to occur between alveoli and blood. During heavy exercise, where cardiac output might increase five- to sixfold, it follows that total lung flow increases five- to sixfold as well. The velocity of flow, however, increases only about two- to threefold because of the redistribution of flow and perfusion of capillaries that were not utilized at rest. Pulmonary transit time during maximal exercise is thus approximately 0.35 seconds and still adequate for complete gas exchange, except in unusual circumstances.

Although the pulmonary vessels are supplied with sympathetic and parasympathetic nerve fibers, they respond only minimally to autonomic stimulation. Pulmonary circulation is thus primarily passive in nature, and is determined by changes in blood pressure in the pulmonary artery. An important exception is the potent reflexive vasoconstrictor response that occurs when there is a local decrease in alveolar P_{O_2} (see Box 3.7). This **hypoxic vasoconstrictor effect** is unique to the pulmonary vessels and the opposite of that which normally occurs in other blood vessels, where local hypoxia results in vasodilation. Since the vasoconstrictive effect only occurs in the immediate vicinity of those alveoli with low P_{O_2}, the reflex serves to direct blood flow away from this region to non-hypoxic alveoli. The mechanism is also thought by some to be a residual effect of that controlling pulmonary circulation in the fetus, where pulmonary arteries are greatly constricted and less than 20% of the cardiac output passes through the lungs. With the first breath at birth, the hypoxia is suddenly alleviated and there is a rapid increase in pulmonary flow. Whatever the reason for this reflex, it can prove to be a problem with ascent to altitude. Now, instead of a local vasoconstriction, there is a generalized vasoconstriction of all pulmonary arteries and arterioles that elevates pulmonary arterial pressure, increasing the work of the right heart and elevating pulmonary capillary hydrostatic pressure. If the latter effect is severe enough, it will result in the net movement of plasma out of the pulmonary capillaries, a condition known as **pulmonary edema**.

Gas transport and exchange Although some gases (such as nitrogen) are slightly soluble in blood and others (such as carbon monoxide) are very soluble, the most important gases that are transported and exchanged are oxygen and carbon dioxide. Gas exchange between the blood and the lung depends upon the differences in partial pressure (see Box 3.8) at the two sites. The partial pressures of alveolar gases are illustrated in Figure 3.17. Approximately 21% of the volume of the atmospheric air consists of oxygen and approximately 79% is nitrogen, with trace amounts of carbon dioxide and other gases. At sea level, the P_{O_2} of atmospheric air is thus approximately 160 mmHg and the P_{N_2} is approximately 600 mmHg. However, these pressures will change when this air is drawn into the lungs. This is because water vapor and carbon dioxide

BOX 3.7

The threshold for the hypoxic vasoconstrictor reflex appears to be a P_{O_2} of approximately 70 mmHg. A decline in alveolar P_{O_2} from its normal value of 100 to 70 mmHg has little effect but, when it drops below this value, there is a marked vasoconstrictor effect. It is also known that it is the P_{O_2} of the alveolar gas, and not that in the pulmonary artery, that triggers this response (45).

BOX 3.8

Molecules of gas are in constant random motion, so that at any one time some are colliding with the walls of their container and exerting outward pressure on it. The greater the number of molecules that are present, the greater the number of collisions. Thus, the pressure that a gas exerts is directly dependent on its **concentration**. In a mixture of gases, the concentration of a given gas is expressed as a percentage of the total volume. The pressure exerted by each gas is known as its **partial pressure** and the total pressure exerted by the mixture is equal to the sum of the partial pressures of each component gas (Dalton's law). The total pressure exerted by the weight of the gas in the atmosphere is termed the **barometric pressure**. At sea level, the barometric pressure is approximately 760 mmHg and, since the concentration of oxygen is approximately 21%, it follows that the P_{O_2} equals 21% of 760, or 160 mmHg.

Figure 3.17 The fractional concentration (*F*) and partial pressure (*P*) for oxygen, nitrogen, carbon dioxide, and water vapor in ambient and alveolar air at sea level (barometric pressure 760 mmHg).

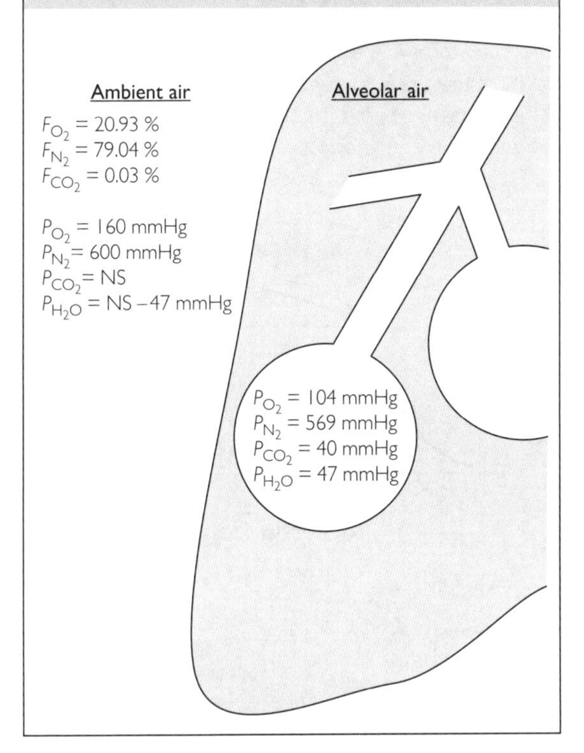

have been added to the alveolar gas. At the end of each expiration, the air remaining in the bronchi, trachea, and throat is the last to leave the alveoli and is therefore saturated with water vapor and high in carbon dioxide. With the next inspiration, this air is then drawn back into the lungs, diluting the amount of oxygen that reaches the alveoli. The volume of gas that remains in the airways at the end of each expiration is known as respiratory **dead space**. The volume of the dead space depends upon body size, but on average is about 150 mL (see Box 3.9). Because of this dead-space effect, the P_{O_2} of alveolar gas will average about 100 mmHg, the P_{CO_2} is about 40 mmHg, the P_{N_2} is about 570 mmHg, and the P_{H_2O} is 47 mmHg, because it is saturated with water vapor. When tidal volumes are relatively consistent, these values remain quite constant and do not fluctuate with each inspiration and expiration, as they would in the trachea and larger bronchioles. This is because the extensive branching of the smaller bronchioles provides a spatial buffering effect, so that it would take several seconds before

BOX 3.9

Dead space in milliliters is approximately equal to an individual's body weight in pounds. Thus, with an inspiration of 600 mL (a normal resting tidal volume for an individual of this size), a 170 lb athlete would move only about 430 mL (600–170) of fresh ambient air into the alveoli. In this example, 28% of each inspired volume remains in the dead space. During exercise, the physiological consequence of dead space is greatly diminished because there is a large increase in inspired volume and only a slight increase in dead space (due to slight dilation of the bronchioles). For example, during heavy exercise, inspired volume might now be 3000 mL, whereas dead space might increase to only about 200 mL or about 6% of each breath.

rapid changes in breathing patterns (e.g., hyperventilation or breath-holding) become apparent.

Oxygen Since the P_{O_2} of alveolar gas averages about 100 mmHg, oxygen diffuses from alveoli to blood as the blood flows through the pulmonary capillaries, until the P_{O_2} of the blood becomes 100 mmHg as well (see Figure 3.18 depicting transit time). At this P_{O_2}, the hemoglobin in the red blood cells combines readily

Figure 3.18 The change in partial pressure of oxygen (P_{O_2}) in blood as it flows through a typical pulmonary capillary. Although red blood cells are exposed to alveolar gas for approximately 0.75 seconds (pulmonary capillary transit time), they are normally fully saturated with oxygen within the first 0.3 seconds.

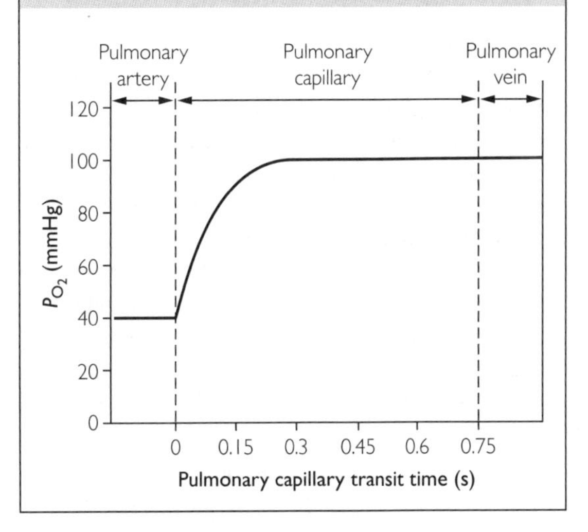

Figure 3.19 (A) The oxyhemoglobin dissociation curve showing the relationship between P_{O_2}, percentage hemoglobin saturation, and blood oxygen content in a typical individual (at pH 7.4, P_{CO_2} 40 mmHg, blood temperature 37°C, and a hemoglobin concentration of 15 g·100 mL^{-1} of blood). Volume % refers to the content in milliliters per 100 mL of whole blood. (B) The O_2 content of arterial blood (P_{O_2} = 100 mmHg) and venous blood (tissue P_{O_2} = 40 mmHg) in the resting condition. Note that arterial blood has a content of approximately 20 vol %, but at the tissue level it can only bind approximately 15 vol %. Thus, for every 100 mL of blood that flows through the tissues, 5 mL of O_2 are dissociated or unloaded. (C) The O_2 content of blood during heavy exercise. Note that the curve has shifted to the right (Bohr effect) as a result of increased CO_2, heat, and H^+ production at the exercising muscles. These changes, combined with the lower muscle P_{O_2} due to exercise, result in a marked decrease in the ability of hemoglobin to bind oxygen and thus an increase in the amount of oxygen unloaded (more than 15 vol %) as blood flows through the muscle.

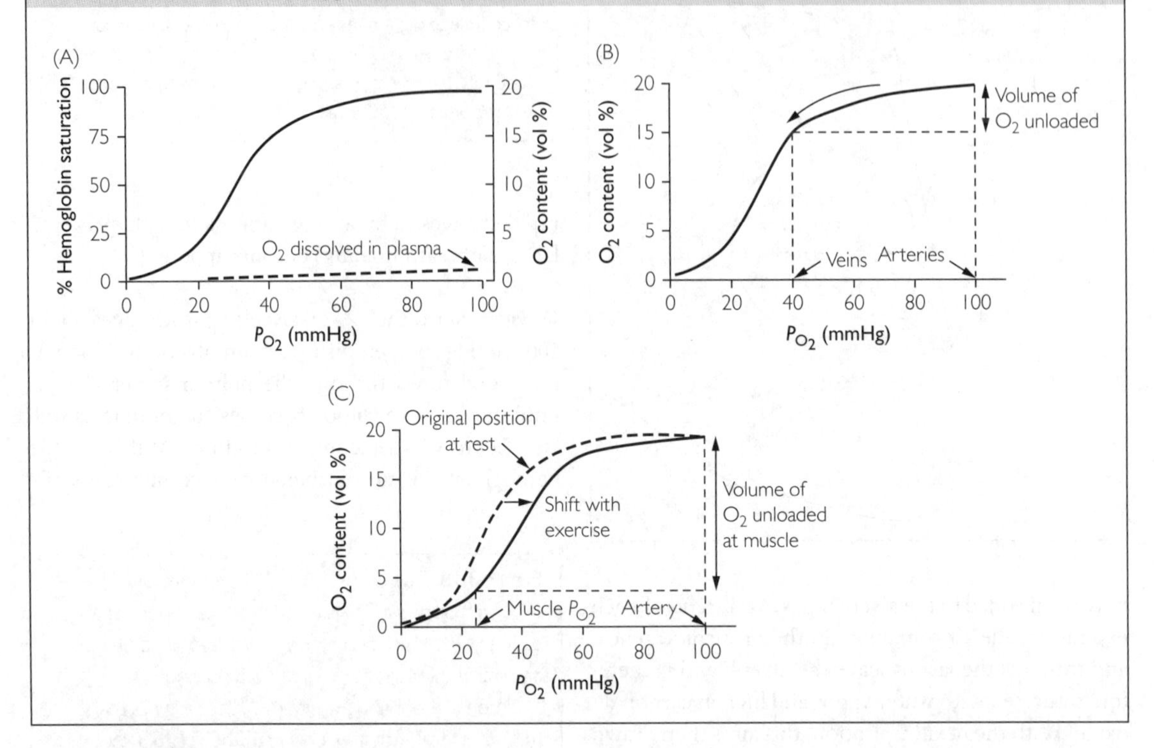

with oxygen to form oxyhemoglobin. The relationship between hemoglobin binding and P_{O_2} is illustrated in the **oxyhemoglobin dissociation curve** depicted in Figure 3.19. Notice that, as P_{O_2} increases from 0, there is a steep rise in hemoglobin saturation and the oxygen content of the blood. This continues in an almost linear fashion until a P_{O_2} of approximately 50 mmHg, at which point the curve becomes flatter. At a P_{O_2} of 100 mmHg, the hemoglobin is approximately 97% saturated, so that the oxygen content of the blood (the amount bound to hemoglobin plus the small amount dissolved in plasma) is approximately 20 vol %.[4] This P_{O_2} and oxygen content is maintained as the blood passes through the pulmonary vein to the left heart and then out through the aorta to the arteries and arterioles. At the capillaries, the P_{O_2} in the blood will decline to equilibrate with that in the surrounding tissue: about 40 mmHg for most resting tissues. Reading from Figure 3.19, we can see that, at this P_{O_2}, the hemoglobin is approximately 75% saturated and the oxygen content of the blood will be reduced to approximately 15 vol %. This content of 15 vol % will be returned by the veins to the right heart and then to the lungs to begin the cycle again. Thus, the oxygen extraction or **arteriovenous oxygen difference** [(a-v) O_2 difference] in resting tissue averages about 5 vol %, or 5 mL of oxygen per 100 mL of blood flow. It should be noted that, in tissues such

[4] Blood content of a gas is usually measured in volumes %; i.e., the quantity of gas, in milliliters, per 100 mL of blood.

as kidney and skin, P_{O_2} is much higher and thus extraction lower, whereas in tissues such as brain and heart extraction is considerably greater. On average, however, a sample of mixed venous blood in individuals with a normal hemoglobin concentration would have an oxygen content of approximately 15 vol %. (Mixed venous blood represents an average of all the blood returning to the right heart and the sample would have to be taken from the right ventricle or pulmonary artery.)

The shape of the oxyhemoglobin dissociation curve—quite flat at the top, but much steeper in the middle and lower portions—has several functional advantages. The significance of the flat portion is that relatively large decreases in alveolar P_{O_2}, as might occur in certain pathological conditions or on ascent to higher altitudes, have only a slight effect on the amount of oxygen loaded into the blood at the lungs. In contrast, at the steep portions of the curve, a slight decrease in P_{O_2} results in a large reduction in hemoglobin saturation and thus enhanced oxygen extraction by the tissues. This constitutes a very efficient autoregulatory mechanism, as a result of which tissues with higher metabolic rates (e.g., heart and brain) and lower P_{O_2} automatically have higher oxygen extraction. During increasing exercise, P_{O_2} in the active muscles becomes lower and lower, resulting in greater and greater oxygen extraction. Sampling of blood in veins draining a muscle has indicated P_{O_2} levels as low as 15 mmHg (35) during maximal aerobic exercise. Since these samples include blood from capillaries surrounding type 2 fibers, which would have been less active during such exercise, it is probable that P_{O_2} in many active fibers approaches 0.

Hemoglobin's ability to bind with oxygen is affected not only by P_{O_2} but also by the temperature of the tissue and local concentration of carbon dioxide and H^+. An increase in temperature, P_{CO_2}, or $[H^+]$ reduces this binding ability and thus results in a rightward shift of the oxyhemoglobin dissociation curve (Figure 3.19). This is termed the **Bohr effect** and means that, at any given P_{O_2}, more oxygen is extracted or unloaded. During exercise, muscle temperature increases, as does CO_2 and H^+ production, causing greater O_2 extraction at the exercising muscles. The consequences of this rightward shift are most apparent at the unloading or steep portions of the curve, with little or no effect on oxygen loading at the lungs. This is because loading occurs at the flatter part of the curve, and because alveolar P_{CO_2} actually *decreases* during heavy exercise. Thus, during heavy and maximal exercise, loading of oxygen at the lungs is not compromised, while unloading is greatly enhanced and the muscle extracts virtually all of the oxygen that is delivered to it.

Once oxygen enters the muscle cell, it combines with **myoglobin**. Myoglobin is a muscle protein with a structure similar to that of hemoglobin, but with a greater oxygen-binding capacity. It occurs in greatest concentration in type 1 fibers and in lowest concentration in type 2X fibers. The myoglobin-O_2 dissociation curve is considerably steeper than that for hemoglobin and, as a result, as the P_{O_2} in the vicinity of the mitochondria decreases with exercise, oxygen is rapidly released from myoglobin to maintain the oxidative process. Myoglobin thus acts as an oxygen buffer, ensuring a more even distribution of oxygen throughout the exercising fiber than would occur with diffusion alone. In addition, the extra oxygen that is bound to myoglobin in the resting state ensures an immediate supply at the sudden onset of exercise, before blood flow has had time to increase.

Carbon dioxide Because carbon dioxide is about 20 times more soluble in water than is oxygen, a significant amount is transported dissolved in the plasma. A small amount of this dissolved carbon dioxide will combine with water to form carbonic acid (H_2CO_3), a portion of which will dissociate in the form of H^+ and bicarbonate, HCO_3^-:

$$CO_2 + H_2O = H_2CO_3 = H^+ + HCO_3^-$$

In plasma, the first reaction occurs relatively slowly, while the second (the dissociation step) occurs very rapidly. Both reactions are readily reversible and proceed until chemical equilibrium occurs. About 9% of the total CO_2 that is generated in the body is transported in plasma in the dissolved state and about 1% as H_2CO_3.

The remaining 90% is transported as products of reactions that occur in the red blood cells. Mature red blood cells contain the enzyme **carbonic anhydrase**, a potent catalyst for the reaction by which CO_2 combines with water to form carbonic acid. In the presence of carbonic anhydrase most of the CO_2 that has diffused into the blood cell combines with the water in the cell at a rate that is several hundred times more rapid than in plasma and, in the process, generates large quantities of H_2CO_3. This H_2CO_3 immediately dissociates into H^+ and HCO_3^-. The H^+ ions become buffered by hemoglobin (see Box 3.10) and the

BOX 3.10

In an acid environment, many proteins can act as buffers by becoming proton acceptors and thus neutralizing or taking up H^+. In blood this is accomplished by plasma proteins such as albumin, and in the red cell by the protein hemoglobin. Reduced hemoglobin (hemoglobin that has released its O_2) is a particularly effective buffer and thus, during exercise, where greater oxygen extraction occurs, the pH of the red cell is maintained despite large increases in carbon dioxide transport by venous blood.

HCO_3^- ions diffuse out of the cell into the plasma. As this happens, ionic balance between the red cells and plasma is maintained by the inward diffusion of chloride ions (Cl^-), a process known as the chloride shift. Approximately 70% of the total CO_2 in blood is in the form of HCO_3^- as generated in the red blood cells by these reactions. At the same time that this is occurring, some CO_2 will also bind loosely with hemoglobin within the blood cell to form the compound **carbaminohemoglobin**. This union is facilitated at the tissue level by the freeing up of binding sites as the hemoglobin releases O_2. In contrast, in the lungs the uptake of O_2 by hemoglobin helps to displace CO_2, causing it to be unloaded and removed in expiratory gas. The mutually beneficial interactions between O_2 and CO_2 binding and release by hemoglobin are known as the **Haldane effect**. Approximately 20% of the CO_2 in blood is transported bound to hemoglobin.

In summary, carbon dioxide is transported in blood in three ways: dissolved in plasma, in combination with hemoglobin within red blood cells, and as bicarbonate (Figure 3.20). Approximately 70% of all carbon dioxide transported is in the form of bicarbonate, 20% is combined with hemoglobin, and 10% is in the dissolved form. Each of the chemical reactions involved in these mechanisms is reversible and its direction depends upon the P_{CO_2}. Thus, when P_{CO_2} is high, the reaction proceeds to the right; when it is low, net movement is to the left, causing a release of carbon dioxide from the blood. Since carbon dioxide is produced by the tissues, P_{CO_2} is highest (approximately 46 mmHg at rest) at that site. As blood passes through the capillaries, carbon dioxide diffuses rapidly until its partial pressure matches that of the tissues (46 mmHg). At this P_{CO_2}, the net movement of the reactions is to the right, causing carbon dioxide to be loaded into the blood. When the blood reaches the lungs, it encounters an alveolar P_{CO_2} of 40 mmHg. At this P_{CO_2}, the net movement of all reactions is to the left, causing some of the carbon dioxide to be unloaded.

The relationship between P_{CO_2} and blood carbon dioxide content is depicted in Figure 3.21. Notice that, at a P_{CO_2} of 46 mmHg, blood carbon dioxide

Figure 3.20 A summary of the three ways by which carbon dioxide is transported by the blood: dissolved in plasma, combined with hemoglobin in the red blood cells, and as bicarbonate generated by the production of carbonic acid in the red blood cells and diffused into the plasma. See text for details.

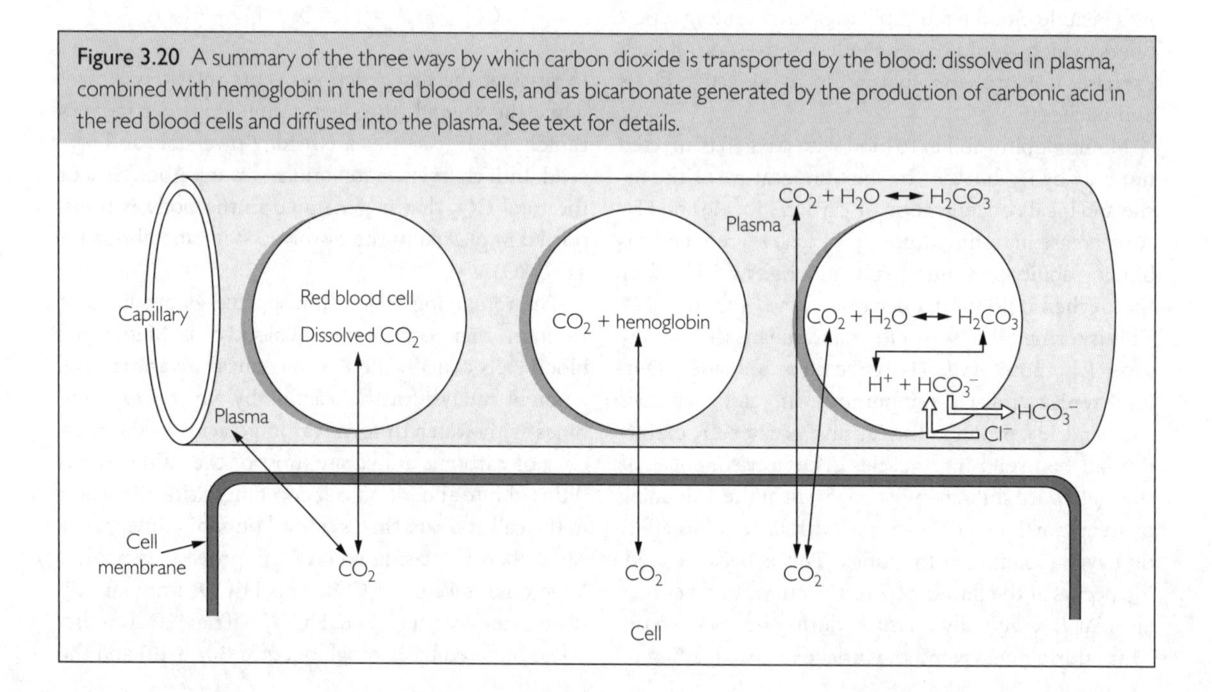

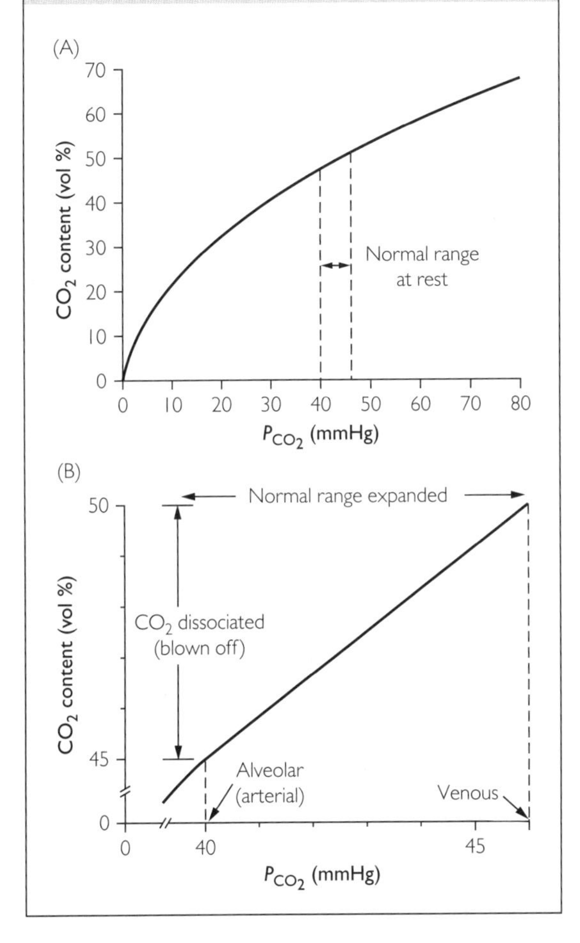

Figure 3.21 (A) The CO_2 association/dissociation curve. Note that between P_{CO_2} levels of approximately 20 and 60 (the normal physiological limits) the slope is almost linear. (B) The CO_2 association/dissociation curve, expanded to show CO_2 content in venous and arterial blood in the resting state. Note that when venous blood (CO_2 content of 50 vol %) flows through the lung (alveolar P_{CO_2} = 40 mmHg), the content drops to 45 vol %. See text for details.

content is approximately 50 vol %. Thus, venous blood contains approximately 50 mL of CO_2 per 100 mL of blood. When this blood is returned to the lungs, where P_{CO_2} is 40 mmHg, its content drops to approximately 45 vol % and the dissociated carbon dioxide is blown off by ventilation, leaving the arterial blood with a P_{CO_2} of 40 mmHg and a carbon dioxide content of 45 vol %. This venous/arterial difference in CO_2 content (5 vol %) is similar to the arterial/venous difference for oxygen; as a result, expired ventilatory volume (VE) is almost identical to inspired ventilatory volume (VI).

Figure 3.22 summarizes the partial pressures of O_2 and CO_2 in the lungs, muscle, and blood at rest and during heavy exercise (shown in parentheses on the figure). Since O_2 is utilized at a rapid rate during exercise, the P_{O_2} of muscle and capillary blood decreases considerably. In this example, P_{O_2} in a typical capillary has dropped from its normal resting value of 40 mmHg to 10 mmHg. At this P_{O_2}, the O_2 content of the blood would be approximately 3 vol %, resulting in an arterial/venous O_2 difference across the muscle of approximately 17 vol % (20 − 3). As this blood is returned to the right heart, it becomes mixed with blood returning from other tissues, so that mixed venous P_{O_2} would be approximately 20 mmHg, with an O_2 content of approximately 5 vol %. Because of the increase in CO_2 production generated by oxidative metabolism, the P_{CO_2} in the muscle might be as high as 60 mmHg. At this P_{CO_2}, the CO_2 content of a typical muscle capillary would be approximately 58 vol % (Figure 3.21). Upon its return to the lung, this blood encounters an alveolar P_{CO_2} of 40 mmHg or less. At exercise intensities up to approximately 75% of $\dot{V}O_{2max}$, alveolar P_{CO_2} is maintained at 40 mmHg by well-controlled increases in ventilation that are directly proportional to the increase in CO_2 production. As detailed in the following paragraph, at exercise intensities above this point, ventilation begins to increase disproportionately to metabolic CO_2 production, causing alveolar P_{CO_2} to drop below 40 mmHg. In this example, alveolar P_{CO_2} has declined to 35 mmHg. At this P_{CO_2}, the CO_2 content of the blood would be approximately 40 vol % and the venous/arterial CO_2 difference 18 vol % (58 − 40). The venous/arterial CO_2 difference (18 vol %) is thus slightly greater than the arterial/venous O_2 difference across the muscle (17 vol %) and expired ventilatory volume is slightly greater than the inspired volume. Notice that alveolar P_{O_2} has also increased slightly (to 110 mmHg) in the exercise condition. This is a result of an increase in tidal volume, reducing the effect of dead space, and the fact that alveolar P_{CO_2} has decreased slightly.

Ventilation during exercise When an individual begins exercise, both tidal volume (V_T) and frequency of breathing (*f*) increase immediately, with the increase in V_T being proportionally greater than the increase in *f*. If the exercise is progressive in nature, these changes tend to mirror the increase in exercise intensity, so

Figure 3.22 The partial pressures of O_2 and CO_2 throughout the system and pulmonary circuits at rest and during heavy exercise (shown in parentheses). RA and LA indicate right and left atria and RV and LV right and left ventricles. See text for further explanation.

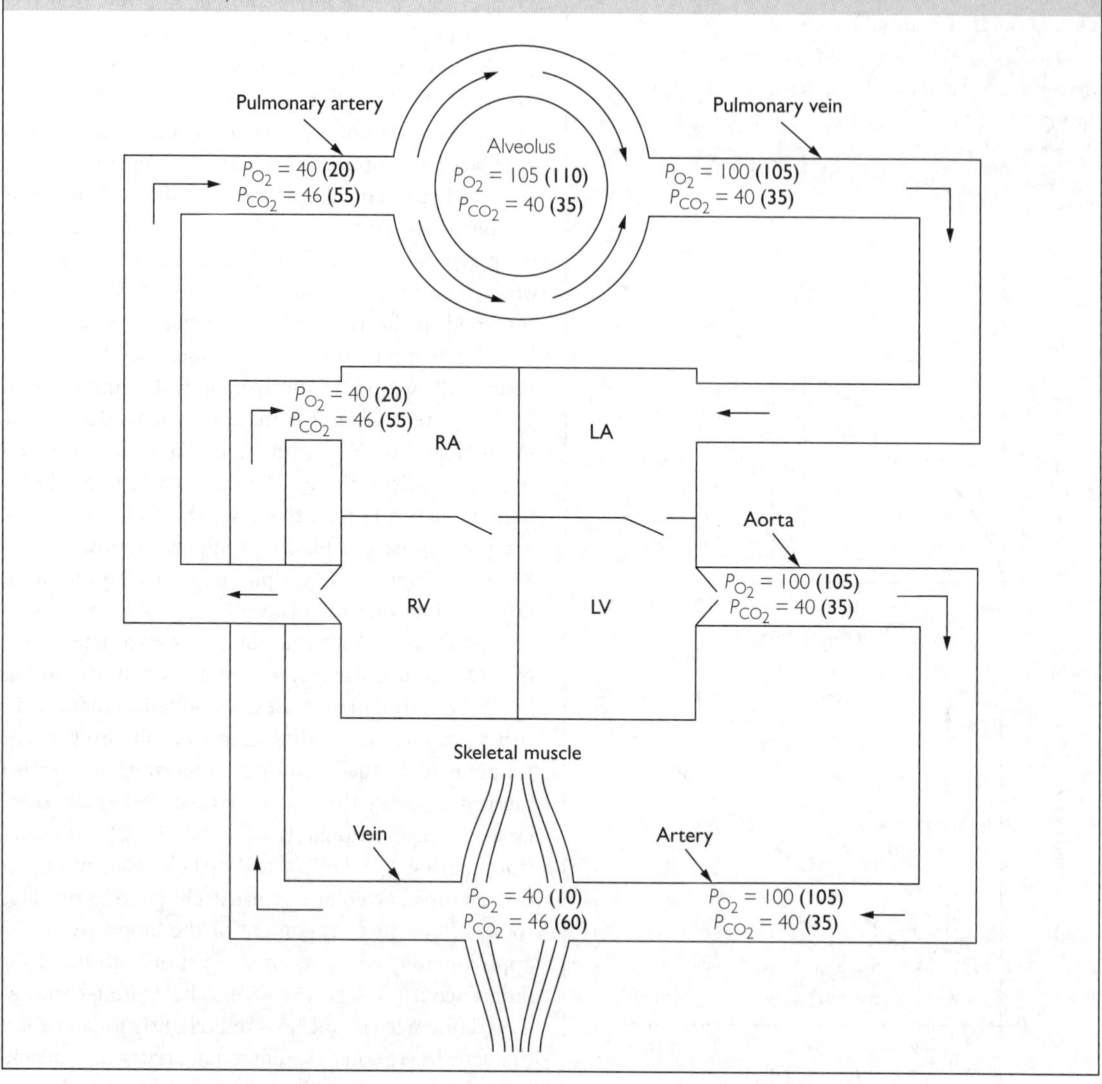

that, until about 65–75 % $\dot{V}O_{2max}$, ventilation increases almost linearly with oxygen consumption (Figure 3.23). For example, ventilation in an endurance athlete might increase from ≈7.2 L·min^{-1} at rest (V_T = 600 mL and f = 12 breaths per minute) to ≈72 L·min^{-1} (V_T = 3000 mL and f = 24 breaths per minute) at an exercise intensity that requires 60% of his $\dot{V}O_{2max}$. This fivefold increase in V_T with only a twofold increase in breathing frequency is a more effective means of increasing ventilation because it minimizes the effects of dead space on alveolar P_{O_2}. However, as exercise intensity approaches maximum, further increases in ventilation are achieved by increases in f, with little change (and perhaps a slight decrease) in V_T.

As exercise intensity increases beyond 65–75% of $\dot{V}O_{2max}$, ventilation begins to increase in an exponential or alinear fashion. The point at which this departure from linearity first occurs is known as the **ventilatory threshold**, or T_{vent}. The mechanisms that stimulate the exaggerated ventilatory response from this point onwards are complex and controversial. Since, in most situations, there is a good

Figure 3.23 Typical changes in ventilation that occur as a resting subject undergoes progressive exercise to exhaustion. The point where ventilation begins to increase in non-linear fashion is known as the ventilatory threshold (T_{vent}).

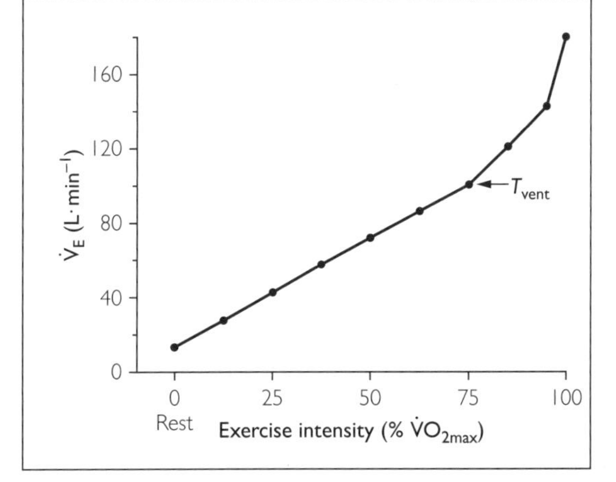

correlation between the appearance of the ventilatory threshold and the **lactate threshold** discussed in Chapter 2, it is generally accepted that at least part of the extra ventilatory drive occurs in response to the appearance of lactic acid in the blood. The buffering of this lactic acid by sodium bicarbonate generates excess CO_2 that acts as a further stimulus for increased ventilation:

$$\text{Lactic acid} + NaHCO_3 = \text{Na lactate} + H_2CO_3 \rightarrow H_2O + CO_2$$

In addition, as exercise intensity approaches maximum, some unbuffered H^+ will dissociate from both lactic and carbonic acid to stimulate ventilation as well. It is unlikely that the release of lactic acid from muscle is the only cause for the ventilatory threshold, since some of the extra ventilatory stimulus is believed to arise from increases in circulating catecholamines and potassium and spillover from descending neural traffic from the motor cortex to the muscles, all of which tend to increase exponentially at about the same point.

Control of ventilation during exercise The signals that stimulate the respiratory muscles to contract can be traced back to a cluster of specialized neurons in the medullary portion of the brain stem. There are two separate neuronal networks: a dorsal group that activates inspiration, and a ventral group that activates both inspiration and expiration. Collectively they are known as the **respiratory control center** (Figure 3.24). Output from the respiratory control center is transmitted as neural impulses along the phrenic nerve to the diaphragm and along the spinal nerves to the intercostal and abdominal muscles. Rhythmicity in breathing patterns is caused by regular discharges from the control center modulated by inhibitory feedback from stretch receptors in the airways and in the walls of the lung.

There are a number of sources of input to the respiratory control center. Probably the most important input source is from the respiratory **chemoreceptors**. The chemoreceptors are specialized neurons that are sensitive to changes in P_{O_2}, P_{CO_2}, and H^+ concentration. They are located on the medullary surface of the brain stem (**central chemoreceptors**) and in the carotid and aortic bodies (**peripheral chemoreceptors**). The central chemoreceptors are surrounded by cerebrospinal fluid (CSF) and are very sensitive to changes in its H^+ concentration. An increase in H^+ results in an immediate increase in respiration, while a decrease results in a decrease in respiration. The H^+ concentration of the CSF is controlled primarily by the P_{CO_2} of the arterial blood, since CO_2 diffuses out of the cerebral arteries, across the blood–brain barrier and into the CSF. Some of this CO_2 generates carbonic acid, which ionizes to release H^+. Because the protein content of the CSF is less than that of blood, it is less effective as a buffer and increases in blood P_{CO_2} thus have a greater effect on the H^+ concentration of the CSF than on that in the blood.

The peripheral chemoreceptors are located at the bifurcation of the common carotid arteries, on both sides of the neck, and in the aortic arch. In humans, the carotid body receptors are more important than those in the aorta. The carotid bodies receive a very high blood flow in proportion to their size. As a result, the (a-v) O_2 difference across them is minimal and capillary blood P_{O_2} is almost identical to that of arterial blood. The receptors are sensitive to changes in *arterial* blood chemistry and increase their rate of discharge if there is a decrease in P_{O_2}, an increase in P_{CO_2}, or an increase in H^+ concentration (decrease in pH). Of these variables, within the normal physiological range, they are most sensitive to changes in P_{CO_2} and demonstrate little or no response until P_{O_2} decreases to about 65 mmHg. Because they are in direct contact with blood, their response is more rapid than that of the central chemoreceptors.

In the resting state, ventilation is regulated almost exclusively by feedback from the chemoreceptors in

Figure 3.24 A schematic summary of the afferent and efferent pathways of the respiratory control center. Input to the center comes from the cerebral cortex, central chemoreceptors, and peripheral chemoreceptors. Output goes to the motor neurons of the diaphragm, intercostals, and abdominals. CSF, cerebrospinal fluid.

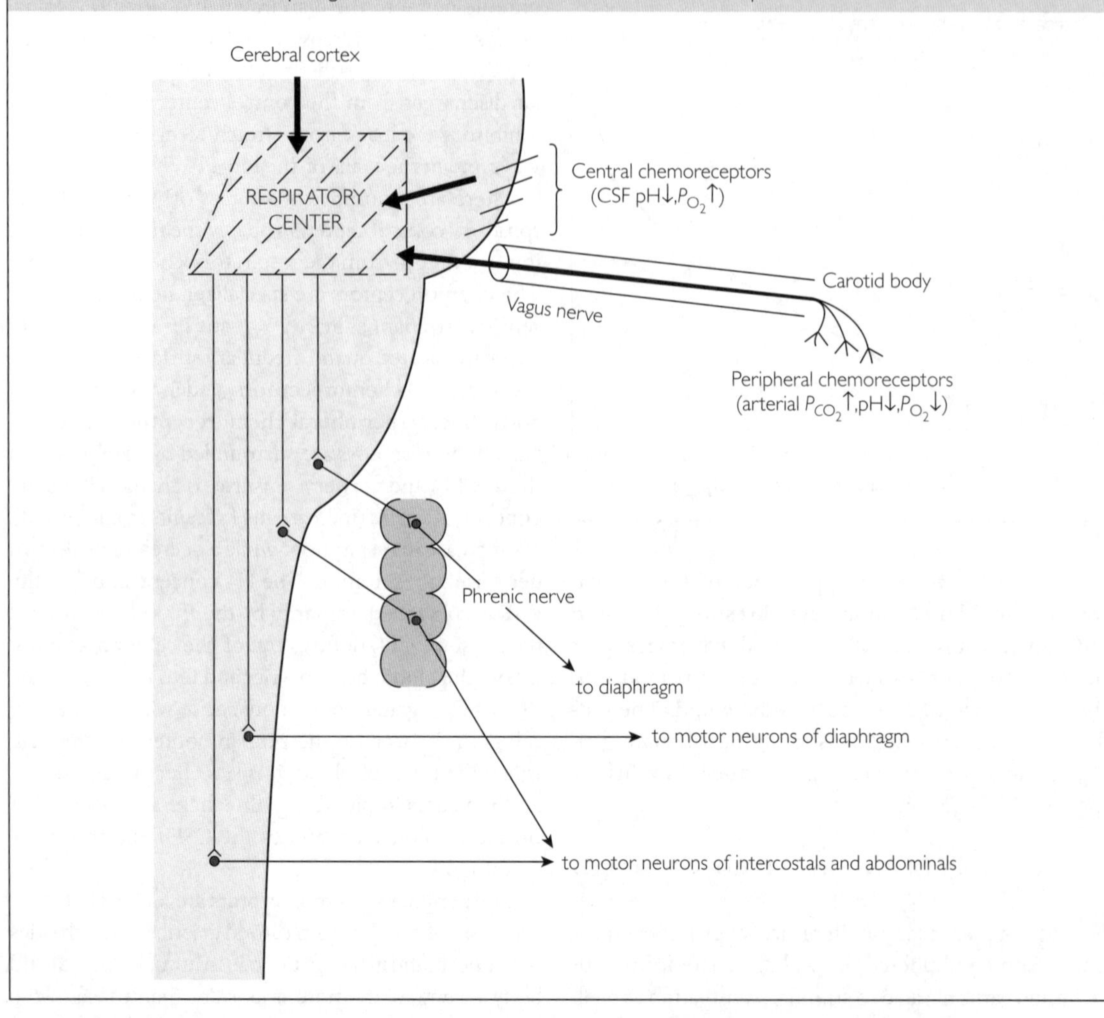

response to changes in arterial P_{CO_2}. Both types of receptors function simultaneously, with proportionally more input coming from the central chemoreceptors. In addition, because of the rapidity with which the peripheral receptors respond, slight increases in P_{CO_2} are met with an almost immediate increase in ventilation. Thus, breathing is regulated by our rate of CO_2 production and not by our need for O_2. We breathe in order to remove or blow off CO_2 and, under normal conditions, this hypersensitivity to CO_2 ensures that ventilation is always adequate to maintain O_2 delivery as well.

During exercise, the respiratory center receives input from several sources in addition to the chemoreceptors. These include neural input from the motor cortex area of the brain, as impulses are sent to the muscles, as well as a form of feedback from the muscles themselves. This latter pathway is thought to involve movement or **proprioceptors** in the muscle that become activated by movement as the muscle contracts. Because these sources of input are neural in nature and directly proportional to the intensity of exercise, they cause abrupt increases in ventilation in response to the sudden onset of exercise or rapid changes in intensity. Thus adequate ventilation is immediately ensured, allowing time for the chemoreceptors to catch up with the changes in CO_2 production caused by increased metabolism. A final source of input during exercise may be triggered by increases in body temperature, but whether or not this is a direct effect on the respiratory center is not known.

The oxygen cost of ventilation Inspiration and (during exercise) expiration are caused by contraction of the respiratory muscles. The act of breathing itself thus consumes oxygen. Oxygen consumption by the respiratory muscles is very low during breathing at rest, but increases exponentially during progressive exercise. This is primarily because of the fact that the respiratory muscles must contract more and more forcefully against the elastic resistance of the lung in order to provide a given increase in tidal volume. In addition, during exercise, there is an increase in breathing frequency. It has been estimated that the oxygen cost of ventilation accounts for less than 4% of the total resting $\dot{V}O_2$, while during maximal exercise it increases to more than 10% of the total exercise $\dot{V}O_2$. To put this into perspective, in a typical resting individual (resting $\dot{V}O_2 = 0.3$ L·min^{-1}), the oxygen cost of breathing would be approximately 12 mL·min^{-1}. In contrast, during maximal exercise (e.g., $\dot{V}O_{2max} = 5.0$ L·min^{-1}), the respiratory muscles would consume more than 500 mL of O_2·min^{-1}. We might assume that this increase steals oxygen from the skeletal muscles and thus limits $\dot{V}O_{2max}$, but, as we will see in the following section (Does Respiratory Performance Limit Exercise Performance?), this is not the case.

Does Respiratory Performance Limit Exercise Performance?

In order for lung function to be considered a weak link in the body's ability to deliver adequate oxygen to the exercising muscles, we would first have to demonstrate that a decline occurs in the oxygen content in the pulmonary artery during exercise. In most athletes this does not happen, even during maximal exercise to the point of exhaustion. We are therefore left to conclude that there is considerable reserve in the lung's capacity to move air, which cannot be maximally taxed during strenuous exercise. This is illustrated in Figure 3.25, which summarizes the typical changes in ventilation, $\dot{V}O_2$, and arterial P_{O_2} that would occur as an athlete undergoes progressive increases in exercise intensity leading to exhaustion in 10 minutes.

Figure 3.25 Changes in ventilation, oxygen uptake ($\dot{V}O_2$), and arterial P_{O_2} as an athlete undergoes progressive exercise leading to exhaustion in 10 minutes. Notice that at $\dot{V}O_{2max}$ (at approximately 9 minutes) he or she has not reached maximal ventilation and that there is minimal decline in arterial P_{O_2}. MVV, maximal voluntary ventilation (see text).

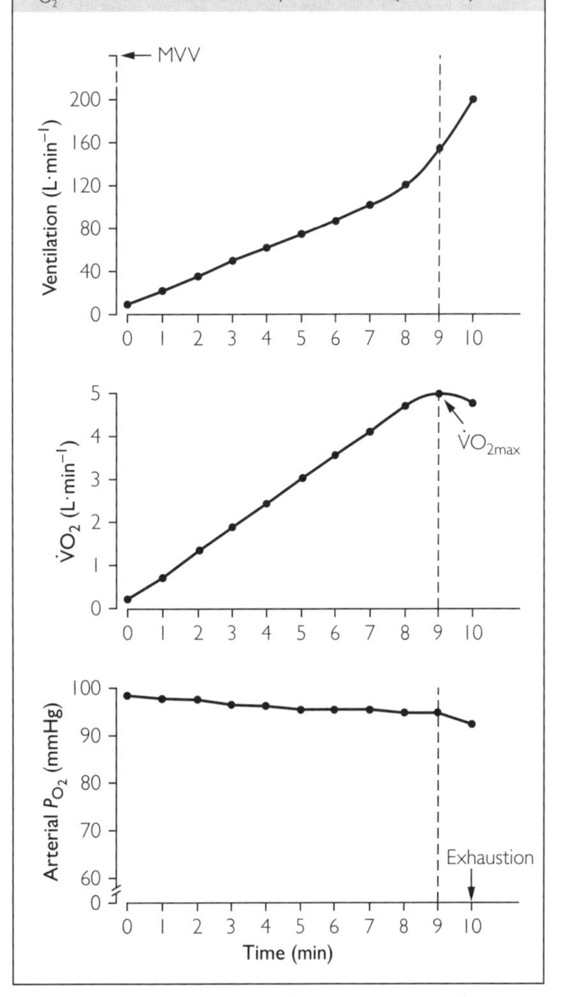

As exercise intensity and $\dot{V}O_2$ increase, there is a proportional linear increase in pulmonary ventilation. This direct relationship continues until the exercise intensity requires approximately 70% of the athlete's $\dot{V}O_{2max}$, at which point ventilation begins to demonstrate the exaggerated or curvilinear response discussed previously. With further increases in intensity, a point is reached at which the oxygen cost of the exercise exceeds the $\dot{V}O_{2max}$ of the athlete, as indicated by the plateau in $\dot{V}O_2$. Although the athlete has reached his or her $\dot{V}O_{2max}$ at this point, he or she is still able to tolerate further increases in exercise intensity by drawing on anaerobic energy delivery processes for perhaps a minute or two before becoming exhausted. Notice that, at the point where $\dot{V}O_2$ begins to plateau,

BOX 3.11

Maximal voluntary ventilation (or maximal breathing capacity) and other lung volumes can be measured by breathing into a recording spirometer. To measure maximal ventilation, the subject is instructed to move as much air as possible in 10 seconds by breathing as deeply and rapidly as possible into the spirometer. The volume achieved in 10 seconds is then multiplied by 6 to arrive at a maximal minute ventilation.

ventilation is far from maximal and is still able to increase another 30–40 $L \cdot min^{-1}$ before exhaustion. In addition, even at exhaustion, ventilation is considerably less than can be achieved in a test of maximal voluntary ventilation (see Box 3.11). The fact that ventilation is submaximal when $\dot{V}O_2$ plateaus and that further increases in ventilation do not elevate $\dot{V}O_2$ indicates that inadequate ventilation does not limit $\dot{V}O_{2max}$.

Further evidence that lung function does not normally limit exercise performance is furnished by examining arterial P_{O_2}. Studies consistently show that during maximal exercise leading to exhaustion, in healthy young subjects and the majority of endurance trained athletes, arterial P_{O_2} remains within 90% of normal resting levels (Figure 3.24). This slight decline is not great enough to exert an effect on performance and can be interpreted as evidence that, even at maximal exercise, adequate oxygen saturation is occurring and the lung is still doing its job.

The exception The exception to this rule occurs in approximately 25% of highly trained endurance athletes ($\dot{V}O_{2max}$ > 65 $mL \cdot kg^{-1}$). In these individuals, a marked decrease in arterial P_{O_2} can be detected, with values dropping to 60–70% of normal (11). The cause of this desaturation of oxygen as the blood flows through the lungs is not known but may be a function of the pulmonary capillaries' inability to properly accommodate the extremely high cardiac outputs and pulmonary arterial pressures that such athletes are capable of generating.

Remember that little or no structural or functional adaptation occurs in the respiratory system as a result of endurance training. In other words, lung volumes and capacities of highly trained endurance athletes tend to be the same as those in age-matched healthy untrained subjects. On the other hand, endurance training results in pronounced increases in exercise SV and cardiac output. As a result, maximal cardiac outputs in such athletes may be 60–70% greater than those in untrained individuals. It has been postulated that, in some individuals, this mismatch between cardiovascular and pulmonary adaptation may result in increased blood flow rates (decreased pulmonary transit times) to the extent that there is not adequate time for blood to become saturated with oxygen. A second possibility is that the increase in pulmonary arterial pressure caused by such large cardiac outputs may result in a transient form of pulmonary edema and that this increase in extravascular lung water causes a partial collapse of some alveoli, thus reducing the surface area available for diffusion.

The Cardiovascular System and Temperature Regulation

Humans belong to a classification of animals known as **homeotherms**, meaning that they must maintain their body temperature within a very narrow range at all times (Figure 3.26). In the resting state, internal body temperature (often called core temperature) is

Figure 3.26 The range of internal or core temperature in humans. Note the narrow range that must be maintained and that deviations above or below this range can be fatal.

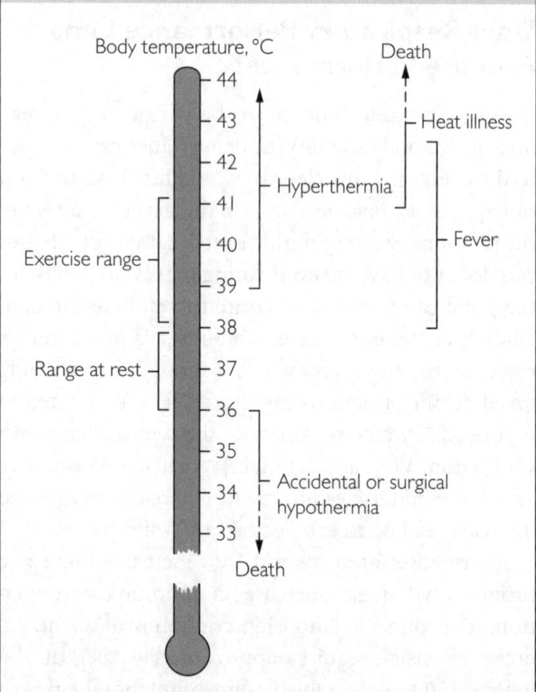

normally about 37°C. A temperature below 36°C would indicate some degree of **hypothermia**, while a temperature greater than 39°C indicates some degree of **hyperthermia**. Notice that a deviation of 6°C from the normal resting set point (or even less, if the condition is prolonged) can be fatal. For example, if for some reason an individual's temperature were to fall below 31°C, within a few hours the heart would begin fibrillating, causing death. Similarly, a rise in core temperature to approximately 44°C or more will eventually result in a breakdown or denaturing of the protein structures of many key enzymes in the body, also resulting in death. This need to maintain a constant body temperature within a few degrees obviously requires highly evolved thermoregulatory mechanisms, with little room for error. It also allows us, however, to function across a wide range of environmental temperatures, from polar regions to the equator, unlike amphibians and reptiles.

Body temperature (the core–shell concept)

While internal body temperature, as in arterial blood, tends to fluctuate very little, a lower and wider range of regional temperatures occurs in the skin and the extremities. For example, in a thermoneutral environment where air temperature is 22°C, core temperature would be maintained at approximately 37°C, while skin or shell temperature might vary from 34°C in the thorax region to 31°C in the fingers or toes. Since heat can only be transferred when a temperature difference, or gradient, exists, these cooler shell temperatures ensure that heat is constantly flowing from the core through the skin to the ambient air. Core temperature is maintained by mechanisms that maximize this rate of heat flow in conditions of high ambient temperature (or during exercise) and minimize it in the cold.

Heat can be lost or gained by **conduction**, **convection**, or **radiation**. Conduction is the transfer of heat that occurs when the body is in direct contact with another body or object. Heat will flow from the warmer body to the cooler body at a rate that is affected by the temperature gradient, the area of contact, and the thermal conductivity of the two bodies. Convection is a form of conduction that is aided by movement of one of the conducting bodies. For example, in still air, heat would be lost from the body by conduction into the ambient air. In windy conditions, heat loss would be accelerated by the addition of convection. Radiation refers to heat transfer between the body and an object that is not in contact. For example, the addition of a block of ice to a large thermoneutral room would result in an increase in radiant heat loss from the skin, without a measurable effect on room temperature. Similarly, exercise on a sunny day could result in excess heat gain by the body through radiation from the sun, pavement, or playing surface. At any given time, heat loss or gain is the summed or net effect of all three physical processes.

A fourth process that contributes to heat loss from the body is **evaporation**. Evaporation is the mechanism by which water changes from its liquid state to water vapor. Since evaporation consumes heat energy, heat is lost to the environment and skin temperature decreases when water or sweat evaporates from the surface of the body. The rate at which this will occur depends upon the vapor pressure gradient between the skin and the air (it slows as relative humidity increases) and the surface area of the skin that is exposed to the environment.

Since heat is a byproduct of the metabolic pathways discussed in Chapter 2, even at rest the body produces a significant amount of heat. Heat production can be indirectly estimated from the rate at which oxygen is consumed (see Box 3.12). In the resting state, where $\dot{V}O_2$ is approximately 0.25 $L{\cdot}min^{-1}$, heat production would be approximately 1.25 $kcal{\cdot}min^{-1}$. This is equivalent to the heat generated by a 90 W light bulb. Unlike a light bulb, however, our skin does not feel hot to the touch because the heat is being lost through a much larger surface area.

During heavy exercise, where $\dot{V}O_2$ might increase to 5.0 $L{\cdot}min^{-1}$ or more, heat production will increase by more than 20-fold! This means that, for core temperature to remain the same, the body must now eliminate 20 times as much heat as at rest. Overheating

BOX 3.12

Studies in the late 1800s used whole-body calorimeters to determine the relationship between oxygen consumption and heat production. It was found that, for each liter of O_2 consumed, approximately 5 kcal of heat were produced. This is known as the **caloric equivalent** for oxygen. More detailed investigations, where different foods were oxidized in a calorimeter, determined that this value varied slightly depending on the food. For example, it was approximately 5.05 $kcal{\cdot}L^{-1}$ for carbohydrate, 4.74 $kcal{\cdot}L^{-1}$ for fat, and 4.46 $kcal{\cdot}L^{-1}$ for protein.

(hyperthermia) is thus a major danger during heavy prolonged exercise, especially in a hot and/or humid environment.

Mechanisms for Heat Loss

Body temperature is controlled by neural signals that originate from an area of the brain known as the **hypothalamus**. The hypothalamus responds to input from two sources: the temperature of the arterial blood flowing to it, and heat and cold receptors located in the skin. Neurons in the anterior portion of the hypothalamus are activated by increases in body temperature, while those in the posterior portion respond to decreases in temperature (Figure 3.27). Output from the anterior hypothalamus occurs via the sympathetic system to both beta-adrenergic and cholinergic receptors in the arterial walls of the skin, as well as to cholinergic receptors in the sweat glands. Output from the posterior hypothalamus is also transmitted via the sympathetic system, but to alpha-adrenergic receptors in the arterial walls of the skin.

When body temperature begins to increase above its normal set point, the warming of the arterial blood is immediately detected by the anterior hypothalamus, causing it to increase its sympathetic output to the blood vessels in the skin. This increase results in vasodilation and an increase in cutaneous blood flow. In a process similar to the one that occurs in the radiator of your car, the warm blood from the interior of the body is cooled as it flows through the skin and returns to the interior to cool metabolically active tissues. This mechanism is effective as long as the skin temperature remains 4 or 5°C cooler than the temperature of arterial blood (the core–shell gradient). However, as more blood flows through the cutaneous vessels, the skin becomes warmer and this gradient is reduced. This information is transmitted to the hypothalamus by the heat receptors in the skin, causing it to respond by sympathetic output to the sweat glands and thus cooling the skin and re-establishing an effective core–shell gradient. The mechanisms for increasing the rate of heat loss thus involve two separate phases: an increased rate of blood flow to the skin and, under more severe conditions, the addition of the sweating response. A small amount of heat is also lost through breathing, by the expiration of warmer air to the environment. While this amount increases with increased ventilation during exercise, it still accounts for only a very small proportion of total heat loss.

During heavy exercise, these adjustments are only partly successful in maintaining core temperature at its normal resting set point, and it will increase with exercise intensity and duration. Figure 3.28 summarizes the changes that would typically occur in muscle (rectus femoris), core (rectal), and average skin temperature as an athlete undergoes 60 minutes of treadmill running at 75% $\dot{V}O_{2max}$ in a 22°C environment. In the pre-exercise state, rectal temperature is 37°C and skin temperature (averaged over several sites) is approximately 33.5°C. Deep muscle temperature of the quadriceps, which are inactive and receiving minimal blood flow, is also less than core temperature, at approximately 33.5°C.

Notice that, as soon as the athlete begins running, there is an immediate increase in muscle temperature to 40°C or more. As exercise continues, it continues to increase gradually to more than 41°C. Skin temperature increases slightly over the first few minutes as a result of increased blood flow but declines thereafter as a result of sweating. Although sweating accelerates the rate of heat loss by widening the core–shell gradient, it is only partially successful in dissipating the heat generated in the exercising muscles and core temperature continues to rise over time, reaching levels in excess of 40°C.

Sweating Thermally initiated sweating occurs as a result of activation of the more than 2–3 million eccrine sweat glands that exist in an adult. They are distributed over the entire body surface, with concentrations in the head and neck regions, the palms, and the soles of the feet. The secretory coil of each gland is embedded in the deep dermal layer of the skin and surrounded by a capillary network. When activated by sympathetic output from the hypothalamus, fluid is drawn from the interstitial space, secreted by the sweat glands and delivered through the sweat ducts to the skin by pulsatile contractions of smooth muscle-like cells.

With increased blood flow to the sweat glands, the composition of sweat in the secretory coil approaches that of plasma but, because of the selective permeability of their cell membranes, it contains little or no protein. As this fluid moves through the ducts that conduct it to the skin, a high proportion of the electrolytes (especially Na^+ and Cl^-) is osmotically re-absorbed into the interstitial space, so that by the time that the fluid is released to the skin their concentrations are considerably lower than in plasma. With rapid sweating, however, the time available for re-absorption is

Figure 3.27 Mechanisms by which the hypothalamus regulates body temperature. The anterior portion of the hypothalamus controls heat dissipation (shown as dashed arrows) while the posterior portion controls heat conservation (shown as solid arrows). They receive input from the temperature of the blood and from heat and cold receptors in the skin. Output is via the sympathetic system to blood vessels and sweat glands in the skin and via the spinal cord to muscle motor neurons, which initiate the shivering reflex.

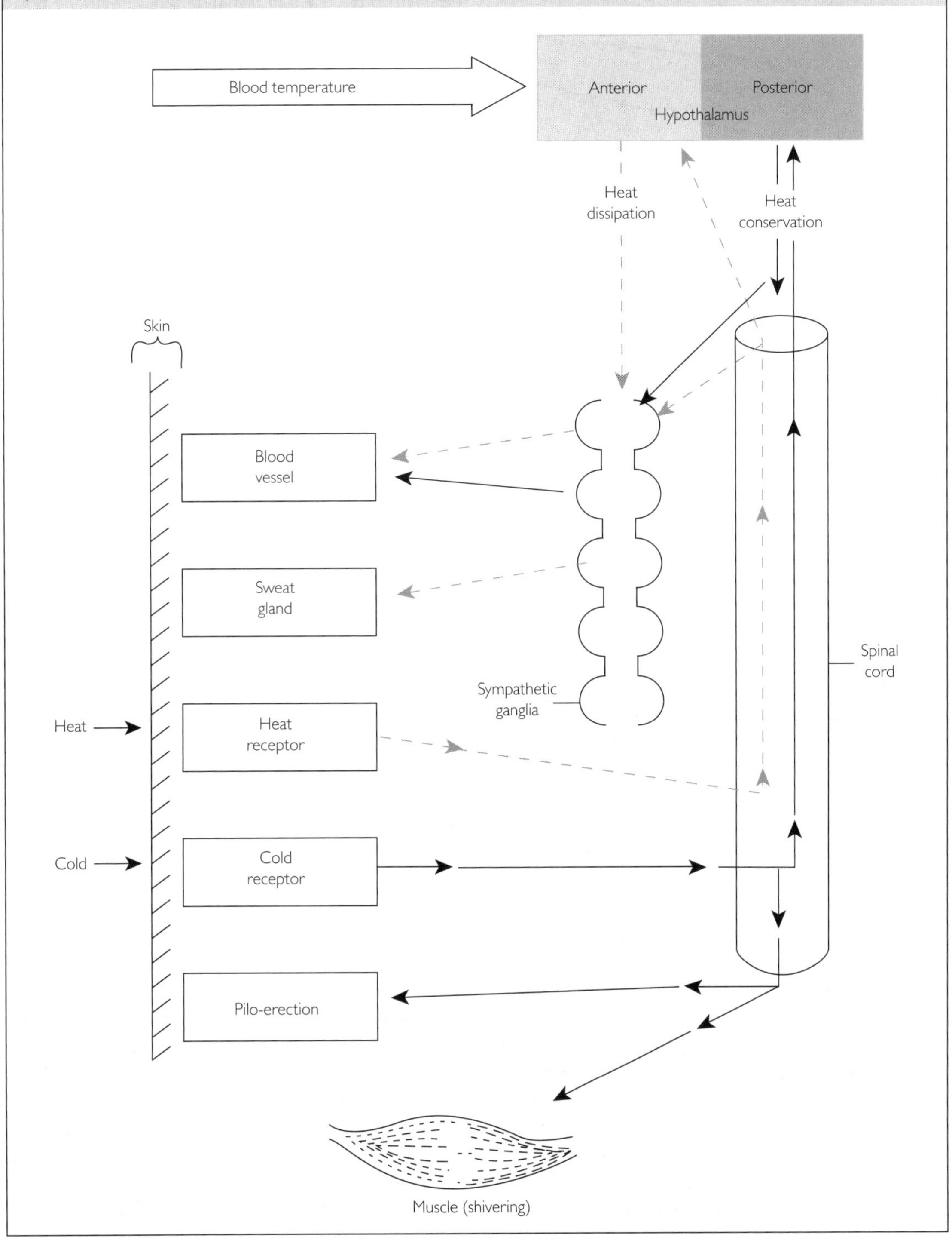

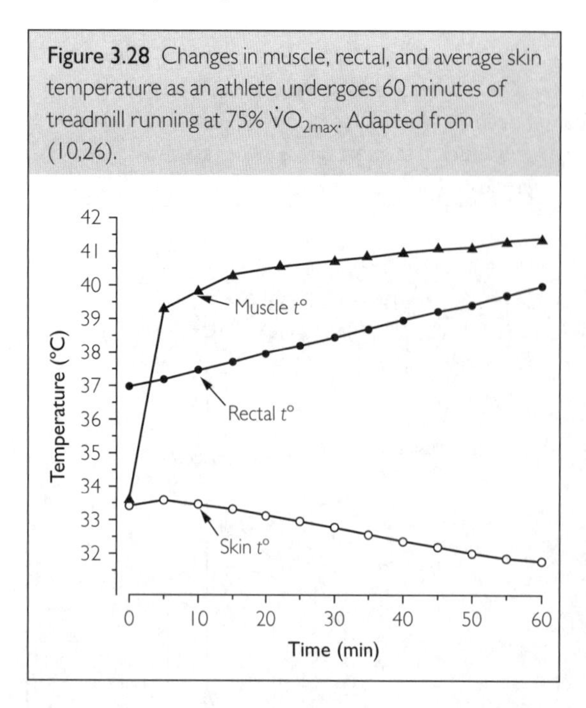

Figure 3.28 Changes in muscle, rectal, and average skin temperature as an athlete undergoes 60 minutes of treadmill running at 75% $\dot{V}O_{2max}$. Adapted from (10,26).

decreased and sweat electrolyte concentrations do not decrease to the same extent, although they still remain hypotonic to those in plasma. In addition, because the muscle contractions within the glands are largely dependent on anaerobic metabolism, lactate concentrations in sweat are also very high. Even so, sweat is still more than 99% water and therefore an excellent evaporatory coolant, with the evaporation of each milliliter of water removing about 0.58 kcal of heat from the surface of the skin. The rate at which evaporation will occur is dependent upon the temperature and relative humidity of the ambient air, the temperature of the skin, the tonicity of the sweat, and the velocity of ambient air flow over the skin.

Sweating occurs most profusely in the head region, followed by the neck and torso, and is lowest in the extremities. Sweating rates during prolonged heavy exercise in a normal environment are typically in excess of 1.0 $L \cdot h^{-1}$ and have been recorded as high as 3.7 $L \cdot h^{-1}$ in an elite marathoner running in a hot and humid environment (3). At such high rates of production, much of the sweat does not fully evaporate and simply drips off the body, providing little cooling. Even with fluid intake, it is inevitable that water loss at these levels will reduce the interstitial water compartment and eventually plasma volume. Procedures for minimizing fluid and electrolyte loss during exercise in the heat are discussed in Chapter 11, which deals with nutrition.

Thermoregulatory demands on the cardiovascular system During heavy prolonged exercise, blood flow to the skin may rise to 2–3 $L \cdot min^{-1}$ for the purpose of heat elimination. This represents a demand for cardiac output *in addition* to that required by the muscles. Although the additional demand normally poses no problem at low to moderate exercise intensities (where cardiac outputs are submaximal), when exercise intensity requires maximal or near-maximal cardiac output, skin blood flow cannot increase enough to compensate for the increase in heat production. It is also probable that vasodilation occurs to some extent in the cutaneous veins. Although this has the advantage of slowing skin blood flow and allowing the loss of more heat, it does so at the expense of decreasing the rate at which blood from the skin is returned to the heart. The result is a small but significant effect on ventricular filling rate and thus SV. Again, at low and moderate exercise intensities this can be offset by an increase in heart rate to maintain cardiac output. This cardiovascular drift occurs during prolonged exercise in even a normal temperature environment, but is exaggerated during exercise in the heat.

The water that is lost through sweating is derived from the interstitial fluid compartment. In a typical 75 kg male athlete, the interstitial fluid compartment would contain a total of approximately 12 L of water. The 3–3.5 L of water that makes up the plasma is separated from the interstitial compartment by the porous walls of the capillaries. As a result, any reduction in the volume of the interstitial water is accompanied by an almost proportional reduction in plasma volume. The loss of 1 L of sweat thus results in approximately a 220 mL reduction in plasma volume. As exercise continues, this progressive decline in blood volume is manifested by a decline in SV, requiring the cardiovascular control centers to further increase heart rate in an attempt to maintain cardiac output. The body is able to buy more exercise time through these adjustments but, eventually, as heart rate approaches its maximum and diastolic filling time is compromised, SV will begin to decrease to a point where cardiac output falls as well and exercise can no longer be maintained at the same intensity.

Exercise in the cold When the non-exercising body encounters lower than normal ambient temperatures, even the slightest decline in core temperature activates the posterior hypothalamus, causing it to transmit impulses to the alpha-adrenergic receptors in the arterioles of the skin (Figure 3.27). These impulses result in vasoconstriction and the shunting of blood away from the skin, thus decreasing the rate at which heat is lost through the shell. If this mechanism is insufficient, the further decrease in core temperature will trigger the shivering response. Shivering is caused by rhythmic involuntary contractions of skeletal muscle and can increase heat production by three- or fourfold. These adjustments, along with appropriate levels of insulating clothing, normally can greatly extend the time that even extremely cold temperatures can be tolerated. The exception, however, occurs in incidents of cold-water immersion. Because water conducts heat away from the body approximately 25 times more rapidly than air at the same temperature, survival time is extremely short.

Training or competition in cold ambient temperatures is not normally a problem because of the large increase in heat production generated by muscular exercise. For example, Nordic skiers wearing only 1.5 Clo Units[5] still find it necessary to sweat profusely when competing in temperatures as cold as −25°C. With appropriate clothing and access to shelter immediately following exercise, hypothermia is very rare among winter sports athletes or during any type of large-muscle exercise in the cold. Again, however, the exception would occur with swimming in cold water, as might be encountered in events like the triathlon.

Cardiovascular Adaptations to Training

Repeated bouts of endurance exercise (training) cause a number of adaptive changes within the cardiovascular system that will enhance oxygen transport and exercise performance. These include structural changes in the heart that improve its capacity for pumping blood, an increase in the number of capillaries that supply blood to the skeletal muscles, and an increase in blood volume itself.

The Heart

Structural adaptations Echocardiographic assessment of the hearts of elite endurance athletes shows that, on average, internal ventricular diameters are approximately 10% greater and ventricular walls approximately 18% thicker than those of non-athletic controls (13). These enlargements occur symmetrically in both right and left ventricles (18), causing total heart volumes to be approximately 30% larger in the athletes. Since such differences tend to disappear when athletes undergo a period of detraining (12,14,33), it is apparent that they are, at least in part, the result of training per se and not simply inherited traits. Although longitudinal studies of the effects of training in previously sedentary subjects usually show smaller degrees of cardiac enlargement than those reported in cross-sectional comparisons, it is generally accepted that chronic endurance training results in an increase in cardiac chamber size and wall thickness. The functional significance of these adaptations is an increase in SV, both at rest and during exercise. On their own, however, these structural changes can account for only a portion of the increase in SV that occurs during heavy exercise.

The increase in internal ventricular volume is apparently caused by an increase in the number of sarcomeres per cardiac muscle fiber and/or remodeling of connective tissue (7). The stimulus for this adaptation is believed to be the *stretch* that is imposed on the walls of the ventricles by greater than normal ventricular filling. This increased filling, sometimes referred to as a **volume overload**, is caused by the increased venous return to the heart that occurs during endurance exercise. In skeletal muscle, stretch is known to be a potent stimulus for increasing muscle length by the addition of sarcomeres to the ends of the fibers. Using animal models, it has been shown that, while both the intensity and the duration of the stretch affect the magnitude of this response, duration appears to be the more important variable (2).

The increase in ventricular wall thickness is caused by an increase in the cross-sectional area of the cardiac muscle fibers. As in the case of skeletal muscle fibers that have been subjected to mechanical overload, the stimulus for this adaptation is thought to be the increased contractile force that occurs in the ventricular fibers during exercise. Part of this increased contractility occurs in response to the greater end-diastolic volumes brought about by increased venous return to the heart (the Frank–Starling mechanism; see Box 3.1). Part of it is also due to the elevated peripheral resistance (blood pressure) that the ventricles encounter as they attempt to eject blood into the aorta or pulmonary

[5] A Clo Unit is the insulation value provided by a single thin layer of clothes, as would be worn in still air at an environmental temperature of 22°C.

artery. As also discussed previously, this **pressure overload** varies with the type of exercise, being lowest with endurance exercise and highest with resistance exercise. As a result, it is not surprising that several studies have noted greater wall thickness in bodybuilders and weight lifters than in endurance athletes.

Heart rate Chronic endurance training causes a reduction in resting heart rate, a condition known as **bradycardia**.[6] This adaptation is also evident during submaximal exercise, particularly when we compare the pre- and post-training heart rates seen at the same absolute exercise intensity. Although heart rates at the same relative exercise intensity are generally unaffected by training, there may be a slight reduction in maximal heart rate if the training period has been prolonged. The mechanism responsible for training bradycardia was originally thought to be an increase in parasympathetic discharge rate to the heart (vagal tone) and, to a lesser extent, decreased sympathetic output. It is now known that the mechanisms are more complex and also involve a slowing of intrinsic heart rate that is independent of autonomic control.

Whatever the mechanism(s), bradycardia is probably the most predictable adaptation to any exercise training program. Several months of training typically reduce previously sedentary individuals' resting heart rates by 5–10 bpm, while the heart rates of highly trained athletes are typically 20–25 bpm lower than those of sedentary controls. The functional significance of a reduced heart rate is that the extended diastolic phase allows more time for ventricular filling to occur. Ventricular filling rate is most rapid early in diastole, but continues gradually until the beginning of systole. Thus, any decrease in heart rate results in an increase in ventricular filling. The relationship, however, is not linear. For example, a 20% reduction in heart rate could be expected to increase end-diastolic ventricular volume only by about 4%, if all other factors remained constant.

Blood Volume

Endurance training also stimulates an increase in total blood volume. This is the result of a two-phase process: an expansion of plasma volume and an expansion of erythrocyte volume. The increase in plasma volume is a rapid response that can be detected within the first few days of beginning a training program, whereas the increase in erythrocyte volume does not become evident until after 2 or 3 weeks. Since plasma volume and erythrocyte volume increase in proportion to each other, after a month or so, hemoglobin concentration and hematocrit differ little between the trained and untrained states. Most short-term (1–3-month) longitudinal training studies indicate that, on average, total blood volume expands by about 8–10% (37). In contrast, cross-sectional comparisons between elite endurance athletes and sedentary controls indicate that the athletes tend to have 35–45% greater blood volumes relative to body weight (17).

The mechanisms responsible for this training-induced increase in blood volume are not yet fully understood but may be related in part to the increase in body temperature that occurs during exercise. As discussed previously in this chapter (see Figure 3.14), plasma volume is determined by the balance that exists between capillary hydrostatic pressures and capillary osmotic pressures. It has been shown that heat acclimatization, on its own, can result in increased synthesis of the plasma protein albumin by the liver, as well as a decrease in its degradation rate (38). This increase in total circulating protein elevates capillary osmotic pressure, causing greater re-absorption from the interstitial space and an expansion of the plasma volume. It is thus thought that the exercise-induced elevation of body temperature may be a stimulus for the increased plasma volume that occurs with training.

Since the erythropoiesis process requires a minimum of 6 days before final maturation of a red blood cell, it is not surprising that erythrocyte expansion is usually not apparent until after 2 weeks of training. Erythrocyte production rate is controlled primarily by the concentration of the hormone erythropoietin (EPO), which is very sensitive to any decline in tissue P_{O_2}. It is well known that chronic exposure to moderate or high altitude induces an elevation in circulating EPO concentration and subsequently in red blood cell count and hemoglobin concentration (44). In such instances, it is generally accepted that the increase in EPO production rate is caused by reduced P_{O_2} in the kidneys and liver. Whether or not exercise, on its own, induces enough hypoxemia to stimulate functionally significant increases in EPO is debatable, since most

[6] Bradycardia is technically defined as a heart rate of less than 60 bpm. The term is commonly used, however, to refer to any reduction in resting or exercise heart rate that occurs as a result of exercise training.

studies show that resting plasma EPO concentrations are the same in trained and untrained subjects.

The training-induced increase in blood volume enhances venous return to the heart and ventricular filling pressure during exercise. As a result, it is probably the major contributor to the increased SV and cardiac output adaptations that occur with training. In a cross-sectional comparison it has been shown that, during cycle ergometry, ventricular filling rate was approximately 71% greater in trained endurance athletes than in untrained controls (15). An expanded plasma volume also provides important thermoregulatory benefits by improving the athlete's ability to maintain high skin blood flow and sweat rates during exercise in a hot or humid environment.

Muscle Capillarization

Endurance training provides a stimulus for an increase in the number of capillaries in the trained muscle. In one of the first longitudinal investigations of peripheral adaptations, Ingjer (20) analyzed muscle biopsy samples of the quadriceps, before and after 6 months of heavy training. The capillary density (number of capillaries per square millimeter) was approximately 26% greater following training and the number of capillaries per fiber was 29% greater, indicating the addition of capillaries in the trained muscle. A number of subsequent studies have reported similar results (23,41).

The mechanism(s) by which these new capillaries develop (a process known as **angiogenesis**) is not yet known but may involve sprouting or splitting of existing capillaries. It is also not known whether this angiogenesis is an adaptation to the reduced P_{O_2} that occurs in muscle during exercise (muscle hypoxia) or to the mechanical stress imposed on the capillary wall by the large increases in blood flow that occur during training (34). The muscle hypoxia hypothesis arose from earlier studies which reported increased capillary density in animals exposed to chronic altitude hypoxia, as well as findings of greater than normal muscle capillary density in high-altitude natives. Later studies have, however, indicated that even severe hypoxia, on its own, does not stimulate capillary growth in human subjects (19,25), although training at altitude increases capillary density to a greater extent than the same amount of training under normoxic conditions (19,28). It also appears that angiogenesis is affected by training intensity, being more pronounced following a high-intensity program than a program of lower intensity but longer duration (22,43).

An increase in the number of capillaries reduces the average diffusion distance between each muscle fiber and its surrounding capillaries and increases the surface area available for exchange between muscle and blood. As a result, the rate at which O_2 and fuel sources such as glucose and fatty acids enter the fiber and byproducts such as CO_2 and lactic acid can be removed from the fiber during exercise is accelerated. In addition, since velocity of blood flow is inversely related to the total cross-sectional area of the capillary bed, transit time across the muscle for a given blood flow is prolonged, thus increasing the time available for diffusion. The increased capillary density that occurs with endurance training is accompanied by an almost parallel (46) increase in mitochondrial volume density (discussed in Chapter 2) and together they greatly enhance the maximal aerobic capacity of the muscle.

The capillary density of skeletal muscle is also affected by changes in fiber size. For example, if the muscle were to atrophy, the capillary density would increase, despite the absence of change in total capillary numbers. Although average fiber size tends to change very little with endurance training, with strength training there is normally a large increase in fiber area and volume. In such cases, the increase in fiber volume has been shown to exceed any increase in angiogenesis and there is actually a reduction in capillary density. The increase in contractile protein that occurs with strength training also causes a reduction in the volume density of mitochondria in the fiber, resulting in a net reduction in oxidative capacity relative to muscle mass.

Heat Acclimatization

Endurance training improves an athlete's ability to tolerate the increase in body temperature that occurs with exercise. The expanded blood volume helps to maintain venous return to the heart when large increases in skin blood flow are necessary to dissipate heat. This helps to preserve SV during prolonged exercise and/or in instances where blood volume has declined as a result of high sweat rates.

Exposure to a hot environment will, on its own, improve the effectiveness of the body's thermoregulatory mechanisms. Within 2 or 3 days, the hypothalamus becomes more sensitive to elevations in temperature and initiates cutaneous vasodilation and sweating at a lower core temperature. After approximately 1 week,

Figure 3.29 Rectal temperatures in a trained subject ($\dot{V}O_{2max}$ = 5.1 L·min^{-1}) and untrained subject ($\dot{V}O_{2max}$ = 3.5 L·min^{-1}) after 10 minutes of cycle ergometry at the same absolute power outputs ($\dot{V}O_2$) and at an additional $\dot{V}O_2$ of 4.5 L·min^{-1} for the trained subject (shaded bars).

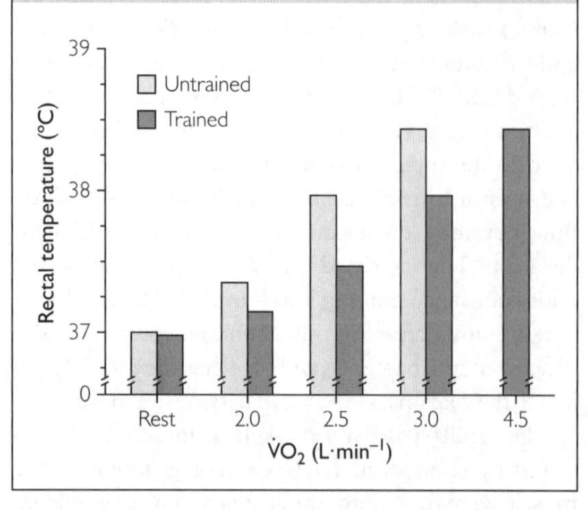

the more effective sweating mechanism allows the individual to maintain a normal core temperature with a lower skin blood flow. A certain degree of acclimatization is also achieved through training itself, even when performed in a cool or temperate environment. Apparently the daily elevation in core temperature that occurs while the athlete is exercising provides a stimulus for acclimatization. As a result, when exercising at the same absolute power output, the increase in core temperature is considerably less in a trained athlete than in an untrained subject (Figure 3.29). Since heat production is the same for both individuals at the same absolute $\dot{V}O_2$, the fact that rectal temperature is lower in the trained subject indicates a higher heat dissipation rate. This is achieved by more profuse sweating and greater skin blood flow for a given rate of heat production (32). Although heat acclimatization can be gained either by simple exposure to a hot environment on its own or by daily exercise on its own, the process is most effective for competition purposes when the athlete combines the two and trains in a hot and/or humid environment.

Summary of Adaptations to Training

Endurance training is a stimulus for a number of cardiovascular adaptations that together result in an increase in exercise SV and maximal cardiac output. The cardiac muscle fibers in all four chambers become slightly longer by adding more sarcomeres, thus increasing the volume of each chamber and the heart's capacity for accommodating greater venous return. In addition, these fibers (especially those of the ventricles) undergo a slight hypertrophy, thus becoming stronger and capable of more forceful contractions. Endurance training also results in an expansion of the athlete's total blood volume, which increases ventricular filling pressure. This mechanism combines with the above structural adaptations to greatly enhance ventricular filling and, to a somewhat lesser extent, ventricular emptying rate, even at very high exercise heart rates. At rest and during submaximal exercise, ventricular filling is further augmented by the longer diastole resulting from the acquired bradycardia. Finally, the reduced systolic and mean blood pressures observed following training indicate a lower peripheral resistance at the same absolute exercise intensity than before training. This is thought to be due to a more effective sympathetic vasodilatory response in the muscle vasculature and/or increased capillary density, and serves to further augment SV for a given cardiac contractile force.

These adaptations to training are frequently referred to as **central adaptations** and, while their magnitude is affected by both the intensity and duration of the training stimulus, duration appears to be the more important of the two variables. The major **peripheral adaptation** to endurance training is an increase in the number of functional capillaries in the muscle. This reduces the diffusion distance between capillary and muscle fiber and slows the rate at which blood flows through the muscle during exercise, thus ensuring greater gas exchange for a given blood flow. Peripheral adaptations appear to be affected more by training intensity than by the duration of the stimulus. The central adaptations result in a greater delivery of oxygen to the exercising muscle, while the peripheral adaptations result in a greater unloading of oxygen. Together, they produce an increase in the athlete's maximal aerobic power ($\dot{V}O_{2max}$) and maximal aerobic capacity (ability to perform exercise at a high percentage of $\dot{V}O_{2max}$ before becoming fatigued).

Maximal Aerobic Power ($\dot{V}O_{2max}$)

As discussed in Chapter 1, maximal aerobic power, or maximal oxygen consumption ($\dot{V}O_{2max}$), refers to the maximal amount of oxygen that can be taken from the

BOX 3.13

In 1870, German physiologist Adolph Fick deduced that cardiac output (Q_c) could be calculated by determining the amount of oxygen that is taken up by the body and the amount that is returned to the heart in the venous circulation: $Q_c = \dot{V}O_2/(a\text{-}v)\ O_2$. Written another way, $\dot{V}O_2 = Q_c \times (a\text{-}v)\ O_2$. Thus, $\dot{V}O_{2max} = \max Q_c \times \max (a\text{-}v)\ O_2$.

atmosphere and utilized by the muscles during exercise. It can be measured at the mouth as the difference between the amount of oxygen that is inhaled and the amount that is exhaled and expressed as a volume (V) per minute. As outlined in Chapter 11, it is usually measured during a progressive test to exhaustion on a standardized exercise ergometer. Because it is affected by the mass of the active muscles, its magnitude will vary with different modes of exercise. $\dot{V}O_{2max}$ is normally measured in athletes using an ergometer that most closely simulates the muscular activity in their sport. For example, for athletes in a sport where running is the major component, a treadmill will be used; for cyclists, a cycle ergometer; for rowers, a rowing ergometer; and so on.

According to the Fick principle (see Box 3.13), an increase in $\dot{V}O_{2max}$ will occur if there is an increase in maximal cardiac output or an increase in maximal (a-v) O_2 difference (oxygen extraction). Although both of these changes are known to occur as a result of training, most studies indicate that the increase in maximal muscle oxygen extraction is usually considerably less than the increase in maximal cardiac output. In addition, the percentage increase in $\dot{V}O_{2max}$ correlates more closely with the percentage increase in cardiac output than with the increase in (a-v) O_2 difference. Consequently, it is commonly accepted that, while both central and peripheral adaptations contribute to the training-induced increase in $\dot{V}O_{2max}$, central factors (i.e., increased maximal cardiac output) are the more important variables.

How much can $\dot{V}O_{2max}$ increase with training?

$\dot{V}O_{2max}$ values for world-class endurance athletes (e.g., distance runners) are normally 70–80 $mL \cdot kg^{-1} \cdot min^{-1}$, while that of the average healthy young male is normally 40–50 $mL \cdot kg^{-1} \cdot min^{-1}$. With several years of dedicated training, can the average individual aspire to become a world-class distance runner? Unfortunately, the answer is no. Training studies with previously sedentary subjects indicate that the upper limit for improvement is about a 20% increase in $\dot{V}O_{2max}$. In more physically active subjects, such as physical education students, the upper limit is closer to 10%, with similar increases in maximal cardiac output. It is thus apparent that heredity accounts for a large portion of the exceptional cardiovascular attributes of champion athletes.

Maximal Aerobic Capacity

Since maximal aerobic capacity refers to the total amount of ATP that can be generated by oxidative metabolic processes (Chapter 1), it is often indirectly quantified as the maximal amount of work that can be performed aerobically in a given time, or at a given percentage of $\dot{V}O_{2max}$, before fatigue. In endurance events, it is thus obvious that any increase in maximal aerobic capacity will manifest itself as an increase in athletic performance. With endurance training, it is also well known that the improvement in maximal aerobic capacity is proportionally greater than the improvement that occurs in $\dot{V}O_{2max}$ (Figure 3.30)

Training increases not only the athlete's $\dot{V}O_{2max}$ but also the capacity to tolerate and sustain exercise at higher intensities (percentages of $\dot{V}O_{2max}$) than before training. The practical implications that such an

Figure 3.30 Typical changes in maximal aerobic power ($\dot{V}O_{2max}$) and maximal aerobic capacity (MAC) that would occur when an untrained subject undergoes 6 months of endurance training.

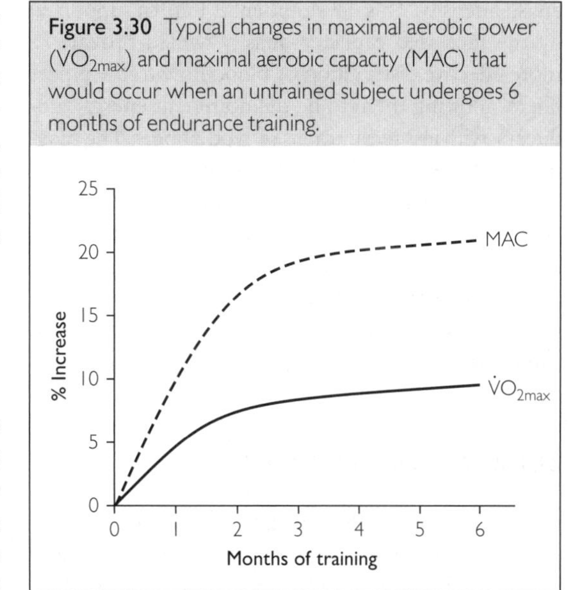

adaptation have on performance are illustrated in the following example: an untrained individual decides that she will compete in a 10 km run that will take place in 4 months' time. She begins training by running for 30 minutes, four or five times per week. After a few weeks, she realizes that, with the same degree of effort, she is now running further in 30 minutes than she did at the beginning. This progress continues up until the time of the event and she is able to complete the 10 km with ease and in a creditable time. Over the training period her $\dot{V}O_{2max}$ would have increased (perhaps 5–10%), but, more importantly, her maximal aerobic capacity would have increased to a greater extent (perhaps 15–20%). Whereas initially she was able to sustain continuous running only at a pace that required about 60% of her $\dot{V}O_{2max}$, after 4 months of training she is able to sustain a pace that requires about 75% of $\dot{V}O_{2max}$. As will be discussed in Chapter 6, had she also added some high-intensity interval training to her program her performance would have been even more impressive.

The capacity to tolerate submaximal exercise at higher intensities following endurance training is largely related to the elevation that occurs in the lactate threshold (Chapter 2). This increase in the lactate threshold can be attributed primarily to the increased muscle mitochondrial and capillary densities that occur with such training. The enhanced oxidative enzyme activity that accompanies the increased mitochondrial numbers permits the oxidation of more pyruvate (and thus less lactic acid production) for a given absolute exercise intensity. This development is in turn supported by the increased capillary density, which provides a more effective distribution of blood flow to the type 1 fibers, allowing them to be the dominant energy producers, with less recruitment of type 2 fibers. The magnitude of these peripheral adaptations (often 30–40%) is considerably greater than the improvement that occurs in $\dot{V}O_{2max}$ and correlates well with the improvement in maximal aerobic capacity. It is thus generally accepted that peripheral factors are more important in determining the athlete's ability to perform prolonged submaximal exercise at a high intensity.

Summary of Key Points

- During exercise, the principal role of the cardiovascular system is to deliver oxygen to the exercising muscles and to remove carbon dioxide. It does this by increasing muscle blood flow in proportion to the increase in ATP production rate. Muscle blood flow is increased in two ways: by increasing the cardiac output of the heart and by increasing the proportion of this cardiac output that is directed to the muscle. The former is accomplished by increases in both heart rate and SV, and the latter by vasodilation of the arterioles leading to the muscles, in conjunction with vasoconstriction of the arterioles in many other parts of the body. These functions are controlled by complex interactions between the autonomic nerves to the heart and those that control the diameters of the blood vessels.
- The increase in heart rate is caused by a reduction in the activity of the parasympathetic nerves to the heart, as well as an increase in the activity of the sympathetic nerves. The increase in SV is caused by greater ventricular filling during diastole and by more forceful contractions of the ventricles during systole. Vasodilation of the blood vessels leading to the exercising muscles occurs in response to regional decreases in P_{O_2} and increases in various metabolic byproducts. In addition, there is a generalized increase in sympathetic stimulation to all blood vessels, resulting in further vasodilation of those that supply the heart and the muscles, but vasoconstriction elsewhere. As a result of these adjustments, as much as 85% of the total cardiac output can be directed to the exercising muscles during maximal dynamic exercise.
- In addition to its rate of blood flow, the amount of oxygen that a muscle receives is dependent upon the amount of oxygen in the arterial blood and the proportion that is extracted as the blood flows through it. The oxygen content of arterial blood is determined by its hemoglobin concentration and by the lung's ability to increase ventilation and saturate the blood as it passes through the pulmonary capillaries. The lung has an extremely large reserve capacity and, during maximal exercise, is capable of increasing pulmonary ventilation more than 20-fold. As a result, only in unusual cases is it incapable of adequately saturating the arterial blood with oxygen. The respiratory system is thus not normally considered a weak link to exercise performance.
- The amount of oxygen extracted from the blood as it flows through the capillaries of the muscle

is determined by the amount of oxygen that is released (or dissociated) from the hemoglobin. The dissociation of oxygen and hemoglobin is autoregulated by the P_{O_2} in the muscle. At rest only about 25% of the oxygen bound to hemoglobin is released, but during exercise this amount increases markedly, approaching 100% at maximal exercise intensity. Since arterial oxygen content declines very little during intense exercise and the muscle is capable of extracting almost all the oxygen that is delivered to it, it is apparent that the major factor that determines the upper limit for aerobic performance is the athlete's maximal cardiac output.

- The cardiovascular system also performs an important thermoregulatory function by controlling the rate of blood flow to the skin. During exercise, the increase in skin blood flow accelerates the rate at which heat is removed from the body, thus slowing the rise in body temperature. The increase in skin blood flow, however, represents a demand for cardiac output that is in addition to that required by the muscles and often cannot be met during heavy prolonged exercise or exercise in a hot or humid environment. In such instances, the result is a decrease in exercise SV, which can be further exacerbated by the decline in plasma volume caused by excessive sweating.
- Endurance training causes a number of cardiovascular adaptations that improve oxygen transport and exercise performance. These include increased atrial and ventricular volume, increased ventricular wall thickness, and increased total blood volume. These central adaptations result in a greater SV and thus greater maximal cardiac output. In addition, endurance training provides a stimulus for an increase in the number of capillaries in the trained muscles. This peripheral adaptation improves gas exchange by reducing the diffusion distance between the capillaries and the fibers and by slowing blood transit time across the muscle. As a result of these adaptations, there is an increase in the athlete's $\dot{V}O_{2max}$ and an even greater increase in maximal aerobic capacity. The extent to which central adaptations occur appears to be more dependent upon increased duration of the training stimulus, whereas peripheral adaptations appear more sensitive to increased intensity.

4

Muscle Physiology

Ultimately, performance depends on the ability of muscles to generate force and power, and the ability of the nervous system to activate the muscles appropriately for a given task. The purpose of this chapter is to provide an understanding of muscle physiology. Here we will consider how muscles are controlled by the nervous system, and how the nervous system and muscles work together to produce what are commonly called strength and power performance.

Muscle Ultrastructure

The structure of muscle is illustrated in Figure 4.1. Muscles are composed primarily of cells called muscle

Figure 4.1 Muscle structure. Muscles are composed mainly of muscle fibers, which are in turn composed mainly of myofibrils (20). Myofibrils are made up of thick myosin and thin actin filaments, as well as a cytoskeleton (not shown). From Billeter and Hoppler (6), © John Wiley & Sons.

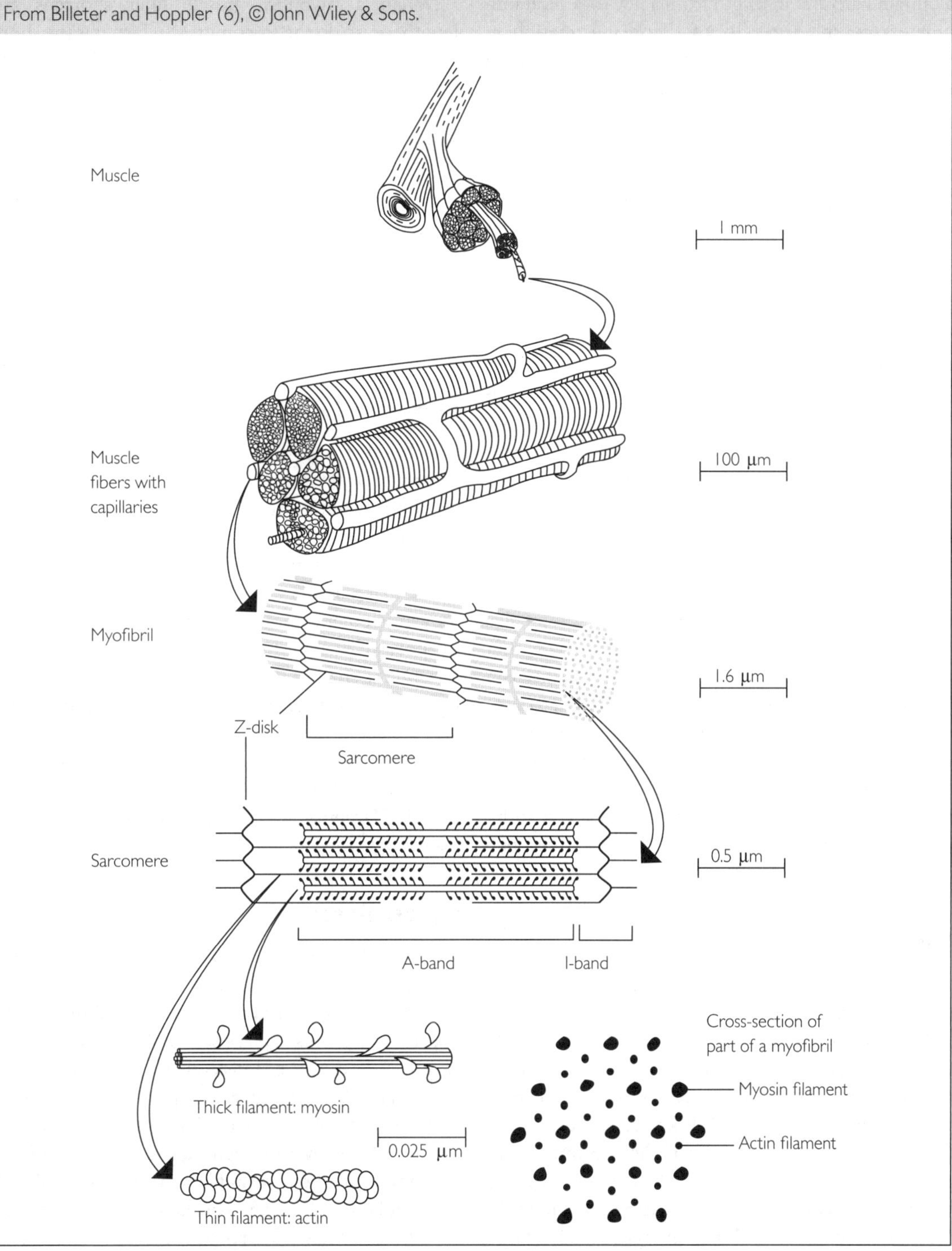

fibers. Muscle fibers range in thickness from about 50–100 μm (a thick fiber would have the diameter of a human hair) and in length from a few to several millimeters (consult Appendix A for more information on units of measurement). Each muscle fiber contains a few to several thousand myofibrils, each of which is about 1–2 μm in diameter but usually has the same length as the muscle fiber. Myofibrils make up about 85% of the contents of a muscle fiber, the remaining 15% being largely composed of sarcoplasmic reticulum (SR), mitochondria, glycogen granules, and fat droplets.

Within each myofibril are a large number of myofilaments, of which there are two types: thick myosin filaments and thin actin filaments. The myosin filaments, so-named because they are composed of the protein myosin, are about 16 nm thick (1 μm = 1000 nm; 1 million nm = 1 mm) and 1.6 μm long. The actin filaments, composed primarily of the protein actin, are about 6 nm thick. Whereas the length of myosin filaments is similar across species, the length of actin filaments varies in different species of animals; for example, actin filaments are approximately 0.95 and 1.27 μm long in frog and human muscle, respectively (98). The myosin molecules that comprise the myosin filaments have heads that project from the shaft of the filament. The myosin heads can act as cross-bridges (CBs) that bind to sites on the actin filaments. It is the interaction between myosin CBs and actin filaments that produces the force of muscle contraction. Note also that the myosin heads project in different directions (six to be exact, as will be shown) around the circumference of the myosin filament.

Sarcomeres

Seen from the side (longitudinal view), it is evident that the myosin and actin filaments within a myofibril are organized into functional units called sarcomeres (Figures 4.1 and 4.2). Each sarcomere is bordered by two Z-disks. Anchored to and projecting inward from the Z-disks are the thin actin filaments. Located centrally in the sarcomere are the thick myosin filaments. They are held in place by cytoskeletal or accessory proteins not shown in Figures 4.1 and 4.2. As shown in Figure 4.2B, a large number of sarcomeres is needed in series to span the length of a myofibril and hence the whole muscle fiber.

When a myofibril/muscle fiber shortens and lengthens, so do the sarcomeres; in fact, it is the shortening of the sarcomeres, brought about by the interaction between the myosin and actin filaments, that causes the myofibril/muscle fiber to shorten during contraction. The working length range of the sarcomere depicted in Figure 4.3 would be about 2 μm (based on frog muscle). For 5000 sarcomeres in series as shown in Figure 4.2B, the collective **working range** would be 10 000 μm (10 mm). As will be discussed, the force produced by a sarcomere varies at different sarcomere lengths, a variation known as the **force–length relationship**.

Hexagonal Array

The bottom right portion of Figure 4.1 shows the cross-sectional view of part of a myofibril. The cross-section is taken in the region of the A-band (equivalent to the length of the myosin filaments), which is where actin and myosin filaments overlap. Note that six actin filaments surround each myosin filament, an arrangement known as a **hexagonal array**. The purpose of the multi-directional projection of CBs around the shaft of the myosin filament, shown at the bottom left of Figure 4.1, is now evident; the CBs in fact project in six directions so that a myosin filament can form CBs with each of the six actin filaments that surround it. The hexagonal array, including actin and myosin filaments and myosin CBs, is shown schematically in Figure 4.4.

Cytoskeleton

Reference has been made to cytoskeletal or accessory proteins. These proteins form the **cytoskeleton** that provides a framework or scaffold to hold the myosin and actin filaments in place within the sarcomeres (see Box 4.1). In fact, the Z-disks that have been referred to as part of a sarcomere are part of the cytoskeleton. The cytoskeleton is illustrated in Figure 4.5. Figure 4.5A shows the cytoskeleton of a myofibril, with the actin and myosin filaments removed. The Z-disks and the M-lines are labeled. Figure 4.5B shows a longitudinal view of parts of the cytoskeleton, with additional proteins labeled.

Figure 4.2 Sarcomeres in series. (A) Two sarcomeres are shown, including Z-disks (also called Z-lines) and thin actin and thick myosin filaments. One end of each actin filament is anchored to the Z-disk. The myosin filaments are stacked in the middle of the sarcomere by accessory proteins (not shown). (B) Many sarcomeres are connected in series to span the length of a myofibril. For a sample myofibril/muscle fiber 10 mm long, 5000 sarcomeres in series would be needed, assuming a sarcomere length of 2 μm.

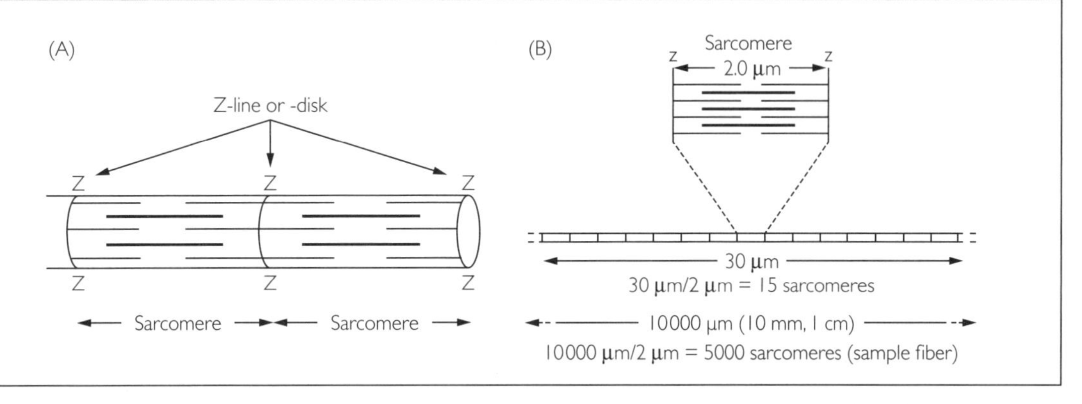

Figure 4.3 Working range of a sarcomere. For actin and myosin filaments of the lengths shown, the working range would be 3.5 – 1.5 = 2.0 μm. In fact, the long length shown is slightly too long, because, as will be shown later, there must be at least some overlap of the actin and myosin filaments for force generation. For 5000 sarcomeres in series, the collective working range would be 10000 μm (10 mm). The term "optimal" length refers to the length at which sarcomere force is greatest. This will be discussed in more detail in the section Force–Length Relationship.

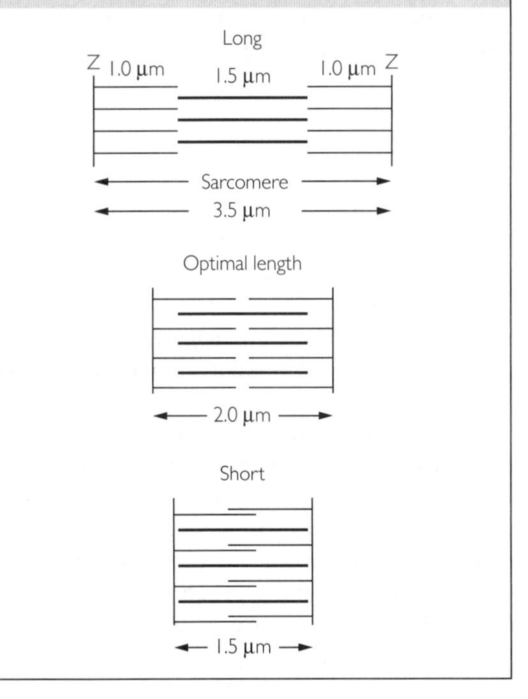

Figure 4.4 Hexagonal array. A schematic representation of one myosin and six actin filaments in cross-section. Since six actin filaments surround each myosin filament, the arrangement is called a hexagonal array. Note that myosin CBs project to each actin filament.

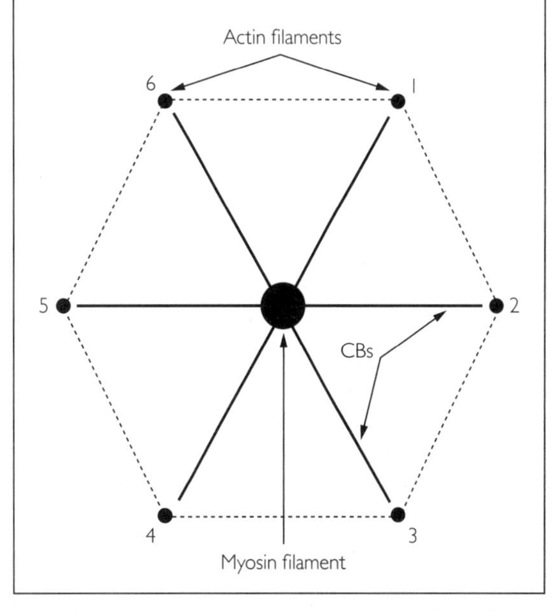

BOX 4.1 ROLE OF THE CYTOSKELETON

The cytoskeleton plays important roles in exercise, training, and performance. One example is that the cytoskeleton is a site of the muscle damage caused by unaccustomed exercise, which can lead to delayed-onset muscle soreness. A second example is that the cytoskeleton is a site of storage of elastic energy which enhances performance of activities like running and jumping. The roles of the cytoskeleton in performance will be considered in more detail in subsequent parts of the text.

Contractile Mechanism

Cross-Bridge Power Stroke

The basis for muscle contraction is the interaction between the myosin heads or CBs and the actin filaments. Under the right conditions, the CBs can bind to specific sites on the actin filaments and pull on them, producing force. If the pulling force exceeds any opposing resistance, the actin filaments are pulled toward the middle of the myosin filament; that is, the sarcomere shortens. The CB action, called a power stroke or working stroke, is shown in simplified form in Figure 4.6A (see also Box 4.2). For CB power strokes to be effective, all CBs must pull toward the middle of the sarcomere. This is achieved by the orientation of the myosin molecules that make up the thick filament. The thick filament is made up of about 500 myosin molecules, which consist of a tail and a head or CB. [In fact, the myosin molecule consists of two myosin heavy chains intertwined and two heads (see Figure 4.7B); for simplicity, the myosin is shown with a single head in Figure 4.6 and several other figures.] The molecules are arranged so that all of the tails are pointing toward the middle of the filament (Figure 4.6B). This is crucial because in the power stroke, myosin heads pull toward their tails. The arrangement as shown in Figure 4.6B ensures that, on both sides of a sarcomere, the actin filaments are pulled toward the center of the sarcomere.

CB power strokes always attempt to cause muscle shortening. Whether they succeed depends on the force opposing muscle contraction. If the collective force of active CBs exceeds an opposing force, the muscle shortens, producing a concentric or shortening contraction. If the opposing force just matches the CB force, there is no change in muscle or sarcomere length, producing an isometric contraction. Finally, if the opposing force exceeds the force produced by active CBs, the muscle is forcibly lengthened, producing an eccentric or lengthening contraction (see Box 4.3).

Cross-Bridge Cycle

During contraction, CBs go through a series of states known as the CB cycle. As illustrated in Figure 4.7 and discussed in Boxes 4.4 and 4.5, CBs bind to actin, produce a power stroke, detach from actin, bind to actin again, and so on. Note that ATP plays two roles in the CB cycle: (1) the binding of an ATP molecule to the CB allows the CB to detach from actin at the end of a power stroke (transition from states 1 to 2) and (2) energy from ATP splitting into ADP and P_i, and the release of the latter products from the CB, is coupled to the power stroke (transition from states 4 to 1) (42).

There are some notable features of CB cycling. In fast, unopposed muscle shortening (concentric contraction) there may be a few hundred CB cycles per second and the actin and myosin filaments may slide by each other at speeds up to 15 $m \cdot s^{-1}$! During contraction, the CBs do not cycle synchronously, like rowers in a boat, but asynchronously, with CBs in various states of the CB cycle (39). This ensures that at all times some CBs are actively pulling on the actin filaments. The number of CB cycles per second and the maximum shortening velocity depend on the nature of the myosin molecules; that is, on muscle fiber type. Fiber types will be considered in detail in Chapter 5.

Control of CB Cycling

Role of calcium ions There must be a way of turning CB cycling on when contraction is desired and turning cycling off when relaxation (no contraction) is needed. Calcium ions (Ca^{2+}) ultimately control whether or not CBs are cycling (29). When Ca^{2+} is present in high concentration among the actin and myosin filaments, CB cycling occurs. If Ca^{2+} is removed, CB cycling stops. How the presence of Ca^{2+} can initiate CB cycling is illustrated in Figure 4.8. The thin myofilament is mainly composed of the protein actin, but also includes the regulatory proteins tropomyosin and troponin. The actin filament consists of two twisted strings of spherical actin molecules. Adhering to the actin strings are many threadlike

Figure 4.5 Cytoskeleton. (A) A schematic representation of the cytoskeleton of a myofibril, with the actin and myosin filaments removed. The Z-disks and M-lines have been labeled. The projections from the Z-disks and M-region allow connections to the corresponding structures of adjacent myofibrils. From Wang and Ramirez-Mitchell (112), © Wang K and Ramirez-Mitchell R, 1983. (B) Longitudinal view of sarcomeres from two adjacent myofibrils, with some of the associated cytoskeletal proteins, as labeled. From Billeter and Hoppler (6), © John Wiley & Sons.

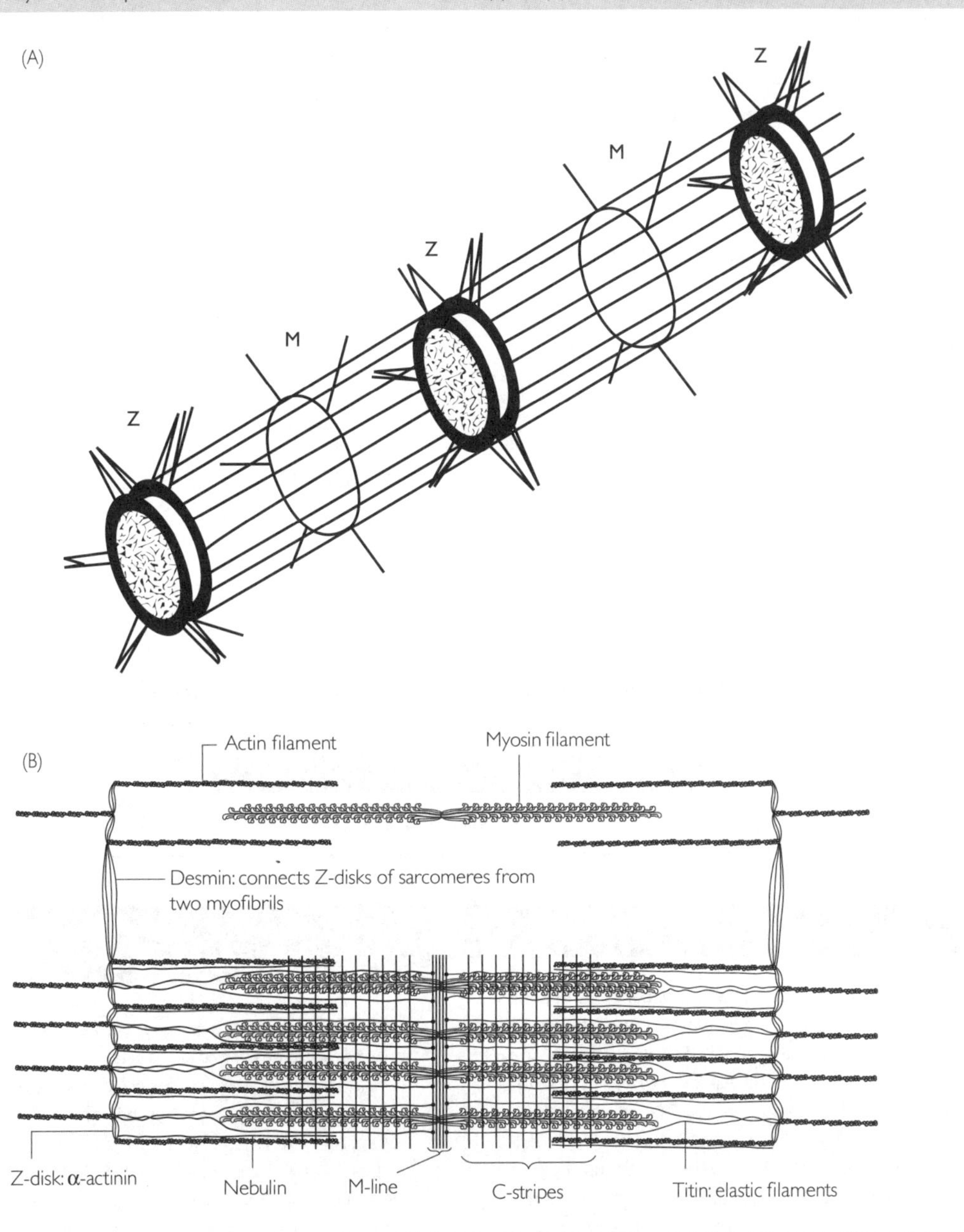

Figure 4.6 (A) Schematic representation of the CB power stroke. From top to bottom: CB detached from actin filament; CB binds to a site on the actin filament; CB executes a power stroke by pulling the actin filament and Z-disk toward mid-point of sarcomere and myosin filament. Actual shortening of the sarcomere occurs during a shortening (concentric) contraction. Note that the part of the CB that binds to actin maintains its orientation relative to the actin-binding site (42). The power stroke is achieved by a bending action between the two parts of the CB. (B) Orientation of myosin molecules on the two halves of the myosin filament. The heads (CBs) of myosin molecules pull toward their tails; thus, all CBs pull toward the center of the sarcomere, as indicated by the arrows. Note that in this schematic the number of CBs on the myosin filament has been greatly reduced for clarity. Also recall that CBs are distributed around the myosin filament in a hexagonal array (Figures 4.1 and 4.4).

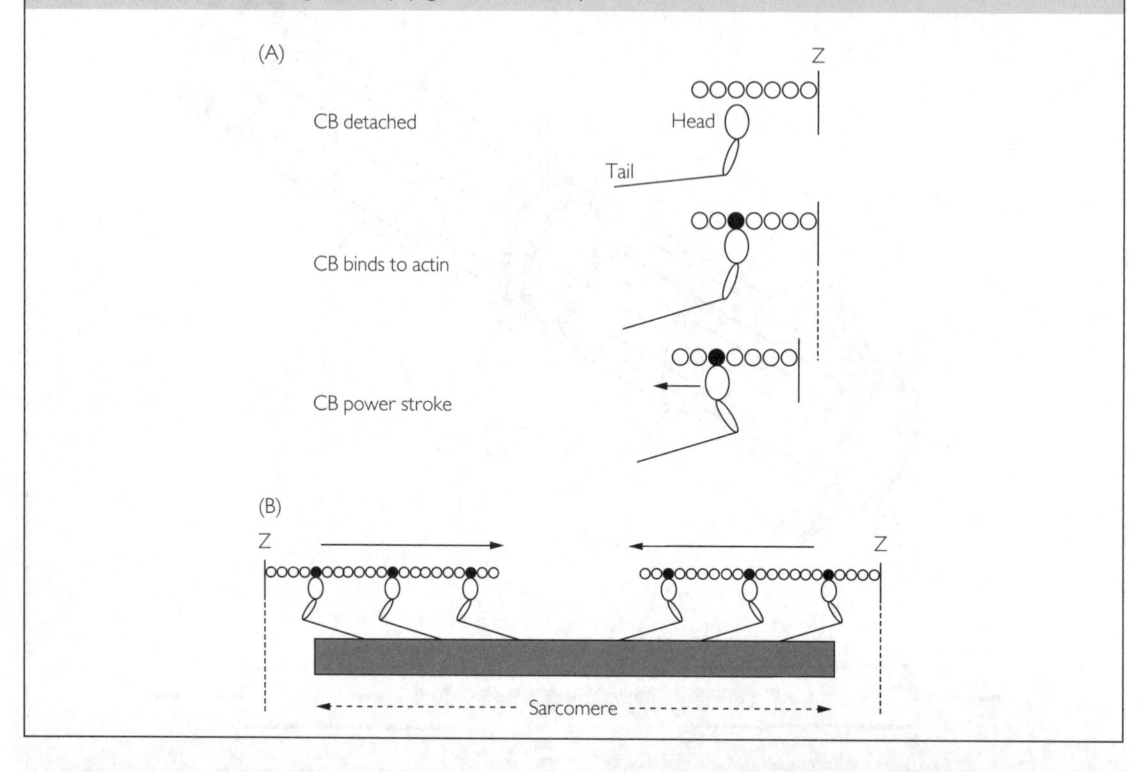

BOX 4.2 POWER STROKE STATISTICS

A CB power stroke, acting alone and with minimal opposing load, could move an actin filament through a displacement of about 5–10 nm (92,108). With larger loads the power stroke is smaller and slower (73,91). Assuming a power stroke of 10 nm and one CB acting alone, 100–200 power strokes would be required to shorten a half of a sarcomere 1 μm. However, hundreds of CBs would be performing power strokes, so a single CB would not have to perform that many power strokes. The force produced by a single power stroke in an isometric contraction is about 5–10 pN (1 trillion pN = 1 N) (15,91,92). Thus, a force of 1 N (0.1 kg) would require the simultaneous pull of 100 billion CBs!

BOX 4.3 TERMINOLOGY RELATED TO MUSCLE CONTRACTION

In the strictest sense, isometric and eccentric contractions would appear to be contradictions in terms, because "contraction" literally means "shortening." For this reason, some have advocated the term "action" instead of "contraction" to apply to active muscle shortening and lengthening, as well as isometric activity (57). There has also been criticism of the terms "concentric" and "eccentric" as applied to shortening and lengthening contractions, respectively (25). However, the terms "contraction," "concentric," and "eccentric" continue to be commonly used in muscle physiology, and will be used in this book to avoid confusion.

Figure 4.7 (A) CB cycle. In state 1, a power stroke has been completed and the CB is in a rigor state, still bound to actin. A new ATP molecule binds to the CB, causing it to detach from actin, forming state 2. The CB then undergoes a recovery stroke, putting it into the pre-power stroke position (state 3). The completed recovery stroke initiates the splitting of the ATP molecule into ADP and P_i, and the CB binds to actin (state 4). The power stroke proceeds with the release of ADP and P_i. The CB has returned to state 1. ATP, adenosine triphosphate; ADP, adenosine diphosphate; P_i, inorganic phosphate. Adapted from Mesentean *et al.* (80). (B) For simplicity, the myosin molecule is usually depicted as having one head; in fact, it has two heads as shown schematically. The possible functions of the double head are discussed in Box 4.4.

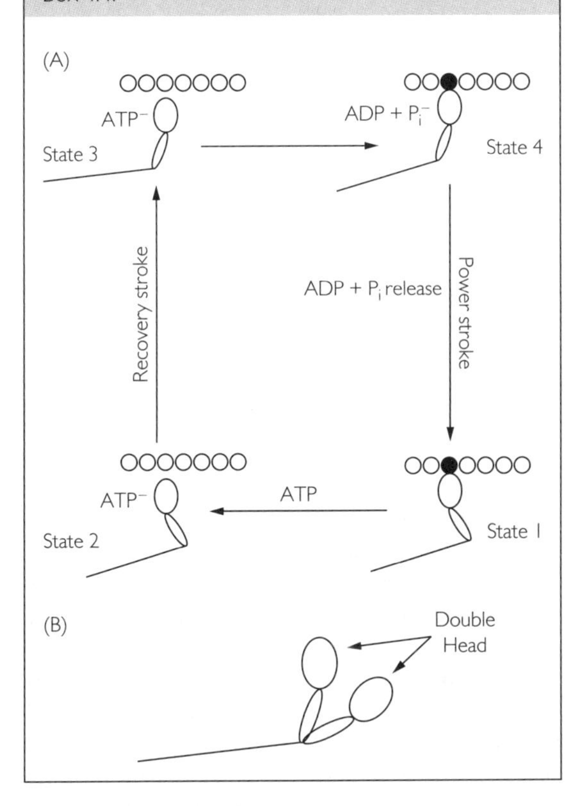

tropomyosin molecules, which are joined end to end along the length of the actin filament. Near the junction of adjacent tropomyosin molecules is a three-component troponin complex (see Box 4.6) that, for simplicity, is shown as a sphere in Figure 4.8.

When the muscle is not contracting (at rest or relaxed), tropomyosin covers the actin-binding sites for myosin CBs, thereby preventing CB cycling. However, when Ca^{2+} is present, it binds to one part of the troponin complex (Box 4.6), causing a conformational change in the troponin–tropomyosin positioning, so that tropomyosin is pulled aside, exposing the actin-binding sites (Figure 4.8). These sites can now bind with myosin CBs and CB cycling begins. If Ca^{2+} is removed and is no longer bound to troponin, the conformational change reverses and tropomyosin again covers the actin-binding sites, leading to cessation of CB cycling and relaxation.

BOX 4.4 ARE TWO HEADS BETTER THAN ONE?

In illustrations of the CB power stroke and CB cycle, myosin molecules are usually shown, for simplicity, as having a single head. In fact, myosin molecules have two heads (39), as illustrated schematically in Figure 4.7B. Different possible roles for the double head have been advanced. One possible role is that the two heads bind to actin quickly one after the other, each contributing to the power stroke (108). Another possible role is that one head optimizes the orientation of the second head for performing the power stroke (108). It has also been proposed that whereas only one of the two heads is attached to actin at any time in isometric and concentric contractions, both heads of myosin interact simultaneously with actin when a stretch is applied to a muscle during an isometric contraction, producing an eccentric contraction (28). This would largely explain why eccentric contractions are stronger than isometric contractions, as will be discussed in more detail in an upcoming section of the chapter.

BOX 4.5 WEAKLY AND STRONGLY BOUND CBS

In the CB cycle (Figure 4.7A), after ATP has split into ADP and P_i, the CB initially binds weakly to actin (transition from states 3 to 4). As ADP and P_i are released from the CB, the CB becomes strongly bound to actin and the power stroke proceeds (transition from states 4 to 1) (42).

Control of Ca^{2+} To start a muscle contraction, there must be a way of delivering Ca^{2+} to the actin and myosin filaments; to stop a contraction, there must be a way of removing the Ca^{2+}. The delivery of Ca^{2+} begins with the initiation of a muscle action potential (MAP) in the muscle fiber membrane, called the sarcolemma.

Figure 4.8 Role of Ca^{2+}. The thin filament consists of the proteins actin, tropomyosin, and troponin. With Ca^{2+} absent, tropomyosin covers binding sites (for myosin heads) on actin molecules, preventing actin-myosin binding and thus CB cycling (preventing transition from state 3 to state 4 in Figure 4.7). If Ca^{2+} is present and binds to sites on troponin, a conformational change occurs in which tropomyosin is pulled away from the actin-binding sites, allowing myosin CBs to bind to actin and begin CB cyling. If Ca^{2+} is removed, tropomyosin will slide over the binding sites again and CB cycling will stop (i.e., relaxation). From Brooks *et al.* (10), © McGraw-Hill Education.

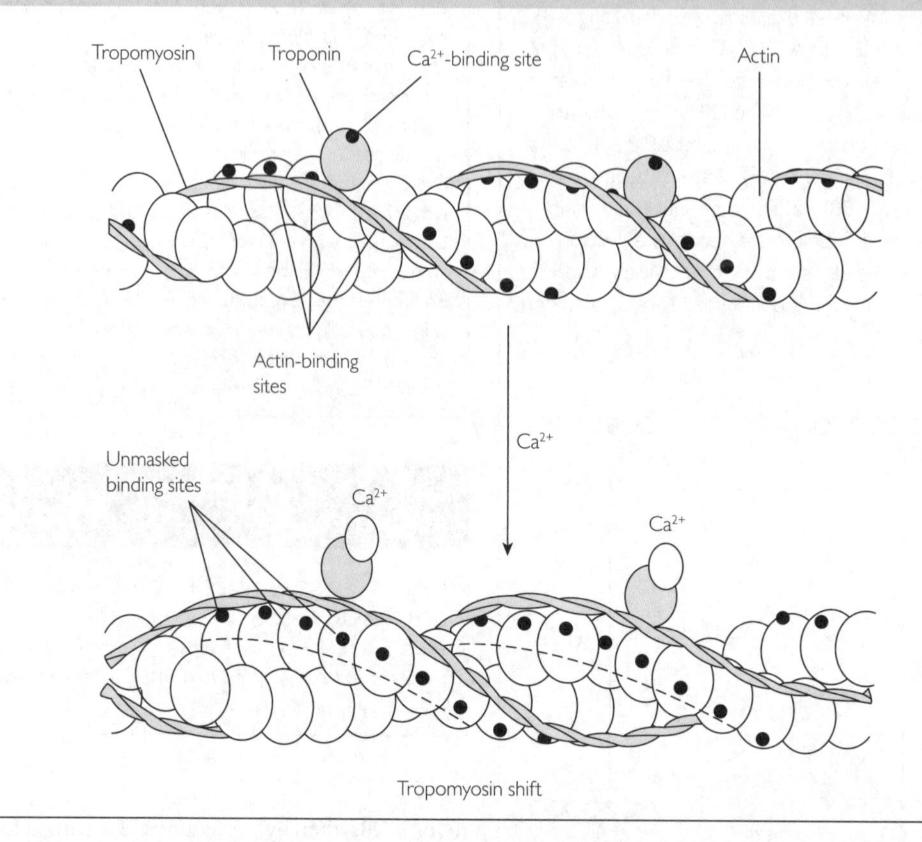

This is achieved when a nerve impulse traveling down the axon of a motor nerve cell (motoneuron) reaches the junction between the nerve cell and the muscle fiber, called the **neuromuscular junction** (NMJ) (Figure 4.9). The NMJ consists of the end terminal of the axon and the motor endplate, a specialized area of the sarcolemma beneath the terminal. After a nerve impulse has invaded the terminal, a neurotransmitter, acetylcholine, is released from the terminal and excites a MAP in the sarcolemma. (The process by which acetylcholine is released and excites a MAP is called **neuromuscular transmission**.) The MAP travels away from the NMJ in both directions to the ends of the muscle fiber. An important feature of the sarcolemma is inward channels called transverse tubules (T-tubules) that allow the MAP to penetrate to the center or core of a fiber, which may be 100 μm in diameter (Figure 4.9).

The sarcolemma, especially the T-tubular portion, is linked to a separate tubular network called the sarcoplasmic reticulum (SR) (Figure 4.10). The functions of the SR are to store Ca^{2+} when the muscle is relaxed, release Ca^{2+} in response to a MAP, and reuptake or sequester Ca^{2+} once the effect of the MAP has subsided. The storage of Ca^{2+} is achieved in two ways. First, the

BOX 4.6 THREE COMPONENTS OF TROPONIN

Troponin (Tn) has three interacting components: (1) TnC, which binds to Ca^{2+}, (2) TnI, which can inhibit myosin CBs from binding to actin, and (3) TnT, which binds the troponin complex to tropomyosin. When Ca^{2+} binds to TnC there is a conformational change among the three components with the result that tropomyosin is pulled aside, allowing myosin CBs to bind to actin [for a more detailed review, see (29)].

Figure 4.9 A nerve impulse traveling down a motor nerve axon reaches the neuromuscular junction, initiating a MAP in the sarcolemma of the muscle fiber. The sarcolemma has inward channels (transverse tubules) that allow the MAP to penetrate to the center of the muscle fiber.

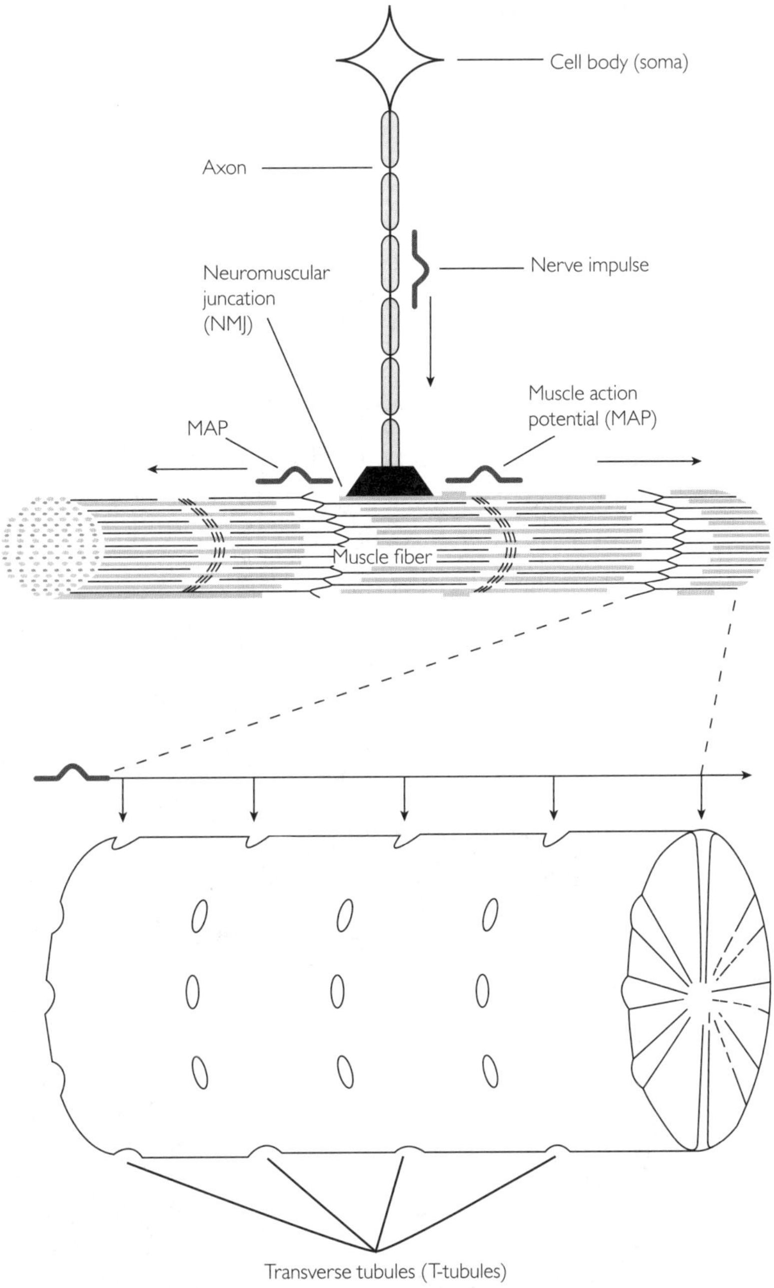

SR membrane has an active transport system, the Ca^{2+} pump that takes up Ca^{2+} so that the concentration of Ca^{2+} is very high inside the SR but very low outside in the sarcoplasm. Secondly, the SR has Ca^{2+} channels that are effectively closed when the muscle is at rest, ensuring that Ca^{2+} pumped into the SR cannot easily leak out. The Ca^{2+} release function of the SR is achieved in response to a MAP. The MAP, as it passes down the T-tubules, alters the linkage between the SR and T-tubules (see how the SR is positioned adjacent to the T-tubules in Figure 4.10),

Figure 4.10 A segment of a muscle fiber oriented vertically. The sarcoplasmic reticulum (SR) envelopes each myofibril and is positioned adjacent to the transverse tubules (T-tubules). As a MAP passes along the sarcolemma and down the T-tubules, it excites the SR to release Ca^{2+}. The Ca^{2+} binds to troponin and CB cycling begins. When the effect of the MAP subsides, the SR sequesters the Ca^{2+} that, by no longer binding to troponin, leads to relaxation. Note that the number of myofibrils in the muscle fiber has been reduced for clarity. Adapted from Krstić (58), © Springer, 1978.

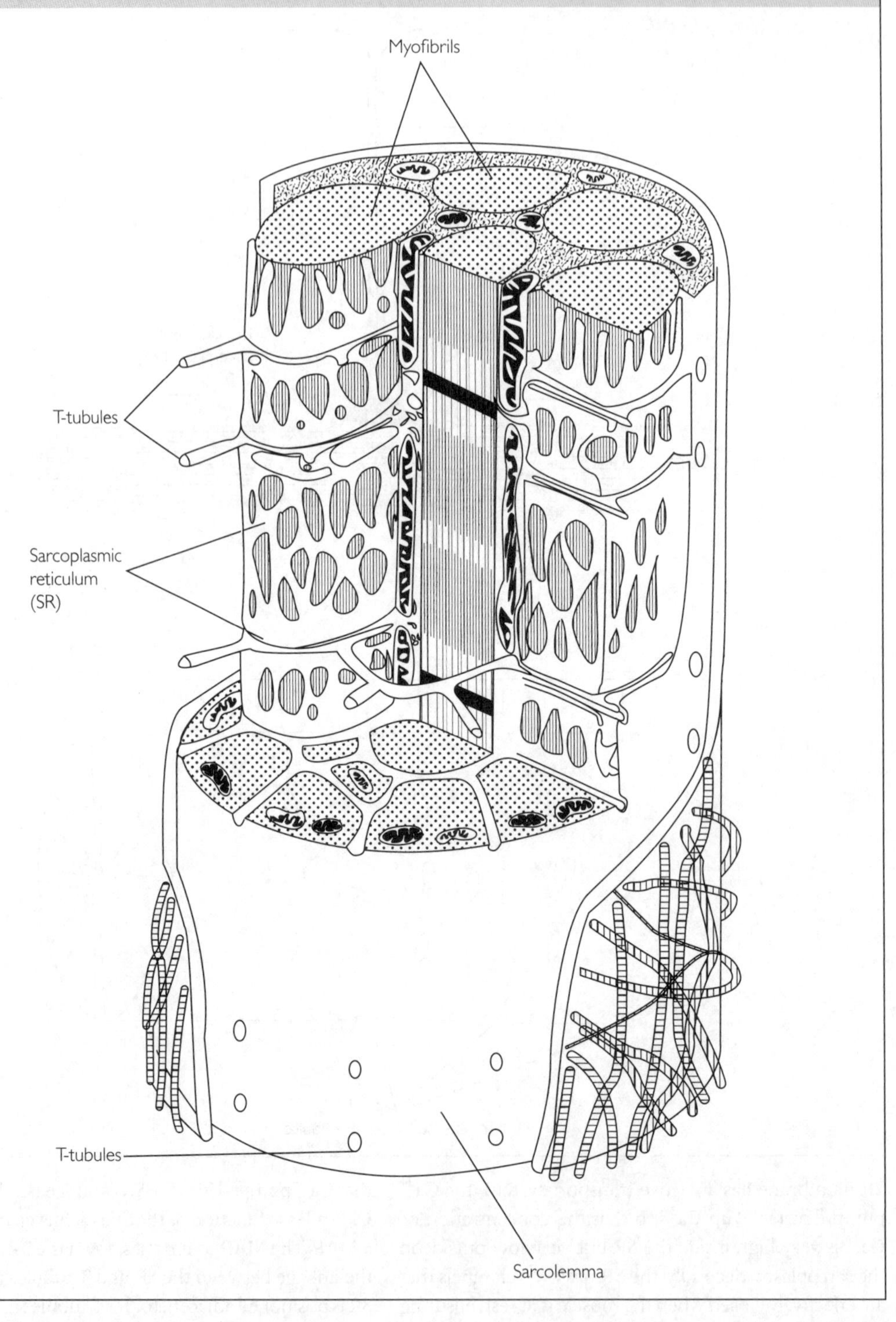

Figure 4.11 Summary of excitation–contraction coupling. A MAP travels along the sarcolemma and down the transverse tubules (T-tubules), causing the SR to release Ca^{2+}, which, by binding to troponin, initiates CB cycling. When the effect of the MAP has subsided, the SR sequesters the Ca^{2+}, resulting in relaxation. Adapted from Vander *et al.* (109), © McGraw-Hill Education.

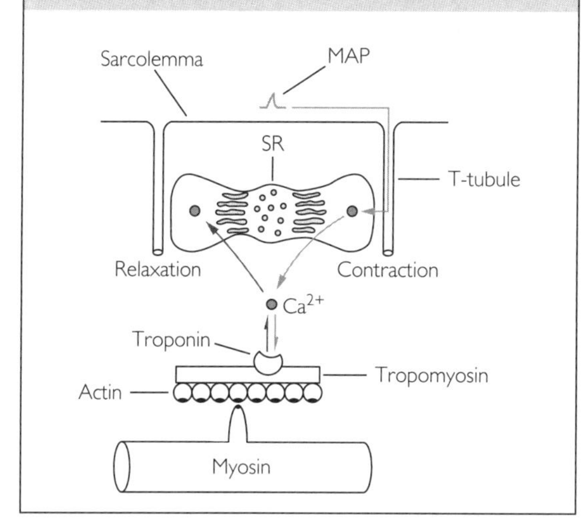

Figure 4.12 Twitch contraction. Top: a nerve impulse travels down the branches of the motoneuron's axon, triggering a MAP and twitch in each muscle fiber. Bottom: collectively, the activated fibers produce a motor unit twitch. The contraction and relaxation phases of a twitch correspond to the periods of Ca^{2+} release and reuptake, respectively.

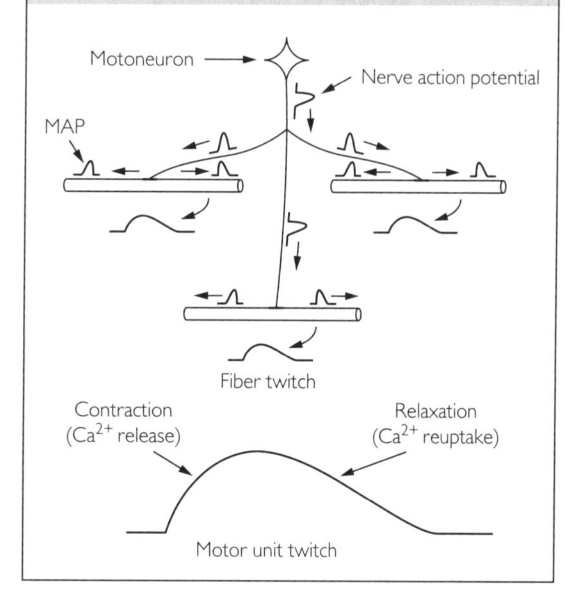

causing the SR's Ca^{2+} release channels to open, allowing Ca^{2+} to diffuse out of the SR and bind with troponin, initiating CB cycling (i.e., muscle contraction). The reuptake of Ca^{2+} back into the SR is done by the Ca^{2+} pump, which is stimulated by the increased concentration of Ca^{2+} outside the SR. The sequestering or reuptake of Ca^{2+} is aided by the dissipation of the MAP's effect on the Ca^{2+} release channels, causing the latter to close, restricting efflux of Ca^{2+} from the SR once it has been pumped back in. The process by which the MAP triggers CB cycling, known as **excitation–contraction coupling**, is summarized in Figure 4.11.

Twitch, Summation, and Tetanus

Twitch

A twitch is the contractile response of a muscle fiber to a single MAP (Figure 4.12, top). Collectively, twitches evoked in all muscle fibers innervated by one motoneuron produce a motor unit twitch (Figure 4.12, bottom) (motor units will be considered in detail in Chapter 5). If all of the muscle fibers (i.e., all motor units) in a muscle were activated to produce a twitch, the collective response would be a *whole muscle* twitch. Single twitches are rare in normal function; however, twitches serve to explain fundamental properties of muscle contraction. For this reason, in research single twitches are often evoked artificially by electrical stimulation. A twitch has a contraction or rising phase, and a relaxation or falling phase. The rising phase is caused by the brief period of Ca^{2+} release after a MAP, whereas the falling or relaxation phase corresponds to the reuptake of Ca^{2+} by the SR after the effect of the MAP has subsided. Note in Figure 4.12 that the duration of the rising phase is shorter than the falling phase, indicating that Ca^{2+} release (opening of the SR's Ca^{2+} release channels) is more rapid than Ca^{2+} reuptake (Ca^{2+} pump).

By convention and for practical reasons, twitch contractions are usually evoked and studied with the two ends of the whole muscle or muscle fiber fixed at constant length; that is, under isometric conditions. A twitch contraction has a brief duration, usually measured in milliseconds; for example, the human quadriceps muscle has twitch duration of a few hundred milliseconds. A twitch is also relatively weak, producing only about 10% of the force attained in a high-frequency tetanic isometric contraction (see section on Tetanus). For this reason, the twitch is sometimes considered more a measure of excitation–contraction coupling than force-generating capacity. A muscle twitch is very sensitive to changes in the state of a muscle, including potentiation, fatigue, temperature, and muscle length, factors to be discussed subsequently in the text.

Summation

If nerve impulses, and therefore MAPs, occur at long enough time intervals, the result will be two or more twitch contractions (Figure 4.13A). However, if subsequent MAPs occur before the effect of the previous one has subsided, that is, before all of the Ca^{2+} has been taken back up into the SR, the effect of the second MAP is partially added to that of the first, and the force of the contractile response increases. This additive effect of MAPs placed close enough together (i.e., at a high-enough frequency) is called **summation**. Figure 4.13B shows the effect of two trains of nerve impulses, which lead to two trains of MAPs (not shown). The higher-frequency train on the right produces greater summation and therefore greater force.

Figure 4.13 Summation. (A) On the left the MAPs are not close enough in time for summation to occur, so two separate twitches are produced. At right, the MAPs are close enough; that is, the second twitch begins before the first has completely relaxed. The result is increased force or summation. (B) The magnitude of summation is dependent on the closeness (frequency) of the nerve impulses and thus MAPs comprising a train. Increased frequency brings greater force of contraction. Each of the two contractions shown is called a tetanus (plural = tetani). See also Figure 4.14.

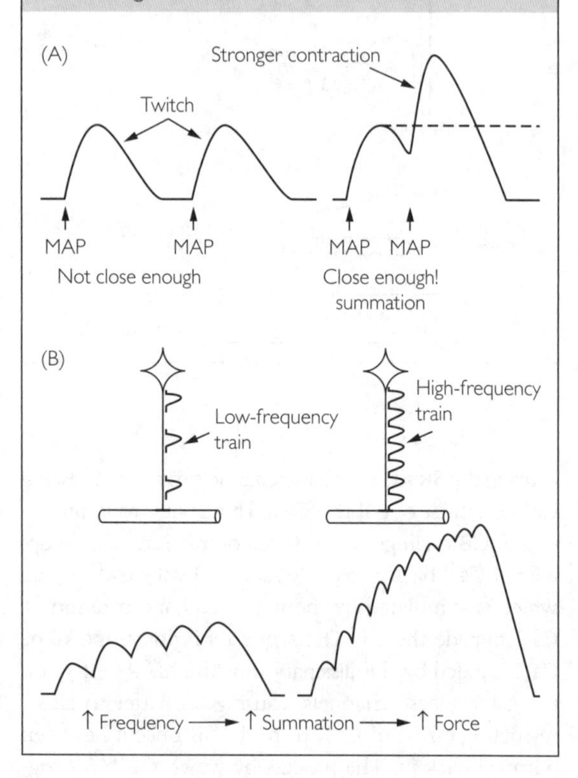

Tetanus

The contractile response to a train of MAPs causing summation is called a tetanic contraction or tetanus (Figures 4.13B and 4.14). A low-frequency tetanus may have partial relaxation phases, so it is said to be "unfused." A high-frequency tetanus shows no relaxation phases; therefore, it is called "fused" or "smooth." The force of a twitch and a high-frequency tetanus can be compared as a twitch/tetanus ratio (Figure 4.14). The twitch/tetanus ratio can be affected by several factors, including postactivation potentiation, temperature, muscle length, and fatigue; these factors will be covered in subsequent sections of the text.

Force–Frequency Relationship

Motoneurons can activate their constituent muscle fibers with a range of frequencies, and in so doing can greatly vary the force produced by the muscle fibers (Figure 4.15). The relationship between frequency and force is known as the **force–frequency relationship**. "Frequency" refers to the number of MAPs per second induced in muscle fibers. The force–frequency relationship is one of two principal ways by which the nervous system can vary the force of muscle contraction. Note in Figure 4.15 that at low frequencies a

Figure 4.14 Tetanus. A tetanus or tetanic contraction is the contractile response to a train of MAPs. Low-frequency tetani may be unfused; that is, partial relaxation phases may be seen. High-frequency tetani are fused; that is, there are no relaxation phases so the contraction has a smooth, unrippled appearance. The force of a high-frequency tetanus can be compared to that of a twitch as a twitch/tetanus ratio. In this example, the frequency of the chosen tetanus is that producing the greatest force. Thus the twitch/tetanus ratio is 0.1 or 10%.

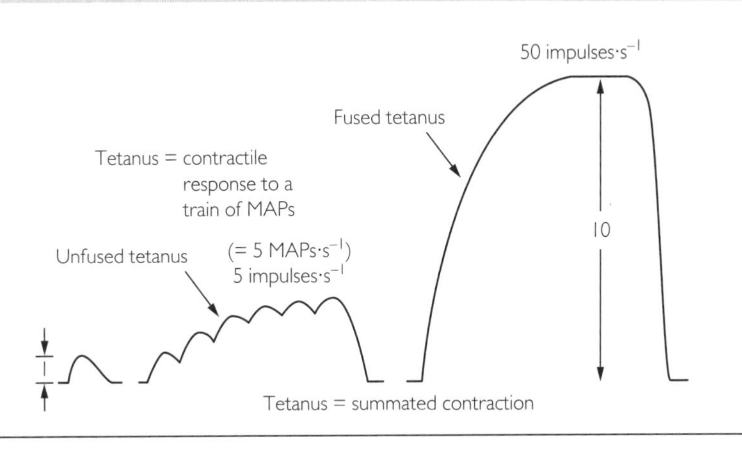

small change in frequency causes a large change in force, whereas at the highest frequencies a change in frequency has little effect on force. This is referred to as the plateau region of the force–frequency relationship.

Figure 4.15 Force–frequency relationship. The inset shows a twitch contraction and tetani at 10 and 50 Hz. The main figure shows how these and other frequencies can be plotted to produce a force–frequency relationship. Note how force increases sharply with increased frequency in the low-frequency range, but then plateaus at higher frequencies. Hz = hertz, the international unit for frequency; for example, 50 Hz corresponds to 50 MAPs·s $^{-1}$, which in turn corresponds to a stimulation frequency of 50 stimuli·s $^{-1}$.

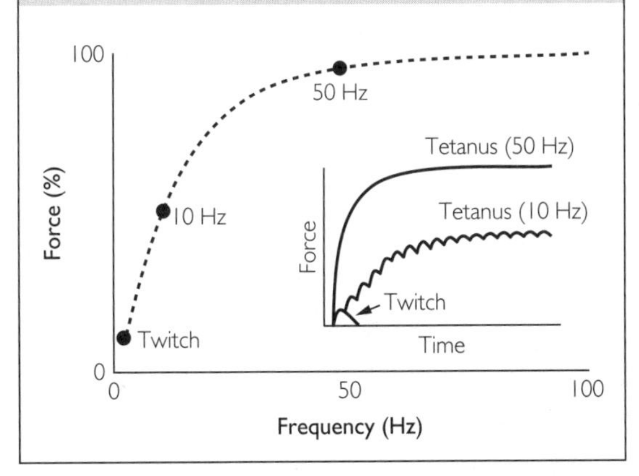

Mechanism of Summation

What is the mechanism of summation; that is, why does force increase as frequency of stimulation (MAPs) increases? There are two mechanisms. First, a greater frequency of stimulation causes a greater release of Ca^{2+} from the SR (Figure 4.16), which results in more troponin molecules being bound by Ca^{2+}, leading to more CB cycling (see Figure 4.11 for a review). More active CBs produce greater force. The second mechanism is related to what is known as the **series-elastic component** (SEC) (40). The SEC refers to elastic structures situated between the force-generating CBs in a muscle and the bones to which a muscle is attached (Figure 4.17A). The elastic structures include the muscle fiber's cytoskeleton, connective tissue between the muscle and tendons, and the tendons themselves. Before the force developed by CBs can be fully transmitted to the bones, these in-series elastic structures must be fully stretched out. In a twitch contraction, the CB cycling does not last long enough for the SEC to be fully taken up; therefore, the external force on the bone is less than the potential CB force (Figure 4.17B). In contrast, the duration of CB cycling in a tetanus is long enough for the SEC to be fully taken up, so that

Figure 4.16 Relationship between amount of Ca^{2+} released (i.e., Ca^{2+} concentration, or $[Ca^{2+}]$) among the actin and myosin filaments and the resulting force production. Note that a given change in $[Ca^{2+}]$ has a large effect on force at low $[Ca^{2+}]$, but little effect at high $[Ca^{2+}]$. The similarity between the force–$[Ca^{2+}]$ relationship shown here and the force–frequency relationship shown in Figure 4.15 is not coincidental; increased frequency of stimulation (MAPs) causes increased Ca^{2+} release from the SR.

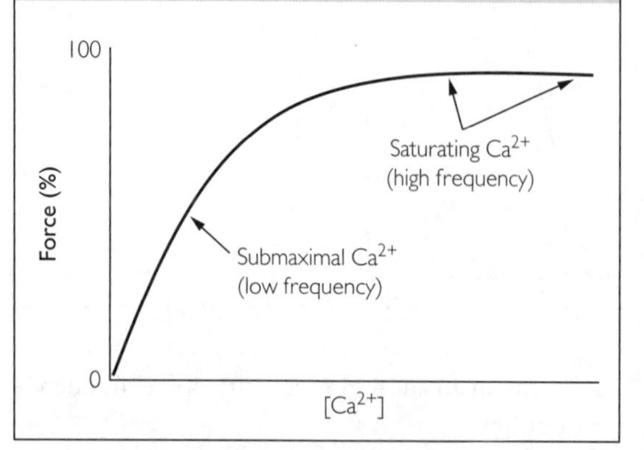

all of the potential CB force is realized as external force on the bone (Figure 4.17B). The two mechanisms responsible for summation of contraction are summarized in Figure 4.18. A tetanus is stronger than a twitch because in a tetanus there is both a greater number of CBs cycling due to greater Ca^{2+} release, and greater take-up of the SEC due to a longer period of CB cycling.

Contraction Types

Definitions

Reference has been made to concentric, isometric, and eccentric contractions. These three contraction types are illustrated in Figure 4.19 (see also Box 4.7). In concentric contractions, muscle force exceeds the opposing force and muscle shortening occurs. In isometric contractions, muscle force equals the opposing force, so there is no change in muscle length.

Figure 4.17 Series-elastic component (SEC). (A) Concept of SEC. CB cycling is the force generator in the muscle. Before CB force can be transmitted through to the bones to which the muscle is attached, elastic elements between the CBs and the bones must be stretched out fully. Otherwise, CB force will be lost taking up the SEC. The SEC may include the cytoskeleton and the tendons. (B) In a twitch contraction the duration of CB cycling is too brief to fully take up the SEC, so potential CB force is not fully transmitted as external force on the bones. In a tetanus, CB cycling lasts long enough for the SEC to be fully taken up; therefore, external force equals potential force.

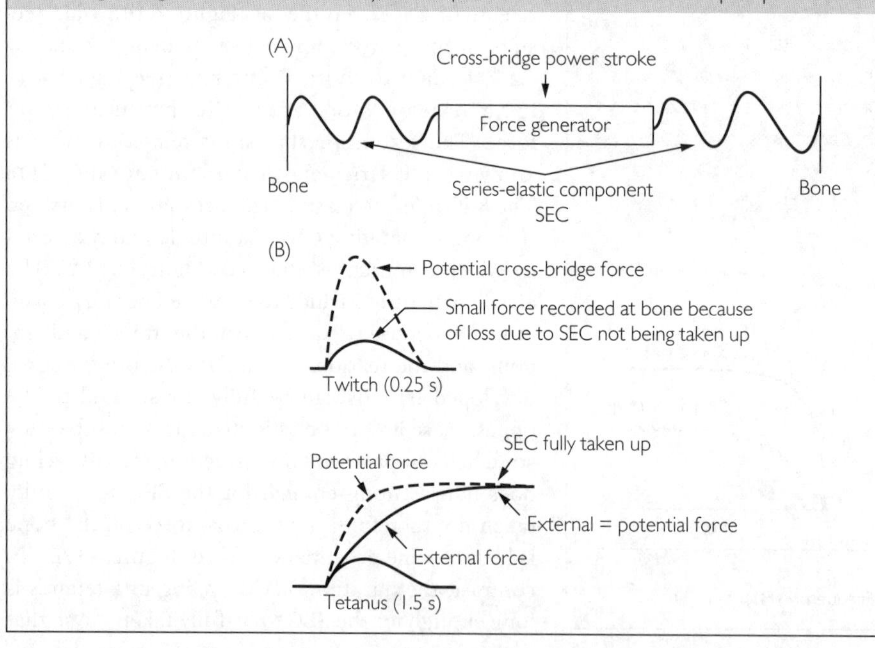

Figure 4.18 Summary of mechanisms explaining summation of contraction; that is, why a tetanus is stronger than a twitch. First, in a tetanus there is greater potential CB force because more CBs are cycling as a result of greater Ca^{2+} release from the SR. Secondly, there is the time factor: the longer tetanus allows the SEC to be fully taken up, which means that all CB potential force is realized as external force transmitted to the bone.

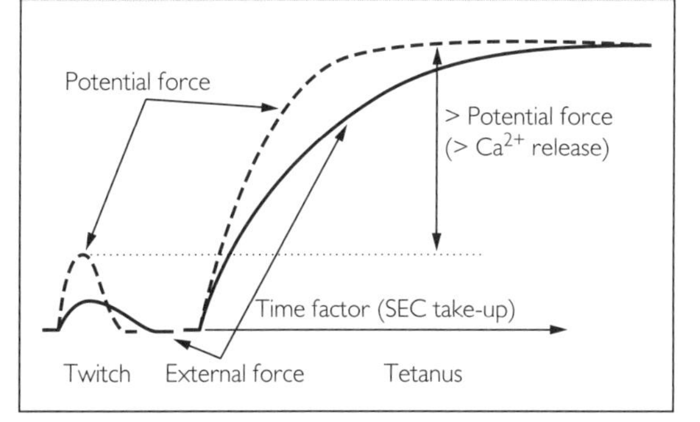

In eccentric contractions, the opposing force exceeds muscle force, and the muscle lengthens. "Concentric" literally means "toward the center;" that is, the two ends of the muscle move towards its center as in muscle shortening. Conversely, "eccentric" [initially introduced with the spelling "excentric;" (71)] means "away from the center," as in muscle lengthening. Eccentric contractions should be distinguished from passive lengthening of a relaxed (i.e., non-contracting) muscle. "Isometric" means "same measure," but for contractions the meaning is same or constant muscle length.

Effect of Contraction Type on Force

Anyone who has done repetitions of a weight-lifting exercise will have noticed that the concentric phase (raising the weight) of each repetition feels more difficult than the eccentric phase (lowering the weight), even though the force required for the two phases is the same (if done at a relatively slow, near-constant velocity). Similarly, it is easier to walk down a flight of stairs (eccentric) than up (concentric). The same applies to Figure 4.19; it is easier to lower the object with an eccentric contraction than to raise it with a concentric contraction. It should be noted in Figure 4.19 that the subject's nervous system has deliberately under-activated the elbow flexors for the eccentric contraction, so that the object's weight exceeds contraction force, causing the object to be lowered. Holding the object still with an isometric contraction would feel in between—harder than eccentric but easier than concentric. These perceptions are a consequence of muscles having their highest and lowest force-generating capacity (strength) in eccentric and concentric contractions, respectively. The superiority of eccentric contractions in force production has been demonstrated in many studies. The force of concentric and eccentric contractions is affected by the velocity of shortening and lengthening, respectively, known as the **force–velocity relationship** (as shown in the section Force–Velocity Relationship). But first it must be explained why eccentric contractions in general are stronger than isometric contractions, and why concentric contractions are weaker than isometric contractions.

The explanation for the difference in strength among the contraction types should be based on the CB power stroke and the CB cycle (Figures 4.6 and 4.7), although non-CB mechanisms may also be involved (94). The force of a contraction is a result of the number of active CBs (i.e., those interacting with actin filaments), and the force exerted by each CB while attached to the actin filament. Therefore, there are two possible explanations for the greater force of eccentric contractions. First, compared to isometric contractions at the same (maximal) level of activation, eccentric contractions have a greater number of attached CBs at a given point in time. The greater number of CBs yields greater force. The greater number of attached CBs may be partly due to the fact that during an eccentric contraction, increased strain on one head (CB) of a myosin molecule facilitates attachment of the second (partner) head (70); thus, both heads of single myosin molecules (Figure 4.7) are attached simultaneously to actin (Figure 4.20 and Box 4.4). This is in contrast to isometric and concentric contractions, in which only one head of each myosin molecule is attached at any time (28). In an eccentric contraction, the number of attached CBs may be up to twice the number attached in isometric contractions (69). A second mechanism is that in eccentric contractions, CBs do not complete CB cycles but instead become suspended between states 3 and 4 (Figure 4.7). CBs are forcibly detached in state 4 during the eccentric

Figure 4.19 Contraction types. Concentric: muscle force exceeds opposing force or load, and muscle shortens. Isometric: muscle force equals opposing force; thus, muscle length does not change. Eccentric: muscle force is less than opposing force; therefore, the muscle lengthens.

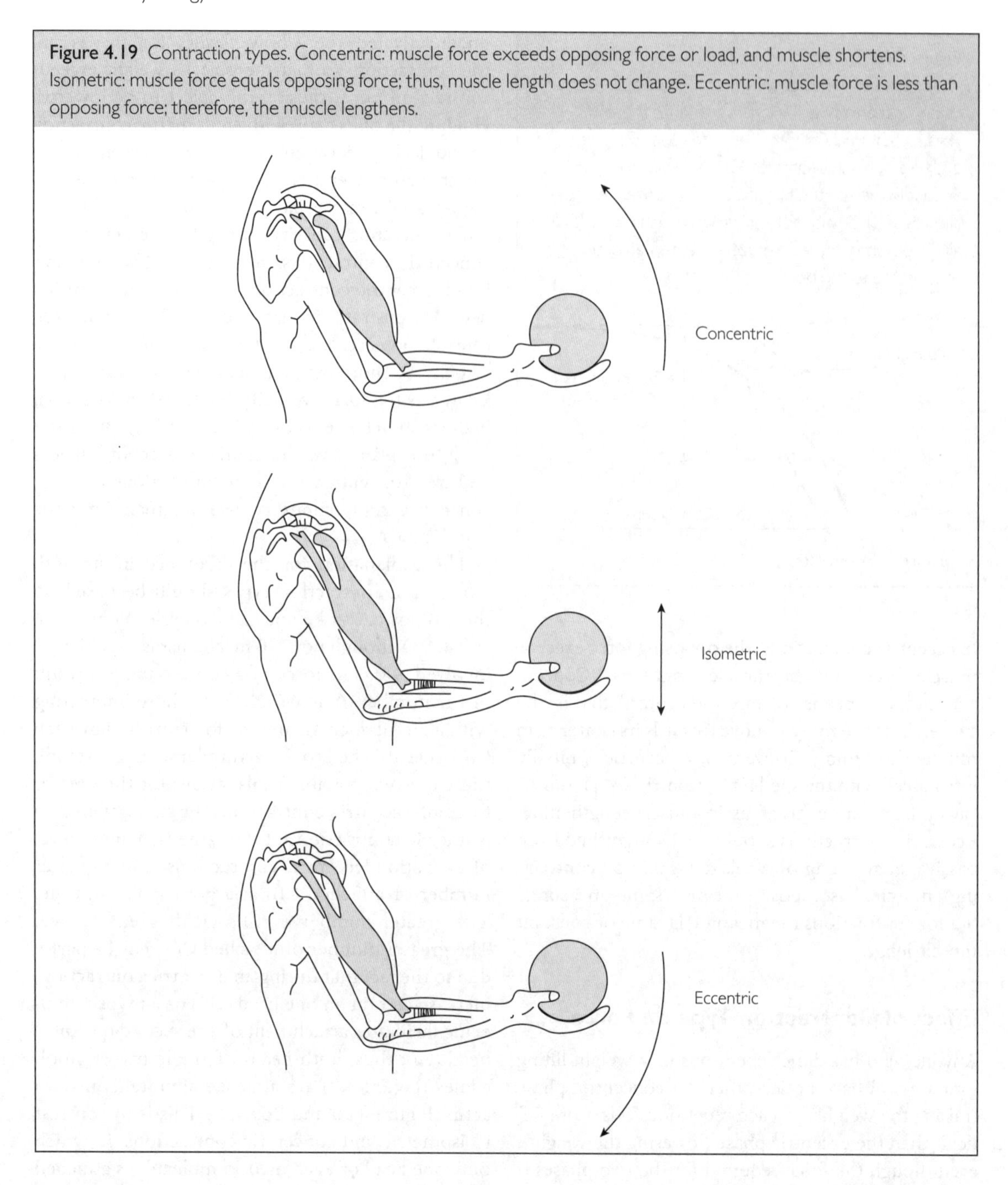

contraction but rapidly re-attach to actin, increasing the number of attached CBs at any time (28). Another effect of suspended CB cycles is that the amount of ATP needed to maintain force is reduced (46) because CB cycles are not completed (Figure 4.7). In addition to the increased number of attached CBs, the force per CB may also increase (21), but this may only be important at the beginning of an eccentric contraction. As more CBs are attached as the contraction proceeds, the force per CB may

BOX 4.7 TERMINOLOGY IN TYPES OF MUSCLE CONTRACTIONS

Additional terms have been used to describe muscle contractions. The term "isotonic" means "same tension or load." At one time "isotonic" was commonly used to describe muscle contractions that were not isometric, such as weight lifting and calisthenics. These activities are now commonly described as comprising concentric and/or eccentric contractions. Taking the literal meaning of isotonic (constant force), concentric, isometric, or eccentric contractions could, under some conditions, be "isotonic" (i.e., have a constant force).

The term "isovelocity" refers to concentric and eccentric contractions done at constant velocity. Isovelocity contractions are rare in normal activities; however, "isokinetic" dynamometers allow concentric and eccentric contractions to be tested at various constant joint angular velocities, and it is inferred that constant joint velocities correspond to constant velocity of shortening and lengthening of contracting muscles. Indeed, the terms "isokinetic" and "isovelocity" are usually considered to be synonymous.

Some activities consist of an eccentric contraction followed, after a very brief isometric phase, by a concentric contraction. Such coupled eccentric–concentric contractions are referred to as "stretch-shortening cycles" (57). Exercises involving the stretch-shortening cycle (or SSC) are often referred to as "plyometrics," although "plyometric" (also spelled "pliometric") is also a synonym for eccentric (25). Although not commonly used, a synonym for concentric or "shortening" contraction is "miometric" (25). Finally, the terms "static" and "dynamic" are sometimes applied to muscle contractions. "Static" is a synonym for isometric, whereas "dynamic" contractions are either concentric or eccentric.

Figure 4.20 Schematic representation of one mechanism by which eccentric contractions are stronger than isometric and concentric contractions. (A) In an isometric contraction (or concentric contraction, not shown), only one head (CB) of a myosin molecule is attached to the actin filament at a given time. (B) In an eccentric contraction, increased strain on one head facilitates the attachment of the second head. The result is an increase in the number of simultaneously attached CBs, which increases force. See discussion in Box 4.4.

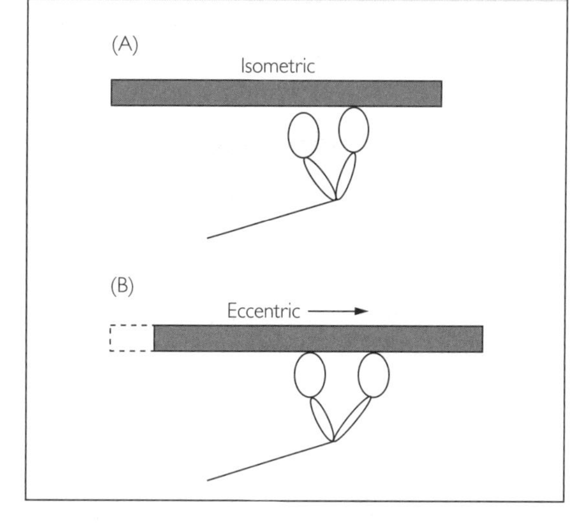

decrease back to the level seen in isometric contractions (28).

How can the reduced force of concentric contractions be explained? During muscle shortening against a load less than the load producing an isometric contraction, there is a decrease in the number of CBs bound to actin filaments at any moment, which reduces muscle force to match the load (92,96). Secondarily, there may be a small decrease in the force per attached CB, particularly when the load is small and corresponding shortening velocity is high (92). Concentric force may also be reduced because some attached CBs may exert a drag or braking effect if they remain attached to actin too long as the actin filament slides by (22,29,45) (Figure 4.21).

Force–Velocity Relationship

The velocity of shortening and lengthening affects the force of concentric and eccentric contractions, respectively. For concentric contractions, force decreases as velocity increases, whereas for eccentric contractions force increases (Figure 4.22).

Measurement

There are two basic methods of measuring the force–velocity relationship. In the first method, velocity is controlled or set, and the force attained at each set

Figure 4.21 The drag action of some CBs in concentric contractions. Three CBs are shown at three different stages of the CB cycle (recall that CBs cycle asynchronously). The CB at the far right is about to begin a power stroke. The middle CB has just completed a power stroke. The CB at far left has remained bound to actin too long, and thus exerts a drag or braking effect that opposes the force produced during the power stroke.

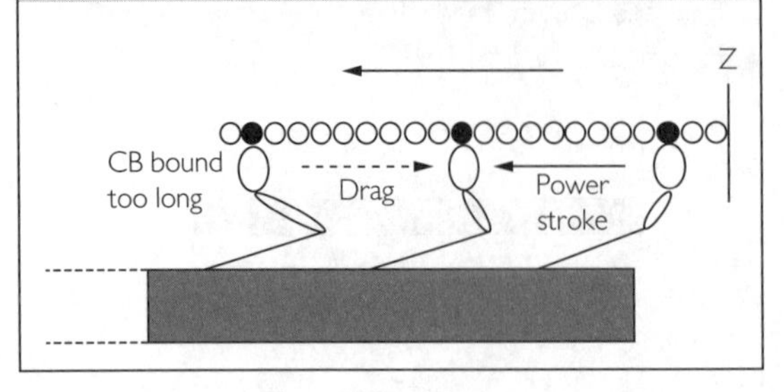

velocity is measured. This is the method used with isokinetic dynamometers. Since their introduction in the late 1960s, these dynamometers have been widely adopted for testing human subjects; thus, for many the concept of the force–velocity relationship is based on the "set velocity, measure force" method (method 1, Figure 4.23, left). However, the earliest measurements of the force–velocity relationship, in both humans and animal preparations, were done with the "set load, measure velocity" method (method 2, Figure 4.23, right). It can be argued that method 2 is the most applicable to human performance, because in the activities of daily living and in sports, various loads, including body mass, must be moved at various velocities, sometimes at maximum possible velocity. In addition, method 2 can be used to test bilateral, multi-joint movements, whereas the isokinetic dynamometers used in method 1 are typically designed to test unilateral (one limb at a time), single-joint movements.

With reference to Figure 4.23, two terms should be explained further. ISO_{max} refers to the force of a maximal isometric contraction (the acronym P_o is also used to denote maximum isometric force). V_{max} refers to maximum unresisted shortening velocity. V_{max} can be calculated from an equation derived from measurements obtained with method 2, or can be measured directly using a slack test, in which the term V_o is used.

Figure 4.22 An illustration of the force–velocity relationship. Force and velocity are given in arbitrary units. Symbols represent measurements of force at the various velocities. The line drawn through the symbols is the force–velocity relationship. Note that by convention lengthening velocities are given negative values. It can be seen that concentric force decreases as velocity increases. The velocity at which force just reaches 0 is V_{max}, the maximum unresisted or unloaded shortening velocity. In contrast, eccentric force increases as velocity of lengthening increases, but then a plateau is reached in which force increases no further despite increased velocity. In isometric contractions there is neither shortening nor lengthening; therefore, isometric force is plotted at 0 velocity.

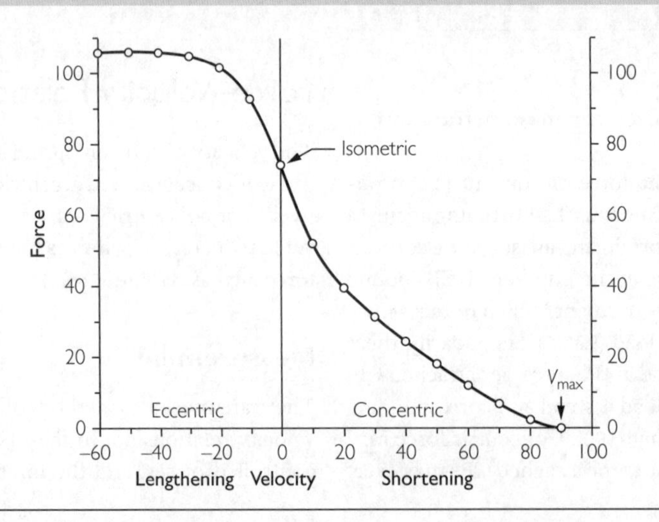

Figure 4.23 Two methods of measuring the force–velocity relationship. In method 1, velocity is set and force is measured. This method was used in the previous figure (Figure 4.22). In method 2, various loads are set and the attained velocity is measured. Load corresponds to force. Note that when velocity is set, it is on the horizontal axis (method 1), whereas when measured, velocity is on the vertical axis (method 2). For simplicity, only the concentric part of the force–velocity relationship is shown. ISO_{max}, maximum isometric force; V_{max}, maximum unresisted shortening velocity.

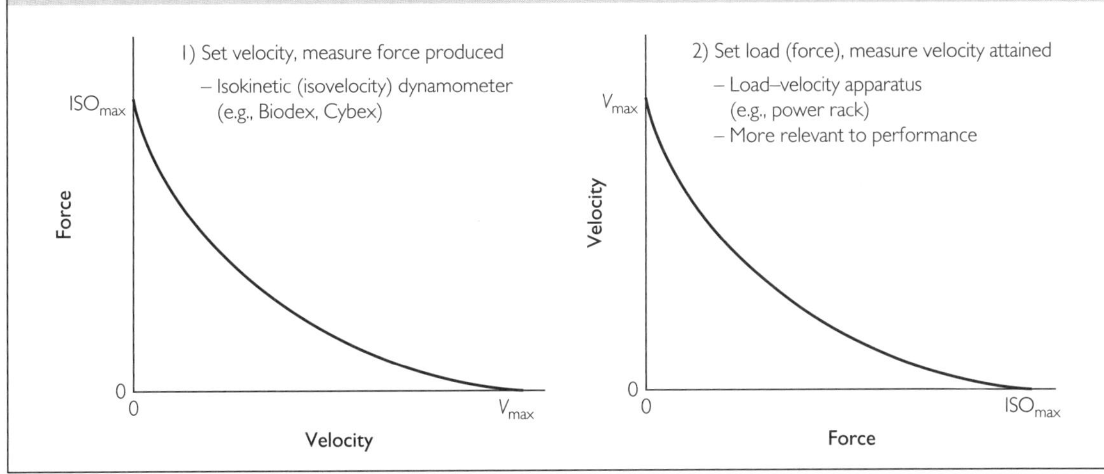

For simplicity, in this book the term V_{max} will be used regardless of how maximum shortening velocity is measured.

Decreased Concentric Force as Velocity of Shortening Increases

We have given possible explanations for concentric contractions being weaker than isometric contractions: (1) reduced number of attached CBs, (2) a small decrease in force per CB, and (3) increased number of CBs that exert a drag or braking effect. The simplest way to explain the decrease in concentric force with increased shortening velocity is that the three explanations given above are amplified at faster shortening velocities. With progressive decreases in imposed load and resultant increases in shortening velocity, there is a progressive decrease in the number of attached CBs (92), the force per CB (92), and an increased number of CBs exerting a drag or braking effect (22,29,46).

Increased Eccentric Force as Velocity of Lengthening Increases

In experiments with isolated muscle, the force of eccentric contractions can increase, with increased velocity of lengthening, up to two times isometric force before reaching a plateau (72). It will be recalled that, compared to isometric contractions at the same (maximal) level of activation, eccentric contractions have a greater number of attached CBs at a given point in time. The greater number of CBs yields greater force (70). With increased velocity of lengthening, the number of attached CBs can increase to close to two times the isometric value (70).

It is clear from Figure 4.22 that eccentric force does not continue increasing as velocity increases; a plateau in force is reached. At the point where the plateau begins, there is no further increase in the number of attached CBs (69).

Force–Velocity Relationship and Performance

The influence of the force–velocity relationship on performance is best considered in the context of the second method of measurement (method 2, see Figure 4.23); that is, when various loads are set and the velocity achieved measured. In Figure 4.24 it is shown that for a given load (mass of load, *M*), the greater the accelerating force (*F*), the greater the acceleration (*a*), and hence velocity (*V*) or speed attained. Conversely, the greater the load (*M*), the

Figure 4.24 The load–velocity relationship and performance. The velocity (V) or speed attained with a load with a mass (M) depends on the acceleration (a) achieved by the application of a force (F) against the load. The greater the applied force, or the smaller the load, the greater the acceleration (a) and therefore the velocity. This is illustrated in the load–velocity relationship. Only the concentric part of the load–velocity relationship is shown.

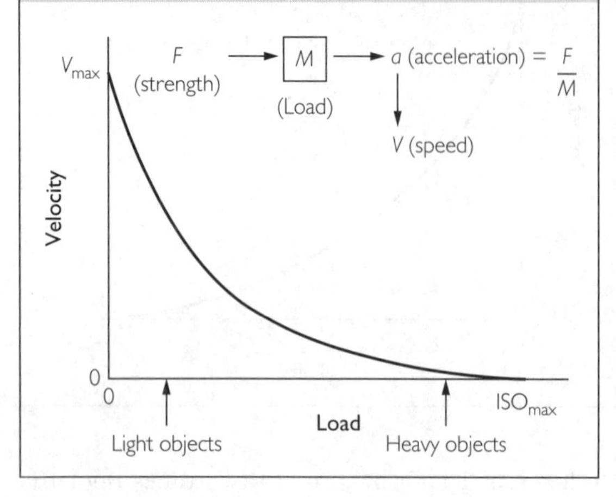

Figure 4.25 The position of some activities on the force/load–velocity relationship is shown. These are estimated positions, and it should be noted that in many movements the velocity varies; that is, the load is accelerated through the range of motion. 1 RM means one repetition maximum, the heaviest weight that can be lifted once (concentric phase) in a conventional weight-lifting exercise. Note that an Olympic lift (snatch or clean and jerk) must be done at a higher velocity than a 1 RM to be successful; therefore, the load must be lighter.

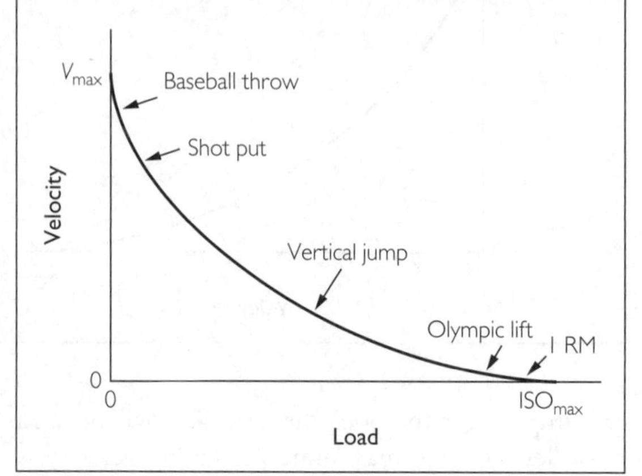

smaller the acceleration and velocity attained for a given applied force. It is important to note that the force applied to a given load can vary widely, depending on the intent (to move the load slowly or quickly) of the individual and his/her strength, rate of force development, and speed capability. In the load–velocity relationship depicted in Figure 4.24, it is assumed that the intent is to move the various loads as quickly as possible.

In Figure 4.25 the positions of a few activities on the force–velocity relationship are estimated. In viewing this figure, it is best to think of loads opposing movement rather than force. As already explained, the force applied to a load can vary depending on various factors, one being the intent to move the loads slowly or quickly. It is important to note that in all of the activities shown in Figure 4.25, with the possible exception of the 1 RM test, the goal is to move the load as fast as possible; the actual velocity attained is directly proportional to the applied force and inversely related to the magnitude of the load. It seems reasonable to suppose that for a given activity, for example the shot put, the smaller the load (weight of shot) as a percentage of the athlete's maximum strength, the farther the shot should be able to be thrown, because the release velocity will be greater. This concept is illustrated in Figure 4.26. The athlete with the greater strength should be able to throw the shot put farther, assuming that the two athletes are similar in other respects (e.g., technique, fiber-type distribution, motivation, etc.) In fact, however, these other respects may not be equal, with the result that two athletes might show force–velocity relationships as illustrated in Figure 4.27 (left panel). Athlete A's greater isometric and low velocity concentric strength may be related to A's greater muscle mass, whereas B's greater high-velocity strength, may be the result of a higher percentage of fast-twitch muscle fibers or superior neural activation related to high-velocity performance. However, it is possible that one athlete may be superior to another across the whole force–velocity relationship (Figure 4.27, right panel).

What about the eccentric force–velocity relationship and performance? The eccentric force–velocity relationship has, by comparison to the concentric

Figure 4.26 The shot put has a mass of 7.26 kg. In theory, the smaller this mass is as a percentage of the isometric maximum force (ISO_{max}) the higher the velocity attained with the shot put. Thus, an athlete with an ISO_{max} corresponding to 100 kg would attain a higher velocity with the shot put (i.e., throw it farther) than an athlete with an ISO_{max} of 75 kg. V_{max}, maximum shortening velocity (with zero load).

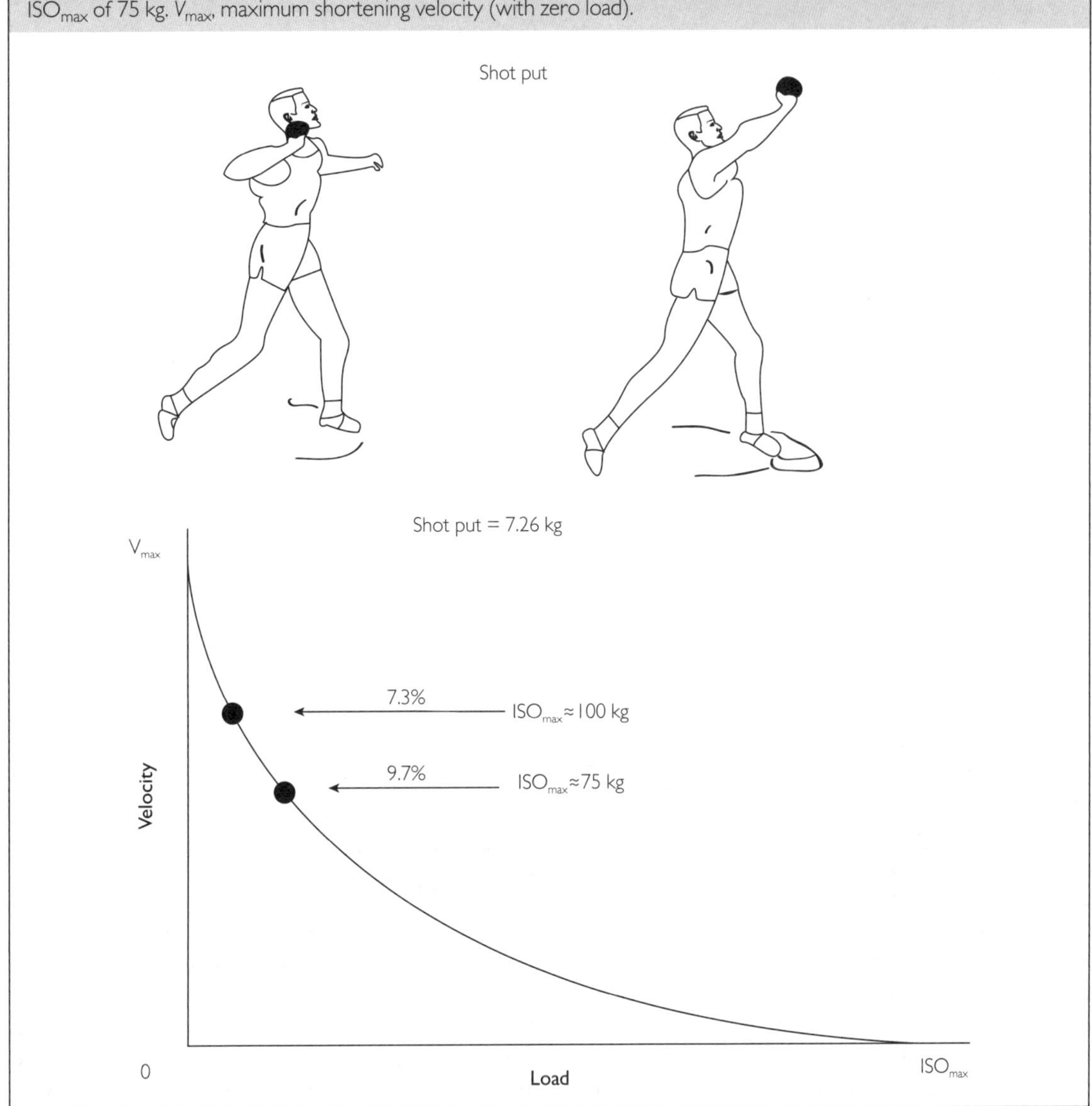

relationship, received less attention; it has been referred to as the "dark side" of the force–velocity relationship (71). Eccentric contractions may play their biggest role in performance when they are coupled to concentric contractions in what is known as the stretch-shortening cycle (SSC). The SSC will be considered in detail later in the chapter. At this point the power–velocity relationship will be considered.

Power–Velocity Relationship

Power can be defined as the rate of doing work (power = work/time) or as the product of force and velocity (power = force × velocity). This latter equation for power is best for explaining the connection between the force– and power–velocity relationships. If force has been measured at a series of velocities,

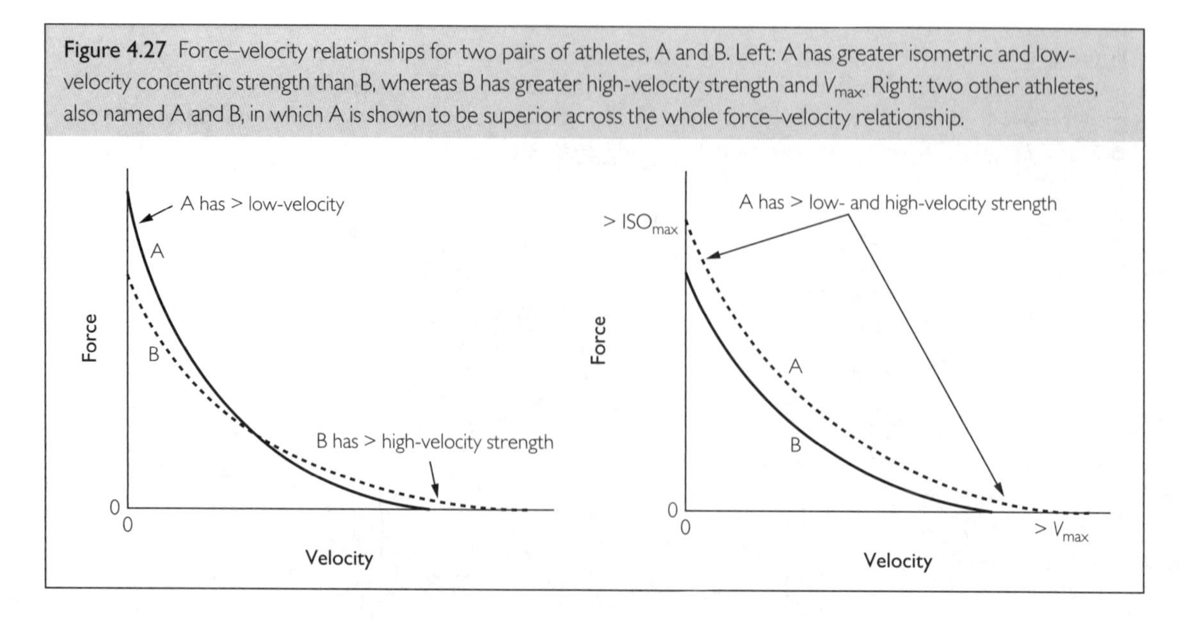

Figure 4.27 Force–velocity relationships for two pairs of athletes, A and B. Left: A has greater isometric and low-velocity concentric strength than B, whereas B has greater high-velocity strength and V_{max}. Right: two other athletes, also named A and B, in which A is shown to be superior across the whole force–velocity relationship.

the power can be calculated as power = force × velocity. This approach is illustrated in Figure 4.28. From the values of force and velocity in the table of the figure, power has been calculated and plotted in reference to the right-hand vertical axis of the graph. Note the inverted U shape of the power–velocity relationship. At a velocity of zero (i.e., isometric contraction) power must also be zero regardless of the force produced, because any force multiplied by zero velocity must equal zero. Through an initial range of shortening velocity (concentric contractions), power increases toward a maximum value because the increase in velocity exceeds the decrease in force. Beyond the maximum power value, power decreases because the decrease in force exceeds the increase in velocity. At the velocity where no force can be generated, power has again decreased to zero. In the example shown in Figure 4.28, maximum power occurred at a velocity corresponding to about 45% of V_{max}. For those who associate power with fast-velocity movements, it may be surprising to learn that maximum power often occurs at less than 50% of maximum velocity (V_{max}).

What about power generated by eccentric contractions? In Figure 4.29 an example of both eccentric and concentric power is given. In contrast to the inverted U shape of the concentric power–velocity relationship, the eccentric one rises virtually linearly as lengthening velocity increases. This is because force increases and then plateaus as velocity increases; thus, the product of the two increases across the velocity range depicted. Note, in Figure 4.29, that lengthening (eccentric) velocities are given as negative, to distinguish them from positive shortening (concentric) velocities. This is why the right-hand axis in Figure 4.29 gives power as positive (+, concentric) or negative (−, eccentric) (see Box 4.8).

Power–Velocity Relationship and Performance

Whenever force is produced at a particular velocity (other than zero), power has also been produced (power = force × velocity). Although power is often associated exclusively with fast movements, even the slowest movements exhibit power. In fact, it is possible that an athlete may have a greater maximum speed (V_{max}) but a smaller peak power than another athlete (Figure 4.30). The explanation for this power paradox is that peak power occurs at relatively low velocities. In many sport actions, however, movements must be done at high velocities; that is, at velocities greater than those at which peak power occurs. Thus, an athlete (B in Figure 4.30) with a high V_{max} will produce greater power at high velocities despite having a lower peak power.

An example of the power–velocity relationship in sprint cycling is shown in Figure 4.31. Two

Figure 4.28 Using method I (set velocity, measure force, Figure 4.23), force is measured at a series of velocities (see table). The force values obtained are plotted to produce a force–velocity relationship as shown. The power attained at each velocity is calculated (force × velocity) and also plotted. In this example the highest power value ($Power_{max}$) occurred at a velocity corresponding to 44.4% of maximum unresisted shortening velocity (V_{max}). This is within the range (30–50%) usually observed. In the figure, power is given in arbitrary units; however, the international unit for power is the watt (W).

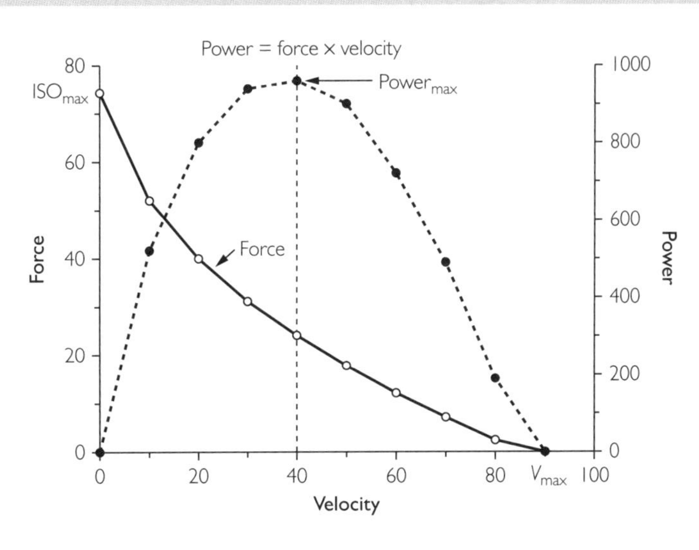

Velocity	Force	Power (force × velocity)
0	74	0
10	52	520
20	40	800
30	31.2	936
40	24	960 ← 44.4% V_{max}
50	18	900
60	12	720
70	7	490
80	2.4	192
90	0	0

subjects did brief, maximal effort sprints at a series of velocities on an isokinetic cycle ergometer. Power was calculated from the torque (force) values obtained. The subject with a higher percentage of fast (type 2) fibers in one muscle of the thigh (vastus lateralis) was able to generate a greater peak power at a higher velocity.

Stretch-Shortening Cycle

In the section on Contraction Types, reference was made to a combined eccentric–concentric contraction referred to as a **stretch-shortening cycle** (SSC) (55). Actually, the SSC also includes a brief isometric contraction linking the eccentric and concentric contractions. The duration of this isometric phase, which may be as brief as 10–15 ms, is also referred to as the coupling time between the eccentric and concentric phases of the SSC (55). It is important to emphasize that the stretch in the SSC is an *active* (eccentric contraction) rather than a *passive* stretch of a relaxed muscle. It is also important to note that the three phases of the SSC (eccentric, isometric, concentric) occur in the same muscle(s). For example, squatting down and immediately jumping

Figure 4.29 A plot of both eccentric and concentric power in relation to velocity. To distinguish between lengthening and shortening, velocity, work, and power of eccentric contractions are given a negative sign, whereas the same values for concentric contractions are given a positive sign (2). Therefore, the power axis (right) gives values of either concentric (+) or eccentric (−) power. The concentric power values are the same as those shown in Figure 4.28.

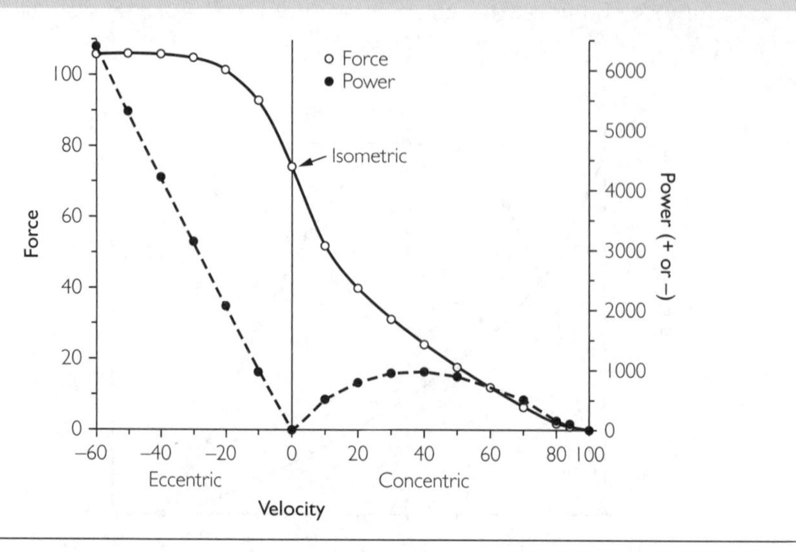

upward would constitute a SSC. The major muscles involved would be the hip and knee extensors and the ankle plantarflexors.

SSC Potentiation

The SSC is of interest and the subject of considerable research because when a concentric contraction is done immediately after an eccentric contraction, the force, power, and speed of the concentric contraction are increased compared to an isolated concentric contraction (110). This enhancement of concentric performance can be referred to as SSC potentiation. A schematic representation of SSC potentiation is shown in Figure 4.32. Note that the largest SSC potentiation occurs early in the concentric contraction following the eccentric contraction. SSC potentiation is present in many common activities such as walking, running, jumping, kicking, and throwing. SSC potentiation may also be used in many common exercises such as push-ups, chin-ups, and most weight-training exercises. The influence of the SSC on bench-press performance is illustrated in Figure 4.33. SSC potentiation must also be considered in strength and power testing (Box 4.9).

Possible Mechanisms of SSC Potentiation

High initial force level In viewing Figures 4.32 and 4.33, the most obvious mechanism of SSC potentiation is the high force level transferred from the end of the eccentric contraction and brief isometric phase to the subsequent concentric phase (101,102). This initial force is greater than what could be achieved in an isolated concentric contraction. The high initial force should allow greater power and speed to be

BOX 4.8 PLOTTING THE ECCENTRIC POWER–VELOCITY RELATIONSHIP

It could be argued that the eccentric (−) power values should have extended below the zero on the power axis of Figure 4.29, in the same way that eccentric velocity extended to the left of zero on the velocity axis; however, for simplicity, both eccentric and concentric power have been plotted on the same upward scale. Because eccentric contractions are associated with negative work and power, in strength training eccentric contractions are sometimes referred to as "negatives."

Figure 4.30 The power paradox. Force– and power–velocity relationships for two athletes, A and B. A has greater isometric strength (force at zero velocity) than B, whereas B has a higher V_{max} than A (see also Figure 4.27, left). Despite having a lower V_{max} than B, A has greater peak power because peak power occurs at relatively low velocities, where A can generate greater force. However, it is also clear from the graph that at higher velocities (where many performance actions occur) B can produce greater power than A.

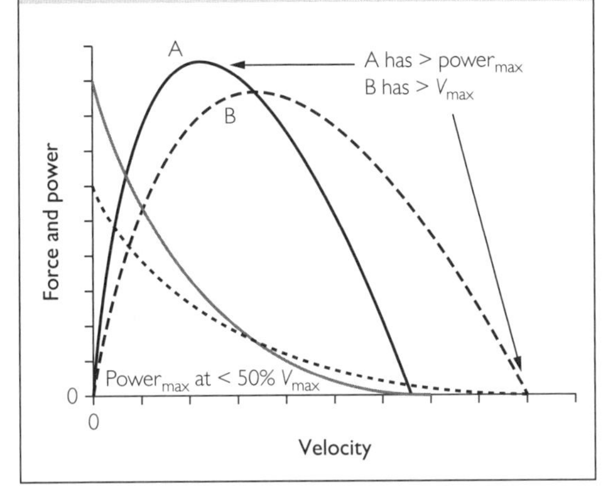

Figure 4.31 Power–velocity relationship in sprint cycling. Subjects did brief maximal effort sprints at a series of set velocities on an isokinetic cycle ergometer. Power was calculated from the torque values obtained at the various velocities. An equation was used to extrapolate power values beyond those obtained from torque measurements. Subject A, who had a higher percentage of fast-twitch (FT, also known as type 2) fibers in vastus lateralis (a head of quadriceps) than subject B, had a greater peak power and the peak power occurred at a higher crank velocity. Adapted from McCartney *et al.* (77).

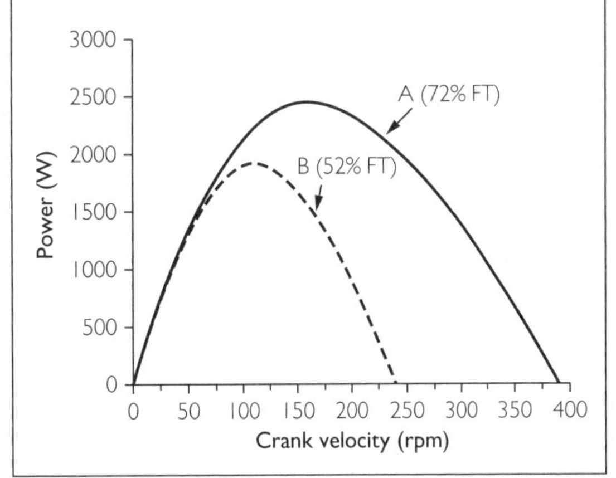

developed, particularly at the onset of the concentric phase of the SSC (110,113). The eccentric and isometric force phases of the SSC would also eliminate the time that would otherwise be needed to fully activate the muscles (61) and therefore build up force in an isolated concentric contraction (7) and take up the SEC (see below).

Take-up of SEC The SEC was discussed in connection with summation of contraction (see the section on Mechanism of Summation, and Figure 4.17). When a muscle is activated, there is a time period between appearance of MAPs propagating along the muscle fibers of a muscle (electrical activity) and the onset of muscle force (mechanical event). This time period is called the electro-mechanical delay (EMD) (13). The major part of the EMD is the time needed to take up the SEC sufficiently so that force begins to develop at the muscle's attachments to bone. The SEC has to be fully stretched (taken up) before all of the force generated by myosin CBs can be transmitted through to the bone(s) to which a muscle is attached. This is a particular challenge in isolated fast concentric (shortening) contractions against a small load that begin with the muscle relaxed (43), because there is limited time to build up force and fully take up the initially slack SEC (110), and the SEC will always be tending to become slack as the shortening contraction proceeds. The process of taking up the SEC could decrease rate of force development and the ultimate attained force, as illustrated in Figure 4.34. In the SSC, however, the SEC is already taken up during the eccentric and brief isometric phases, so that at least at the start of the concentric phase the SEC does not have to be taken up. As referred to above, the benefit of prior SEC take-up would be especially important for fast concentric contractions, where the EMD can be about 40 ms (43), and the time for full take-up of the SEC even longer.

Storage of elastic energy Reference has already been made to the SEC, the elastic structures within a muscle. The elastic structures include the tendon,

Figure 4.32 Schematic illustration of SSC potentiation. The force of an isolated concentric (CON) contraction, referenced to time, is compared to that of a concentric contraction preceded by an eccentric (ECC) contraction. The vertical line separating the ECC and CON phases of the SSC is the brief isometric (ISO) phase or coupling time. It can be seen that the force (and hence work and power) of CON is greatly enhanced (potentiated) by the preceding ECC, particularly in its early phase (see vertical double-headed arrows).

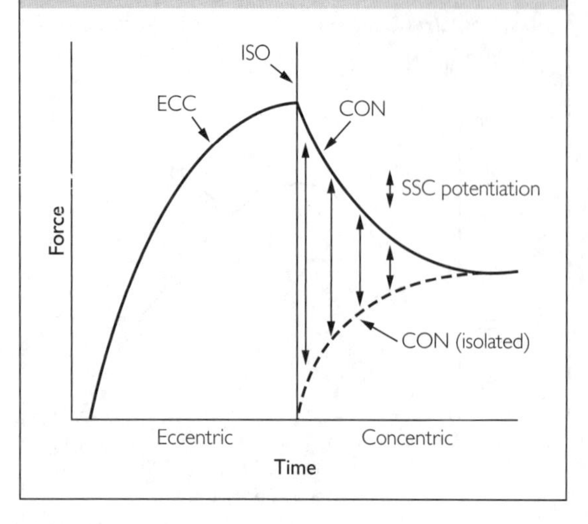

Figure 4.33 Effect of SSC potentiation on bench-press performance. Left: when subjects were allowed to lower (eccentric phase) the weight and immediately lift it up (concentric phase), their maximum single lift (one repetition maximum, 1 RM) was approximately 15% greater than when the lift was done without the SSC (isolated concentric contraction). Right: recordings of force exerted on the bar during the concentric (CON) phase of a lift with and without the SSC. Note that most of the force potentiation occurred during the first 250 ms of the concentric phase. Adapted from Wilson *et al.* (113).

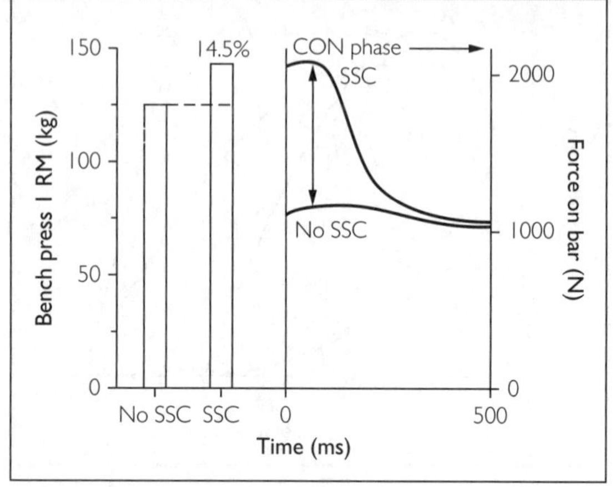

aponeuroses, cytoskeleton within muscle fibers (94,106), myofilaments, and the myosin CBs themselves (5) (Figure 4.35). The elastic structures have the potential to store elastic energy during the eccentric phase of the SSC. The elastic energy can then be released during the subsequent concentric phase. Most of the elastic energy is stored in tendons and aponeuroses (5). An example of how much a tendon can be stretched and energy stored is shown in Figure 4.36. Storage of elastic energy is particularly important in fast activities like jumping and throwing (40), in which the release of elastic energy can

BOX 4.9 CONTROLLING FOR SSC POTENTIATION IN PERFORMANCE TESTING

SSC potentiation should be considered when testing jump height in athletes, because the SSC can significantly affect jump height (see Figure 4.42). If SSC potentiation is to be allowed, athletes should be instructed to squat down quickly to a specified level and then immediately jump as high as possible. If SSC potentiation is not to be used, the athlete should be instructed to lower slowly into the squat position, and then hold the position until instructed to jump as high as possible. A common method of testing strength is the one-repetition maximum (1 RM); that is, the heaviest weight that can be lifted (concentric phase) for one repetition. Like jump tests, SSC potentiation can affect the 1 RM (see Figure 4.33). Usually, care is taken to eliminate SSC potentiation in a 1 RM, so that a measurement of maximum concentric strength can be made without the influence of varying degrees of SSC potentiation. In competitive power lifting, one role of the referee is to ensure that SSC potentiation is not used in the squat and bench-press lifts. The referee does this by requiring the athlete to hold the isometric phase after lowering the weight in the eccentric phase. The referee then gives the signal for the concentric phase to begin. The referee should be consistent in setting the duration of the isometric phase, because variation in the duration can affect the performance of the concentric phase (see Figure 4.43).

Figure 4.34 SEC and the SSC. In an isolated concentric (CON) contraction, time must be spent taking up the SEC. This, together with the time needed to activate the muscle and build up force, reduces rate of force development and the ultimate force attained. In contrast, in the eccentric (ECC) and brief isometric phase (depicted as the vertical line between the ECC and CON phases) the muscle is already activated and the SEC is taken up. The result is a large enhancement of the initial part of the CON phase.

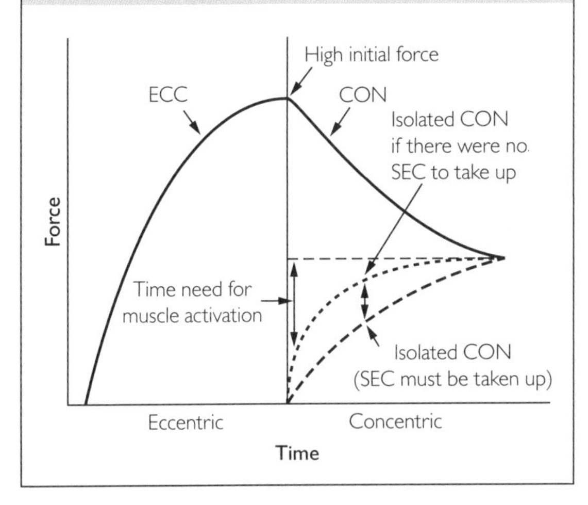

Figure 4.35 Schematic illustration of storage of elastic energy during the eccentric (stretch) phase of the SSC. Stretch occurs in myosin CBs (A), the cytoskeleton, for example the titin elastic filaments (B), and muscle tendon (C). The actin and myosin filaments also have compliance and can be stretched (not illustrated).

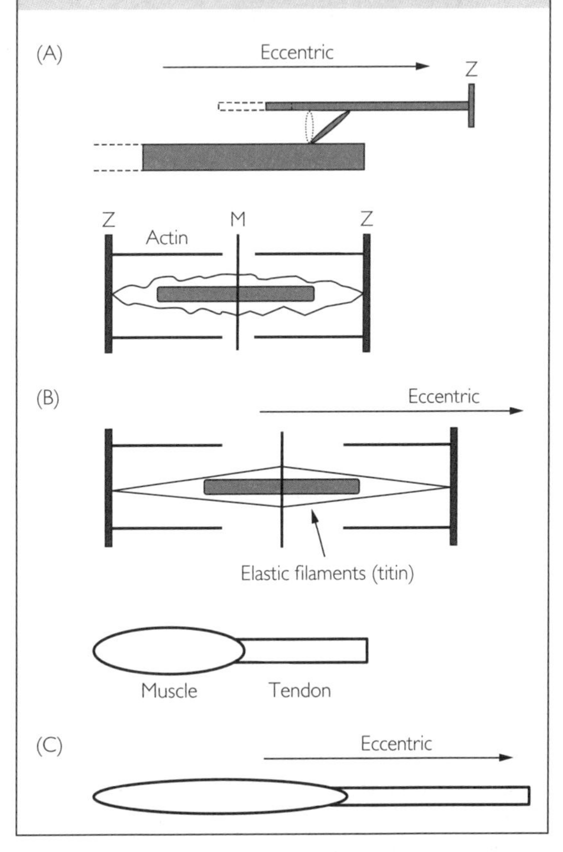

provide a catapult action (41), attaining speeds that could not be achieved by the contractile mechanism alone (40).

Reflex potentiation When a muscle is forcibly stretched while contracting, as in an eccentric contraction, a stretch-sensitive receptor within the muscle, called the muscle spindle, is also stretched. The muscle spindle is responsible for the stretch reflex; that is, a reflex muscle contraction in response to stretch. The role that the stretch reflex could have on SSC potentiation is illustrated in Figure 4.37. The stretch reflex evoked in an eccentric contraction potentiates the activation of the muscle by other parts of the nervous system (9,19). A more complete activation of the muscle yields greater force and power. The stretch reflex is more pronounced in rapid than slow stretches (86); thus, reflex potentiation's contribution to SSC potentiation may be greater in fast movements like running and jumping (19) (see Box 4.10).

Factors Affecting the Magnitude of SSC Potentiation

Submaximal versus maximal eccentric contractions at a given velocity For a given velocity of eccentric contraction, a greater force of eccentric contraction would result in a greater initial force of the subsequent concentric contraction, which should increase SSC potentiation (Figure 4.38).

Force–velocity relationship of eccentric contractions According to the force–velocity relationship of eccentric contractions, the force of maximal eccentric contractions increases up to a certain velocity before reaching a plateau (Figures 4.22 and 4.29). The increased eccentric force at higher

Figure 4.36 Storage of elastic energy in the patellar (quadriceps) tendon. A maximal isometric knee extension (MVC) at a knee joint angle of 100° resulted in an initial shortening (concentric) phase, which caused the patellar tendon to stretch 33 mm, representing an energy storage of 15 J. A more prolonged isometric contraction followed the brief shortening phase. Based on Kubo et *al.* (59).

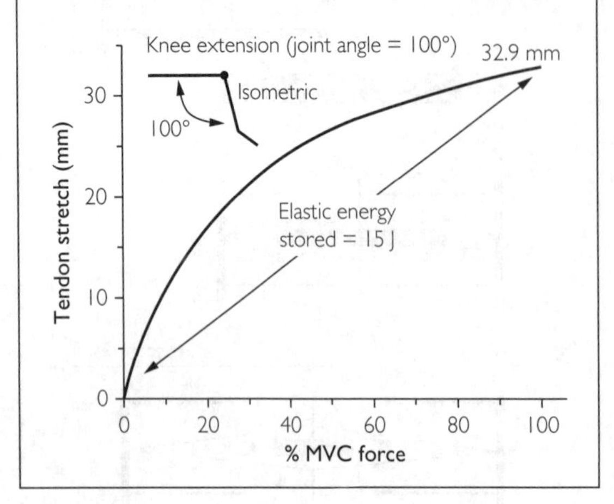

BOX 4.10 WHICH MECHANISM OF SSC POTENTIATION IS MOST IMPORTANT?

The relative importance of the various mechanisms of SSC potentiation has been a matter of debate in the scientific literature (110). The main issues appear to be over the importance of reflex potentiation (e.g., 19,57,110) and storage of elastic energy (e.g., 7,8,103). Some activities may emphasize a particular mechanism over others (47), and muscles acting together may differ in the main mechanism employed (48).

velocities would provide a greater initial force at the start of the concentric phase, thereby increasing the force of concentric contraction (Figure 4.39). Figure 4.40 gives an example of the effect of velocity on SSC potentiation. SSC potentiation raises the concentric force–velocity relationship; that is, greater force is developed at a given shortening velocity. The elevation of the concentric force–velocity

Figure 4.37 Schematic representation of reflex potentiation in the SSC. In the eccentric phase of the SSC, the stretch of muscle spindles (stretch-sensitive sensory receptors) in the muscle excites motoneurons in the spinal cord, which in turn excite the muscle fibers they innervate. This stretch reflex assists (i.e., potentiates) other nerve centers acting to activate the muscle. A more complete activation of the muscle leads to greater force and power output.

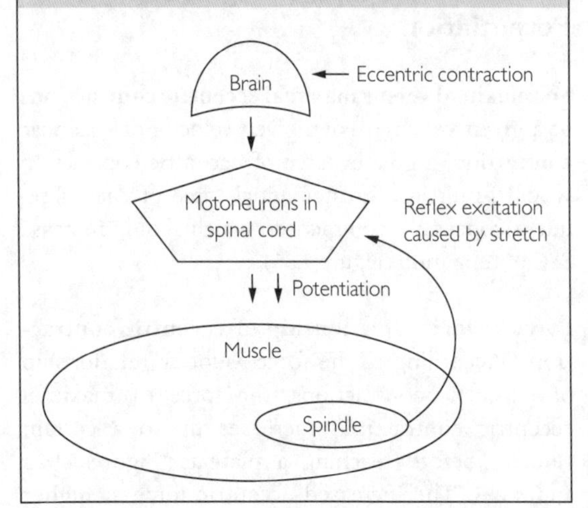

Figure 4.38 Comparative effects of submaximal (submax) and maximal (max) effort eccentric (ECC) contractions at a given velocity on the force produced in a subsequent maximal concentric (CON) contraction at a given velocity. The maximal eccentric contraction provides a greater initial force at the start of the maximal concentric contraction, and therefore greater concentric force (SSC potentiation). However, even the submaximal eccentric contraction enhances concentric force relative to an isolated maximal concentric contraction.

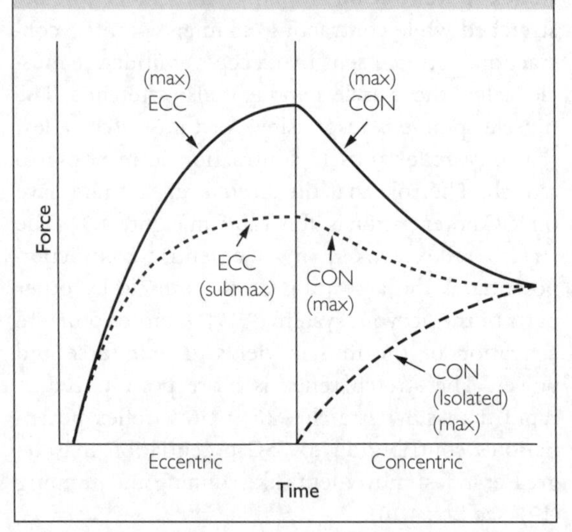

Figure 4.39 Schematic representation of the effect of velocity of maximal eccentric (ECC) contractions on the enhancement of concentric (CON) force. At faster velocities, eccentric contractions are stronger and can therefore provide a greater initial force at the start of the concentric phase, resulting in greater SSC potentiation (see vertical double-headed arrows). In contrast, isolated concentric contractions are weaker at higher velocities; thus, concentric contractions have more to gain with SSC potentiation at faster velocities, whereas faster eccentric contractions have more to give.

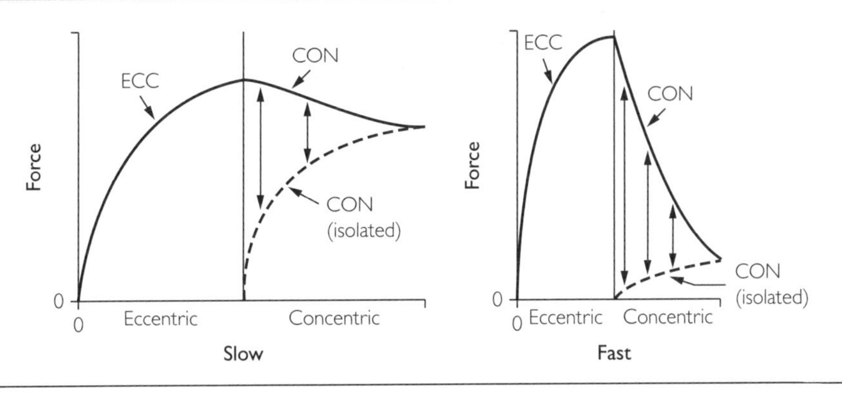

Figure 4.40 Effect of velocity of maximal eccentric contractions on the SSC potentiation of maximal concentric contraction force of ankle plantarflexors. On an isokinetic dynamometer, maximal concentric contractions were done in isolation and immediately after eccentric contractions at the two indicated joint angular velocities. SSC potentiation is given as the percentage increase in torque above that attained by isolated concentric contraction. There was greater SSC potentiation at the higher of the two velocities. It is also notable that at both velocities, there was greater potentiation in women than men. Based on the data of Svantesson and Grimby (100).

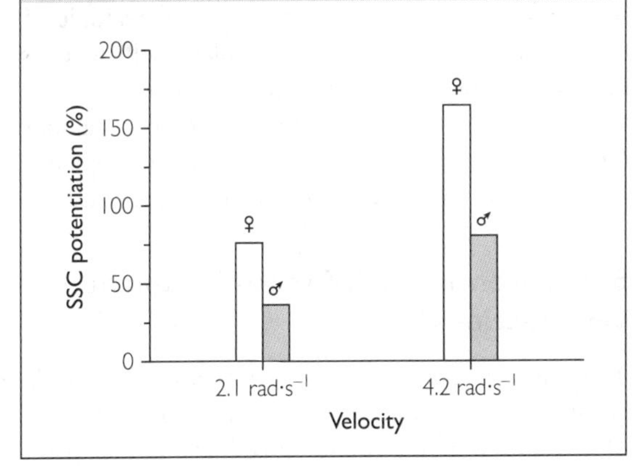

relationship is relatively greater at higher velocities (12,101) (Figure 4.40).

However, there is a level of eccentric force beyond which there is a decline in SSC potentiation. As shown in Figure 4.41, concentric force and work (average force × displacement) decrease at higher eccentric velocities, even when a high eccentric force is maintained. This pattern has been observed in single muscle fibers (103) and in single and multi-joint movements (102,103).

The decline in concentric force and work begins when the eccentric peak force exceeds approximately 1.8 times the isometric maximum force (ISO_{max}) in isolated muscle fibers and approximately 1.25 times ISO_{max} for intact human muscles (103). This phenomenon may also partially explain the impaired vertical jump performance that occurs when the drop phase of drop jumps is done from too great a height (Figure 4.42). Inhibition of muscle activation on landing may also partially explain the decline in jump height at the highest drop jump heights (56).

Duration of coupling time Recall that coupling time is the duration of the brief isometric phase between the eccentric and concentric phases of the SSC. The coupling time is typically very brief (e.g., 15 ms). If the coupling time is prolonged, SSC potentiation is reduced (55). The simplest explanation for the reduction is that the high eccentric force degrades to a

Figure 4.41 Effect of magnitude of eccentric (ECC) work (average force × displacement of single contractions) on concentric (CON) work in human elbow flexors. Work (relative work) is expressed relative to the concentric work after a brief isometric contraction (100%). Eccentric work was varied by increasing lengthening (joint angular) velocity from 30 to 180°·s^{-1}. All concentric contractions were done at 90°·s^{-1}. The greatest enhancement of concentric work occurred after eccentric contractions at 90°·s^{-1}. At higher eccentric velocities concentric work decreased even though high eccentric work was maintained at the higher velocities. Based on the data of Takarada *et al.* (103).

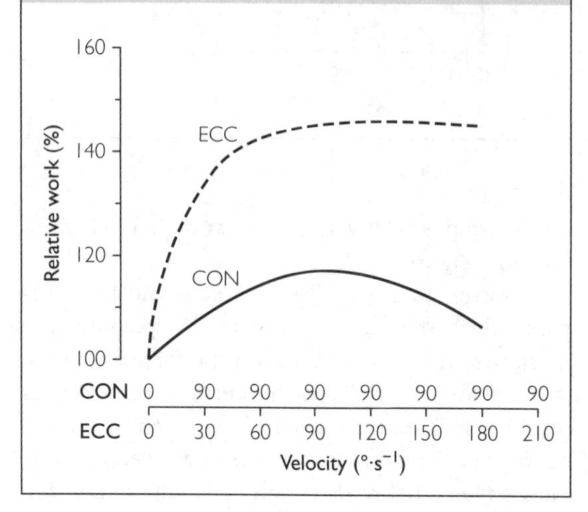

Figure 4.42 Effect of SSC potentiation on vertical jump height. Height attained in a squat jump (SJ) done without the SSC is compared with a counter-movement jump (CMJ) in which the subjects, from an initial standing position, rapidly squatted (eccentric phase) and then immediately jumped upward (concentric phase). A faster and stronger eccentric phase was achieved by having the subjects jump down from a height (DJ = drop jump from the heights indicated). Note that up to a DJ height of 40.4 cm, the faster and stronger eccentric phase produced a higher jump height. At the highest DJ height (69.9 cm), however, jump height began to decline, producing a pattern similar to that in Figure 4.41. Based on the data of Asmussen and Bonde-Petersen (3).

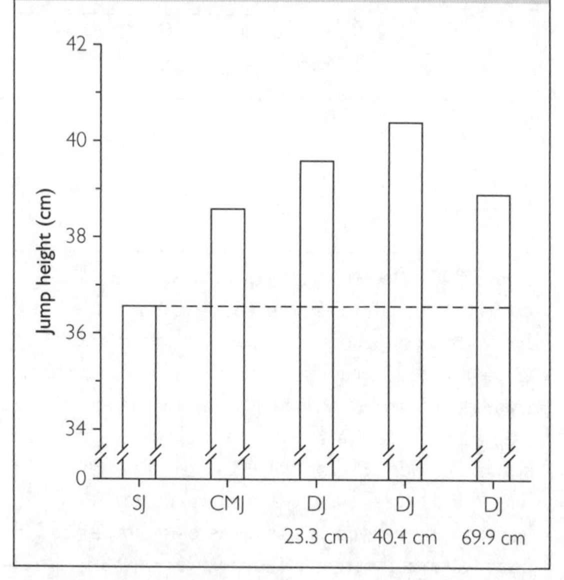

lower isometric force during the prolonged coupling time, which is the isometric phase (55) (Figure 4.43). An example of the influence of coupling time on SSC potentiation is shown in Figure 4.44. Figure 4.44 also suggests, and it has been demonstrated (101,111), that an isometric contraction without a prior eccentric contraction can increase subsequent concentric force, albeit to a lesser extent than the SSC.

SSC and Efficiency

Although SSC potentiation is usually associated with enhancing performance in activities like jumping and throwing, it may play an equally important role in increasing efficiency, particularly in walking and running. Muscular efficiency is the ratio of mechanical energy output (e.g., speed of walking) to the metabolic energy input. Efficiency is often expressed as a percentage: (energy out/energy in × 100). The efficiency of exercise is always less than 100%. The wasted energy is heat, which explains why muscle and core body temperature increase during exercise. An example of the role that SSC potentiation can play in efficiency is illustrated in Figure 4.45. The amount of ATP needed to maintain force is reduced in eccentric contractions (46); that, along with recoil of stored elastic energy, contributes to the increased efficiency of the SSC.

SSC Potentiation Without Eccentric Contractions?

The stretch associated with SSC potentiation typically consists partly of an eccentric contraction and partly of tendon stretch. Another possibility is that most

Figure 4.43 Effect of coupling time duration on SSC potentiation. With a long coupling time (isometric phase, ISO), the relatively high eccentric (ECC) force degrades to a lower isometric force before the concentric (CON) phase begins. With a short coupling time, the concentric phase starts at a higher force level.

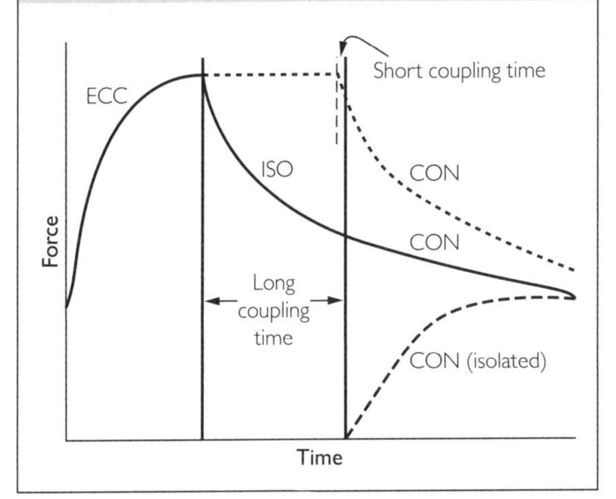

or all of the stretch is in the tendon whereas the muscle contraction is isometric. These two possibilities are illustrated in Figure 4.46 (see also Box 4.11). Figure 4.47 shows an example of the SSC potentiation that can be achieved when tendon stretch is combined with an isometric contraction. The mechanisms for SCC potentiation with isometric contraction are similar to

Figure 4.44 Effect of coupling time duration on bench-press performance. SSC (i.e., lowering weight before lifting up) with brief coupling time increased the bench-press lift by 14.4% above that attained with no SSC. With long coupling times of 0.6 and 1.3 s, SSC potentiation was reduced. Based on the data of Wilson *et al.* (113).

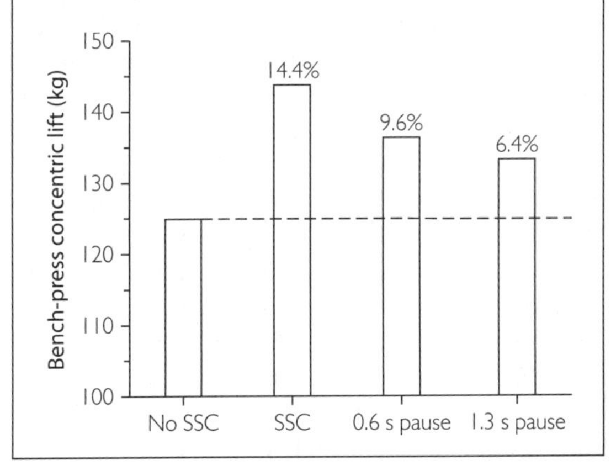

Figure 4.45 Effect of SSC on efficiency. The energy cost (energy in) of repetitive squatting with and without the SSC was compared. Subjects squatted at various speeds (repetition frequency) to generate different mechanical power (energy out). In squatting without the SSC, subjects squatted down but sat briefly to eliminate SSC potentiation. In contrast, subjects squatted and immediately stood up to utilize SSC potentiation. As expected, metabolic rate increased as mechanical power increased. Note, however, that the metabolic rate at any given power output (e.g., 150 W) was less when the squatting used the SSC; in other words, the efficiency (energy out/ energy in) was greater with the SSC. Based on the data of Asmussen and Bonde-Petersen (4).

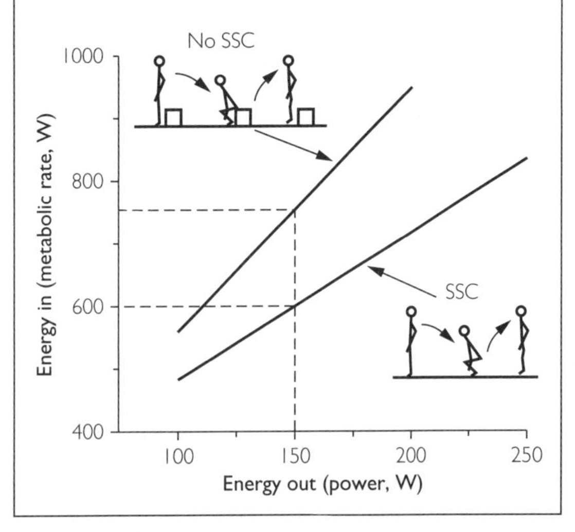

Figure 4.46 Two types of SSC. (A) Muscle at rest. (B) Stretch of the muscle/tendon unit consists of an eccentric (ECC) contraction and tendon stretch. (C) All of the stretch is in the tendon whereas the contraction is isometric (ISO). Both types of SSC can potentiate the force of concentric contractions.

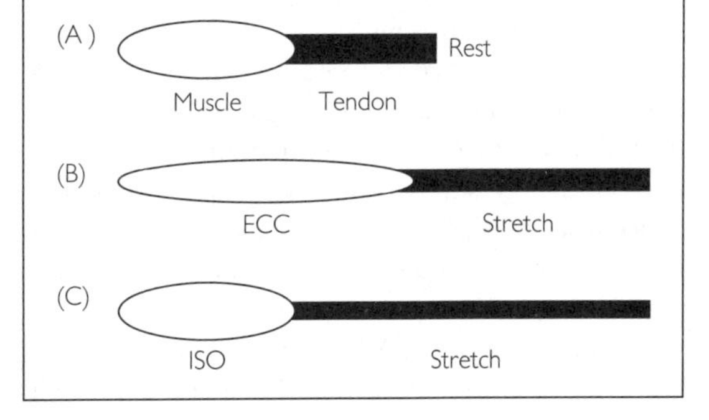

BOX 4.11 THE KANGAROO: THE MASTER OF SSC POTENTIATION?

With their hopping ability, kangaroos might be considered the masters of SSC potentiation. The largest species of kangaroo, the red kangaroo, can stand up to 2 m tall and attain a body mass of 135 kg. It can cover a distance of over 9 m in a single leap, hop over obstacles 3 m high, and achieve speeds up to 65 $km \cdot h^{-1}$ in short bursts. Even more impressive is the efficiency of the kangaroo's hopping. In one study of kangaroos, the energy cost of hopping *decreased* as speed increased from 7 to 23 $km \cdot h^{-1}$. This is in contrast to running animals like the wild dingo dog that preys on kangaroos; in four-legged running animals, and in running humans, the energy cost of running increases as speed increases. The great efficiency of hopping allows the kangaroo to travel at speeds up to 40 $km \cdot h^{-1}$ for several kilometers (16). Interestingly, the SSC potentiation used by kangaroos consists of a stretch primarily of the tendon; the muscle contraction is isometric as shown in Figure 4.46C (5). The tendons of the kangaroo are well designed for this type of SSC; for example, a 40 kg kangaroo has an Achilles tendon 1.5 cm in diameter and 35 cm in length (16).

Figure 4.47 SSC potentiation of velocity and force in human plantarflexor muscles in subjects performing a horizontal jumping task. The task was maximal effort ankle plantarflexion (concentric contraction) either without (No SSC) or with (SSC) a stretch before the plantarflexion. The stretch was found to consist of an isometric contraction combined with elongation of the tendon. There was no eccentric contraction during the stretch phase. SSC potentiation increased both the velocity and force attained in the jump task. The No-SSC condition was not an isolated concentric contraction because an isometric contraction at 40% of maximum strength was briefly held before the concentric phase. Based on the data of Kawakami et *al.* (53).

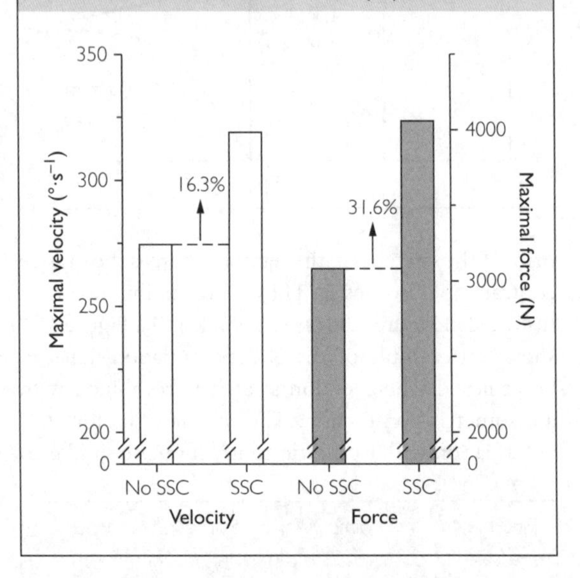

those with eccentric contraction: high initial force at the start of the concentric phase, take-up of the SEC, and storage of elastic energy. However, reflex potentiation based on muscle lengthening would not be as significant because the muscle does not lengthen during the stretch phase.

Isometric–Concentric Potentiation

In the section on Stretch-Shortening Cycle it was found that potentiation of concentric contraction force could occur following a stretch of the muscle–tendon unit (MTU), consisting of either an eccentric contraction combined with tendon stretch, or an isometric contraction combined with tendon stretch (Fig 4.46). It was also shown that in a SSC involving an eccentric contraction, a prolonged coupling time, resulting in a degrading of the eccentric contraction to an isometric contraction, still produced some potentiation of concentric force (Figures 4.43 and 4.44). Now it will be shown that an isometric contraction, even when it is not preceded by an eccentric contraction or an increase in length of the MTU, can still potentiate the force of concentric contraction. This potentiation will be referred to as isometric–concentric (ISO-CON) potentiation. Figure 4.48 illustrates the comparison between the concentric potentiation resulting from prior eccentric and isometric contraction. Although a maximal isometric contraction does not produce as much force as a maximal eccentric contraction, the isometric contraction can still result in significant potentiation of concentric force. An example of the amount of potentiation produced by ISO-CON in comparison to SSC potentiation (ECC-CON) is shown in Figure 4.49. In this example maximal isometric contractions were done prior to maximal concentric contractions, and compared with isolated maximal concentric contractions. As with eccentric contraction (Figure 4.38), lower-intensity isometric contraction (i.e., a smaller percentage of ISO_{max}) would be expected to produce

Figure 4.48 Isometric–concentric (ISO-CON) potentiation is compared to SSC potentiation involving an eccentric (ECC) phase. Although the isometric force is not as great as the ECC force, the isometric force allows the concentric (CON) phase to begin at a higher level of force compared to an isolated concentric contraction.

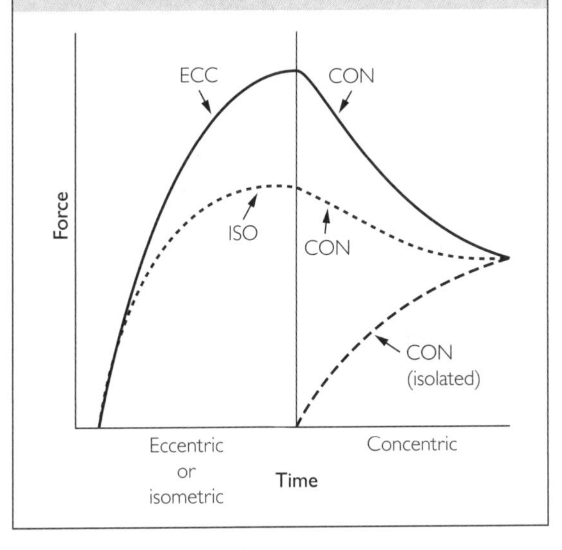

Figure 4.49 Comparison of ISO-CON potentiation with SSC potentiation involving an eccentric (ECC) contraction (ECC-CON). Subjects did maximal ankle plantarflexions (CON) at a joint angular velocity of 240°·s^{-1}. The plantarflexions were done as isolated concentric contraction, concentric contraction following a maximal isometric contraction (ISO-CON), and concentric contraction following a maximal eccentric contraction (ECC-CON) at 240°·s^{-1}. Both ISO-CON and ECC-CON produced marked potentiation of concentric torque. Based on the data of Svantesson *et al.* (101).

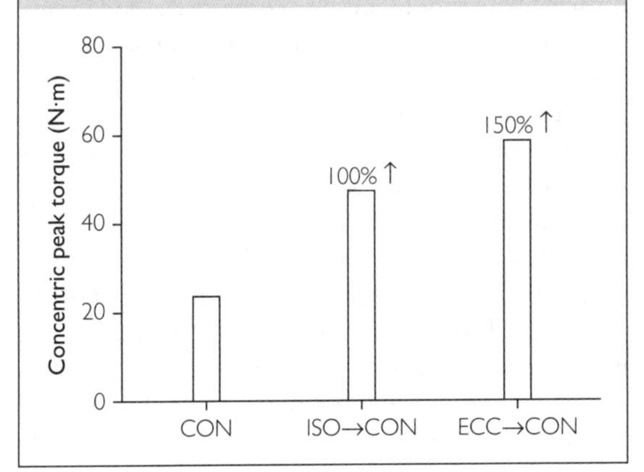

Figure 4.50 ISO-CON potentiation. Subjects did maximal ankle plantarflexions (CON) at joint angular velocities of 120 and 240°·s^{-1}. The plantarflexions were done as isolated concentric contraction (CON) and concentric contraction following a maximal isometric contraction (ISO-CON). As expected from the force–velocity relationship, isolated concentric torque was greater at 120 than at 240°·s^{-1}. In contrast, concentric torque after isometric contraction differed little between the two velocities. Note also that the ISO-CON potentiation expressed as a percentage increase in concentric torque was much greater at the higher velocity. Based on the data of Svantesson *et al.* (101).

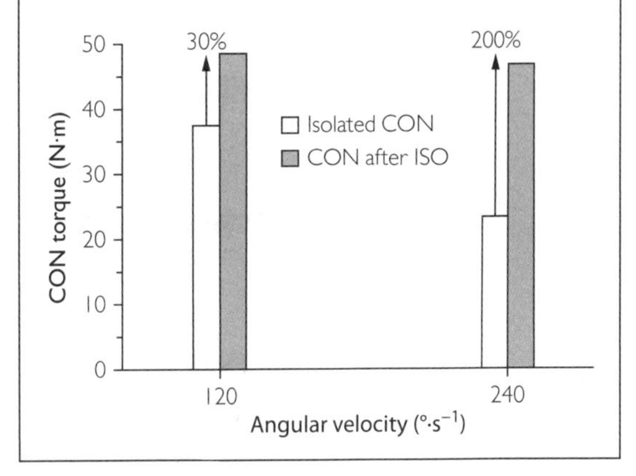

less concentric potentiation (49,89). Like SSC potentiation, ISO-CON potentiation raises the concentric contraction force–velocity relationship, with a relatively greater effect at higher concentric velocities (Figure 4.50).

It is important to distinguish between ISO-CON potentiation and SSC potentiation involving an isometric contraction combined with tendon stretch (Figure 4.51). In this type of SSC, the contraction is isometric contraction but forced lengthening increases, by stretching the tendon, the length of the MTU as a whole, resulting in a change in joint angle. In contrast, in ISO-CON potentiation there is no change in the length of the MTU, and thus there is no change in joint angle. In fact, the isometric phase begins with a shortening phase (concentric contraction) as the SEC is taken up by stretching the tendon, before settling into an isometric contraction (Figure 4.51C). An example of a brief initial concentric contraction before a longer isometric contraction in a nominal isometric contraction at a fixed joint angle is shown in Figure 4.36.

Figure 4.51 Comparison between what occurs in SSC potentiation with tendon stretch and isometric contraction, and ISO-CON potentiation. (A) Muscle at rest. (B) Stretch of the MTU consists of an isometric (ISO) contraction and tendon stretch, causing elongation of the MTU and a change in joint angle. This is a type of SSC potentiation. (C) Overall length of the MTU is fixed, and so is the joint angle. As contraction begins, there is a brief shortening (concentric) phase as the tendon is stretched to take up the SEC. This is followed by a longer isometric phase before the beginning of concentric contraction. This is an example of ISO-CON potentiation.

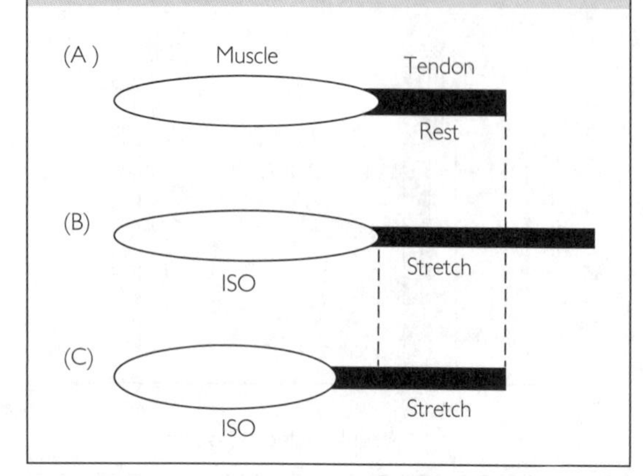

Mechanisms of ISO-CON Potentiation

ISO-CON potentiation shares some of the same mechanisms of SSC potentiation. A maximal isometric phase produces full muscle activation, a high initial force, and complete take-up of the SEC prior to the concentric phase. The stretch of tendons and resultant storage of elastic energy during isometric contraction (Figure 4.36) may also increase the force of the concentric phase, particularly at higher shortening velocities (101). ISO-CON potentiation is usually not as great as SSC potentiation (Figures 4.47 and 4.48), largely because maximum isometric force is not as great as maximum eccentric force (Figure 4.47); nevertheless, ISO-CON potentiation can play an important role in performance (Box 4.12). In fact, ISO-CON potentiation is sometimes used as the control in studies of SSC potentiation (Figure 4.47). To quantify ISO-CON potentiation, an isolated concentric condition (i.e., muscle initially relaxed before the concentric phase) must be included for comparison (Figures 4.48 and 4.49).

BOX 4.12 ISO-CON POTENTIATION AND PERFORMANCE

ISO-CON potentiation is common in sports performance. For example, in the jump ball at the start of a basketball game, the players use an isometric contraction to hold the squat position before jumping upward (concentric phase). This would not be a case of maximal ISO-CON potentiation because the isometric contraction would not be maximal (89), only strong enough to support the athlete's body weight. In power lifting's squat and bench-press lifts, the athlete must support the weight with an isometric contraction before lifting the weight upward (concentric). Like the jump squat, the isometric contraction is submaximal, so the ISO-CON potentiation is not maximal but still significant (Figure 4.44). Therefore, in strength testing protocols that eliminate SSC potentiation, there may still be some ISO-CON potentiation if the test involves an isometric phase. Fortunately, the extent of ISO-CON potentiation is much easier to standardize than the extent of SSC potentiation. The alternative is to eliminate, where possible, both ISO-CON and SSC potentiation (Figures 4.44 and 4.49).

ECC-ISO Force Enhancement

SSC (ECC-CON) and ISO-CON potentiation increase the force and power of concentric contractions. The force of an isometric contraction can also be increased by a preceding eccentric contraction. This ECC-ISO force enhancement (FE) has been termed stretch-induced **residual force enhancement** of isometric force (23). "Stretch-induced" refers to an eccentric contraction (lengthening contraction) rather than a passive stretch of a relaxed muscle. A schematic illustration of ECC-ISO FE is shown in Figure 4.52. ECC-ISO FE has been demonstrated in myofibrils, single muscle fibers, whole-muscle preparations, and *in vivo*, the latter including experiments in human muscles (36). Importantly, in relation to human performance, ECC-ISO FE has been observed in subjects performing voluntary contractions of small, single-joint muscles (63,104) and large multi-joint muscle groups (30). Additionally, ECC-ISO FE can be produced when the coupled voluntary eccentric and isometric contractions are maximal or submaximal in intensity (36). In submaximal contractions, ECC-ISO FE increases with increased intensity of contraction (88).

The magnitude and duration of ECC-ISO FE can be substantial. For example, a FE of approximately

Figure 4.52 Schematic illustration of ECC-ISO force enhancement (FE). First, a reference or control isometric contraction (ISO alone, solid line) is done at a particular muscle length as shown in B, producing a force as shown in A. Next, an isometric contraction (ISO before lengthening, dashed line, shown in A) is done at a shorter length than "ISO alone," as shown in B (dashed line). Then the muscle is forcibly stretched, changing the "ISO before lengthening" into an eccentric contraction (ECC). The eccentric contraction ends when the stretch attains the length of the "ISO alone" (B), at which point a second isometric contraction (ISO after lengthening) is done (A). Force decreases from the eccentric level to a *steady-state* level of isometric force that is greater (FE) than that of the "ISO alone" done at the same length. In this example of ECC-ISO FE, the stretch was done on the descending limb of the force–length relationship; that is, at lengths greater than optimal length (L_o) for isometric force, as shown in C. This explains why the force of "ISO before lengthening" is greater than the force of the "ISO alone." In contrast, if the stretch had occurred on the ascending limb of the force–length relationship, the force of the "ISO before lengthening" would have been smaller than the "ISO alone;" nevertheless, there still would have been ECC-ISO FE, but to a lesser extent.

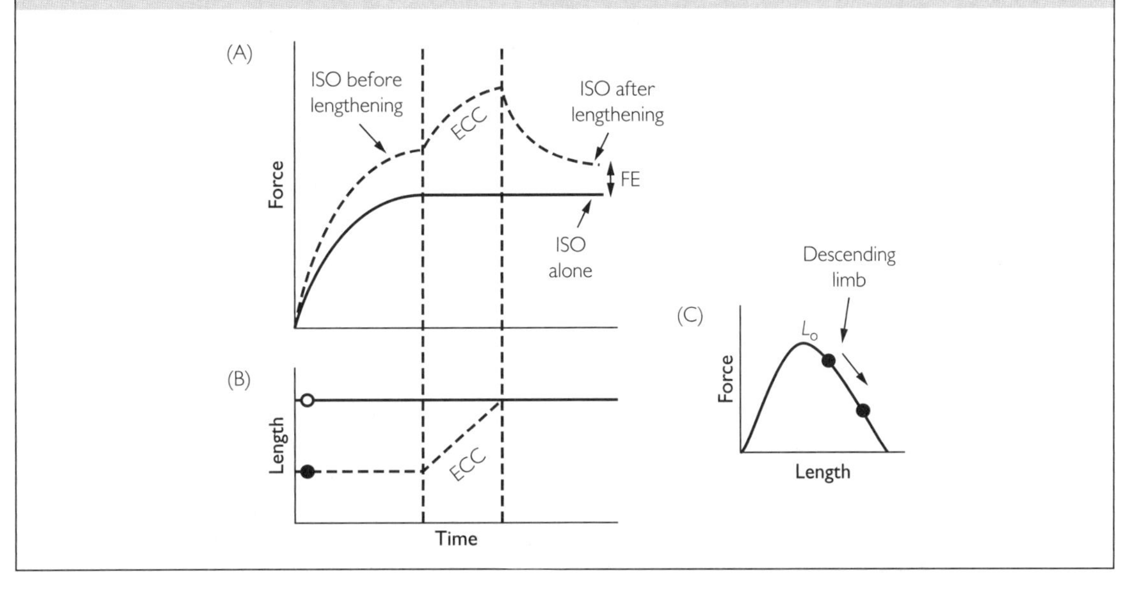

15%, sustained for at least a few seconds, has been observed in the thumb adductor muscle (63). As a second example, ECC-ISO FE averaged approximately 8% over a period of 3 s in a leg-press movement (involved hip and knee extensors and ankle plantarflexors) (30).

Factors Affecting the Magnitude of ECC-ISO Force Enhancement

A number of factors influence the magnitude of ECC-ISO FE (36). Within a certain range, a larger increase in muscle length during the eccentric phase (stretch) produces a greater FE; in contrast, the speed of stretch has a minimal if any effect on the magnitude of FE (34). For a given change in muscle length during stretch, FE is greater when the final length after stretch is at a relatively long muscle length. If a brief relaxation occurs during the isometric contraction after lengthening and then the isometric contraction is restored, FE is abolished (82). If a concentric contraction is done before the eccentric contraction the magnitude of ECC-ISO FE is reduced (Box 4.13).

Characteristics of human subjects may influence the magnitude of ECC-ISO FE. For example, subjects with shorter twitch contraction times, less fatigue resistance, and greater postactivation potentiation exhibited greater FE in submaximal contractions (87). Since shorter twitch contraction times, less fatigue resistance, and greater postactivation potentiation are associated with a higher percentage of fast, type 2 muscle fibers (31), fiber-type distribution in muscles may affect the magnitude of ECC-ISO FE.

BOX 4.13 EFFECT OF A CONCENTRIC CONTRACTION BEFORE THE ECCENTRIC CONTRACTION ON ECC-ISO FORCE ENHANCEMENT

Experimentally, ECC-ISO FE is usually produced as shown in Figure 4.52: the muscle is initially made to contract *isometrically* (isometric contraction before lengthening), and then the muscle is forced into an eccentric contraction by an imposed stretch to a specified length, after which the isometric contraction resumes, showing the FE. Thus, the sequence is ISO-ECC-ISO. However, if during the initial isometric contraction the muscle is allowed to shorten (concentric contraction) before being stretched (eccentric contraction) to the specified muscle length, the magnitude of ECC-ISO FE is reduced. The sequence would be ISO-*CON*-ECC-ISO. A greater amount of shortening during the concentric contraction produces a greater reduction in ISO FE after the eccentric contraction (36).

Active and Passive Components of ECC-ISO Force Enhancement

ECC-ISO FE consists of two components. The active component acts during the isometric contraction after eccentric contraction but disappears when the isometric contraction ends and the muscle is relaxed. The passive component not only acts during the isometric contraction after eccentric contraction, but persists after the isometric contraction ends, increasing the passive force of the relaxed muscle at a given muscle length. "Passive" force refers to the resistance of a relaxed (no CB cycling) muscle to stretch. (Passive force will be considered in more detail in a subsequent section called the Force–Length Relationship.) A schematic representation of the active and passive components of ECC-ISO FE is shown in Figure 4.53. The magnitude of the passive component is revealed when the muscle relaxes after the "ISO after lengthening." The passive component can make a substantial contribution to the total ECC-ISO FE. For example, in voluntary contractions of human adductor pollicis, the passive component can contribute up to approximately 25% of the total FE of the isometric contraction after the eccentric contraction (63). The passive component, which can last for more than 25 s after the end of the isometric contraction (38), is most pronounced at relatively long muscle lengths where passive force is relatively large in a relaxed muscle that has not undergone an immediately prior ECC-ISO sequence (63). Interestingly, the enhancement of passive force in the relaxed muscle may be abruptly reduced or even abolished if the muscle is allowed to shorten briefly from the set length before being restored back to the set length. An amount of shortening equal to the amount of the stretch that produced the enhanced passive force is needed to abolish the passive component (38).

Mechanisms of ECC-ISO Force Enhancement

The mechanism responsible for ECC-ISO FE has not been unequivocally established (36). For the active component (present during the isometric contraction after eccentric contraction) of ECC-ISO FE, one proposed mechanism is a redistribution of sarcomere lengths along the length of myofibrils in muscle fibers (Figure 4.2) after the eccentric contraction, with the result being an average sarcomere length that is closer to optimal for force generation at a particular whole muscle length (or muscle fiber length) (81). Another possible mechanism involves the CB action during contraction. The increased number of recruited CBs during the eccentric contraction (28) may persist into the *steady-state* phase of the subsequent isometric contraction (37), which would enhance the isometric force. Also contributing to the enhanced isometric force would be a decreased CB detachment rate during the isometric contraction after eccentric contraction (97), resulting in an increased number of attached CBs at a given time point.

The passive component of ECC-ISO FE likely involves the protein **titin**. Titin is a large protein in the form of elastic filaments that span the distance between the Z-disk and M-line in a sarcomere (Figures 4.5 and 4.35B). In a relaxed muscle, titin is attached only to the Z-disk and M-line. In an eccentric contraction, however, part of the titin filament attaches to the actin filament, causing the titin filament to become stiffer. It has been proposed that the attachment of titin to actin and resultant increased stiffness persists during the subsequent isometric contraction and for a time after

Figure 4.53 Schematic illustration of active and passive components of ECC-ISO FE. First, a reference or control isometric contraction (ISO alone, solid line) is done at a particular muscle length as shown in B, producing a force as shown in A. Next, an isometric contraction (ISO before lengthening, dashed line, shown in A) is done at a shorter length than "ISO alone," as shown in B (dashed line). Then the muscle is forcibly stretched, changing the "ISO before lengthening" into an eccentric contraction (ECC). The eccentric contraction ends when the stretch attains the length of the "ISO alone" (B), at which point a second isometric contraction (ISO after lengthening) is done (A). Force decreases from the eccentric level to a *steady-state* level of isometric force that is greater (FE) than that of the "ISO alone" done at the same length. In this example of ECC-ISO FE, the stretch was done on the descending limb of the force–length relationship; that is, at lengths greater than optimal length (L_o) for isometric force, as shown in C of Figure 4.52. This explains why the force of "ISO before lengthening" is greater than the force of the "ISO alone." The ECC-ISO FE consists of active + passive components during the "ISO after lengthening." When the muscle relaxes, the active component is abolished but the passive component persists.

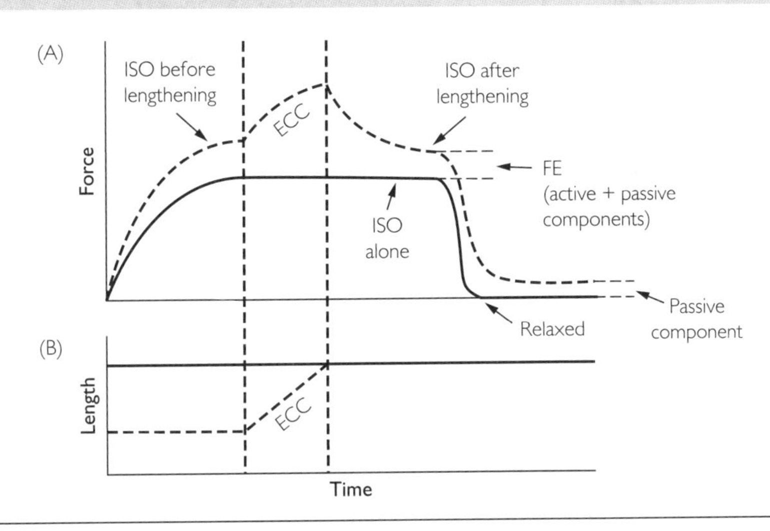

the isometric contraction ends. The increased stiffness may be partly responsible for the enhanced isometric force after eccentric contraction and entirely responsible for the elevated passive force at a given muscle length when the isometric contraction ends and the muscle is relaxed (no myosin–actin interaction) (67).

Implications of ECC-ISO Force Enhancement for Performance

Perhaps the classic example of ECC-ISO FE affecting performance is a gymnast lowering into an iron cross on the rings. A cross is an isometric hold position that must be maintained for at least 2 s. If the gymnast lowers into the cross with an eccentric contraction using the same muscles that will perform the cross, ECC-ISO FE should make it easier to hold the cross (isometric contraction) compared to pulling up into the cross position (Figure 4.54). ECC-ISO FE also affects performance in weight-lifting exercise. If a weight is lowered with an eccentric contraction and then held with an isometric contraction, the force capability of the isometric contraction is briefly increased. The enhanced isometric capability provides a stronger base for the subsequent lifting phase (concentric contraction). This might be referred to as ECC-ISO-CON potentiation. Check Figure 4.44, which shows that the lifting (concentric) phase of the bench press is enhanced following a 0.6 and 1.3 s isometric phase that in turn followed an eccentric phase. The eccentric phase enhanced the isometric phase, which then enhanced the concentric phase, allowing a heavier weight to be lifted compared to an isolated

Figure 4.54 An example of ECC-ISO FE's effect on performance. The illustration shows a gymnast lowering into an iron cross on the rings. The iron cross is an isometric (ISO) hold position that in competition must be maintained for at least 2 s. At the top, the gymnast is lowering into the cross with an eccentric contraction (ECC) of the muscles activated in holding the cross isometrically. The eccentric phase should enhance the force of the subsequent isometric phase, making it easier to hold the cross in comparison to pulling up into the cross without a prior eccentric contraction.

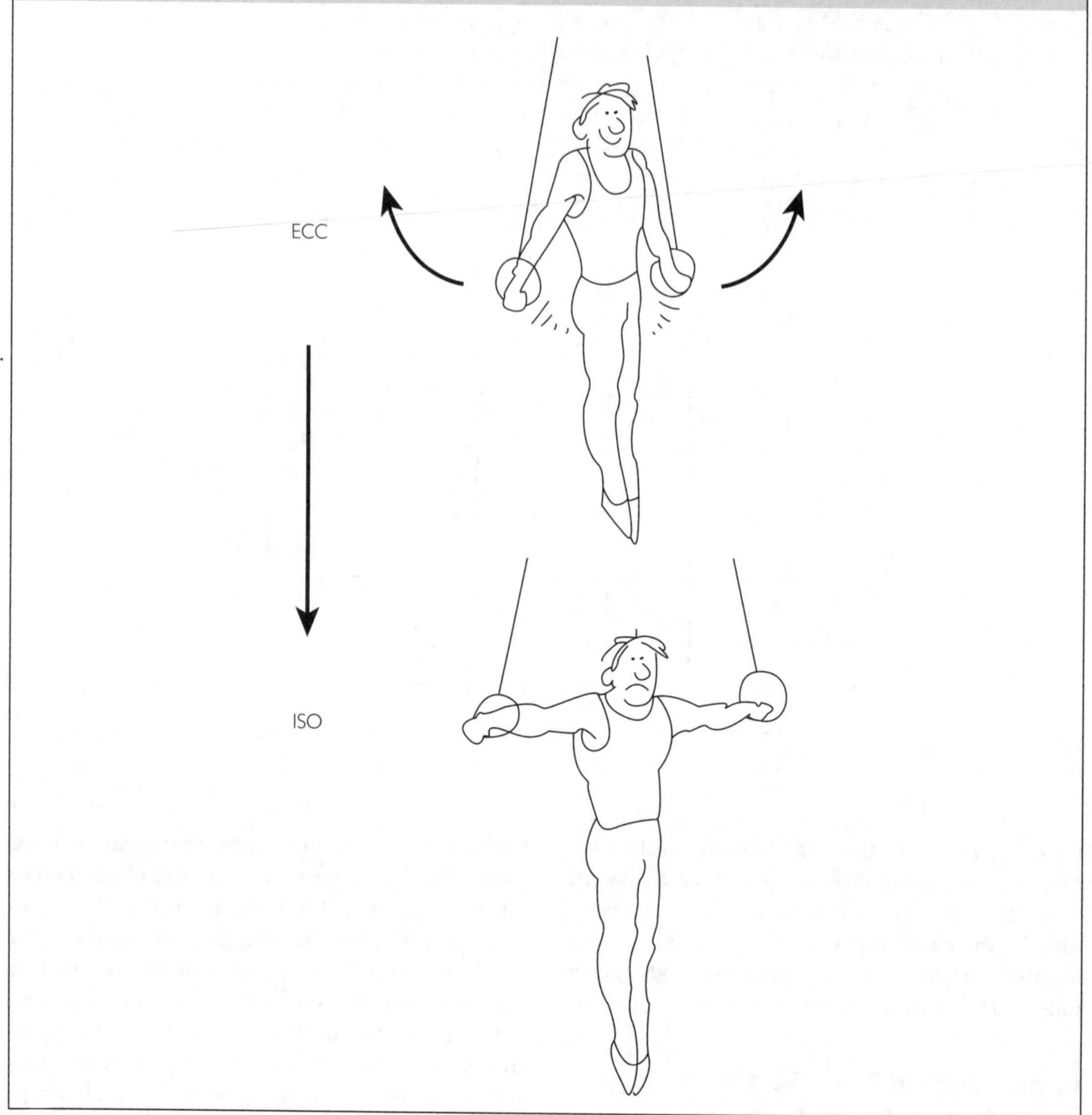

concentric contraction. ECC-ISO-CON potentiation must be considered in performance testing (Box 4.14).

ECC-ISO FE may play at least one other role in performance. Unaccustomed eccentric exercise can lead to muscle damage and delayed-onset muscle soreness (or DOMS). The site of the damage is considered to be the cytoskeleton (Figure 4.5), of which the protein titin is a part. The increased passive force produced by titin during and after eccentric contractions, particularly at long muscle lengths, may attenuate the damage caused by exercise that includes eccentric contractions (67).

BOX 4.14 ECC-ISO FORCE ENHANCEMENT AND PERFORMANCE TESTING

Attention has been drawn to the importance of controlling for SSC (ECC-CON) potentiation in performance testing. In weight-lifting strength tests, the most reliable control was eliminating SSC potentiation altogether by imposing a prolonged (a few seconds) coupling time (isometric phase) between the eccentric and concentric phases. However, even if the coupling time is prolonged for a few seconds, there is lingering ECC-ISO FE that in turn enhances concentric performance (Figure 4.44). How can ECC-ISO FE be controlled? The longer the isometric phase, the smaller the ECC-ISO FE; also, fatigue increases as the duration of the isometric phase increases. Therefore, the best control is to standardize the duration of the isometric phase before the concentric phase. Controlling for ECC-ISO FE is also an issue in the squat and bench-press lifts in power-lifting competition. If a referee required one competitor to hold the isometric phase for 4 s but another only 2 s, the shorter isometric phase would provide an advantage. It might be possible to eliminate ECC-ISO FE by requiring a very long (several seconds) isometric phase, but then fatigue would be a confounding factor in the test. Another approach in weight-lifting tests would be to eliminate an eccentric or isometric phase. This would require equipment that allowed the test to be an isolated concentric contraction.

CON-ISO Force Depression

In contrast to the enhancement of isometric force after an eccentric contraction, a prior concentric contraction causes a depression of subsequent isometric force compared to the force of an isolated isometric contraction at the same muscle length. An example of CON-ISO force depression (FD) is shown in Figure 4.55. In animal muscles, CON-ISO FD has been demonstrated in myofibrils (51), and in single muscle fibers and whole muscles (96). In humans CON-ISO FD occurs with either electrically stimulated or voluntary contractions of small and large muscles (17,18,65,96,104). CON-ISO FD expressed absolutely (e.g., FD expressed in newtons of force) increases with increasing activation (increased level of electrical stimulation or voluntary effort); however, the relative (%) FD is similar for maximal and submaximal contractions (18,64,99).

Factors Affecting the Magnitude of CON-ISO Force Depression

The magnitude of CON-ISO FD is proportional to the amount of work (force × displacement) done during the concentric contraction; thus, any combination of increased force and/or displacement (amount of shortening) will increase the magnitude of CON-ISO FD (34). In contrast to ECC-ISO potentiation, which is little affected by the speed of eccentric contraction, CON-ISO FD is greater at slower speeds (96) because, based on the force–velocity relationship (Figure 4.22), concentric force is greater at slower speeds. CON-ISO FD is more pronounced at lengths greater than optimal length for isometric force (Figure 4.55) but has also been shown at lengths shorter than optimal length (34,78).

Mechanisms of CON-ISO Force Depression

The mechanism(s) that causes CON-ISO FD is not known, but mechanisms have been proposed and debated. One proposed mechanism is a sarcomere length *non-uniformity* that develops in the series of sarcomeres along the length of myofibrils in muscle fibers (Figure 4.2) during concentric contraction. The non-uniformity persists during the subsequent isometric contraction. The result is an average sarcomere length that is farther from optimal (Figure 4.55) compared to the average sarcomere length of an isolated isometric contraction at the same whole muscle length (82). Another possible mechanism for CON-ISO FD is an inhibition of CB attachment during concentric contraction as a result of stress imposed on actin filaments, resulting in reduced force of the subsequent isometric contraction (51).

Implications of CON-ISO Force Depression for Performance

In Figure 4.54, it was shown that if a gymnast on the rings lowered into an iron cross with an eccentric contraction, the subsequent isometric contraction (the cross) would, because of ECC-ISO FE, be easier to hold compared to an isolated isometric contraction. In contrast, if the gymnast pulled up into the cross using a concentric contraction, the cross would,

Figure 4.55 Schematic illustration of CON-ISO force depression (FD). First a reference (control) isometric contraction (ISO alone, solid line) is done at a particular muscle length as shown in B, producing a force as shown in A. Next, an isometric contraction (ISO before shortening, dashed line, shown in A) is done at a longer length than "ISO alone," as shown in B (dashed line). Then the muscle is released, allowing it to shorten, changing the "ISO before shortening" into a concentric contraction (CON). The concentric contraction ends when the shortening attains the length of the "ISO alone" (B), at which point a second isometric contraction (ISO after shortening) is done (A). Force increases from the concentric level to a *steady-state* level of isometric force that is smaller (CON-ISO FD) than that of the "ISO alone" done at the same length. In this example of CON-ISO FD, the shortening was done on the descending limb of the force–length relationship; that is, at lengths greater than optimal length (L_o) for isometric force, as shown in C. This explains why the force of "ISO before shortening" is smaller than the force of the "ISO alone." In contrast, if the shortening had occurred on the ascending limb of the force–length relationship, the "ISO before shortening" would have been greater than the "ISO alone;" nevertheless, there still would have been CON-ISO FD.

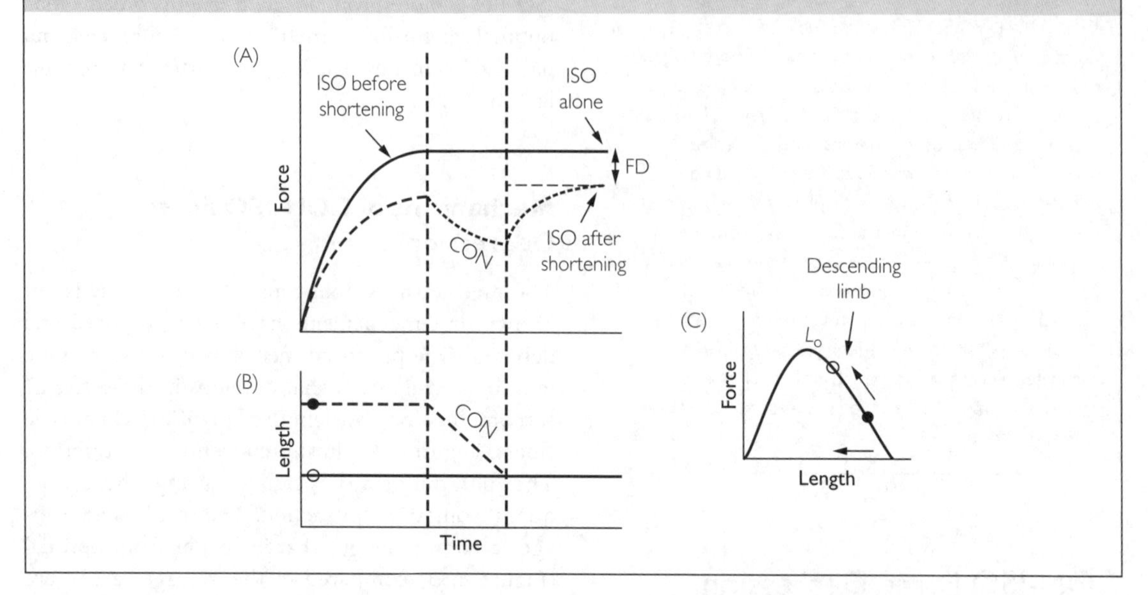

because of CON-ISO FD, be more difficult to hold compared to an isolated isometric contraction. Any situation in which a weight, including body weight, has to be raised or lifted (concentric) and then held in position (isometric) would be affected by CON-ISO FD (Figure 4.56). In activities involving submaximal contractions, CON-ISO FD would require the nervous system to increase the activation of the muscle in an isometric contraction after concentric contraction, to compensate for the reduced force-generating capacity of the isometric contraction. Doing an eccentric contraction before the concentric contraction has little effect on the magnitude of CON-ISO FD (Box 4.15).

Concentric Effect on Concentric Force

Isometric force is reduced by a preceding concentric contraction (CON-ISO FD), but a concentric contraction also causes a reduction in its own force as shortening proceeds (18,78). For example, a concentric contraction done at a set constant velocity undergoes a progressive loss of force during shortening, which can be called CON FD. A faster concentric contraction produces a greater CON FD during shortening (Figure 4.57). Remember, however, that a slower concentric contraction, due to its greater force (Figure 4.57), produces greater work (force × displacement) and therefore greater CON-ISO FD. CON FD

Figure 4.56 Two examples of CON-ISO FD. The top shows the maximum weight that can be held for a few seconds in an isometric contraction alone (20 kg) compared to an isometric contraction after concentric contraction (16 kg). The CON-ISO FD would be 20%. The bottom shows that if a subject pulls up (concentric) and then holds the position shown (isometric) as long as possible, CON-ISO FD would reduce the maximum holding time compared to being able to start the test in the hold position without prior concentric contraction.

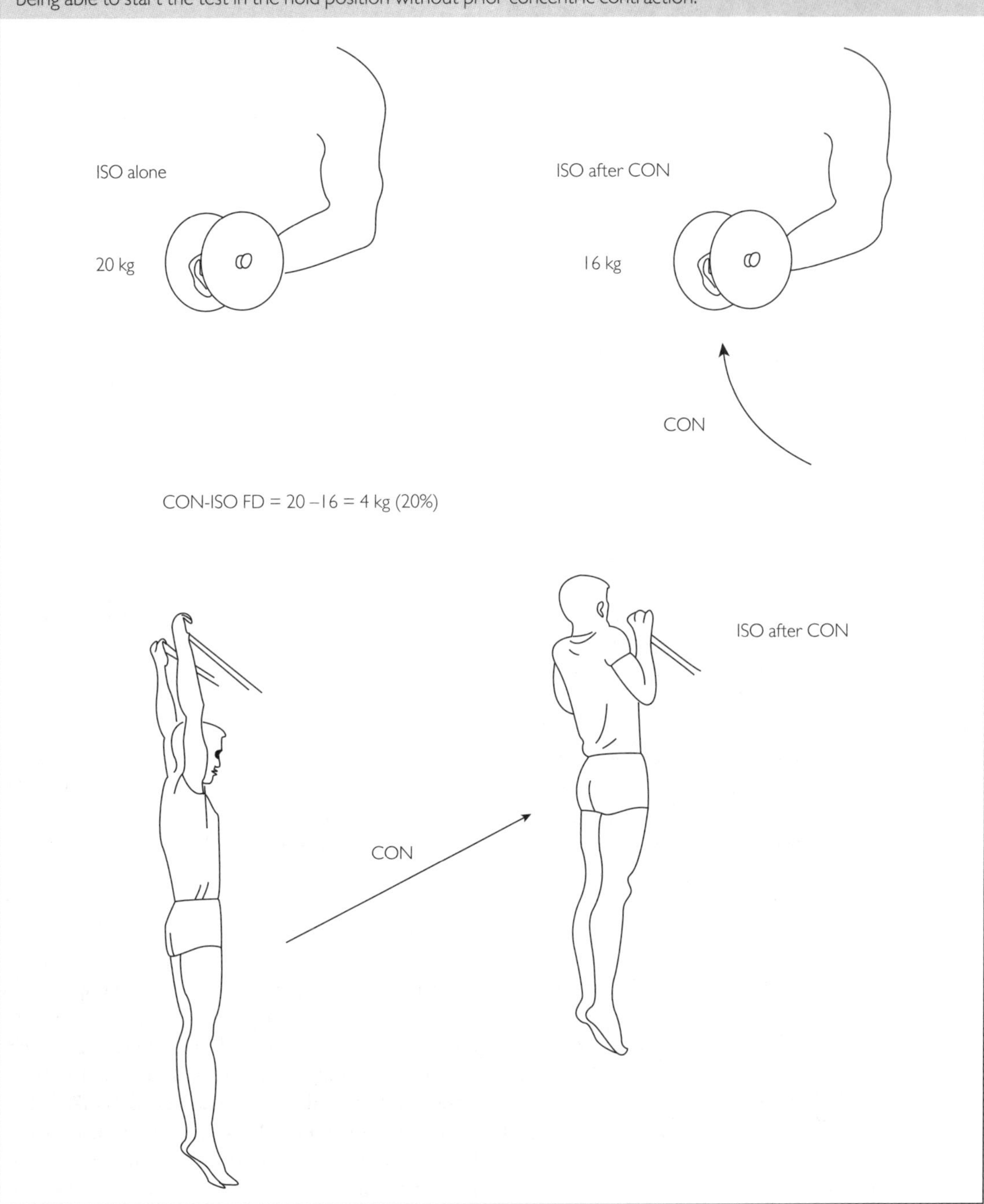

BOX 4.15 DOES ECCENTRIC CONTRACTION BEFORE THE CONCENTRIC CONTRACTION AFFECT THE MAGNITUDE OF CON-ISO FORCE DEPRESSION?

In the section on ECC-ISO FE, it was pointed out (see Box 4.13) that a concentric contraction before the eccentric contraction reduced the magnitude of ECC-ISO potentiation. In contrast, an eccentric contraction before concentric contraction has much less, if any, effect on CON-ISO FD (37,66).

can also affect the load–velocity relationship. A greater amount of shortening in concentric contraction lowers the load–velocity relationship; that is, the velocity attained with a given load is decreased (Figure 4.58). If the velocity is reduced with a given load, power is also reduced because power equals force (load) × velocity. The FD during concentric contraction is likely related to a progressive decrease in the number of attached CBs during the shortening (78).

CON FD affects any performance that involves concentric contractions; for example, crank power in cycling can be significantly reduced in the later part of each concentric phase (79). In the pull-up exercise shown at the bottom of Figure 4.56, there would be a progressive decrease in the force-generating capacity from the beginning to the end of the concentric phase. Then, as discussed, there would be CON-ISO FD if the top position is held. In weight lifting, each concentric phase of a repetition would be influenced by CON FD (top of Figure 4.56). As shown in Figure 4.58, CON FD and its potential effects, such as decreased velocity and power when lifting loads, would be greater in exercises with a greater range of movement; that is, a

Figure 4.57 Schematic example of FD during concentric contractions (CON). First, isometric force (ISO) was measured at various lengths as shown. Then, at the longest length, the muscle was made to contract isometrically and then released to perform an isovelocity slow (CON slow) or fast (CON fast) contraction. Note the progressive decline in force during concentric contraction, with a greater decline in the CON fast. The progressive FD was not due to shortening onto a weak part of the force–length relationship, because the isometric force–length relationship was relatively flat through the length range of muscle shortening. Also, the CON, whether slow or fast, were done at constant velocity (isovelocity) throughout the shortening; therefore, the progressive FD was not caused by the muscle speeding up. Adapted from the data of de Ruiter et *al.* (18).

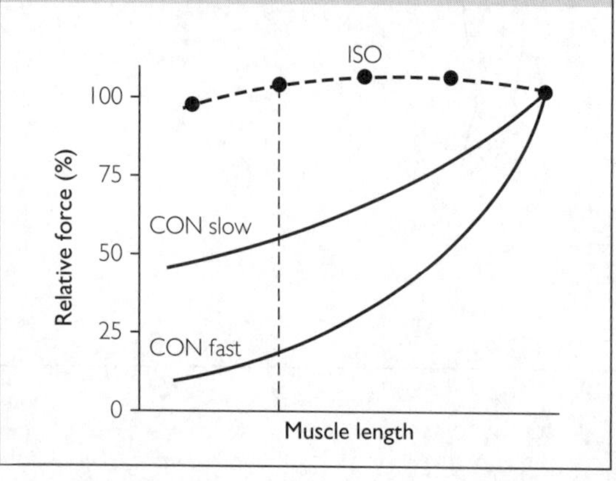

Figure 4.58 Schematic illustration of the effect of the amount of shortening in a concentric contraction on the velocity attained with different loads ranging from no load (V_{max}) to the load that just prevented shortening (ISO_{max}). A greater amount of shortening (horizontal dashed line) to the length at which velocity was measured (vertical dashed line) caused a greater decrease in the velocity that could be attained with the various loads. Both shortening distances would have produced some CON-ISO FD (Figure 4.55), but it was greater after the greater shortening distance (ISO FD). A decreased velocity with given loads would also decrease power because power is equal to force (load) × velocity (Figure 4.28).

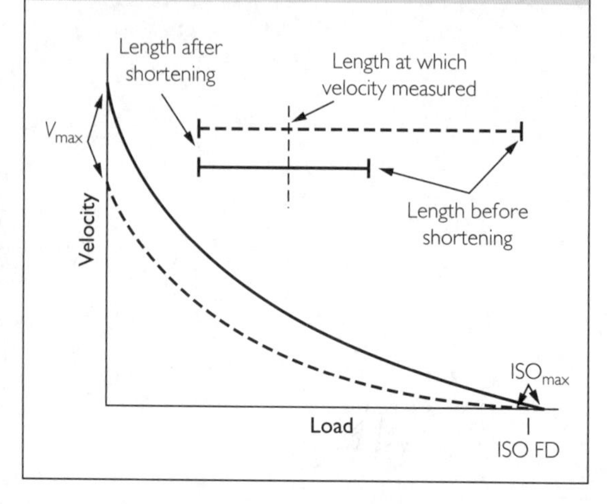

greater amount of shortening during concentric contraction. But earlier in the chapter it was shown that the force of concentric contraction can be increased by a prior eccentric contraction (SSC potentiation) (Figure 4.32). Thus, SSC potentiation can offset the effect of CON FD. An analysis of muscle performance therefore requires not only an identification of the types of contraction involved (concentric, isometric, eccentric) and the velocity of contraction, but also the contractile history, which is the sequence of contractions (ECC-CON, ISO-CON, ECC-ISO, CON-ISO, etc.). All of these forms of contractile history affect human performance, including actions referred to as "explosive" (24), such as jumping, throwing, and kicking.

Figure 4.59 Force–length relationship of a muscle. Active force, the force produced by the contractile mechanism, is zero at a very short length, increases to a maximum value (denoted as optimal length, L_o) as the muscle is lengthened, but then decreases to zero again at a very long length. Passive force is the resistive force offered by a *relaxed* muscle to stretch. The length at which passive force first rises above zero is called the resting length. Total force is the sum of active and passive force.

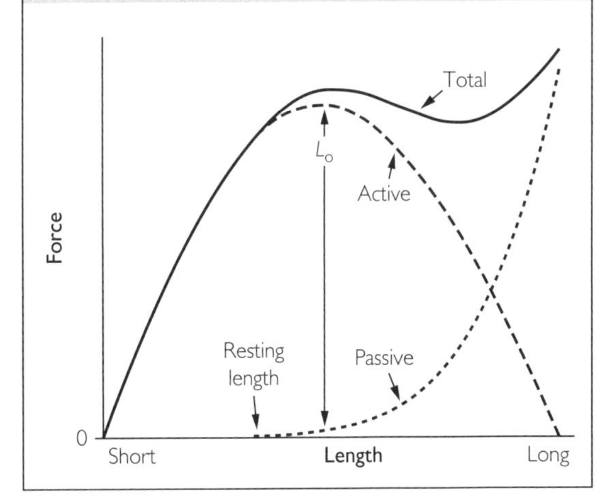

Force–Length Relationship

When a muscle contracts, the force it produces is affected by the length of the muscle at the time of contraction. This is referred to as the force–length relationship.

Passive, Active, and Total Force

A force–length relationship is produced by fixing a muscle at various lengths. Two tests are done, as shown in Figure 4.59. First, the *passive* force is measured at a series of muscle lengths. Passive force is the resistive force produced by a relaxed (non-contracting) muscle as it is set to longer lengths. Note in Figure 4.59 that at the shortest lengths the passive force is zero. The length at which passive force rises above zero is called the **resting length**. At progressively longer lengths, passive force increases exponentially.

The second test is to stimulate the muscle to produce maximal isometric contractions at the same series of muscle lengths at which passive force was measured. The force produced by stimulation is called the **total force**. Up to the length at which passive force begins to develop (resting length), the total force is entirely **active force**; that is, force produced solely by the contractile mechanism. Beyond resting length, the total force includes both passive and active force (Figure 4.59). If the tests are done at long enough lengths, the total force will consist entirely of passive force because the force produced by the contractile mechanism will have decreased to zero. Active force is calculated by subtracting passive force from total force. The length at which maximum active force occurs is called **optimal length** (L_o). The part of the force–length relationship at lengths shorter than L_o is called the **ascending limb**; the part at lengths greater than L_o is called the **descending limb** (bottom of Figure 4.60). Various muscles operate primarily on the ascending, descending, or both limbs of the force–length relationship and this affects the relationship between L_o and resting length (Figure 4.60, Box 4.16).

In isolated animal nerve/muscle preparations, active force is evoked by electrical nerve stimulation. In human experiments, electrical nerve stimulation can also be used, but an alternative is to have subjects do maximal voluntary isometric contractions. The active force–length relationship is usually determined using isometric contractions, but as will be shown, the force–length relationship of isometric contractions can be affected by contractile history, and the force–length relationship may differ for concentric and eccentric contractions compared to isometric contractions.

Figure 4.60 Top: relationship between resting and optimal (L_o) length in three frog leg muscles. Note how this relationship affects total force in relation to active force. Based on Rassier *et al.* (98), The American Physiological Society/CC-BY. Bottom: gastrocnemius, which operates on the ascending limb of its force–length relationship, has a resting length shorter than L_o. In contrast, semitendinosus, which operates on the descending limb of its force–length relationship, has a resting length longer than L_o. Copyright © 1999, The American Physiological Society.

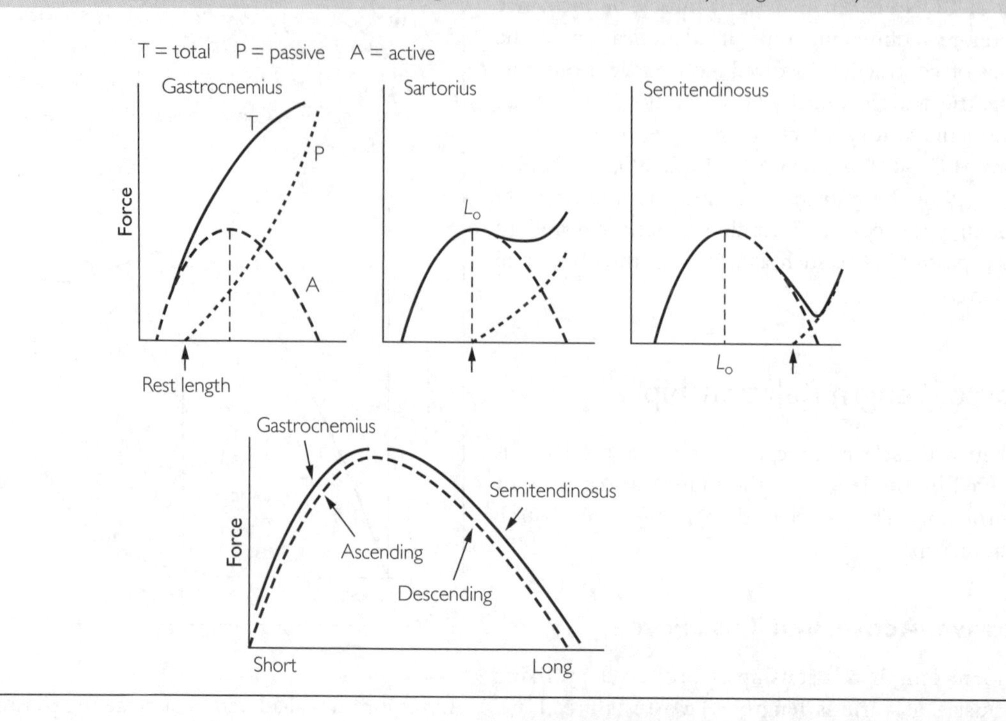

Mechanisms of Variation in Passive and Active Force

Passive force is the resistive force offered by a muscle (e.g., the cytoskeleton; Figures 4.5 and 4.35) and its tendons (Figure 4.36) to stretch. As the MTU is stretched to longer lengths, its resistive (passive) force increases exponentially (Figure 4.59). The magnitude of passive force at a given length would depend on factors such as the relationship between resting length and L_o (Figure 4.60, Box 4.16), and the properties of the muscle and tendon.

The mechanism of variation in active force of a whole muscle is determined by the force–length relationship of each of the thousands of sarcomeres (for a review, see Figure 4.1) that are linked in series along the length of myofibrils (for review, see Figures 4.2 and 4.3). The force–length relationship of each sarcomere is determined by the lengths of the actin and myosin filaments, and the extent of overlap between the actin and myosin filaments, which in turn dictates the

BOX 4.16 RELATIONSHIP BETWEEN OPTIMAL LENGTH AND RESTING LENGTH

Resting length in Figure 4.59 occurs at a length shorter than L_o. This is not always the case. In some muscles the resting and optimal lengths occur at the same length; in other muscles, resting length is longer than optimal length. The relationship between resting length and optimal length depends in part on the part of the force–length relationship used by the intact muscle in normal function. An example for three leg muscles in the frog is shown in Figure 4.60. Gastrocnemius, which operates on the ascending limb of its force–length relationship (i.e., from short lengths up to L_o), has a resting length shorter than L_o. In contrast, semitendinosus, which operates on the descending limb of its force–length relationship (i.e., at L_o and longer lengths), has a resting length longer than L_o. In sartorius, which operates on both ascending and descending limbs of the force–length relationship, resting length and L_o occur at the same muscle length. The relationship between resting length and L_o ensures that some, but not excessive, passive force will be present in the part of the force–length relationship used by the muscle.

Figure 4.61 Myosin (1.6 μm) and actin (0.95 μm) filament lengths in frog muscle are shown, expressed in micrometers (μm). Z-disk thickness is given as 0.1 μm. Note that filament lengths given here are more precise than those given in Figure 4.3, which were given to the nearest 0.5 μm. The bare zone is the small (0.2 μm) mid-portion of the myosin filament that contains no myosin CBs. In the illustration the sarcomere length is set at 3.6 μm. At this length active force should be zero because there is no overlap between actin and myosin filaments. Note that the sarcomere length includes half the thickness of adjacent Z-disks. Sarcomere lengths and Z-disk thickness taken from Rassier *et al.* (98).

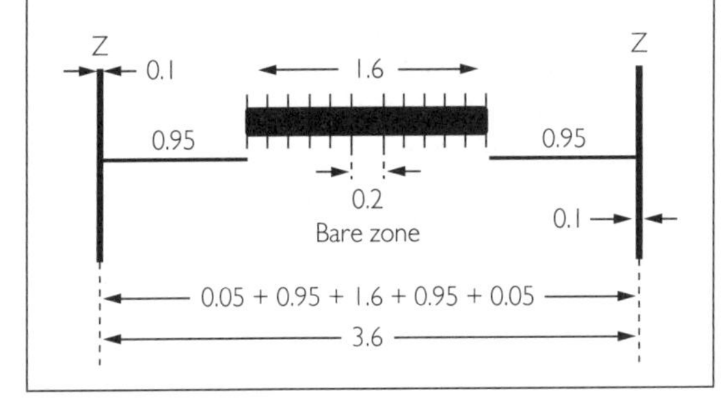

interaction between myosin CBs and actin-binding sites. The lengths of the actin and myosin filaments in frog muscle are shown in Figure 4.61.

The frog sarcomere force–length relationship is shown in Figure 4.62. L_0 is actually a 0.2 μm plateau, the result of the 0.2 μm **bare zone** in the middle of the myosin filament. Thus, at sarcomere lengths of 2.0–2.2 μm active force is the same because the same number of CBs can interact with the actin filaments. At lengths greater than 2.2 μm active force decreases because there is less overlap between actin and myosin until at 3.6 μm, where force has fallen to zero (no overlap of actin and myosin; see also Figure 4.61). At sarcomere lengths less than 2.0 μm, actin filaments begin to overlap with each other, causing a disruption of CB binding to actin and a loss of force (see also Figure 4.63). As actin/actin overlap increases, there is more CB disruption and therefore more force loss. At a sarcomere length of 1.7 μm (length of myosin filament, 1.6 μm, plus two half thicknesses of Z-disks, 0.1 μm; see also Figure 4.62), the Z-disks begin to compress the myosin filaments, causing a steep decrease in force. At 1.27 μm, force has dropped to zero. It should be noted that the plateau of L_0 seen at the sarcomere level (Figure 4.62) is not seen at the muscle-fiber or whole-muscle level (Figures 4.59 and 4.60) because of sarcomere length non-uniformity (see Box 4.17).

Figure 4.62 The sarcomere force–length relationship of frog muscle. For five points on the relationship, the corresponding depictions of actin/myosin overlap are shown. Note the plateau of optimal length (L_0) from 2.0 to 2.2 μm, corresponding to the 0.2 μm bare zone without CBs in the middle of the myosin filament. Based on Rassier *et al.* (98), The American Physiological Society/CC-BY. Copyright © 1999, The American Physiological Society.

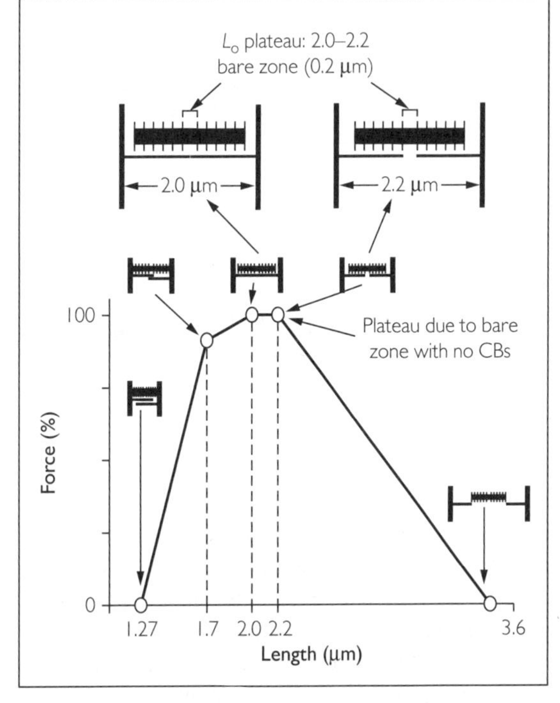

Interspecies Differences in the Sarcomere Force–Length Relationship

The standard textbook depiction of the sarcomere force–length relationship is based on frog muscle, because frog muscle was first used to establish the sarcomere force–length relationship. Frog muscle was chosen because it is very stable in experimental conditions. Does human muscle have the same sarcomere force–length relationship as frog muscle? The key factor is filament lengths. Apparently, many species ranging from frog to human have similar myosin filament lengths but different actin filament lengths (Box 4.18).

Figure 4.63 Schematic depiction of how CB action could be disrupted when actin filaments overlap with each other at less than optimal sarcomere lengths. (A) All CBs can effectively bind to actin. (B) At short sarcomere lengths, actin filaments overlap with each other, preventing some CBs from binding to the appropriate actin filament. This reduces the number of effectively attached CBs, with a resultant loss of force. Note that in this schematic the number of CBs on the myosin filament has been greatly reduced for clarity.

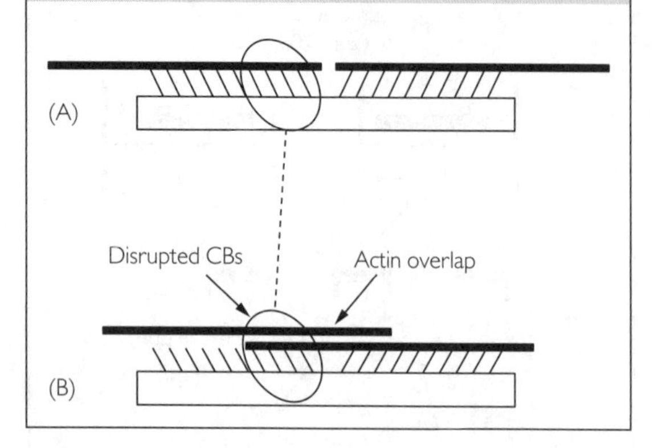

Frog and human filament lengths are compared in Figure 4.64. Note that both frog and human muscle have the same myosin filament length (1.6 μm), but human actin filament length is greater (1.27 vs 0.95 μm). The effect of the difference in actin filament length on the sarcomere force–length relationship is shown in Figure 4.65. First, human sarcomeres can produce force over a greater length range (2.97 vs 2.33 μm). Secondly, the optimal length (L_0) plateau starts at a longer sarcomere length (2.64 vs 2.00 μm). Thirdly, the length range from the start of the L_0 plateau to the length (1.7 μm) at which the Z-disks begin to abut the ends of the myosin filaments is greater in human sarcomeres (0.94 vs 0.3 μm). On the other hand, other features of Figure 4.65 can be explained by similar frog and human myosin filament lengths. First, the length of the L_0 plateau is the same (0.2 μm; see also Figure 4.64). Secondly, there is a similar slope of force decrease beyond

BOX 4.17 WHY DOES A MUSCLE FIBER OR WHOLE MUSCLE NOT HAVE THE FLAT PLATEAU REGION SEEN IN SARCOMERES?

The flat plateau region of the sarcomere force–length relationship corresponds to the bare zone on the myosin filament (Figures 4.62 and 4.65). In contrast, in the whole-muscle or muscle-fiber force–length relationship, the flat plateau is replaced with a smoothed summit (Figures 4.59 and 4.60). The explanation for the latter may be that, for a given muscle or muscle fiber length, sarcomere lengths are non-uniform either longitudinally or in cross-section. The non-uniform sarcomere lengths cause a smoothing of the plateau region (98).

BOX 4.18 VARIATION IN ACTIN FILAMENT LENGTH AMONG SPECIES

Other species (e.g., rat, rabbit, cat, and monkey) have actin filament lengths intermediate between humans and frogs. The reason for the difference is not known; however, it can be noted that longer actin filaments, by increasing the length range (working range; Figure 4.3) over which a single sarcomere can produce force, reduce the number of sarcomeres in series that are needed along the length of a myofibril (and muscle fiber) to achieve a given length range of force production (see Figure 4.2).

Figure 4.64 Comparison of frog and human actin and myosin filament lengths (μm). Myosin filament length is the same in frog and human but the human actin filament is longer. Note the effect that the actin filament has on the sarcomere length at which actin and myosin filaments just fail to overlap. Filament lengths from Rassier *et al.* (98).

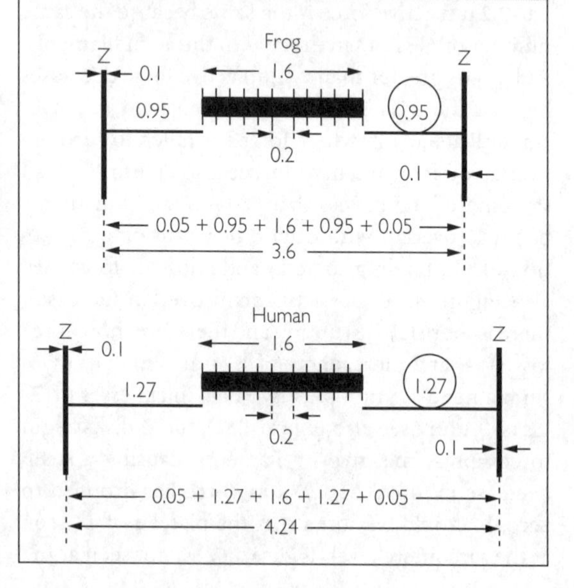

Figure 4.65 Comparison of sarcomere force–length relationship in frog and human muscle. The longer human actin filament (1.27 vs 0.95 μm) causes optimal sarcomere length (L_0) to occur at a greater length (2.64 vs 2.00 μm). On the other hand, the length of the L_0 plateau (0.2 μm) is the same because human and frog have the same myosin filament length (1.6 μm) and the bare zone is the same length (0.2 μm). Filament lengths from Rassier *et al.* (98).

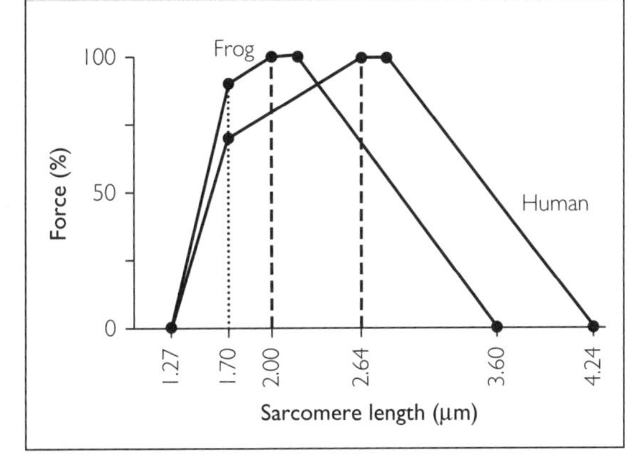

L_0, and the same length range (1.4 μm) over which force decreases to zero. Thirdly, the sarcomere length (1.7 μm) at which the Z- disks begin to abut the ends of the myosin filaments is the same.

Figure 4.66 Force–length relationships for opposing (antagonistic) muscles acting at the ankle joint, derived from a group of subjects performing maximal isometric contractions at a series of joint angles. Soleus (Sol) produces plantarflexion (ankle extension) force whereas tibialis anterior (TA) produces dorsiflexion (ankle flexion) force. A joint angle of 60° indicates a long muscle length for Sol but a short length for tibialis anterior; the reverse applies for the angle of 135°. An ankle joint angle of 90° is the neutral position. Both muscles operate on the ascending limb and plateau region of their force–length relationships. Note that the force for each muscle is normalized with respect to the maximum force at the plateau (100%). The absolute force is about four times greater in soleus than tibialis anterior. Based on the data of Magnaris (74).

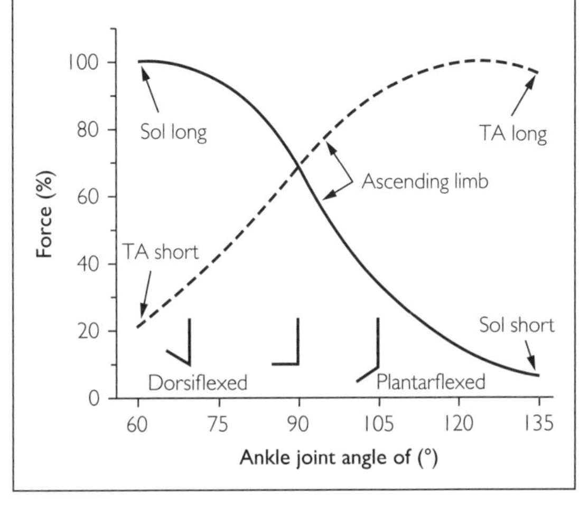

Working Range of Force–Length Relationship in Human Muscle

It has been pointed out (see Figure 4.60) that different frog leg muscles operate, when intact in the body, on different parts of the active force–length relationship. The same is true for intact human muscles (68). The part of the force–length relationship used by an intact muscle during normal function can be referred to as its working range or alternatively, the **expressed section** of the force–length relationship (115). Figure 4.66 shows the working range for two antagonistic muscles that act on the ankle joint. Both soleus, an ankle plantarflexor, and tibialis anterior, the principal ankle dorsiflexor, operate on the ascending limb and plateau region of their respective force–length relationships. The working range of the isometric active force–length relationship used by some other human muscles is shown in Figures 4.67 and 4.68. In a given muscle there may be inter-individual differences in the portion of the force–length relationship used in a particular movement (Box 4.19).

The force–length relationship's working range in some muscles has also been estimated for some specific activities. For example, whereas vastus lateralis (one head of quadriceps) operates on the ascending limb, plateau, and descending limb of its force–length relationship in isolated knee extension through a full range of motion (35), this muscle operates mainly on the descending limb when pedaling a bicycle (27). Gastrocnemius (an ankle plantarflexor like soleus shown in Figure 4.66) uses a relatively small portion of the ascending limb of

Figure 4.67 Working range of isometric active force–length relationship in selected human upper-limb muscles. All of the muscles use the part of the force–length relationship near optimal length (L_o). The elbow flexors (biceps, brachialis, and brachioradialis) use a large part of the ascending limb. Based on the data of Murray *et al.* (83).

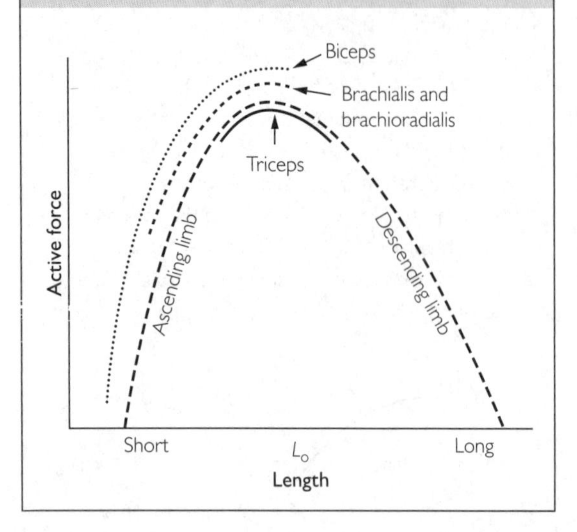

Figure 4.68 Working range of isometric active force–length relationship in selected human lower limb muscles. All of the muscles use the part of the force–length relationship near optimal length (L_o). Triceps surae includes soleus and gastrocnemius. Force–length relationships based on references 35, 44, 74, and 115.

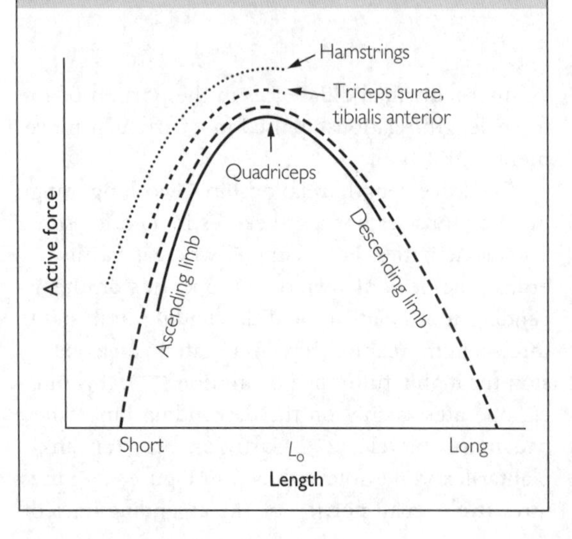

BOX 4.19 INTER-INDIVIDUAL DIFFERENCES IN THE WORKING RANGE OF MUSCLES

Depending on the muscle, there may be small or large inter-individual differences in the working range of the force–length relationship. For example, the working range for rectus femoris was determined in a group of subjects. They were evenly distributed in using predominantly the ascending limb, descending limb, or both limbs of the force–length relationship. In contrast, most of the same subjects used predominantly the ascending limb in gastrocnemius, but there were a few outliers who used the descending limb or both ascending and descending limbs. The greater inter-subject differences in rectus femoris were attributed to its relatively short tendon and long muscle fibers (115).

its force–length relationship in walking but a larger part of the ascending limb and also the plateau region in jumping (27).

Strength Curves

For intact muscles, changes in muscle length are brought about by changes in joint angle (e.g., Figure 4.66). Muscles generate **torque** (also called a **moment of force**) as they contract at different joint angles and therefore at different lengths. Torque is equal to the force produced by the muscle, transmitted through its tendon, multiplied by its **moment arm** (defined in the Moment Arm subsection). The muscle's moment arm may change as the joint moves through a range of motion. Therefore, the variation in torque produced at various joint angles is not only determined by a muscle's force–length relationship, but may also be affected by changes in the muscle's moment arm at different joint angles. The pattern of change in torque through a range of motion is called an **angle–torque relationship** or **strength curve** (62).

The concept of a strength curve is illustrated in Figure 4.69. The formation of a strength curve usually assumes that the involved muscles have been maximally activated, either by electrical stimulation or maximal voluntary contractions, to produce maximal isometric contractions at a series of joint angles

Figure 4.69 Concept of a strength curve. Strength (torque) varies through a joint range of motion. The joint position of greatest strength is called the summit or peak of the strength curve.

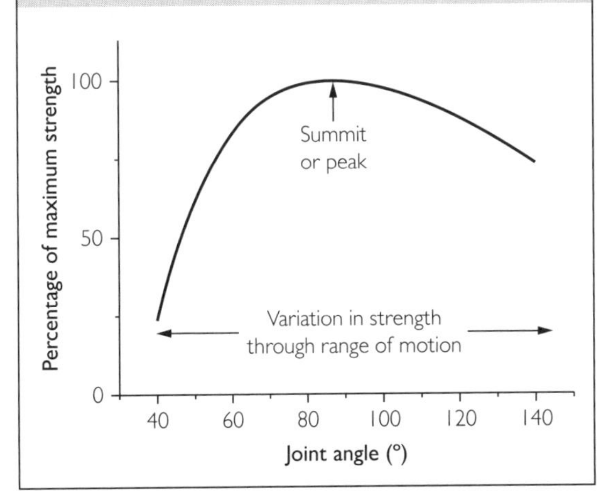

through a range of motion at a joint. Strength curves can also be generated with concentric or eccentric contractions; the effect of contraction type and velocity on the shape of strength curves will be considered later. Strength curves can have various shapes; two examples are shown in Figure 4.70.

Figure 4.70 Sample strength curves for knee extensor (top) and flexor (bottom) muscles. Curves were derived from isometric force measurements made at a series of joint angles. Subjects were seated for knee extension and lying prone for knee flexion. Adapted from Houtz *et al.* (44). Copyright © 1957, The American Physiological Society.

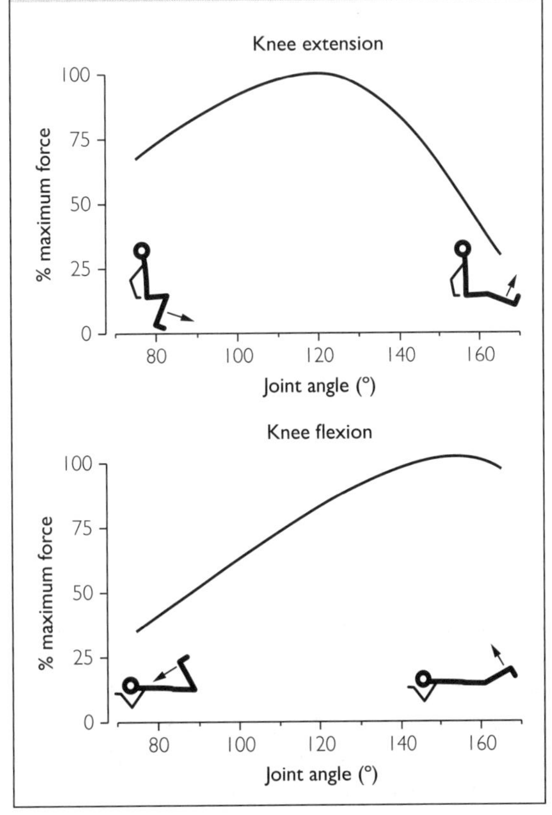

Moment Arm

As noted, the shape of a strength curve can be influenced by the muscle's force–length relationship and changes in the muscle's moment arm. The concept of muscle moment arm is illustrated in Figure 4.71, and a simple depiction of torque generated by brachial biceps is shown in Figure 4.72. The moment arm of a muscle changes at different joint angles (Figure 4.73) and the magnitude of the moment arm and the amount by which it changes through a range of motion vary among different muscles. For example, compared to triceps, brachial biceps has a larger maximum moment arm (4 vs 3 cm) that occurs at a larger joint angle (100 vs 60°) (Figure 4.74). Even for muscles that are synergists, there can be large differences in moment arm magnitude and pattern of change, as shown for example in elbow flexors (Figure 4.75). In elbow flexors (84) and probably other muscles there are inter-individual differences in the absolute values of moment arms and in the joint angle of the peak moment arm.

Figure 4.71 Muscle moment arm (ma) is defined as the perpendicular distance between the line of force (*F*) of the muscle and the joint center of rotation. Muscle force multiplied by the moment arm equals the torque (moment of force) produced by muscle contraction (torque = $F \times$ ma).

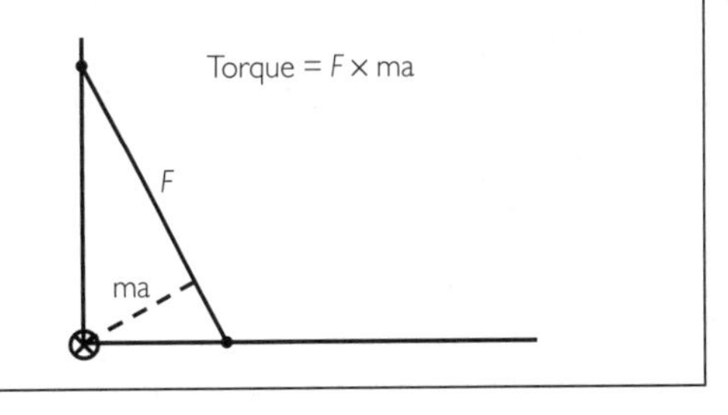

Figure 4.72 Torque produced by brachial biceps at an exemplar elbow joint angle.

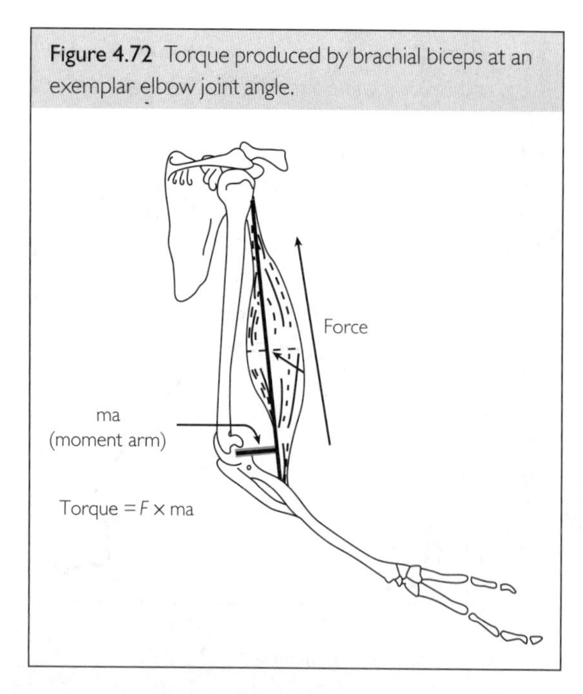

Figure 4.74 Maximum moment arm and pattern of change are compared between biceps and triceps. The maxima for triceps (3 cm) and biceps (4 cm) occur at notably different joint angles (60 vs 100°, respectively). Based on Murray *et al.* (85). © 1995 Published by Elsevier Inc.

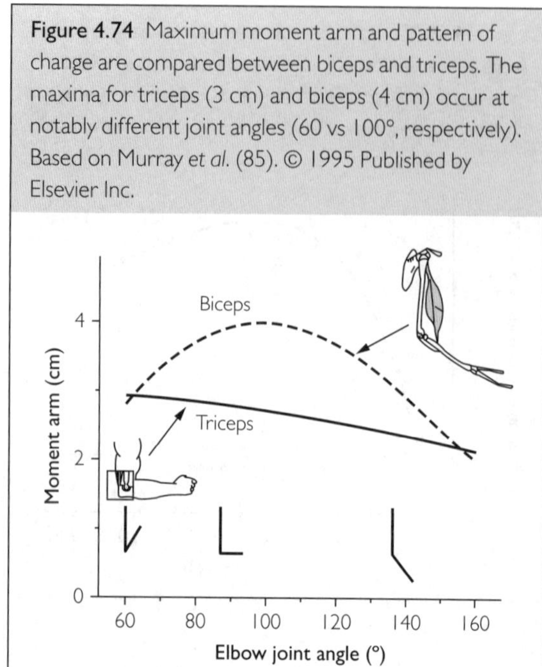

Interaction between Force–Length Relationship and Moment Arm

The shape of a strength curve is dictated by the interaction of two main factors: (1) the force–length relationship of the involved muscle(s) and (2) the magnitude and pattern of change in moment arm of the involved muscle(s). Figure 4.76 shows in a relatively simple example the interaction between the soleus muscle's force–length relationship and moment arm to produce an ankle plantarflexion strength curve. Note that whereas muscle force increases as the muscle becomes longer (i.e., joint angle becomes smaller), the moment arm becomes smaller. Thus the effect of the changing moment arm is to shift the summit of the strength curve to the left; that is, to a larger joint angle (shorter muscle length) relative to the optimal length for the force–length relationship. In contrast, the force–length relationship of soleus tends to shift the summit of the strength curve to the

Figure 4.73 Change in muscle moment arm through a range of movement. On the right the change in joint angle causes the moment arm to get larger.

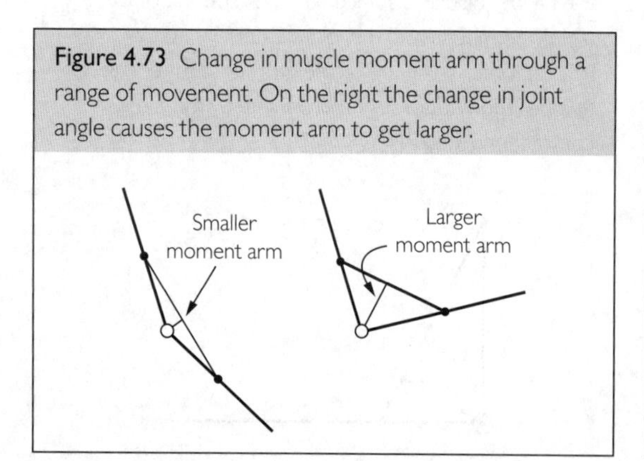

Figure 4.75 Moment arm changes for the three elbow flexors. Note the difference in the maximum moment arm and pattern of change. Elbow joint angle is referenced to 180° as full extension. Based on the data of Murray *et al.* (85).

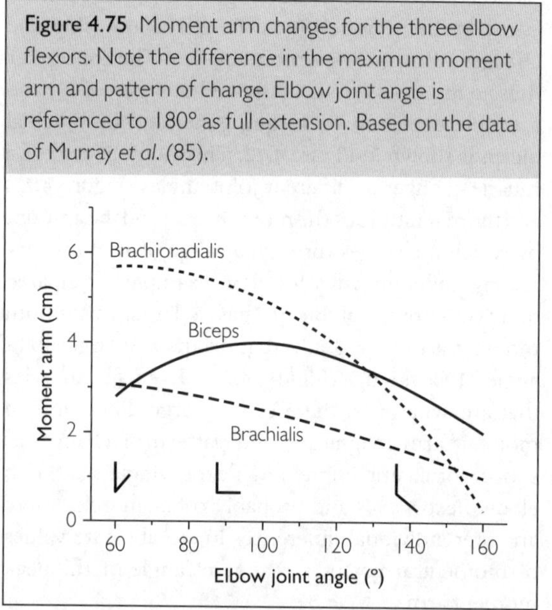

Figure 4.76 Interaction between the active force–length relationship (FLR) and moment arm of the soleus muscle to produce a strength curve (torque). Soleus is one of two large muscles (the other is gastrocnemius) that generate a plantarflexion (extension) force at the ankle joint. Note that a joint angle of 135° denotes the shortest muscle length (i.e., full plantarflexion). The moment arm does not undergo a large change through the range of movement; therefore, the force–length relationship of the soleus muscle makes the greatest contribution to the shape of the strength curve. Nevertheless, the pattern of change in the moment arm acts to shift the strength curve to the left (larger joint angles) relative to the force–length relationship. Based on the data of Magnaris (74).

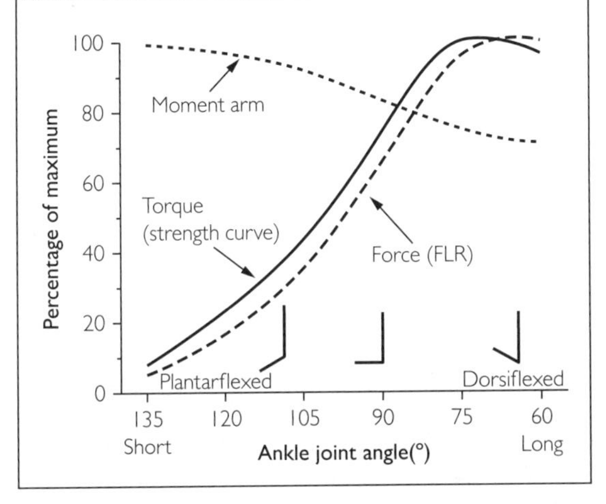

BOX 4.20 MOMENT ARMS ARE GREATER IN CONTRACTING THAN RELAXED MUSCLES

It is now recognized that moment arms are larger when muscles are contracting in comparison to being relaxed. In Figure 4.76 the moment arm of soleus was measured as subjects did maximal isometric contractions. The pattern of change in the moment arm through the range of motion was similar for the contraction and relaxed states; however, the moment arm of soleus was about 25% greater when measured during a maximal contraction compared with the muscle relaxed (75). The moment arm of quadriceps (patellar tendon) is greater in maximal concentric and eccentric contractions compared to the relaxed state, but in addition the amount of increase with contraction varies through the range of motion (105).

right (smaller joint angle, longer muscle length). In this example the force–length relationship of soleus dominates the shape of the plantarflexion strength curve, because the force–length relationship changes more than the moment arm through the range of motion.

The moment arm influences the strength curve not only by its pattern of change through a joint range of motion, but also by its absolute values (e.g., measured in centimeters). For example, at a particular joint angle, a moment arm (ma) of 5 cm (0.05 m) multiplied by a force (F) of 2500 N would produce a torque of 125 N·m (torque = F × ma). In Figure 4.76, the moment arm of soleus (Achilles tendon) was measured as the subjects did a maximal isometric contraction. If the moment arm had been measured with the muscle relaxed, the moment arm would have been smaller (see Box 4.20) and the influence of the moment arm on the torque generated and the shape of the strength curve would have been underestimated.

A second example of the interaction between muscle moment arms and force–length relationships is shown in Figure 4.77. In this case the moment arms and force–length relationships of the three elbow flexors interact to produce the elbow flexor strength curve. Generally, the moment arms of biceps, brachialis, and brachioradialis (Figure 4.77A) act to shift the summit of the strength curve (Figure 4.77C) to the left (smaller joint angle); conversely, the muscles' force–length relationships act to shift the summit to the right (larger joint angle) (Figure 4.77B). The strength curve's summit at a joint angle in the middle of the range of motion (≈120°) is the result of the interaction between moment arms and force–length relationships.

Significance of Small Muscle Moment Arms

Most muscles have small moment arms compared to those of the loads they must support or move. An example is shown in Figure 4.78. To support a 10 kg load in the hand, the elbow flexors must generate a force equivalent to 80 kg. Thus, muscles must generate very large forces to produce even moderate torque at a joint. Another potential consequence of the large forces is muscle or tendon strain and

Figure 4.77 Influence of moment arms (A) and the force–length relationships (B) of biceps, brachialis, and brachioradialis on the shape of the elbow flexor strength curve (C). The moment arms tend to shift the summit of the strength curve to a smaller joint angle whereas the force–length relationships tend to shift the summit to a larger joint angle. As a result of the interaction between moment arms and force–length relationships, the summit of the strength curve takes an intermediate joint angle. Based on the data of Murray *et al.* (83,85) and Tsunoda *et al.* (107).

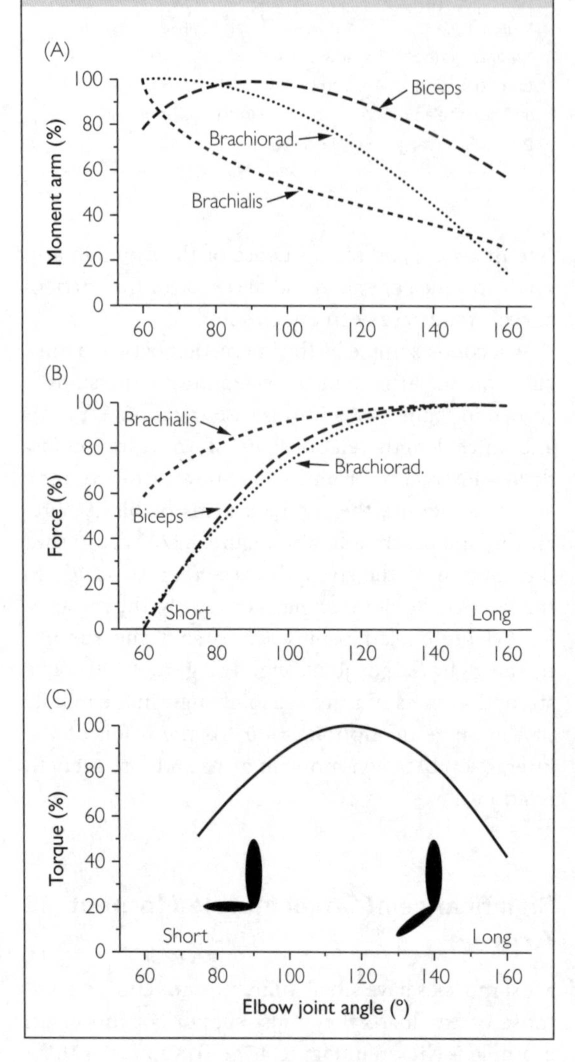

injury. The question could then be asked: why are muscle moment arms so small if they decrease external torque generation and pose a risk of injury? The answer is that small compared to large moment arms provide a greater range and speed of angular motion for a given amount of muscle shortening (Figure 4.79).

Figure 4.78 At a joint angle of 90°, the average moment arm for the elbow flexors (only the biceps is shown here) is about 4 cm (see Figure 4.75). By contrast, the 10 kg load held in the hand has a moment arm of 32 cm. Thus, to support the 10 kg load, the elbow flexors must generate a force equivalent to 80 kg (≈800 N). Many muscles operate against a similar mechanical disadvantage.

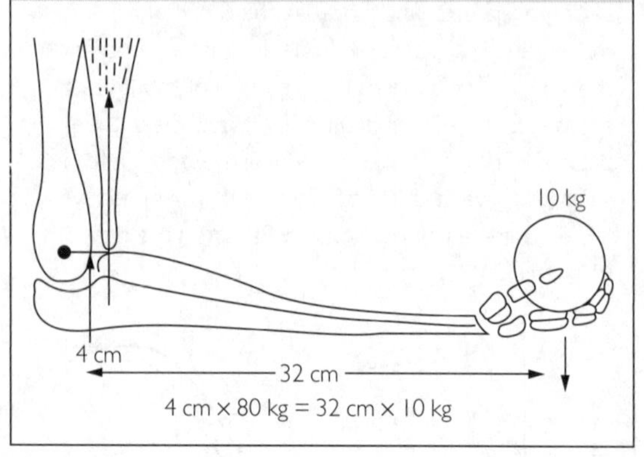

Figure 4.79 Effect of moment arm length on range and speed of angular motion. Muscle A has a small moment arm, but when the muscle shortens a given amount, the joint goes through a large range of motion. Muscle B with its larger moment arm will be able to generate greater torque, but for the same amount of muscle shortening as in A, the angular displacement will be less. Assuming that the shortening speed of muscles A and B are the same, the joint angular velocity produced by muscle A will be greater.

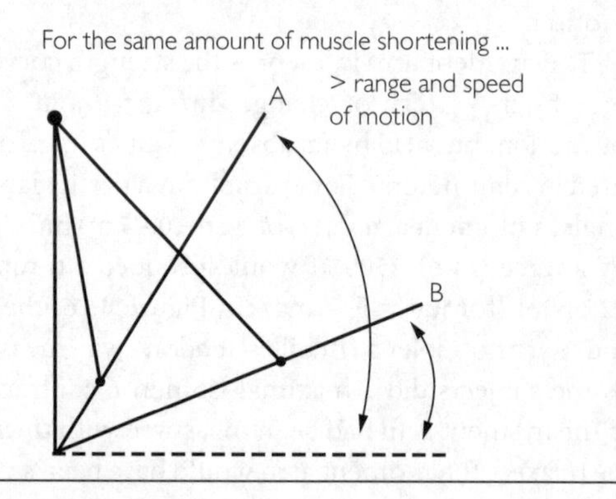

Other Factors Influencing the Shape of Strength Curves

Other factors can affect the shape of strength curves, mainly by influencing the force–length relationship. These factors include training, injury, and fatigue. The effects of training and injury will be considered in more detail in Chapter 8, and the major discussion of fatigue will be in Chapter 5. However, a few examples will be presented here. In regard to training, training that is restricted to a particular range of movement and therefore a particular range of muscle length may alter the strength curve to enhance performance (force) in that range of movement. For example, isometric strength training at one joint angle may cause a disproportionate increase in strength at the trained angle because of neural adaptations that make it easier to fully activate the muscle(s) at the trained angle (Figure 4.80). This adaptation alters the shape of the strength curve. A strength curve could also be altered by adaptations within the muscles that affect the muscle's force–length relationship. Fatigue can alter the shape of a strength curve by affecting the force–length relationship. A fatigued muscle displays a greater loss of force at short muscle lengths (Figure 4.81). Greater fatigue at short muscle lengths is likely related to impaired excitation–contraction coupling (see Figure 4.11). A strength curve could be altered by injury as well (Figure 4.82). In addition to a general loss of strength, an injury could change the shape of the strength curve. For example, pain may be present at a point on the strength curve that reduces voluntary activation. Alternatively, injury could induce scar tissue that would affect the muscle's force–length relationship.

Figure 4.81 Effect of fatigue on the shape of the quadriceps (knee extensor) strength curve. Fatigue causes a disproportionate loss of force at short muscle lengths. Adapted from Rassier (95). © John Wiley & Sons.

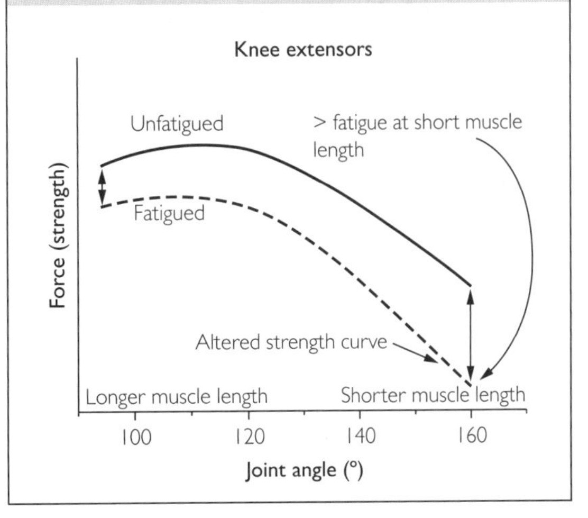

Figure 4.80 Effect of isometric strength training at one joint angle on the shape of a strength curve. The altered strength curve could have been the result of a neural adaptation to training, but other mechanisms are possible (see text). Based on the data of Kitai and Sale (54).

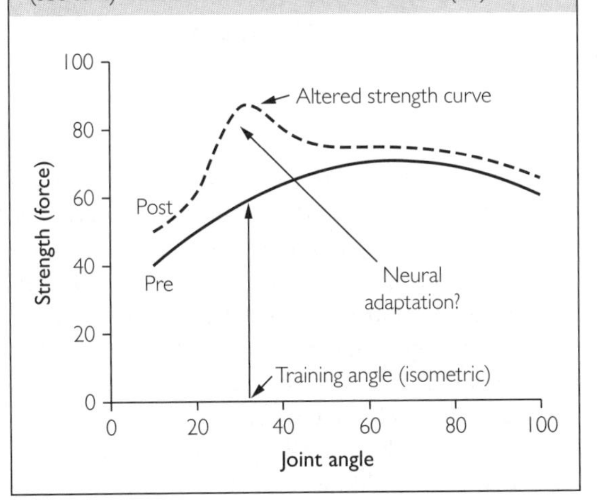

Figure 4.82 Effect of injury on the shape of a strength curve. In this example, strength curves were generated by having the subject do maximal concentric contractions on an isokinetic dynamometer. In addition to an overall loss of strength, an injury may cause an amplified deficit at a particular point on a strength curve.

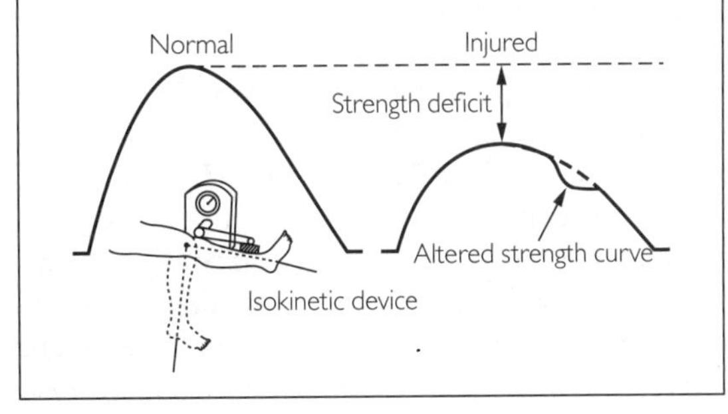

Figure 4.83 Schematic illustration of how ECC-ISO FE and CON-ISO FD could affect the isometric (ISO) force–length relationship and, consequently, the shape of a strength curve. The force–length relationship for isometric contraction without prior eccentric (ECC) or concentric (CON) contraction is shown as a solid line. The effect of FE and FD on the isometric force–length relationship will vary depending on the factors known to affect FE and FD, as covered earlier in the chapter.

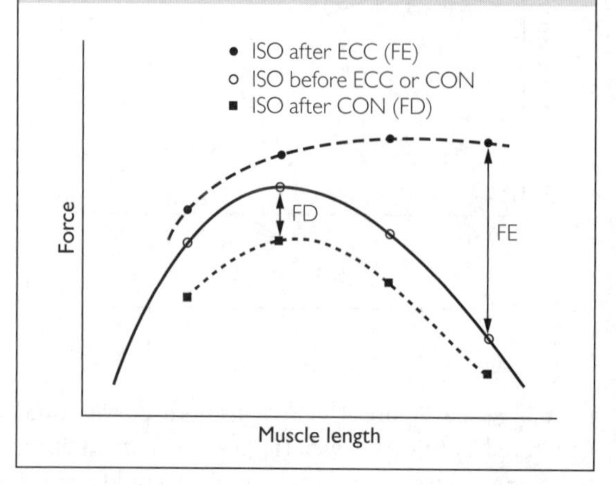

Figure 4.84 Effect of contraction type and velocity on the shape of the thumb adductor (adductor pollicis) strength curve. The isometric (ISO) curve was based on brief, maximal isometric contractions done at a series of joint angles through the range of motion. The concentric (CON) curves were based on single concentric contraction (one slow, one fast) that began as an isometric contraction at a relatively long muscle length at a joint angle of 30°. The muscle was then released to shorten (CON, see arrow) at a constant velocity through the range of motion. The eccentric (ECC) curves are based on single eccentric contraction (one slow, one fast) that began as an isometric contraction at a relatively short muscle length at a joint angle of 0°. The muscle was then forcibly lengthened (ECC, see arrow) at the same constant velocities as the concentric contraction through the range of motion. The difference in force according to contraction type and velocity was expected, but note the marked difference in the shape of the strength curves. Based on the data of Lee and Herzog (63,64).

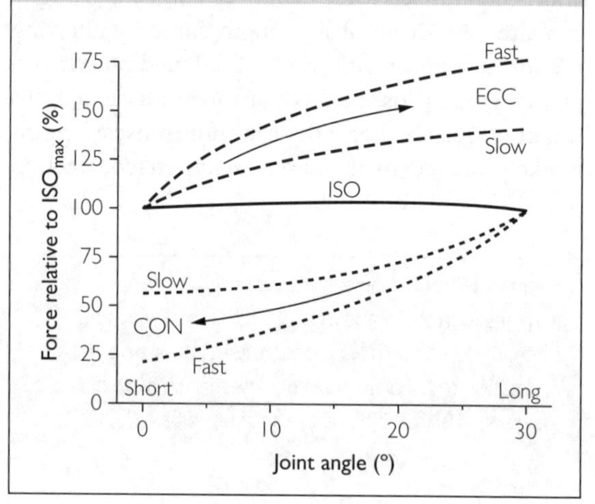

Contractile history In previous sections of this chapter the effect of prior eccentric and concentric contractions on isometric force were covered: ECC-ISO FE and CON-ISO FD. Thus, eccentric and concentric contraction before isometric contraction should affect the isometric force–length relationship and consequently the shape of an isometric strength curve (Figure 4.83). Recall that, as covered earlier in the chapter, the magnitude of ECC-ISO FE and CON-ISO FD varies according to factors such as the amount of stretch (eccentric contraction) or shortening (concentric contraction). These factors can therefore have a highly variable effect on the isometric force–length relationship.

Contraction type and velocity There is a progressive depression of concentric force (CON FD) as a muscle shortens (see Figure 4.57). The pattern of change in concentric force through a range of muscle lengths is much different from that of the isometric force–length relationship. Therefore, CON FD would be expected to affect the shape of a strength curve, as illustrated for a hand muscle in Figure 4.84.

Compared to the relatively flat isometric strength curve, the curves for both concentric and eccentric contraction are much different in shape. The figure also shows the effect of contraction velocity. The effect of contraction type on the shape of the knee extensor strength curve is shown in Figure 4.85. For concentric contraction, the peak of the strength curve at a relatively long muscle length (larger joint angle), followed by a decline in torque as the muscle shortened, could be the result of the CON FD

Figure 4.85 Effect of contraction type on the shape of the knee extensor strength curve. The concentric (CON) and eccentric (ECC) curves were based on single contractions done at a relatively slow joint angular velocity (30°·s^{-1}). The knee joint angle is referenced to 0° = full extension. Thus, a smaller joint angle indicates a shorter muscle length. The concentric contraction started at 90° and progressed to 10°, whereas the eccentric contraction started at 10° and progressed to 90°, according to the arrows. The difference in the shape of the concentric and eccentric strength curves is apparent, as well as the difference in the angle at which the peak torque occurred. Note that the peaks of the concentric and eccentric strength curves are set to 100%. Based on the data of Aagaard *et al.* (1).

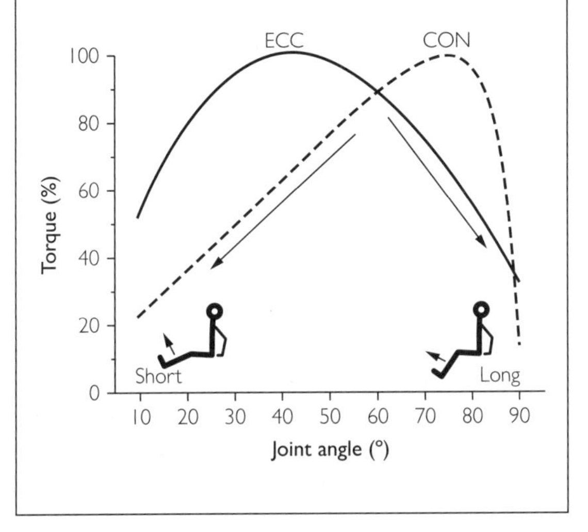

Figure 4.86 Effect of concentric (CON) contraction velocity on the shape of the knee extensor strength curve. Concentric contractions were done at the joint angular velocities shown. The knee joint angle is referenced to 0° = full extension. Thus, a smaller joint angle indicates a shorter muscle length. Torque is referenced to the maximum attained at 30°·s^{-1}. Peak torque decreased as velocity increased according to the force–velocity relationship, but note how the peak torque occurred at smaller joint angles at higher velocities. Based on the data of Kawakami *et al.* (52).

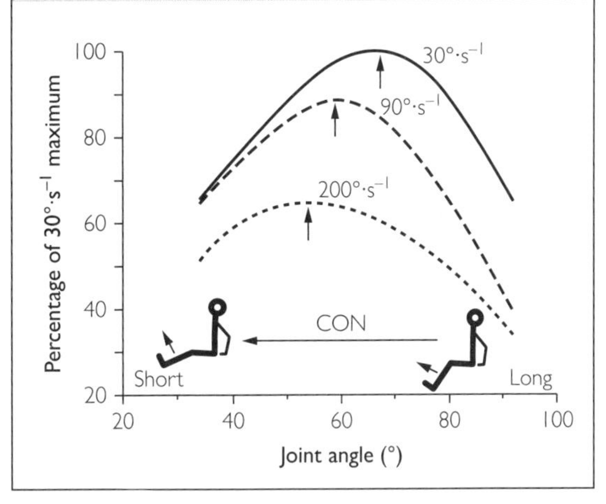

also shown in Figure 4.84. The effect of contraction velocity on the shape of the knee extensor concentric strength curve is shown in Figure 4.86. At higher velocities, the peak of the strength curves shifted to a smaller joint angle (shorter muscle length). The shift was attributed to less stretch of the patellar tendon at higher velocities because of the smaller contraction force. The result of variable stretch of the tendon was that optimal muscle length shifted to smaller joint angles at higher velocities (52).

Strength Curves and Strength Measurement

In the laboratory, strength is often measured by having subjects do maximal voluntary isometric contractions. Typically, tests are done at one selected joint angle in a range of motion. The existence of strength curves requires that the setting of the joint angle be as precise as possible, otherwise differences in force measured might simply reflect different points on the strength curve. When the muscle or muscle group tested has a strength curve like the one in Figure 4.87, the selected angle is often on the summit of the strength curve. An advantage of this selection is that the slightly flattened summit results in little difference (error) in measured force if there is a small error in setting the joint angle. In contrast, if the selected angle is on the slopes of the strength curve, the same joint angle setting error would cause greater variability in force measurement. Unfortunately, some strength curves do not have conveniently flat summits, so special care must be taken to set joint angles precisely (Figure 4.87). Even with careful measurement there is the problem that there is inter-individual variation in the shape of strength curves (Box 4.21).

Figure 4.87 To reduce error in isometric strength measurements, it is important to test each subject at the same joint angle. The same would apply for repeated testing in the same subject. The joint angle selected is usually on the summit of the strength curve. For the strength curve illustrated, this choice reduces measurement error, because a small error in setting the joint angle has little effect on the force measured. In contrast, the same joint angle setting error on the upward or downward slope of the curve would cause greater variation in the force measurement. Other muscles (inset) may not have a conveniently flat summit, so particular care must be taken in setting joint angles.

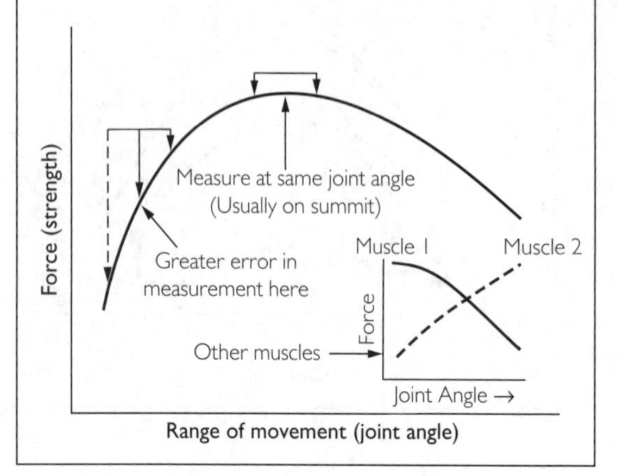

Figure 4.88 Influence of body positioning (hip angle) on the isometric knee flexor strength curve. Compared to the prone position (hip extended), testing in the seated position (hip flexed) causes the summit of the knee flexor strength curve to shift to a smaller (more flexed) joint angle. Based on the data of Houtz et al. (44).

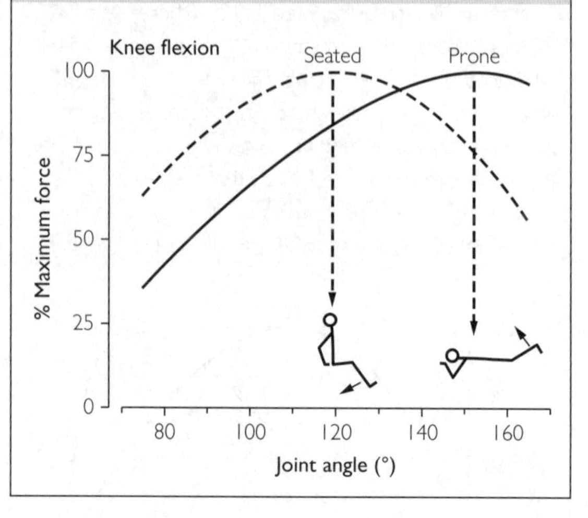

BOX 4.21 INTER-INDIVIDUAL VARIATION IN STRENGTH CURVES

There may be inter-individual variation in strength curves in untrained subjects with an apparently homogeneous activity background (60). Variation is likely to be even greater among athletes who train a particular muscle group in different ways. Therefore, if sport scientists conducted isometric testing at only one joint angle in a range of motion, interpretation of the results could be misleading. Different groups of athletes could have peaks of strength curves at different joint angles. Within a group of athletes, a prescribed training program or an injury could shift the joint angle at which the strength curve peaks. Because of all of these potential variations, obtaining useful isometric strength data from athletes requires testing at a series of joint angles throughout the range of motion.

Influence of body positioning When a muscle acts at two joints, it is important to standardize the positioning of both joints. For example, the peak of the knee flexor strength curve occurs at different joint angles depending on the hip angle (Figure 4.88). The knee flexor (hamstring) muscles act not only to flex the knee joint, but also to extend the hip. Thus, the hip angle (extended or flexed) will affect the force–length relationship of the knee flexors. An extended hip shortens the knee flexors at any knee joint angle; the shortening shifts the optimal muscle length to a larger knee joint angle compared to when the hip is flexed (Figure 4.88). As a second example, the knee extensor strength curve (Figures 4.70 and 4.81) is also affected by hip angle. One head of quadriceps, rectus femoris, crosses both hip and knee joints; consequently, hip angle affects its force–length relationship. With the hip extended, optimal length of rectus femoris occurs at a larger joint angle (greater extent of knee extension). The result, compared to when the hip is flexed, is that the peak of the knee extension strength curve is shifted to a larger joint angle (35).

Isokinetic testing With the introduction of isokinetic dynamometers in the late 1960s (90), it became possible to test the strength of concentric contractions through a full range of motion at various set velocities (e.g., Figure 4.86). In time, these dynamometers evolved to have the capability to make the same measurements with eccentric contractions (e.g., Figure 4.85). One advantage of isokinetic dynamometers is that a strength curve can be produced with a single contraction through a range of motion, as opposed to isometric testing in which contractions have to be performed at a series of joint angles to obtain a strength curve. A second advantage of an isokinetic dynamometer is that it automatically adjusts to inter-individual variations in the shape of a strength curve. The dynamometer mechanism matches any force applied to it in order to maintain the set joint angular velocity.

Establishing strength curves with single concentric or eccentric contractions must take into account factors such as fatigue if slow contractions are tested. In fast contractions, part of the range of motion may be traversed before the limb accelerates sufficiently to catch up to the resistance mechanism of the dynamometer at the set speed (33). Also, in fast contractions, momentum of the accelerating limb as it reaches the speed set on the dynamometer may affect the shape of a strength curve by producing an impact artifact (114).

Figure 4.89 There may an advantage to having the resistance pattern of a training exercise closely match the strength curve; that is, provide a matching variable resistance. In this way, as strength changes through a range of motion, resistance would change in a similar way; therefore, the exercise would be equally challenging throughout the range of motion. In contrast, if a strength exercise offers the same resistance throughout the range of movement (non-matching constant resistance), or a resistance that varies but does not match the strength curve (non-matching variable resistance), it may provide either not enough or too much resistance through parts of the range of movement.

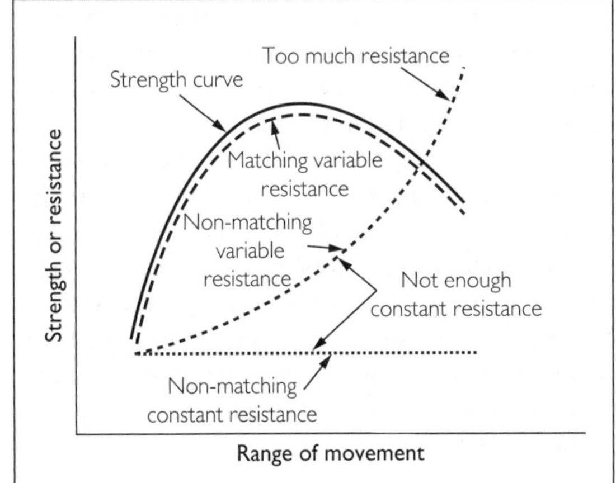

Strength Curves and Training Equipment

When a piece of training equipment is used through a range of motion, the equipment offers a pattern of resistance; the resistance may remain constant or it may vary. Theoretically, it may be advantageous to have a pattern of resistance that closely matches the strength curve through the range of motion (Figure 4.89). A pattern of matching resistance would make the exercise feel equally difficult throughout the range of motion. Therefore, the training stimulus would be similar (optimal?) throughout the range of motion. On the other hand, a non-matching resistance pattern might feel too easy at some points in the range of motion and too difficult at others.

An isokinetic device has the advantage of providing a matching resistance to any training movement, provided the movement is not done too quickly. At any point in the range of motion, the device automatically imposes resistance that precisely matches the strength curve. Furthermore, isokinetic devices adjust automatically to changes in strength curves brought about by injury (Figure 4.82), fatigue (Figure 4.81), and training (Figure 4.80). Limitations to isokinetic devices, as a training mode, are their expense and their restriction primarily to unilateral, single joint actions. In addition, an isokinetic (constant velocity) resistance is less specific to sports performance than an isotonic (constant load) resistance, as discussed in relation to the force–velocity relationship (see Figure 4.23).

Weight-training equipment is most commonly used for strength training. Basic equipment includes barbells and dumbbells (so-called free weights), and an assortment of weight-stack or plate-loading machines. This equipment provides resistance patterns that may or may not closely match strength curves. An example is the standing dumbbell lateral raise (Figure 4.90). As the dumbbell is raised the resistance it imposes increases and is maximal when the upper

Figure 4.90 The contrast between the strength curve for shoulder abduction and the resistance pattern (curve) offered by the dumbbell lateral raise exercise. The progressive increase in the resistance as the dumbbell is raised is the result of the increasing moment arm of the dumbbell as it is raised, becoming greatest when the upper arm is horizontal. The depicted isometric strength curve is based on the data of Clarke *et al.* (14). It is clear that there is a poor match between the strength curve and pattern of resistance. Note that the resistance curve (pattern) depicted applies to a relatively slow action with minimal acceleration and deceleration.

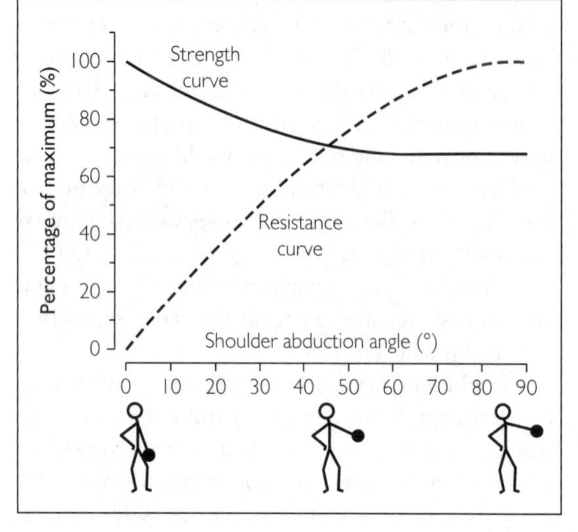

arm is horizontal. In contrast, the shoulder abductor muscles are strongest when the arm is vertical at the trainee's side, and weaken by about 25% as the arm is raised to the horizontal position. Thus, there is a mismatch between the dumbbell's resistance pattern and the muscles' strength curve. One way to almost eliminate this mismatch would be to have the trainee lie on his/her opposite side as the dumbbell is raised. In this posture, the dumbbell would provide its greatest resistance when the shoulder abductors are strongest. A pulley machine with a handle at floor level would also provide a resistance pattern that more closely matches the strength curve. Another approach is to use a "cheating style." In the dumbbell lateral raise example (Figure 4.90), the trainee could accelerate the weight at the beginning of the movement when the abduction strength is greatest. The momentum produced would make it easier to complete the range of motion where the strength is reduced. With practice, a trainee could learn to apply just the right amount of acceleration at the beginning of the movement to make it possible to complete the movement. On the other hand, excessive acceleration would eliminate any resistance at the end of the movement, and might pose the risk of injury.

Some weight-training equipment manufacturers have produced variable-radius or cam weight-training machines that vary the resistance through a range of motion, with the presumed goal of providing resistance patterns that closely match the corresponding strength curves. Technically, it should not be difficult to design machines that closely match *average* strength curves for many joint actions, because strength curves based on isometric and dynamic (concentric and eccentric) contractions are readily available in the scientific literature (several examples have been given in this chapter). Perhaps surprisingly, the attempts by equipment manufacturers to produce cams that provide matching resistance patterns have had only limited success (11,26,32,50,93). Even if equipment were produced that successfully provided matching resistance for average strength curves, there would still be the challenge of accommodating all of the factors that can affect the shape of a strength curve: contraction type and velocity, training background, fatigue, and injury. The many factors affecting strength curves may detract from the theoretical advantages of average matching variable resistance (Box 4.22).

BOX 4.22 IS THE THEORETICAL ADVANTAGE OF A MATCHING RESISTANCE PATTERN SUPPORTED BY EVIDENCE?

In one experiment (76), one group of subjects trained on a knee-extension weight machine that, because of a variable-radius cam, provided a resistance pattern closely matching the isometric strength curve for knee extensors. A second group trained on the same machine, but the variable-radius cam was replaced with a circular sprocket that provided a constant resistance through the range of motion. After 10 weeks of training both groups made similar increases in isometric strength at all joint angles tested, indicating that the variable-radius cam did not produce a superior training response.

Summary of Key Points

- Muscles are composed primarily of muscle fibers that in turn are made up primarily of myofibrils. A cytoskeleton organizes myofibrils longitudinally into sarcomeres. Within each sarcomere are overlapping thick myosin and thin actin filaments.
- The myosin molecules of thick filaments have heads that can form cross-bridges (CBs) with the actin filaments, resulting in muscle contraction.
- During contraction CBs undergo CB cycles, each consisting of attachment to actin, a power stroke and detachment. ATP provides energy for the power stroke and allows detachment of the CB from actin at the end of the power stroke.
- Within sarcomeres CBs are oriented so that their power strokes attempt to cause sarcomere and thus muscle shortening by pulling actin filaments toward the center of the sarcomere. A shortening (concentric) contraction will occur if the collective CB force exceeds the load imposed on the muscle. If CB force just matches the load, an isometric contraction without a change in muscle length occurs. If the load exceeds CB force the muscle is forcibly lengthened, producing a lengthening (eccentric) contraction.
- CB cycling is initiated when Ca^{2+}, released from the sarcoplasmic reticulum (SR) surrounding each myofibril, binds to the regulatory protein troponin that is part of the actin filament. Relaxation occurs when Ca^{2+} is taken back up into the SR.
- Ca^{2+} release is triggered by a muscle action potential (MAP) that travels along the muscle fiber membrane (sarcolemma) and down into the fiber via the transverse tubules. When the effect of the MAP subsides the SR's pump causes Ca^{2+} reuptake into the SR. The MAP originates at the neuromuscular junction in response to a nerve impulse.
- A twitch, the contractile response to a single MAP, has contraction and relaxation phases corresponding to Ca^{2+} release and reuptake, respectively. Single twitches are rare in normal function that instead consists of tetanic contractions (tetani) resulting from a train of several MAPs. A low frequency of MAPs produces an unfused tetanus whereas a high frequency produces a fused (smooth) tetanus.
- Tetani exhibit summation, producing greater force than twitches. Summation increases with increased frequency until a plateau frequency is reached with no further increase in force. At the plateau of the force–frequency relationship tetanic force can be about 10 times greater than twitch force. Tetani are stronger than twitches, and high-frequency tetani are stronger than low-frequency because more Ca^{2+} is released and the SEC is taken up to a greater extent.
- With maximal activation eccentric contractions are stronger than concentric contractions with isometric contractions in between. Eccentric force is superior to concentric and isometric contraction mainly due to a greater number of attached CBs. Concentric force is inferior both because of a reduced number of attached CBs and because some CBs may exert a braking effect on sliding actin filaments.
- Concentric force decreases as shortening velocity increases until no force can be produced, whereas eccentric force increases to a plateau value as lengthening velocity increases.
- The force–velocity relationship can be measured with the "set velocity-measure force" method or the "set load-measure velocity" method. The latter has more application to athletic performance. In various sports athletes may attempt to achieve maximum possible velocities with light (e.g., baseball pitching) or larger (e.g., body mass in vertical jump) loads.
- Stronger eccentric contractions at faster velocities result from a greater number of attached CBs. Weaker concentric contractions at faster velocities result from fewer attached CBs, decreased force per CB and an increased number of CBs exerting a braking effect.
- Values of force (load) and velocity from the force–velocity relationship can be used to calculate the power–velocity relationship (power = force × velocity). The concentric power–velocity relationship has an inverted U shape with zero power at maximum isometric force (velocity = 0) and at maximum unloaded shortening velocity (V_{max}) (force = 0). From the slowest velocities power increases until velocity has reached 30–50% V_{max} (peak power) and then decreases. The eccentric power–velocity relationship is

almost linear with increased velocity, with power increasing as a result of both increased velocity and force.

- Power is produced with any combination of force and velocity. Although power is sometimes associated only with fast movements, peak power usually occurs at less than 50% of V_{max}.
- The SSC is a coupled ECC-CON contraction of the same muscle(s) with a brief isometric phase (coupling time) between the eccentric and concentric phases. The SSC can enhance concentric force, speed, and power (SSC potentiation) compared to an isolated concentric contraction. The preceding eccentric contraction provides a greater initial force at the start of the subsequent concentric contraction. Also contributing to SSC potentiation are take-up of the SEC, storage and recoil of elastic energy and reflex potentiation. SSC potentiation can occur at slow movement velocities but is most pronounced in fast movements such as sprinting, jumping and throwing.
- SSC potentiation generally increases as the force of the eccentric contraction increases but the highest eccentric forces may cause a decrease in subsequent concentric force.
- The isometric coupling time between eccentric and concentric phases is usually brief (e.g., 15 ms); if it is prolonged SSC potentiation is reduced because the eccentric force degrades to a lower isometric force before the concentric phase.
- The stretch associated with SSC potentiation typically consists partly of an eccentric muscle contraction and tendon stretch of the MTU. Another possibility is that most or all of the stretch is in the tendon whereas the muscle contraction is isometric. This non-eccentric SSC potentiation is important even if the isometric contraction does not provide the high initial force seen with eccentric contraction.
- Isometric contraction can also potentiate concentric contraction even if there is no lengthening of the whole MTU. An initial concentric phase stretches the tendon before becoming an isometric contraction, which is then followed by a second concentric phase. Compared to an isolated concentric contraction, ISO-CON potentiation greatly increases concentric force and power.
- A prior eccentric contraction can increase the force of a subsequent isometric contraction (ECC-ISO potentiation), making it easier to hold a load after it has been lowered. In contrast, a prior concentric contraction decreases the force of a subsequent isometric contraction (CON-ISO FD), making it more difficult to hold (isometric contraction) a load after it has been raised with a concentric contraction.
- When a concentric contraction is done at a set velocity, there is a progressive FD as shortening proceeds (CON FD). For example, crank power in cycling decreases at the later part of each concentric phase.
- The force of muscle contraction is affected by muscle length, called the force–length relationship. There is both a passive (stretch of relaxed muscle) and active (contracting muscle) force–length relationship. Total force is the sum of passive and active force. Passive force increases exponentially as length increases. Active force increases from a short muscle length to an optimal length (L_o) and then decreases at longer lengths. The active force–length relationship is determined by the extent of actin and myosin filament overlap at different sarcomere lengths, which affects how many myosin CBs can interact with actin filaments.
- A strength curve is the variation in torque through a range of joint motion. The shape of a strength curve is the result of the interaction between the involved muscles' force–length relationships and their joint moment arms. The shape of a strength curve can also be affected by contraction type and velocity, fatigue, training history, contractile history (e.g., ECC-ISO potentiation), and body positioning. Strength curves must be considered in strength testing to avoid unwanted variability in test results. Commonly used strength training equipment may or may not provide resistance patterns that closely match strength curves.

5 Neuromuscular Bases for Performance

In Chapter 4 we described how muscles contract, and how contraction force is influenced by factors such as contraction type, contraction velocity, muscle length, and contractile history. The first part of Chapter 5 (Motor Unit Activation) will explain how the nervous system controls muscle contraction as reflected in the activation of motor units. The second part of the chapter (Neuromuscular Fatigue) will focus on exercise-induced fatigue and its effects on neural and muscle function. The third part (Factors Affecting Strength, Power, and Speed Performance) will consider factors such as muscle size and muscle temperature that affect strength, power, and speed performance.

Motor Unit Activation

Motor Units

Several parts of the brain as well as spinal reflexes control the force of muscle contraction, but the final common pathway for muscle control is the **motor unit** (MU) (141), which consists of a motor nerve cell, also called a motor neuron or motoneuron, and the muscle fibers it innervates (87) (Figure 5.1). A motoneuron consists of a neuronal cell body (also called a soma) situated in the ventral horn of the spinal cord or within nuclei of the brain stem, dendrites that extend from the soma, and a motor axon that extends from the soma and is part of a motor nerve serving a muscle. Within the muscle, the axon divides into several branches, each innervating a single muscle fiber.

Large exercise muscles have MUs containing a few hundred muscle fibers. The number of muscle fibers per MU is called the **innervation ratio**. For example, biceps brachii has about 300 000 muscle fibers organized into about 750 MUs. The *average* innervation ratio would be 400 (300 000/750 = 400) but there can be a large *range* in innervation ratios within a muscle. The muscle fibers of a MU may intermingle with the fibers of as many as 10–25 other MUs (48) (Figure 5.2).

Figure 5.1 Schematic illustration of a motor unit (MU). A MU consists of a motor nerve cell, also called a motor neuron or motoneuron, and the muscle fibers it innervates. A motoneuron consists of a cell body or soma that is typically situated in the ventral (anterior) horn of the spinal cord, and a motor axon that is part of a motor nerve serving a muscle. In the muscle the axon divides into several branches, each branch innervating a single muscle fiber. The junction between each axonal branch and a muscle fiber is the neuromuscular junction (NMJ; see also Figure 4.9). The illustrated components of the MU and the spinal cord are not to scale but sample dimensions are given. The illustrated MU has only three muscle fibers for simplicity. Typically, MUs in exercise muscles such as biceps brachii or quadriceps femoris have a few to several hundred muscle fibers.

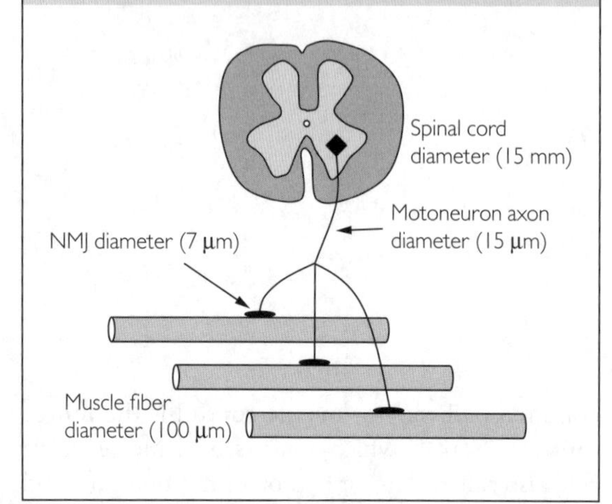

Motor Unit Types

MUs are classified primarily on the basis of the contraction speed and fatigue-resistance of their muscle fibers. For example, MUs have been divided into S (slow, very fatigue-resistant), FR (fast, fatigue-resistant), and FF (fast, fatigable) groups (49). The corresponding muscle fiber types would be type 1, 2A, and 2X, respectively. The terms fast-twitch and slow-twitch are commonly used for the two main MU types, based on the isometric twitch contraction time (short contraction time = fast). The contraction speed of a MU's muscle fibers can also be based on their maximum unloaded shortening velocity (V_{max}). Figure 5.3 summarizes the names given to different MU types and the predominant muscle fiber type of each MU type.

Figure 5.2 Schematic illustration of how the muscle fibers of different MUs intermingle with each other. Two MUs are shown. For simplicity, each MU is shown with only two muscle fibers. Typical exercise muscles have MUs with at least a few hundred muscle fibers.

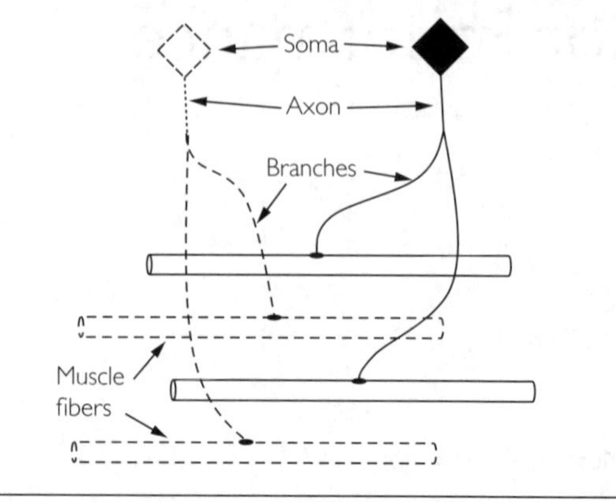

The names type 1, 2A, and 2X are based on the myosin heavy chain (MHC) isoforms contained within muscle fibers (38) that determine contractile speed. While it is convenient to think of muscle fibers as being pure, containing only one MHC

Figure 5.3 Nomenclature used to classify MU types and their constituent muscle fibers. Slow-twitch (ST), also called slow, very fatigue-resistant (S) MUs, contain type 1, also called slow-oxidative (SO), muscle fibers. Fast-twitch (FT) MUs are divided into two subtypes: fast, fatigue-resistant (FR) MUs containing type 2A fibers, also called fast-oxidative-glycolytic (FOG) fibers; and fast, fatigable (FF) MUs containing 2X fibers, also called fast-glycolytic (FG) fibers.

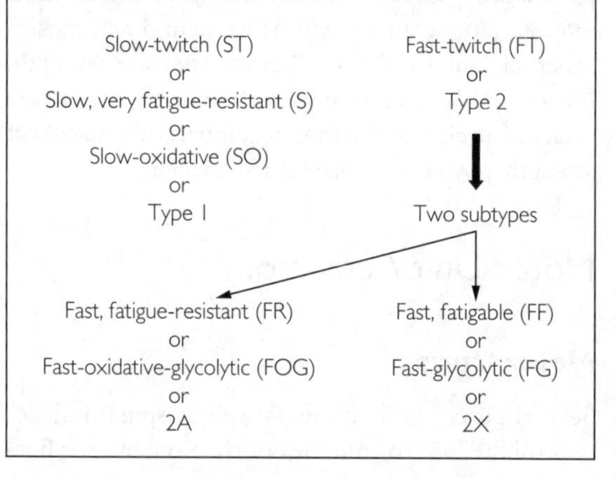

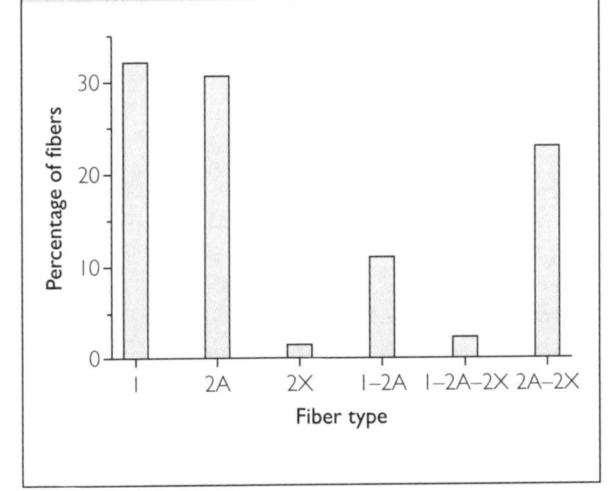

Figure 5.4 Distribution of pure (1, 2A, 2X) and hybrid muscle fibers in human vastus lateralis. Values are the averages from six young men and six young women. About 36% of the fibers were hybrids. Data from Williamson *et al.* (257). Copyright © 2001, The American Physiological Society.

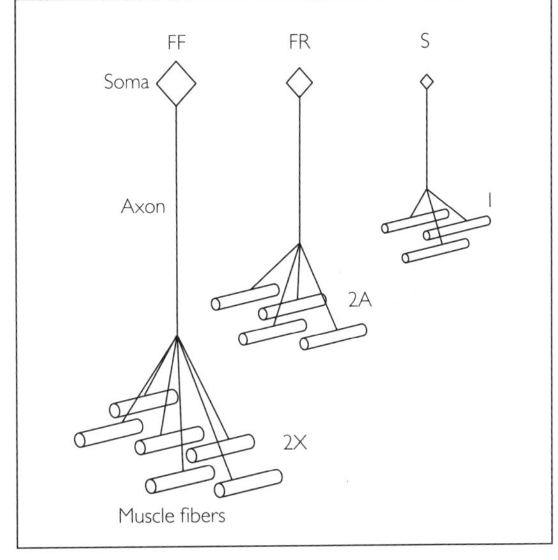

Figure 5.5 Size matters. Schematic representation of the three types of MU: (1) fast, fatigable (FF) MUs with 2X muscle fibers; (2) fast, fatigue-resistant (FR) MUs with 2A fibers; (3) slow (S) MUs with type I fibers. The figure illustrates how on *average* the MU types differ in size. FF units have the largest motoneuron somata (plural of soma), the largest-diameter motor axons, and the largest number and size of muscle fibers. The S units are smallest in all respects, while the FR units are closer in size to FF units. The shown number of muscle fibers per MU is less than typical for clarity.

isoform (1, 2A, 2X), muscles actually contain many **hybrid** fibers containing more than one myosin isoform (Figure 5.4). Hybrid fibers may make up 20–35% of the total number of fibers within a muscle (143,160,188,206,257). Note in Figure 5.4 that there are more hybrid fibers containing 2X myosin than pure 2X fibers; consequently, whereas pure 2X fibers may account for only 0–10% of the total number of fibers in a muscle (44,143,159,188,185,206,257), the 2X MHC isoform may account for 20–25% of the whole muscle MHC content in different muscles and subject groups (6,185,257). Thus, although the depiction of MU types in Figures 5.3 and 5.5 is a useful guide, it should be recognized that a large proportion of MUs contain fibers with both exclusively one MHC isoform and hybrid fibers. Moreover, within each type of MU there will be a variable distribution of pure and hybrid fibers, which has an important influence on the MU's functional characteristics.

Functional Characteristics of Motor Unit Types

A key distinction among S, FR, and FF MUs is size. FF MUs tend to have the largest motoneuron soma size, nerve axon diameter, innervation ratio, and muscle fiber cross-sectional area (CSA), followed by FR and then S MUs (Figure 5.5). Correspondingly, FF MUs (followed by FR and S MUs) have the highest recruitment threshold, the fastest axonal nerve conduction velocity, and the greatest contraction force, respectively. These and other functional characteristics are discussed in more detail below.

Isometric force Maximum tetanic isometric force of a MU is determined by its innervation ratio, average fiber CSA, and average fiber specific force. FF and FR produce greater isometric force than S MUs primarily because of their larger innervation ratios but also because of greater fiber area and specific force (Figures 5.6 and 5.7). In human muscle, however, type 2 fibers are not always bigger than type 1 (Box 5.1).

A greater fiber area translates directly into greater isometric force. Specific force is the force generated per unit CSA (e.g., newtons per square centimeter, or $N{\cdot}cm^{-2}$). The greater specific force of type 2 than type 1 fibers is the result of both a greater fraction of

Figure 5.6 MUs in human triceps brachii. Average ratio values for FF and FR MUs are expressed as ratios to those of S MUs (= 1.0). Tetanic force is determined from innervation ratio, fiber area, and fiber specific force, with innervation ratio making the largest contribution. Innervation ratios adapted from Enoka and Fuglevand (90), assuming a muscle total of 300 000 muscle fibers. Fiber (cross-sectional) area and specific force from Harridge *et al.* (121). Tetanic force ratios were calculated from ratios for innervation ratio, fiber area, and specific force.

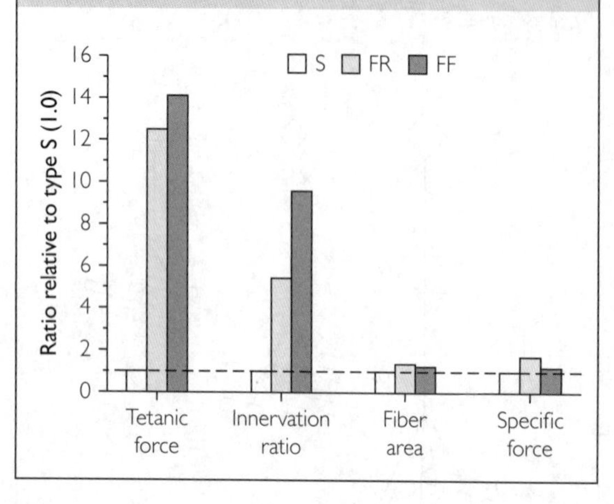

BOX 5.1 ARE TYPE 2 FIBERS LARGER THAN TYPE 1 IN HUMANS?

The scheme depicted in Figure 5.6 shows muscle fiber CSA ranked in the order 2X > 2A > 1. This rank order is often observed in animal muscles but not always in human muscles. In Figure 5.6 the rank order for fiber area in human triceps brachii was 2A > 2X > 1 from data obtained in seven men. In two large groups of untrained men, the rank order in vastus lateralis was 2A > 2X > 1, whereas in two large groups of women the rank order was 1 > 2A > 2X (227,232); that is, the reverse of the order shown schematically in Figure 5.5. Even when 2X fibers are not the largest, F MUs still produce the greatest isometric force because of their greater number of muscle fibers [see innervation ratio of FF (2X) MUs in Figure 5.6]. In women, the greater specific force of type 2X compared to type 1 fibers (Figures 5.6, 5.7) would compensate for the former's smaller fiber area.

attached myosin cross-bridges (CBs) during maximal activation and a greater force per attached CB (147).

Speed of activation Before muscle fibers contract they must be activated by their parent motoneurons to initiate CB cycling. The motoneurons of faster MUs

Figure 5.7 Specific force and maximum shortening velocity (V_{max}) in 19 type 1, 10 type 2A, and 10 type 2X muscle fibers from human vastus lateralis. Mean (average) values are given with standard deviation bars. V_{max} is expressed in muscle fiber lengths per second ($L \cdot s^{-1}$). Note that fiber-type differences are more distinct in V_{max} than in specific force. Data from Bottinelli *et al.* (44).

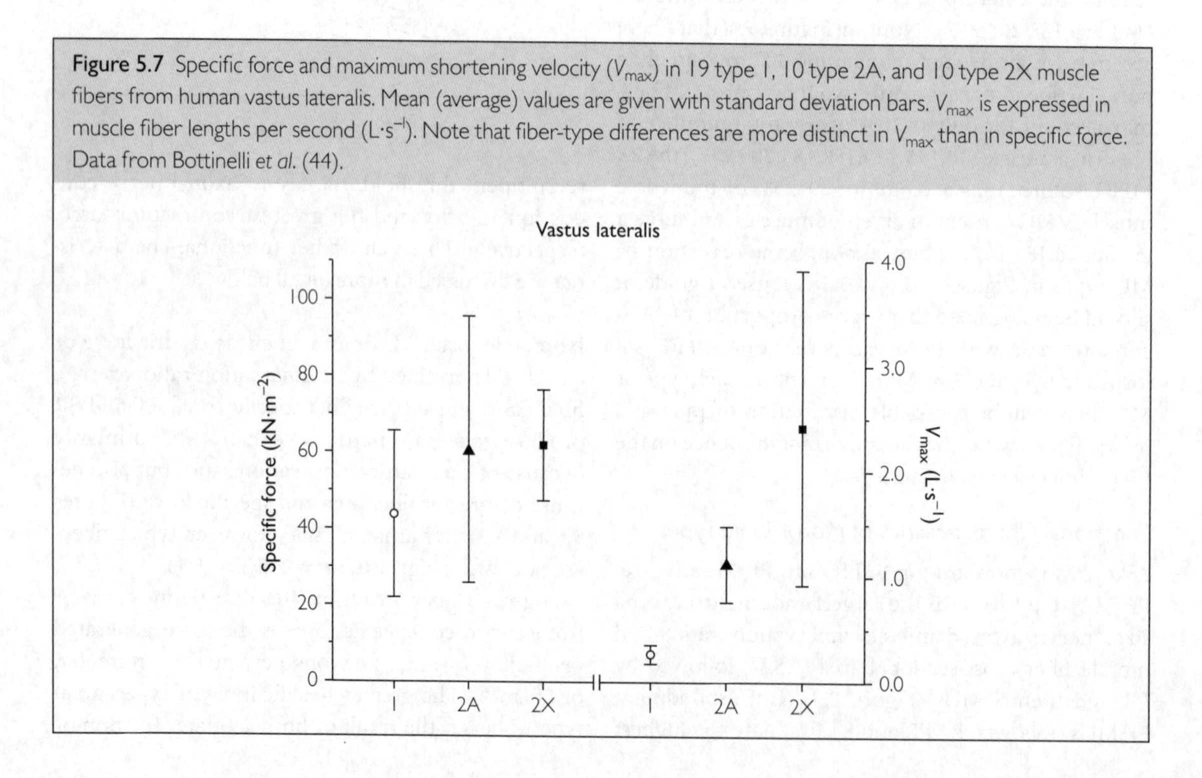

Figure 5.8 Speed of activation. The fastest FF MUs containing type 2X muscle fibers have larger-diameter axons with a faster nerve-impulse conduction velocity (meters per second, m·s^{-1}) compared to the slowest S MUs containing type 1 fibers. Velocities of 30 and 60 m·s^{-1} would correspond to axon diameters of approximately 6 and 12 μm, respectively (165). Note also that 2X fibers have a faster muscle action potential (MAP) conduction velocity.

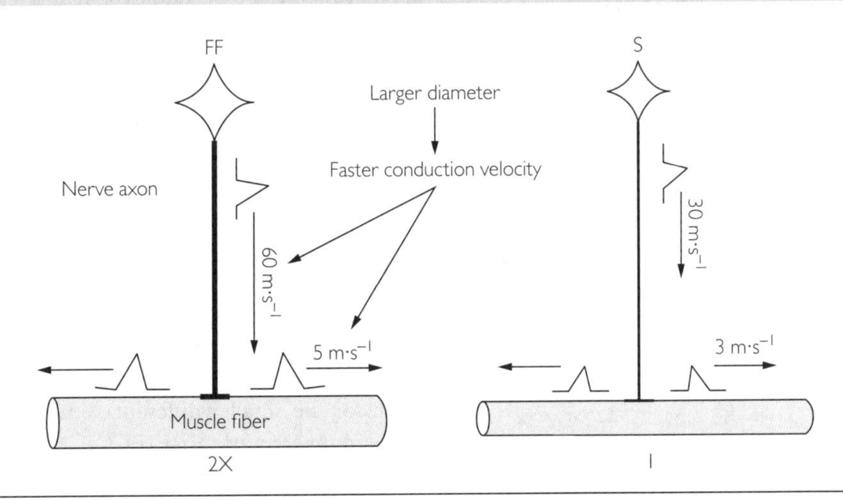

provide a faster speed of activation by a faster nerve-impulse conduction velocity due to larger-diameter nerve axons, and neuromuscular junctions (NMJs) are designed to release more neurotransmitter in response to each nerve impulse, to initiate more quickly a muscle action potential (MAP) in the muscle fiber membrane (219) (Figure 5.8).

Twitch contraction time Once a MAP has been initiated in a muscle fiber, several factors contribute to the shorter (faster) isometric twitch contraction time of fast fibers (Figure 5.9) (219), including a fiber membrane that responds more quickly to transmitter release and propagates the MAP faster, a sarcoplasmic reticulum/transverse tubular network that facilitates faster release and reuptake of Ca^{2+}, and troponin designed for rapid release of Ca^{2+} (58).

Maximum shortening velocity (V_{max}) V_{max} is the maximum *unloaded* shortening velocity of a muscle fiber. There is a large difference in V_{max} across fiber types (Figures 5.7 and 5.10). Type 2X fibers may have an average V_{max} about four or ten times faster than type 1 fibers depending on how V_{max} is measured (Box 5.2). Hybrid fibers have V_{max} values intermediate between those of 2X, 2A, and 1 fibers (Figure 5.10). Since most MUs contain various mixtures of pure and hybrid fibers and there is variation even within pure fiber types (Figures 5.7 and 5.10), there is a continuum of

Figure 5.9 Schematic illustration of twitch contractions in the three pure fiber types. Hybrid fibers would show intermediate twitch responses in the same way they show intermediate values for V_{max} (see Figure 5.11). Note that *isometric* twitches are shown; therefore, the 2X fiber's fast-twitch actually refers to a short-duration isometric twitch. The 2X fiber's short rising (contraction time) and falling (relaxation time) twitch phases correspond partly to more rapid release and reuptake of Ca^{2+}, respectively (for a review, see Figure 4.12). The greater force seen in the 2X fiber twitch is largely due to the greater amount of Ca^{2+} released in response to a MAP (219). Inset: shows how the duration of a twitch contraction time is divided into a contraction time (CT)—that is, time to peak twitch force—and a relaxation time (RT).

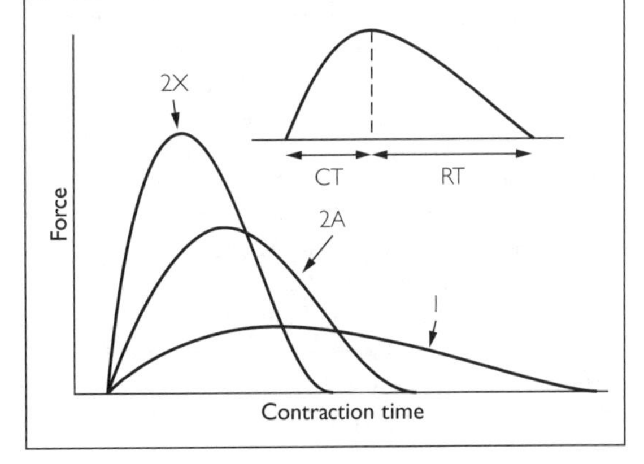

Figure 5.10 Maximum shortening velocity (V_{max}) in pure and hybrid fiber types from human vastus lateralis. Muscle fibers are typed as I, I–2A, 2A, 2A–2X, and 2X. Data (mean ± SD) are plotted with the average (mean) for type I fibers set to 1.0. Adapted from data of Bottinelli *et al.* (45).

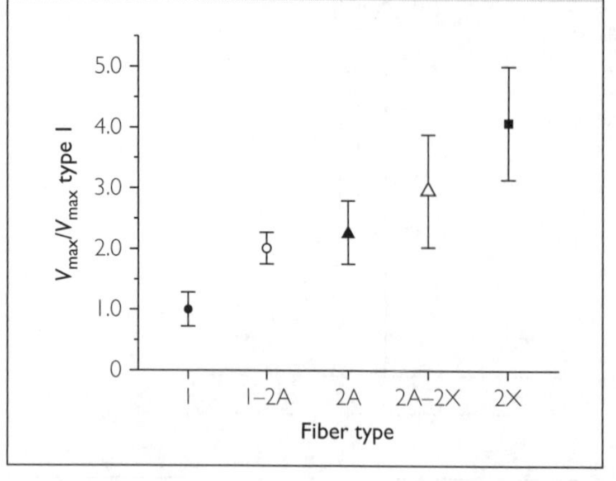

MU V_{max} (and other characteristics discussed below) across the MU population of a muscle, from the slowest MUs with mainly type 1 fibers to the fastest MUs with predominantly 2X fibers. V_{max} is determined by the speed (rate) of CB cycling, which is affected by the rate of ATP breakdown, the displacement produced by each CB power stroke, the duration of CB attachment to actin, and how quickly ADP is released during the CB cycle (184). The myosin isoform dictates the speed of these steps.

BOX 5.2 TWO WAYS OF MEASURING MAXIMUM SHORTENING VELOCITY

In Chapter 4, two methods of measuring maximum shortening velocity were described. One method is a direct measurement, using what is called a slack test. The acronym V_0 is used for this measurement. The second measurement is an extrapolation of a measured force–velocity relationship; in this case the acronym is V_{max}. In Chapter 4 it was stated that, to avoid confusion, V_{max} would be used to denote maximum shortening velocity regardless of the method of measurement. This point is being reviewed here because the method of measuring maximum shortening velocity affects the magnitude of differences in maximum shortening velocity among fiber types. In Figure 5.7, in which maximum shortening velocity was measured directly (V_0), although the acronym V_{max} was used in the figure legend, the average V_{max} (V_0) was about eight times greater in 2X than in I fibers. In contrast, in Figure 5.10 average V_{max} was only about four times greater in 2X than in I fibers. In this case the extrapolation method was used to measure maximum shortening velocity. Regardless of the method of measurement, there are large differences in average V_{max} across fiber types, as well as a variation in V_{max} within each fiber type (Figures 5.7 and 5.10).

Force–velocity relationship Figure 5.11 shows the effect of fiber type on the force–velocity relationship. The greatest effect is on V_{max} and concentric force and the smallest effect is on eccentric force. Previously it was explained that a type 2X fiber's greater V_{max} results from a greater rate of CB cycling whereas its greater ISO_{max} results from a greater fiber area and specific force, the latter due to greater number of attached CBs per unit CSA. But what is the explanation for the smaller difference among fiber types in maximum eccentric force (ECC_{max}) than ISO_{max}? In all fiber types, greater ECC_{max} than ISO_{max} is attributed mainly to a greater number of attached CBs per unit CSA. Therefore, if a large number of available CBs is already attached in type 2 fibers in an ISO_{max} as noted

Figure 5.11 Schematic representation of force–velocity relationships for type I, 2A, and 2X fibers. ISO_{max} refers to maximum isometric force. Maximum shortening velocity (V_{max}) corresponds to the point where the force drops to zero on the velocity axis. See also Figures 5.7 and 5.10 for values of V_{max}. Note that the largest difference between fiber types is in V_{max}, whereas the smallest difference is in eccentric force. Adapted from the data of Linari *et al.* (147).

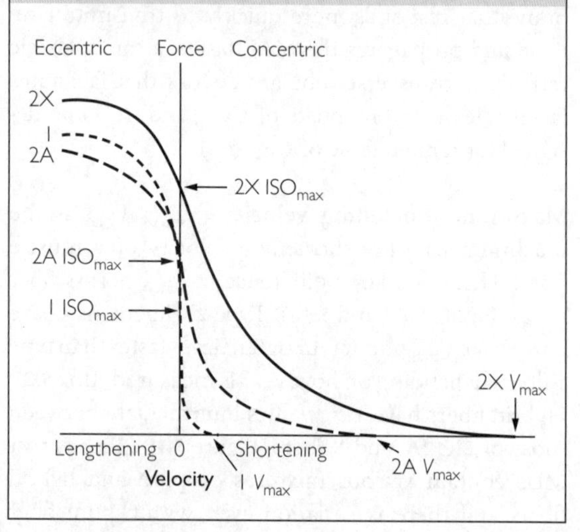

previously, there are fewer additional CBs that can be recruited in an ECC_{max}, thus reducing the amount by which ECC_{max} can exceed ISO_{max} in type 2 fibers. In contrast, type 1 fibers have a smaller number of attached CBs in an ISO_{max} and thus there is a larger number of additional CBs that can be recruited in an ECC_{max}. This increases the amount by which ECC_{max} can exceed ISO_{max} in type 1 fibers (219), producing a smaller difference between type 2 and 1 fibers in ECC_{max} than ISO_{max}.

An additional advantage for type 2 fibers is that their concentric force–velocity relationship has a smaller curvature (219), which allows a greater percentage of V_{max} to be attained with a given load. This effect is best illustrated when the force–velocity relationship of type 1 and 2 fibers is plotted with their ISO_{max} and V_{max} normalized to the same values on the force and velocity axes (Figure 5.12).

Figure 5.12 Schematic representation of how the curvature of the concentric force–velocity relationship differs between type I and 2X fibers. To highlight the effect of curvature, ISO_{max} and V_{max} for the two fiber types have been normalized (given the same values on the axes). For a given load (L) the smaller (less) curvature in type 2X fibers allows a higher percentage of V_{max} to be attained.

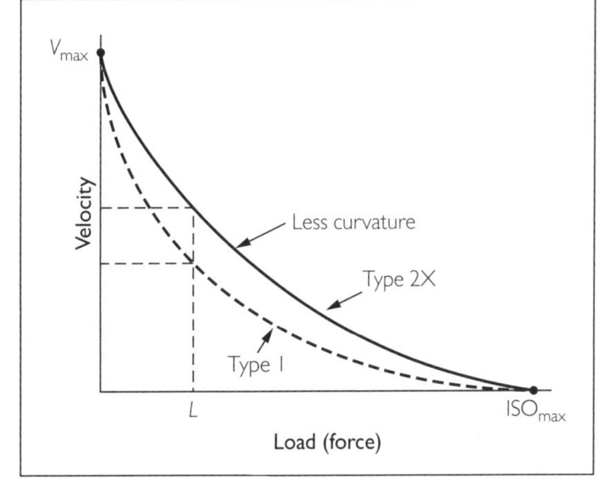

Power–velocity relationship Compared to type 1 fibers, type 2 fibers produce much greater concentric peak power, the peak power occurs at a higher velocity, and power is produced over a greater range of velocities (Figure 5.13). Greater concentric force, a higher V_{max}, and a smaller curvature of the force–velocity relationship contribute to type 2 fibers' greater power. Since power = force × velocity, the differences among fiber types in power are greater than for force and velocity alone (compare Figures 5.11 and 5.13). In contrast, the eccentric power–velocity relationship is similar for all fiber types (Figure 5.13) because the difference in eccentric force is small across fiber types (Figure 5.11). In addition, whereas the low V_{max} of type 1 fibers greatly diminishes their concentric power, it has no effect on their eccentric power.

Isometric rate of force development Similar to twitch contractions (Figure 5.9), both absolute and relative rate of force development (RFD) (Figure 5.14) of maximal tetanic isometric contractions are greater in type 2 than 1 fibers (Figure 5.14).

Metabolic power Metabolic power is the rate at which energy is consumed and is determined by the rate of CB cycling during muscle contraction. Since the rate of CB cycling is greatest in type 2X fibers, followed by type 2A and type 1 fibers, metabolic power is ranked in the same order. For example, metabolic power during brief, maximal intensity exercise is about five times greater in type 2 than type 1 fibers (219). To support their greater rate of energy consumption through splitting of ATP, type 2 fibers have corresponding metabolic pathways for splitting of phosphocreatine (PCr) and anaerobic breakdown of muscle glycogen; both of these pathways rapidly resynthesize ATP. Chapter 2 covers energy metabolism in detail.

Fatigue resistance Whereas type 2 fibers excel in strength, power, and speed, type 1 fibers excel in fatigue resistance or endurance (Figure 5.15). Type 1 fibers' superior fatigue resistance is primarily the result of their capacity for aerobic metabolism (219) (Figure 5.16). There are additional advantages for type 1 fibers. Their lower rate of CB cycling requires a lower rate of ATP consumption, which makes it easier for back-up metabolic pathways, particularly the oxidative pathway, to resynthesize ATP as fast as it is being consumed. The slower CB cycling of type 1 fibers makes them more economical because less ATP is consumed per unit contraction force and per unit time of contraction. Finally, the muscle fiber membranes have type-based properties that allow type 1 fibers to propagate a prolonged succession of MAPs, whereas type 2X fibers can support only a brief but high-frequency train of MAPs (211) (Figure 5.17).

Figure 5.13 Schematic representation of power–velocity relationships in type 1, 2A, and 2X fibers. Because power = force × velocity, the large difference in concentric contraction power ($Power_{CON}$) between type 2 and 1 fibers is due to type 2 fibers' greater force (Figure 5.11), smaller curvature of the force–length relationship (Figure 5.12) and greater V_{max} (Figure 5.11). Note also that the velocity at which peak $Power_{CON}$ occurs increases in the order 2X > 2A > 1. In contrast, all fiber types show the same pattern of eccentric power ($Power_{ECC}$) because the velocity range is the same and the differences in eccentric force are relatively small (Figure 5.11). Power values are relative to the $Power_{CON}$ value for type 1 fibers (= 1). Note the much greater values of $Power_{ECC}$ than $Power_{CON}$. Power–velocity relationships are based on the data of Linari *et al.* (147).

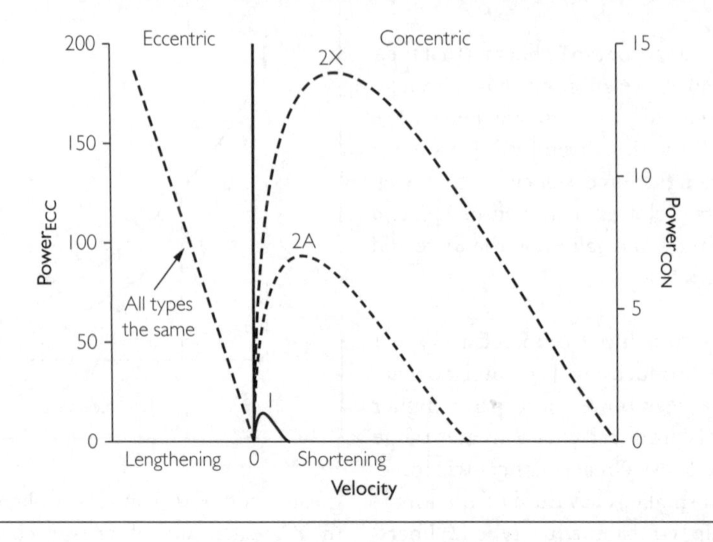

Figure 5.14 Schematic illustration of isometric rate of force development (RFD) in type 1, 2A, and 2X muscle fibers. The scheme assumes that each fiber has been stimulated to produce a maximal tetanic isometric contraction. Note that maximum force for each fiber type has been normalized, that is, set to 100%, to better show the differences in RFD.

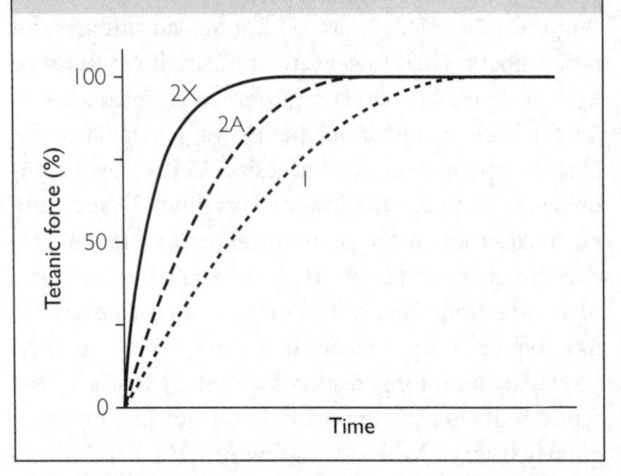

Figure 5.15 Schematic illustration of the fatigability of the different MU types. Force is referenced to the peak force attained by the FF MU. In repeated, stimulated maximal isometric contractions, the S MU with type 1 fibers showed no loss of force over the time period observed, whereas the FF containing type 2X fibers rapidly lost most of its force-generating capacity. The FR MU (2A fibers) showed an intermediate fatigue response. Adapted from the data of Burke *et al.* (49).

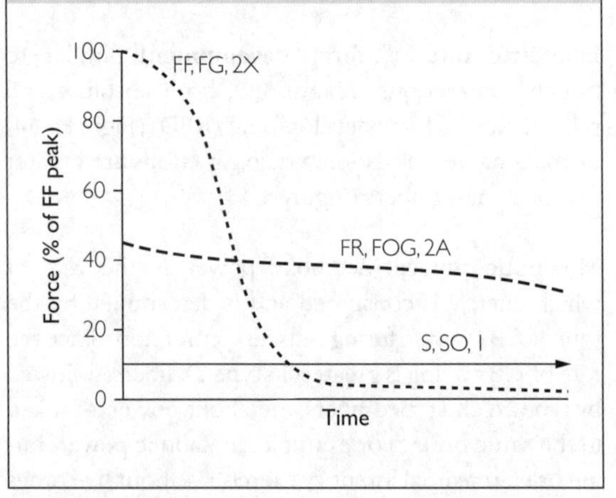

Figure 5.16 Some of the factors affecting fatigability by influencing the rate at which ATP can be resynthesized by the oxidative pathway. Type I fibers excel in oxidative metabolism because of their greater mitochondrial (mitochond.) volume, higher concentration of myoglobin (Mb), enhanced transport of fuel into the fiber, and contact with more capillaries. Smaller fiber diameter facilitates diffusion of oxygen into the fiber, diffusion of carbon dioxide and hydrogen ions out of the fiber, and a more rapid exchange of lactate (La). Chapter 2 provides a detailed discussion of energy metabolism.

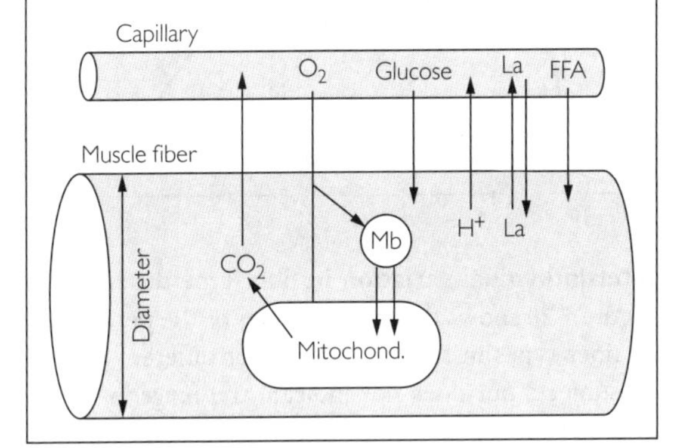

Figure 5.17 Illustration of factors in addition to those discussed in Figure 5.16 that could account for differences in fatigability between types of MUs. The soma of a type S (SO, I) MU responds to a constant level of excitation with a relatively small decrease in firing rate (frequency of nerve impulses) that maintains a high level of summation of muscle contraction. In contrast, the soma of a type FF (FG, 2X) MU responds with a marked decrease in firing rate, which could cause decreased summation and hence a decrease in force. There are also differences at the NMJ. The terminals on type I fibers contain a large pool of transmitter (vesicles of acetylcholine, ACh) that supports a prolonged succession of MAPs. In contrast, the terminals on type 2X fibers allow a greater quantal release of acetylcholine for each nerve impulse, but cannot support a prolonged succession of MAPs.

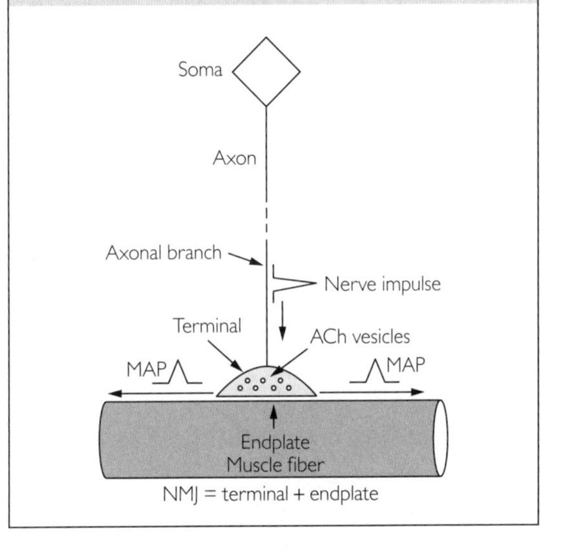

The superior fatigue resistance of S MUs is not confined to the properties of their type 1 fibers. The motoneuron soma and NMJs are adapted, respectively, to maintain long periods of low-frequency nerve impulses (142,230) and repeated neuromuscular transmission (207) (Figure 5.17).

Force–length relationship Compared to type 2 fibers, type 1 fibers produce less passive force at normalized length (equivalent number of sarcomeres in series) (201), and the active force–length relationship operates over a greater range in length (219) (Figure 5.18). The longer, more compliant form of the cytoskeletal protein titin in type 1 fibers likely accounts for their lower passive force (201), whereas longer actin filaments would explain the greater sarcomere length operating range of the active force–length relationship (108).

Distribution of Motor Unit and Fiber Types Within Muscles

With the same muscles humans can perform both prolonged endurance exercise and brief, fast powerful activities; therefore, it is not surprising that most human muscles involved in performance contain a mixture of all MU (FF, FR, S) and fiber types (2X, 2A, 1). And, as noted previously, most MUs contain hybrid as well as pure fibers (Figure 5.4), with functional properties intermediate between those of the pure (2X, 2A, 1) fiber types (Figure 5.10). Thus, MUs in mixed human muscles provide a broad range of functional capability. For example, activities done at various velocities can recruit MUs with the most suitable concentric power–velocity relationship (Figure 5.19). A mixture of fiber types is not as important for generating eccentric power because the different fiber types are similar in eccentric power output (Figure 5.13). Mixed muscles are also important for endurance performance in that the motoneurons of S MUs are designed to be preferentially recruited during relatively low-intensity but

Figure 5.18 Fiber-type differences in the passive and active force–length relationship. Length is normalized to the same number of sarcomeres in series. (A) The active force–length relationship operates over a greater range in length in type 1 fibers because of their longer actin filaments. The greater force of type 2 fibers at optimal length is the result of their greater specific force and possibly greater fiber area. (B) At any given length, passive force is less in type 1 than type 2 fibers, a factor that has been linked to their longer and more compliant titin protein filaments.

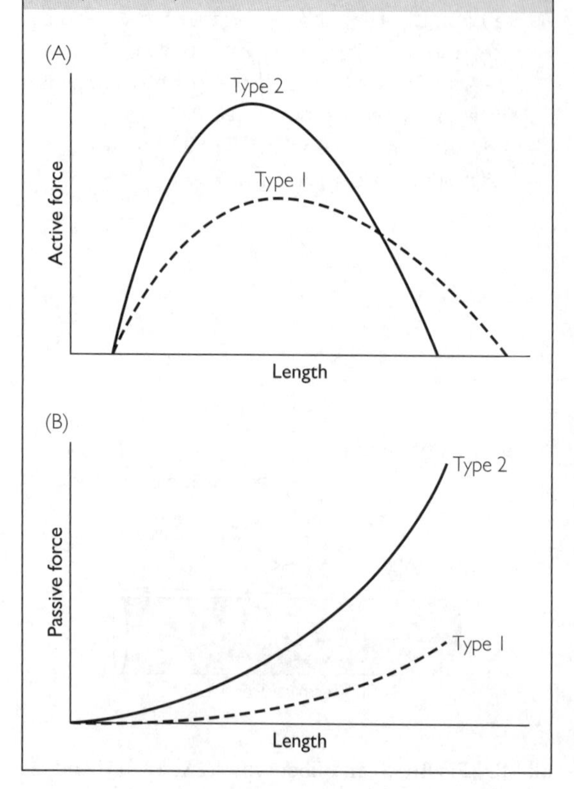

Figure 5.19 Schematic illustration of how a muscle with a mixture of pure fiber types and hybrid fibers can generate concentric power over a large range of shortening velocities because of the variation in the velocity at which peak power occurs.

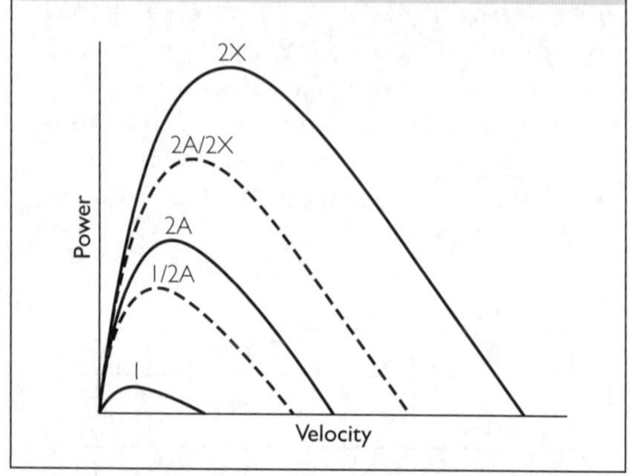

prolonged exercise. If the exercise were to suddenly become more intense (e.g., sprinting up a hill), then FR and FF MUs would be recruited as well to provide the necessary speed and power.

Although human muscles have a mixed distribution of MU and fiber types (Figure 5.20), muscles used for locomotion and maintaining posture possess a greater share of fatigue-resistant S MUs containing mainly type 1 fibers. For example, soleus is about 80% type 1 fibers and may have no 2X fibers, whereas most muscles have equal numbers of type 1 and 2 fibers. Although not shown in Figure 5.20, up to about 30% of the fibers in a muscle could be hybrid fibers (Figure 5.4).

Inter-individual variation in fiber-type distribution
Figure 5.20 shows the *average* values of the percentage of fiber types in several muscles for different groups of subjects but does not indicate the *range* in values among subjects. In a given muscle the range in fiber-type distribution is quite large in sedentary (227) or untrained subjects (232) (Figure 5.21), suggesting a genetic influence on fiber-type distribution. Fiber-type distribution in a large population of untrained subjects exhibits a normal distribution or bell curve (Figure 5.22) that encompasses the extremes seen in power and endurance athletes (Figures 5.22 and 5.23). Thus, the extremes seen in untrained subjects could account for the extremes seen in athletes, both determined by genetic endowment. Can training influence fiber-type distribution? The most common fiber-type conversion (also called transition or shift) observed in longitudinal training studies is conversion from 2X to 2A fibers. There is less evidence for transition from type 1 to 2 fibers or vice versa. It would be unlikely, for example, that an untrained subject with 75% type 1 fibers could achieve 75% type 2 fibers with speed training. However, there is some evidence from cross-sectional studies of athletes suggesting an increased percentage of type 1 fibers in endurance-trained muscles (Figure 5.24), but in cross-sectional studies it is uncertain whether training rather than endowment was responsible for the fiber-type distribution in the trained muscles. The issue of training-induced fiber-type conversion will be considered in detail in Chapter 8.

Figure 5.20 Fiber-type distribution in several human muscles. Typing was done with the histochemical method. Muscles are ordered from left to right according to the percentage of type I fibers. Sol, soleus; Gastr, gastrocnemius; Delt, deltoid; VL, vastus lateralis; LD, latissimus dorsi; Bic, biceps brachii; Tri, triceps brachii; PM, pectoralis major. Data shown are from (121,227,231,256).

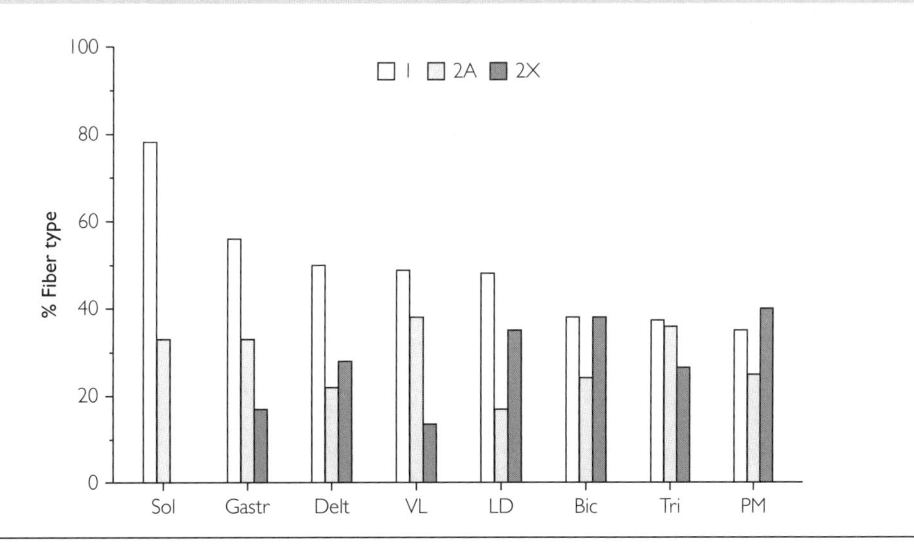

Distinction between fiber-type and motor-unit-type distribution in a muscle If a muscle contains 50% type 1 fibers, is the muscle composed of 50% type 1 MUs? This would be the case if all MUs had the same innervation ratio (number of muscle fibers per

Figure 5.21 Mean (average) and range in fiber-type distribution in vastus lateralis of young women and men. The large number of subjects tested contributed to the large range in distribution. Based on the data of Staron *et al.* (232).

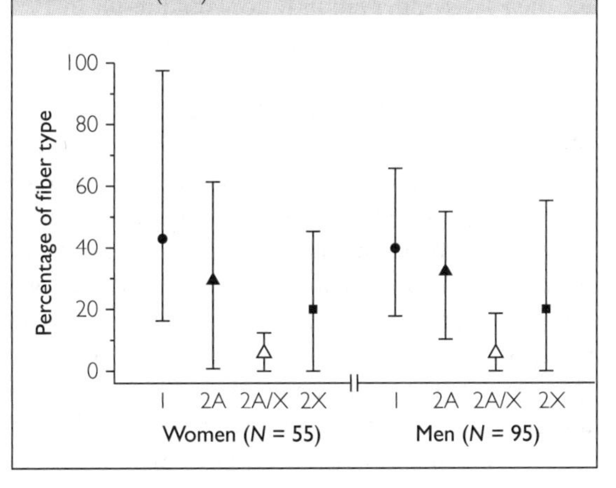

Figure 5.22 The distribution of type I and type 2 fibers in a large population of untrained subjects. The schematic assumes a normal distribution or bell curve and would apply to a muscle in which the average distribution is about 50% type I and 50% type 2 fibers (see Figure 5.20 for examples). The fiber-type distribution observed in elite athletes falls within the range observed in a large population sample (e.g., 227,232).

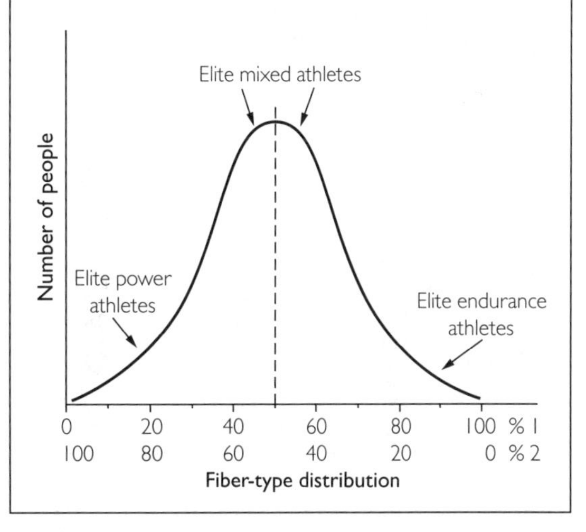

Figure 5.23 Percentage of type I fibers (vastus lateralis) and maximal aerobic power ($\dot{V}O_{2max}$) in runners who compete over distances ranging from 100 to 10 000 m. Longer distances are associated with a higher percentage of type I fibers and a higher maximal aerobic power. Note in particular the two middle distance runners who, although having the same performance, differ slightly in their percentage type I fibers and aerobic power. The runner with a higher percentage of type I fibers and aerobic power is suited for the fastest steady pace but does not have the finishing spurt or kick of the runner with a higher percentage of type 2 fibers. Based on the data of Saltin (215).

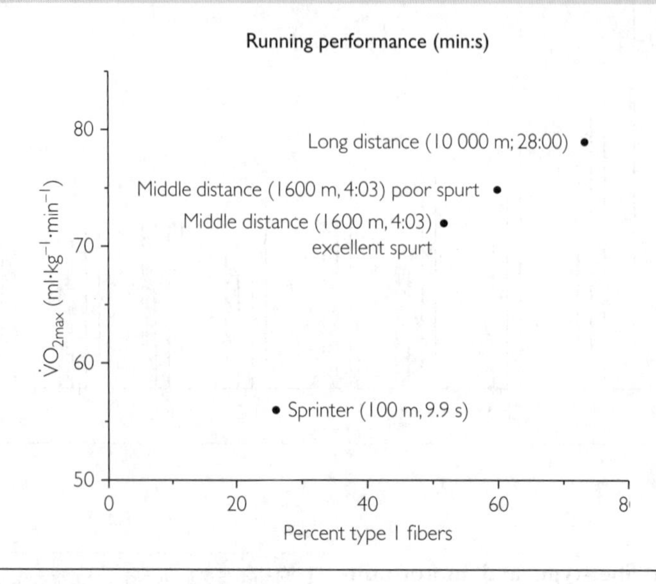

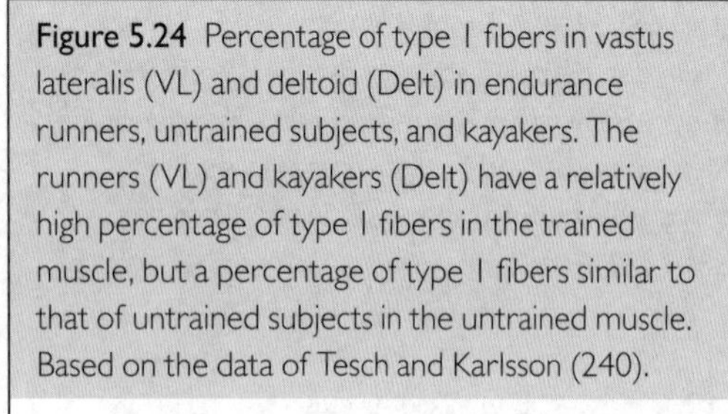

Figure 5.24 Percentage of type I fibers in vastus lateralis (VL) and deltoid (Delt) in endurance runners, untrained subjects, and kayakers. The runners (VL) and kayakers (Delt) have a relatively high percentage of type I fibers in the trained muscle, but a percentage of type I fibers similar to that of untrained subjects in the untrained muscle. Based on the data of Tesch and Karlsson (240).

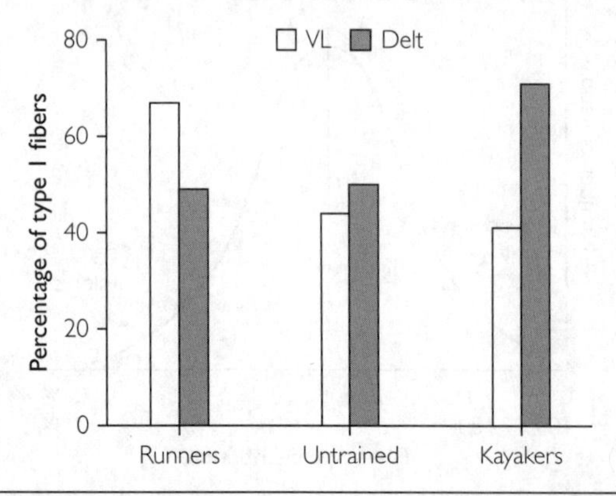

MU). But as covered earlier in the chapter, innervation ratios differ between MU types in the order FF (2X) > FR (2A) > S (1) (Figures 5.5 and 5.6). The difference in innervation ratios across MU types is so large that, for example, if 75% of the MUs in a muscle were type S, this MU type's small innervation ratio would result in the muscle containing only about 35% type 1 fibers. In contrast, if only 5% of the MUs were type FF, their large innervation ratio would result in 20% of the fibers being type 2X. FR MUs might account for 20% of MUs but their 2A fibers represent about 50% of total muscle fibers. The distinction between MU and fiber-type distribution illustrated in Figure 5.25 has implications for performance and training. In performance requiring maximal voluntary effort, failure to recruit the high-threshold FF (2X) MUs may seem trivial if they account for only 5% of the total number of MUs in a muscle. However, if this small number of MUs innervates about 20% of the muscle fibers in the muscle, and if these fibers (2X) have the greatest CSA (Figure 5.6), specific force (Figures 5.6 and 5.7), and power (Figure 5.13), then failure to

Figure 5.25 Comparison of percentage fiber type and percentage MU type in triceps brachii. The total number of MUs and muscle fibers is estimated to be 750 and 300 000, respectively. Based on these estimates, the inset shows the average number of muscle fibers for each MU type (innervation ratio). The percentage of MU types is given horizontally, and the percentage of fiber types is given vertically. For simplicity, the comparison assumes there are no hybrid muscle fibers. Percentages of fiber and MU types from Enoka and Fuglevand (90).

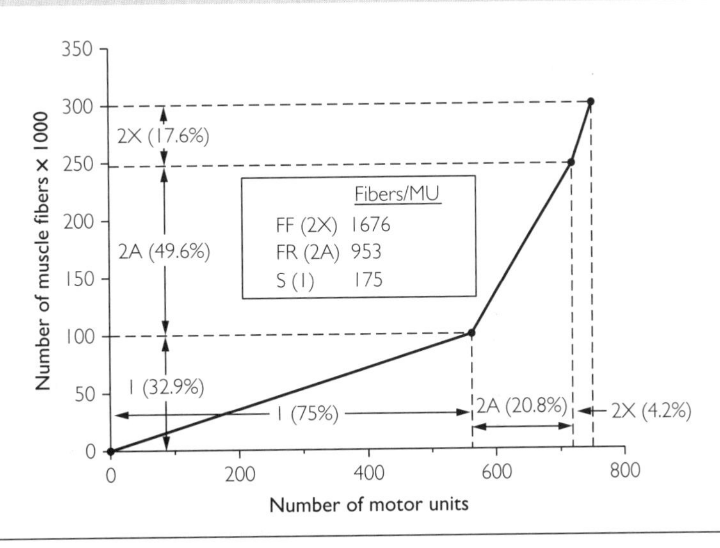

activate these MUs consistently is actually significant. Similarly, when training to increase strength, speed, and power, it would be important to consistently activate the FF MUs.

Gradation of Contraction Force

The nervous system can vary the force of muscle contraction by one or both of increasing or decreasing the number of MUs recruited and increasing or decreasing MU firing rates.

Recruitment

Muscles are composed of several hundred MUs so the most obvious way to vary (grade) contraction force is to recruit a varying number of MUs. MUs are not recruited randomly to increase force but instead are usually recruited in order according to their size in the order S, FR, and FF. For example, in a ramp voluntary isometric contraction in which force is increased from zero to maximal force, the small S (SO, 1) MUs are recruited at low force levels and, as force increases to maximum, FR (FOG, 2A) and finally FF (FG, 2X) MUs are recruited (Figure 5.26). The same pattern occurs when progressively heavier weights are lifted

Figure 5.26 Schematic illustration of increased MU recruitment during a voluntary isometric contraction in which force is increased from zero to maximum. Recruitment of MUs is in order according to the size of the motoneuron soma, indicated by the size of the symbol. Note that S, FR, and FF MUs innervate type 1, 2A, and 2X muscle fibers, respectively. Note also that for simplicity only a small number of MUs is shown; a typical exercise muscle contains several hundred MUs (e.g., Figure 5.25).

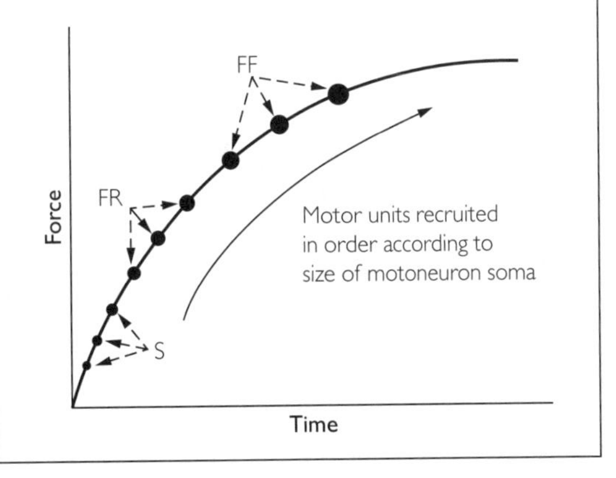

Figure 5.27 Recruitment according to the size principle in weight lifting. A single lift (concentric action) with a light weight (far left) is achieved with the recruitment of S MUs. Lifting a heavier weight (middle) requires recruitment of FR MUs in addition to all of the S MUs. An even heavier weight (right) requires also the recruitment of FF MUs. Note that for simplicity only a small number of MUs is shown; a typical exercise muscle contains several hundred MUs (e.g., Figure 5.25).

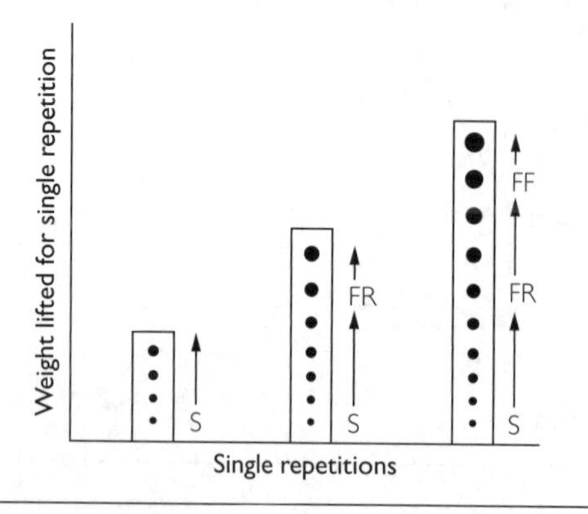

(Figure 5.27) and when exercise intensity is increased during graded exercise such as cycling on an ergometer (Figure 5.28).

Size principle The orderly recruitment of MUs according to size is referred to as the **size principle** of MU recruitment (124). The mechanism of the size principle is directly related to the size of the somata and dendrites of motoneurons. A larger size requires a higher level of excitation to initiate the discharge of nerve impulses (Figure 5.29). MUs are recruited along a continuum from the smallest S MUs to the largest FF MUs (Figures 5.26, 5.27, and 5.29), emphasizing that recruitment is according to size both *within* and *across* MU types (81) (see also Box 5.3).

Recruitment wisdom As intensity of exercise increases, the orderly recruitment of MUs according to size could be considered **recruitment wisdom** because in low-intensity exercise, which is often prolonged and requires endurance, the S MUs composed of fatigue-resistant type 1 fibers will be preferentially recruited. Exercise at progressively higher intensities

Figure 5.28 Recruitment of MUs during exercise with step increases in intensity. Intensity can be expressed as force, load, speed, and power output. Increases in intensity require the recruitment of additional MUs in accordance with the size principle: S > FR > FF. The scheme shown would apply to many types of exercise; for example, step increases in speed or power in running or cycling, respectively. Note that for simplicity only a small number of MUs is shown; a typical exercise muscle contains several hundred MUs (e.g., Figure 5.25).

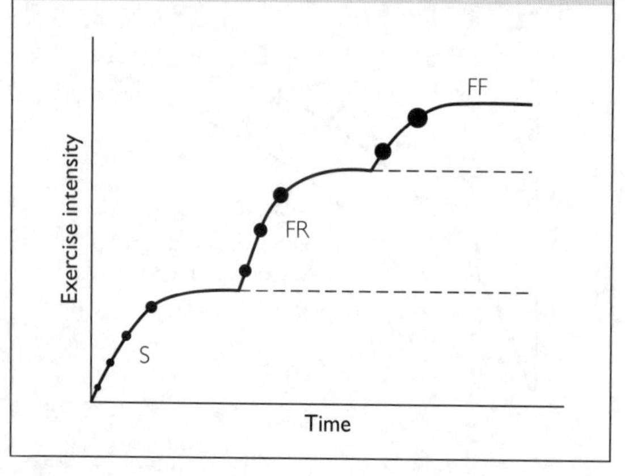

Figure 5.29 Mechanism of the size principle. At left the S MU's smaller motoneuron is easier to excite and therefore preferentially recruited when excitation is relatively small, as would occur with low voluntary effort. At right, the level of excitation has increased and is sufficient to recruit the FF as well as the S MU. The S MU would respond to increased excitation by discharging nerve impulses at a greater frequency, also referred to as an increased firing rate.

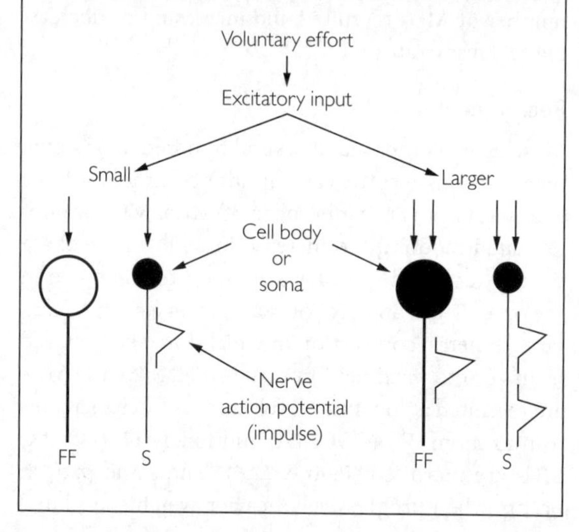

BOX 5.3 DOES THE SIZE PRINCIPLE APPLY ACROSS SYNERGISTS?

Synergists are two or more muscles that perform the same action. The MUs of closely related synergists may be treated by the nervous system as one motor pool (259). For example, soleus and gastrocnemius share a common (Achilles) tendon to plantarflex (extend) the ankle joint and their MUs are activated as a single motor pool (229).

would require recruitment of FR and finally FF MUs in addition to the already recruited S MUs. High-intensity exercise usually calls for some combination of increased force, velocity, and power. This combination is well provided by the 2A and 2X muscle fibers of FR and FF MUs, respectively (Figure 5.30). The percentage of maximum force or intensity at which a MU is recruited is called its **recruitment threshold**. Thus, S MUs are called low-threshold MUs, whereas FR and FF MUs are called medium- and high-threshold MUs, respectively.

In addition to increasing intensity, increased recruitment of MUs can occur if exercise at a low- or moderate-intensity is continued until fatigue requires a maximal effort to complete the exercise; for example, sustaining a submaximal isometric contraction as long as possible (Figure 5.31) or doing as many repetitions as possible with a weight that is less than the maximum single lift (Figure 5.32). The progressive recruitment of MUs is also observed in prolonged, submaximal endurance exercise (Figure 5.33).

Thus, the key factor affecting MU recruitment is the extent of voluntary effort. For an exercise to produce maximal MU recruitment, the exercise must require maximal or near-maximal effort. This is achieved either by brief fatiguing or non-fatiguing exercise at maximal or near-maximal intensity, or by submaximal intensity exercise continued to the point where the

Figure 5.30 Recruitment wisdom and the size principle. The small somata of S MUs generally guarantee that they will be preferentially recruited in low-intensity exercise. This is recruitment wisdom because lower threshold S MUs contain fatigue-resistant type 1 fibers best suited for prolonged exercise. In high-intensity exercise requiring greater force, velocity, and power, the larger and higher-threshold FR and FF MUs, composed of type 2A and 2X fibers, respectively, would be recruited in addition to the S MUs. In the schematic example shown, the low-intensity exercise is a sustained isometric contraction at a low percentage of maximum force. The high-intensity exercise is an isometric contraction increasing to maximum force. Note that for simplicity only a small number of MUs is shown; a typical exercise muscle contains several hundred MUs (e.g., Figure 5.25).

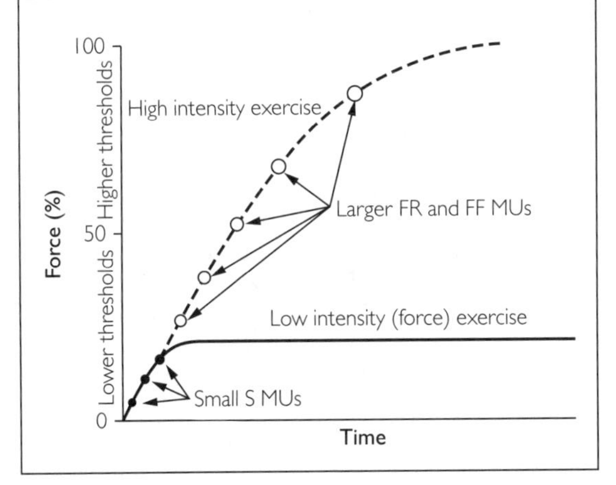

Figure 5.31 Schematic depiction of recruitment of MUs during a brief, relatively non-fatiguing maximal voluntary isometric contraction (100% MVC) and during a submaximal (50% MVC) contraction sustained for about 60 s, at which time maximal effort was required to maintain the 50% MVC force because of increasing fatigue. Note the increasing MU recruitment as fatigue developed in the sustained submaximal contraction, finally attaining the level of recruitment observed with the MVC.

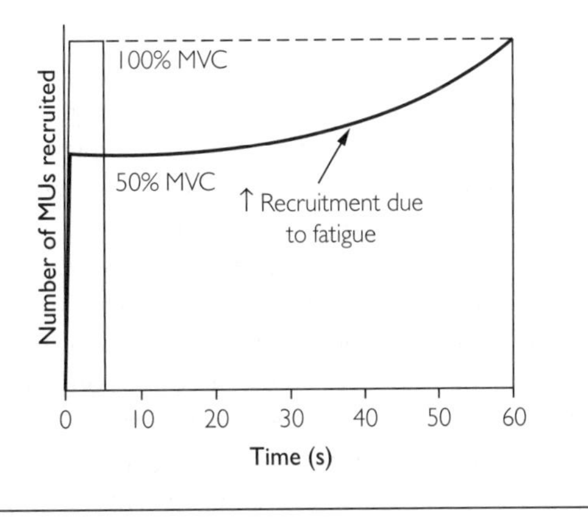

Figure 5.32 Schematic depiction of recruitment of MUs during a single maximal weight lift, called a one-repetition maximum (1 RM), and when doing as many repetitions as possible with a weight that is some percentage of the 1 RM. In the example shown, the selected sub-1 RM weight allowed 15 repetitions to be completed; therefore, this weight would be referred to as the 15-repetition maximum (15 RM). Note that in the 1 RM all possible MUs were recruited. In contrast, not all MUs were recruited in the early repetitions of the 15 RM set, but as fatigue developed more effort was required to complete repetitions. The increasing effort caused more MUs to be recruited, finally attaining the same number of recruited MUs as occurred in the 1 RM.

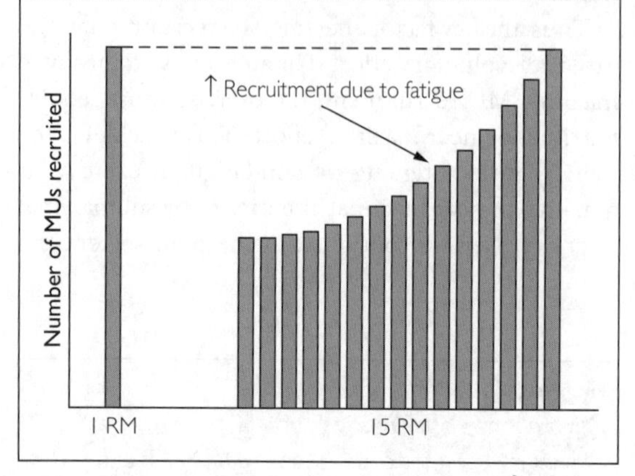

Figure 5.33 MU recruitment, as reflected by glycogen depletion in muscle fibers, during prolonged, fatiguing cycling exercise at 75% of maximal aerobic power ($\dot{V}O_{2max}$). S and FR MUs were recruited at the start of exercise as indicated by glycogen depletion in type 1 and 2A muscle fibers. With increasing fatigue, higher-threshold MUs containing hybrid 2AX and finally 2X fibers were eventually recruited, indicated by the onset of glycogen depletion (vertical arrows and dashed lines). Based on the data of Vollestad *et al.* (250).

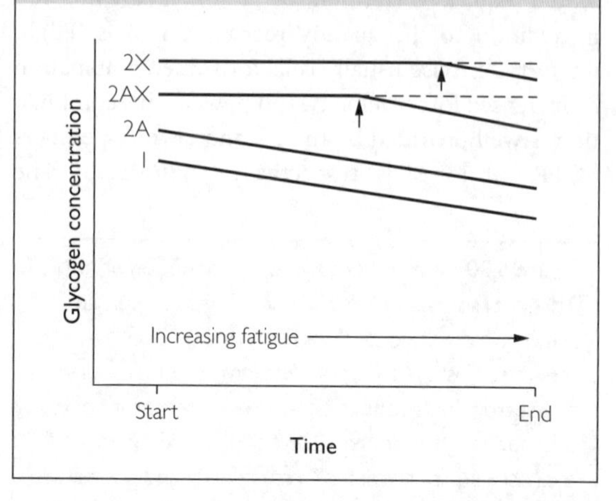

intensity can no longer be maintained. Many athletes and coaches are aware of the role that intensity plays in recruiting MUs and therefore different muscle fiber types; fewer are aware of the role played by fatigue. This issue will be raised again when strength and power training is covered in Chapter 8.

Is the size principle ever violated? If all motoneurons receive the same level or excitation, then the size principle would be in effect according to the size of their somata. To violate the size principle, higher-threshold MUs would have to receive *biased* excitatory input; that is, more than an equal share of excitation (Figure 5.34). MUs receive several types of excitatory and inhibitory input that can act to either reinforce or oppose the size principle, so the tools for violating the size principle exist (123,199). There are conditions in which violation of the size principle might enhance performance. For example, jumping over S MUs to preferentially activate FR and FF MUs might benefit

Figure 5.34 Schematic illustration of how the size principle could be violated. If the S and FF MUs receive homogeneous or *unbiased* excitatory input (left) as indicated by the number of arrows, recruitment according to the size principle is assured. If the excitatory input is *biased* in favor of the FF MU, there may be a reversal of recruitment order (FF first) or the FF MU may be exclusively recruited.

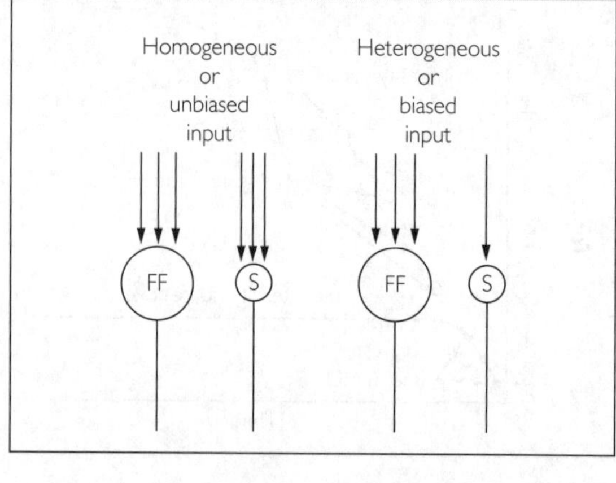

explosive or ballistic actions. In multifunctional muscles, it may be beneficial, when executing a specific sport movement pattern, to recruit FR and FF MUs in one part of the muscle while leaving S MUs unrecruited in another part that would contribute little to the action. Possible violations of the size principle in regard to contraction velocity, contraction type, and movement pattern have been a matter of debate and will be discussed in subsequent sections.

Motor Unit Firing Rate

In addition to recruiting different numbers of MUs, contraction force can be graded by varying the firing rate (also called discharge rate or rate coding) of individual MUs. Firing rate is expressed as **nerve impulses per second** or **hertz (Hz)**, the international unit for frequency. The discharge of each nerve impulse from a motoneuron produces a MAP. As covered in Chapter 4 (for a review, see Figures 4.13–4.15), higher MU firing rates produce stronger muscle fiber contractions according to the force–frequency relationship (Figure 5.35A). By varying firing rate, a MU can vary force over about a 10-fold range, as shown by the force–frequency relationship (Figure 5.35B).

Firing rates of different motor unit types The firing rate needed for summation of contraction would be higher in MUs with shorter twitch contraction times. FF MUs with their shorter twitch contraction times (Figure 5.10) would therefore require higher firing rates than S MUs (Figures 5.36 and 5.37). Do FF MUs actually fire at higher rates than S MUs? The evidence is equivocal (81) but some evidence (114,115,177,186) supports the scheme shown in Figure 5.38.

Synchronous vs asynchronous motor unit firing rates In some experiments with human muscles, electrodes are placed over motor nerves to **synchronously** stimulate the axons of MUs, which is in contrast to the **asynchronous** firing of MUs in voluntary contractions. At relatively low frequencies, asynchronous firing produces smoother and stronger *whole*-muscle contractions than synchronous stimulation (Figure 5.39). With low-frequency synchronous stimulation there are relaxation phases in the contraction, whereas in asynchronous firing of MUs each MU's relaxation phase is filled by the contraction phase of other MU's, producing a smooth, fused contraction of the muscle as a whole (Figure 5.40). Regarding contraction force at low frequencies, the marked relaxation phases between stimuli with synchronous stimulation allow the series-elastic component to become slack. With each subsequent stimulus, some CB cycling is wasted taking up the series-elastic component again,

Figure 5.35 (A) Schematic illustration of a MU including the motoneuron's soma and axon. In the muscle the axon branches to supply the MU's muscle fibers. By varying its firing rate, the MU can vary contraction force. Force can range from that of a single twitch or series of twitches to a high-frequency tetanus. (B) The MU uses the force–frequency relationship to vary contraction force over about a 10-fold range in this example.

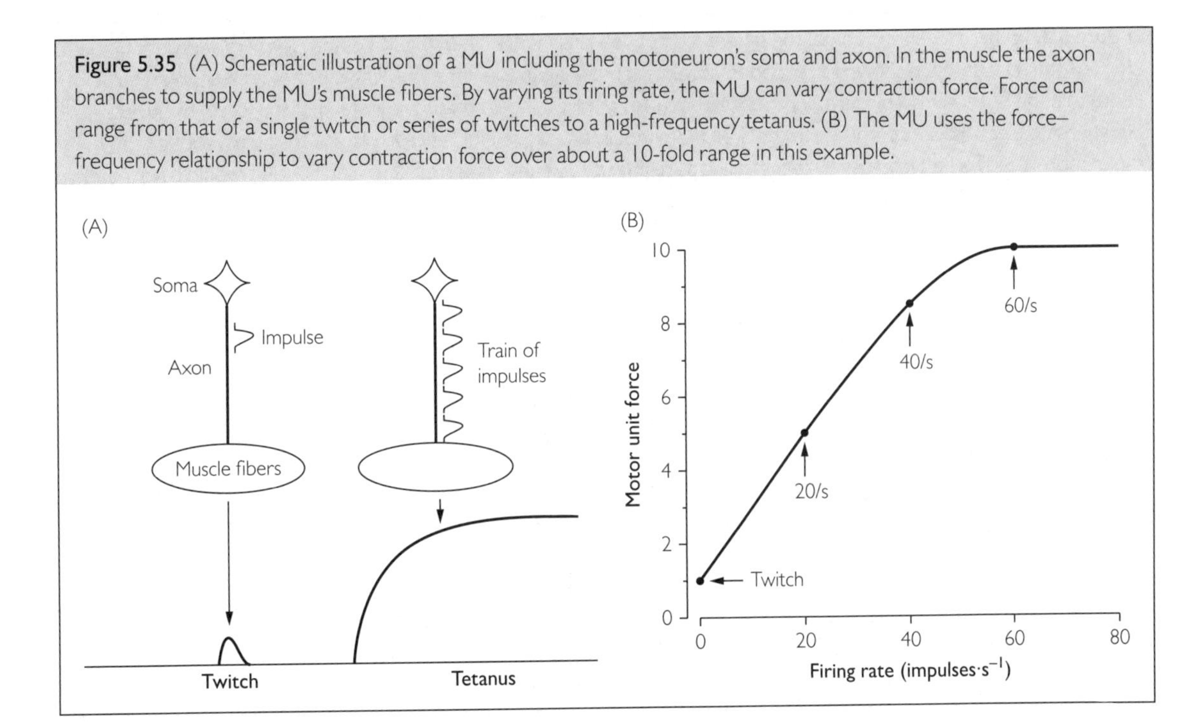

Figure 5.36 Why MUs with shorter twitch contractions need higher firing rates to achieve summation of contraction. (Top) For the S MU the interval between stimuli (vertical arrows) is short enough so there is summation of the long-duration twitch. (Middle) The same interval between stimuli, corresponding to a given MU firing rate, fails to cause summation in the FF MU's short-duration twitch. (Bottom) A shorter interval between stimuli (i.e., a higher MU firing rate) is needed to produce summation in the 2X fiber. The inter-stimulus intervals required for summation of FR (2A) MUs would be intermediate between those required for S and FF MUs.

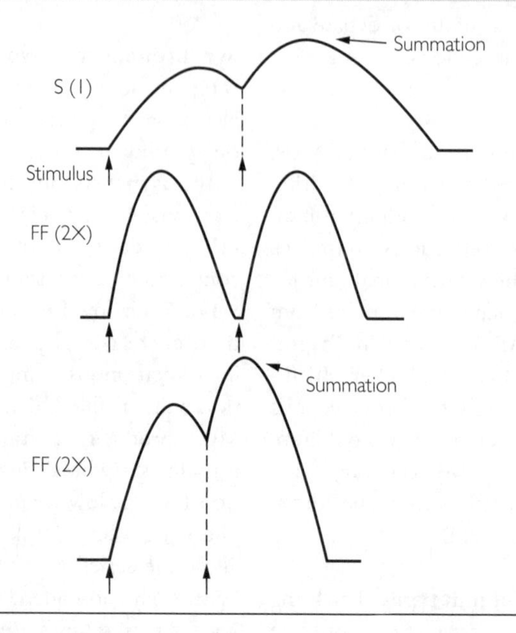

Figure 5.37 Schematic illustration of the difference between FF and S MUs in the force–frequency relationship. On the left the relationship is shown in relation to absolute force. As expected the FF MU generates greater force, but note that the FF requires a higher firing rate (60 impulses·s^{-1} or 60 Hz) to reach the plateau of the force–frequency relationship (see arrows) compared to the S MU (30 impulses·s^{-1}). The difference in plateau frequency (firing rate) is seen more clearly when force level is normalized as relative (%) values, shown on the right. FR MUs would have force–frequency relationships intermediate between FF and S MUs.

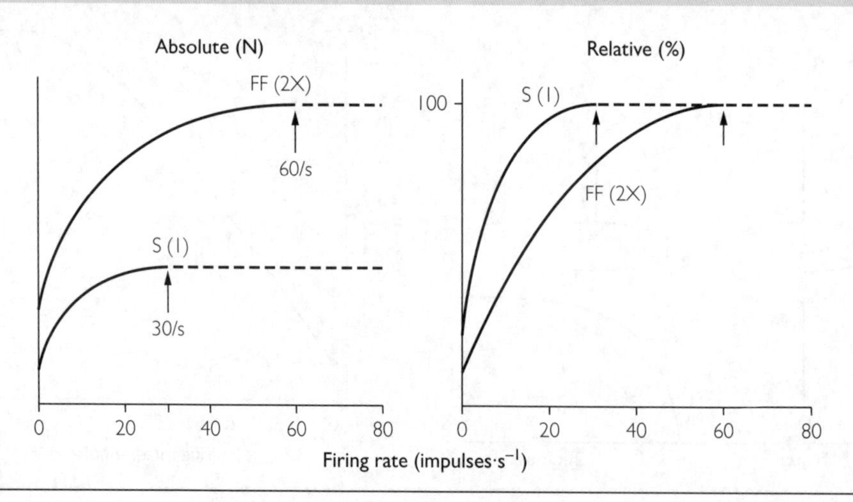

Figure 5.38 Schematic illustration of how MU firing rates would increase as contraction force increased from zero to maximum. In accordance with the size principle of MU recruitment, the S unit is recruited first, followed by the FR and FF MUs. Note that the minimum and maximum firing rates of each type of MU correspond to their twitch contraction times: shorter contraction times require higher firing rates for summation.

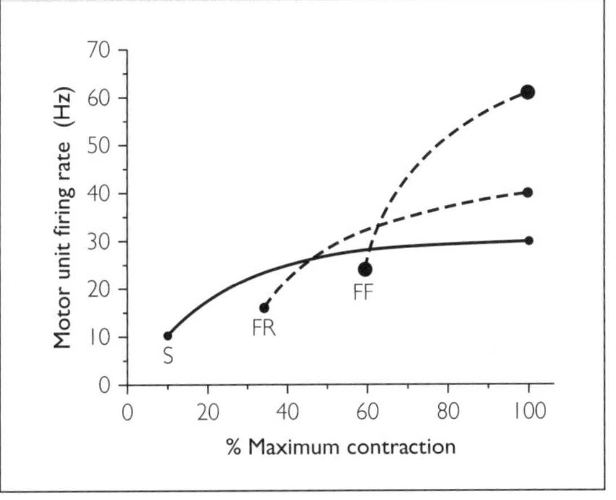

Figure 5.40 Mechanism by which asynchronous firing of MUs at low frequencies produce more fused and thus smoother contractions than synchronous stimulation. MUs A and B are firing at a rate that individually produces unfused contractions. By firing asynchronously the net effect (A + B) of the two MUs is a more fused contraction with smaller relaxation phases. In a typical voluntary contraction 50–100 or more MUs would be firing asynchronously, resulting in the completely fused contraction shown in Figure 5.39.

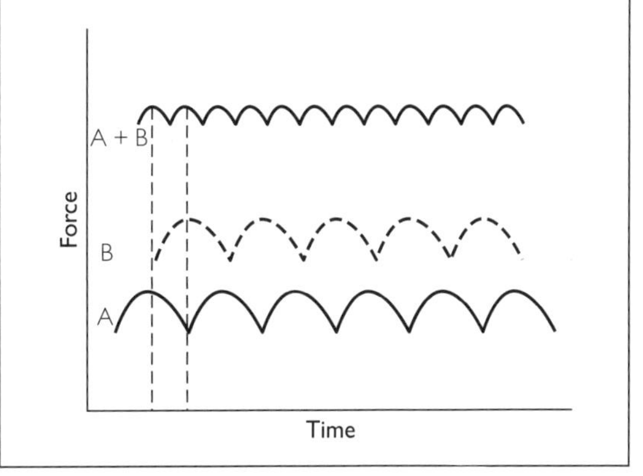

Figure 5.39 Comparison of synchronous versus asynchronous activation of MUs within a muscle. (Left) Stimulating electrodes (Stim.) synchronously activate all of the motor axons supplying the muscle at a frequency of 10 Hz, producing an unfused (i.e., partial relaxation between stimuli) and relatively weak tetanus. (Right) The same number of MUs (axons) firing asynchronously at about the same frequency in a voluntary contraction (Vol.) produce a fused contraction of greater force.

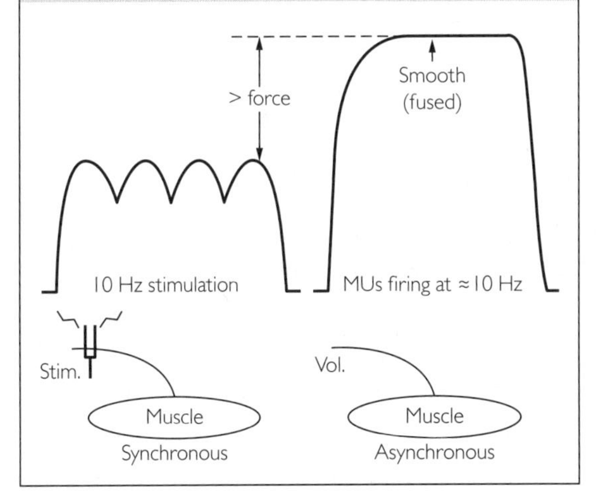

reducing potential force production. In contrast, asynchronous stimulation eliminates the slackening of the series-elastic component, allowing greater force to be developed. At high frequencies, however, contractions are fully fused even with synchronous stimulation, so asynchronous stimulation offers little benefit, as illustrated with their respective force–frequency relationships (Figure 5.41). Note that the force–frequency relationship is shifted to the left with asynchronous vs synchronous stimulation (Figure 5.41) so that a greater percentage of maximum force is attained at a relatively low frequency. This increases efficiency because less CB cycling and hence energy consumption is needed to maintain a given force, thus delaying the development of fatigue.

Coordination of Recruitment and Firing Rate

The relative roles of MU recruitment and firing rates in grading contraction force vary across muscles (Box 5.4). In small hand muscles virtually all MUs are recruited at forces less than 50% of maximum (67,146), indicating that (increasing or decreasing) firing rate is the main if not sole mechanism for grading force at high force levels (170). In large muscles such as biceps (146), deltoid (67), and tibialis anterior (248), MUs

Figure 5.41 The effect of asynchronous and synchronous stimulation on the isometric force–frequency relationship. Asynchronous stimulation corresponds to asynchronous firing of MUs in a voluntary contraction, whereas synchronous stimulation typically occurs when the motor nerve supplying a muscle is stimulated with electrodes (Figure 5.39). Compared to synchronous stimulation, asynchronous stimulation shifts the force–frequency relationship to the left, resulting in higher percentages of maximum force at lower frequencies. Note also that with asynchronous stimulation maximum force is attained at a lower frequency (vertical dashed lines). Based on the data of Lind and Petrofsky (149).

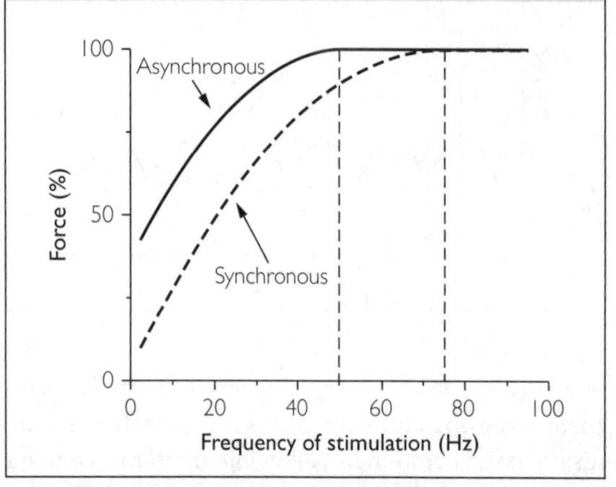

may continue to be recruited until force has reached 80–90% of maximum, but more than 50% of the MUs are recruited at forces less than 50% of maximum. In biceps, for example, 80% of the MUs are recruited at forces less than 50% of maximum (146). In these larger muscles variation in firing rate is predominant in grading forces from 50 to 100% of maximum. In the soleus muscle, however, not only are MUs recruited up to 95% of maximum force, but recruitment is distributed evenly over the whole range of contraction force (186), indicating that recruitment and firing rate play approximately equal roles in grading force over the full force range. The relative roles of MU recruitment and firing rates in different muscles are illustrated in Figure 5.42.

Influence of rate of force development, contraction type, and contraction velocity This discussion and illustration of the relative roles of recruitment and firing rates are based on performance of ramp isometric contractions lasting at least a few seconds in which force rises linearly from zero to maximum or near maximum. However, if the ramp contraction is done quickly to the same force level—that is, with a high RFD–MU recruitment thresholds decrease, indicating that firing rate plays a greater role in force gradation (Figure 5.43). Less is known about the relative roles of recruitment and firing rates in concentric and eccentric contractions.

BOX 5.4 WHAT WOULD ACCOUNT FOR DIFFERENCES AMONG MUSCLES IN THE RELATIVE ROLES OF RECRUITMENT AND FIRING RATES IN GRADING CONTRACTION FORCE?

Differences could be partly related to variation in MU type distribution. The equal roles of recruitment and altered firing rate in soleus (186) may be the result of it being composed almost entirely of S MUs innervating type I muscle fibers. Type I fibers have long twitch contraction times and therefore reach the plateau of their force–frequency relationship at low firing rates (Figure 5.37), leaving little scope for increasing force by increasing firing rate. As a consequence, to increase force, MUs have to be successively recruited throughout the force range. Muscles such as deltoid and biceps contain a mixture of S, FR, and FF MUs. Because of their shorter twitch contraction times, the FR and FF MUs recruited at higher force levels have a greater range of firing rates to increase force (Figure 5.37). In addition, FR and especially FF MUs generate greater force than S MUs (Figures 5.6 and 5.11) so fewer of these MUs have to be recruited at higher force levels to produce a given increment in force. The difference between a small hand muscle and large muscles like deltoid and biceps in the relative roles of recruitment and firing rates may be related to muscle size and complexity of motor control (67). A small hand muscle might contain only approximately 100 MUs; thus, recruitment of each MU causes a relatively large increment in force. This would pose difficulty in making fine force gradations during strong contractions when relatively large MUs would be recruited. Finer gradation would be achieved by altering MU firing rates of already recruited MUs to move up and down the force–frequency relationship (Figures 5.35 and 5.37). In contrast, biceps and deltoid contain approximately 750 and 1000 MUs, respectively. Recruitment of additional MUs at high force levels on top of an already large number of recruited MUs would add only a small increment in force, allowing relatively fine gradations of force to be made. Also, these large muscles would not require the precise control needed for controlling small hand muscles.

Figure 5.42 The relative roles of MU recruitment and firing rates in grading (varying) the force of muscle contraction. In tibialis anterior, deltoid, and biceps, MUs are recruited up to 80–90% of maximal voluntary contraction (MVC) but about 80% of the MUs are recruited by 50% MVC. In soleus, MUs are recruited up to 95% MVC and recruitment thresholds are evenly distributed from 0 to 100% MVC.

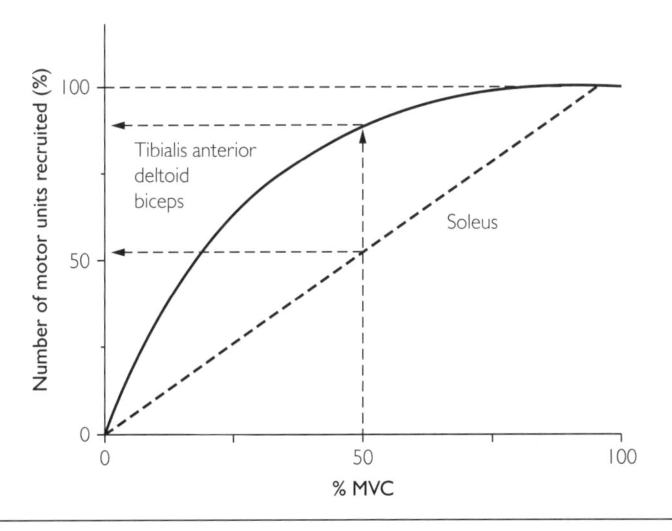

Figure 5.43 Effect of RFD on MU force thresholds. The figure shows isometric contractions increased to the same target force either slowly (slow ramp contraction) or quickly (fast ramp contraction). Note that the two monitored MUs maintained the same recruitment order according to size but both the absolute and relative (percentage of maximum attained force) force thresholds of both MUs decreased in the fast ramp contraction. Based on the data of Desmedt and Godaux (74).

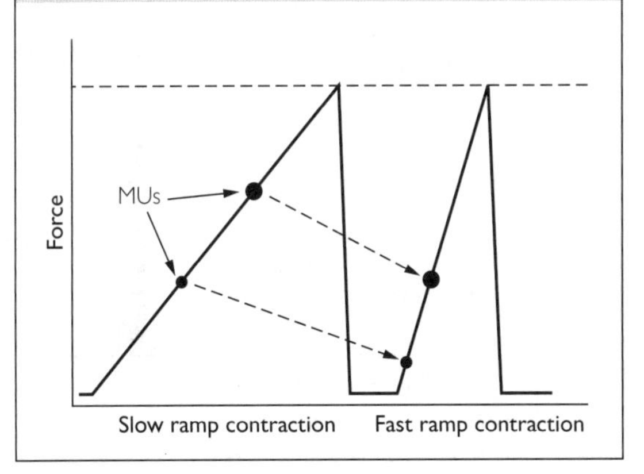

Motor Unit Activation

MU recruitment and altered MU firing rates combine to activate muscles during exercise. The combined effect of recruitment and firing rates is called **motor unit activation**. When an athlete is instructed to perform a maximal voluntary contraction with a muscle (or muscle group), the athlete is attempting to recruit all of the muscle's MUs (Figures 5.26 and 5.27) and have them firing at rates high enough to reach the plateau of their force–frequency relationships (Figures 5.35 and 5.37). If successful, full or complete MU activation, also referred to as 100% MU activation, would be achieved. However, despite maximal effort, an athlete may not be successful in achieving 100% MU activation and specific tests would be needed to measure the degree of success. Therefore it cannot be assumed that a maximal voluntary contraction equals 100% MU activation. The following sections will cover how MU activation can be monitored and quantified, the degree of success in achieving 100% MU activation, and the influence of several factors affecting MU activation, including contraction velocity, contraction type, movement pattern, and muscle length.

Monitoring and Quantifying Motor Unit Activation

Electromyography Electromyography (**EMG**) is the most common method of monitoring MU activation. EMG records the MAPs produced when a MU discharges nerve impulses to its constituent muscle fibers. *Surface* EMG can measure the total amount of MU activation in a muscle (81) whereas *intramuscular* or *subcutaneous* electrodes can monitor the activity of single MUs (Figure 5.44). There are technical challenges in using intramuscular electrodes to record single MU activity (55,81), such as electrode movement especially in concentric and eccentric contractions (68). Consequently, much of what is currently known about individual MU activity has come from recordings during isometric contractions.

Magnetic resonance imaging One application of magnetic resonance imaging (MRI) can detect what muscles or parts of muscles have been active during an exercise and indicate the degree of usage (5,60,239). MRI correlates highly with EMG in assessing the extent of MU activation (5). A difference between the EMG and MRI methods is that EMG recording occurs *during* exercise whereas MRI assessment is made a short time (e.g., 2–3 min) *after* exercise (e.g., Figure 5.45). An advantage of MRI over EMG is that MU activation can be measured in deep muscles that would be inaccessible to EMG, and can more easily indicate what muscles or parts of muscles are activated in different exercises (239). A limitation of MRI is that its application is mainly confined to limb and neck muscles; it would be more difficult to apply it to muscles like pectoralis major and latissimus dorsi.

Metabolic monitoring The extent of MU activation can be determined from the extent of depletion of ATP, PCr, and glycogen from muscle fibers (e.g., 113,138). This method cannot monitor individual MUs but can indicate what type of MU has been active based on the fiber types exhibiting depletion after exercise. Figures 5.33 and 5.46 show examples of metabolic monitoring.

Extent of Motor Unit Activation in Maximal Voluntary Effort

Measurement A true maximum contraction would be one in which all available MUs are recruited and firing at a rate high enough to achieve maximum summation of contraction and therefore maximum

Figure 5.44 Illustration of the two main methods of electromyography (EMG). One method uses surface electrodes on the skin overlying a muscle to record from as many MUs as possible in a muscle (only one MU is shown for simplicity). The second method uses inserted intramuscular (or subcutaneous) electrodes to record activity of individual MUs. A single MU MAP is actually a composite of the MAPs produced by the muscle fibers of the MU.

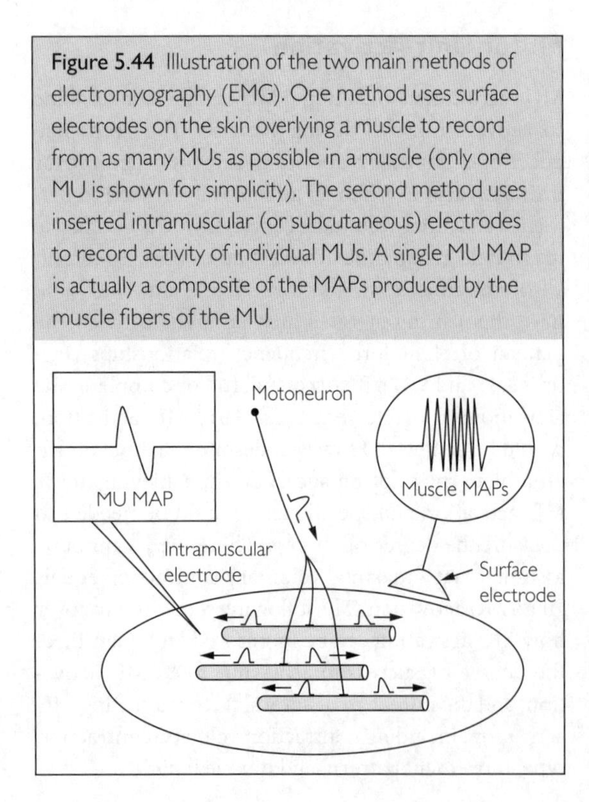

Figure 5.45 Use of magnetic resonance imaging (MRI) to monitor muscle activation of muscles involved in neck extension. Results for three of nine monitored muscles are shown: trapezius (Trap), longissmus capitis and cervicis (LSC), and semispinalis capitis (SEC). The percentage of active muscle CSA is shown before and after neck-extension exercise (five sets of 10 RM). Note the difference in activation of the three muscles. Based on the data of Conley et *al.* (60).

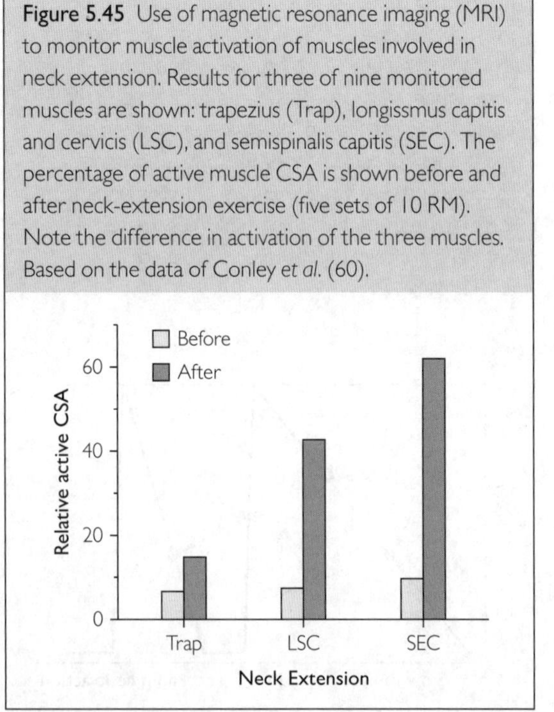

Figure 5.46 Effect of exercise intensity on the activation of type 1, 2A, and 2AX muscle fibers of vastus lateralis. Intensity is expressed as a percentage of MVC of the knee extensors. Seven 1 s contractions were done at each intensity (1 s contract, 1 s rest). Activation (recruitment) of muscle fibers was quantified as the decrease in the phosphocreatine/creatine (PCr/Cr) ratio, indicating that PCr was being depleted. At the lowest intensity (39% MVC) only the type 1 fibers showed a marked decline in the PCr/Cr ratio, in accordance with the size principle of MU recruitment. At the intermediate intensity (72% MVC) type 1 fibers showed a greater decline in the ratio compared to 39% MVC and the type 2A fibers showed a marked decline in the ratio compared to rest. Only at the highest intensity (87% MVC) did the 2AX fibers show a marked decline in the PCr/Cr ratio compared to rest, indicating that these fibers were significantly activated only at the highest intensity. Based on the data of Beltman *et al.* (34).

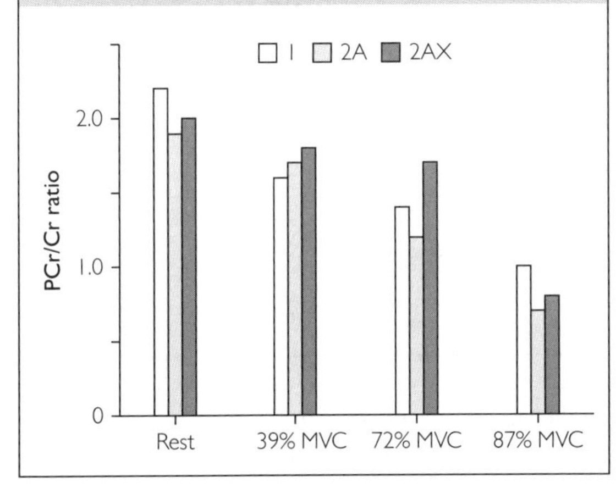

possible force. This would constitute 100% MU activation. How successful are people in achieving 100% MU activation? Various methods of measuring the extent of MU activation up to 100% could be used. EMG might seem to be an obvious method, but while EMG will show the activation associated with a maximal voluntary contraction, it cannot indicate whether or not 100% MU activation has been attained (Figure 5.47). Another apparently straightforward method would be to compare the force of a maximal voluntary contraction with that produced by maximal electrical high-frequency stimulation (Figure 5.48). This method is simple in principle but has limitations. Maximal tetanic stimulation is painful and poorly tolerated. In its application to many muscles, stimulation intense enough to maximally stimulate the muscle(s) of interest may also stimulate other muscles including antagonists. This would complicate the comparison of tetanic force with MVC force. It is also difficult with tetanic stimulation to replicate the pattern of activation in a MVC. Finally, maximal tetanic stimulation could pose the risk of injury such as dislocation of the patella when the knee extensors are stimulated. However, when the intensity of high-frequency (e.g., 50 Hz) tetanic stimuli is increased to the maximum tolerable level, tetanic stimulation has been used to assess MU activation (e.g., 254).

A commonly used method to assess the extent of MU activation is the **interpolated twitch technique (ITT)**. The ITT has been applied using stimulation of motor nerves (32) or less commonly with transcranial magnetic stimulation of the brain (246). The ITT has been used primarily to test the extent of MU activation

Figure 5.47 Schematic illustration of a surface EMG recording of a ramp isometric contraction in which force (left) increases to maximum in a few seconds. (Middle) The raw or unprocessed EMG recording shows that MU activation increased as force increased. (Right) The EMG recording can be processed to permit easier quantification of the EMG. EMG indicates whether MU activation is increasing or decreasing but cannot by itself determine if 100% MU activation has been attained.

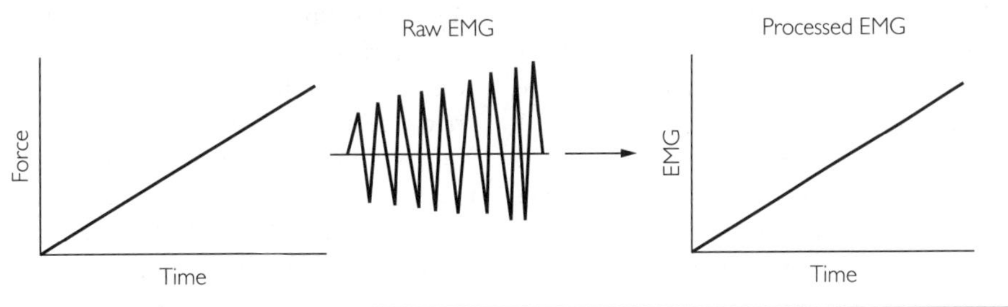

Figure 5.48 Schematic comparison of force produced by a MVC and a maximal electrically evoked isometric tetanus for the measurement of the extent of MU activation. For example, if the MVC force was 80% of the evoked tetanic force, the MU activation would be 80%.

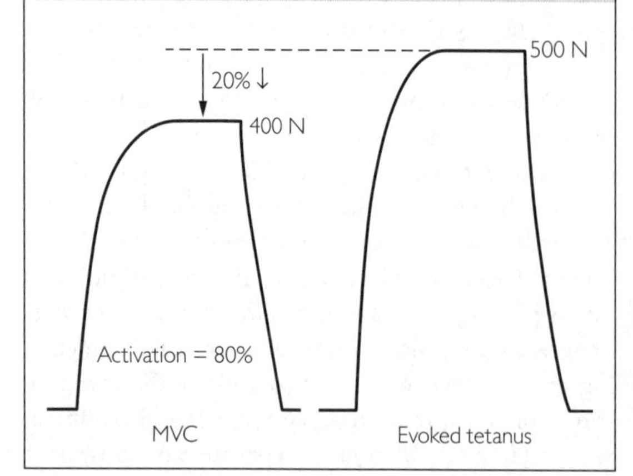

Figure 5.49 Principle of the interpolated twitch technique (ITT). The schematic illustration shows, from left to right, a maximal isometric twitch evoked at rest by a single maximal electrical stimulus (vertical arrow), an isometric MVC during which a second maximal stimulus is delivered, evoking an interpolated twitch, and a maximal isometric twitch evoked at rest after completion of the MVC. The difference between the force of the interpolated and resting twitch is used to determine the extent of activation. A smaller interpolated twitch relative to the resting twitch indicates greater activation (or less inactivation). Note that the resting twitch evoked after the MVC is larger (potentiated) than the twitch evoked before the MVC despite the same level of stimulation. The potentiated twitch is most often used to determine the extent of activation.

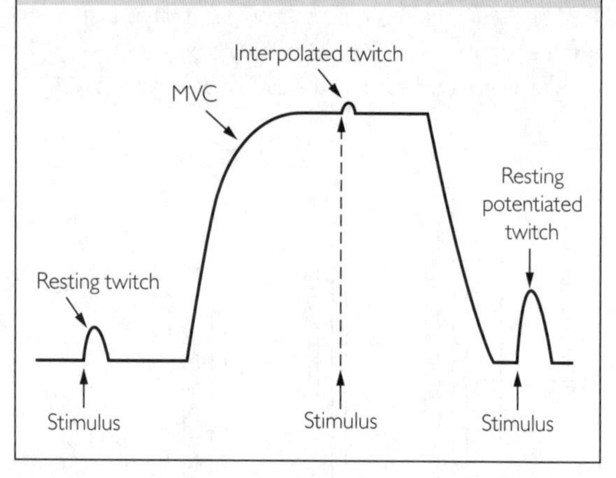

in isometric contractions but has also been applied to concentric and eccentric contractions (15,16,103,181). The ITT compares the force or torque of an electrically evoked maximal resting muscle twitch to that produced by a twitch when it is evoked during a MVC (Figure 5.49). The force of the interpolated twitch, expressed as a percentage of the resting twitch force, indicates the proportion of the muscle not activated in the MVC (96). Since there is evidence that the interpolated twitch itself is potentiated, the potentiated twitch (Figure 5.49) is most commonly used to determine the extent of activation (96).

Despite being simple in principle the ITT's validity and related technical issues have been debated (32, 64,65,66,96,128,144,225,236,237). Nevertheless, the ITT remains a useful method for establishing whether an individual can fully activate a muscle or muscle group. Contrary to the belief that there is a vast reserve of strength that can only rarely if ever be exploited, healthy *untrained* individuals can regularly achieve a high degree of muscle activation (Figure 5.50). It should be noted, however, that the ITT is mainly applied to unilateral, single-joint, isometric, and isovelocity concentric and eccentric actions. It would be difficult to use the ITT to assess percentage MU activation in bilateral, multi-joint training exercises such as an isoinertial (weight-lifting) squat or bench press, nor could the ITT be used to measure percentage MU activation in explosive actions like jumping. It is possible and perhaps likely that the extent of MU activation would be less in these more complex actions. Thus, the application of the ITT to sport-specific movements and training exercises is limited. On the other hand, the ITT applied in single-joint tests may be useful, for example in monitoring percentage MU activation in athletes following injury and during rehabilitation (e.g., 130).

Effect of Contraction Velocity

Motor unit thresholds It was noted that MU thresholds decrease when isometric contractions are done with a greater RFD (see Figure 5.43). In isometric contractions done with very high RFD, called ballistic contractions despite being isometric with no movement, the force thresholds of recruited MUs decrease to zero because all the MUs begin firing and thus their muscle fibers begin producing MAPs before force begins to develop. Stronger ballistic contractions recruit

Figure 5.50 The extent of MU activation (percentage activation) assessed in four muscle groups using the interpolated twitch technique: knee extensors (KE), elbow flexors (EF), ankle plantarflexors (PF), and ankle dorsiflexors (DF). The muscle groups were tested in the same group of subjects. The potentiated twitch evoked after an isometric MVC (see Figure 5.50) was used to determine activation. In addition, two closely placed stimuli (called a doublet) rather than one stimulus was used to evoke the resting twitch. Based on the data of Behm *et al.* (33).

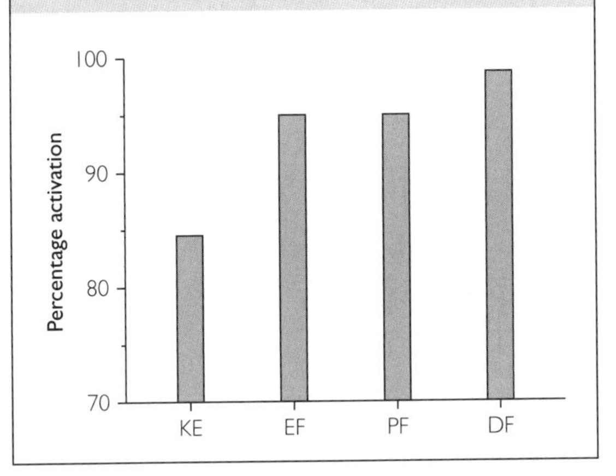

additional MUs but these also have force thresholds of zero (Figure 5.51). This would also apply to ballistic concentric contractions such as throwing (e.g., shot put), jumping, or moving a load as quickly as possible (75). The distinction between the fastest ramp contractions and ballistic contractions is that ballistic contractions are so fast that the MU activation associated with them cannot be adjusted by voluntary or reflex control once the contraction has been initiated (75).

Motor unit firing rates In relatively slow ramp isometric contractions, firing rate increases as force increases (Figure 5.52) to achieve greater summation of contraction according to the force–frequency relationship (Figure 5.53). In contrast, in ballistic *isometric* contractions firing rate is initially high (Figure 5.52) to achieve high RFD (Figure 5.53) and then decreases rapidly (Figure 5.52). Note also that the maximum required firing rate is greater for maximum RFD than force (Figures 5.52 and 5.53). The same pattern has been observed in ballistic *concentric* contractions (75). Higher MU firing rates are needed to attain maximum force and power in concentric contractions, and more so at higher concentric velocities (Figure 5.54).

Figure 5.51 Effect of isometric RFD on MU force thresholds. From the left, slow and fast ramp isometric contractions were done to a target force in the times indicated. Note that the force recruitment thresholds for the two MUs shown (symbols) decreased in the fast ramp contraction, with a greater decrease in the higher threshold MU. The two contractions at the right were ballistic contractions that reached the target force in 0.15 s. The first ballistic contraction was done to the same force as the two ramp contractions. Note that the thresholds of the MUs decreased to zero. The ballistic contraction at far right was done to a higher force. Note that an additional MU was recruited to produce the greater force, but its threshold was also zero. Based on the data of Desmedt and Godaux (74).

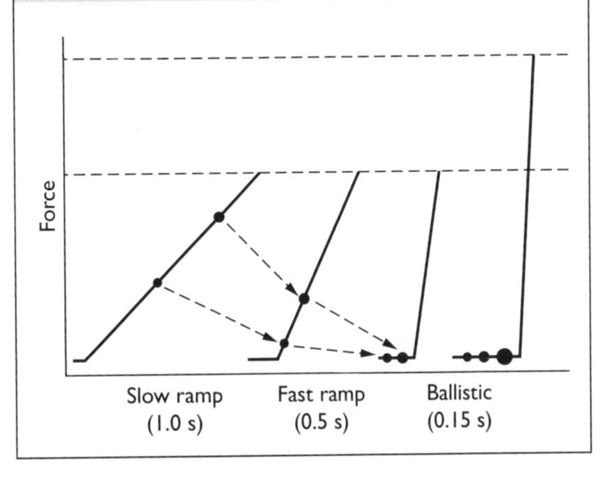

Consequently, the force–frequency relationship is shifted to the right in fast concentric contractions, so that a higher frequency is needed to reach the force plateau (Figure 5.55).

Previously it was shown that higher threshold MUs need higher firing rates because of their muscle fibers' shorter twitch contraction times (Figures 5.36–5.38). There is some evidence that this is true for ballistic as well as slower contractions (74) (Figure 5.56).

Grading the force of ballistic contractions Similar to isometric and slow contractions, both increased recruitment of MUs (Figure 5.51) and increased firing rates (Figure 5.55) are used to increase the force of ballistic contractions.

Is there selective recruitment of fast, high-threshold motor units (type FF) in ballistic

Figure 5.52 Comparison of MU firing rates in an isometric slow ramp contraction reaching the target force in approximately 1 s and an isometric ballistic contraction in which the same target force is reached in approximately 0.1 s. The same MU is recorded in both contractions. (Top) In the slow contraction the MU begins to fire at 10 Hz at the beginning of the contraction and attains 50 Hz as the target force is reached. (Bottom) In the ballistic contraction the MU fires at its highest rate (60 Hz) initially, followed by a rapid decrease. To attain the same target force, the higher peak firing in the ballistic contraction was needed to produce a high RFD. The inset shows how firing rates are calculated from interspike intervals in the MU recording. MU firing rate is calculated as the reciprocal of the time interval between two successive MU MAPs, each called an interspike interval (ISI). Because MUs typically fire trains of nerve impulses, changes in MU firing rate correspond to changes in the ISI between any two successive MAPs. Based on the data of Desmedt and Godaux (74).

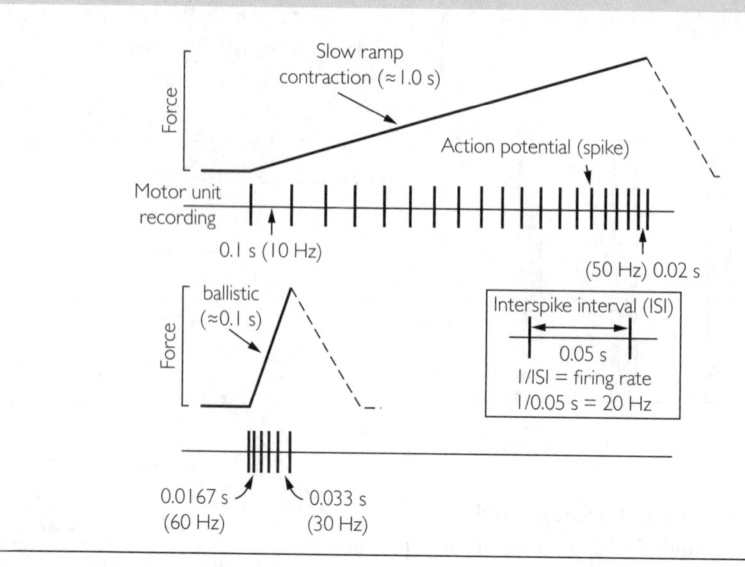

Figure 5.53 The effect of stimulation frequency on isometric force and RFD. Note in this example that twice the frequency (100 vs 50 Hz) was needed to attain maximum RFD compared to maximum force, explaining why MUs must fire at high rates to achieve high RFD. Based on the data of Miller et *al.* (169).

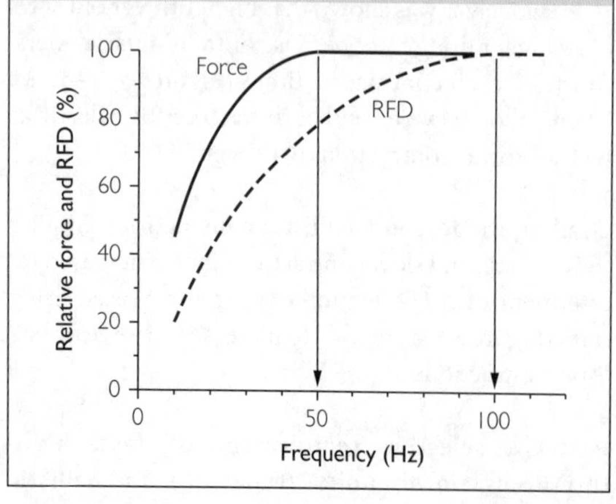

contractions? The concentric force– and power–velocity relationships differ markedly between type 1 and type 2 muscle fibers (Figures 5.11, 5.13, and 5.19). Furthermore, in muscles with a mixture of fiber types, maximum shortening velocity (V_{max}) is determined by the fastest fibers (137). Therefore, the nervous system might be considered wise to alter the size principle-based recruitment order and recruit the fast MUs first or even exclusively when executing fast, explosive actions. Although there is conflicting evidence (74,81,114), the majority of evidence argues against a reversal of recruitment order or selective recruitment of fast MUs (90). Nevertheless, it is worth considering possible advantages and disadvantages of exclusively recruiting fast MUs in explosive actions. One advantage would be the energy saved by not activating MUs that might contribute little if anything to the force and power of an explosive action. In addition, the slow relaxation times of type 1 fibers (Figure 5.9) might make their participation a hindrance

Figure 5.54 Effect of frequency of stimulation on the concentric contraction force– and power–velocity relationships. Velocity was not taken out to V_{max}. (Top) In the force–velocity relationships, isometric force (at zero velocity) was similar at the two frequencies, but as concentric velocity increased, force was greater at the higher frequency. (Bottom) In the corresponding power–velocity relationships, higher frequency produced greater power and peak power occurred at a higher velocity. Based on the data of de Hann (64).

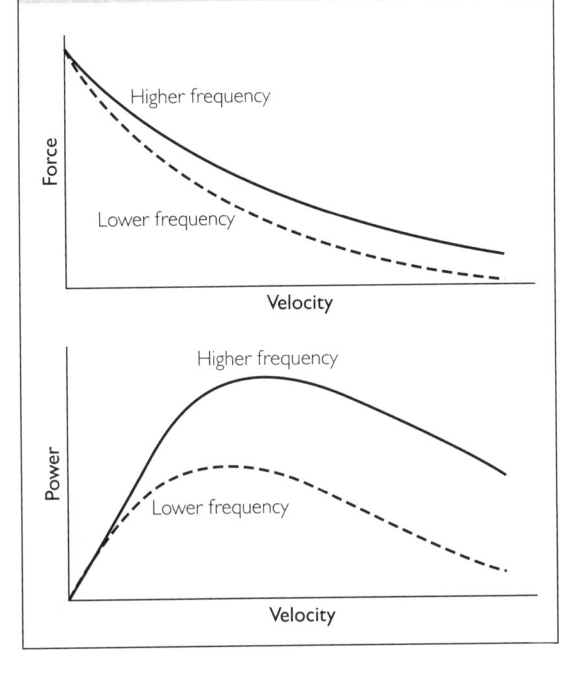

Figure 5.55 The force–frequency relationships for isometric and fast concentric contractions are compared. Force–frequency relationships are derived from a series of contractions at various frequencies. In the fast concentric contraction the plateau of the force–frequency relationship begins at a higher frequency (rightward shift), corresponding to MU firing rates. The lower plateau force of the concentric contraction would be expected based on the force–velocity relationship. Based on the data of de Haan (64).

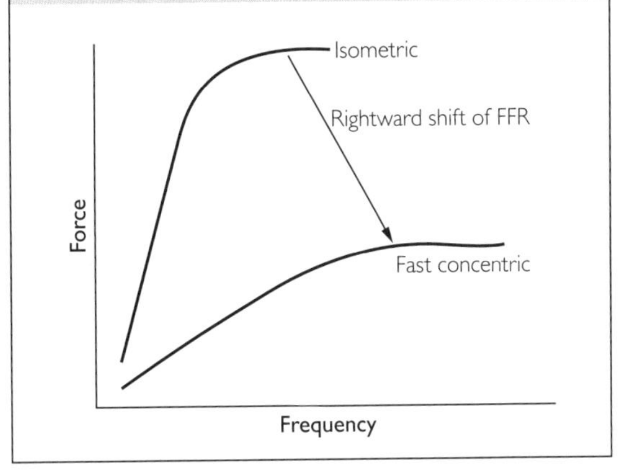

in rapid alternating movements (see Box 5.5). But there would be disadvantages. One is that close to half or more of the muscle fibers in many muscles are type 1 (Figure 5.20) so a large portion of many muscles would be inactive in an explosive action and consequently would be dead weight. This would be a particular disadvantage if the athlete had to accelerate his or her body weight, as in jumping. Also to be considered is that many explosive actions start at zero velocity and then accelerate rapidly. Type 1 fibers could contribute to the beginning of these actions, and it is known that low-threshold, presumably type S, MUs exhibit the same initial high MU firing rates as high-threshold MUs in ballistic contractions, leading to increased RFD (74). This would explain why type 1 fibers are active in a 30 s sprint run, as indicated by glycogen depletion (113). It is also likely that

Figure 5.56 Schematic illustration of firing rates in a low-threshold S and a high-threshold FF MU. In the slow ramp contraction both MUs increase firing rates but the FF MU had higher initial and final firing rates. In the ballistic contraction both MUs begin with firing rates higher than occurred in the ramp contraction and again the FF MU had higher firing rates.

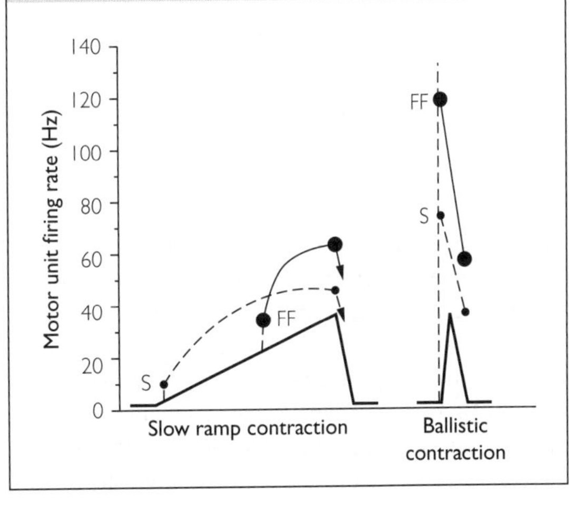

BOX 5.5 SELECTIVE ACTIVATION OF FAST MUSCLES IN A GROUP OF SYNERGISTS

The majority of evidence argues against selective recruitment of fast MUs *within* a muscle in explosive actions. However, one potential advantage of selective fast MU recruitment would be in rapid alternating and cyclic movements, where slowly relaxing type I fibers might hinder performance. In a group of synergistic muscles, there is some evidence that the synergist with the higher percentage of type 2 fibers will be preferentially if not exclusively activated in rapid cycling. In ergometer cycling at a set load, as crank velocity increased to 170 revolutions per minute (rpm), gastrocnemius became progressively more activated whereas its synergist, soleus, became less activated (82). Gastrocnemius' higher percentage of type 2 fibers (Figure 5.20) would allow it to contribute more power, whereas reduced activation of the slower soleus would reduce the interference caused by a slow relaxation phase (Figure 5.9). It is not known whether there would be reduced type I fiber activation in gastrocnemius.

some low-threshold MUs contain, in addition to type 1 fibers, hybrid type 1–2A fibers that could contribute to fast contractions (Figures 5.10 and 5.19). Finally, many explosive actions involve the stretch-shortening cycle and in fact depend on stretch-shortening cycle potentiation (covered in Chapter 4) to enhance the force and power of fast concentric contractions. In contrast to the large differences in concentric force and power between type 1 and 2 muscle fibers, they are similar in producing eccentric contraction force and power (Figures 5.11 and 5.13). Thus, in fast eccentric contractions there would be nothing to be gained and much to be lost if only type 2 fibers were activated. It appears then that the nervous system would be wise *not* to exclusively recruit fast MUs in explosive actions, and the balance of evidence supports this wisdom.

The value of recruiting S motor units in generating power at different velocities At relatively low velocities type 1 fibers can still contribute significantly to power. Cycling is a good example. The optimal pedaling rate for power output is approximately 120 rpm (218). At 60 rpm type 1 fibers could provide up to approximately 70% of the total power; at 120 rpm the contribution would drop to approximately 30% (218). These relative contributions would vary depending on the muscles' fiber-type distribution; that is, the relative numbers of type 1 and 2 fibers. Since slower MUs contain hybrid fibers with a range of force– and power–velocity relationships (Figure 5.19), and pedaling rates of 60–120 rpm are within the range commonly used by cyclists, it would be counterproductive to derecruit S MUs in this range. However, at even higher pedaling rates up to 170 rpm, it might be advantageous to reduce the activation of type 1 fibers because their contribution to power would be very small and their slow relaxation might impede high pedaling rates (see Box 5.5).

Effect of Contraction Type

Since eccentric contractions are stronger than concentric contractions (Figure 5.11), eccentric contractions require less MU activation to produce a given *absolute* force. For example, in doing arm curls there would be greater EMG, indicating greater MU activation, when raising (concentric) than lowering (eccentric) a weight (e.g., 5) (Figure 5.57A). Lifting heavier weights would increase MU activation in both concentric and eccentric contractions, but MU activation would be greater in concentric contraction (37,88). In contrast, concentric and eccentric contractions done at the same *relative* (as a percentage of maximum) force such as maximal contractions (100% of maximum *possible* force) should result in the same MU activation. However, there is often less MU activation in maximal eccentric than in concentric contraction (Figures 5.57B and 5.58) (80,88), indicating greater difficulty in fully activating muscles in eccentric contraction (see Box 5.6). Reduced activation in maximal eccentric contraction reduces maximum possible force, seen when voluntary contractions are compared to those produced by muscle stimulation (15,35,83,254) (Figure 5.58). In addition, maximum eccentric force is greater than maximum isometric force in stimulated contractions but may not be greater in voluntary contractions (15,35,83,254). Reduced MU activation in maximal eccentric contraction results from a combination of reduced MU recruitment and decreased MU firing rates (68).

Is there preferential recruitment of fast motor units in eccentric contraction? Observed selective recruitment of high-threshold, presumably fast MUs in gastrocnemius during eccentric contractions in some task-oriented movements with moderate loads in a

Figure 5.57 Schematic comparison of MU activation (MUA) in concentric and eccentric contraction. (A) Greater EMG, indicating greater MU activation, occurs when a weight is raised with concentric contractions than when it is lowered with eccentric contraction. (B) Maximal concentric and eccentric contractions done at a set velocity on an isovelocity (isokinetic) dynamometer. Lower EMG in eccentric contraction indicates reduced ability to fully activate the muscle compared to concentric contractions.

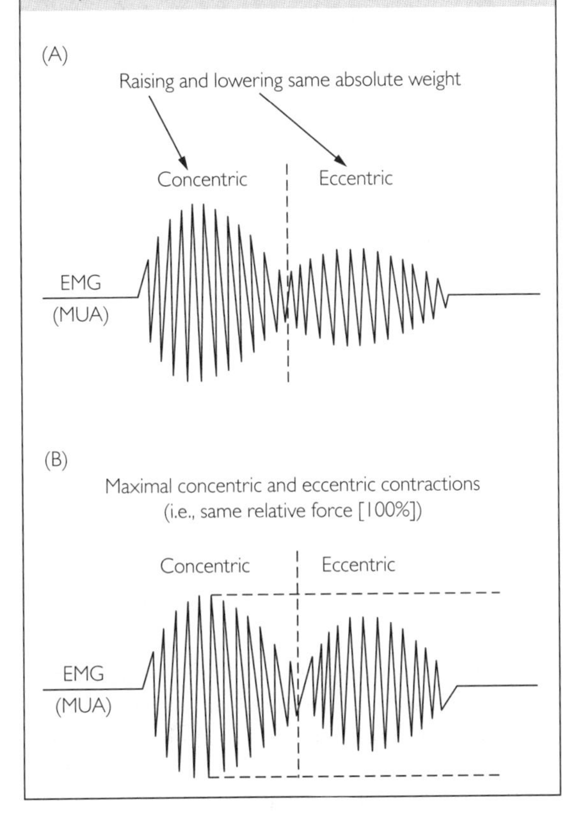

Figure 5.58 Schematic illustration of the expected force–velocity relationship if MU activation (EMG) is similar and complete (100%), and what is usually observed because MU activation is reduced in maximal eccentric contractions compared to maximal isometric and concentric contractions. The expected force–velocity relationship is based on that produced by electrical muscle stimulation.

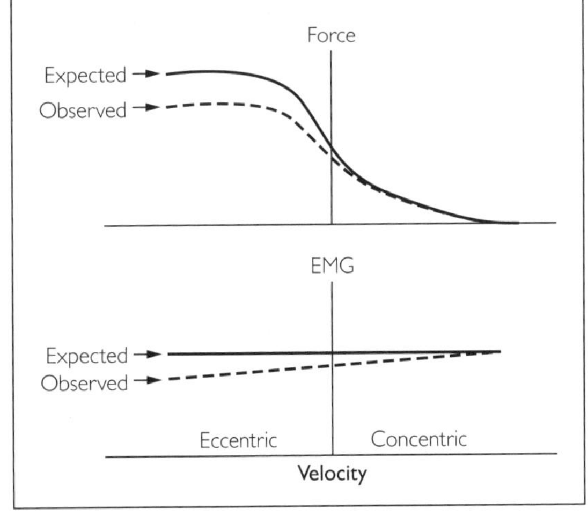

restricted range of velocities (180) has raised the issue of whether selective activation of type 2 muscle fibers occurs generally in eccentric contractions that are part of sport actions and training (55). The majority of studies using single MU recording (55,80,81,90) and a study using the metabolic monitoring method (35) indicate that selective activation of type 2 fibers does not occur in eccentric contractions at a large range of intensities and velocities, including maximal contractions. In fact, in maximal eccentric contractions in which it is difficult to fully activate MUs, it is the type 2 fibers that show less activation than type 1 fibers (35), as would be expected if the size principle were operating. Also, because type 1 fibers are similar to type 2 fibers in producing eccentric force and power (Figures 5.11 and 5.13), it would be a disadvantage if the nervous system jumped over type 1 fibers and exclusively or preferentially activated type 2 fibers, particularly in the many muscles in which up to half the muscle fibers are type 1 (Figure 5.20). Regardless of whether the size principle is upheld or violated in eccentric contractions, there is evidence that higher nerve centers initiate and control eccentric contractions differently from concentric contractions

BOX 5.6 WHY IS IT MORE DIFFICULT TO ACTIVATE MUSCLES IN MAXIMAL ECCENTRIC VS MAXIMAL CONCENTRIC OR ISOMETRIC CONTRACTION?

The mechanism or mechanisms are unknown, but it has been suggested that large eccentric forces could potentially damage muscles and joints; therefore, depressed activation in maximal eccentric contractions could be the result of some protective mechanism (68). Strength training can increase activation in maximal eccentric contractions (1).

(80,88), a fact that has implications for specificity of training (covered in Chapter 8).

Effect of Muscle Length and Strength Curves

Not only must adjustments in MU activation be made for the same force produced by different contraction types because of different maximum possible force (Figure 5.57A), but adjustments must also be made for the same force produced at different muscle lengths because maximum possible force varies according to the force–length relationship. MU activation must be greater at lengths greater and less than optimal length (L_o) to compensate for smaller maximum force (Figure 5.59). Additionally, the small twitch/tetanus ratio at short lengths requires higher MU firing rates to reach the plateau of the force–frequency relationship (Figure 5.60). What applies to the force–length relationship also applies to strength curves (Figure 4.69), which are influenced by both the force–length relationship and changes in muscle moment arm.

Figure 5.60 Illustration of how muscle length affects the force–frequency relationship. Compared to optimal length (L_o), higher frequencies are needed to attain the plateau of the force–frequency relationship at lengths shorter than L_o; that is, the force–frequency relationship is shifted to the right. At lengths greater than L_o, lower frequencies are needed compared to L_o, shifting the force–length relationship to the left. Note that the twitch force (force at frequency 0) is also affected by muscle length. Twitch force is a lower percentage of maximum tetanic force at $< L_o$; that is, the twitch/tetanus ratio is smaller. Note that force is normalized; maximum force is set to 100% for the three lengths tested. *Absolute* tetanic force would be greater at L_o.

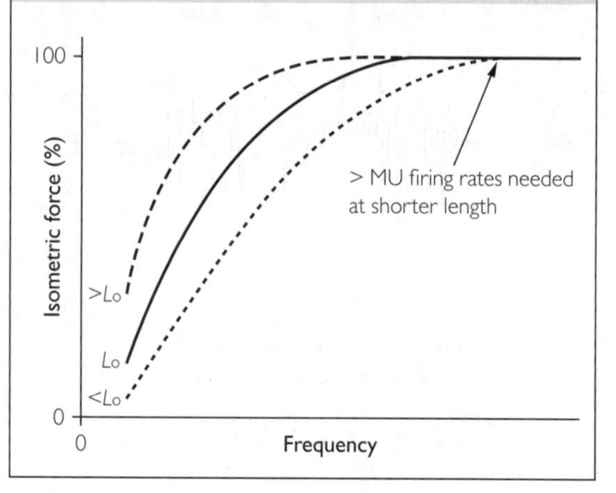

Figure 5.59 Effect of the isometric force–length relationship on the MU activation needed to produce a given force. (Top) A typical force–length relationship. The dashed horizontal line indicates particular muscle lengths less than and greater than L_o at which maximal force is the same. (Bottom) MU activation for a given force (dashed line) should be greater at lengths $< L_o$ and $> L_o$ than at L_o because a higher percentage of maximum force must be produced.

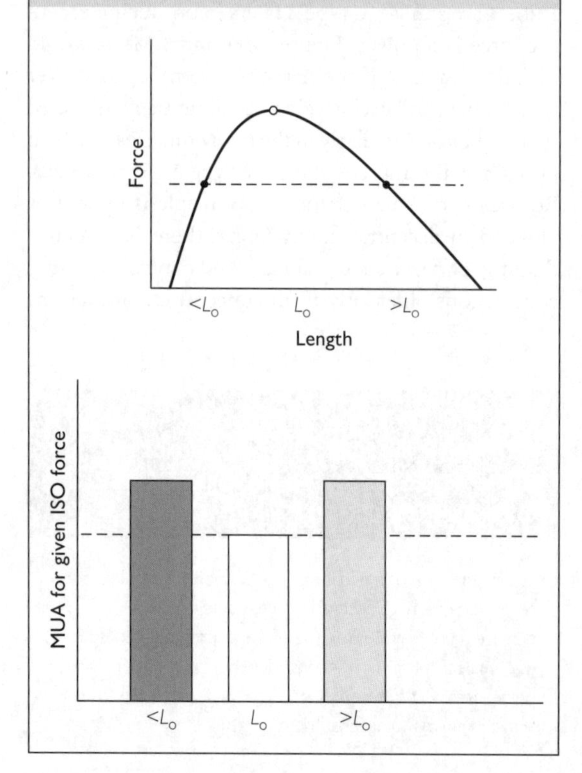

The higher MU firing rates needed to reach the plateau of the force–frequency relationship at short muscle lengths might suggest greater difficulty in achieving 100% MU activation, but evidence is mixed on this point (11,17,30,73,104). The mixed evidence may be that, in addition to the force–frequency relationship, variations in sensory feedback could affect MU activation at different muscle lengths/joint angles, for example reducing activation at short lengths because force capacity is too small (139) or reducing MU activation at long lengths to protect muscles and ligaments (130). Most research on the effect of muscle length on MU activation has involved unilateral, single joint actions; however, a study of the bilateral, multi-joint, isometric leg press at different knee joint angles showed a marked variation in MU activation in several muscles (118). These results emphasize the importance of specificity in strength testing (Chapter 11) and strength training (Chapter 8).

Effect of Movement Pattern

Many muscles are multi-functional in that they can perform more than one movement at a joint. For example, pectoralis major can contribute to shoulder joint flexion and extension, horizontal flexion, and adduction. Some MUs within pectoralis major and entire parts of pectoralis major may be better suited to contribute to a particular joint movement. Pectoralis major has two parts, one that arises from the clavicle (clavicular portion) and one that arises from the sternum (sternal portion). The clavicular portion is relatively more active in flexion whereas the sternal portion is more active in extension and adduction. Both portions are similarly active in horizontal flexion. These differences can be seen in common weight-training exercises (Figure 5.61). In the same exercises the anterior deltoid (composed of anterior, medial, and posterior heads) was most active in the horizontal and decline bench-press exercises (25). In other muscles with multiple parts or heads, such as quadriceps (each of the four heads is given its own name) and triceps brachii, there is varied activation of the heads depending on the movement executed (239).

Selective recruitment of MUs even occurs within one part or head of a muscle. For example, MUs laterally located in the long head of biceps are preferentially recruited in elbow flexion but not supination, whereas medially located MUs are preferentially recruited in supination (238). Thus, there are task groups of MUs (150) that receive input from higher nerve centers to execute specific tasks. Is the size principle violated when higher threshold MUs (FR or FF) are recruited in one task group while low-threshold MUs (S) in another task group are left unrecruited? The answer seems to be that the size principle operates *within* a task group but may be violated *across* task groups (Figure 5.62). In large muscles with distinct parts, such as pectoralis

Figure 5.61 Activation of sternal and clavicular portions of pectoralis major (PM) in decline, horizontal and incline bench-press exercises and the overhead (vertical) press. The maximum observed activation of sternal and clavicular portions is set to 100%. The relative activation of the two portions of pectoralis major varies markedly in the different exercises. Based on the data of Barnett *et al.* (25).

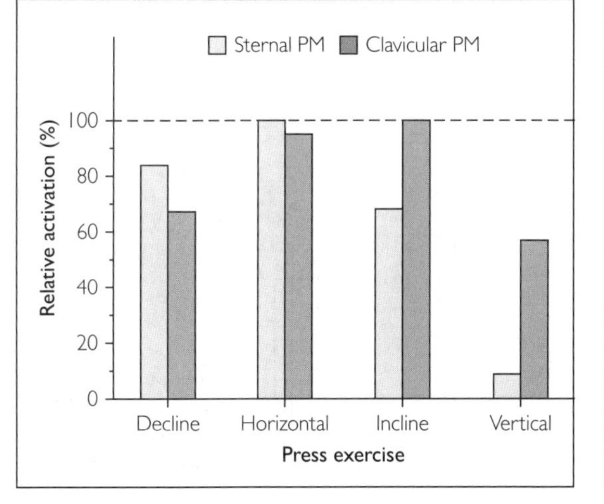

Figure 5.62 The size principle within and across MU task groups. The example is MU recruitment in the long head of biceps. (Top) In elbow flexion the laterally located flexion group of low- and high-threshold MUs is recruited (filled symbols) according to the size principle as force increases but both *low-* (arrow) and high-threshold MUs in the medially located supination group are inactive (unfilled symbols), an apparent violation of the size principle *across* task groups. (Bottom) In forearm supination, the pattern is reversed. Based on the data of ter Haar Romeny *et al.* (238).

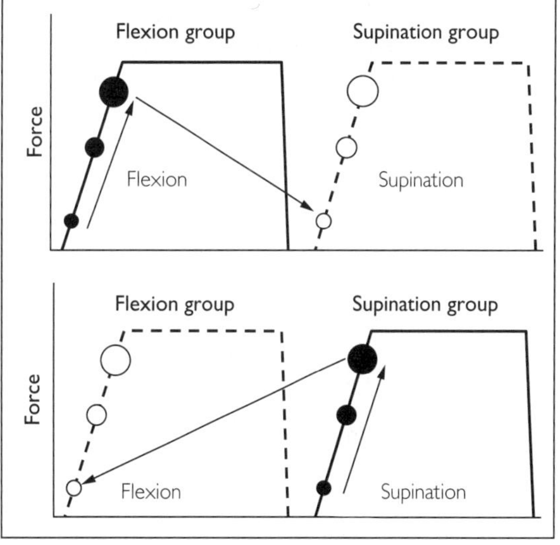

major and deltoid, whole populations of low- and high-threshold MUs will be activated in one joint action while another population is almost silent.

Neuromuscular Fatigue

Definition of Fatigue

Fatigue can be defined as an exercise-induced decrease in maximum force and/or maximum power-generating capacity (Figure 5.63). There is no power developed in isometric contractions because velocity and displacement are zero, but with fatigue rate of force development and relaxation decrease along with force. Both force and velocity can decrease in concentric contractions, thus decreasing power (8). Fatigue also increases the curvature of the force–velocity relationship (134), which will be discussed in more detail below. In eccentric contractions such as lowering a weight, fatigue increases rather than decreases velocity because the weight becomes a higher percentage of maximum eccentric force. The increased velocity attenuates the decline in eccentric power as force declines.

Figure 5.63 Effect of fatigue on isometric, concentric, and eccentric contractions. No power is generated in isometric contractions because no work is done (displacement = 0) and velocity is zero, but fatigue decreases rate of force development (RFD) and rate of force relaxation (RFR) along with force. Both force and velocity decrease in concentric contractions, leading to reduced power. Repeated eccentric contractions done on an isovelocity device result in decreased force and power but velocity remains constant. In eccentric contractions done to lower a weight at a controlled velocity, fatigue would actually cause an increase in lengthening velocity that, paradoxically, would tend to increase power, attenuating the decrease in eccentric power.

Isometric: ↓ force + ↓ RFD + ↓ RFR

Concentric: ↓ force + ↓ velocity

↓ power

Eccentric: ↓ force + ↕ velocity?

↓ power

Fatigue vs Fatigue Failure

Fatigue is sometimes defined as the failure to continue exercising at a set exercise intensity but the fatigue *process* is usually underway before failure occurs. For example, as many repetitions as possible are done with a weight equivalent to 80% of the one-repetition maximum (1 RM). Ten repetitions were completed but there was failure to complete the eleventh repetition. The fatigue process, in this case the decrease in maximal concentric force-generating capacity, started at the beginning of exercise and continued until concentric force fell below the level needed to lift the weight, at which point failure to complete the eleventh repetition occurred (Figure 5.64). Thus, a distinction should be made between the fatigue process and one of its consequences, the failure to continue at a set intensity (102). Examples of fatigue and fatigue failure in isometric and isokinetic concentric exercise are shown in Figures 5.65 and 5.66, respectively.

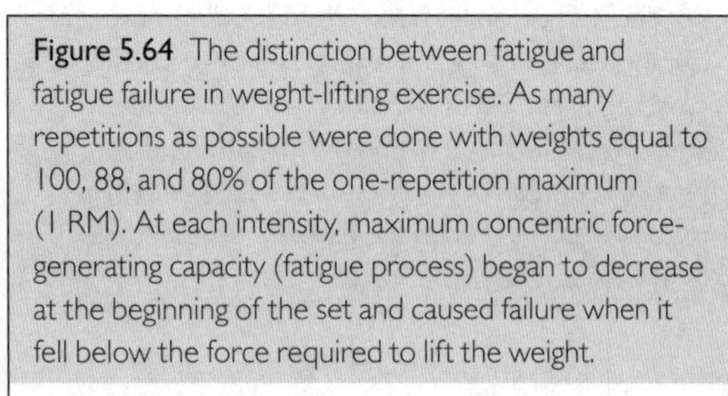

Figure 5.64 The distinction between fatigue and fatigue failure in weight-lifting exercise. As many repetitions as possible were done with weights equal to 100, 88, and 80% of the one-repetition maximum (1 RM). At each intensity, maximum concentric force-generating capacity (fatigue process) began to decrease at the beginning of the set and caused failure when it fell below the force required to lift the weight.

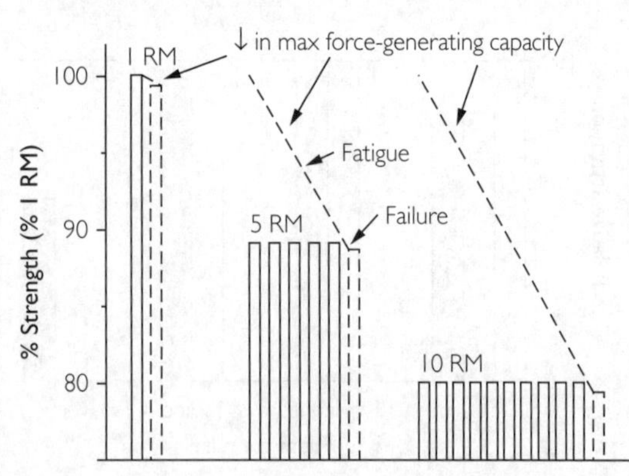

Figure 5.65 The distinction between the fatigue *process* and fatigue *failure* in isometric exercise. (Left) An isometric MVC is continued with maximal effort for 1 minute, after which it is stopped. Maximum force generating capacity starts declining almost immediately into the contraction and falls to 50% of the initial maximum force after 1 minute. In this type of contraction the decline in force directly monitors the fatigue process. (Right) A contraction at 50% of maximal force can be sustained for 1 minute before maximum force-generating capacity (dashed line) falls below the 50% level, at which point failure occurs. Note that although fatigue failure did not occur until 1 minute into the contraction, the fatigue process began as soon as the contraction started.

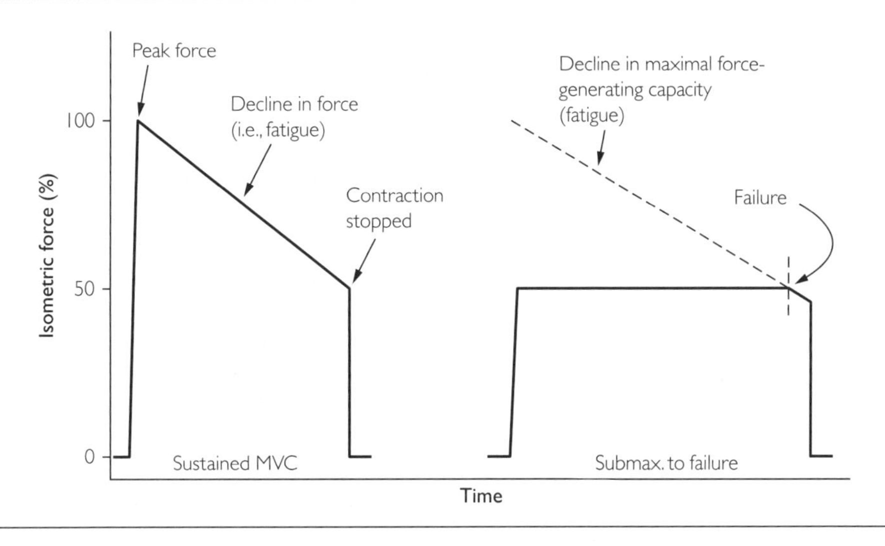

Figure 5.66 Fatigue in isokinetic exercise. An athlete does a knee-extension fatigue test consisting of 50 maximal concentric contractions at a set velocity of 3.14 rad·s^{-1} (180°·s^{-1}) on an isokinetic dynamometer. Note the progressive decline in maximum force- and power-generating capacity (fatigue) during the 50 contractions.

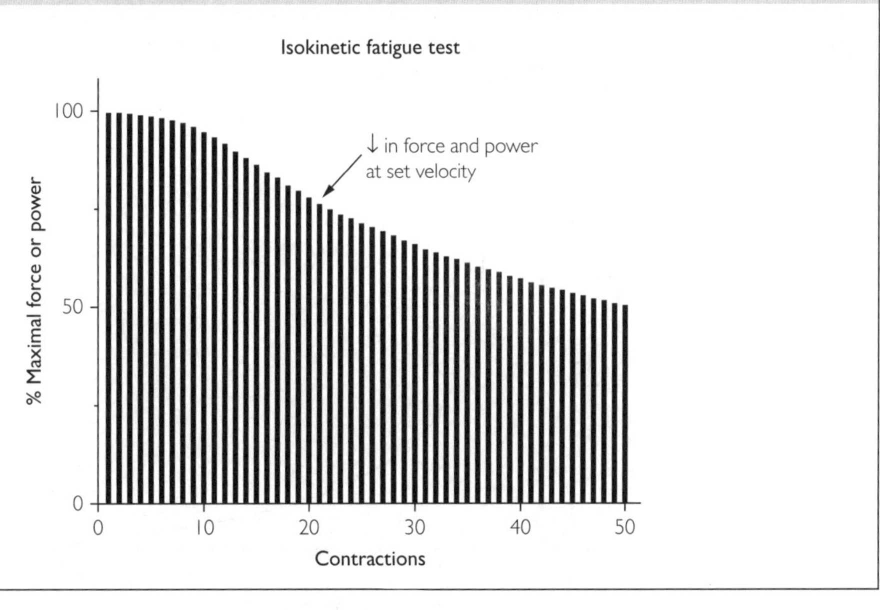

Exercise Intensity and Time to Fatigue Failure

Regardless of how exercise intensity is expressed, as intensity increases fatigue develops more rapidly and failure occurs sooner (Figures 5.67, 5.68, and 5.69). At each relative intensity there is a large range in time to failure (endurance) due to differences in fiber-type distribution and other factors. For example, in a group of untrained subjects, the average isometric knee-extension endurance at 50% MVC was approximately 1 minute (Figure 5.67), but the range was 38–102 seconds (162). In athletes the type of training used could also affect endurance.

Mechanisms of Fatigue

Fatigue can occur at any point between the voluntary command issued from the brain to contract a muscle down to the level of the myosin CB. Thus, many changes brought about by exercise could contribute to fatigue. As a result scientists in many different disciplines (e.g., physiology, biochemistry, psychology) have conducted research on fatigue (4). Regardless of the varied mechanisms that contribute to fatigue, they ultimately cause fatigue by affecting myosin CB function (check Chapter 4 for a review of the CB cycle). Since muscle force and power depend on the number of cycling CBs, the force per CB, the rate of activation and deactivation of CB cycling, and the rate of CB cycling, these aspects of CB function are the bottom lines of fatigue (Figure 5.70). Figure 5.71 illustrates how three of these could affect isometric force and concentric force and power.

Overview of Sites and Mechanisms of Fatigue

There are many possible sites and mechanisms of fatigue (Figure 5.72). In the brain there could be decreased volitional drive to the motor cortex. The motor cortex would then decrease excitation of the motoneurons, which would result in reduced MU recruitment (MU drop-out) and decreased MU firing rates. In addition, the motoneurons might become less excitable, further increasing the possibility of MU drop-out and decreased firing rates. There could also be reflex inhibition or reduced reflex excitation that would also act to cause MU drop-out and decreased MU firing rates. There may be neuromuscular transmission failure that would cause muscle fiber drop-out. Excitation–contraction coupling might be

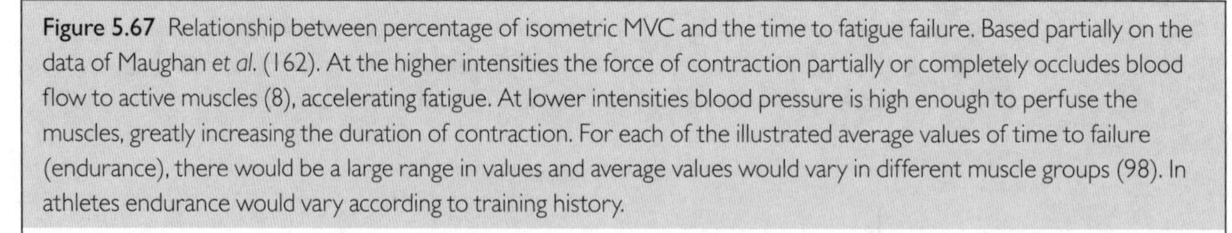

Figure 5.67 Relationship between percentage of isometric MVC and the time to fatigue failure. Based partially on the data of Maughan *et al.* (162). At the higher intensities the force of contraction partially or completely occludes blood flow to active muscles (8), accelerating fatigue. At lower intensities blood pressure is high enough to perfuse the muscles, greatly increasing the duration of contraction. For each of the illustrated average values of time to failure (endurance), there would be a large range in values and average values would vary in different muscle groups (98). In athletes endurance would vary according to training history.

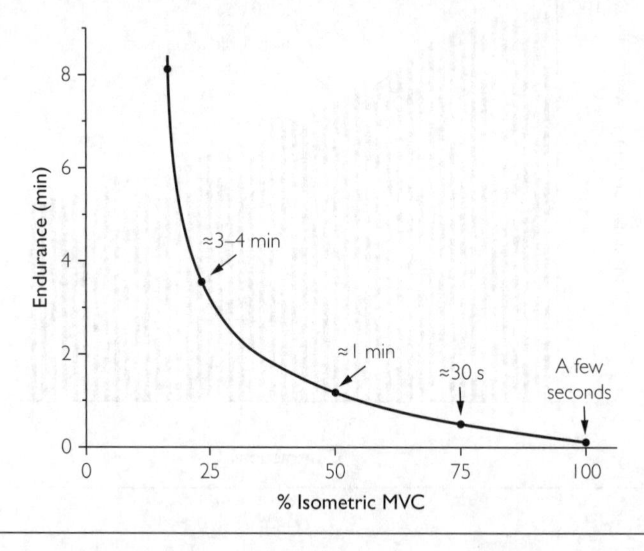

Figure 5.68 Relationship between percentage of the 1 RM and the number of repetitions completed in the bench-press exercise. The illustrated average values of repetitions to failure are from a group of male university physical education students (at McMaster University). For each of the illustrated average values of repetitions to failure (endurance), there would be a large range in values and average values would vary in different exercises (muscle groups). In athletes, endurance would vary according to training history.

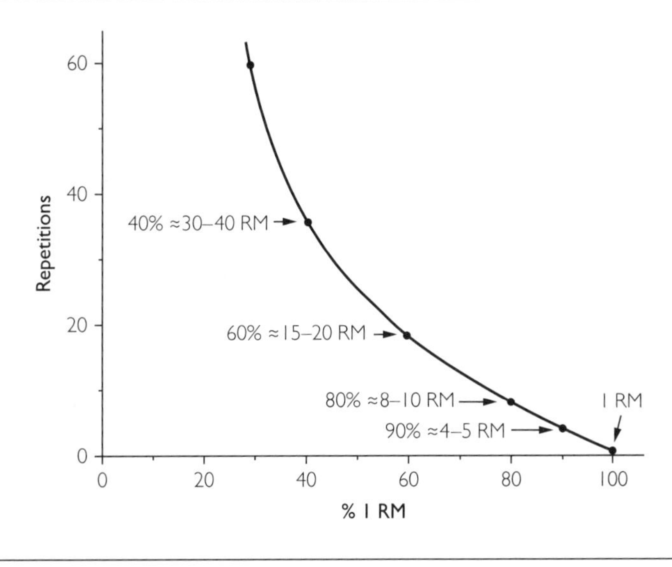

impaired, which would decrease the force of contraction. Finally, there are metabolic factors such as fuel depletion and metabolite accumulation that have been implicated in fatigue, perhaps having both a direct effect at the level of CBs and an indirect effect such as reducing the number of active CBs by, for example, reflex inhibition and impaired excitation–contraction coupling. Because there are so many possible

Figure 5.69 Relationship between percentage of maximal aerobic power ($\dot{V}O_{2max}$) and time to exhaustion (failure). Values can vary according to the test (running, cycling, rowing, etc.) and athlete characteristics (e.g., muscle fiber-type distribution, training history).

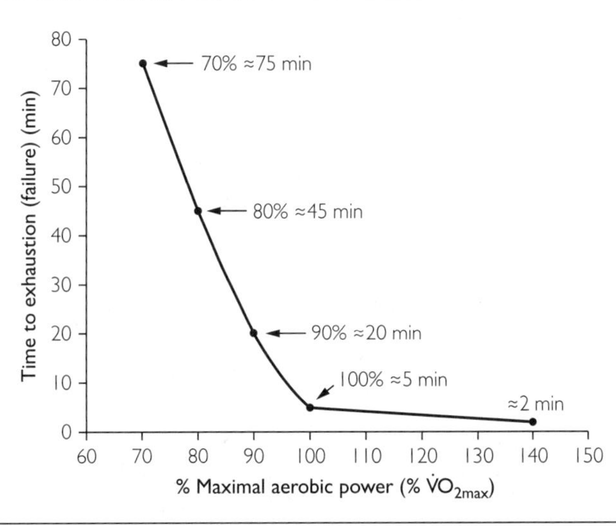

Figure 5.70 The bottom lines of fatigue. There are many possible mechanisms of fatigue but they ultimately affect aspects of myosin CB function, the so-called bottom lines of fatigue. Note how changes in CB function affect muscle contraction.

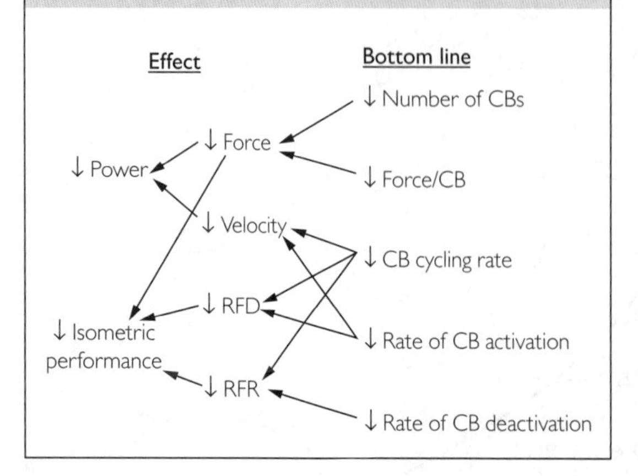

mechanisms of fatigue, it is difficult to include them all in one scheme, such as that shown in Figure 5.72. For example, fatiguing exercise can cause disturbances in electrolytes [e.g., potassium (K^+)] that could contribute to fatigue (8). The scheme also does not indicate how the relative importance of various mechanisms differs with different intensities and durations of exercise (8,134), or that some illustrated mechanisms, such as the roles of H^+ accumulation (7,51,95,253) and afferent feedback from muscles (10,161) have been the subject of considerable controversy and debate (7,95). Adding to the complexity is that various mechanisms may interact with others to amplify fatigue (95).

The multiple potential mechanisms of fatigue have been grouped in different ways, for example central and peripheral fatigue mechanisms (8). Here, mechanisms will be grouped as neural and muscular.

Neural Mechanisms

The neural mechanisms of fatigue ultimately result in failure to fully activate the muscle fibers of MUs; in other words, failure to achieve full or 100% MU activation. MU activation or activation failure consists of a combination of not maintaining recruitment of all possible MUs and not having them fire at rates sufficient for full summation of contraction (43,116,142). In sustained maximal contractions MU activation may be 100% at the beginning of contraction but with fatigue high-threshold FF MUs may quickly drop-out and/or reduce firing rates (Figure 5.73). In sustained submaximal contractions to fatigue failure, MU activation increases as fatigue develops but full activation of high-threshold FF MUs may not be attained (Figure 5.74). Since FF MUs contain a large number of muscle fibers, their drop-out or reduced firing rate represents a significant loss of force.

Figure 5.71 Schematic illustration of how fatigue could affect maximum isometric force (ISO_{max}) and concentric force and power. Changes brought about by fatiguing exercise, such as decreasing the number of active CBs, the force per CB, and the rate of CB cycling, affect myosin CB function. Decreased CB number and force would decrease ISO_{max} and concentric force while decreased rate of CB cycling would decrease maximum shortening velocity (V_{max}) and high-velocity concentric force and power, and increase the curvature of the force–velocity relationship.

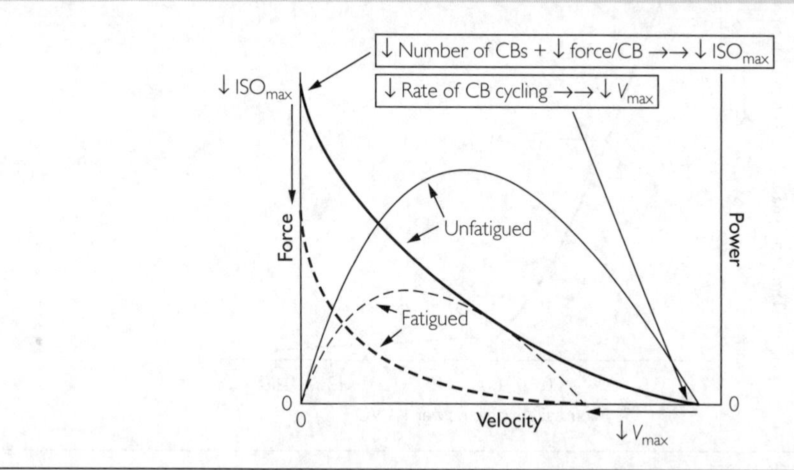

Figure 5.72 Illustration of many possible sites and mechanisms of fatigue and their effects on CB function.

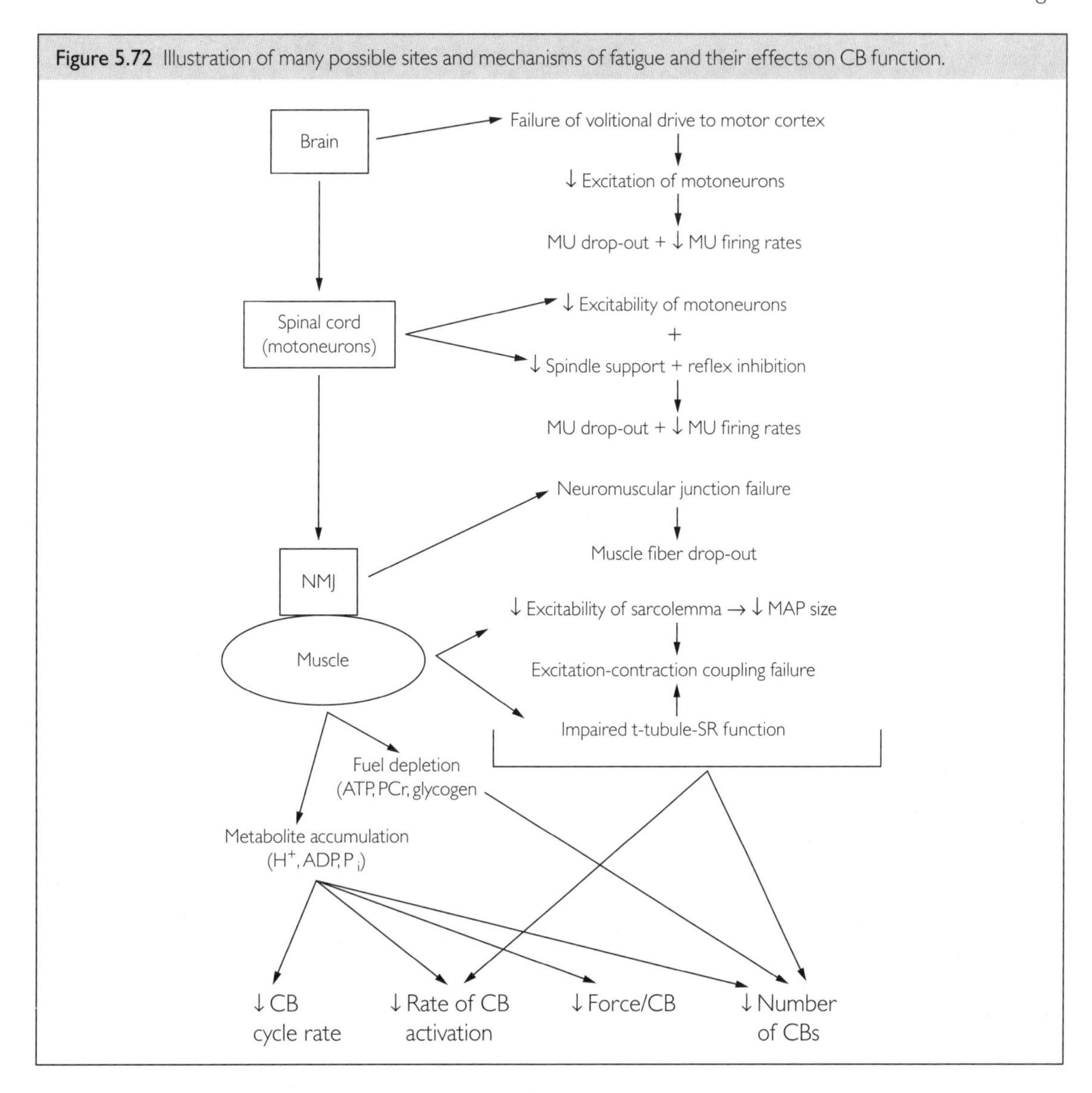

Activation failure can occur in a wide variety of activities; for example, during 5 hours of intermittent running at submaximal intensity there was a progressive decrease in the ability to fully activate muscles in brief maximal voluntary contractions (198). Activation failure could result from reduced excitation of the motor cortex, reduced excitation of motoneurons by the motor cortex, reflex inhibition of motoneurons, reduced excitability of motoneurons, and impaired neuromuscular transmission.

Reduced excitation from higher centers Sensory feedback from exercising muscles, triggered by metabolite accumulation giving rise to sensations of discomfort or muscle pain, would make it difficult to sustain the effort needed to maintain or increase excitation of the motor cortex, that in turn would decrease excitation of motoneurons (10,102).

Altered reflex inputs to motoneurons Muscle sensory receptors react to metabolites (89) such as increased H^+, P_i, and ADP that could induce reflex inhibition of motoneurons, making it more difficult to maintain MU recruitment and firing rates. These are the same muscle receptors that influence excitation of the motor cortex (102), as discussed under Reduced excitation from higher centers.

In strong muscle contractions, muscle spindles within muscles are excited by their own small (called gamma) motoneurons that through the stretch

Figure 5.73 Schematic illustration of MU drop-out and decreased firing rate in a sustained maximal voluntary isometric contraction. (Left) A subject rapidly increases force to maximum and attempts to sustain it. After a few seconds force begins to decrease (failure to hold) and the subject is told to relax. (Right) Recordings of two high-threshold MUs (FF 1, FF 2) during the contraction above. Both MUs initially fire at a high rate to build up force rapidly but then firing rates rapidly decline. FF 1 ceases to fire after approximately 2.5 seconds. FF 2 continues to fire but the rate continues to decrease. Inset: shows that decreased firing rate could decrease summation of contraction according to the force–frequency relationship. MU firing patterns based on the data of Borg *et al.* (43) and Grimby *et al.* (116).

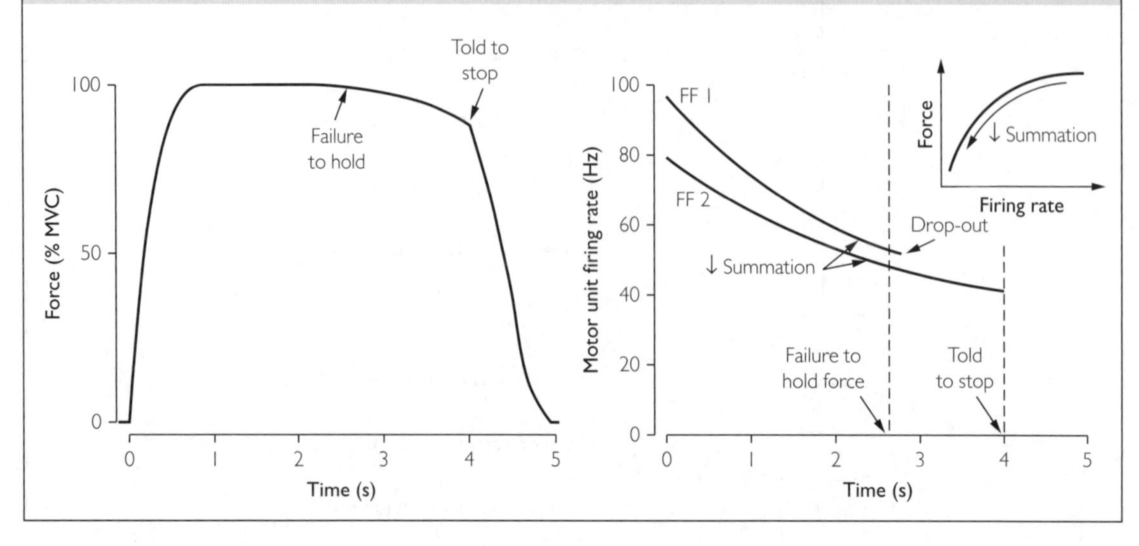

Figure 5.74 Schematic illustration of the increase in MU activation during a submaximal isometric contraction sustained to fatigue failure. As fatigue develops within active muscle fibers, there is a compensatory increase in MU recruitment and firing rates to maintain the force of contraction. There may be activation failure; that is, inability to recruit all MUs or to get them firing at a sufficient rate for maximal summation of contraction.

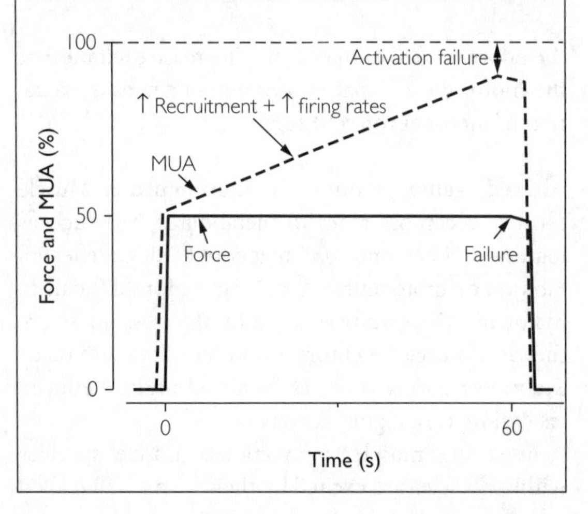

reflex pathway excite the motoneurons (called alpha) innervating muscle fibers, adding to the excitation of motoneurons provided by the motor cortex (Figure 5.72). This muscle spindle support (Figure 5.72) decreases during a sustained contraction because the gamma motoneurons innervating the muscle spindles progressively decrease their firing rates, reducing excitation of the alpha motoneurons (102).

Decreased motoneuron excitability In addition to receiving less excitation during fatiguing exercise, motoneurons themselves may become less excitable (Figure 5.72). Even if a constant level of excitation is maintained, motoneurons exhibit a progressive decrease in motoneuron firing rate called **adaptation** (142,230). Adaptation is most pronounced in high-threshold FF MUs, making them more susceptible to firing rates falling too low for full summation of contraction (43) (Figure 5.73).

Failed neuromuscular transmission The link between the motoneuron and muscle fibers is the NMJ, by which neuromuscular transmission induces MAPs in the muscle fiber membrane (sarcolemma) to cause contraction. Failure to initiate MAPs could result from reduced transmitter release at the NMJ or inability of the sarcolemma to propagate MAPs (43). With failure to initiate MAPs the fiber could not contract (fiber drop-out) to contribute force (Figure 5.72). Decreasing MU firing rates largely protects against neuromuscular transmission failure (failure to initiate MAP) because there is more recovery time between transmissions (8). Asynchronization of MU firing rates also reduces the risk of neuromuscular transmission failure because lower firing rates are needed to produce a given force (Box 5.7). However, as fatigue progresses a greater decrease in firing rate is needed to prevent neuromuscular transmission failure, and in FF MUs that are more susceptible to neuromuscular transmission failure partly because of their higher firing rates, firing rates may fall too low to maintain full summation (43) (Figure 5.72).

Muscular Mechanisms

Fatigue mechanisms in muscle fibers range from impaired excitation–contraction coupling to metabolic factors such as fuel depletion and metabolite accumulation.

Impaired excitation–contraction coupling Several steps in excitation–contraction coupling (Figure 5.75; see also Chapter 4) could be affected by fatigue (7,8,197). Increased extracellular K^+ from repeated MAPs would decrease the amplitude of the MAP as it propagated along the sarcolemma and especially down the transverse tubules. This would reduce the activation of the transverse tubules' voltage-sensing proteins that in turn would cause reduced opening of the Ca^{2+}-release channels, resulting in less Ca^{2+} being released per MAP. Even more important for inhibiting the Ca^{2+}-release channels would be accumulation of Mg^{2+} as ATP is broken down, especially in fast type

Figure 5.75 Summary of the steps in excitation–contraction coupling leading to CB cycling and the key step in relaxation of contraction. All of these steps can be adversely affected by fatigue in some exercise conditions. SR, sarcoplasmic reticulum; t-t, transverse tubules.

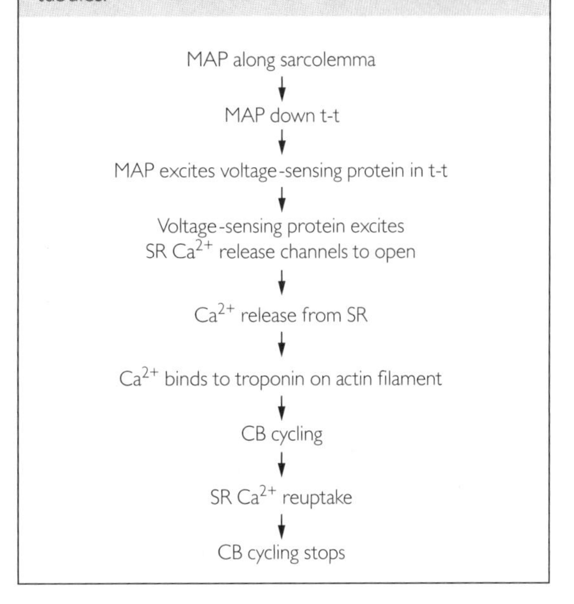

2 fibers. Increased P_i from the breakdown of PCr enters the sarcoplasmic reticulum and, by binding with Ca^{2+}, reduces the Ca^{2+} available for release. The slowing of relaxation with fatigue is partly related to the effect of increased ADP, P_i, and H^+, and decreased ATP/ADP·P_i ratio near the sarcoplasmic reticulum Ca^{2+}-reuptake pumps in the sarcoplasmic reticulum. In summary, all of these fatigue-induced alterations in excitation–contraction coupling reduce force by reducing the amount Ca^{2+} released per MAP, thus reducing the number of cycling CBs. In relation to excitation–contraction coupling, slowed relaxation of contraction is caused by slowing of the sarcoplasmic reticulum Ca^{2+} pumps.

Metabolic factors In brief but high-intensity exercise such as sprinting there is rapid depletion of **ATP** and **PCr**, especially in type 2 fibers (e.g., Figure 5.76), and a corresponding increase in **ADP** and $\mathbf{P_i}$ (8). Depletion of ATP, which is the immediate energy source for CB cycling, and depletion of PCr, which provides the most rapid pathway for resynthesizing ATP, combine to decrease the number of cycling CBs (Figure 5.72). Increased ADP decreases maximum

BOX 5.7 ASYNCHRONOUS FIRING OF MUS ATTENUATES SOME MECHANISMS OF FATIGUE

Earlier in this chapter it was shown that, compared to synchronous stimulation, in asynchronous firing of MUs a lower firing rate is needed to produce a given percentage of maximum force (Figure 5.41). Lower firing rates in fatiguing contractions reduce the probability of impaired neuromuscular transmission and excitation–contraction coupling.

Figure 5.76 Depletion of ATP and PCr of vastus lateralis after 25 seconds of maximal-effort cycling against an isokinetic (isovelocity) resistance at a crank velocity of 120 rpm. Concentration is referenced to the Pre (before exercise) PCr value of type 2X/2A fibers set to 100%. Fiber type influenced the initial concentration of ATP and PCr and the extent of depletion. Mechanical power output, referenced to the right vertical axis, decreased to approximately 60% of the peak value. Based on the data of Karatzaferi *et al.* (138).

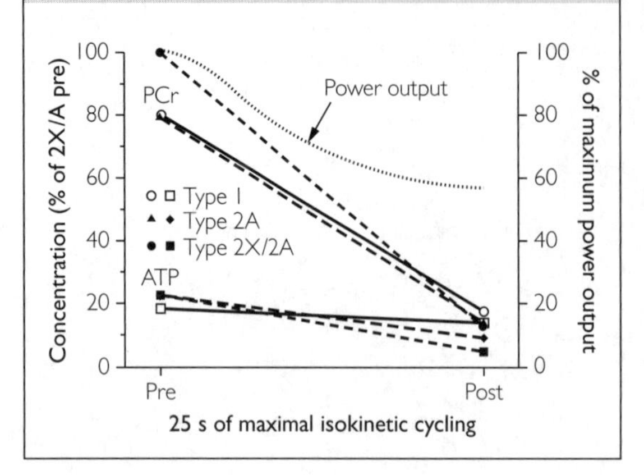

Figure 5.77 Muscle glycogen depletion during prolonged aerobic exercise. Endurance-trained men cycled to exhaustion (failure) at an intensity equivalent to 95% of the lactate threshold (≈70% $\dot{V}O_{2\,peak}$). The exercise was done twice, once after consuming a high-carbohydrate diet to produce a high pre-exercise level of muscle glycogen (HG), and once after consuming a low-carbohydrate diet to produce a low pre-exercise level of glycogen (LG). Subjects were able to exercise longer with a higher initial muscle glycogen concentration. Values are expressed as a percentage of the HG pre-exercise level. Based on the data of Baldwin *et al.* (20).

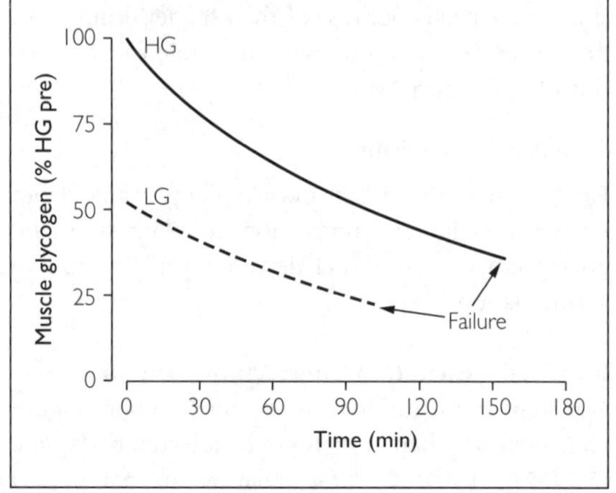

shortening velocity (V_{max}) (8) by decreasing the rate of CB cycling (Figure 5.72). Increased P_i reduces Ca^{2+} sensitivity of the myofibrillar proteins, thus reducing the number of active CBs, and also the force per CB (8) (Figure 5.72).

Muscle Glycogen

Muscle glycogen can be depleted to low levels during prolonged endurance exercise (Figure 5.77) and repeated high-intensity exercise (Figures 5.78 and 5.79). In prolonged exercise glycogen is initially depleted more in type 1 and type 2A muscle fibers but as fatigue failure approaches there is progressive depletion in type 2X fibers (Figure 5.33). In repeated high-intensity exercise, all fiber types are active and use glycogen but type 2 fibers use glycogen at a faster rate (Figure 5.19) and to a greater extent because of their greater metabolic power (Figure 5.76).

The mechanism by which glycogen depletion causes fatigue in aerobic exercise is not clear (8). It had been thought that glycogen depletion would lead to inadequate provision of substrate for aerobic metabolism and resynthesis of ATP, but it has been shown that a lack of glycogen availability causes little disturbance of levels of ATP, PCr, and metabolic intermediates (20). Low muscle glycogen is associated with impaired excitation–contraction coupling but again the mechanism is unknown (8). Glycogen depletion during prolonged exercise is concurrent with other changes that could cause fatigue, such as reduced blood glucose (hypoglycemia) from liver glycogen depletion and increased body temperature (hyperthermia), which could impair muscle activation, among other mechanisms (183).

The marked glycogen depletion in repeated, high-intensity exercise, especially in type 2 muscle fibers, might cause fatigue by reducing the effectiveness of the second fastest (breakdown of PCr is the fastest) way of rapidly resynthesizing ATP. After several previous bouts with brief recovery periods, the duration or power output of a high-intensity bout of exercise might be limited because, with reduced availability of glycogen (23), there must be more reliance on PCr breakdown and aerobic metabolism for ATP resynthesis (101). In an ice hockey game, for example, players

Figure 5.78 Muscle glycogen depletion following intermittent high-intensity exercise. Subjects did six 1 minute bouts of exercise at an intensity corresponding to 150% $\dot{V}O_{2max}$. There was a 10 minute rest period after each bout. After six bouts totalling 6 minutes of exercise, muscle glycogen had decreased 63%, or to 37% of the pre-exercise value. Values shown are for *whole*-muscle glycogen, representing an average of the depletion in type 1 and 2 muscle fibers. There was greater glycogen depletion in type 2 fibers. Based on the data of Gollnick *et al.* (109).

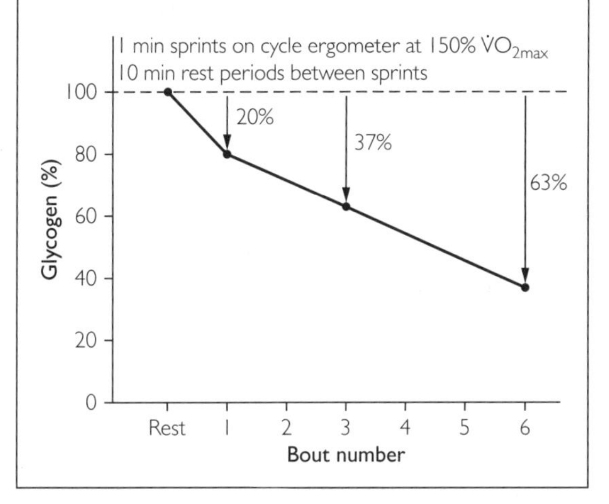

would have to shorten their shifts as muscle glycogen became depleted. As mentioned in connection with prolonged exercise, glycogen depletion could also adversely affect excitation–contraction coupling. During prolonged intermittent exercise with rest periods (e.g., many team sports), carbohydrate ingestion improves performance by glycogen sparing resulting from blood glucose uptake, accelerated glycogen resynthesis during rest periods, and the effect of increased blood glucose on effort perception and muscle activation (195).

In intense exercise like sprinting, glycogen is rapidly broken down into **lactic acid**, which takes the form of **lactate** and **hydrogen ions** (H^+). Accumulation of H^+ reaches the highest levels after maximal exercise lasting 1–4 minutes. In track running, for example, this would correspond to 400–1500 m. In 100 and 200 m sprints the rate of lactic acid production is high but the sprints do not last long enough to produce a large H^+ accumulation (41). In distance events (10 000 m, marathon) there is enough time for H^+ to accumulate but the rate of accumulation is very

Figure 5.79 Muscle glycogen depletion during two weight-training routines with the same volume but differing intensities, as shown. The two routines produced a similar total glycogen depletion but the high-intensity routine produced the depletion with fewer repetitions. Based on the data of Robergs *et al.* (209).

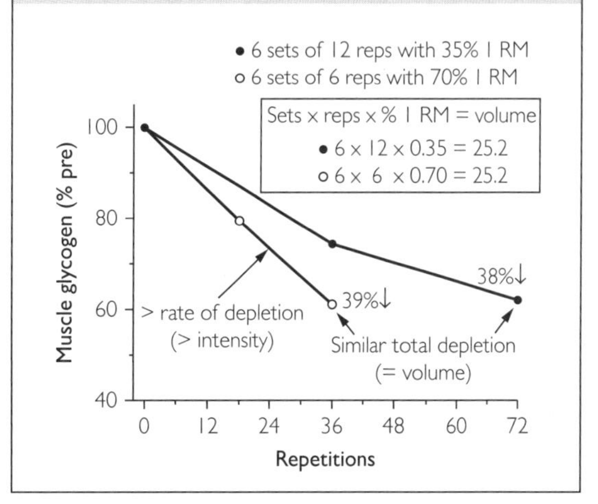

low. Accumulation of H^+ increases the acidity level, quantified as a *lowering* of muscle pH (e.g., 42). The importance of increased H^+ as a cause of fatigue is controversial (51,253). The main effect of H^+ may be reducing the sensitivity of the myofibrillar proteins to Ca^{2+}, which could reduce the number of cycling CBs (8), but low pH may also reduce the force per CB (95). Induced *acidosis* (decreasing pH) impairs performance whereas induced *alkalosis* (increasing pH) improves performance (51), pointing to accumulation of H^+ as a contributor to fatigue. Finally, high-intensity interval training increases muscle buffering capacity (e.g., 255) and delays fatigue development in high-intensity exercise, suggesting that H^+ contributes to fatigue and that increased buffering capacity is one adaptive mechanism by which fatigue can be delayed.

Summary of Fatigue Mechanisms

Simplified summaries of the mechanisms of fatigue are shown in Figures 5.80 and 5.81. Note in viewing the figures that some mechanisms (e.g., NMJ failure, H^+) have been the subject of debate. In addition, indications of when particular mechanisms become more or less important are estimates based on available evidence. Although fatigue is viewed negatively, it may have benefits (Box 5.8).

Figure 5.80 Summary of sites and causes of fatigue in events lasting up to 60 seconds. As lines become more solid, relative importance of a site or cause increases. E-C coupling, excitation–contraction coupling; MUA, MU activation.

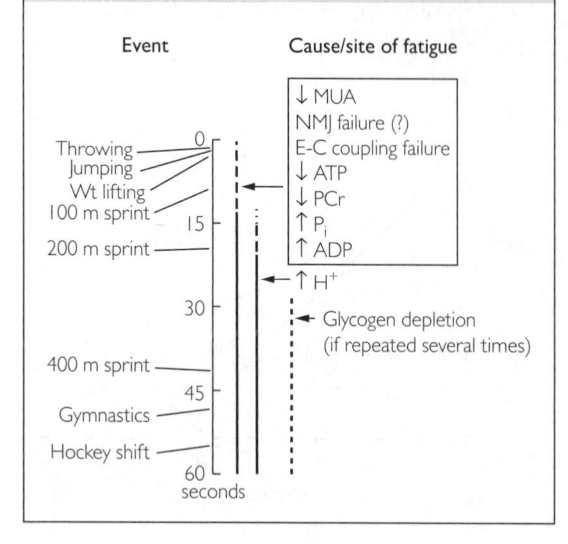

Figure 5.81 Summary of sites and causes of fatigue in events lasting up to 120 minutes. As lines become more solid, relative importance of a site or cause increases. E-C coupling, excitation–contraction coupling; MUA, MU activation.

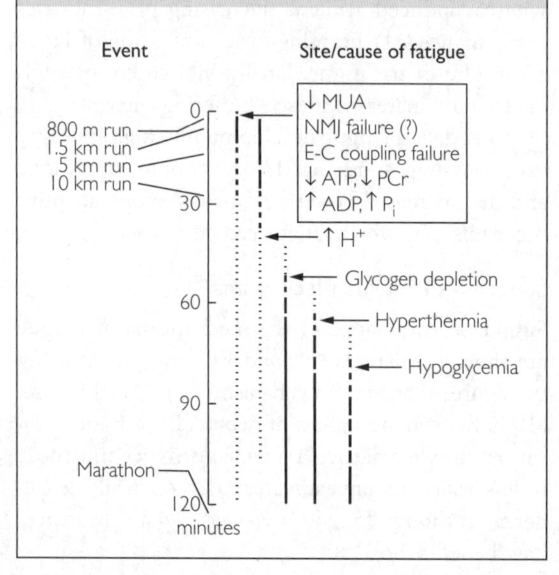

BOX 5.8 WHY DO WE FATIGUE?

Fatigue is an impediment not only to athletic performance but also to tasks of daily living. What benefits does fatigue provide? Fatigue may provide a functional reserve (10) or act as a safety mechanism (3) to prevent muscle exhaustion, particularly in fatigue-susceptible type 2 muscle fibers (116), that could lead to muscle damage. In brief, high-intensity exercise, ATP can be depleted to relatively low values in type 2X/2A (Figure 5.76) and likely also in 2X fibers. Complete depletion of ATP could cause a rigor state not unlike rigor mortis (stiffness of death). If fatigue did not occur in prolonged endurance exercise, there could be the risk of severe hyperthermia and hypoglycemia. Thus, it is fortunate that training can delay but not prevent fatigue.

Factors Affecting Fatigability

Contraction Type

In repeated maximal voluntary and electrically stimulated contractions, fatigue occurs more rapidly in concentric than in eccentric contractions, with isometric contractions in between (8,71,193). The energy cost is higher for repeated maximal concentric than isometric (8) or eccentric contractions (193); thus, there is greater probability of metabolic factors and impaired excitation–contraction coupling contributing to fatigue.

Force– and Power–Velocity Relationships

Fatigue decreases force across both the eccentric and concentric force–velocity relationships (Figure 5.82). Compared to concentric contractions, in fatigue eccentric contraction exhibits a greater *absolute* but much smaller *relative* (percentage) decrease in force. Decreased V_{max} and increased curvature of the force–velocity relationship contribute to the large decline in concentric force. Increased curvature depresses force at intermediate velocities or, alternatively, lowers the velocity attained with a given load (Figure 5.12). The mechanism of the increased curvature is unknown, but is highly correlated with the slowing of relaxation that occurs with fatigue, which could result from impaired Ca^{2+} reuptake into the sarcoplasmic reticulum or to a slowing of the rate of CB cycling (134).

Power is the product of force and velocity and is also influenced by curvature of the concentric force–velocity relationship; therefore, the negative effect of fatigue on concentric power is greater than on concentric force, V_{max}, and curvature alone (Figure 5.83B). Fatigue not only decreases peak power, it decreases both

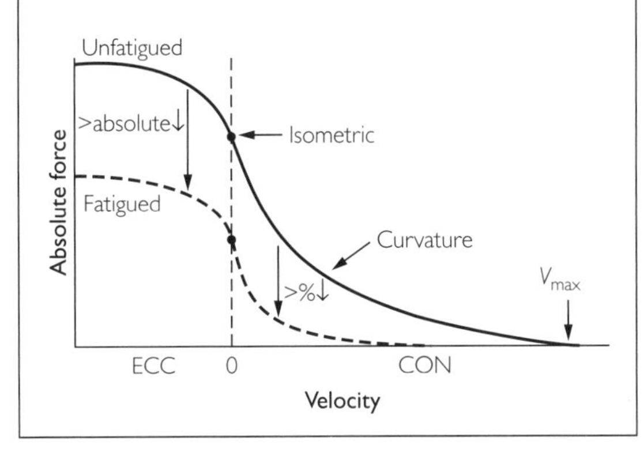

Figure 5.82 Schematic illustration of the effect of fatigue on the force–velocity relationship. Fatigue was induced by a series of concentric contractions. Fatigue produced a greater absolute decrease in eccentric (ECC) than concentric (CON) force but the relative (%) decrease was greater in concentric contractions. Isometric force exhibited an intermediate response. Note how fatigue increased the curvature of the concentric force–velocity relationship, which contributed to the greater relative decrease in concentric force. Based on the data of de Ruiter et *al.* (71).

the velocity at which peak power occurs and the velocity range over which power can be produced (Figure 5.83A). The large decrement in power is responsible for the rapid performance decline in repeated power activities such as sprinting and jumping.

Fiber-Type Distribution

In a given muscle, individuals with a higher percentage of S MUs (105) and thus type 1 fibers (Figure 5.15) generally have greater resistance to fatigue (more endurance) (Figure 5.84). However, variation in muscle activation, motivation, and activity history can also affect performance in fatigue tests. The poor correlation sometimes observed (e.g., 163) between fiber-type distribution and fatigability could be related to these factors and also to the methods for determining fiber-type distribution.

Training

A primary training goal is to delay fatigue. Training can increase both the *absolute* intensity (force, power, speed) and *relative* intensity (percentage of maximum absolute intensity) that can be sustained for prolonged periods (Box 5.9). For example, a trained athlete might be able to exercise at a relative intensity of approximately 85% $\dot{V}O_{2max}$ for 1 h whereas an untrained individual can only manage approximately 50% $\dot{V}O_{2max}$ (Figure 5.85). The athlete has greater *relative* endurance.

Training can also increase *absolute* endurance by increasing $\dot{V}O_{2max}$ (Figure 5.85). Thus, the endurance-trained athlete can do prolonged exercise at a higher percentage (relative) of a higher (absolute) $\dot{V}O_{2max}$.

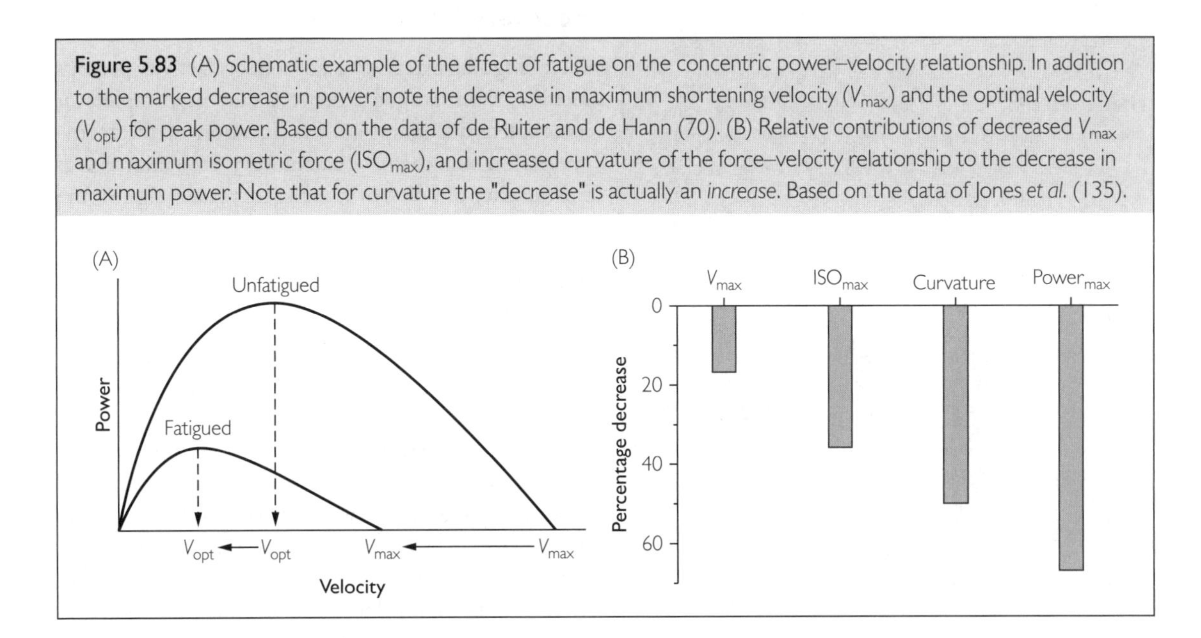

Figure 5.83 (A) Schematic example of the effect of fatigue on the concentric power–velocity relationship. In addition to the marked decrease in power, note the decrease in maximum shortening velocity (V_{max}) and the optimal velocity (V_{opt}) for peak power. Based on the data of de Ruiter and de Hann (70). (B) Relative contributions of decreased V_{max} and maximum isometric force (ISO_{max}), and increased curvature of the force–velocity relationship to the decrease in maximum power. Note that for curvature the "decrease" is actually an *increase*. Based on the data of Jones et *al.* (135).

Figure 5.84 Effect of fiber-type composition on fatigability. (A) Subjects did 50 maximal knee-extension concentric contractions at 3.14 rad·s^{-1}. The subject with a higher percentage of type 2 fibers in vastus lateralis showed a greater decline in torque (i.e., greater fatigue). Based on the data of Thorstensson and Karlsson (243). (B) Subjects with a range of percentages of type I fibers in vastus lateralis did a leg-press isometric endurance test with 50% of isometric MVC. Subjects with a higher percentage of type I fibers exhibited greater endurance. Based on the data of Hulten et *al.* (131).

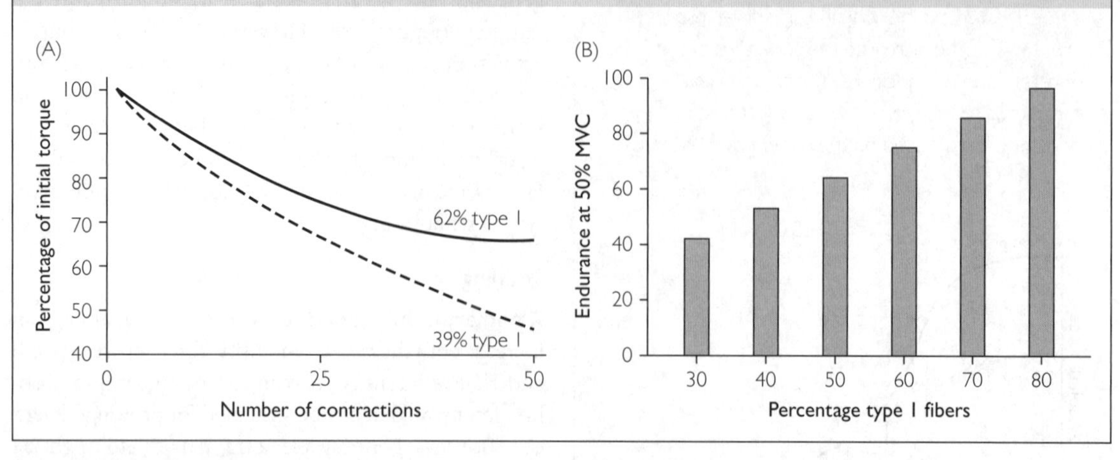

It should be emphasized that endurance sports are tests of absolute rather than relative endurance (Figure 5.85, Box 5.9).

For simplicity, runners A, B, and C (Figure 5.85, Box 5.9) were matched in relative endurance, which actually varies among endurance athletes. It is therefore possible that two marathon runners could have the same best race time (e.g., 2:10); one might have a slightly higher $\dot{V}O_{2max}$, whereas the other can run at a slightly higher percentage of $\dot{V}O_{2max}$. Regardless of how a time is achieved, a marathon race remains a test of absolute endurance. The same is true of other endurance sports such as swimming, rowing, speed skating and canoeing.

Figure 5.85 Effect of training on resistance to fatigue (endurance). (Left) Endurance-trained subjects can sustain a higher percentage of $\dot{V}O_{2max}$ in prolonged exercise lasting 1 and 2 hours. Based on data reviewed by Bassett and Howley (26). (Right) Exemplar $\dot{V}O_{2max}$ values for trained and untrained subjects, showing that a given *relative* intensity (% $\dot{V}O_{2max}$) in trained subjects is also a greater *absolute* intensity ($\dot{V}O_2$). Based on the data of Powers et *al.* (200).

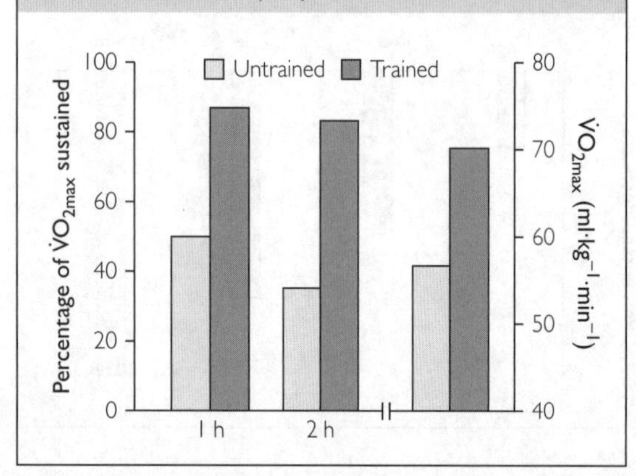

Recovery from Fatigue

After fatiguing exercise, rest or reduced activity permits recovery from fatigue. Recovery may be complete or incomplete depending on the duration of the recovery period, and the needed recovery time increases as the **number** of bouts of exercise increases (e.g., Figure 5.87).

Rate of Recovery is Correlated with Rate of Fatigue

Fatigue develops rapidly in a single bout of brief, high-intensity exercise, but recovery is also rapid (8). For example, a 100 m track sprinter can recover completely in hours if not minutes, whereas a marathon runner requires days if not weeks to fully recover; in both cases complete recovery means being able to repeat the performance. Figure 5.88 shows an example of the same level of fatigue (same decrease in maximal isometric strength) produced over a short or long period of exercise. Although the extent of fatigue was the

BOX 5.9 THE DISTINCTION BETWEEN ABSOLUTE AND RELATIVE ENDURANCE

In a **relative endurance** test exercise is continued as long as possible at a given percentage of maximal strength, power, speed, $\dot{V}O_{2max}$, etc. For example, an isometric contraction at 50% of maximum strength is sustained as long as possible (Figure 5.84B). In contrast, an **absolute endurance** test would require sustaining a force of 200 N, which would represent a variable percentage of maximum strength across subjects, as long as possible. Figure 5.86 compares the results of relative and absolute endurance tests administered to three distance runners, A, B, and C, with $\dot{V}O_{2max}$ values of 60, 70, and 75 ml·kg^{-1}·min^{-1}, respectively. The runners had the same relative endurance, the time run at 80% $\dot{V}O_{2max}$ (Figure 5.86, left). However, in the absolute endurance test in which the runners ran at a speed corresponding to a $\dot{V}O_2$ of 60 ml·kg^{-1}·min^{-1}, runner A could only run for approximately 15 minutes because 60 ml·kg^{-1}·min^{-1} was 100% of $\dot{V}O_{2max}$. At the other extreme, 60 ml·kg^{-1}·min^{-1} was only 80% of runner C's $\dot{V}O_{2max}$, so C achieved the same time as in the relative endurance test, also done at 80% $\dot{V}O_{2max}$. Runner B, for whom the absolute intensity was 86% $\dot{V}O_{2max}$, ran for a little under 2 hours. Distance running events are tests of *absolute* rather than *relative* endurance (right side of Figure 5.86). On average, a 2.25 hour (2:15) marathon requires a running speed equivalent to a $\dot{V}O_2$ of 60 ml·kg^{-1}·min^{-1} (26). To achieve the 2.25 hour marathon time, 60 ml·kg^{-1}·min^{-1} should not exceed approximately 80% of the runner's $\dot{V}O_{2max}$. The required $\dot{V}O_2$ exceeded the 80% limit in runners A and B, so they could not maintain the 2.25 hour pace. Runner C was able to maintain the pace because the limit was not exceeded.

same, recovery from the long-duration exercise was much slower.

Mechanisms of Recovery

The mechanisms of recovery are a reversal of the mechanisms of fatigue. As described below, some recovery mechanisms are rapid while others are slower.

Muscle activation Recovery of the ability to fully activate muscles in a brief maximal contraction occurs quickly, within minutes (102). Recovery of activation must be at least as rapid as the recovery of voluntary contraction force. For example, almost complete recovery of activation and force occurred within a few minutes after a 2 minute sustained maximal isometric

Figure 5.86 The distinction between *relative* and *absolute* endurance. Three distance runners, A, B, and C, have maximal aerobic power ($\dot{V}O_{2max}$) values of 60, 70, and 75 ml·kg^{-1}·min^{-1}, respectively. (Left) When asked to run as long as possible at a treadmill speed corresponding to 80% of their $\dot{V}O_{2max}$, all three runners ran for the same time; that is, they had the same relative endurance. (Right) On another occasion the three runners ran as long as possible at a speed corresponding to a $\dot{V}O_2$ of 60 ml·kg^{-1}·min^{-1}; that is, at the same absolute intensity. The percentage values above the bars indicate the percentage of $\dot{V}O_{2max}$ at which the athlete had to run. For runner A, the required $\dot{V}O_2$ was 100% of $\dot{V}O_{2max}$ so failure occurred after a relatively short time. For athlete C the intensity of 60 ml·kg^{-1}·min^{-1} was 80% of $\dot{V}O_{2max}$; therefore, this athlete ran at the same relative intensity as in the relative endurance test at the left and, not surprisingly, the time in the two tests was similar. Runner B had to run at 86% $\dot{V}O_{2max}$ and could continue for a little under 2 h, not as long as when the run was done at 80% $\dot{V}O_{2max}$ as shown at the left. Runner C would have completed a marathon in approximately 2.25 hours (2:15). See Box 5.9 for further discussion.

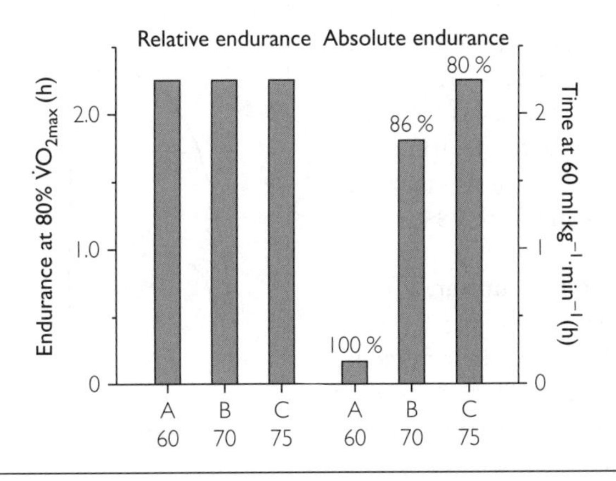

Figure 5.87 Effect of inter-set rest period duration on recovery in weight training. American football players attempted to complete, in both the leg-press and bench-press exercises, three successive sets of 10 repetitions with the heaviest weight that could be lifted for 10 repetitions (10-repetition maximum, 10 RM) in the first set. This was done on two occasions, once with 3 minute and once with 1 minute rest periods between sets. Three-minute rest periods allowed completion of three sets of 10 repetitions, suggesting complete or at least sufficient recovery. With 1 minute rest periods there was a progressive decrease in the number of repetitions, indicating incomplete recovery that was cumulative. Based on the data of Kraemer (145).

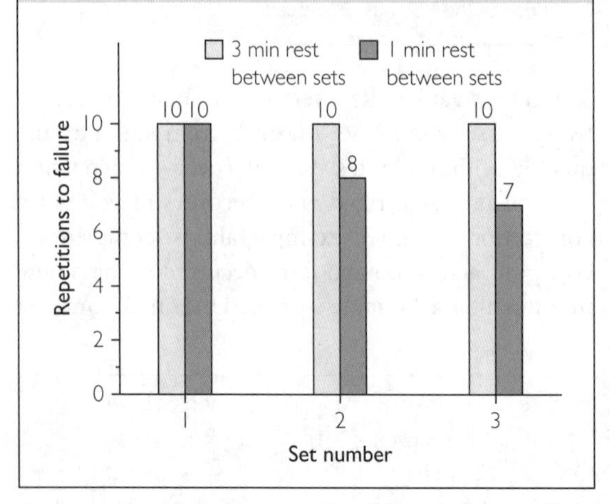

Figure 5.88 Effect of short- and long-duration ankle dorsiflexion exercise on recovery from fatigue. Fatigue to the same level of isometric MVC was induced by either a sustained 2 minute MVC ("Short") or intermittent MVCs ("Long") over a period of 15–20 min, in which the product of force and time was five to seven times that of "Short." Recovery was rapid and complete 15 minutes after "Short" but was relatively slow and incomplete after 15 minutes for "Long." Based on the data of Baker *et al.* (19).

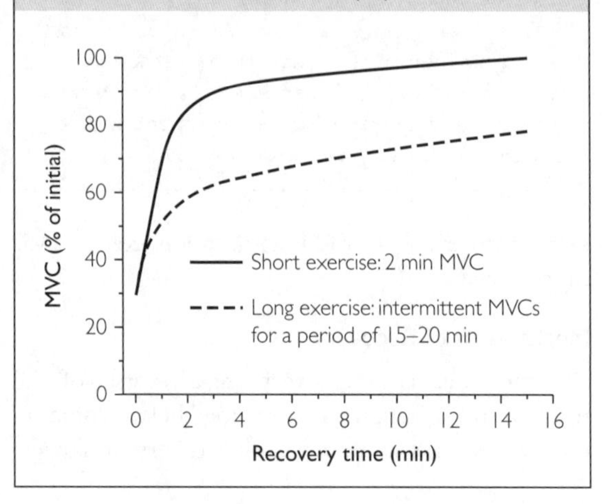

contraction ("Short" in Figure 5.88). In fact, recovery of activation was also rapid after the "Long" exercise shown in Figure 5.88; the longer recovery of force was due to other factors (discussed below).

Neuromuscular transmission and MAP propagation As discussed, there is often no evidence of neuromuscular transmission failure or failure of MAP propagation in fatiguing exercise because of the decrease in MU firing rates. If failure were to occur, recovery must be at least as rapid as activation, because activation depends on preserved neuromuscular transmission and MAP propagation, as was the case in the fatiguing exercise depicted in Figure 5.88.

Excitation–contraction coupling Impaired excitation–contraction coupling results in reduced Ca^{2+} release in response to each MAP, which in turn reduces the Ca^{2+} available to activate CB cycling. A feature of impaired excitation–contraction coupling is that force is more depressed when a muscle is stimulated at low vs high frequencies (Figure 5.89). The recovery of low-frequency tetanic force can take hours as opposed to minutes for high-frequency force (Figure 5.90). The delayed recovery of low-frequency tetanic force is called

Figure 5.89 Schematic illustration of a fatigue-induced reduction in high- and low-frequency isometric force. The muscle was stimulated at the frequencies indicated. Note the relatively greater reduction in low-frequency fatigue (LFF).

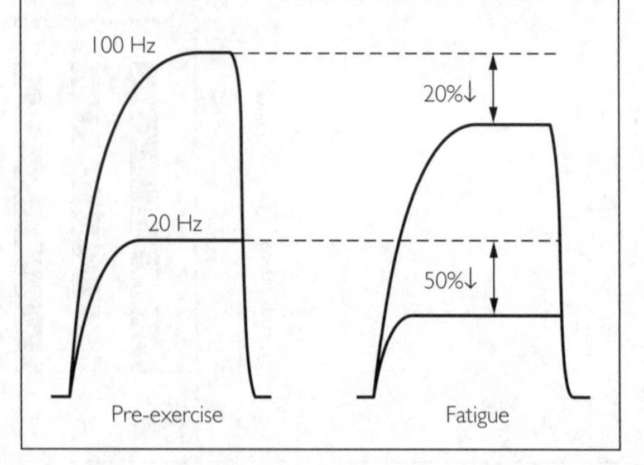

Figure 5.90 During the recovery period after fatiguing exercise, muscle force in response to 20 and 100 Hz stimulation is tested several times over 24 hours. Note the delayed recovery of force evoked at the lower frequency, which is referred to as low-frequency fatigue (LFF) or prolonged low frequency force depression (PLFFD). Based on the data of Edwards *et al.* (86).

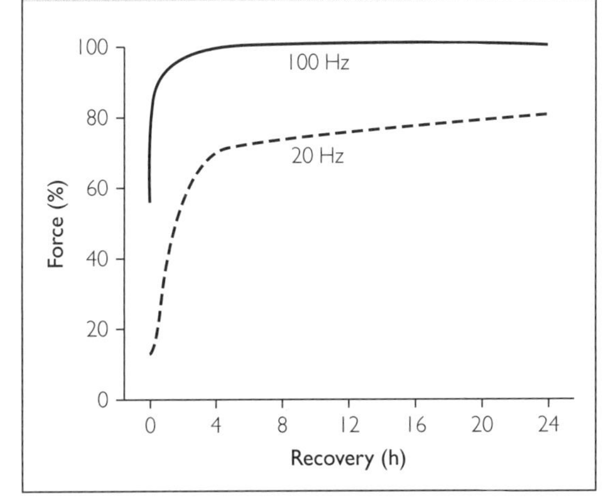

Figure 5.91 Schematic illustration of the effect of fatigue on the force–frequency relationship. The plateau of the force–frequency relationship for the unfatigued and fatigued muscle has been normalized (set to 100%) to highlight the effect of fatigue on the shape of the force–frequency relationship. Note that fatigue depresses force more at low than high frequencies, producing a downward and rightward shift of the force–frequency relationship. (Inset) Relationship between Ca^{2+} available to activate CB cycling and contraction force. At high Ca^{2+} concentrations [Ca^{2+}], corresponding to high frequencies, an increase or decrease in Ca^{2+} release (arrows) has little effect on force because Ca^{2+} is at a saturated level (plateau region of force–frequency relationship). In the low [Ca^{2+}] range corresponding to low frequencies, force is much more sensitive to reduced Ca^{2+} release.

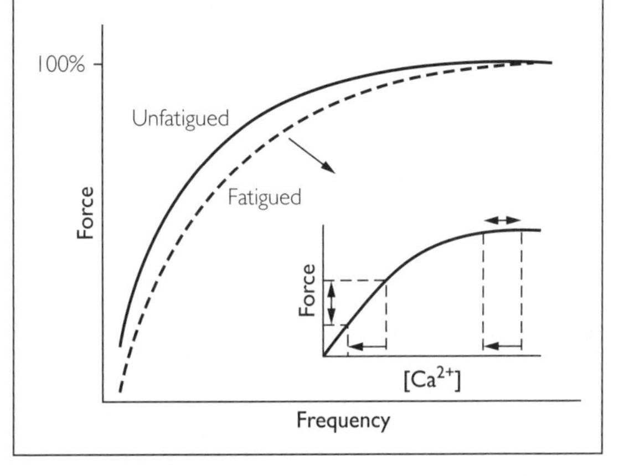

low-frequency fatigue (LFF) or **prolonged low-frequency force depression** (PLFFD) (8). LFF/PLFFD is greater in fast, type 2 muscle fibers (208).

The mechanism of LFF/PLFFD is a prolonged reduction of Ca^{2+} release, which has a greater effect at low than high frequencies (Figure 5.91). It is unclear why reduced Ca^{2+} release lasts so long (8,22). It persists long after some metabolic factors that might affect excitation–contraction coupling acutely, such as ATP and PCr depletion and ADP, P_i, and H^+ accumulation, have recovered completely (e.g., 19,86). However, low muscle glycogen is associated with LFF/PLFFD (261) and recovery of glycogen can take several hours.

LFF/PLFFD may develop in athletes engaged in intense, prolonged training sessions repeated several days per week. Apparent weakness and increased perception of effort in doing light exercise may be a sign of LFF/PLFFD because when MUs are firing at low rates the force is depressed. Paradoxically, if the athlete attempts a brief maximal effort, in which MU firing rates are high, there would be no or little sign of impairment (Figure 5.90). In the absence of reduced maximal strength or power, a coach without knowledge of LFF/PLFFD might dismiss an athlete's complaint of weakness or chronic fatigue as psychological or malingering when in fact the athlete could be suffering from LFF/PLFFD. LFF/PLFFD may be responsible for the "dead-legs" phenomenon about which many athletes complain when their training involves a large volume of running, jumping, or skating. Again, if these athletes are subjected to brief strength and power tests (e.g., jump, lift), there may be little or no indication of impairment. The presence of LFF may be one sign of overtraining (see Box 5.10).

Metabolic factors After brief, intense exercise, ATP and PCr recover to pre-exercise levels relatively quickly and more quickly in type 1 than type 2 muscle fibers (Figure 5.92). Similarly, ADP and P_i (19,41) recover relatively quickly to pre-exercise levels. In contrast, muscle pH (H^+) takes longer to recover. The first half of pH recovery is fairly rapid (107) but then slows, requiring in some cases up to 40–60 minutes to return to baseline (Figure 5.93).

Restoring depleted muscle glycogen can require several hours to a few days depending on the intensity

BOX 5.10 LOW-FREQUENCY FATIGUE IN AN ATHLETE

A kayak paddler begins to complain of upper body weakness and fatigue when doing low-intensity bouts of paddling and doing warm-up sets in her weight-training program. However, when her coach administers a short kayak sprint test and 1 RM tests in her weight-training exercises, the results show no decline from previous tests. Based on the test results the coach is tempted to dismiss the athlete's complaint; however, the coach knows a sport scientist at the local university and contacts her about the problem. In the absence of any known medical problem, the sport scientist recommends a laboratory test in which isometric contractions of the elbow flexors are evoked at frequencies of 15 and 50 Hz. The ratio of 15 Hz:50 Hz force is found to be well below normal, indicating low-frequency fatigue (LFF). In contrast, the athlete's voluntary isometric peak force (MVC) is quite high. Thus, the lab tests confirm the field observations and indicate that the athlete is experiencing LFF. Based on the test results, the coach and scientist discuss ways to adjust the volume and frequency of training to minimize the risk of LFF.

and duration of exercise (196,221). At one extreme, following a marathon run muscle glycogen may not have fully recovered after 7 days on a high-carbohydrate diet (Figure 5.94). At the other extreme, after nine sets of a weight-training exercise muscle glycogen

Figure 5.92 Changes in ATP (squares) and PCr (circles) in vastus lateralis after a 30 second sprint on a cycle ergometer and after 4 minutes of recovery. In type 1 fibers (unfilled symbols) recovery of ATP and PCr to pre-exercise levels was complete by 4 minutes but had recovered only to approximately 70% in type 2 fibers (filled symbols). Based on the data of Casey *et al.* (52).

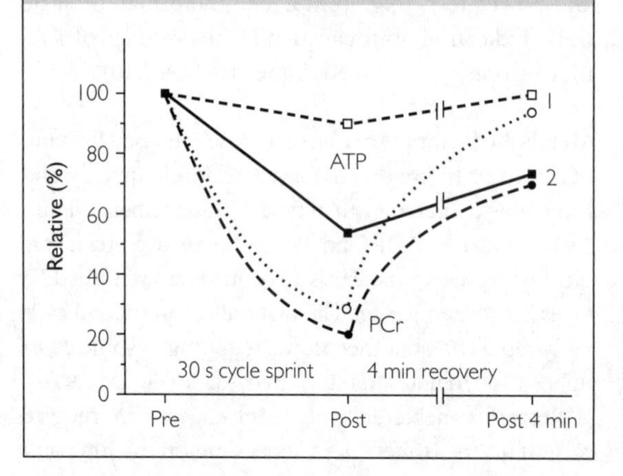

Figure 5.93 Schematic illustration of the increase in muscle H^+ concentration (dashed line) during 2 minutes of maximal exercise (e.g., 800 m run) and the time course of the return of H^+ to pre-exercise levels. Based on the data of Hermansen and Osnes (125), which showed a decrease of muscle pH from approximately 7.0 to 6.4 after 2 minutes of maximal exercise.

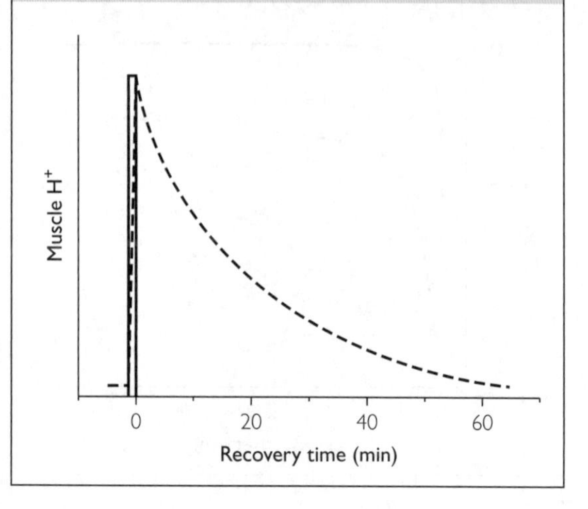

Figure 5.94 Time course of glycogen repletion in lateral gastrocnemius during recovery from a marathon run (42.2 km). Trained male runners, who achieved personal best times (≈3 hours) in the run, had consumed a relatively high-carbohydrate diet before and after the race. During the race glycogen depleted to approximately 12% of the pre-exercise value and had recovered to approximately 70% after 7 days of rest. Based on the data of Sherman *et al.* (223).

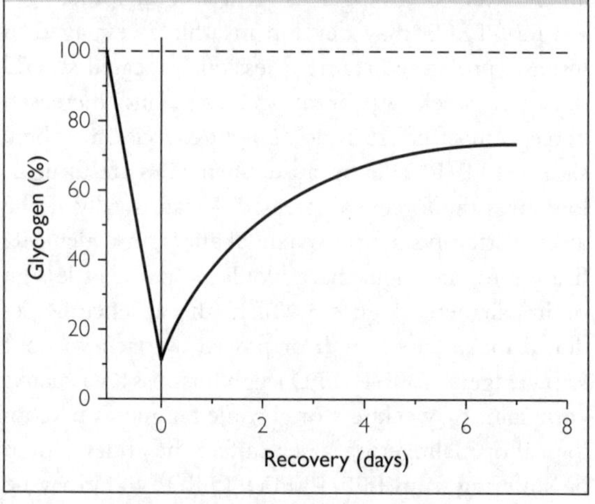

Figure 5.95 Glycogen repletion in vastus lateralis after resistance exercise. Subjects did nine sets of six repetitions of a knee extension exercise with 70% of the 1 RM, with 30 second rest periods between sets. After exercise, glycogen had depleted to approximately 70% of the pre-exercise value. During recovery, glycogen was restored to approximately 85 and 90% of the pre-exercise value after 2 and 6 hours, respectively. Based on data of Pascoe *et al.* (191).

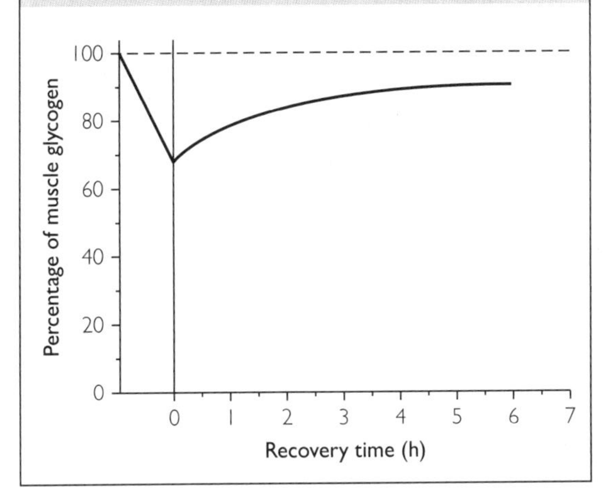

Figure 5.96 Time course of glycogen repletion in vastus lateralis during recovery from high-intensity cycle ergometer exercise. Subjects did 1 minute bouts of exercise at a power output corresponding to 140% $\dot{V}O_{2max}$, alternated with 3 minute rest periods. The session ended when a bout could not continue beyond 30 seconds. The average total exercise time, excluding rest periods, was approximately 10 minutes. Glycogen depleted to approximately 30% of the pre-exercise value at the end of exercise, and recovered to approximately 70 and 100% of the pre-exercise value after 12 and 24 hours, respectively. Based on the data of MacDougall *et al.* (153).

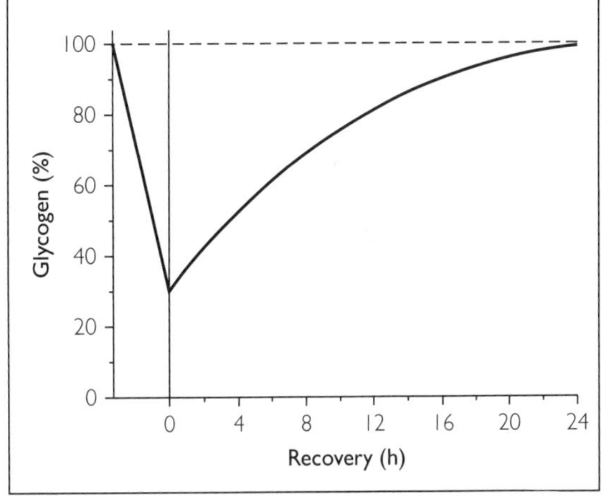

can be restored to approximately 90% of the pre-exercise value in 6 hours (Figure 5.95). In between these extremes, after a total of approximately 10 minutes (excluding resting time) of high-intensity intermittent exercise, muscle glycogen was fully restored in 24 hours (Figure 5.96). Glycogen is resynthesized faster in type 2 muscle fibers (14,192,196); therefore, recovery of glycogen to pre-exercise levels would be relatively fast after repeated, intense exercise because it is the type 2 fibers that are extensively depleted.

Linking Mechanisms of Recovery to Recovery of Performance

Since recovery mechanisms have different time courses, some aspects of performance can recover completely at a time when others have not. For example, strength, the ability to produce brief, maximal contraction, recovers faster than endurance, the ability to sustain submaximal contractions (212) (Figure 5.97). Following brief, maximal contractions (e.g., "Short" in Figure 5.88), recovery of strength, activation, ATP, ADP, PCr, and P_i all have a similar time course, suggesting they may be linked (19). In contrast, after repeated maximal contractions (e.g., "Long" in Figure 5.88) and sustained submaximal contractions, the slower recovery of strength and endurance is more temporally linked with recovery of muscle pH and excitation–contraction coupling (19,212).

Fully restoring muscle glycogen after prolonged, endurance exercise lasting more than 90 minutes would contribute to being able to replicate the performance (122), but there is more to recovery from a prolonged endurance event than restoring muscle glycogen, such as repairing muscle damage (Box 5.11). On the other hand, daily non-exhausting endurance training sessions lasting approximately 1 h may not require complete recovery of muscle glycogen to be able to complete training sessions and maintain the same level of perceived exertion (224).

A distinction should be made between recovery from acute fatigue and recovery from training sessions. For example, following a weight-training session or a session of sprint training, muscle glycogen will be restored within 24 h (Figures 5.95 and 5.96), as will other fatigue mechanisms such as muscle activation and muscle pH. Yet daily strength and power

Figure 5.97 Strength vs endurance recovery. (Left) At left an isometric MVC is continued for a set time and force decreases by 50%. At right a second contraction is done after a recovery period. The force at the beginning of the second contraction was the same as the first, indicating complete strength recovery, but the rate of decrease in force was greater, indicating incomplete endurance recovery. (Right) At left a brief MVC is done followed by a submaximal (50% MVC) contraction sustained as long as possible. At right, after a recovery period, the MVC attained the same force level as the first MVC, indicating complete strength recovery but the submaximal contraction was not sustained as long, a sign of incomplete endurance recovery.

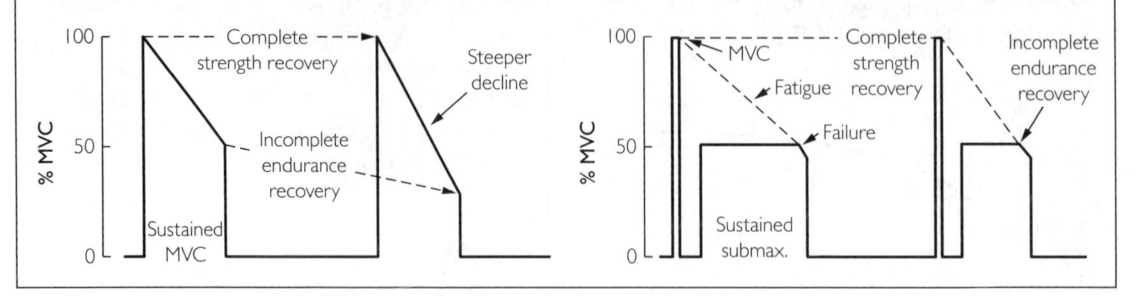

training of the same muscle group is not well tolerated even by highly trained athletes. Thus, additional factors are involved in recovery from training sessions or some competitions, such as repair of muscle damage (Box 5.11) and the time course of muscle protein synthesis. Recovery from training sessions will be discussed in more detail in Chapter 8.

Effect of Training on Recovery

A goal of training is delaying fatigue but it has the added benefit of speeding recovery. Examples include accelerated glycogen (126) and PCr (166) resynthesis, and increased buffering capacity (255). Increased muscle capillarization with training (discussed in Chapter 8) also accelerates recovery. Genetic endowment expressed in various ways could also affect resistance to and recovery from fatigue (Box 5.12).

BOX 5.11 INTERACTION BETWEEN EXERCISE-INDUCED FATIGUE AND MUSCLE DAMAGE

A prolonged endurance event like a marathon run causes not only fatigue but also muscle damage (251). Muscle damage most likely occurs in prolonged activities with an eccentric contraction component (50). Some effects of fatigue could also be attributed to muscle damage, such as loss of strength (136) and LFF (136). Muscle damage also delays muscle glycogen resynthesis (77). Thus, the prolonged loss of strength after a marathon run (222) and the LFF after prolonged exercise (261) likely reflect both mechanisms of fatigue and muscle damage. For example, both fatigue and muscle damage impair excitation–contraction coupling, resulting in LFF.

Muscle damage can also occur after high-intensity intermittent exercise and resistance exercise (e.g., 106); therefore, delayed recovery from this type of exercise could also reflect a combination of fatigue and muscle damage.

Factors Affecting Strength, Power, and Speed Performance

Several factors that influence strength, power, and speed performance have already been covered: in Chapter 4 we discussed the influence of the force– and

BOX 5.12 ROLE OF GENETIC ENDOWMENT IN RESISTING FATIGUE

It is recognized that the superior performance of elite endurance athletes is the result of both training and genetic endowment. Determining the relative importance of training and endowment is difficult. One way to establish the existence of a genetic factor is to test *untrained* muscles in elite athletes and non-athletes. In one study the wrist flexors of elite distance runners exhibited superior fatigue resistance that was associated with a higher level of ATP and lower P_i/PCr ratio, indicating greater oxidative capacity, and a smaller decrease in muscle pH. By testing a small muscle group the effect of the runner's greater cardiac output was minimized (189). Inheriting a high percentage of type I fibers would contribute to increased fatigue resistance.

power–velocity relationships, contractile history (e.g., stretch-shortening cycle potentiation, and isometric force enhancement and depression), muscle length, and strength curves; in this chapter we have covered the effects of neural activation and fatigue. Some additional factors affecting strength and power performance will be covered in this section.

Muscle Size

The most obvious factor affecting strength and power is muscle size, usually expressed as **muscle cross-sectional area (CSA)**. CSA can be measured as **anatomical CSA (ACSA)** or **physiological CSA (PCSA)**. ACSA is measured as the area of a transverse section of the muscle, whereas PCSA is the summed CSA of the muscle's muscle fibers. In relating muscle force to muscle size, PCSA is the most valid measure because ultimately the goal is to measure the summed CSA of all active muscle fibers. In a muscle where all fibers are parallel to the tendon's line of pull and all fibers pass through the point of measurement, ACSA is equal to PCSA (Figure 5.98, left). However, many muscles have a pennate fiber arrangement in which fibers are attached obliquely to the tendon's line of pull. By allowing more fibers to be attached to the tendon, the pennate arrangement causes PCSA to exceed ACSA (Figure 5.98, right) (see also Box 5.13). The pennate muscle shown schematically in Figure 5.98 is called **unipennate** because fibers are attached on one side of the tendon. Other muscles called **bipennate** have fibers attached on both sides of the tendon, like a feather. Still other muscles are **multipennate**. The difference between ACSA and PCSA could be particularly pronounced in bipennate and multipennate muscles.

PCSA and ACSA are usually correlated with *isometric* force but have also been correlated with weight lifting 1 RM (168) and concentric force and power at various velocities (214,218). To establish true muscle force, torque measurements must be corrected for moment arms (158) (review Figures 4.71 and 4.72) and activation level. In pennate muscles a correction must be made for fiber pennation angle (158) (Figure 5.98, right).

The correlation between CSA and muscle force (strength) is positive and fairly high but not perfect (Figure 5.99), partly because of the methods used to measure force and CSA (24,91,100) and partly because factors other than CSA could affect strength, e.g., intersubject differences in muscle activation (47,158) and muscle specific force (discussed in the section on

Figure 5.98 Distinction between anatomical cross-sectional area (ACSA) and physiological cross-sectional area (PCSA). ACSA is measured as the area of a transverse section (horizontal dashed line) of the muscle. PCSA is the summed CSA of the muscle's muscle fibers. (Left) Muscle fibers are arranged in parallel with the line of pull of the tendon (vertical arrow). In this case ACSA = PCSA. (Right) Muscle fibers are attached at an angle (obliquely) to the line of pull of the tendon, a fiber arrangement referred to as pennate (or pinnate). The pennate arrangement allows a greater number of muscle fibers to be attached to the tendon, producing a large PCSA. Because of the pennate orientation of the muscle fibers, the PCSA is greater than the ACSA.

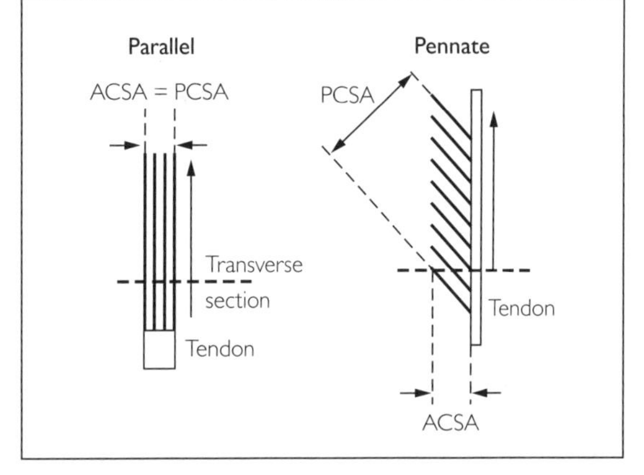

BOX 5.13 ADVANTAGES AND DISADVANTAGES OF PENNATE MUSCLE FIBER ARRANGEMENT

One advantage of the pennate fiber arrangement is that it increases the force of a muscle's contraction by allowing more fibers to attach to the tendon and thus increasing PCSA. A disadvantage is that the pennate arrangement of fibers causes them to pull obliquely to the tendon's line of pull (Figure 5.98, right) so that some of the fibers' contraction force is lost; however, this loss is more than compensated for by the increased number of muscle fibers attached to the tendon. An additional advantage of the pennate fiber arrangement, for example the bipennate arrangement, is that a given absolute amount of fiber shortening causes a greater absolute amount of tendon (and whole muscle) shortening (164). Thus, in a shortening contraction the velocity of whole-muscle shortening is greater than muscle-fiber shortening velocity.

Specific Force), the latter affected by fiber-type distribution (Figures 5.6 and 5.7). Thus, subjects could have the same strength but different CSA or vice versa (see boxes in Figure 5.99).

Specific Force

Specific force, also called **specific tension**, is the ratio of a muscle's force to its PCSA. Specific force is expressed in kilonewtons per square meter (kN·m^{-2}) or newtons per square centimeter (N·cm^{-2}). Specific force is traditionally determined from isometric force measurements made at optimal muscle length but isokinetic concentric force (9) and weight-lifting 1 RM (168) have also been used. Here the focus will be on isometric force. Measurement of specific force in animal muscles has produced values of 20–30 N·cm^{-2} (58,91), which are within the range found in human muscles (Figure 5.100). Specific force measurements in humans require corrections for muscle moment arm and fiber pennation angle and, if voluntary rather than stimulated (158) contractions are used, corrections for variation in agonist activation and antagonist co-activation must be made (91,179).

There is a large inter-subject variation in specific force. In the quadriceps, for example, the reported range was from approximately 20 to 40 N·cm^{-2} in 27 young men (91). In this example most of the factors that might cause variation in specific force were carefully controlled; therefore, it is difficult to explain the large inter-subject variation in specific force. There is evidence that human type 2 muscle fibers have greater specific force than type 1 fibers (219) but fiber-type distribution likely does not account for all variation in specific force.

Body Size (Mass)

Since muscle CSA is positively correlated with body mass, particularly lean (also called fat-free) body mass (e.g., 9), a positive correlation between strength and body mass (or lean body mass) would be expected and has been found (46,168,194). The most striking example of the correlation between body mass and strength

Figure 5.99 Schematic relationship between muscle CSA and isometric muscle force (strength). Each symbol represents the results for one subject. If the correlation between force and CSA were perfect ($r = 1.0$), all symbols (data points) would lie on the regression line. But only three of 11 points lie on the regression line; the remainder of the results are scattered on either side of the line, producing a correlation of <1.0. Despite a good overall correlation between force and CSA, there are discrepancies. In one highlighted box two subjects have the same CSA but different force values. A second box shows three subjects with similar force but different CSA. Possible factors responsible for the discrepancies are discussed in the text.

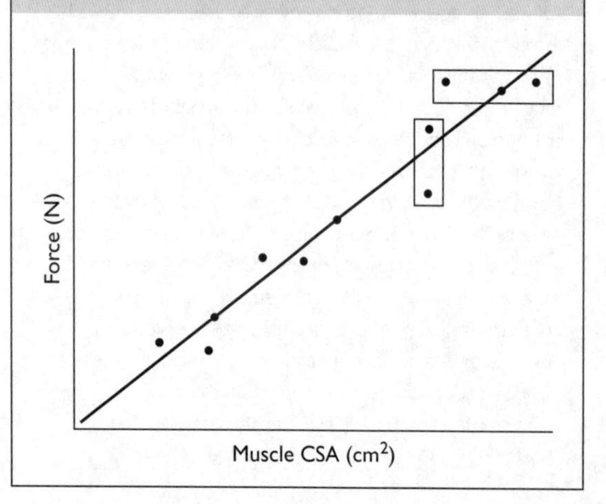

Figure 5.100 Specific force in human muscles. Values are mean ± standard deviation (SD). *N* is the number of young men who were subjects for each muscle. LG, lateral gastrocnemius; Sol, soleus; TA, tibialis anterior; Quad, quadriceps. In LG and Quad, subjects did maximal voluntary isometric contractions; force was adjusted for agonist activation and antagonist co-activation. In Sol and TA, isometric force was produced by maximal tetanic stimulation. Based on the data of Morse et *al.* (179), LG; Magnaris et *al.* (158), Sol and TA; and Erskin et *al.* (91), Quad.

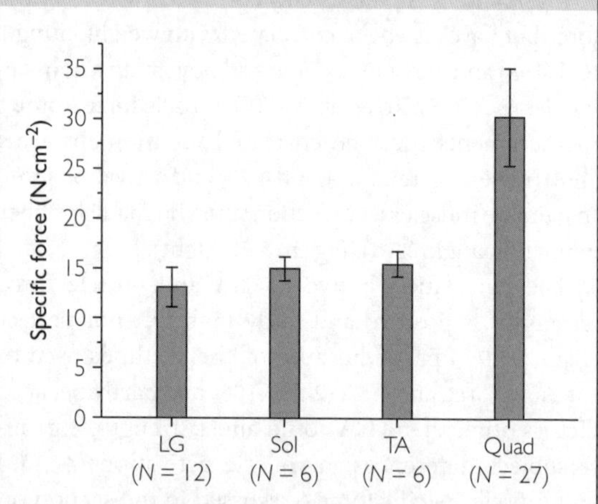

is seen in competitive weightlifters (46) because of their similar training history and, except for the highest weight class, similar body composition. Note in the top part of Figure 5.101 the very high correlation between lift total and weight class. Contributing to the high correlation were the large range in body mass and lift total. The high correlation between body mass and lift total in weightlifters (Figure 5.101) has also been observed in elite powerlifters who compete in the squat, bench press, and deadlift (46).

Absolute Strength vs Strength/Mass Ratio

Strength measurements can be expressed as **absolute** strength or **relative** strength. Absolute strength is expressed in units of force (N), torque (N·m), or, in weightlifting tests, the unit of mass (kg). Relative strength is expressed in relation to another measure such as body mass, lean body mass, or muscle CSA. When considering athletic performance, the most relevant expression of relative strength is the **strength/body mass ratio**, or

Figure 5.101 Men's and women's weight-lifting results at the 2012 Olympics. A and B show the winning lift totals (for snatch, and clean and jerk) plotted against weight class. Note the high *positive* correlation (*r*) between *absolute* strength (lift total) and weight class (body mass, BM). C and D show the strength/body mass ratio; that is, lift total divided by weight class (BM). Note the high *negative* correlation between the strength/body mass ratio and body mass. See text for further discussion.

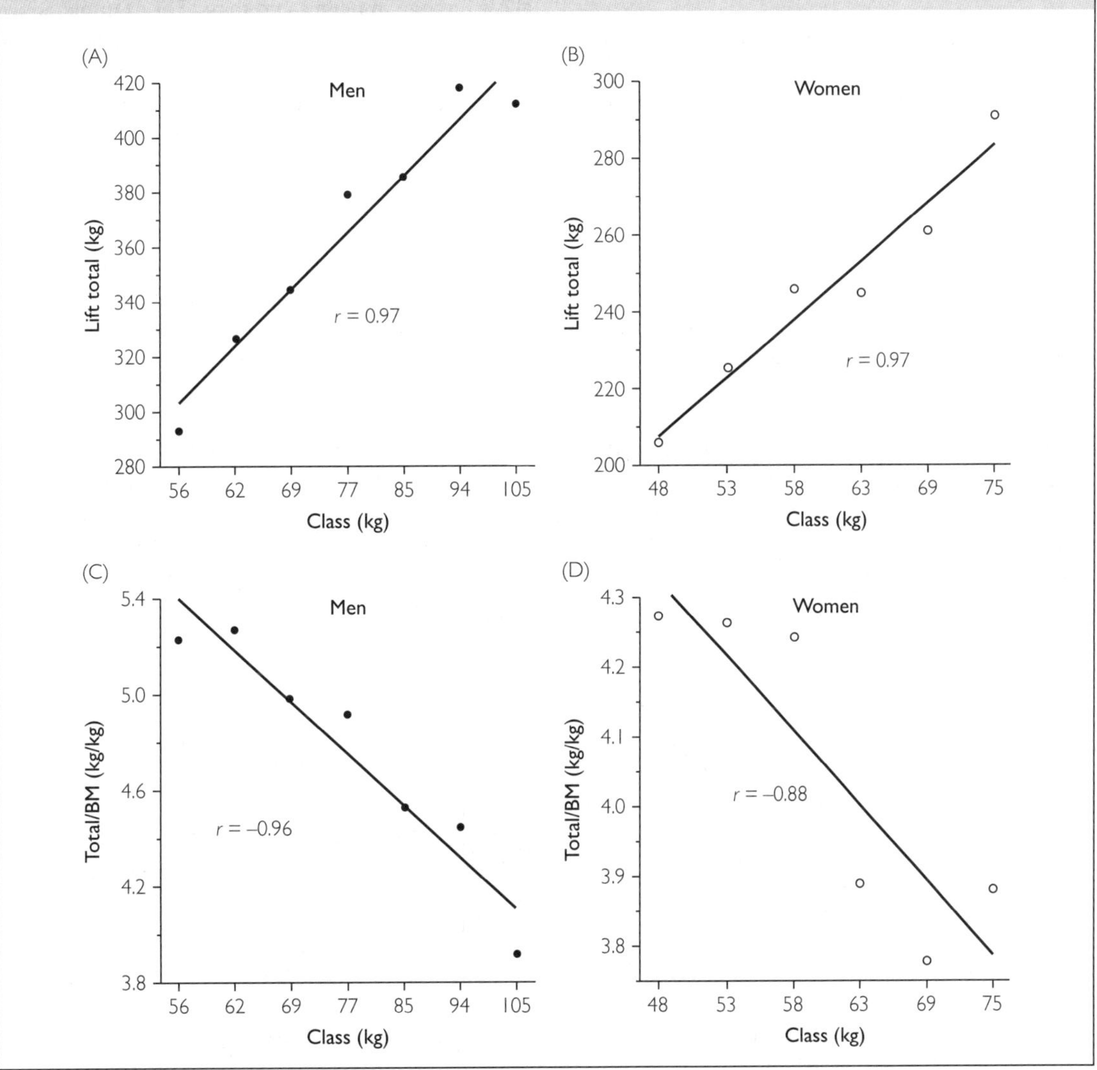

BOX 5.14 INFLUENCE OF GENDER ON THE CORRELATION BETWEEN STRENGTH/MASS RATIO AND BODY MASS

The negative correlation between body mass and the strength/mass ratio is seen in women and men separately (Figure 5.101, bottom) but may not be present if the data from male and female athletes are combined (e.g., 194). The negative correlation between body mass and strength/mass ratio is abolished with the combined data (Figure 5.102B). The positive correlation between absolute strength and body mass is preserved but weaker (Figure 5.102A). The gender influence is probably related to gender differences in body composition and regional differences in the distribution of body mass (e.g., 168).

more simply, the **strength/mass ratio**. In contrast to absolute strength, the strength/mass ratio is *negatively* correlated with body mass; that is, as body mass increases the strength/mass ratio decreases (Figure 5.101, bottom) (see Box 5.14 for influence of gender).

Body Mass and Performance

Because *absolute* strength correlates positively with body mass, sports and athletic events requiring primarily absolute strength would tend to be dominated by tall, large athletes (e.g., football linemen, weight throwers in track and field). At the other extreme, activities requiring mainly a high *relative* strength (strength/mass ratio) are dominated by relatively short, light athletes (e.g., gymnasts, figure skaters). It would be very difficult for a lean 105 kg athlete to hold an iron cross on the rings (Figure 4.54), regardless of absolute strength. Between the cited extremes are many activities that require a high level of both absolute and relative strength. In team sports, for example, absolute strength and body mass help in physical contact with opponents, whereas a high strength/mass ratio helps to accelerate the athlete's own body mass in running, jumping, and skating. The many athletic events requiring a blend of absolute strength and strength/mass ratio are populated by athletes of average height and body mass, although some sports allow a range in height and mass. For example, in ice hockey one athlete may excel because of body mass and absolute strength, while another smaller player may excel because the higher strength/mass ratio provides faster acceleration in skating. In competitive weight lifting, expressing performance as the absolute weight lifted would favor the heaviest lifters, whereas expressing performance as weight lifted per kilogram of body mass would favor the lightest lifters. Fairness is achieved by having weight classes, in which athletes with similar body mass compete with each other (Figure 5.102). Weight classes are used in judo, boxing,

Figure 5.102 Men's and women's weight-lifting results at the 2012 Olympics. The data for men and women presented separately in Figure 5.101 are combined here. (A) The positive correlation (r) between absolute strength (lift total) and body mass (weight class) is maintained in the combined data but the correlation is not as high as in the separate data. (B) Combining the male and female data abolishes the negative correlation (NS, not statistically significant) between the strength/mass ratio (total/BM) and body mass (BM) seen in the women and men separately. Both men and women compete in the 69 kg class (see box). See Box 5.14 for further discussion.

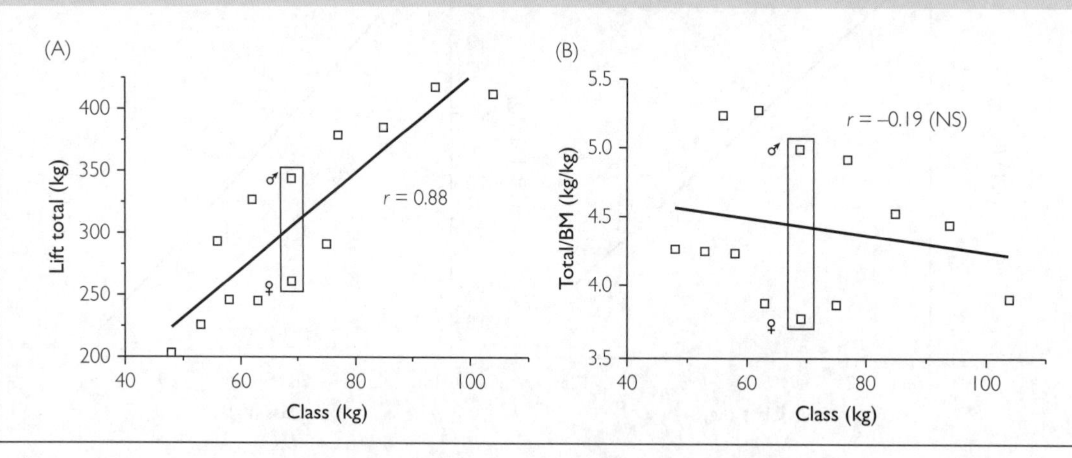

and wrestling to prevent extreme variation in absolute strength. The correlations between body mass and absolute strength and the strength/mass ratio must also be considered in strength and power testing (Box 5.15).

Why Does the Strength/Mass Ratio Decrease as Body Mass Increases?

That absolute strength increases as body mass increases is intuitive, but it is not as clear why the strength/mass ratio decreases as body mass increases. The quick explanation is that in a larger person with greater body mass, the associated increase in muscle mass is greater than muscle CSA. Since strength is directly proportional to CSA rather than muscle mass, the greater increase in mass than in CSA decreases the strength/mass ratio. Figure 5.103 provides a more detailed explanation.

BOX 5.15 SOME FIELD TESTS OF STRENGTH ARE BIASED IN FAVOR OF SMALL PEOPLE

Expressing strength as a strength/mass ratio might be viewed by some as an equalizer, as being a fair way to compare strength among individuals. But it has now been pointed out that strength/mass ratios are higher in smaller people (Figure 5.101). For this reason, some field tests of strength, such as number of completed dips and chin-ups, and the number of repetitions achieved with a weight set as a percentage of body weight, are biased in favor of smaller people because these tests assess strength relative to body weight. This point is often overlooked in classifying people as strong or weak based on these tests.

Figure 5.103 Explanation of why the strength/mass ratio decreases as body mass increases, based on a comparison of two cubes of muscle. For the comparison specific force and muscle density are given as $10\ N{\cdot}cm^{-2}$ and $1.0\ g{\cdot}cm^{-3}$, respectively. One cube is two times higher than the other. It follows that the surface area of each face of the higher cube would be four times greater. Since strength (force) is proportional to area, the larger cube would generate four times the force (absolute strength) of the small cube. However, the larger cube has eight times the volume and thus eight times the mass of the small cube. To obtain the strength/mass ratio of the larger cube, the four times greater force would be divided by the eight times greater mass. The result is that the larger cube would have half (4/8) the strength/mass ratio of the smaller cube. Note that the specific force (strength/area) of the two cubes is the same.

Strength/CSA = 10 N/cm^2 Density = 1 g/cm^3

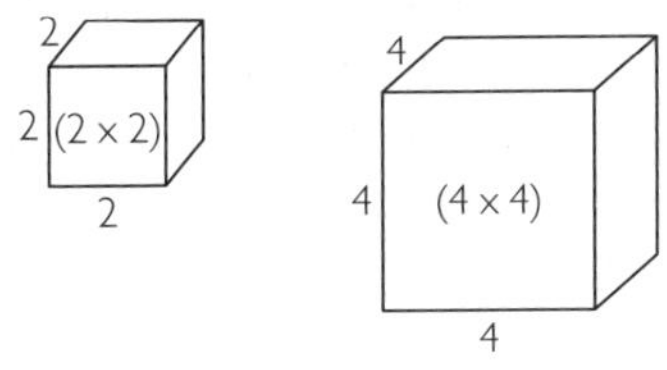

	Small cube (2 × 2)	Large cube (4 × 4)	Ratio
Length	2 cm	4 cm	2/1
CSA	4 cm^2 (2 × 2)	16 cm^2 (4 × 4)	4/1
Mass	8 g (2 × 2 × 2 × 1)	64 g (4 × 4 × 4 × 1)	8/1
Strength	40 N	160 N	4/1
Strength/mass	5 N/g (40 N/8 g)	2.5 N/g (160 N/64 g)	1/2 $\left[\frac{4}{1}/\frac{8}{1}\right]$
Strength/area	10 N/cm^2 (40 N/4 cm^2)	10 N/cm^2 (160 N/16 cm^2)	1/1

Muscle Fiber-Type Distribution

Compared to type 1 fibers, type 2 muscle fibers generate much greater concentric force and power, and have a higher maximum shortening velocity (V_{max}) (Figures 5.12, 5.13, and 5.14). Therefore, athletes with a higher percentage of type 2 fibers would excel in power and speed activities (e.g., Figure 5.23).

Muscle Temperature

Resting Muscle Temperature

In room conditions with an air temperature of approximately 20°C, resting muscle temperature (T_m) is approximately 34–36°C, about 1–3°C below core temperature (T_c) of 37°C. T_m is higher in deeper-lying than in superficial muscles and higher in the deeper parts of superficial muscles (140,217). Resting T_m in superficial muscles could be as low as 30–32°C in an inactive, unprotected athlete standing on an outdoor field on a cold, wet, windy day. If the athlete were suddenly pressed into action, low T_m would impair performance, as discussed in this section.

Active and Passive Warm-Up

Before a training session or competition most athletes do an **active warm-up** that typically consists of a bout of submaximal aerobic exercise, stretching, and sport-specific drills (97). During 10–15 minutes of submaximal aerobic exercise like running or cycling, T_m in active muscles will quickly increase by about 2–3°C and then plateau, indicating little value in prolonging this part of the warm-up beyond 15–20 minutes (12) (Figure 5.104). Passive, static stretching will have no effect on T_m but sport-specific drills will maintain or even elevate T_m further. At conclusion of warm-up, T_m decreases relatively slowly if the athlete is in a warm (≈20°C) environment or wears protective clothing (Figure 5.104). Muscles inactive during exercise show a smaller (140) or no (12) increase in T_m even if the exercise increases T_c and sweating occurs. For example, lower-body aerobic exercise (e.g., cycling) will have a limited effect on upper-body T_m. If training or competition involves the upper body, sport-specific warm-up activities are needed to increase upper body T_m.

Passive warming is achieved with warm-water baths or heated blankets or apparel. For example, immersing the leg in a water bath at 44–46°C for approximately 30 minutes will increase T_m to the same level

Figure 5.104 Schematic illustration of the increase in quadriceps muscle temperature during 15 minutes of submaximal (≈50–60% $\dot{V}O_{2max}$) aerobic exercise and the time course of temperature decrease after exercise. Based on the data of Saltin *et al.* (216) and Kenny *et al.* (140).

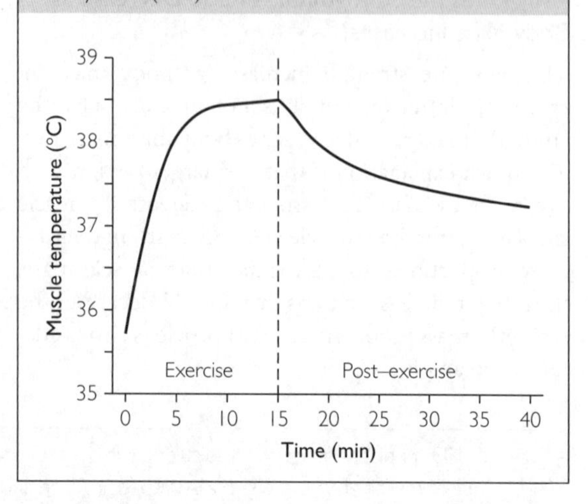

(e.g., 39°C) achieved with exercise (61). To simulate the cold exposure an athlete might experience, passive cooling is done with cold water baths at temperatures as low as 1°C sustained for 30–45 min, which can lower T_m to 20–30°C (36,57,61).

Isometric Contractions

Increasing T_m above a normal resting level increases RFD and RFR more than peak force (36,62,69) (Figure 5.105). Similarly, cooling has a greater effect on RFD and RFR than peak force (36,62,69).

Force– and Power–Velocity Relationships

Changes in T_m have a greater effect on the concentric than eccentric force–velocity relationship (204) (Figure 5.106). Increasing T_m increases concentric force and V_{max}, and reduces the curvature of the force–velocity relationship (70). Reduced force–velocity relationship curvature produces greater force at intermediate velocities or, alternatively, higher velocities are attained with given loads (review Figure 5.12). Fatigue has the opposite effect on the concentric force–velocity relationship (Figure 5.82), as does lowering T_m. The positive effects of increased T_m on the concentric force–velocity relationship translate to greater power and an increase in the velocity at which peak power occurs (Figure 5.107).

Figure 5.105 Schematic illustration of the effect of muscle temperature on isometric peak force (PF), RFD, and RFR. Subjects did brief maximal voluntary contractions and then relaxed as quickly as possible. Relatively low resting and high post-warm-up muscle temperatures are shown. The indicated effects would be smaller for a smaller temperature range (e.g., 36–39°C). Note the rank-order effect of the temperature effect: RFR > RFD > peak force. Based on the data of Asmussen *et al.* (13).

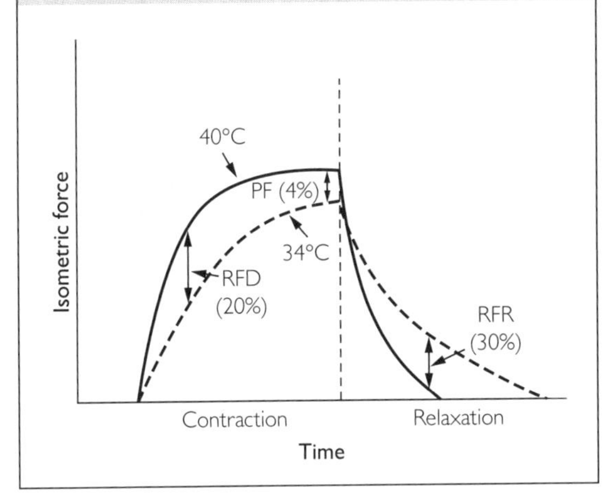

Figure 5.106 Schematic illustration of the effect of muscle temperature on the force–velocity relationship. Note the greater effect on the concentric side of the force–velocity relationship. With cooling, maximum shortening velocity (V_{max}) is decreased and the curvature is increased. Based on the data of de Ruiter and de Hann (70).

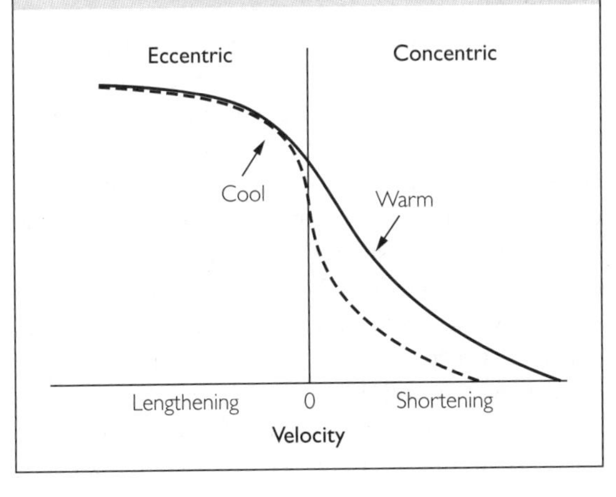

Figure 5.107 Schematic illustration of the effect of muscle temperature on the concentric power–velocity relationship. Peak power is affected more than maximum shortening velocity (V_{max}) because power is the product of force and velocity. Note that temperature also shifts the velocity at which peak power occurs. Based on the data of de Ruiter and de Hann (69).

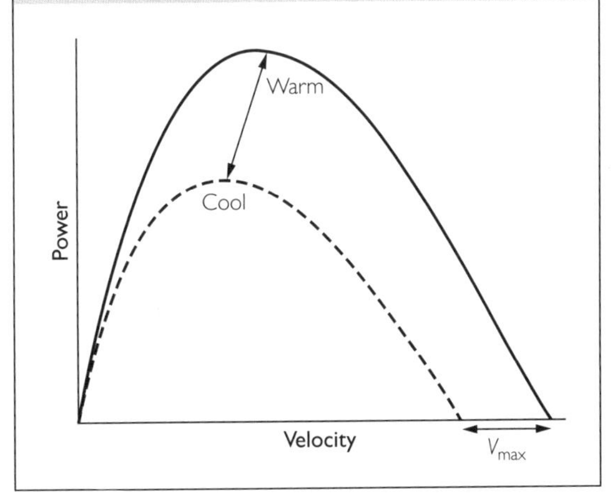

Performance

The effects of T_m on isometric contractions and the force– and power–velocity relationships result in significant effects on strength and power performance (174,217). Figure 5.108 shows examples in which T_m above or below the normal resting range can positively and negatively affect performance, respectively. With warm-up T_m can be increased up to 38–39°C, increasing force, speed, and especially power (e.g., 174). Also important is the large impairment with low T_m that could occur in some outdoor settings if an athlete is not protected from the cold. Thus, protecting an athlete from low T_m is as important as raising it a few degrees above normal resting values. After warm-up has elevated T_m, it is important to maintain it until performance begins (93). Similarly, the increased T_m produced by warm-up and maintained by a training session or competition should not be allowed to decrease during prolonged rest periods such as between periods in hockey, quarters in football, and halves in soccer (see Box 5.16).

Mechanisms by which Increased T_m Enhances Performance

Elevated T_m increases maximal contraction force by increasing the average force of attached myosin CBs.

Figure 5.108 Effect of muscle temperature on vertical jump height (top), maximum cycling speed (middle), and maximum cycling power (bottom). Core and muscle (vastus lateralis) temperature were increased and lowered by exercise and cold-water immersion of the legs, respectively. In the control condition, a normal core temperature of 37°C was associated with an average muscle temperature of 36.2°C. Vertical dashed lines indicate the expected temperature range in resting muscle at a room temperature of 20°C. Based on the data of Bergh and Ekblom (36).

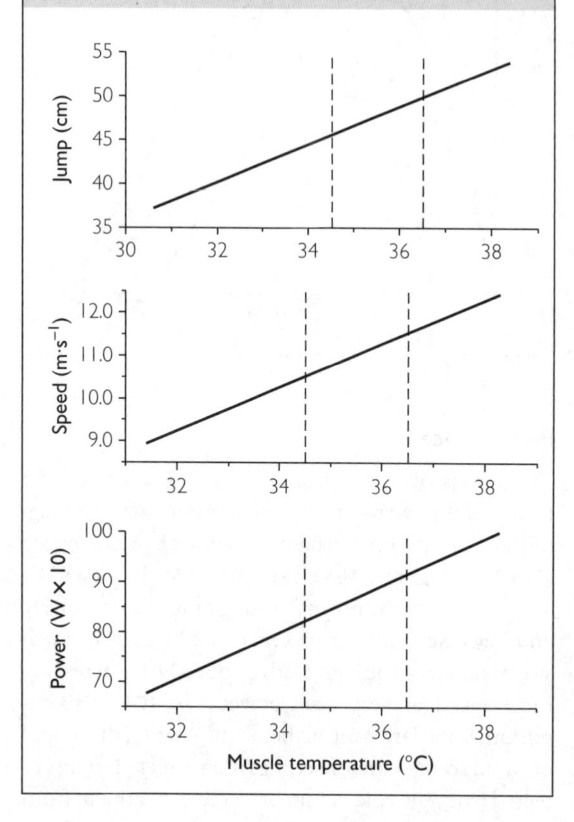

The number of attached CBs does not increase, but a greater proportion of them are in the high-force state of the CB cycle (59,63,148,203) (check Figure 4.7 to review the CB cycle). Increased RFD and V_{max} and decreased curvature of the concentric force–velocity relationship result from the increased rate of CB cycling brought about by increased myosin ATPase activity (203) that in turn increases the rate of ATP turnover (112). Increased nerve (39,175) and muscle fiber action potential (92,112) conduction velocity promote a more rapid release of Ca^{2+} from the sarcoplasmic reticulum (72), leading to a faster onset of CB cycling. Rate of relaxation increases

BOX 5.16 THE BENEFIT OF RE-WARM-UP

In many team sports there are 15–20 minute rest periods between periods of play. A study of soccer players found that the increased T_m attained during the pre-game warm-up was sustained during the first half. However, during the 15–20 minute half-time intermission of passive rest, T_m decreased by 1.5°C and was not restored to the first-half level at 5 minutes into the second half. The same T_m decrease was also associated with impaired performance of an intermittent sprint test. In contrast, players who spent the last 7–8 minutes of half-time intermission doing a re-warm-up of low- to moderate-intensity exercise were able to elevate T_m to the level attained at the end of the first half. In addition, there was no impairment of sprint test performance. Based on these results it was recommended that the first approximately 7 minutes of intermission be spent on giving tactical advice and allowing for rehydration and mental recovery, then spending the final 7–8 minutes doing a re-warm-up. The re-warm-up will allow the players to perform better in the first part of the second half (174). The same approach can be applied to any sport in which the athlete has prolonged rest periods before going back into action. It is important to select exercise intensities that minimize the risk of causing fatigue.

from increased rate of CB cycling, a faster dissociation of Ca^{2+} from troponin, and faster reuptake of Ca^{2+} by the sarcoplasmic reticulum (70). The relatively small effect of increased T_m on eccentric contractions is largely attributed to incomplete CB cycles; CBs undergo repeated strain and release without going through the state of the CB cycle, involving P_i release and ATP breakdown, that is affected by T_m (204).

Effect of Core Temperature (T_c) on Strength and Power

Typical warm-ups increase T_c as well T_m. If done in a comfortable, ambient environment the increase in T_c should have no effect on strength and power in brief maximal efforts. However, increasing T_c to high levels (39–40°C), either by passive exposure to heat stress or by exercising in a hot environment, may in some conditions *decrease* strength (178,241). Hyperthermia (elevated T_c) can reduce the ability to activate muscles (56,99) by increasing brain temperature (182). The concurrent increase in T_m would also make it more difficult to fully activate muscles because the faster contractile response would shift the force–frequency relationship to the right (Figure 5.109), requiring

Figure 5.109 Effect of muscle temperature on the force–frequency relationship. Force has been normalized for each temperature. Compared to the normal resting muscle temperature (T_m), cooling and warming increases and decreases twitch contraction time, respectively. For a given interval between stimuli (vertical arrows), which corresponds to a given frequency and MU firing rate, there is less summation of contraction when T_m increases. Therefore, MUs must fire at higher rates to achieve full summation of contraction. Note that lowering T_m has the opposite effect.

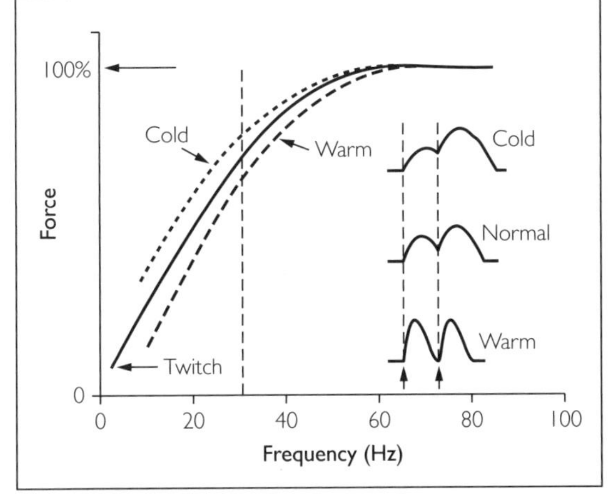

higher MU firing rates to achieve full summation of contraction (245). However, this effect is partially offset by the leftward shift of the force–frequency relationship caused by contractile slowing associated with fatigue. Postactivation potentiation, discussed in the section of that name, also causes a leftward shift of the force–frequency relationship.

High-velocity concentric force is less affected by hyperthermia than isometric and low-velocity concentric contractions (99), perhaps because activation is less impaired (99) and because high-velocity concentric contractions benefit more from the increased CB cycling rate caused by increased T_m. In fact, exposure to a hot, humid environment that increases T_m may increase the peak power attained in one or two sprints (21) but the concomitant increase in T_c may impair performance of multiple sprints (79). Although the focus here has been effect of T_m and T_c on strength and power, it is important to recognize that elevated T_m and especially T_c can have a large impact on endurance performance (see Box 5.17).

Warm-Up: More Than Just Increasing T_m and T_c

Warm-up activities can enhance performance in ways independent of increased T_m and T_c. For example, a warm-up bout of continuous exercise of sufficient

BOX 5.17 EFFECT OF T_m AND T_c ON ENDURANCE

In the Fatigue section of this chapter, brief reference was made to the role of hyperthermia (elevated T_c) in impairing prolonged submaximal exercise performance (see Figure 5.81). During prolonged exercise T_c steadily increases and more rapidly in a hot, humid environment, reaching values of 39–40°C. Fatigue develops more rapidly, shortening the time to fatigue failure (152,190). Hyperthermia can impair prolonged exercise performance by several mechanisms (56) that will not be reviewed here, although one mechanism, impaired muscle activation, was discussed in the text in relation to strength and power performance.

Increasing T_m without increasing T_c can also decrease endurance or increase the rate of fatigue in repeated isometric and concentric contractions (69,72). In sustained submaximal isometric contractions, the optimal starting T_m for endurance can be as low as 27°C, several degrees below the normal resting range of T_m (57,85,242). The efficiency (economy) of isometric contraction is reduced at higher T_m because the increased rate of CB cycling accelerates fatigue by increasing the rate of ATP turnover. In dynamic exercise, however, the effect of increased T_m on efficiency depends on the velocity of concentric contractions. In cycling for example, increased T_m may reduce efficiency when pedaling at 60 rpm but increase it at 120 rpm (94).

Due to the deleterious effect of hyperthermia in prolonged endurance exercise and repeated high-intensity exercise, various pre-cooling methods have been tested for their effectiveness in delaying or preventing the high T_c that hastens fatigue (171,202,210), particularly if the exercise is to be performed in a hot, humid environment (202,226,252). Pre-cooling is definitely effective for prolonged, submaximal exercise, but there is a dilemma in applying pre-cooling to high speed/power activities such as multiple sprints. Pre-cooling may impair power at the beginning of intermittent, high-intensity exercise but delay fatigue later on, whereas pre-warming will have the opposite effect (53,228). T_m does not have to decrease very much before there is a decrement in sprint performance (Box 5.17, Figure 5.108). Therefore, deliberate cooling to lower T_c must be used cautiously in repeated high speed/power activities to avoid lowering T_m to the point where speed and power are adversely affected (78).

intensity and duration increases the endurance of a subsequent bout of intense exercise lasting approximately 2–10 minutes (132) (Figure 5.110). The increased endurance is associated with an increased aerobic contribution to the total energy turnover during the subsequent exercise. It is important to note that the common 15–20 minutes of low-intensity exercise (e.g., 50–60% $\dot{V}O_{2max}$) typical of general warm-ups (97) will probably not produce the beneficial shift in aerobic metabolism even if T_m increases (Figure 5.104). The required intensity for the aerobic shift is above the lactate threshold but the optimal intensity and duration of the aerobic shift type of warm-up and the optimal rest period between warm-up and performance have not been determined (18,132). The optimal prescription is likely sport-specific (40).

Postactivation Potentiation

Almost any type of **pre-activity**, including weightlifting exercise, sport-specific drills, and various warm-ups, activate a mechanism causing **activity-dependent potentiation** (155) or **postactivation potentiation** (PAP) (213). Here, the general term PAP will be used. PAP has been traditionally demonstrated as an increase in evoked isometric twitch force following activity (Figure 5.111). The principal mechanism of PAP is the **phosphorylation** of a myosin **regulatory light chain** (RLC) that is a part of the neck of the myosin CB (260). RLC phosphorylation (P-RLC) is caused by Ca^{2+} released from the sarcoplasmic reticulum during muscle contraction. In addition to binding to troponin to initiate CB cycling, Ca^{2+} activates an enzyme that phosphorylates the RLC, which in turn alters the orientation of the CB, moving it closer to the actin filament, resulting in an increased probability of the CB binding to actin in response to Ca^{2+} (154,155,234) (Figure 5.112). Put another way, P-RLC increases the sensitivity of myosin–actin interaction (CB cycling) to a given amount of Ca^{2+} (234). The result is increased force at a given concentration of Ca^{2+}. There may be additional mechanisms that contribute to PAP. Like

Figure 5.110 Effect of warm-up on endurance. In warm-up and no warm-up conditions, subjects cycled to exhaustion at intensities corresponding to 100, 110, and 120% $\dot{V}O_{2max}$. The warm-up was 6 minutes of cycling at 75% $\dot{V}O_{2max}$, followed by a 10 minute rest period. Warm-up increased endurance at all three intensities. Based on the data of Jones *et al.* (133).

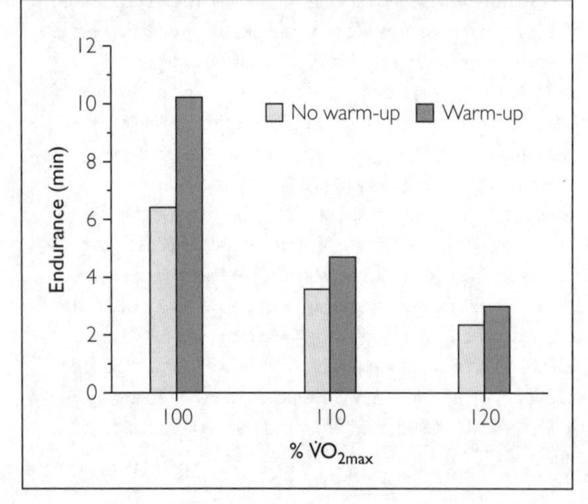

Figure 5.111 Schematic illustration of postactivation potentiation (PAP), indicated as an increase in maximal isometric twitch force. (A) A series of isometric twitch contractions is evoked in rapid succession. Twitch force increases (is potentiated) during the series. This type of potentiation is called **staircase** or **treppe**. (B) A twitch is evoked before and after an isometric MVC or evoked maximal isometric tetanic contraction (tetanus). This type of twitch potentiation is called **post-tetanic potentiation** (PTP) if the pre-activity is an evoked tetanus, and PAP if a MVC is used. Note the difference in the timescale of the twitches versus the MVC or tetanus. Part B of the figure is from Sale (213), ©2002 The American College of Sports Medicine, with permission.

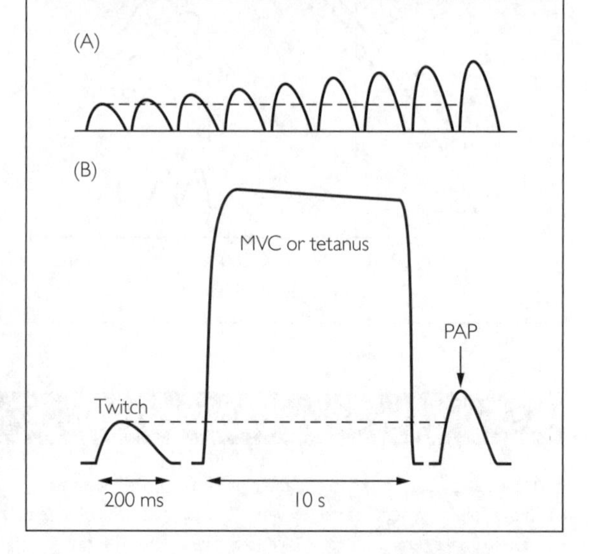

Figure 5.112 Mechanism of PAP. Myosin molecules are composed of two myosin heavy chains, each consisting of tail, neck, and head regions. For simplicity only one myosin heavy chain is shown in the figure. Attached to the neck region is a regulatory light chain (RLC). At rest (unpotentiated) the neck is oriented so that the head is farther from the actin filament, reducing the probability of myosin–actin interaction. Ca^{2+} released during contraction activates an enzyme that phosphorylates the RLC (P-RLC). The P-RLC alters the orientation of the neck region, moving the head closer to the actin filament and therefore increasing the probability of CB cycling for a given amount of Ca^{2+} released. The result is potentiation of twitch force as illustrated in Figure 5.111. If necessary, check Chapter 4 to review the CB cycle and excitation–contraction coupling.

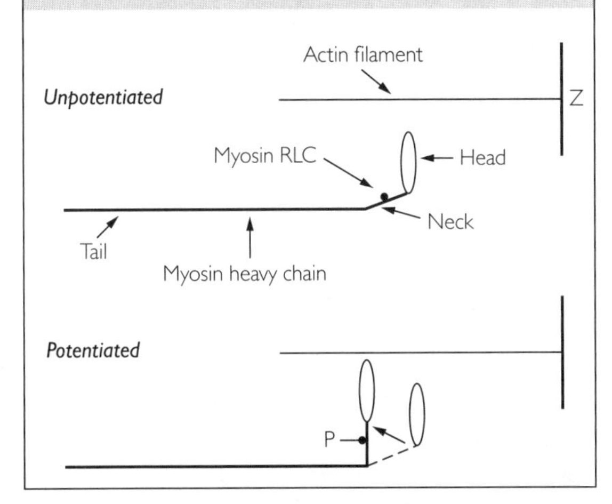

Figure 5.113 Effect of PAP on the isometric force–frequency relationship. The figure shows the force–frequency relationship in a control condition and after PAP has been induced. The lowest frequency corresponds to a twitch contraction. Since frequency determines the amount of Ca^{2+} released and therefore its concentration ($[Ca^{2+}]$), low and high frequencies correspond to low and high $[Ca^{2+}]$, respectively. The amount of Ca^{2+} released in a twitch or low-frequency tetanus is relatively small (i.e., low $[Ca^{2+}]$). Therefore, by increasing sensitivity to Ca^{2+}, PAP increases twitch and low-frequency force. In contrast, at high frequencies corresponding to high $[Ca^{2+}]$, saturating Ca^{2+} levels ensure a maximum number of cycling CBs; therefore, increasing sensitivity to Ca^{2+} (PAP) has no effect on force. ISO_{max}, maximum isometric force. If necessary, check Chapter 4 for a review of excitation–contraction coupling and the force–frequency relationship.

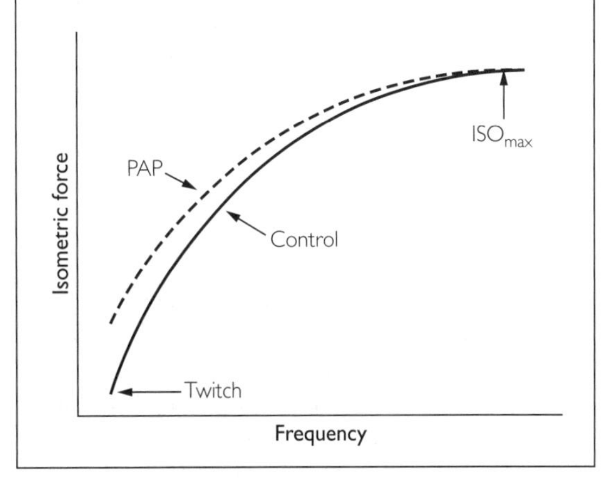

P-RLC, these mechanisms affect excitation–contraction coupling (155). For simplicity, in this discussion P-RLC will be considered the sole mechanism of PAP.

By increasing sensitivity to Ca^{2+}, P-RLC increases isometric twitch and low frequency tetanic force. However, high-frequency tetanic force (maximum isometric force) is not affected by P-RLC because Ca^{2+} is at saturating concentrations where increased sensitivity to Ca^{2+} has no effect (234). The result is a leftward shift of the isometric force–frequency relationship without an increase in maximum isometric force (Figure 5.113).

Factors Affecting Magnitude of Postactivation Potentiation

The magnitude of P-RLC and resulting PAP depends mainly on the intensity and duration of the pre-activity (28,249) (Figure 5.114). Increased intensity ensures that more muscle fibers are active and undergoing P-RLC. Increased duration allows more time for P-RLC to develop and, if fatigue develops, more muscle fibers will be activated and undergo P-RLC. A disadvantage of excessive duration is that fatigue could depress contractile force (PAP) (Figure 5.115) even if P-RLC continues to increase (111,234) (Figure 5.116). Figure 5.116 shows that a distinction should be made between the PAP mechanism, P-RLC, and the resulting contractile response, PAP. PAP but not P-RLC is affected by fatigue. In fact, fatigue from prolonged pre-activities could cause initial twitch depression rather than potentiation (PAP) despite a high level of P-RLC (111).

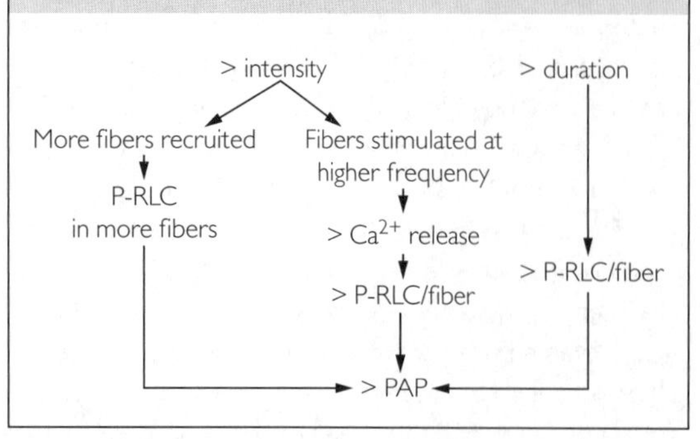

Figure 5.114 Muscle contractile activity that induces the PAP is called the pre-activity. Increasing the intensity and duration of the pre-activity increases both the number of muscle fibers that undergo P-RLC and the extent of P-RLC in each fiber.

Fiber type In response to the same pre-activity that activates both fiber types, P-RLC and thus PAP is greater in type 2 than type 1 muscle fibers (234); consequently, subjects with a higher percentage of type 2 fibers exhibit greater PAP (Figure 5.117). The greater PAP in type 2 fibers results from greater activity of the enzyme responsible for P-RLC (234).

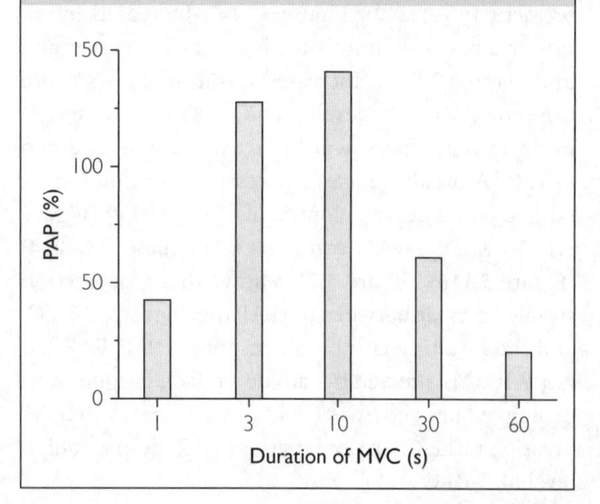

Figure 5.115 Effect of duration of a pre-activity, in this case isometric MVC, on PAP (twitch potentiation) of ankle dorsiflexors. Twitches were evoked before and after (within ≈2 seconds) the conditioning contractions. Note the large increase in PAP from 1 to 3 seconds, a much smaller increase from 3 to 10 seconds and decreases in PAP for 30 and 60 seconds. Based on the data of Vandervoort *et al.* (249).

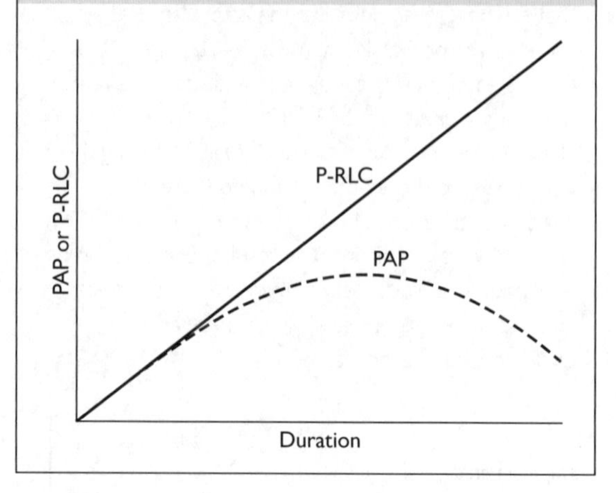

Figure 5.116 Schematic illustration of the effect of duration of a high-intensity pre-activity on PAP (twitch potentiation) and the mechanism of PAP, P-RLC. P-RLC is shown increasing in proportion to activity duration. PAP tracks P-RLC up to a certain duration then falls off and eventually declines as a result of increasing fatigue.

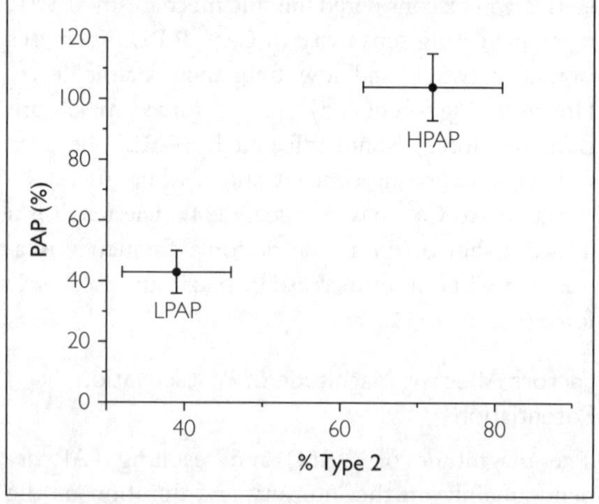

Figure 5.117 Influence of fiber-type distribution in vastus lateralis on magnitude of PAP in the whole quadriceps muscle. Maximal isometric twitches were evoked before and 5 seconds after a 10 second isometric MVC. A group of subjects (HPAP) with an average of 72% type 2 fibers had an average PAP of 104%. A group (LPAP) with 39% type 2 fibers had a PAP of 43%. Average values are given with standard deviation bars. Based on the data of Hamada *et al.* (120).

Muscle length PAP is greater at short than long muscle lengths (157,234,249,260). At long muscle fiber lengths Ca^{2+} sensitivity is already high because interfilament spacing is small; therefore, increasing Ca^{2+} sensitivity by P-RLC has little additive effect. In contrast, at short fiber lengths interfilament spacing is large so that increasing Ca^{2+} sensitivity by P-RLC has a greater effect, producing more PAP (Figure 5.118).

Pre-activity contraction type Single eccentric, isometric, and concentric contractions of similar (brief, e.g., 6 second) duration and *relative* intensity (e.g., maximal effort) produce the same magnitude of PAP as indicated by *isometric* twitch potentiation (27), suggesting that all contraction types cause the same degree of P-RLC. However, because repeated eccentric contractions produce less fatigue than concentric and isometric contractions (193), for the same degree of P-RLC, repeated eccentric contractions would exhibit greater and more sustained PAP, and a slower decay of PAP (Figure 5.119).

Figure 5.118 Mechanism for greater PAP at short muscle fiber length. (Top) At shorter fiber length there is greater space between actin and myosin filaments, resulting in reduced probability of myosin CBs interacting with actin (i.e., reduced Ca^{2+} sensitivity). (Bottom) Phosphorylation (P–) of RLC (bottom; see also Figure 5.112) orients the CB closer to the actin filament, offsetting the effect of the large interfilament spacing at short length. In contrast, the interfilament spacing is already small at a longer fiber length and thus Ca^{2+} sensitivity is already high. As a result, phosphorylation of RLC has little additive effect; that is, there is little PAP.

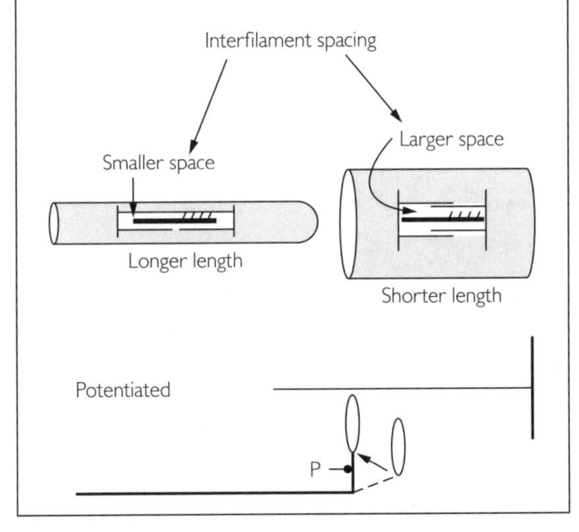

Potential Role of Postactivation Potentiation in Strength and Power Performance

PAP does not increase maximum isometric force of evoked high-frequency tetani or maximal voluntary contractions, but can increase RFD (2,28,157) (Figure 5.120) by increasing the rate of CB attachment to actin and transition to the strong binding state of the CB cycle (155,156,235). In maximal concentric contractions PAP does not increase V_{max} (233,234) but, by the same mechanism that increases RFD, reduces the curvature of the load–velocity relationship (29,157,234), allowing greater velocity to be attained with intermediate loads. The increased velocity with given loads increases power and shifts peak power to a higher

Figure 5.119 Effect of contraction type on the interaction between PAP and fatigue. Twelve young adults (six women) did 10 sets of five maximal isokinetic contractions with the knee extensor muscles at a joint angular velocity of $60°\cdot s^{-1}$. In the brief rest period (≈6 seconds) after each set, maximal twitch contractions were evoked to monitor PAP. PAP was also monitored for a 10 minute recovery period after the fatiguing sets of contractions. On separate days in random order, the sets were done with concentric and eccentric contractions. In the figure the twitch torque before the first set of contractions is equivalent to zero. Note the much greater PAP during the sets of eccentric contractions compared to concentric contractions, and the greater PAP after eccentric contractions during the first 4 minutes of recovery. Based on the data of Staples and Sale (unpublished, McMaster University).

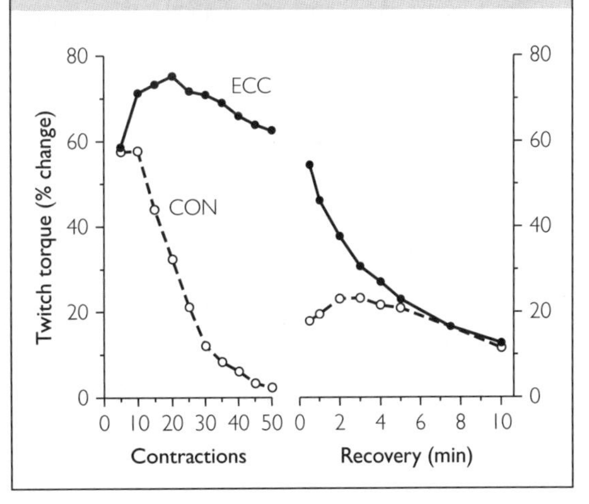

Figure 5.120 Schematic illustration of the effect of PAP on isometric RFD of a twitch and high-frequency tetanus. Note that in the high-frequency tetanus PAP does not increase the peak force attained at the final force plateau but does increase RFD. Increased RFD increases the force attained in the early phase of the contraction (vertical dashed line).

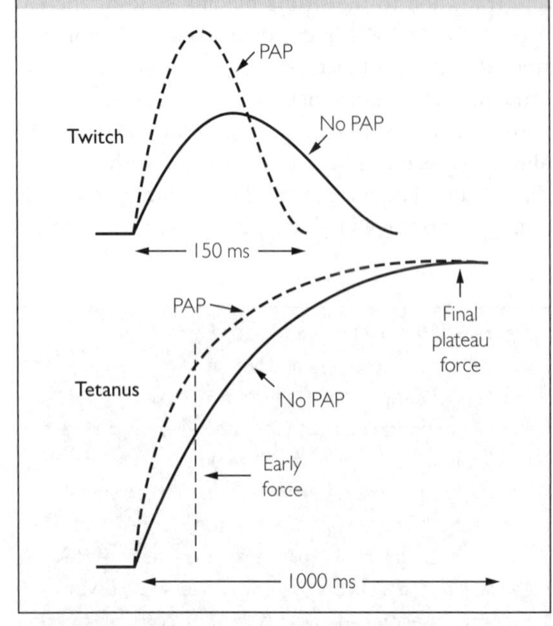

Figure 5.121 Schematic depiction of the effect of PAP on the concentric load–velocity relationship. PAP does not increase maximum unloaded (zero load) shortening velocity (V_{max}) or maximum isometric force (ISO_{max}), but reduces the curvature of the load–velocity relationship, which increases the velocity attained with intermediate loads. Since power = load × velocity, power is increased and peak power occurs at a higher velocity.

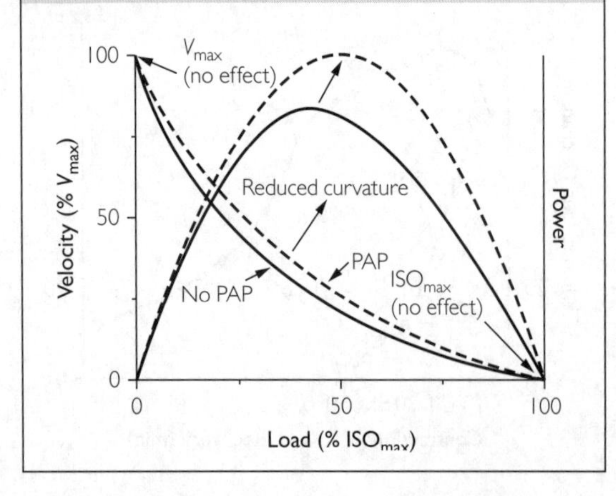

velocity (Figure 5.121). In contrast, PAP has little effect on eccentric contractions (54).

Interaction Between Postactivation Potentiation and Fatigue

PAP can increase RFD, reduce the curvature of the concentric load–velocity relationship, and increase power, all of which should enhance strength and power performance (213). Unfortunately, pre-activities that induce PAP also produce fatigue (155,205), which has effects opposite to those of PAP (Figure 5.122). Furthermore, more intense and prolonged pre-activities both amplify the PAP mechanism and increase fatigue (Figure 5.123). On completion of the pre-activity, the rate of decay in PAP may differ from the rate of recovery from fatigue, and these rates would be affected by the intensity and duration of the pre-activity (e.g.,Figure 5.123).

Figure 5.122 Interaction between PAP and fatigue. A pre-activity such as a warm-up routine activates the mechanism of PAP but also causes fatigue by various mechanisms. PAP and fatigue have opposing effects, as shown. The net effect of opposing mechanisms is uncertain because the magnitude of fatigue and PAP varies with variation in the intensity and duration of the pre-activity, and with the difference in the time course of recovery from fatigue versus the decay in PAP. ISO_{max}, maximum isometric force; ISO RFD, maximum isometric rate of force development; V_{max}, maximum unloaded shortening velocity; LVR, concentric load–velocity relationship.

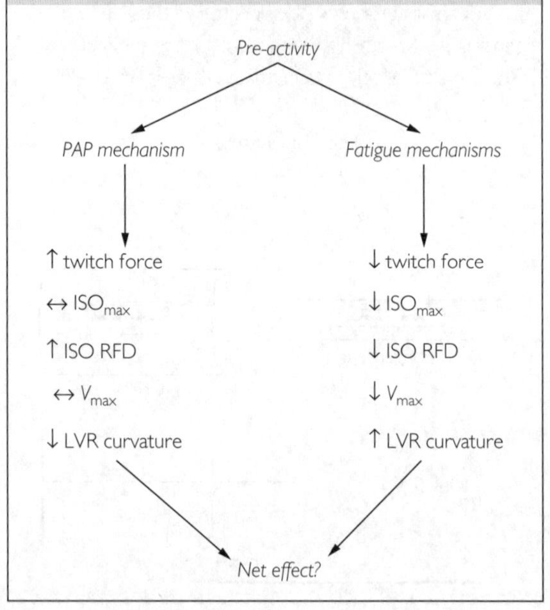

Figure 5.123 Interaction of PAP and fatigue. Isometric twitch force was measured after isometric maximal voluntary contractions of knee extensors sustained for 10 or 60 seconds on different occasions. The 60 second MVC produced both greater P-RLC and metabolic changes related to fatigue. Twitch force is expressed as percentage change from the pre-MVC twitch. Twitch force above zero indicates PAP (twitch potentiation). PAP was maximal immediately after the 10 second MVC and then declined as shown. In contrast, twitch force was depressed immediately after the 60 second MVC due to fatigue but became potentiated after approximately 1.5 minutes, and by 4 minutes PAP was greater than after the 10 second MVC, indicating the greater P-RLC produced by the 60 second MVC. Based on the data of Houston and Grange (129).

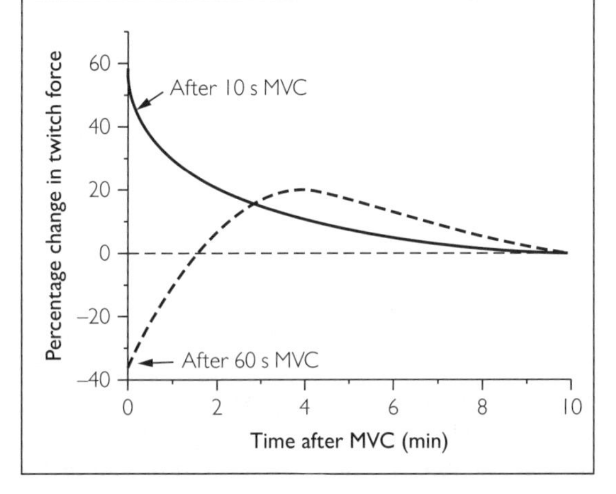

Figure 5.124 Effect of a 60 second isometric MVC of knee extensors on subsequently measured P-RLC, twitch force, and the force of a 1 second MVC. The pre-60 second MVC values (not shown) equal 100% (horizontal dashed line) on the vertical axis. The mechanism of PAP (P-RLC) was at its highest level a few seconds after completion of the 60 second MVC, then declined as shown. In contrast, twitch force, a common indicator of PAP, was initially depressed but at 4 minutes of recovery was potentiated. The force of the 1 second MVC was also initially decreased but had recovered close to the pre value after 4 minutes. The figure shows that twitch force fails to indicate the level of P-RLC when the conditioning activity produces considerable fatigue. Note also that twitch force can be potentiated at a time when MVC force has not fully recovered. Based on the data of Grange and Houston (111).

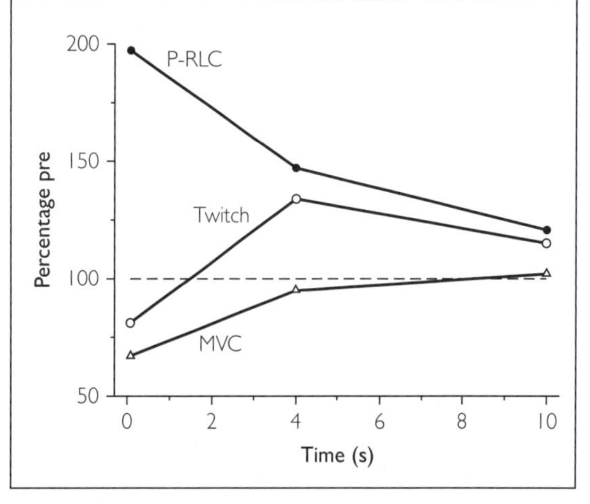

The interaction between PAP and fatigue makes it difficult to predict whether a particular pre-activity will have a positive, negative, or no effect on subsequent performance. Adding to the difficulty is that the magnitude of twitch potentiation, a common marker for the presence of PAP, may not correspond to changes in P-RLC, the PAP mechanism. For example, twitch force may be depressed rather than potentiated at the time when P-RLC is greatest (Figures 5.123 and 5.124). Alternatively, twitch potentiation may be maximal immediately after a pre-activity whereas positive effects such as increased RFD and reduced load–velocity relationship curvature are delayed (28,29). And the presence of twitch potentiation may coincide with incomplete recovery of maximal voluntary contractions (Figure 5.124); thus, the presence of twitch potentiation is not a guarantee that there is recovery of, let alone potentiation of, maximal strength and power (31). Finally, pre-activities may also increase muscle temperature (T_m), which has positive effects similar to those of PAP (Figures 5.106 and 5.107).

Acute Effect of Postactivation Potentiation on Strength and Power Performance

Many studies have tested the effect of various pre-activities, such as weight-lifting exercise, on power performance such as sprinting and jumping. While the often-stated aim of these studies has been to test the value of PAP, the previous discussion makes it clear that the success or lack of success of a particular pre-activity is the result of interaction among the magnitude of P-RLC induced, the rise in T_m attained, and the extent of fatigue incurred during the pre-activity (see Box 5.18). Also affecting the outcome is the recovery time between the end of the pre-activity and the beginning of performance. Given the large variation in the type of pre-activity, its intensity and duration, and the duration of recovery periods that have been tested, it is perhaps not surprising that the studies collectively have produced equivocal results (for reviews see 76,84,127,151,156,244,258). Figure 5.125 illustrates how variation in one factor, recovery time after a particular pre-activity, could affect subsequent performance.

It may be reasonable to assume that most pre-activities produce some P-RLC, increased T_m, or fatigue; nevertheless, actual measurements of these factors are needed to determine which factors changed the most during the pre-activity, and which were dominant at the end of the recovery period when performance occurred. Few studies have included such measurements. One study used evoked twitch contractions to confirm the presence of PAP 4 minutes after a set of squats, the time at which vertical jump height increased (172). Another study monitored T_m and used twitch contractions to monitor PAP and fatigue after a traditional and a shorter warm-up, and found that the shorter warm-up was more successful in increasing 30 second sprint cycle power primarily by producing less fatigue. Since some fatigue was evident even after the short warm-up, it was suggested that an even shorter warm-up might have produced even better results (247). This study illustrates the value of including measurements of the factors acting during and after a pre-activity, to determine their roles in performance enhancement or impairment.

BOX 5.18 PROPER USE OF THE TERM "PAP"

There has been a trend to expand the meaning of the terms "activity-dependent potentiation" and "postactivation potentiation" (PAP) to include any mechanism by which a pre-activity might enhance performance. The expansion of the term PAP in this way is considered inappropriate (155), partly because it causes confusion. For example, altered neural activation and changes in muscle fiber pennation angle have been included as mechanisms of PAP (244), whereas in muscle physiology the terms "activity-dependent potentiation" and "postactivation potentiation" are attributed solely to the contractile effects of P-RLC (155,213). This latter attribution is recommended for clarity because, as discussed in the text and shown in Figure 5.125, factors in addition to PAP determine the success or failure of a pre-activity in enhancing power performance.

Figure 5.125 Schematic illustration of the interaction among the mechanism of PAP (P-RLC), muscle temperature (T_m), and fatigue. A pre-activity increases T_m and P-RLC, both having a positive effect on subsequent performance. In contrast, fatigue developed during the pre-activity has a negative effect. In the recovery period after the pre-activity, performance will be enhanced if recovery from fatigue is faster than the decrease in T_m and P-RLC, *provided* that the recovery time is long enough. The equivocal results of various studies may be partly accounted for by the selection of various recovery times, resulting in impairment, no change, or enhancement of performance.

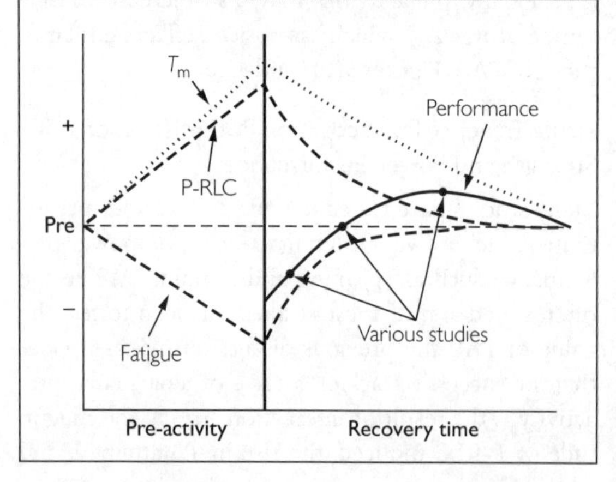

Strategies for Applying Postactivation Potentiation to Performance

The first goal of a pre-activity is to be specific by activating the same muscles that will be used in performance (119). The second goal is to activate as many muscle fibers as possible, especially the type 2 fibers that exhibit the greatest PAP. This is achieved with a high-intensity pre-activity. The third goal is to minimize fatigue. Excessive fatigue developed during the pre-activity and insufficient subsequent recovery are the main causes of a pre-activity's failure to improve performance (247,258). One challenge is to select a high-intensity pre-activity of sufficient duration to induce enough PAP without producing too much fatigue. The second challenge is to select the optimal recovery period that allows dissipation of fatigue without too much decay of PAP (Figure 5.125). Unfortunately, there is no simple formula to calculate the optimal duration of pre-activities and recovery periods. Trial and error is currently the only option. It may also be necessary to make adjustments for individual athletes based on fiber-type distribution and training history (see below).

Training history will have an influence on the effectiveness of a particular pre-activity (119,176,187). Strength training would tend to amplify the PAP induced by a pre-activity by causing preferential hypertrophy of type 2 fibers and by facilitating a more consistent activation of high-threshold MUs. However, strength training could also lead to more rapid fatigue during a pre-activity and prolong recovery by decreasing muscle oxidative capacity. On balance, athletes with strength training experience show greater performance improvement following a pre-activity consisting of sets of weight-lifting exercise (258). Endurance training, by increasing fatigue resistance and speeding recovery, would shift the interaction between PAP and fatigue in favor of PAP (167). Endurance training may also increase the content of fast myosin RLCs in type 1 fibers, which would increase the magnitude of P-RLC (220).

It should also be recognized that the performance *after* the pre-activity could produce both PAP and fatigue, particularly if the performance consists of repeated actions (110,173), and the repeated actions could themselves enhance performance, such as increasing height in a series of vertical jumps (e.g., 117).

A potential solution to the opposing effects of PAP and fatigue is to use eccentric contractions as the pre-activity. Single maximal eccentric contractions produce the same PAP and presumably P-RLC as maximal concentric and isometric contractions (27), but with repeated contractions eccentric contractions produce much less fatigue (193), resulting in greater and sustained PAP that persists at a higher level in the subsequent recovery period (Figure 5.119). A pre-activity consisting of several high-intensity eccentric contractions would maximize P-RLC and its benefits with a minimum of fatigue. Recovery from any fatigue would be rapid, allowing performance to occur sooner after the pre-activity when there has been little decay of P-RLC (Figure 5.119). A potential disadvantage of repeated high-intensity eccentric contractions is that they can produce muscle damage and soreness. However, a gradual introduction of eccentric contractions over several sessions will permit an adaptation that progressively reduces the extent of damage and soreness. A second disadvantage of eccentric contractions is that they may not be contraction-type specific to performance; however, they can be specific to muscle group and movement pattern. Thus, repeated eccentric contractions are a pre-activity worthy of testing for effectiveness in enhancing performance.

Summary of Key Points

- A MU is a motoneuron and its few hundred to several hundred constituent muscle fibers. A typical exercise muscle contains several hundred MUs.
- MUs are typed as S (slow, very fatigue-resistant), FR (fast, fatigue-resistant), and FF (fast, fatigable) based primarily on the contraction speed and fatigue-resistance of their muscle fibers, referred to as type 1, 2A, and 2X, respectively.
- In addition to pure type 1, 2A, and 2X fibers, MUs contain *hybrid* fibers that have functional characteristics intermediate between those of pure fibers.
- Faster MUs produce greater isometric force, isometric RFD, V_{max}, and concentric force and power. Fast and slow MUs are more similar in eccentric force and power.
- The superior fatigue resistance of S MUs is due to properties of both their motoneurons and muscle fibers.
- By containing a mixture of MU types most muscles can contribute to both prolonged endurance exercise and brief, fast, powerful activities.
- In muscles the distribution of different MU and fiber types varies among individuals, providing an advantage for either power or endurance performance. Differences in distribution are largely due to genetic endowment but training can induce some fiber-type transitions.
- Muscle contraction force can be graded by recruiting different numbers of MUs and varying their firing rates.
- In contractions of increasing strength, MUs are recruited according to the size principle in the order S, FR, and FF, causing S MUs to be preferentially recruited in prolonged, low-intensity activities requiring endurance. Additional

recruitment of FR and FF MUs occurs in high-intensity activities requiring strength and power.

- As force increases the firing rates of all MUs increase to reach the plateau of their respective force–frequency relationships. Faster MUs require higher firing rates because of their shorter twitch contraction times.
- In most muscles recruitment is the primary mechanism for grading force at low force levels whereas altering firing rate is more dominant at high force levels.
- The combination of MU recruitment and firing rates is called MU activation. Complete or 100% MU activation would occur if all MUs were recruited and firing at rates sufficient to reach the plateau of their force–frequency relationships. Untrained people can achieve a high percentage of MU activation in unilateral, single-joint movements but likely lower activation in the bilateral, multi-joint movements common in sport actions and training exercises.
- As RFD and velocity of contraction increase, MU thresholds decrease but there is little evidence of preferential activation of fast MUs. At onset of ballistic contractions, MU firing rates are initially high and then decrease rapidly. In slowly developing contractions MU firing rates increase as force increases to maximum but firing rates are not as high as in ballistic contractions.
- Because of greater maximum force, for a given force MU activation is less in eccentric than in isometric and concentric contractions. In maximal contractions it is more difficult to achieve full activation in eccentric than in isometric and concentric contractions. There is little evidence of preferential activation of fast MUs in eccentric contractions.
- For a given force there is greater MU activation at weaker than stronger points on strength curves and at muscle lengths farther from optimal length.
- In muscles that contribute to more than one joint action, for each action there are task groups of MUs recruited according to the size principle while other task groups may remain unrecruited. Thus, there may be violation of the size principle *across* but not *within* task groups of MUs.
- Fatigue is the exercise-induced decrease in maximum force and/or maximum power-generating capacity. Decreased force and velocity contribute to decreased power.
- The fatigue process begins almost at the onset of exercise and in time will cause failure to maintain a given exercise intensity.
- The higher the exercise intensity, the shorter the time to fatigue failure.
- Fatigue can occur at any point between the voluntary command issued from the brain to contract a muscle down to the level of the myosin CB. Ultimately, the bottom lines of fatigue are affected aspects of CB function.
- Neural mechanisms of fatigue include reduced activation of the motor cortex, reduced excitation of motoneurons by the motor cortex, reduced excitability of motoneurons, altered sensory inputs that reduce excitation of motoneurons, and impaired neuromuscular transmission.
- Muscular mechanisms include metabolic factors such as fuel depletion and metabolite accumulation that collectively can adversely affect all aspects of CB function either by direct effects on CBs or indirectly by impairing neural activation and excitation–contraction coupling.
- The relative importance of various fatigue mechanisms is affected by the intensity and duration of exercise.
- Several factors affect fatigability:
  - fatigue occurs more rapidly in repeated maximal concentric than in eccentric or isometric contractions;
  - force and power are affected more in concentric than eccentric contractions;
  - individuals with a higher percentage of S MUs and thus type 1 fibers generally have greater resistance to fatigue (more endurance);
  - several training adaptations can increase resistance to fatigue.
- After exercise, rest or reduced activity permits recovery from fatigue.
- The more rapidly fatigue occurs, the more rapid is the recovery, even if the extent of fatigue (force decline) is the same.
- The mechanisms of recovery are a reversal of the mechanisms of fatigue. Some recovery mechanisms are rapid while others are slower:
  - the ability to fully activate muscle is relatively rapid, as is recovery from any impairment of neuromuscular transmission;

  - impaired excitation–contraction coupling can persist for many hours, particularly decreasing force in response to low MU firing rates;
  - ATP, ADP, PCr, and P_i recover quickly, within minutes, whereas muscle pH may take up to an hour to return to pre-exercise values;
  - restoring depleted muscle glycogen can require several hours to a few days depending on the intensity and duration of exercise, and the extent of glycogen depletion.
- Since recovery mechanisms have different time courses, related aspects of performance can recover at different rates.
- Recovery from a training session involves more than reversal of mechanisms associated with acute fatigue. Recovery also includes repair of muscle damage and the time course of muscle protein synthesis, which can persist long after neural activation and muscle metabolic factors have fully recovered.
- Adaptations to training both delay fatigue and accelerate recovery.
- Muscle size, quantified as the summed PCSA of all of a muscle's fibers, is directly correlated with maximal strength.
- The lack of a perfect correlation between PCSA and strength results partly from the large inter-individual variation in muscle specific force, the ratio of a muscle's force to its PCSA.
- There is a positive correlation between body mass and *absolute* strength, whereas *relative* strength (strength / body mass ratio) is negatively correlated with body mass. Thus, sports requiring primarily absolute strength are dominated by taller, heavier athletes whereas sports requiring primarily relative strength are dominated by shorter, lighter athletes.
- Athletes with a higher percentage of type 2 fibers should excel in specific force and especially in concentric force and power at higher velocities.
- Increased muscle temperature can enhance power performance by increasing RFD and rate of relaxation, increasing maximum shortening velocity, and increasing the velocity attained with given loads. Exposure to cold that lowers muscle temperature has the opposite effects.
- Exercise combined with high environmental temperatures could raise core body temperature, resulting in decreased power, especially repeated power performance.
- Pre-activities such as weight-lifting exercise, sport-specific drills, and various warm-up activities induce PAP, which has the potential to enhance strength, power, and speed performance. Since pre-activities also produce fatigue, it is difficult to predict how the interaction between PAP and fatigue will affect performance. Training history and fiber-type distribution will affect the interaction between PAP and fatigue. The most successful pre-activity is intense enough to induce PAP in all muscle fibers and brief enough to avoid excessive fatigue.

PART II
Training for Different Sports and Activities

6

Training for Endurance Sports

Introduction

As reviewed in Chapter 2, **endurance sports** can be classified as those in which oxidative metabolism is the dominant energy delivery pathway. They thus include activities that require sustained efforts lasting from approximately 2 minutes to several hours. At the shorter end of the continuum (e.g., events that require a maximum effort in the approximate 2–6 minute range), maximal energy delivery from the anaerobic pathways is also an integral component. Such activities are typically referred to as **middle-distance events** and include the 800 and 1500 m track events, 200 m swimming, rowing, and paddling, and most speed-skating events (Table 6.1). Thus, in addition to training to promote adaptations that will maximize the rate of energy delivery from oxidative pathways, athletes in these sports must also concentrate on achieving adaptations that will increase the rate of energy delivery from the anaerobic pathways.

Although the key energy-delivery pathways for **team sports** are anaerobic, replenishment of high-energy phosphate stores and removal of lactate and H^+ following bursts of high-intensity activity occur through oxidative processes. It is not surprising then that endurance-trained athletes recover more quickly after brief maximal intensity efforts than sprint-trained athletes (25). Since the pattern of most team sports is one of brief bursts of maximal sprint-type efforts followed by lower-intensity recovery intervals, the athlete who recovers most quickly is best able to repeat these bursts as the game progresses. As further discussed in Chapter 7, it is thus important to include an endurance training component in training for most

Table 6.1 Examples of events that are classified as endurance sports.

Middle-distance events (aerobic/anaerobic)	Distance events (aerobic)
Track: 800 m, 1500 m	Track: 3000 m–marathon
Rowing	Triathlon
Paddling	Cross-country skiing
Swimming: 200 m, 400 m	Swimming: 800 m, 1500 m
Speed skating: 1000–5000 m	Speed skating: 10 000 m
Cycling: sprint, pursuit	Cycling: road racing

team sports. The same is true for combative sports such as boxing, wrestling, and judo.

When designing a training program for an athlete in a given sport, key variables that must be considered include the intensity of the training stimulus, the duration of the training stimulus, and the optimal amount of recovery time following each training session (frequency of training). In training for endurance sports, intensity is conventionally quantified relative to the athlete's maximal aerobic power (% $\dot{V}O_{2max}$) and, as illustrated in this chapter, it is a simple procedure to derive an approximate target heart rate (HR) that corresponds to any given intensity. Since the adaptations that occur with training are usually specific to the muscles involved in that exercise, it is obvious that the training stimulus (mode of exercise) must be specific to the sport. For example, runners normally train by running, rowers by rowing, swimmers by swimming, and so on. The possible exception applies to cardiovascular adaptations that can be partially transferred to other modes of exercise. For example, the increase in heart volume and wall thickness that occurs in the cross-country skier who trains by running in the summer months would be beneficial with the resumption of skiing as well.

Training Intensity vs Training Volume

Coaches have long debated the relative importance of the intensity vs the duration of the training stimulus in preparing their athletes for competition. Each factor is important, but each stimulates different adaptations and should be emphasized at different stages in the preparation plan.

In the simplest of terms, all forms of endurance training can be considered as falling under one of two classifications: either **continuous training** or **endurance interval training**. When the goal is a high-intensity stimulus, the most effective program will be one that uses high-intensity intervals interspersed with recovery periods. With this form of training, the athlete is able to accumulate a considerably greater total exercise time than could be achieved in a single continuous bout at the same intensity. When the goal is to increase the duration of the stimulus, the most effective program will be one that utilizes continuous training at relatively lower intensities. Such training is usually also the most effective mode for increasing the total volume of the training stimulus[1] because it is less fatiguing.

Exercise Intensity and Time to Fatigue

An athlete's ability to sustain continuous exercise depends upon the exercise intensity. This principle is illustrated in Figure 6.1, which schematically depicts time to fatigue for a well-trained, highly motivated athlete exercising at different percentages of $\dot{V}O_{2max}$. In this instance, fatigue refers to the athlete's ability to continue exercising at that particular intensity or power output. The data points that appear above 100% of $\dot{V}O_{2max}$ have been derived by extrapolation from cycle ergometry fatigue tests in the authors' laboratory and the assumptions of a linear relationship between power output and oxygen cost and a $\dot{V}O_{2max}$ of 5.0 $L{\cdot}min^{-1}$. Although based on cycling, the curve generally applies to any large-muscle-group exercise such as running,

[1] The volume of training is the product of the intensity, frequency, and duration of the stimulus and is usually quantified in units of distance covered per unit of time (e.g., miles per week).

Figure 6.1 The relationship between exercise intensity and time to fatigue. Based upon data from the authors' laboratory. See text for further explanation.

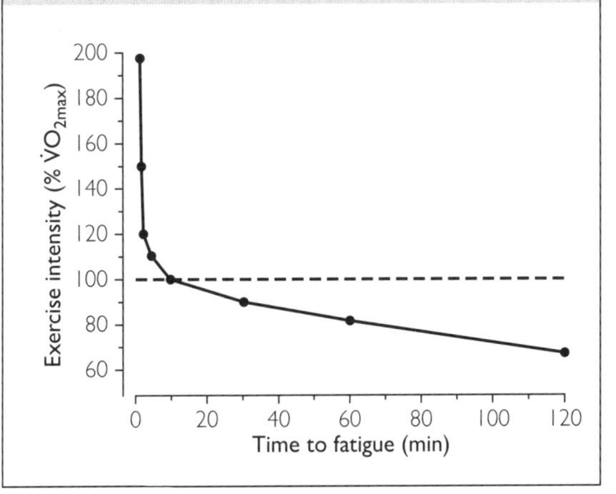

swimming, or cross-country skiing. For example, a runner could expect to be able to maintain a running pace equivalent to 150% of $\dot{V}O_{2max}$ for approximately 40 seconds, 120% for approximately 2 minutes, 110% for approximately 4 minutes, and 100% for approximately 8–10 minutes. Further extrapolating from Figure 6.1, the same runner could maintain a pace that requires 90% of $\dot{V}O_{2max}$ for approximately 30 minutes, and 82% for approximately 60 minutes.

Heart Rate as an Index of Exercise Intensity

When an athlete performs a progressive exercise test to exhaustion, heart rate (HR) increases almost linearly with the increase in oxygen consumption. Consequently, for any given athlete, we can estimate the relative intensity of the exercise if we know that individual's exercise HR relative to his/her maximal HR. The relationship, however, is not a simple 1:1 conversion since the athlete's HR at rest must be factored in as well. For example, even though HR would be maximal during exercise at 100% $\dot{V}O_{2max}$, during exercise at, say, 75% of $\dot{V}O_{2max}$ the person's HR would not be 75% of maximal HR. It would, however, be equal to his/her resting HR *plus* 75% of the *difference* between his/her maximal HR and his/her resting HR. This method of equating exercise intensity to a target HR was first described by Karvonen in the mid-1950s and is often referred to as the **Karvonen index**.

The following example will illustrate how the Karvonen index can be used to determine a target training HR for a given athlete. Suppose a coach wants an athlete to train at 80% of $\dot{V}O_{2max}$ for a given duration. If the athlete's resting HR is 60 bpm and maximal exercise HR is 190 bpm, the target HR for training would be $60 + (190 - 60)\cdot(0.80)$ or 164 bpm. The athlete then adjusts the pace upwards until he/she reaches an exercise HR of 164 bpm. Remember that at any given exercise intensity it takes approximately 2–3 minutes for HR to level off. In addition, since HR at any given $\dot{V}O_2$ tends to drift upwards as the training session progresses and body temperature rises, the procedure should be performed at the beginning of the training session. Because athletes frequently relate better to a given pace (e.g., the time required to cover a standard distance on the track, in the swimming pool, etc.) than to a given HR, it is a simple matter to transpose the target HR to a target pace. For example, when running on a track at a HR of 164 bpm, the above athlete covers 400 m in 95 seconds. His/her target training pace would be 95 seconds per lap on a 400 m track.

Determining Resting and Maximal Heart Rate

With recent developments in technology, there is a wide selection of inexpensive and accurate electronic HR monitors available to the athlete. These typically consist of an elastic strap that encircles the chest, with a sensor positioned on either side of the heart or a wrist sensor that detects radial arterial pulse. The sensors record the electrical activity that occurs with each ventricular contraction and transmit it digitally to a wristwatch-like receiver that converts and displays it as a minute HR. The athlete's resting HR for any given day would be the value that appears before any exercise or warm-up activity.

Although maximal HR is most accurately determined with a progressive test to exhaustion in a laboratory setting, it can also be determined in the field. To do this, the athlete runs (swims, rows, etc.) at the maximal pace that can be maintained for 3 minutes. The HR recorded immediately (within 10 seconds) following the exercise bout can be considered his/her maximal HR. An alternative method is to have the athlete perform several maximal 30 second sprint bouts alternating with 60 seconds of recovery. With this method, maximal HR is usually achieved by the end of the third exercise interval.

HR can also be counted manually by palpating the radial artery (at the wrist) or the carotid artery (at the forward side of the throat) with the finger tips. Because of the increase in exercise stroke volume, these pulse waves are particularly easy to detect immediately following exercise. When palpating the carotid pulse, athletes should make only light contact with the artery. Too much pressure with the finger tips can activate the baroreceptors and introduce an artifact by causing a reflex-enhanced slowing of HR. Since HR begins to decrease following exercise, the maximal HR must be calculated immediately. Counting the pulses over the first 10 seconds following maximal exercise and then extrapolating to a minute value has been shown to give a figure within 1 or 2 bpm of the true maximal exercise HR. Since maximal HR remains quite constant over the course of a training season, the athlete will have to determine it only two or three times per year.

Finally, as reviewed in Chapter 3, an approximate estimate of maximal HR can be derived by subtracting the athlete's age from 220 bpm. However, since the standard deviation for this method exceeds 10 bpm, this estimate is generally not precise enough for training prescription for the serious high-performance athlete.

Factors That Limit Aerobic Capacity and Sport Performance

As discussed in Chapters 2 and 3, performance in endurance sport events is ultimately determined by the maximal rate at which the athlete's muscles can generate ATP from the oxidation of glycogen and fat. This, in turn, is determined by the maximal rate at which oxygen can be delivered to the mitochondria and the rate at which the mitochondrial enzymes can drive the reactions in these pathways. Oxygen-delivery rate is a direct function of the athlete's cardiac output and the proportion of this blood flow that can be directed to the exercising muscles (**central factors**), while oxygen-extraction or -utilization rate is a function of muscle capillarization and mitochondrial enzyme kinetics (**peripheral factors**).

Although exercise physiologists have long debated the issue of whether it is central or peripheral factors that impose the greater limitation on aerobic power and sport performance, they are so interdependent that it would be misleading to separate the two and to single one out over the other. It is, however, generally accepted that central factors play a more dominant role in determining an athlete's maximal aerobic power ($\dot{V}O_{2max}$), while peripheral factors are more dominant in limiting maximal aerobic capacity (the maximal amount of work that can be performed aerobically in a given time). These conclusions are based largely on a number of studies in the training literature that indicate that the percentage improvement in $\dot{V}O_{2max}$ (usually 10–20%) correlates most closely with the percentage increase in maximal cardiac output, while the relative improvement in aerobic work capacity (often more than 100%) correlates most closely with the increase in mitochondrial enzyme activity and muscle capillary density (11, 19). It is also apparent that peripheral adaptations, on their own, would not be particularly effective unless accompanied by an increase in total oxygen delivery to the muscle. Therefore, the ultimate goal for the endurance athlete is to achieve maximal adaptation in *both* central and peripheral factors.

Training to Improve Central Factors

For the athlete, the two most important central adaptations are an increase in the volume of the heart and an increase in total blood volume. The increase in heart volume is the result of an increase in ventricular chamber size and a small but significant increase in ventricular wall thickness (13). The increase in blood volume is a function of expanded plasma and red cell volume and serves to enhance venous return and ventricular filling pressure during exercise. When combined with the changes in heart volume and ventricular contractile force, these adaptations result in relatively large increases in exercise stroke volume and maximal cardiac output.

The extent to which the enlargement of internal ventricular volumes can be attributed to an increase in the number of sarcomeres per cardiac fiber or to simple remodeling of collagen (3) is not known, but it is clear that a major stimulus for these adaptations is the mechanical loading imposed by stretch of the ventricular walls due to greater venous return. The magnitude of this stretch stimulus is reflected by the changes in ventricular end-diastolic volume. Figure 6.2 depicts the relationship between exercise intensity and end-diastolic volume in trained athletes. Notice that, with increased venous return, end-diastolic volume increases almost linearly with exercise intensity until approximately 70% $\dot{V}O_{2max}$. Thereafter, even though

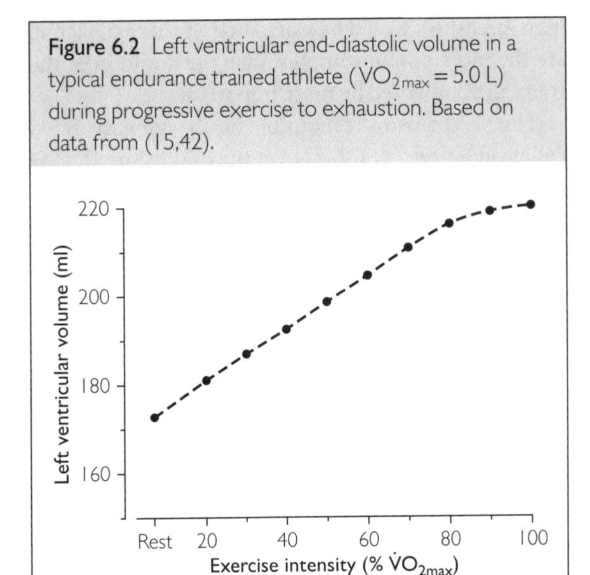

Figure 6.2 Left ventricular end-diastolic volume in a typical endurance trained athlete ($\dot{V}O_{2max}$ = 5.0 L) during progressive exercise to exhaustion. Based on data from (15,42).

ventricular filling pressure continues to rise, the time available for ventricular filling declines as maximal HR is approached and the relationship between end-diastolic volume and exercise intensity tends to flatten out.

Although ventricular end-diastolic volume still continues to increase with exercise intensity from 70 to 100% $\dot{V}O_{2max}$ (46) (unlike the response in untrained individuals), this flattening of the response is such that the magnitude of the stretch stimulus at maximal exercise is not substantially larger than at 70 or 75% $\dot{V}O_{2max}$. It is also apparent that the duration of the stretch stimulus is an important variable in determining the extent to which ventricular volume increases. This conclusion is based on a number of studies that indicate larger ventricular chambers in the hearts of longer-distance endurance athletes, such as cyclists and cross-country skiers, than in team-sport athletes and sprinters (34,35, 36). Since endurance athletes typically train at lower intensities but for considerably longer durations than other athletes, it is probable that the **duration** of the stimulus is a crucial component. In addition, it has been shown that, in skeletal muscle of animals, the duration of the stretch stimulus has a greater effect on increasing muscle length by the addition of new sarcomeres than does the magnitude of the stimulus (1).

It thus appears that the most effective form of training for the purpose of increasing internal ventricular volume of the heart is continuous exercise at approximately 70–75% of $\dot{V}O_{2max}$ for an extended duration. The exact intensity should be just below the athlete's lactate threshold.[2] For example, if an athlete's lactate threshold occurs at 75% of $\dot{V}O_{2max}$, the optimum training intensity for this form of training would be 74% of $\dot{V}O_{2max}$. At this intensity, it is possible for athletes to train for 60 minutes or more in a single bout without feeling exhausted. Training at higher intensities (e.g., 90% of $\dot{V}O_{2max}$) would slightly increase the magnitude of the stretch stimulus but would significantly shorten the duration that the athlete could tolerate the exercise and would thus be less effective.

The increase in ventricular wall thickness is the result of increased cross-sectional area (hypertrophy) of the cardiac muscle fibers. As with skeletal muscle fibers, when cardiac fibers are required to contract more forcefully than they normally do, they adapt by adding more actin and myosin filaments to the periphery of each myofibril, thus increasing the size and strength of the fiber. During endurance exercise, ventricular fibers must contract more forcefully to accommodate the increase in end-diastolic volume, or **preload** (the Frank–Starling mechanism, discussed in Chapter 3), as well as the elevated peripheral resistance, or **afterload**, that they encounter as they attempt to eject blood into the aorta. Thus, both preload and afterload are important stimuli for increasing ventricular wall thickness. The magnitude of the afterload stimulus is reflected by the **mean arterial blood pressure** (see Box 6.1) against which the heart must contract during exercise. As shown in Figure 6.3, during progressive dynamic exercise mean arterial blood pressure increases linearly with exercise intensity until approximately 70–75% of $\dot{V}O_{2max}$, at which point it tends to flatten out. As a result, as with ventricular end-diastolic volume (discussed above), training at

BOX 6.1

Mean arterial blood pressure is the average pressure exerted on the walls of the arteries over each cardiac cycle. Since the duration of diastole exceeds that of systole, mean pressure is closer to diastolic than to systolic pressure. Mean pressure is conventionally estimated as being equal to the diastolic pressure plus one third of the pulse pressure (the difference between systolic and diastolic pressures).

[2] As discussed in Chapter 2, the lactate threshold is defined as the exercise intensity above which there is a non-linear accumulation of lactic acid in the blood. The method for measuring the lactate threshold is presented in Chapter 11.

Figure 6.3 Typical blood-pressure response in healthy subjects undergoing progressive dynamic exercise to exhaustion. From (27). © John Wiley & Sons.

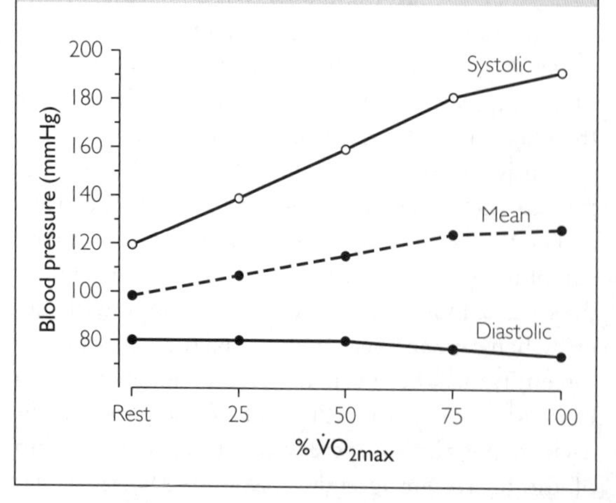

higher intensities does not appreciably increase the magnitude of the stimulus. The exception, of course, is with static or resistance exercise, where peripheral resistance is much higher and directly related to the force of the contraction. Thus, for the endurance athlete, *the most effective form of training for stimulating an increase in ventricular wall thickness is also continuous exercise at approximately 70–75% of* $\dot{V}O_{2max}$.

The mechanisms that cause an increase in total blood volume are not fully understood, but involve both an expansion of plasma volume and an expansion of erythrocyte volume. Increased plasma volume is a rapid response that can be detected within a few days of beginning a training program, whereas increased erythrocyte volume is not apparent until after 2 or 3 weeks. It is believed that most of the increase in plasma volume can be attributed to an increased synthesis of albumin, which causes greater re-absorption of plasma from the interstitial space. Although there is also a slight increase in total body water due to lower urine output, this is not enough to account for the expanded plasma volume that occurs. Increases in albumin and plasma volume have been shown to be related to both elevations in core temperature (7,41) and the acute reductions in plasma volume that occur with increased filtration pressures across the muscle capillary bed (18). It is difficult, however, to determine which is the more potent stimulus since they are interdependent (i.e., when an individual exercises, core temperature increases in proportion to exercise intensity and duration). If hemodynamic factors are the more potent stimulus, then training at very high intensities would be warranted. If thermal factors are the more potent stimulus, then the duration of the training bout would be more important.

The mechanisms responsible for the increase in red cell production are also not yet fully understood. One possibility is that this increase may be directly attributable to the expanded plasma volume (decreased hematocrit) and represents the body's attempt to restore its normal hematocrit. Another possibility is that it is caused by the increase in erythropoietin that is stimulated by the hypoxia occurring during heavy exercise. In either case it is probable that endurance interval training would provide a more effective stimulus for augmenting red cell production than would continuous moderate-intensity training.

It thus appears that increases in plasma volume occur in response to both high-intensity interval training and to prolonged continuous training, whereas increases in red cell production are more likely attributable to high-intensity interval training. We can therefore conclude that, *in order to achieve maximal increases in total blood and red cell volumes, both high-intensity endurance interval training and moderate-intensity (long-duration) continuous training are warranted.*

Time Course for Central Adaptations

With the exception of increases in plasma volume, central adaptations to training occur gradually (i.e., over several months). This conclusion is based on a number of long-term training studies indicating that when previously untrained individuals begin an endurance training program, $\dot{V}O_{2max}$ continues to increase each week for up to about 2.5–3 months, after which little improvement occurs. Since maximal cardiac output is the major determinant of $\dot{V}O_{2max}$, it is generally accepted that these improvements are attributable to increases in heart size and total blood volume. With complete cessation of training (detraining), $\dot{V}O_{2max}$ remains the same for approximately 1–2 weeks and then begins to decline gradually at about the same rate at which it increased during the training period.

Training to Improve Peripheral Factors

Of the many peripheral training adaptations that can occur at the muscle level, the performance of the endurance athlete is probably most affected by an

BOX 6.2

The sequence of molecular events by which an exercise stimulus upregulates those pathways in a muscle cell that cause it to increase its normal rate for mitochondrial and enzymatic synthesis is still largely unknown. The process is thought to begin with the activation of certain mitogen-activated protein kinase (MAPK) signaling pathways in the cytosol. These pathways involve sequential activation of their transduction kinases, with their phosphorylation being the signaling mechanism for each subsequent link in the chain (45). Some of these signal transduction pathways can interact with each other when a kinase in one is also able to phosphorylate a kinase in a different pathway (a process known as cross-talk). The activation of AMP-dependent protein kinase (AMPK) is considered by some investigators to be one of the main triggers for the sequence of events that leads to an increase in mitochondrial synthesis (38). Other possible pathways include that activated by calcineurin (calcium/calmodulin-dependent protein kinase or CaMK) and that by p38 mitogen-activated protein kinase (p38 MAPK).

Whatever the pathway, the message is eventually communicated into the muscle cell nuclei by a final signaling protein. This protein then acts as a transcription factor or activates another transcription factor, eventually causing DNA to be transcribed into the appropriate messenger RNA (mRNA), which is then translocated back into the cytosol. Through the translation process this mRNA is then decoded by the subunits of the ribosome to produce the specific amino acid chain that will form the final proteins which will be incorporated into the mitochondrial structure or its enzymes.

increase in mitochondrial density and enzyme activity and by an increase in the number of capillaries (angiogenesis). Within each fiber, the increase in the number of mitochondria leads to a parallel increase in the total fiber concentration of mitochondrial oxidative enzymes. This results in a greater oxidative reaction rate for a given concentration of pyruvate or free fatty acids, reduced lactate production, and increased lactate disposal at a given exercise intensity. The increase in capillary density reduces the diffusion distance between each capillary and fiber and thus accelerates the rate at which oxygen and fuel sources enter the fiber and byproducts such as carbon dioxide and lactate are removed from it. In addition, greater capillarization will prolong blood flow transit time across the muscle, thus increasing the time available for diffusion.

The synthesis of new mitochondria requires the combined expression of mitochondrial proteins from both the nuclear and mitochondrial genomes (21,22) and occurs only in those muscle fibers that are recruited during training. The stimulus or signal that initiates the process is still under investigation, but is thought to be related to the degree to which calcium release and reuptake occurs with muscle contraction, the deficit between ATP demand and mitochondrial ATP supply (21), or to low oxygen concentrations (hypoxia) (see Box 6.2) in the active mitochondria (23,30,44). Since the magnitude of each of these possible stimuli is directly related to exercise intensity, it is apparent that the most effective form of training will be a program that incorporates high-intensity intervals. In support of this conclusion, it has been shown that brief high-intensity sprint interval training can elevate oxidative enzyme maximal activity by approximately 40% (6,28). However, since cross-sectional studies reveal higher oxidative enzyme activity in muscles of endurance-trained athletes than in sprinters or team sport athletes (9), we must conclude that the duration of the high-intensity stimulus is also of importance.

As discussed in Chapter 3, exercise-induced angiogenesis is thought to be the result of sprouting or branching of existing capillaries (37). The stimulus for this is probably the hypoxia that occurs in the muscle during exercise and/or the mechanical stress that occurs on the capillary wall as a result of large increases in blood flow (see Box 6.3). Again, in either case, the stimulus would be more potent during high-intensity exercise. However, since oxygen extraction and muscle blood flow have theoretically already reached maximal levels at $\dot{V}O_{2max}$, there is little to be gained by training at intensities that exceed 100% $\dot{V}O_{2max}$. It is therefore not surprising that a recent study shows no greater increase in capillarization and proliferation of endothelial cells following interval training at 150% $\dot{V}O_{2max}$ than at 90% $\dot{V}O_{2max}$ (24).

Thus, unlike most central factors, adaptations at the muscle level require a high-intensity exercise stimulus. Since the duration of the stimulus is also an important component, the most effective form of training will be repeated intervals at intensities that approach 100% $\dot{V}O_{2max}$. When exercise is performed at this intensity for 2–3 minutes and then followed by 2–3 minutes of low-intensity recovery, athletes are able to complete 8 to 10 exercise bouts in a given training session.

BOX 6.3

Support for the theory that muscle hypoxia may be an important stimulus for increasing muscle mitochondria and capillary density is provided by a study that was conducted in the authors' laboratory (30). In that investigation, 10 healthy young males underwent 8 weeks of unilateral endurance training on a cycle ergometer, in which one leg was trained while the subject breathed an inspirate of 13.5% oxygen, while the other leg was trained for the same duration and at the same absolute intensity but with the subject breathing normal ambient air (20.9% oxygen). Muscle biopsy samples taken from the quadriceps before and after the training program indicated that greater increases in oxidative enzyme activity and capillary density occurred in the hypoxically trained leg than in the normoxically trained leg. Since muscle oxygen content would have been lower during the training bouts performed while breathing the low-oxygen inspirate, it appears that muscle hypoxia is at least partially responsible for these peripheral adaptations.

Training at intensities that exceed $\dot{V}O_{2max}$ is probably no more effective and generates greater muscle lactacidosis, which markedly reduces the total exercise duration that can be tolerated.

In summary, *the most effective form of training for stimulating mitochondrial and oxidative enzyme adaptations and for increasing muscle capillary density is interval training at approximately 100% of the athlete's $\dot{V}O_{2max}$. Repeated intervals of 2–3 minutes at such intensities, interspersed with 2–3 minute recovery intervals, are probably the most effective way to increase the total duration of the stimulus*. Because significant quantities of lactic acid are still produced in the muscles at these intensities, and only a portion is removed during the recovery intervals, the athlete begins each successive exercise interval with progressively higher muscle and blood lactate concentrations, leading ultimately to fatigue. By increasing the duration of the recovery intervals to 4 and 5 minutes as the training session progresses, athletes are able to repeat more 2–3 minute exercise intervals and thus extend the total duration of the stimulus.

Monitoring HR at the end of each exercise and recovery interval is an effective way of ensuring that intensity and recovery time are optimal. HR at the end of each exercise interval should be a few beats per minute lower than maximal HR and (for a 20–30-year-old athlete) should decline to less than 120 bpm at the end of the recovery interval. If HR has not declined to 120 bpm, the recovery interval should be extended. Because of the accompanying lactacidosis, most athletes find this form of training quite stressful and considerable motivation is required to complete the later-stage intervals as the session progresses. In addition, this form of training rapidly exhausts muscle glycogen stores to the extent that they may not be fully restored prior to the next training session. While most studies show that athletes on a high-carbohydrate diet can completely replenish muscle glycogen within 24 hours of a single high-intensity interval training session (29,42), this is probably not the case if they attempt several successive days of such training (8). For this reason, it is preferable that this type of endurance interval training be performed every second day or, if performed 2 days in succession, be followed by a day of complete recovery.

An additional advantage of training at intensities greater than 90% $\dot{V}O_{2max}$ is that it ensures that the type 2 motor units are activated. As reviewed in Chapter 5, recruitment of the different muscle fiber types is dependent upon the required force of contraction of the exercise. Classic studies of the glycogen depletion patterns in type 1 and type 2 fibers indicate that, at exercise intensities below this level, only type 1 fibers are recruited (16). As exercise intensity approaches 100% $\dot{V}O_{2max}$, more and more of the type 2A and type 2X units are recruited, thus ensuring that the peripheral adaptations described will occur in them as well. Adaptation within this population of fibers is crucial to success in the middle-distance endurance events, since they rely heavily on the utilization of type 2A fibers in addition to type 1 fibers.

Time Course for Peripheral Adaptations

While most central adaptations occur over several months of training, peripheral adaptations are evident after only a few weeks. Increases in oxidative enzyme activity and (presumably) mitochondrial density have been shown to occur within 1 or 2 weeks of high-intensity interval training (6,39) and to reach maximal levels after about 6 weeks (21). Increases in capillary density have been noted after 4 (but not 2) weeks (24) of training. In addition, it appears that 7 weeks of such training is no more effective in stimulating increased capillary growth than 4 weeks (24). An increase in resting muscle glycogen concentration is also one of the first adaptations to occur and is evident after only 1 week of training (6,17).

Since most mitochondrial proteins turn over quite rapidly, with a half-life of approximately 1 week (21), these adaptations are quickly lost when the training stimulus is reduced or removed. The rate at which peripheral adaptations decline with detraining is probably a mirror image of the rate at which they occurred (31).

Periodization of Endurance Training

Training programs for most high-performance endurance athletes are designed on a yearly basis. The ultimate objective is to achieve maximal adaptation in all of the physiological factors that affect performance prior to the most important competition(s) for that year. In determining the optimum sequence for the different forms of training stimuli, there are several key points to be kept in mind:

- central adaptations require several months of training to occur; whereas
- peripheral adaptations peak after 4–6 weeks of training;
- central adaptations are most effectively achieved by high-volume continuous training at approximately 70–75% of $\dot{V}O_{2max}$;
- peripheral adaptations are most effectively achieved by high-intensity (95–100% of $\dot{V}O_{2max}$) endurance interval training;
- endurance interval training is more physiologically and psychologically stressful than continuous training and requires much more motivation and recovery;
- once achieved, both central and peripheral adaptations can be maintained with considerably less training volume;
- with complete detraining, peripheral adaptations are lost more quickly than central adaptations;
- once adaptations have been achieved, performance for a given occasion can be enhanced if it is preceded by a sharp reduction in training volume over approximately 7 days (tapering).

Based on these facts, it is apparent that endurance athletes should begin their program 5–6 months prior to competition. The first phase (3.5–4 months) should be devoted to achieving maximal central adaptations and the next phase (6–8 weeks) to achieving maximal peripheral adaptations, while at the same time maintaining the central adaptations that have been gained. The first phase of the program would thus begin with an emphasis on continuous, high-volume, submaximal intensity training, followed by an emphasis on high-intensity endurance interval training in the second phase, and a short tapering phase just prior to competition (Figure 6.4). The reverse approach (i.e., training for peripheral adaptations first, followed by training for central adaptations) would be far less efficient since it would require more total high-intensity

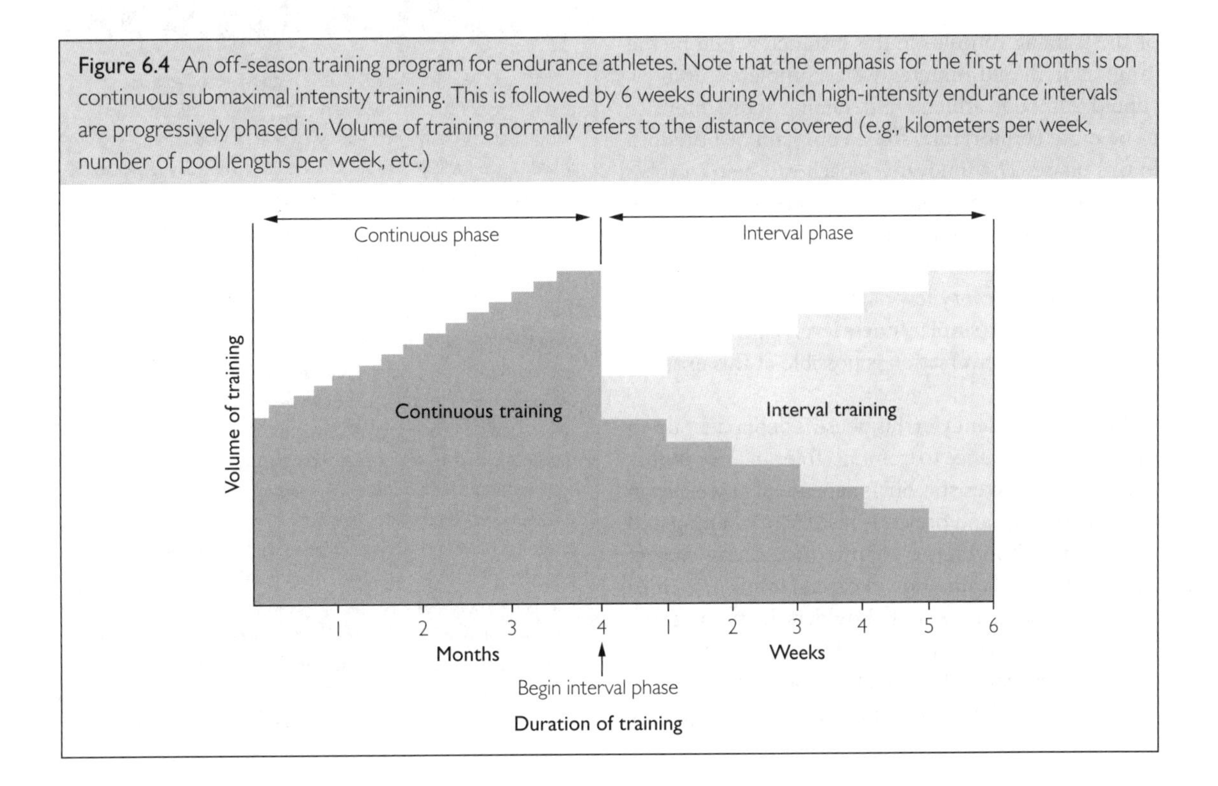

Figure 6.4 An off-season training program for endurance athletes. Note that the emphasis for the first 4 months is on continuous submaximal intensity training. This is followed by 6 weeks during which high-intensity endurance intervals are progressively phased in. Volume of training normally refers to the distance covered (e.g., kilometers per week, number of pool lengths per week, etc.)

training to maintain the peripheral gains achieved throughout the longer duration required to achieve the central adaptations. The result would be a higher risk of injury or overtraining.

Continuous Training Phase

During this phase, the majority of each training session should be devoted to training at a pace that requires approximately 70–75% of the athlete's $\dot{V}O_{2max}$, using HR as an indicator of intensity. The duration of each exercise session should increase gradually over the course of the 3.5–4 months of the phase. For example, athletes might begin with 40 minutes of continuous training in each session of the first week, followed by 45 minutes per session in week 2, 50 minutes in week 3, and so on. By week 10, each session would be 90 minutes long. Ninety minutes of training per session is a reasonable final goal, since this duration still allows most athletes on a high-carbohydrate diet to restore muscle glycogen completely within 24 hours following each session. This is not the case when the exercise exceeds 90 minutes.

The term "continuous" need not be taken literally for this phase of training. For example, an athlete might choose to break the 90 minutes of exercise into three 30 minute blocks, alternating with 2 or 3 minute periods for stretching or hydration. The effectiveness of the training stimulus in this instance would be the same as if the 90 minutes were completed uninterrupted. Although athletes find training at this intensity to be quite comfortable, some complain of boredom in this phase. The innovative coach can help to solve this problem by incorporating variety. For example, when training for sports such as running, cycling, or cross-country skiing, the training route and terrain could be altered every few days. Training with a partner or group with similar fitness levels can add a social element since conversation is possible at this exercise intensity.

Training **frequency** in this phase is dependent upon the athlete's tolerance to training. Tolerance is a highly individual characteristic, but is dependent as well upon the athlete's age and previous history of training and competition. Too high a volume of training can result in minor inflammation in certain joints (e.g., hips and knees in runners and shoulders in swimmers). These symptoms of repetitive-strain or overuse injury should be interpreted as an indication that frequency should be reduced. Other indications that frequency may be too high are a general feeling of fatigue or an increase in susceptibility to colds or other common infections. As a guide, the authors recommend a minimum frequency of 4 days per week during this phase. All athletes should be able to tolerate this, while some will be able to tolerate 5, 6, or even 7 days per week. It should be recognized, however, that the central adaptations to training 6 or 7 days per week are similar to, or only marginally greater than, those achieved with 4 or 5 days per week. Training twice per day (as was once popular in sports such as swimming and rowing) is no more effective than once per day (10) and may even be counter-productive, since it reduces recovery time between training sessions.

The relative importance of strength training for the endurance athlete is discussed in more detail in Chapter 8. At this point, suffice it to say that performance in most middle-distance sports will benefit from the addition of some resistance training during this phase (see Box 6.4). An increase in sport-specific muscle strength can manifest itself at the start of most events and over the first 8 to 10 strides or strokes in speed skating, running events, and rowing. In swimming, an increase in leg strength is also advantageous during turns. The objective is to increase the strength and power (rate

BOX 6.4

For several years in the 1970s, the authors acted as training and fitness consultants for the men's and women's Canadian national cross-country ski teams. At that time, coaches and athletes generally felt that traditional strength training would be of little or no benefit in an endurance sport of this nature. Some even felt that it could have a negative effect on performance because it would cause athletes to bulk up and increase the body mass to be carried throughout the race. Tests in our laboratory indicated that, although most team members had good to very good lower-body (especially hip and knee extension) strength scores, their upper-body (arm and shoulder extension) scores were quite ordinary. Athletes were prescribed an optional strength training program of 4 different exercises selected to increase the strength of the arm and shoulder extensors. The program was carried out 3 days per week along with their normal summer running program. Retesting of those athletes who participated in the program indicated strength gains in the 30–50% range and all felt that it improved skiing performance, especially when double-poling or hill-climbing. The following year, all team members added an upper-body resistance training component to their training.

of force development) of certain muscle groups with minimal hypertrophy or increase in body weight. For most sports, this can be achieved by as little as 10 or 15 minutes of resistance training, two or three times per week. In that time, athletes can complete two or three 10 RM sets (see Chapter 8) of two or three sport-specific exercises. Although some studies (2,20,26) suggest that gains in muscle size and strength may not be optimal when such training is combined with endurance training, this is of secondary importance here, since the goal is to achieve relatively moderate changes. The issue of possible interference effects occurring when strength and endurance training are performed concurrently is also further discussed in Chapters 8 and 9.

Interval Training Phase

This phase begins with the gradual integration of high-intensity intervals into three or four training sessions per week. Exercise intensity should be in the 95–100% range of $\dot{V}O_{2max}$ and each bout should be 2–3 minutes in duration, with a similar recovery interval. Recovery involves only low-intensity exercise (e.g., walking for a runner). As previously indicated in this chapter, HR at the end of each interval should be within a few beats per minute of maximal and less than 60% of maximal at the end of each recovery interval (see Figure 6.5). Athletes can expect it to become more difficult to maintain their target pace as the workout continues, but if they have difficulty completing the first few intervals

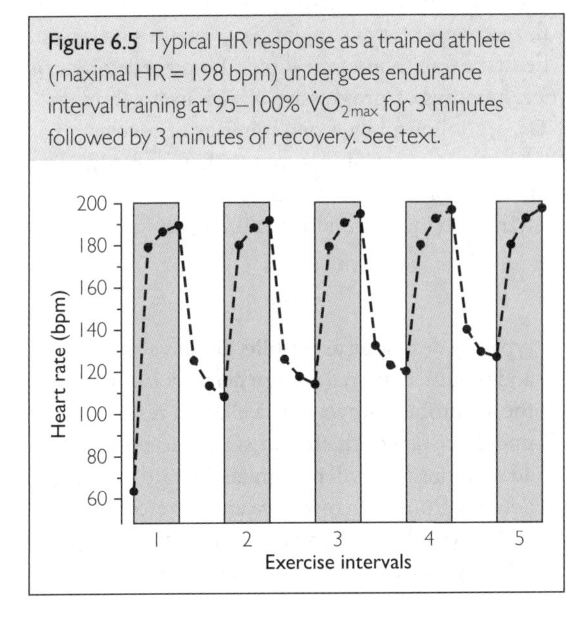

Figure 6.5 Typical HR response as a trained athlete (maximal HR = 198 bpm) undergoes endurance interval training at 95–100% $\dot{V}O_{2max}$ for 3 minutes followed by 3 minutes of recovery. See text.

and HR is maximal at the end of these, it is probable that the intensity exceeds 100% of $\dot{V}O_{2max}$ and the pace should be re-calculated. In addition, as the training session continues, HR during the recovery intervals will not decline as rapidly as following the first few intervals and they should extend the recovery interval in order to reach their target recovery HR.

Although muscle and blood lactate concentrations with this form of training are not as high as with sprint interval training (28), they increase progressively with each interval and our laboratory has recorded fingertip capillary blood concentrations in excess of 15 $mM \cdot L^{-1}$ after completion of a typical training session. As a result, this can be a very uncomfortable form of training, requiring considerable self-discipline and motivation on the part of the athlete. Unlike continuous training, high-intensity interval exercise also results in a sharp elevation in the various stress-response hormones. Immediately following an interval training session, blood concentrations of epinephrine and norepinephrine can be more than 15-fold (4,32) and cortisol and glucagon more than fourfold greater than their normal resting levels and may remain elevated for several hours. This response, sometimes referred to as the fight-or-flight response, greatly increases blood pressure, HR, plasma glucose concentration, and muscle blood flow. It is probably a holdover from our evolutionary past that helped us to survive in times of confrontation or extreme physical danger and is now invoked only when we call on our body to perform maximal exercise. Chronic elevations of these hormones can have a suppressive effect on appetite, sleep patterns, and immune function (Chapter 9). Consequently, adequate rest and recovery are important components of this phase and the number of interval training sessions should never exceed four per week.

A reasonable progression would be to begin this phase with six intervals per session in the first week, seven in the second and so on, until a target of 10 intervals (a total of 30 minutes of stimulus) is reached. For most athletes, three interval training sessions per week will be optimal. These should be on alternate days, separated by a day of moderate-duration (e.g., 60 minutes) continuous training. These 2 or 3 days per week of continuous training provide enough stimulus to maintain the central adaptations gained in the previous phase. Should the athlete attempt four interval training sessions per week, a day of complete rest (no training) should follow the back-to-back sessions.

For middle-distance athletes, especially those whose events are completed in the 2–4 minute range, the above program should be modified to include sprint-type intervals (at intensities greater than 100% of $\dot{V}O_{2max}$) in the last few weeks before the taper. For the shorter-distance athletes, this intensity will be similar to, or slightly above, race pace and helps to instill a good pace sense. Intervals should be approximately 45 seconds in duration, followed by approximately 3 minutes of recovery. Two weeks of such training have been shown to be adequate for increasing muscle glycolytic enzyme activity (33) and anaerobic power output (5). In addition, as discussed in Chapter 7, the extremely high muscle and blood lactate concentrations that are generated during this form of training (28) may be a stimulus for the increased muscle buffering capacity that is found in sprint-trained athletes (12,14,40).

Sample Programs

Examples of a typical week's program for mid-continuous and mid-interval training phases are shown here. Note that these are generic programs that would be slightly modified based upon the duration of the athlete's particular event. Also, training volume is expressed in units of time (minutes) instead of distance. Some coaches may wish to convert these to units of distance (e.g., kilometers, pool lengths, etc., to be covered per training session).

Mid-Continuous Phase (e.g., week 8)

In each training session, athletes will complete a total of 80 minutes of continuous training at a pace that requires approximately 75% of $\dot{V}O_{2max}$. This pace will have been predetermined based on each athlete's target HR or will be calculated during the session using a HR monitor. This latter method of determining pace is often necessary in sports such as cycling or cross-country running or skiing, where a variety of training routes may be used and terrain may vary considerably. The training session will be performed in two 40 minute bouts, separated by an approximate 3 minute break for stretching and/or rehydration. The first exercise bout will be preceded by approximately 3 minutes of low-intensity (less than 50% $\dot{V}O_{2max}$) warm-up exercise.

Athletes will train 5 days per week (e.g., Monday, Tuesday, Wednesday, Friday, and Saturday). On three of the training days (e.g., Monday, Wednesday, and Saturday), they will also perform two or three sets of resistance training. This will involve two or three sport-specific exercises for the prime-mover muscle group(s), to be performed in 10 RM sets, each followed by approximately 3 minutes of recovery.

Mid-Interval Training Phase (e.g., week 3)

During this phase, athletes will perform interval training 3 days per week (e.g., Monday, Wednesday, and Friday) and continuous training 2 days per week (e.g., Tuesday and Thursday). In each interval training session, they will complete eight 3 minute intervals at 95–100% $\dot{V}O_{2max}$ separated by low-intensity (less than 30% $\dot{V}O_{2max}$) recovery intervals. Each recovery interval should last for a minimum of 3 minutes and be long enough for the athlete to reach a recovery HR of less than 60% of their maximal HR. This normally requires increasing the duration of the recovery intervals as the session continues. For example, by the time athletes complete the sixth or seventh exercise interval, recovery intervals might be 4 minutes or longer. The first exercise interval should be preceded by 5 or more minutes of progressive warm-up (to about 75% $\dot{V}O_{2max}$), followed by 2 or 3 minutes of recovery. For the continuous training sessions they will complete 60 minutes of exercise at approximately 75% $\dot{V}O_{2max}$, as in the previous continuing training phase.

Summary of Key Points

- **Endurance sports** are those in which oxidative metabolism is the dominant energy-delivery pathway. They thus include sports that require maximal efforts over a duration of approximately 2 minutes to several hours. Races that are completed in approximately 2–6 minutes are typically described as **middle-distance sports** and require maximal energy delivery from the anaerobic pathways as well. As a result, middle-distance athletes must train to promote adaptations that will maximize the rate of energy delivery from both oxidative and anaerobic pathways. Although the most important energy-delivery pathways for **team sports** are anaerobic, recovery from sprint-type activity involves oxidative pathways. Since team sports typically involve brief bursts of maximal sprint-type activity separated by lower-intensity recovery intervals, the athlete who recovers most quickly is best able to repeat such bursts as the game progresses. Consequently, it is important to include an endurance training component as well when training for most team sports.

- Since many of the adaptations that occur with endurance training are confined to the muscles involved in the exercise, the training stimulus must be specific to the sport (runners train by running, swimmers by swimming, etc.). Both the intensity and the duration of the training stimulus are important, but each stimulates different adaptations and should be emphasized at different stages of the training program. There are typically two types of endurance training: **continuous training** at submaximal intensity (e.g., 75% $\dot{V}O_{2max}$) or **endurance interval training** at maximal (100% $\dot{V}O_{2max}$) or near-maximal intensity. When the goal is to increase the duration of the training stimulus, the most effective form is continuous training. When the goal is a high-intensity stimulus, the most effective program will be one that uses short high-intensity intervals interspersed with recovery periods. With this form of training, the athlete is able to accumulate a much greater high-intensity stimulus than could be achieved in a single continuous bout of exercise at the same intensity.
- The athlete's exercising HR, relative to his resting and maximal HR, can be used as an indicator of exercise intensity. For example, if an athlete wishes to train at 75% of $\dot{V}O_{2max}$, his/her target HR would be equal to his/her resting HR plus 75% of the difference between the resting and maximal HR. Exercise HR can be determined directly by an electronic HR monitor or can be estimated by counting the pulse for 10 seconds immediately following exercise and extrapolating to a minute value.
- Performance in an endurance sport depends directly on the maximal rate at which the athlete's muscles can generate ATP from the oxidation of glycogen and fat. This rate is determined by the maximal rate at which oxygen can be delivered to the mitochondria and the rate at which it can be extracted and utilized by mitochondrial enzymes to drive these pathways. The former is a direct function of the athlete's cardiac output and the proportion that can be directed to the exercising muscles (**central factors**), while the latter is a function of muscle capillarization and mitochondrial enzyme kinetics (**peripheral factors**). Although they are interdependent, central factors are more important in determining an athlete's maximal aerobic power ($\dot{V}O_{2max}$), while peripheral factors are more important in determining maximal aerobic capacity (the maximal amount of work that can be performed aerobically in a given time).
- The key central adaptations to training are an increase in the volume of the heart and an increase in total blood volume. Increased heart volume is caused by an increase in internal ventricular chamber size and (to a lesser extent) an increase in atrial and ventricular wall thickness. These adaptations are most effectively attained by continuous training at approximately 70–75% of $\dot{V}O_{2max}$ for an extended duration. The increase in total blood volume is a function of both expanded plasma and erythrocyte volume. Increases in plasma volume occur in response to both high-intensity interval training and prolonged continuous training, whereas increases in red cell volume are more effectively achieved through interval training.
- The key peripheral adaptations affecting endurance performance are an increase in muscle mitochondrial density and enzyme activity and an increase in the number of capillaries. These adaptations are most effectively achieved by repeated intervals of 2–3 minutes at approximately 100% of $\dot{V}O_{2max}$, interspersed with 2–3 minute recovery intervals. Training at this intensity also requires recruitment of most of the type 2A fibers (and even some 2X fibers), thus ensuring that peripheral adaptations will occur in them. This population of fibers is crucial to success in middle-distance events, which rely heavily on utilization of type 2A fibers in addition to type 1 fibers. By monitoring HR at the end of each exercise and recovery interval, athletes can determine whether intensity and recovery time are optimal. HR at the end of each exercise interval should be within a few beats per minute

of maximal HR and (for a 20–30-year-old athlete) should decline to less than 120 bpm at the end of each recovery interval. As the training session progresses, athletes will probably find that it is necessary to extend the recovery interval in order to achieve their target recovery HR.

- The time course for central adaptations to occur differs from that of peripheral adaptations. With the exception of increases in plasma volume, central adaptations to training occur gradually (i.e., over several months). In contrast, peripheral adaptations are evident after only a few weeks of training and tend to reach maximal levels after 4–6 weeks. Upon cessation of training, central adaptations are preserved for approximately 1–2 weeks and then gradually decline at about the same rate at which they occurred during the training period. Peripheral adaptations are lost within a few weeks of cessation of training. Consequently, endurance athletes should begin their annual training program 5–6 months prior to competition. The first 3.5–4 months should emphasize continuous submaximal training in order to achieve maximal central adaptations, whereas the next 6–8 weeks should be devoted to high-intensity endurance interval training to achieve maximal peripheral adaptations. The interval training phase will also provide enough stimulus to maintain the central adaptations that have been achieved in the first training phase.
- During the continuous training phase, the majority of each training session should be devoted to training at a pace that requires approximately 70–75% of $\dot{V}O_{2max}$, using the athlete's target HR to determine this intensity. Typically, athletes would begin this phase with 40 minutes of continuous training per day in the first week, increasing by 5 minutes each week until they reach a total duration of 90 minutes per session. Training frequency should be a minimum of 4 days per week and as much as 6 days. Since performance in most middle-distance events will benefit from an increase in sport-specific muscle strength, this is also an ideal time to include some resistance training. Two 10 RM sets of two or three sport-specific exercises performed three times per week will prove adequate to increase strength and power of the selected muscle groups with minimal effect on muscle size and body weight.
- The interval training phase should consist of 2–3 minute intervals at 90–100% of $\dot{V}O_{2max}$, separated by similar recovery intervals. Typically, athletes would begin this phase with six intervals per session in the first week, adding one each week until they are performing 10 intervals (a total of 30 minutes of stimulus). For most athletes, three interval training sessions per week will be optimal. These sessions should be performed on alternate days, separated by a day of continuous training (as in the previous phase), and should last for approximately 60 minutes. Athletes whose events are completed in the 2–4 minute range should substitute 45 second sprint-type intervals in the last few weeks before their taper.

7

Training for Anaerobic Events and Team Sports

Introduction

As reviewed in Chapter 2, anaerobic events are those where performance depends directly on the maximal rate at which ATP can be generated by the hydrolysis of phosphocreatine (PCr) and/or anaerobic glycogenolysis. They thus include a wide range of sports, from those requiring maximal efforts over a period as brief as 2 seconds or less (e.g., throwing events) to those requiring maximal efforts for as long as 2 minutes (e.g., 800 m in track). The relative contribution of these two energy-delivery pathways depends upon the duration of the event (Table 7.1). In explosive events such as throwing, jumping, or weight lifting, which require maximal power output over only a few seconds, energy is derived almost exclusively from resting muscle stores of ATP and from PCr hydrolysis. Events that require bursts of maximal activity for approximately 2–15 seconds derive their energy almost equally from PCr hydrolysis and glycogenolysis, while sustained maximal efforts over approximately 15–60 seconds (e.g., 200 and 400 m in track) derive most of their energy from glycogenolysis. In events requiring maximal power output for 2 minutes, approximately 50% of the energy is derived from the anaerobic pathways (predominantly glycogenolysis) and 50% from the oxidation of glycogen.

Table 7.1 Anaerobic events and sports according to energy-delivery pathways.

Predominantly PCr hydrolysis	PCr hydrolysis and glycogenolysis	Predominantly glycogenolysis
Throwing events	Track: 100 m, 200 m, high hurdles	Track: 400 m, low hurdles
Jumping events	American football	Swimming: 50 m, 100 m
Weight lifting	Basketball	Downhill skiing
Baseball	Volleyball	Soccer
Golf	Most gymnastic maneuvers	Ice hockey
Diving	Racquet sports	Rugby
Aerial skiing		Lacrosse
Some gymnastic maneuvers		Combative sports

Team sports typically require bursts of maximal effort separated by lower-intensity intervals or stoppages in play for rule infractions. In some, such as baseball or American football, the duration of the maximal effort is extremely brief (often only 3 or 4 seconds) compared to the duration of the intervals (often several minutes or more) before these efforts are repeated. In others, such as ice hockey or soccer, intense activity may be sustained for up to 60 seconds without a stoppage in play or a slowing in tempo. For some sports, then, PCr hydrolysis will be the dominant energy-delivery pathway while, for others, glycogenolysis will predominate (Table 7.1).

Factors That Limit Performance in Anaerobic Sports

As discussed in Chapter 1, performance in anaerobic sports is determined by four factors:

1. the skill and coordination of the athlete;
2. the maximal strength and rate of force development of the muscle group(s) involved in the sport;
3. the maximal anaerobic power of these muscles; and
4. the maximal anaerobic capacity of these muscles.

The relative importance of each factor depends upon the nature of the sport and the duration over which the muscular effort occurs. This chapter will focus primarily on training procedures for increasing maximal anaerobic power and capacity. Procedures for increasing strength, power, and speed are presented in Chapter 8.

The **maximal anaerobic power** of a muscle, or muscle group, can be defined as the highest rate at which ATP can be generated (e.g., moles of ATP per second) by the hydrolysis of PCr and by anaerobic glycogenolysis. Maximal anaerobic power output can be sustained for only a few seconds (see Figure 7.1) and is determined by enzyme kinetics; that is, the maximal rate at which creatine kinase can hydrolyze PCr to regenerate ATP and the maximal rate at which the glycolytic enzymes can break down glycogen to lactic acid and generate ATP. Recall also from Chapter 1 that, although the maximal ATP-regeneration rate by

Figure 7.1 The typical decline in power output that occurs in an athlete over 30 seconds of maximal exercise on a cycle ergometer (Wingate test). From the authors' laboratory.

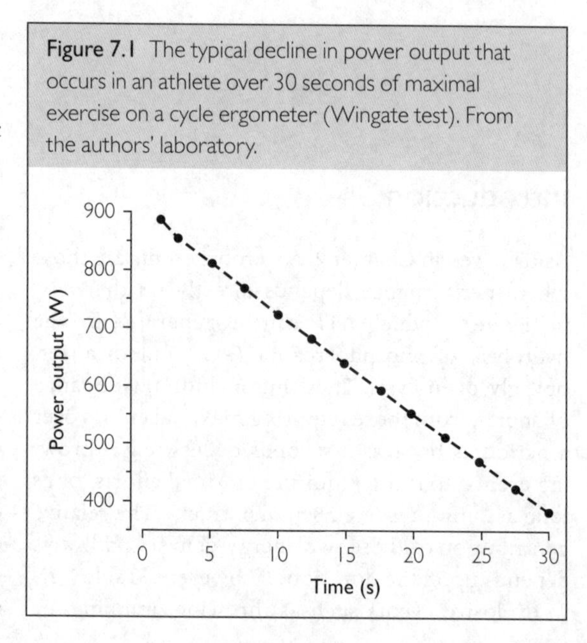

PCr hydrolysis is almost double the rate by glycogenolysis, power output from this source begins to decline after several seconds as muscle PCr stores decrease. In contrast, maximal or near-maximal rates of ATP production from glycogenolysis can be maintained for 40 seconds or more before beginning to decline.

The **maximal anaerobic capacity** of a muscle or muscle group can be defined as the total amount of ATP that can be generated from these two pathways over a given time period (e.g., a maximal effort over 40 seconds). Maximal anaerobic capacity is thus determined largely by the muscles' resting concentration of PCr and the athlete's ability to tolerate increases in lactic acid and other metabolic byproducts.

Following an intense burst of maximal exercise, the restoration of muscle PCr stores and removal of lactate and H^+ are carried out by **oxidative** pathways. Thus, the rate at which the team sport athlete recovers between bursts of activity will be determined largely by aerobic potential. Since recovery rate has a profound effect on late-game performance, it is important to include an endurance-training component in the preparation plan for such sports as rugby, soccer, or hockey (see Box 7.1).

Training to Improve Maximal Anaerobic Power

As documented in Chapter 2, because the activity of creatine kinase is such that the ratio of PCr to creatine remains almost in equilibrium with that of ATP to ADP, it is unlikely that this reaction can be accelerated by training or by an increase in resting muscle concentration of PCr. In contrast, an increase in the activity of the glycolytic enzymes can be expected to increase the maximal rate at which glycogenolysis and thus the generation of ATP can occur.

Sprint interval training is the term often used to describe repeated bouts of maximal efforts over a duration of approximately 5–40 seconds. Such training has been shown to increase the activity of most glycolytic enzymes, including rate-controlling enzymes such as phosphorylase, hexokinase, and phosphofructokinase (7,19,21). In addition, in almost every study that includes a performance measure, such training has been shown to result in an improvement in both peak and maximal short-term power output (9,18,21,24,26). Since changes in glycolytic enzyme activity do not occur, or are minimal, following a program of continuous endurance training (16), it is apparent that a high-intensity stimulus, requiring maximal or near-maximal rates of glycogen breakdown, is a key requirement for their upregulation. We can thus conclude that a program of sprint interval training is the most effective way for an athlete to increase the maximal anaerobic power of the glycogenolytic pathway.

The question then is what sprint interval will maximize increases in enzyme activity. Although increases have been found with repeated intervals as short as 5 seconds (18), most reports in the literature come from studies involving intervals in the 30 second range. Because muscle and blood lactate concentrations reach extremely high values with repeated 30 second intervals (21), an added bonus with such training may be an increase in the athlete's ability to tolerate H^+ and other metabolites, through improvements in ionic regulation and/or muscle buffering capacity (2,12,15,24,27). While such adaptations, on their own, would not be expected to improve maximal anaerobic power, they would improve maximal anaerobic capacity by prolonging the length of time over which such power outputs could be sustained. A second added bonus with this form of training is that it also leads to increases in oxidative enzyme activity (6,21). Such adaptations could be expected to improve the lactate and H^+ removal rate from muscle and the PCr resynthesis rate during recovery, and thus be beneficial in team sports that require repeated high-intensity bursts of activity.

It thus appears that repeated sprint intervals of approximately 30 seconds' duration provide the most effective stimulus for the upregulation of glycolytic enzymes and an increase in maximal anaerobic power. It also appears that *the most effective rest interval for such training is something in the 3–4 minute range.* This duration allows for resynthesis of most of the PCr consumed during the exercise interval (2) as well as removal of some of the lactate and H^+ that have been generated. However, because each successive sprint interval is begun before lactate and other metabolites are completely cleared from the muscle, athletes can expect a progressive decline in total power output with each sprint (Figure 7.2). In practical terms, this means, for example, that if an athlete is running 30 second sprints, the distance covered by the end of each interval will be progressively shorter despite the same maximal effort. Athletes should maintain some degree of low-intensity exercise (e.g., runners would continue walking) during the recovery intervals. This prevents light-headedness and pooling of blood in the

Figure 7.2 Total power output during four successive maximal 30 second efforts on a cycle ergometer (Wingate protocol), each followed by 4 minutes recovery. Values are means ± SD for 12 athletes (21).

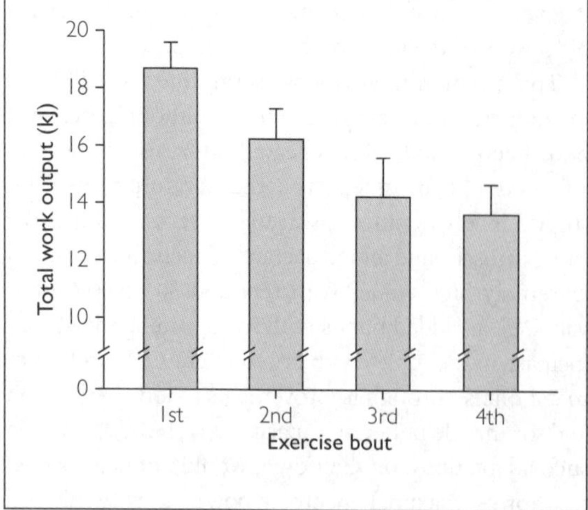

muscles and accelerates lactate removal. For team-sport athletes, this form of training is most effective when incorporated into drills that involve specific skills and simulate actual game conditions.

The optimal number of sprint intervals per training session has not been systematically investigated, but it is probable that the main limiting factor is the individual athlete's tolerance. This form of training is very stressful and uncomfortable, quickly depletes muscle glycogen stores (28,33), and generates extremely high blood lactate levels (21). Some athletes will find it difficult to tolerate more than five or six intervals, while others will be able to tolerate 10 or more. Since motor coordination or form begins to deteriorate as the athlete becomes more and more fatigued, many coaches in sports such as swimming or speed skating prefer not to continue the number of intervals to the point of exhaustion.

Training to Improve Maximal Anaerobic Capacity

Since ATP production rate begins to decline as the muscle becomes fatigued, any adaptation that delays this point will result in greater total ATP production over a given duration. Maximal anaerobic capacity would thus improve as a result of an increase in Na^+/K^+ pump capacity, an increase in muscle buffering capacity, and an increase in resting muscle PCr concentration. Although sprint training seems to have a minimal effect on resting PCr concentration (10,11,17), it has been shown to improve ion pump capacity and muscle buffering (12,23). Thus, in addition to improving maximal anaerobic power, it appears that *repeated bouts of 30 second sprint intervals are an effective means of increasing maximal anaerobic capacity*.

Such training also provides an effective stimulus for increasing resting stores of muscle glycogen (4,15). While the duration of most anaerobic sports or events is too brief for muscle glycogen depletion to become a limiting factor, it could be a limitation in certain team sports such as soccer or ice hockey where games are played back to back without adequate recovery time between games (22). Similarly, not all athletes would benefit directly from those adaptations that improve maximal anaerobic capacity. For example, it is difficult to justify this form of training for athletes whose events are completed in less than 15 seconds or so (jumpers, throwers, 100 m sprinters, 110 m hurdlers, etc.), since there is not enough time for lactic acid and other strong ions to accumulate to the extent that they would limit performance. On the other hand, for 400 and 800 m runners, 100 and 200 m swimmers, and 1000 and 1500 m speed skaters, this form of training is essential.

Frequency of Training

Approximately one third of the muscle's resting glycogen is depleted during the first 30 second interval (3). Although the amount of glycogen used declines with each successive sprint, muscle glycogen stores will obviously be almost completely exhausted after six or seven such intervals. The glycogen resynthesis rate is more rapid following such training than following depletion through prolonged exercise (30) and muscle glycogen can be completely restored within 24 hours in athletes on high-carbohydrate diets (22). Even so, resting muscle glycogen is probably not fully restored after several successive days of sprint interval training, just as it is not restored after successive days of prolonged training (8). This possibility is substantiated by athletes' anecdotal reports of "dead legs" following three or more successive days of interval training. For this reason, such training should not be performed more than four times per week and, if performed two days in succession, should be followed by a day of complete recovery.

Time Course for Adaptations to Sprint Interval Training

Unlike many training responses, adaptations to sprint training occur very quickly. Studies have shown increases in enzyme activity and muscle buffering capacity after only 1 or 2 weeks of such training (5,6,29), in magnitudes comparable to those found following 7 weeks of training (18,21). In practical terms, this means that approximately 3 weeks of sprint interval training should be adequate to achieve optimal increases in enzyme activity. This is good news for the anaerobic athlete since this is a very stressful form of training, which most find unpleasant and which increases the risk of training injury.

It also appears that many of the adaptations to sprint training are not lost as quickly with detraining, or reduced training, as adaptations to endurance training. Although the question has not yet been systematically investigated, there is some evidence that enzymatic adaptations may be retained for over a month following a period of sprint training (31).

Magnitude of Adaptations to Sprint Training

Following an intense sprint interval training phase, fit young athletes can expect to see increases in peak anaerobic power in the 20–30% range and increases in anaerobic capacity (total power output over 30 seconds) in the 20% range[1] (21). In addition, they can expect that the decline in peak power output with subsequent bursts of intense exercise will be considerably less than before training: a key advantage to the team sport athlete.

> **BOX 7.1**
>
> Some team sport athletes and their coaches share the misconception that the addition of an endurance training component might cause them to lose some of their strength and explosive power. While there is some experimental evidence that a prolonged period of continuous submaximal endurance training (as for a marathon) might cause a reduction in type 2 muscle-fiber size (32,33), this apparently does not occur with relatively shorter endurance training programs (2–3 months). In a study conducted in the authors' laboratory (1), 10 healthy young athletes performed 10 weeks of cycle ergometry training with one leg, with the other leg serving as a control. The athletes trained at 75–80% of $\dot{V}O_{2max}$, progressing to 60 minutes per day, 5 days per week, during the last 3 weeks. Measurements of strength and power (low- and high-velocity leg-press strength, evoked twitch properties, and vertical jump power with a single-leg take-off from a force platform) were taken on each leg before and after the training period. In addition, measurements of fiber area and percentage fiber type were taken from needle-biopsy samples of the quadriceps of both legs. Although there was a significant increase in the percentage of 2A fibers and a corresponding decrease in the percentage of 2X fibers in the trained leg, there was no change in fiber areas. Furthermore, the training program had no effect on any of the measures of strength and power. Since the subjects performed a greater volume of endurance training than that normally performed by team-sport and other power athletes, it was concluded that participation in a 10 week endurance training program does not impair muscular strength or power in such athletes.

Periodization of Training for Anaerobic Events and Team Sports

When designing training programs for athletes who compete in anaerobic and team sports it is important to bear in mind the following key points.

- Increases in maximal enzyme activity can occur within 2 or 3 weeks of beginning a sprint interval training program.
- These changes can be maintained for up to several months following cessation of the program.
- This form of training is very physiologically and psychologically stressful and carries a higher risk of muscle injury or overtraining than endurance training.
- Many anaerobic events and most team sports should include some form of muscle-specific strength and power training in the overall program. A minimum of 6 weeks should be allotted for this phase of training.
- Most team sports and combative sports such as boxing, wrestling, and judo should also include an endurance-training component in the overall program. A minimum of 6 weeks should be allotted for this phase of training, prior to the interval-training phase.

[1] These figures apply to anaerobic power and capacity as measured by a 30 second maximal power output test on a cycle ergometer (Wingate test). Consequently, they should not be expected to apply directly to improvements in other exercise modes, such as sprinting, swimming, speed skating, etc.

- For athletes in team sports, sprint interval training is most effective when incorporated into sport-specific drills.
- As in endurance sports, performance on any given occasion can be enhanced by a sharp reduction in training volume immediately prior to competition (tapering).
- While the competitive season for most non-team sports can usually be broken down so that peak performance is required for only three or four crucial competitions, the competitive season for most team sports usually requires that peak performance be maintained over several months, or longer.

Because of these factors, and because the dominant energy-delivery pathways vary with the duration of the event (Table 7.1), periodization of training for the anaerobic athlete is considerably more complicated than for the endurance athlete. For example, for explosive-power athletes such as throwers or jumpers, whose energy is derived exclusively from resting muscle ATP and PCr, training to enhance glycogenolytic enzyme kinetics and maximal anaerobic capacity is unnecessary. For these athletes, then, the training program should focus on enhancing motor skill performance and the strength and power of the sport-specific muscles, in preparation for competitions that occur on perhaps only five or six occasions throughout the year.

For athletes such as the 200 or 400 m sprinter in track or the 50 m swimmer, training must be directed towards attaining the maximal rate of ATP production from glycogenolysis. Typical programs would therefore involve 3 or 4 weeks of sprint interval training, plus 6 weeks or more of training to enhance (or maintain) muscle strength and power prior to each competition for that year.

In contrast, a typical season for a team-sport athlete such as an ice hockey or basketball player might extend over 3 or 4 months (or more, for professional athletes), plus an additional 3–6 weeks for playoff competition. During this time, athletes must compete, and are expected to exhibit peak performance, several times per week. Because a large portion of team practice time must be devoted to learning new systems and specific game-play situations and strategies, developing and maintaining peak physical performance presents a major challenge for the fitness-conscious coach. Since yearly preparation must include an endurance phase and a strength and power training phase as well as a sprint interval phase, the program must obviously begin in the off-season, well before the competitive season opens.

Sample Programs

Sample training programs are given here for a high jumper, a 400 m runner, and an ice hockey player. With slight modifications (so that the specific muscle groups are targeted), they are applicable to almost all anaerobic and power athletes. For example, the program for the high jumper could easily be adapted for long and triple jumpers, discus and javelin throwers, shot putters, and pole vaulters; the program for the 400 m runner is applicable as well to a 200 m runner, a low hurdler, a 50 and 100 m swimmer, or a 500 and 1000 m speed skater; and the program for the ice hockey player could be adapted for athletes in other team sports, such as soccer, rugby, or basketball.

High Jumper

High jumping is an explosive power event. The height that the jumper can achieve depends directly on the power-to-mass ratio of the hip and knee extensors and the plantar flexors of the take-off leg. This event also requires a high level of skill and motor coordination so that the athlete can place his/her body in the optimum position prior to take-off and during the flight phase.

Since the run-up usually involves fewer than 10 strides at a relatively moderate pace and muscle contraction time during the take-off phase is less than 0.2 seconds, the ATP requirements for this event are met exclusively from muscle reserves of ATP and PCr. Training for this event can thus concentrate solely on increasing the strength and rate of force development of certain muscle groups and on skill improvement. High jumpers typically compete fewer than 10 times per year. In this example, we will assume that our athlete uses the flop technique and is emerging from a successful career at the intercollegiate level, with aspirations to qualify for the national team and compete in that year's world championships. The athlete and coach have decided that the athlete's competitive year will consist of three regional competitions, the national team trials, and—hopefully—the world championships. The regional competitions will be considered preparatory and will not involve a tapering process. The program will thus be designed to ensure that performance peaks for the national trials and then again for the world championships.

The athlete will begin the training program 4 months (17 weeks) prior to the first regional competition. The prime objective at this stage is to increase the strength and power of the muscles of the take-off leg. This will require an increase in the size of the fibers in those specific muscles. A secondary objective is to improve overall body and core strength, with minimal change in muscle size, since any increase in the mass of those muscles not directly involved in the activity would be counterproductive.

Pre-Season Phase

Weeks 1–6

The emphasis over the first 6 weeks is on increasing the strength of the hip and knee extensor and plantar flexor muscles. Because of its specificity to the take-off position for jumping, the key exercise at this stage will be the half squat with barbell. This can be performed using free weights or an apparatus that guides the barbell up and down on a track. If free weights are used, one or two spotters should assist with each set. With the barbell on the shoulders, the athlete descends until he or she reaches a knee angle of 90° and then extends the knees and hips until back in the standing position. At this point, the athlete should perform a heel raise to activate the plantar flexors.

Resistance training will be done three times per week. Warm-up will consist of three sets of 10 reps at 50% of the maximum weight that can be lifted (1 RM), with 2 minutes of recovery between sets. The athlete will then complete three sets of six reps at 80% 1 RM, with 2–3 minutes of recovery between sets. At this stage, to avoid possible injury, each rep will be done slowly, with no attempt to perform explosive contractions. After 3 weeks, athletes will progress to four sets of reps to failure at 90% 1 RM (usually three to six).

Because arm swing contributes significantly to jump height, the jumper will also perform overhead presses with dumbbells and forward dumbbell raises. The goal here is to increase the explosive power in these muscles with a minimum increase in size and mass. To accomplish this, these exercises will be performed at only 50% 1 RM at this stage, in three sets of 15 repetitions. This approach will produce minimal hypertrophy in these muscle groups, but will provide a base for the explosive training to be done in the next phase.

At this stage, it is important to strengthen the hip flexors as well, because of their involvement in positioning the body prior to take-off. To do this, the athlete will perform bent-knee sit-ups against a resistance and hanging leg lifts. These will be done in three or four sets of 10–20 reps. To save time, these exercises can be done in a circuit so that, while the jumper is recovering from one set, he can be performing a set of another exercise that involves a different muscle group. While not directly applicable to the event, it is also recommended that for general fitness development or maintenance, the jumper perform 30–40 minutes of moderate-intensity endurance exercise (approximately 75% $\dot{V}O_{2max}$, as outlined in Chapter 6), such as running or cycling, two or three times each week on non-resistance training days.

Weeks 7–12

This phase will begin with the same exercises that were performed in the first 6 weeks, but the concentric portion of the squats and arm exercises will be done as explosively as possible. Because of the heavy weight (90% MVC) that is being lifted, there will be little difference in the actual speed of the movement in the squats, but the *intent* should still be to rise as rapidly as possible. For the overhead presses and forward dumbbell raises, the weights are lowered slowly but lifted as quickly as possible.

After 3 weeks, the jumper will begin adding explosive, high-velocity, and plyometric exercises (see Chapter 8) to the program. These will include jump squats with dumbbells totaling approximately 10% body weight and drop jumps from a 40–60 cm box or platform. The athlete will begin with three sets of 10 repetitions, resting 2–3 minutes between sets and 10–15 seconds between repetitions. These will be phased in so that they are done 1 day per week in week 10, 2 days per week in week 11, and 3 days per week in week 12, on alternate days to the heavy-resistance training. Sit-ups and leg lifts will continue four times per week.

Weeks 13 and 14

The heavy-resistance squats training will be cut back to 2 days per week and the jump squats and drop jumps will be replaced by actual high jumping with a weighted belt, two and then three times per week in week 14. These jumps will be done into the landing pit but without the cross bar. The athlete will use his standard approach and single-leg take-off with arm swing and lift, trying to achieve maximum vertical height each time. It is important that the weight be in the form of a belt (not a vest) and that it not exceed 10% of the athlete's body weight so as not to alter the biomechanics of the movement. The jumper will perform 15–20 jumps per session with 1–2 minutes recovery after each jump.

Weeks 15, 16, and 17

The emphasis during these last 3 weeks will be on jumping mechanics and form (see Box 7.2), and maintenance of strength and power. The heavy-resistance squats training will be cut back further to 1 day per week, with jump training increasing to 4 days per week. On a typical jumping day, about one third of each training session will concentrate on standardizing stride length in the approach run-up, one third on layout position during bar clearance, and one third on jumping for height. For the first two of these components, the bar will be set at a low or moderate height. For the third, the athlete will perform five jumps at 20 cm below his personal best height and five jumps at 5 cm below his personal best, with approximately 2 minutes rest after each jump. The last training session in week 17 will simulate actual competition, with the athlete making no more than eight jumps at progressive heights that he hopes to achieve on the first competition.

BOX 7.2

Since our athlete is an experienced jumper, it is assumed that he has already mastered the motor skills involved in take-off, layout, and landing. Any changes in mechanics will therefore be minimal and based on videos of his performance. For the novice, this phase would obviously constitute a much larger portion of the program.

In-Season Training

During the competitive season, the program will be dictated by the competition schedule. For the regional competitions, it is important that normal training be suspended 3 days before the meet. During this time, the athlete will restrict himself to stretching, practicing approaches, and visualizing jumps, but will do no lifting or forceful jumping. When there are 2 or more weeks between regional competitions, the program will be the same as in the last few weeks of the pre-season phase.

For the national trials and world championships, the ideal situation would be to have a minimum of 4 weeks prior to each competition. If this is the case, for the first 2 weeks he will resume heavy-resistance squat training 3 days per week, as in week 12 of the pre-season. The other two training sessions each week will involve jumping with a weighted belt as in week 14. Two weeks prior to competition, squat training will be cut back to 2 days per week and jumping with the weighted belt will be increased to 3 days.

The week before the competition will be a taper week (see Chapter 9). On day 1 of this week, the athlete will perform squat training. Days 2, 3, and 4 will involve jumping for height, as in week 17. On day 2 he will do 20 jumps, on day 3 he will do 15 jumps, and on day 4 he will do 10 jumps. On the 3 days immediately prior to competition, the athlete will do no forceful jumping.

400 m Runner

As detailed in Chapter 2, most of the energy for this event is derived from the anaerobic glycolytic pathway. Performance thus depends directly on the maximal rate at which the athlete's glycolytic enzymes can generate ATP by breaking down muscle glycogen to lactic acid, and the muscles' ability to tolerate the resultant increasing H^+ and other metabolic byproducts. Training must therefore be directed at improving maximal anaerobic power as well as maximal anaerobic capacity. Both goals can be accomplished through sprint interval training. Assuming that our athlete's best times to date for this event are in the 46–48 second range, we can estimate that, by the end of the race, approximately 20% of the total ATP requirement will be generated by the oxidation of muscle glycogen. Improvements in oxidative enzyme kinetics and maximal cardiac output could thus be expected to have a small but significant effect on total power output as well. In this example, the athlete and coach have decided that the competitive season will include four or five regional competitions, the national team trials, and—hopefully—the world championships. They have also decided that a full tapering process will occur only prior to the national trials and the world championships.

The athlete's program will be centered primarily around sprint interval training and training to enhance strength and power in the muscles of the legs, abdomen, and lower back. The strength and power training phase will begin a minimum of 2 months prior to the first competition of the year and interval training will begin 4 weeks prior to that competition. The athlete's training year can be considered as consisting of three phases: the off-season phase, the pre-season phase, and the competitive-season phase. In this example, the off-season phase is approximately 8 weeks in duration and will begin 3 months prior to the first competition. The pre-season phase will occur over the 4 weeks prior to the first regional competition. Sample programs for each phase are presented in the following pages.

Off-Season Phase

During this phase, the athlete will perform strength and power training 3 days per week and continuous running 5 days per week. This running phase will be at a pace that requires approximately 70–75% of the athlete's $\dot{V}O_{2max}$, using heart rate as an indicator. As documented in Chapter 6, this form of training is the most effective method of stimulating increases in heart volume and total blood volume. The time required for these central adaptations to occur is approximately 2 months. Similarly, increases in muscle strength and power will begin to plateau after approximately 2 months. Since neither this type of running nor strength training is particularly physiologically stressful or fatiguing, athletes can do both on the same training day.

Although many 400 m runners are accustomed to training solely on a track, this is not essential for this phase of the program, and variety can be achieved by altering the running routes and terrain. The program will begin with 30 minutes of continuous running at a target heart rate that corresponds to 70–75% $\dot{V}O_{2max}$ (see Chapter 6 on how to calculate target heart rate). The duration of the session will be increased by 5 minutes each week, so that by the eighth week the athlete will be running for a total of 60 minutes per session. Again, as outlined in Chapter 6, as the duration of the session increases, the athlete may choose to break it into two separate blocks, with 2 or 3 minutes in between for stretching or hydration.

Strength and power training will be done on alternate days (e.g., Monday, Wednesday, and Friday) immediately following the running session for that day. The program will consist of three sets of half squats with a barbell and three sets of hamstring curls on a universal gym or similar apparatus, two sets of high-bench stepping with dumbbells, and three sets of 10–20 resisted bent-knee sit-ups. For the half squats and hamstring curls, the athlete will use a resistance that causes failure after 10–12 reps (Chapter 8), with a minimum of 3 minutes recovery after each set. After 2 weeks, for the half squats, the athlete should attempt to rise explosively on the concentric phase of each lift, while the eccentric phase of the hamstring curls should be done slowly. Since one exercise focuses primarily on the hip and knee extensors and the other on the flexors, it will be most time-efficient if they are alternated, so that the athlete performs a set of one exercise while recovering from the other.

Pre-Season Phase

The emphasis during this phase will be primarily on sprint interval training and maintaining the strength and power gains achieved in the off-season phase. The interval training should provide adequate stimulus to maintain the cardiovascular adaptations gained in the off-season. Sprint interval training will be performed three times per week on alternate days (e.g., Monday, Wednesday, and Friday) and strength and power maintenance training two times per week (Tuesday and Thursday).

To avoid muscle injury in this phase, it is important that the transition to all-out sprinting occur gradually over the first few training sessions in week 1. For example, following a warm-up of approximately 1600 m at moderate intensity (60–70% $\dot{V}O_{2max}$), the athlete will perform sprints of 50, 100, 150, 200, 250, and 300 m, each followed by approximately 3 minutes of low-intensity recovery (walking or jogging). Each interval will begin with a rolling start; that is, the athlete will jog up to the starting point and then accelerate to full sprinting speed. Similarly, deceleration will occur gradually at the end of each interval.

In each session during week 2, the athlete will complete six intervals of 300 m, each followed by approximately 4 minutes of low-intensity recovery. Again, a rolling start will be used for each interval. In week 3, in each session, the athlete will complete seven intervals of 300 m from a standing start, accelerating to maximal effort over the first 10–15 m. In week 4, the first interval of each session will be the full 400 m, followed by seven intervals of 300 m. These will all be run from the starting blocks.

Strength-maintenance training (Tuesday and Thursday) will include 3–10 RM sets of the half squat and hamstring curl exercises done in the previous phase. On one of these days, these sets will be preceded by 30–40 minutes of continuous running, as

BOX 7.3

It is not unusual for 400 m runners to compete in the 200 or 800 m event in addition to their specialty. Their success in these events will be determined at least partially by their muscle-fiber-type percentages. Athletes who have inherited a higher-than-normal percentage of type I fibers will probably also be good 800 m runners, whereas those with a higher percentage of type 2 fibers can also be expected to perform well in the 200 m event. With minimal modifications, the program described would be suitable preparation for the 200 m as well, but would not be the ideal program for the 800 m event (see Chapter 6). If, however, the runner wishes to compete in the 800 m (but as a lower priority), this pre-season phase should be extended for an additional week so that he/she can practice pacing for the longer event.

in the pre-season phase, and on the other by practice starts from the starting blocks.

Competitive-Season Phase

The objective here is to maintain the adaptations that have been achieved during the previous two phases and to avoid injury to the quadriceps and hamstring muscle groups as the athlete competes in the regional track meets. In this example, our runner will compete in a regional meet (usually on a Saturday) every 1 or 2 weeks.

In the week before each regional meet, there will be 2 days of interval training (Monday and Wednesday) as in the pre-season phase but with a 50% reduction in training volume (i.e., 4 × 300 m). Similarly, only two sets of strength exercises will be done on Tuesday and Thursday and no training on Friday.

For the national trials, the runner will perform 3 weeks of preparatory training, as in the last 3 weeks of the pre-season phase, followed by 1 week of tapering (Chapter 9).

Hockey Player

Ice hockey is a high-intensity intermittent body-contact sport. Although there are frequent substitutions and brief stoppages in play for rule infractions, a typical 60 minute game includes approximately 20 minutes of total playing time for forwards and approximately 28 minutes for defensemen. In the course of a game, players can expect to play approximately 15–20 shifts, with 4–5 minutes recovery time between shifts. Time–motion studies indicate that, while shifts may range from 40–70 seconds in duration, stoppages in play reduce actual continuous playing time per shift to approximately 35–40 seconds on average (13,20,25). On each shift, players will spend a total of approximately 6 seconds in maximal-intensity skating or fighting for the puck along the boards or for position in front of the net. The remainder of the shift is spent in fast skating. Because of the intensity of play, blood lactate levels are very high throughout the game and muscle glycogen levels become depleted by approximately 60% (14).

Thus, an athlete's performance in ice hockey will be largely determined not simply by his level of skill, but also by the maximal anaerobic power and capacity of the muscles involved in skating and by the strength and power of his upper- and lower-body muscles. Since between-shift recovery requires the replenishment of high-energy phosphate stores and removal of muscle lactate and H^+, a high level of maximal *aerobic* power in the skating muscles is also essential.

The athlete in this example is a 20-year-old Junior A hockey player with aspirations to play professionally in the National Hockey League. His competitive season extends for approximately 5 months, plus a possible 2 or 3 weeks for playoffs. It is preceded by 1 week of training camp and three or four exhibition games. During the season, he will play an average of three games per week and have on-ice practice 2 days per week. When not playing hockey, he is a college student. His training year will consist of three phases: the off-season phase, the pre-season phase and the competitive-season phase. The pre-season phase begins 4 weeks prior to training camp and is preceded by at least 3 months of off-season training.

Off-Season Phase

During this phase, the athlete will perform strength and power training 3 days per week and continuous endurance training 4 or 5 days per week. The endurance training will normally be performed on a cycle ergometer, although running, inline skating, or road cycling can be substituted occasionally for variety. The endurance training will be intense enough to require 70–75% of $\dot{V}O_{2max}$, using heart rate as an indicator. The purpose of this phase is to stimulate an increase in heart volume and total blood volume. Since neither this form of endurance training nor strength training is particularly physiologically stressful or fatiguing, both can be performed on the same day.

Strength training will be performed on alternate days (e.g., Monday, Wednesday, and Friday), using resistances that cause failure within six to 12 reps. For convenience and safety, most of these exercises will be performed on weight-training machines. Upper-body exercises will include three to four sets of six to eight reps of bench press, seated row, overhead press, pull-downs, biceps curls, and palms-up and palms-down wrist curls. Lower-body exercises will include three to four sets of eight to 12 parallel squats, standing calf raises, hamstring curls, and straight-legged dead-lifts. Warm-up for the first set of each exercise will be 10 reps at 60% of the resistance to be used for that exercise. Since each muscle group should have a minimum of 3 minutes of recovery between sets, the various exercises should be performed in an order that allows one muscle group (e.g., extensors or upper-body muscles) to recover while a set is performed with another muscle group (e.g., flexors or lower-body muscles). This approach significantly reduces the total duration of the workout.

The endurance-training phase will begin with 25 minutes of continuous cycling on an upright cycle ergometer, preceded by 3 minutes of low-intensity warm-up. While the motor pattern involved in cycling is not identical to that of skating, it is similar in the way that hip and knee extensors and flexors are recruited. An added advantage is that pedaling resistance and rate are immediately visible and that most newer ergometers also provide a constant read-out of heart rate. The athlete will pedal at 65–70 rpm, adjusting resistance upwards until the target heart rate is reached. The duration of the session will increase by 5 minutes each week until it reaches a total of 60 minutes (ninth week). At this point, the athlete may choose to break each session into two separate blocks, with 2 or 3 minutes in between for stretching or hydration. In addition, on training days that include strength training, he can choose to shorten his cycling time to 45 minutes in order to reduce total workout duration.

Pre-Season Phase

During this phase, the emphasis will be on increasing the maximal anaerobic power and capacity of the leg muscles and maintaining the strength and power gains achieved in the upper- and lower-body muscles during the off-season phase. To accomplish this, sprint interval training will be performed on the cycle ergometer three times per week on alternate days (e.g., Monday, Wednesday, and Friday) and strength and power training two times per week (Tuesday and Thursday).

Although our athlete should ideally perform his sprint training on ice or with a skating ergometer, he is unlikely to have access to such an apparatus or enough ice time to do so. For sprint training on a mechanically braked cycle ergometer, pedaling resistance will be set at approximately 0.075 $kg{\cdot}kg^{-1}$ of body weight. On most electrically braked models, resistance will be at or close to the highest setting, so that the first few seconds of all-out sprint cycling produce a read-out of 800–900 W. Once the setting has been determined, it will remain constant for the entire training phase. Following a low-intensity 3–4 minute warm-up, the athlete will begin pedaling at maximal velocity, applying the correct resistance after approximately 2 seconds. For the first week of this phase, each all-out sprint interval will be 20 seconds in duration, followed by 3–4 minutes of low-intensity recovery against little or no resistance on the ergometer. After week 1, all intervals will be 30 seconds in duration. Training will consist of four sprint intervals per session in week 1, increasing by one interval each week to a total of seven in week 4. Studies show that this is adequate time to achieve optimal increases in muscle buffering capacity and activity of both glycolytic and oxidative enzymes (5,29). This form of training should also help to preserve the central cardiovascular training adaptations gained during the off-season phase.

On the strength- and power-training days, the athlete will perform three sets of each of the exercises that were done in the off-season. He will attempt to perform the concentric phase of each rep as explosively as possible. Although there will be minimal change in the actual speed of movement at these resistances, the goal here is to train the neuromuscular system to increase the rate of force development (see Chapter 8).

Competitive-Season Phase

During the regular season, the athlete will be on ice an average of five times per week (three games and two practices). Assuming that a significant portion of each practice session is devoted to game-specific conditioning drills (see Box 7.4), this should be enough to maintain the gains in anaerobic power and capacity that have been achieved (31). It is not enough, however, for the cardiovascular adaptations and the increases in muscle size and strength that have been gained. To maintain these gains, the athlete should supplement his on-ice time with two sessions per week of continuous cycling and strength and power training as performed in the off-season phase. The goal here should be 40 minutes of endurance training and two 10 RM sets of resistance training per session. The traveling required for out-of-town games makes it a challenge to find the time for these additional off-ice sessions. One solution is to use the cycle ergometers and universal resistance training apparatus many teams have installed in their dressing rooms, following team practice sessions. Another is to use the fitness centers found in many hotels on the nights before road games.

BOX 7.4

There is an almost infinite variety of game-specific conditioning drills for the sport of ice hockey, limited only by the innovative coach's imagination. The most effective will incorporate all aspects of the game: stopping, starting, changing direction, backward skating, short bursts of all-out skating, passing the puck, and shooting. Each high-intensity drill should be 20–30 seconds in duration for a given player, followed by 2–3 minutes of lower-intensity recovery. The drills should be organized in such a manner that, while one group (usually three or five players) is participating in a high-intensity drill, another group is recovering. During recovery, players engage in lower-intensity skill activities such as passing, shooting and positional play.

Summary of Key Points

- The energy-delivery systems for anaerobic events and team sports are PCr hydrolysis and anaerobic glycogenolysis. The relative contribution of these two pathways depends upon the duration of the event.
  - In events where the maximal effort occurs over 2 seconds or less (e.g., jumping or throwing), energy comes exclusively from stored ATP and PCr hydrolysis. While maximal ATP production rate from this pathway is almost double that of glycogenolysis, power output begins to decline after a few seconds as stored PCr begins to decrease.
  - In events that require maximal efforts over 3–15 seconds (e.g., 100 m sprint and the high-intensity bursts in most team sports), energy comes almost equally from PCr hydrolysis and anaerobic glycogenolysis.
  - In those that require maximal efforts over 15–60 seconds (e.g., 200 and 400 m in track), anaerobic glycogenolysis is the dominant system.
  - In events that require maximal power output for 2 minutes (e.g., 800 m in track), approximately 50% of the energy is derived from the oxidation of glycogen and 50% from the two anaerobic pathways.
- The maximal anaerobic power of a muscle or muscle group is the highest rate at which ATP can be generated by the hydrolysis of PCr and the highest rate at which glycogen can be broken down to lactic acid. These are determined directly by the maximal activity of creatine kinase and the enzymes in the glycolytic pathway respectively. Although the enzyme activity of creatine kinase can probably not be increased by training, the activity of those involved in the glycolytic pathway can.
- The maximal anaerobic capacity of a muscle or muscle group can be defined as the total amount of ATP that can be generated by the two anaerobic pathways over a given time period (e.g., 40 seconds). Since ATP production rate begins to decline as muscles fatigue, any adaptation that delays this point will result in greater total ATP production over this time. Maximal anaerobic capacity would therefore increase as a result of an increase in resting muscle concentration of PCr, an increase in Na^+/K^+ pump capacity or an increase in muscle buffering capacity. While training has a minimal effect on resting muscle PCr concentration, it is known to improve ion-pump capacity and muscle buffering.
- Following a burst of maximal exercise, muscle PCr stores are restored and H^+ and other metabolic byproducts removed by oxidative pathways. The rate at which team-sport athletes recover between

bursts of activity is thus determined largely by the muscles' maximal aerobic power. Since recovery rate affects late-game performance in most team sports, it is important to have an endurance-training component for these sports as well.

- In addition to the athlete's skill and coordination, performance in anaerobic events and team sports will also be determined by the anaerobic power and capacity of the muscle group(s) involved in the sport and by the maximal strength and rate of force development of these muscles. The relative importance of each of these factors will depend upon the nature of the sport and the duration over which the maximal effort occurs.
  - For events such as jumping, throwing or weight lifting, it will be most important to train for improvements in muscle strength and rate-of-force development.
  - For events that are completed in 3–15 seconds (e.g., 100 m and high hurdles in track), it is also important to train for increased maximal anaerobic power, but unnecessary to train for increased maximal anaerobic capacity.
  - For longer-duration events (e.g., 400 and 800 m in track, or 100 and 200 m in swimming), it is crucial to train not only for increased anaerobic power but also for increased anaerobic capacity.
- Both maximal anaerobic power and maximal anaerobic capacity are known to improve with sprint interval training: repeated bouts of maximal effort over approximately 30 seconds followed by 3–4 minutes of recovery. Such training stimulates an increase in the activity of both glycolytic and oxidative enzymes, improves muscle buffering capacity and boosts muscle glycogen concentration. After two or three intervals, athletes find this form of training very stressful and uncomfortable since it generates extremely high levels of blood lactate. As a result, some athletes will find it difficult to tolerate more than five or six intervals, while others may be able to tolerate 10 or more. Since muscle glycogen stores are almost completely exhausted after each session, this form of training should not be performed more than four times per week and, if done 2 days in succession, should be followed by a day of complete recovery. For team-sport athletes, this form of training should occur in the form of sport-specific drills that simulate actual game conditions.
- Increases in maximal enzyme activity and muscle buffering capacity can occur within 2 or 3 weeks of beginning a sprint interval training program and can be retained longer than other training adaptations following a reduction in or cessation of the program. As a result, the sprint interval training phase can be limited to the month prior to the competitive season. If, however, the sport also requires high levels of muscle strength and power and/or aerobic power, this phase should be preceded by several months of off-season training.

8

Training for Strength, Power, and Speed

In Chapter 5 it was shown that strength, power, and speed performance is determined primarily by the functional properties of muscles and by the ability of the nervous system to activate them effectively. Consequently, training to increase performance must stimulate both muscular and neural adaptations. The first part of this chapter will consider these adaptations and the second part will present the training principles for accomplishing them. A recurrent theme will be the role of specificity in adaptations and its importance in the design of training programs.

Adaptations to Training

Muscular Adaptations

Muscle Cross-Sectional Area

Since muscle cross-sectional area (CSA) is a major determinant of muscle force (strength) (Figure 5.99), training that results in an increase in CSA (**muscle hypertrophy**) will also result in an increase in strength. Figure 8.1A shows an example of the progressive increase in muscle CSA during a 6 month period of weight training.

Muscle-fiber hypertrophy Increased muscle size is primarily the result of increased muscle-fiber size (Figures 8.1B and 8.2). Fiber size is usually expressed as fiber (cross-sectional) area, as shown in Figures 8.1B and 8.3. Like whole-muscle size, fiber size increases progressively during the first weeks or months of training (Figure 8.1B). Type 2 fibers enlarge more than type 1 fibers (100,105,215,349), probably because they are less used in daily activities and therefore have more scope for training-induced hypertrophy when activated during training. Figure 8.3 shows that in a group of young women who trained for 20 weeks, both the greatest absolute and relative (%) increases in fiber area occurred in type 2 fibers. The same pattern occurs in young men, although it should be noted, as discussed in Chapter 5 (Box 5.1), that untrained women generally

Figure 8.1 (A) Effect of 6 months of weight training on quadriceps cross-sectional area (CSA). Young men did six sets of eight repetitions of a knee-extension exercise using 80% of the 1 RM. Training sessions occurred every second day. Based on the data of Narici *et al.* (261). (B) Effect of 8 weeks of weight training on quadriceps (vastus lateralis) muscle-fiber (cross-sectional) area. Young men trained twice per week using squat, leg-press, and knee-extension exercises. Each exercise was done for three sets of 6–8 RM one day and three sets of 10–12 RM on the other day. Fiber area data are the average of increases in type 1, 2A, and 2X fibers. Based on the data of Staron *et al.* (339).

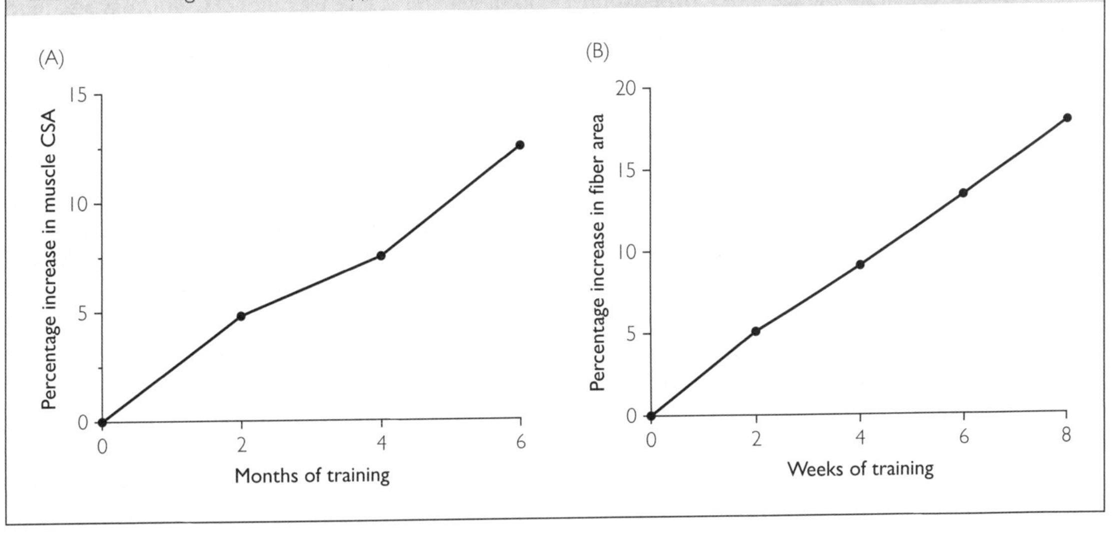

Figure 8.2 Schematic illustration of the hypertrophy process in muscle. (Top) Muscle CSA increases as a result of the increase in muscle-fiber area. (Middle) Muscle-fiber area increases as a result of the increase in both the CSA and number of myofibrils. Myofibrils enlarge to a critical size and then undergo splitting to produce two daughter myofibrils. (Bottom) Myofibrils enlarge as a result of the increased number of actin and myosin filaments. There is no change in the size or arrangement of the filaments. See text for further discussion.

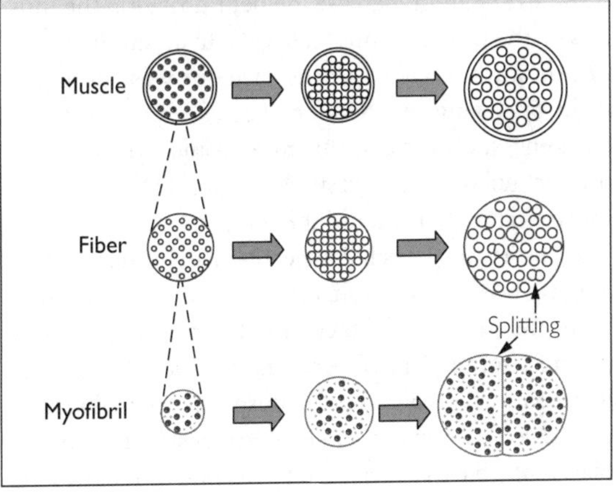

Figure 8.3 Hypertrophy according to fiber type in vastus lateralis after 20 weeks of weight training. Twenty-four young women did three sets (6–8 RM) each of full squats, leg presses, knee extensions, and knee flexions twice per week. Note that in comparison to type 1 fibers, absolute and relative (%) increases in fiber area were greater in type 2 fibers. Based on the data of Staron *et al.* (341).

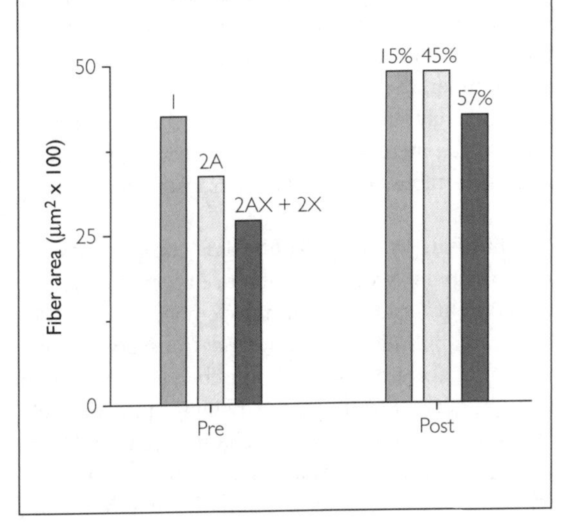

have larger type 1 than type 2 fibers (Figure 8.3), while untrained young men have larger type 2 than type 1 fibers. Nevertheless, both women and men exhibit greater hypertrophy in type 2 fibers in response to training.

Although most studies of muscle-fiber hypertrophy have focused on weight training, sprint and power training can also induce fiber hypertrophy (58,205,229,307,332) and preferential hypertrophy of type 2 fibers (82), although not to the same extent as weight training (159,229,358). For example, 8 weeks of jump training incorporating the stretch-shortening cycle (plyometric training) increased fiber area and also showed greater type 2 fiber hypertrophy (226).

Muscle-fiber hypertrophy does not continue indefinitely with continued training. For example, young men who trained triceps brachii for 6 months attained the same fiber size as bodybuilders and powerlifters who had trained for several years (219). Similarly, an intense 24 week training program failed to increase fiber size in an already highly trained group of female and male bodybuilders (14).

Increased size and number of myofibrils The main contributor to increased fiber size is an increase in the size and number of myofibrils. As illustrated in Figure 8.2, myofibrils increase to a critical size and then split into two daughter myofibrils (215). Myofibrils enlarge as new actin and myosin filaments are synthesized (Figure 8.2). There is no change in the size or arrangement of the filaments (72). This increase in the number of filaments and therefore the number of myosin cross-bridges *in cross-section* results in increased force-generating capacity. Myofilaments are organized into myofibrils by the **cytoskeleton** (Figure 4.5), which transmits force generated by the myofibrils (387). Therefore, as myofibrils increase in size and number, the cytoskeletal framework must adapt quantitatively and qualitatively (179,387) to accommodate the increased size and number of myofibrils and the resultant increased force transmission.

Myofibrillar vs non-myofibrillar components of muscle fibers Myofibrils comprise about 80–85% of the content (volume) of a muscle fiber (12,148,215). The sarcoplasmic reticulum and transverse tubule networks account for 4–5% of fiber volume, as do mitochondria (15,148). The remaining approximately 10% is cytoplasmic volume, including lipid (<1%) and glycogen (15,220). Generally, myofibrillar volume density does not change after a period of strength training (220,369). A decrease in myofibrillar volume density (with a corresponding increase in cytoplasmic volume) has been observed in bodybuilders and weightlifters consuming anabolic steroids (219) but it is not clear whether the drug was responsible for the decrease. High concentrations of muscle glycogen, known to occur with resistance training (215,350), may also tend to increase cytoplasmic volume at the expense of myofibrillar volume. A large increase in muscle glycogen and associated intracellular water content can increase muscle CSA by approximately 3% (268).

Although not shown in Figure 8.2, sarcoplasmic reticulum and T-tubular networks (Figure 4.10) expand to service the increased size and number of myofibrils; therefore, the pre-training volume density of the sarcoplasmic reticulum and T-tubular networks is maintained as hypertrophy occurs (15,215).

Mitochondrial volume density may (213,220) or may not (12,369) decrease after strength training. A decrease is more likely to occur in fibers that have undergone greater hypertrophy, particularly type 2 fibers (220), while no change is more likely in training programs with more sets and repetitions (369).

Hyperplasia While there is consensus that muscle-fiber hypertrophy is the principal contributor to increased muscle size after strength and power training in humans, an increase in muscle-fiber number (**hyperplasia**) may also be a training adaptation that contributes to increased muscle size. Although in humans the final total muscle-fiber number is considered to be attained at birth or soon after (215), mammalian and avian training and loading models show evidence of hyperplasia (23,215). For example, several studies where cats were taught to perform near-maximal unilateral voluntary contractions to receive a food reward show an apparent increase in fiber number in the muscles of the trained limbs (120,117). In addition, in a series of studies where a heavy weight was suspended from one wing of an avian species such as quail or chickens, several weeks of this chronic stretch caused muscle enlargement resulting from both increased muscle-fiber area and number (13,16,23).

Investigating possible hyperplasia in humans is challenging because it is difficult to measure fiber number accurately in living intact muscle. In a thigh muscle (vastus lateralis), the observation that bodybuilders exhibited much greater whole-muscle size than fiber size, compared to control subjects, suggested that hyperplasia may have contributed to the bodybuilders' large muscles (80,198). In contrast, in an upper limb muscle (biceps brachii), muscle-fiber number was similar in untrained men and male bodybuilders, despite a large difference in whole-muscle CSA (more than threefold in some cases) (218). In this study, fiber number was indirectly estimated in a group of elite bodybuilders, a group of intermediate caliber bodybuilders and a group of untrained age-matched controls. Since biceps is traditionally trained by bodybuilders to achieve maximum hypertrophy, the authors hypothesized that, if resistance training causes an increase in fiber number, bodybuilders' biceps muscles should have more fibers than those of the control subjects.

The results indicated that although total fiber number ranged from approximately 172 000 to 419 000 fibers, the average number of fibers was the same for each group (Figure 8.5). Since both groups of bodybuilders had trained their biceps to achieve maximum hypertrophy for a minimum of 6 years but still had the same number of fibers as the untrained controls, it appears that training does not result in a significant increase in fiber number. Within each group there was also a tendency for the subjects with the largest muscles to have a higher than average number of fibers. Thus, although muscle size is primarily determined by the size of the individual fibers, it is also affected by the number of fibers. The massive muscle sizes found in some bodybuilders are therefore the probable result of both large increases in fiber size and their genetically determined larger than normal fiber numbers.

Biceps fiber number, measured by the same method, did not increase after 12 weeks of resistance training, although fiber area and muscle CSA increased (238). Thus, both short-term longitudinal training studies and long-term training represented by cross-sectional studies have not unequivocally demonstrated hyperplasia as an important contributor to whole-muscle hypertrophy in humans.

Figure 8.4 Schematic illustration of a satellite cell. Bottom of the figure shows a segment of a muscle fiber. Top shows a higher magnification at the electron microscopic level. Satellite cells can be distinguished from other myonuclei because they lie *outside* the plasma membrane.

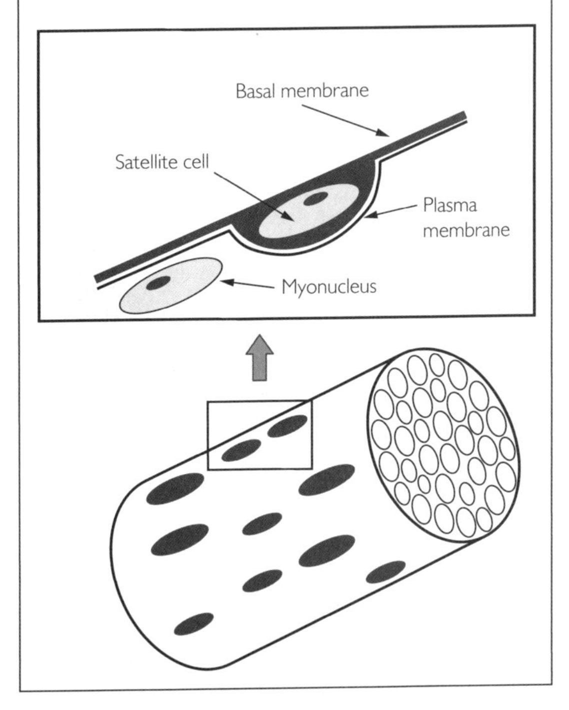

This is not to say that muscle is incapable of generating new fibers. When a muscle fiber is injured or damaged as a result of heavy resistance training, its satellite cells (Box 8.1 and Figure 8.4) become activated

BOX 8.1 ROLE OF SATELLITE CELLS IN MUSCLE HYPERTROPHY

The term satellite cell was first used by Mauro (236) to describe the muscle stem cells that surround each fiber lying inside the basal membrane but *outside* the plasma membrane (Figure 8.4). This location between the two membranes allows them to move up or down the fiber to those sites where they are required. They remain quiescent or dormant until they are activated and enter the fiber to become additional myonuclei for the fiber as it increases in size with maturation or training-induced hypertrophy. They thus ensure that the nuclei-to-cytoplasmic volume ratio is maintained as fibers become larger (166). As they are taken up by the fiber, some of those remaining undergo mitotic division so that their total numbers do not decline. For each hundred myonuclei per fiber there are approximately two to five satellite cells, although this number can vary between different muscles and with the age and physical activity level of the individual (166). A second important function of satellite cells is their role in the regeneration of injured or damaged fibers (40). Because of these functions, it is generally thought that satellite cell activation is essential for muscle hypertrophy (7,167), although not all investigators agree (240).

Figure 8.5 Biceps brachii CSA (top), muscle-fiber area (middle), and muscle-fiber number (bottom) in untrained control subjects (C), intermediate bodybuilders (IBB), and elite bodybuilders (EBB). The greatly enlarged muscle size of the bodybuilders was the result of greatly enlarged muscle fibers rather than a greater number of fibers. Average values are shown with standard deviations. Based on the data of MacDougall *et al.* (218).

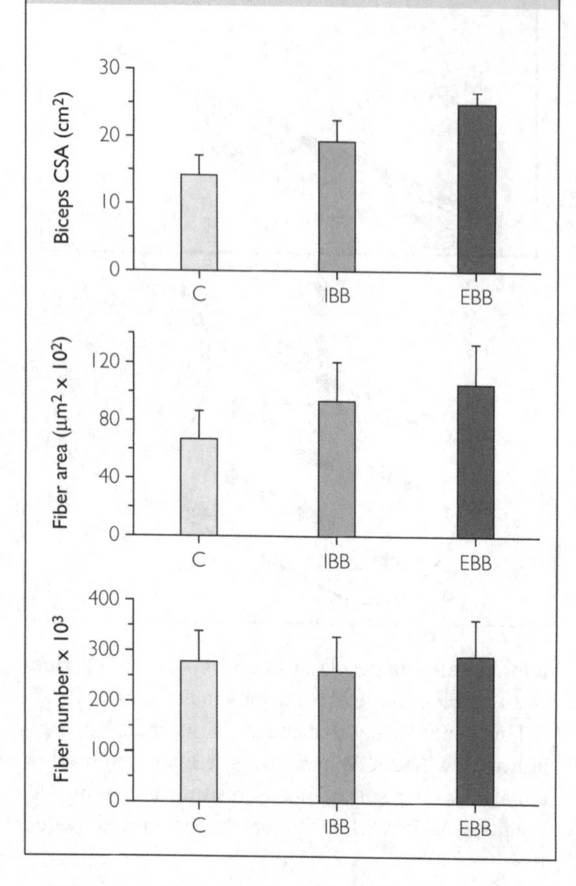

and migrate along the length of the fiber to the site of the injury. They then enter the cell and differentiate to repair the damage by forming new myofibrils in the existing fiber (7), while the debris is removed by macrophages (68). If the damage is too extensive, the satellite cells fuse together to form a multinucleated myotube. The myotube then matures into a new muscle fiber, as occurs during fetal development, and replaces the necrotic fiber (40,322). These new developing fibers probably account for the small population of abnormally small fibers which can be found in the muscles of elite powerlifters and bodybuilders (165,219), because they have been shown to express developmental isoforms for early myogenesis (165).

Since the satellite cells that are activated are only those of the damaged fiber and not recruited from other non-damaged fibers (323), it is unlikely that these new fibers are additional fibers but rather are *replacements* for fibers too damaged to be repaired (215). It thus appears that, when athletes undergo heavy resistance training, there is a constant cycle of damage and repair occurring in the muscles being trained. Milder forms of damage are repaired by incorporation of satellite cells into the existing fibers, while severely damaged fibers are replaced by fusion of satellite cells into myotubes, which develop into new fibers. Otherwise, the finding that at any one time 3% of the fibers in the trained muscles are new developing fibers (166) suggests that years of such training would generate readily detectable massive increases in fiber numbers (215). In fact, large increases in fiber number are not found (Figure 8.5).

Connective tissue As muscle fibers enlarge there is an increase in the absolute but not relative amount of inter-fiber connective tissue (≈10% of muscle CSA), even in greatly hypertrophied muscles (215). Inter-fiber connective tissue (endomysium) surrounds each muscle fiber. Less is known about the effect of training on the thickness of connective tissue (perimysium) enveloping muscle fascicles or the connective tissue (epimysium) enveloping the muscle as a whole (118), although there is evidence that a single bout of exercise evokes a coordinated synthetic response in muscle contractile protein (myofibrillar) and connective tissue (endomysial, perimysial, and tendon collagen) (246).

Stimulus for hypertrophy Various possible stimuli for hypertrophy, and factors that may enhance hypertrophy, are discussed here.

Mechanical tension: a certain minimum level of muscle contraction force, also referred to as mechanical tension or muscle loading, is needed to maintain muscle mass (150). Furthermore, the minimum level of loading must occur frequently, as in daily activities. Prolonged periods of muscle unloading, approximating zero load, such as bed rest, immobilization, and spaceflight, result in **atrophy** and a loss of muscle mass (149). Strength training, which consists of regularly imposing muscle loads greater than those experienced in daily activities, is needed to activate the molecular signaling pathways that stimulate muscle protein synthesis, resulting in muscle-fiber hypertrophy (29,96,137,282,294).

A **minimum threshold** level of contraction force or muscle loading is necessary to induce hypertrophy, because repeated maximal activation of muscle contracting against zero load fails to induce hypertrophy (386). There is some evidence that the minimum *whole-muscle* threshold force is greater than approximately 10% of maximum force (6). For example, 6 second bouts of maximal effort cycling against loads corresponding to approximately 10% maximum strength activate molecular signaling pathways directed toward endurance adaptations rather than muscle hypertrophy (29,114), and thus will produce minimal hypertrophy (10). On the other hand, loads ≥ 15%, if appropriately applied, can effectively produce muscle hypertrophy (6,146).

Athletes and coaches want to identify and apply the **optimal load** for inducing hypertrophy, if that is a goal of training. Whether the range of optimal loading is narrow or broad may depend on whether it is *whole-muscle* force or force at the level of *individual muscle fibers* that is the primary stimulus to muscle protein synthesis leading to hypertrophy. The link between mechanical loading and muscle protein synthesis appears to occur at the level of individual muscle fibers (149). This implies that if training is to induce hypertrophy in the muscle fibers of a motor unit (MU), the MU must be recruited and driven to fire at rates sufficient for maximum or near-maximum force generation, and all of a muscle's MUs must be so activated to achieve maximum *whole-muscle* hypertrophy. In Chapter 5 it was shown that all or most MUs in a muscle can be fully activated by single, brief maximal force contractions or by repeated submaximal force contractions continued to fatigue failure (207) (Figures 5.31 and 5.32). In the former, all MUs are recruited to produce maximum force concurrently in a single contraction, whereas in repeated submaximal contractions to failure, MUs produce their maximum force successively over the time taken to do the contractions. The last recruited high-threshold MUs innervating type 2 fibers generate their maximum force on a background of declining force (due to fatigue) in earlier recruited MUs. If mechanical load at the level of individual muscle fibers is the key stimulus for hypertrophy, and if all fibers in a muscle can be loaded (recruited) either by single maximal contractions or by submaximal contractions repeated until fatigue failure, it follows that there would be a wide range of optimal loading, expressed as a percentage of maximum strength, for inducing hypertrophy. A low correlation between induced hypertrophy and training intensity, expressed as a percentage of single maximum lift (one repetition maximum, 1 RM), would be expected (105). In fact, as will be discussed, a wide range of whole-muscle loading has proven effective, and in some cases equally effective, in producing whole-muscle hypertrophy. Importantly, training to failure with low to moderate loads, representing 20–30% of a single maximal lift (1 RM), produces hypertrophy of the type 2 muscle fibers comprising high-threshold MUs (248,265,370), indicating both that mechanical signaling for hypertrophy occurs primarily at the level of individual muscle fibers, and that submaximal contractions continued to fatigue failure successfully recruit high-threshold MUs innervating type 2 fibers.

In the gym, therefore, an athlete seeking muscle hypertrophy could choose to lift heavy weights for a few repetitions to failure, or relatively light weights for many repetitions (e.g., 20–30) to failure. Traditionally, athletes have trained with heavy loads (high force) and relatively few repetitions. Expressed as a percentage of a single maximum lift (1 RM), conventional hypertrophy loads would be approximately 70–90% 1 RM, or about five to 15 repetitions to failure (5 RM to 15 RM), depending on the exercise and the athlete's training status (18) (Figure 5.68). But relatively light loads, corresponding to 30% 1 RM (≈25 RM), induce a level of myofibrillar protein synthesis that over a 24 hour period post-exercise exceeds the level produced by lifting heavier loads (90% 1 RM, ≈5 RM) (54). Congruently, 10 weeks of training with loads of 30% 1 RM (≈35 RM) and 80% 1 RM (≈10 RM) produce similar increases in whole-muscle and type 1 and 2 muscle-fiber size (248).

Another light-load training method uses blood-flow restriction (BFR), in which pressure cuffs are applied to limbs to impede muscle blood flow during exercise and rest periods between sets of repetitions (6,206–208,210,370). BFR accelerates fatigue and thus increases MU activation during sets of repetitions continued to failure. Like low-load training without BFR, BFR training promotes signaling pathways leading to muscle protein synthesis and muscle hypertrophy (210). The main difference between low-load training (20–40% 1 RM) with and without BFR is that with BFR training failure occurs with fewer repetitions; at higher loads (e.g., 50% 1 RM) there is no difference in the number of completed repetitions because even without applied BFR, muscle contraction force by itself occludes blood flow (370). Furthermore, when BFR is applied to high-load training, there is no augmentation of muscle hypertrophy beyond that seen with high-load training alone (200). Low-load BFR training has produced levels of hypertrophy similar to that attained with traditional high-load training (210,370) and, like low-load training without BFR (248), low-load BFR training increases type 2 fiber area (370), indicating successful activation of high-threshold MUs in this type of training. Alternatives to using BFR to hasten fatigue during low-load training include using short (e.g., 30 second) inter-set rest periods (343) and doing the concentric and eccentric phases of repetitions relatively slowly (320,346,347). Like BRF, both alternatives are effective in producing hypertrophy by accelerating fatigue, leading to recruitment of high-threshold MUs.

It should be noted that although there may be a broad range of mechanical loading that successfully induces muscle hypertrophy, there is still debate over the optimal load within this range (50,321). In addition, the range of optimal loading may be narrower for other adaptations related to increased strength and power. For example, as discussed later in this chapter, high levels of loading may be necessary for optimal tendon and neural adaptations.

Passive stretch: in some animal experimental models passive muscle stretch has also been shown to activate signaling pathways leading to muscle protein synthesis (150) and hypertrophy (23,119,150). This raises the question of whether the type of stretching done by athletes as part of a warm-up routine or to increase flexibility may induce muscle hypertrophy. The effectiveness of such exercises in inducing hypertrophy in human muscles may be limited by the ability to tolerate the magnitude and duration of stretching needed to stimulate it. For example, approximately 30 minutes of intermittent maximally tolerated (without pain) stretch of human soleus muscle failed to increase muscle protein synthesis, whereas intermittent isometric contractions at the same force level (≈30% of maximal voluntary isometric force) and duration did increase muscle protein synthesis (104). The duration of stretching was longer than that occurring in a typical stretch training session. These results suggest that the passive stretch force threshold for inducing hypertrophy is greater than the contraction force threshold, bearing in mind, as discussed, that to be effective low contraction forces must be sustained to fatigue failure. In Chapter 10 (Stretching and Flexibility) it will be shown that one adaptation to regular stretch training is increased stretch tolerance. Nevertheless, it is still unlikely that these greater passive stretch forces are sufficient to induce hypertrophy. This would account for the minimal hypertrophy seen in yoga practitioners even after several years of severe, prolonged stretching.

Muscle damage: a session of heavy resistance training results in structural damage to many of the fibers in the exercised muscles (73,116,338). It is the eccentric or lowering phase of repetitions in weight training that causes the most damage (myofibrillar disruption) (116,201). If the athlete is unaccustomed to such exercise, muscle inflammation and soreness will develop concurrently with the damage, be evident within 24–48 hours and last for several days (73). Although contractile protein damage also occurs in athletes accustomed to strength training, it is less severe after a typical resistance training session and there is little or no muscle inflammation and soreness (115).

Since damage seems to occur primarily in type 2 fibers (162,214) and it is these fibers that show the greatest activation of satellite cells following eccentric exercise (67), the fact that training-induced hypertrophy is also more pronounced in type 2 fibers (367) makes it attractive to speculate that the hypertrophy process may simply be one of *damage and repair*. If the damage to contractile tissue becomes repetitive, as with resistance training, the body attempts to compensate by synthesizing additional and larger myofibrils resulting in a larger and stronger muscle.

Acute hormonal responses: hormones such as testosterone, growth hormone and insulin-like growth factor 1 (IGF-1) can have a potent anabolic effect on skeletal muscle during growth and development, when administered as replacement hormones or taken in addition to normal levels (37,38,46,235). Since blood concentrations of these hormones increase during and immediately following a typical heavy resistance training session, it has been suggested that post-exercise elevations in hormones may be the signal for promoting the protein synthesis that results in muscle hypertrophy (239).

If the training session involves a large number of muscles, hormone concentrations become elevated and peak approximately 15 minutes after exercise but are back to pre-exercise levels within an hour or so (31,371,373). In addition, increases in total and free testosterone are relatively small and similar to normal morning concentrations (31). Because of this, and the fact that the increase in muscle protein synthesis that follows such exercise lasts for 36 hours or more (216,281), the transient elevation in hormones is probably of minimal anabolic importance (207,209,374). This conclusion is supported by the observation that if resistance exercise is done only with a small muscle mass (e.g., biceps brachii), there is little or no increase in testosterone and growth hormone concentration compared to when the biceps exercise is immediately followed by leg exercise. In spite of this, the acute increase in muscle protein synthesis in biceps is similar in both cases (373). Correspondingly, isolated training of biceps produces the same increases in muscle size as training biceps in combination with leg exercise (371).

Muscle-fiber swelling: any athlete familiar with resistance exercise will have experienced the temporary muscle enlargement that occurs after a session of resistance exercise, and may have learned that certain routines, especially those consisting of several sets of several repetitions to failure with short inter-set rest periods, are most likely to create the temporary muscle enlargement referred to as the "pump" by several generations of bodybuilders. To the disappointment of many bodybuilders, the pump is short-lived; the muscle reverts to the pre-exercise size within an hour or less after the exercise session.

The transient muscle enlargement is due to a shift of fluid from the circulation into the muscle. Part of the fluid shift is into the muscle fibers, causing an increase in fiber size called cell swelling (6,206). It has been hypothesized that fiber swelling stimulates a volume sensor that decreases muscle protein breakdown and also activates the signaling pathways leading to muscle protein synthesis (6,206,283,389). The effect would be a net increase muscle protein accumulation that in the course of training could produce muscle-fiber hypertrophy. However, as already noted, the fiber swelling (pump) subsides within an hour post-exercise, whereas muscle protein synthesis can continue for up to 48 and perhaps 72 hours (51,216,281). The question then arises as to the influence of about 1 hour or less of fiber swelling on 2–3 days of post-exercise muscle protein synthesis. As discussed later in the chapter, training with eccentric contractions can effectively induce hypertrophy without fatiguing exercise sessions that produce the transient muscle enlargement. Nevertheless, transient muscle enlargement is at least a sign that the training session, by inducing fatigue, has successfully recruited the type 2 muscle fibers of high-threshold MUs.

Fiber-Type Transitions

A detailed discussion of MU and fiber types is given in Chapter 5 (see also Figures 5.3 and 5.5). Briefly, the main fiber types are the slow-twitch type 1 fibers and the fast-twitch type 2A and 2X fibers. There are also hybrid fibers that contain different proportions of type 1, 2A, and 2X myosin (see Figure 5.4). Muscle fibers can be typed, and therefore possible fiber-type transitions assessed, by histochemical staining for myosin ATPase, by immunohistochemistry, and by single-fiber electrophoresis. It is also possible to measure myosin heavy chain (MHC) 1, 2A, and 2X content in muscle homogenate samples. It has been suggested that, in a training study, at least two of these methods (e.g., histochemical fiber typing plus MHC immunohistochemistry) are needed to assess training-induced fiber-type transitions (106,337).

Chronic stimulation studies in animals have demonstrated fiber-type transitions (also referred to as conversions or switching). Chronic low-frequency stimulation causes a fast-to-slow fiber transition whereas a slow-to-fast fiber transition is produced by repeated, intermittent high-frequency stimulation (227,278,317,381). Endurance exercise and training are typically associated with prolonged periods of relatively low MU firing rates. In contrast, strength and especially power/speed training are associated with repeated brief bursts of high-frequency MU firing rates (review Chapter 5, e.g., Figure 5.52). Thus, endurance (fast-to-slow) and strength/power/speed (slow-to-fast) training may cause fiber-type transitions similar to those seen with stimulation experiments. With endurance training a decrease in type 2X (317) and increases in the number of 2A and type 1 fibers have been observed (227,279) but the increase in type 1 fibers has not been consistent (133).

Strength training The most commonly observed fiber-type transition in response to conventional strength training is an increase in the percentage of type 2A fibers at the expense of type 2X fibers, with no change in the percentage of type 1 fibers (19,39, 100,215,227,320,337,350,381). An example of this common transition is shown in Figure 8.6. The 2X-to-2A transition can occur with a few weeks of training (164,339). Figure 8.6 does not include changes in

Figure 8.6 Effect of resistance training on fiber-type distribution. Thirteen young women and eight young men trained leg muscles with the squat, leg-press, and knee-extension exercises for 8 weeks. On 1 day per week subjects did three sets of 6–8 RM on each exercise; on the second training day per week three sets of 10–12 RM were done. Fiber-type distribution was determined in one head of quadriceps (vastus lateralis). Note that the percentage of type 2A fibers increased at the expense of 2X fibers, whereas there was little change in the percentage of type 1 fibers. Based on the data of Staron *et al.* (339).

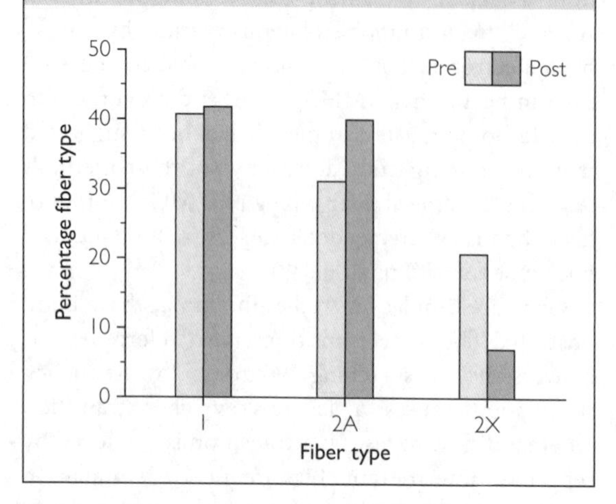

the distribution of hybrid fibers. The percentage of hybrid fibers decreases after a period of training. For example, after a 12 week period of resistance training hybrid fibers decreased from approximately 35 to 10% of the total number of muscle fibers (380). However, the decrease in 2X fibers can be associated with an increase in the number of 2A/2X hybrid fibers as well as an increase in the number of 2A fibers (337).

Sprint training Like strength training, sprint, cycle, and run training produces an increase in the percentage of type 2A fibers (307,381). It is not as clear whether it increases the percentage of type 2 fibers at the expense of type 1 fibers, as suggested by stimulation experiments. Some sprint training studies have shown an increased percentage of type 2 fibers combined with a decreased percentage of type 1 fibers, whereas other studies have not shown this transition or even an increase in the percentage of type 1 fibers (307,381). A season of ice hockey, which would have involved repeated brief but high-intensity bouts of exercise, increased the percentage of type 2A fibers and decreased the percentage of type 2X fibers with no change in the percentage of type 1 fibers (125).

Explosive training Various forms of explosive or ballistic training such as jumping with or without added loads generally either have no effect on fiber-type distribution or show the 2X-to-2A transition also seen in conventional strength training and sprint training (227,229).

For the athlete specializing in speed performance the key question is whether speed training can increase the proportion of the fastest muscle fibers (i.e., 2X). As noted above, the current evidence is equivocal on this question. In fact, the most common finding is a decrease rather than an increase in the percentage of 2X fibers after various forms of speed training. The reason for the contradictory evidence is unknown. It has been suggested that, if speed training has the potential to convert slow (type 1) fibers into fast (type 2A and 2X) fibers, this potential may be compromised if the frequency and volume (e.g., duration and number of sprints, etc.) per training session is relatively high (381). In other words, too much speed training, or speed training combined with endurance training, may be counterproductive. As discussed in the subsection on detraining, the most effective way to increase the percentage of 2X fibers may be to *detrain* after a period of training (19).

Effect of detraining As already discussed, a typical response to strength training is an increase and decrease in the percentage of 2A and 2X muscle fibers, respectively. With detraining, the percentage of 2A and 2X muscle fibers returns to pre-training levels (39,340) (see Figure 8.7). In addition, there is some evidence that, during detraining, the percentage of 2X fibers may increase above or overshoot the pre-training value (20,21). Figure 8.8 provides an example of detraining-induced 2X fiber overshoot. Although the 2X overshoot is not always observed (Figure 8.7), its possible occurrence raises the question of whether detraining or reduced training might be a strategy to enhance speed performance. As shown in Figure 8.7, however, even a greatly reduced training volume may prevent a return of percentage 2X to the pre-training value. It appears that a decrease in percentage 2X is an inevitable consequence of strength training, and likely also power and speed training (227). The 2X overshoot after detraining has been linked to an increase in *unloaded* maximum shortening velocity (V_{max}) and increased

Figure 8.7 Effect of resistance training and detraining or reduced training on the percentage of type 2X fibers in vastus lateralis. Three days per week for 16 weeks three groups of subjects trained leg muscles with the squat, leg-press, and knee-extension exercises. Each exercise was done for three sets of 8–12 repetitions with 75–80% of 1 RM. After the initial 16 weeks of training, one group did no training (Detrain) for 32 weeks. During the same period a second group trained 1 day per week with three sets of each exercise and a third group trained 1 day per week with one set of each exercise. Note that only detraining allowed the percentage of type 2X fibers to recover to the pre-training value. Even markedly reduced training prevented full recovery of the 2X fibers. The percentage of 2A fibers (not shown) exhibited the reciprocal pattern; that is, the initial training increased the percentage of 2A fibers and only detraining allowed a return to the pre-training values. Based on the data of Bickel *et al.* (39).

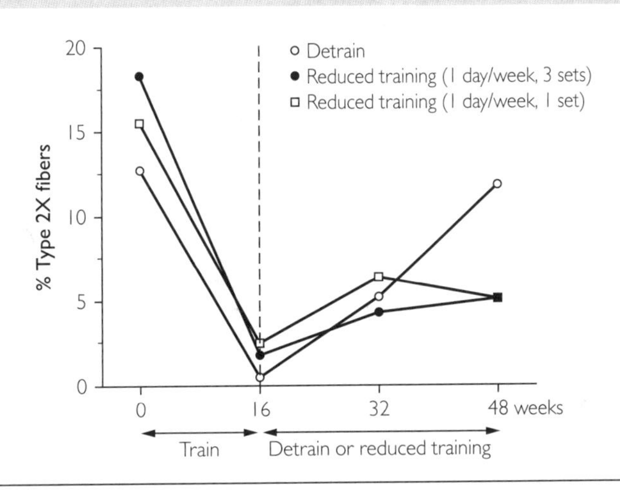

rate of force development of evoked twitch contractions (21); nevertheless, prolonged detraining for the speed athlete is not a viable option because other important adaptations to training, such as increased neural activation, hypertrophy, increased strength, and increased power and speed under *loaded* conditions, are lost with prolonged detraining (21).

Influence of fiber-type transitions on performance The transition from 2X to 2A fibers would seem to be a counterproductive adaptation for strength, power, and speed athletes. Fortunately, there are other compensating training adaptations (19). One is the usually observed greater hypertrophy of type 2 vs type 1 fibers (Figure 8.3) that results in a greater proportion of the trained muscles being occupied by type 2 fibers. Thus, the trained muscle as a whole is stronger and faster despite the reduced percentage of 2X fibers. Note also that the pre-training percentage of 2X fibers is not large (≈15%; see Figures 8.7 and 8.8), so the preferential hypertrophy of type 2 fibers more than compensates for the decline in 2X fibers. For athletes whose performance requires both speed and endurance (e.g., longer sprints, many team sports), the transition from 2X to 2A fibers would improve resistance to fatigue (review Figure 5.9), a benefit that may outweigh the decrease in type 2X fibers. Additional adaptations that compensate for the 2X-to-2A fiber-type transitions include altered muscle architecture, tendon properties, and muscle contractile properties, and increased neural activation (227). These adaptations will be discussed in subsequent sections.

Muscle Architecture

There are two main components of muscle architecture. One is the pennation angle of muscle fibers, as discussed in Chapter 5 (Box 5.13 and Figure 5.98). Pennation angle is affected by muscle length/joint angle and whether the muscle is relaxed or contracted (170). The effects of training on pennation angle are usually assessed in relaxed muscle at a set joint angle. The second component of architecture is muscle-fiber length; like pennation angle, the effects of training on fiber length are measured in relaxed muscle at a standardized joint angle. The methods used to measure pennation angle and fiber length actually measure pennation angles and lengths of muscle fascicles rather than individual muscle fibers (195).

Fiber pennation angle Resistance training typically increases pennation angle as a result of muscle-fiber hypertrophy (41,98,233,326). With increased pennation angle there is less resolved force in the line of pull of the tendon (41,84) (Figure 8.9); however, with the increases in pennation angle observed after training, the loss of resolved force is more than compensated for by the increased physiological CSA resulting from muscle-fiber hypertrophy. The net effect is an increase in *absolute* muscle force (3,169). On the other hand, the decreased resolved muscle force could result in a decrease in *specific force*, i.e., decreased force per unit muscle physiological CSA and also per unit muscle anatomical CSA (154,168,169). An example of the effect of resistance training on pennation angle is given in Figure 8.10.

Figure 8.9 Schematic representation of possible architectural changes in a pennate muscle in response to strength, power, and speed training. (A) Untrained muscle. (B) Training-induced increase in muscle-fiber size causes fiber pennation angle to increase if there is no change in fiber length. (C) Increased fiber length will reduce or eliminate the increase in pennation angle that would tend to occur with increased fiber size.

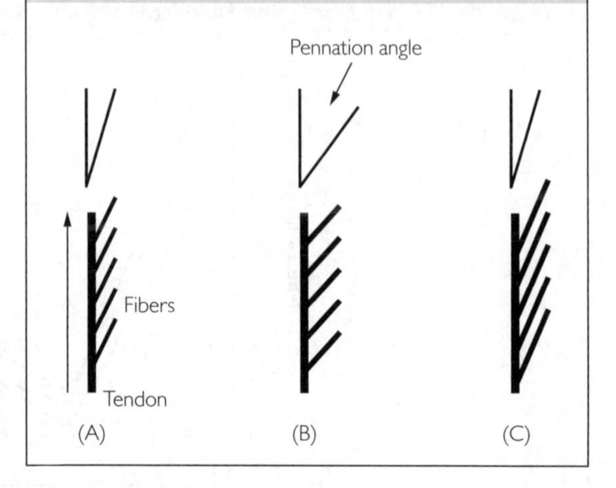

Figure 8.8 Effect of resistance training and detraining on the percentage of type 2X and 2A muscle fibers in vastus lateralis. Three days per week for 3 months, subjects trained leg muscles with the hack-squat, incline leg-press, knee-extension, leg-curl, and calf-raise exercises. Each exercise was done for four or five sets of 6–15 RM, progressing to greater relative loads during the training period. After 3 months of training the percentage of 2A and 2X fibers increased and decreased, respectively. After 3 months of detraining, the percentage of 2X fibers had increased to exceed (overshoot) the pre-training value. Reciprocally, 2A fibers decreased below the pre-training value. Note that the percentage of 2X and 2A fibers are referenced to the left and right vertical axes, respectively. Based on the data of Andersen and Aagaard (20).

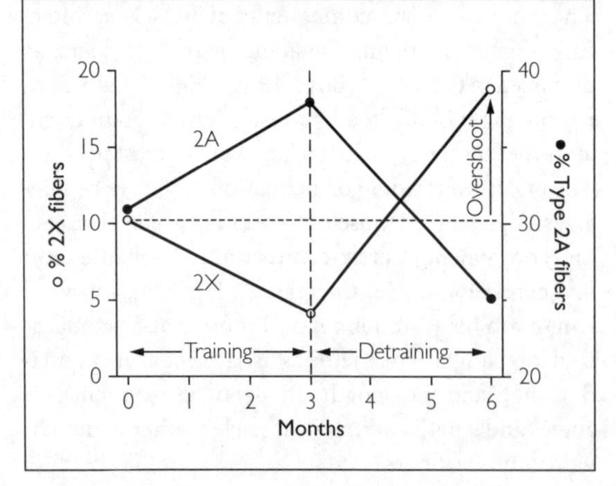

Figure 8.10 Effect of heavy resistance training on muscle-fiber pennation angle in a thigh muscle (vastus lateralis). Eleven men trained 3 days per week for 14 weeks with loads progressing from 10–12 RM to 4–6 RM over the training period. The inset shows that pennation angle increased from an average of 8.0° before training to 10.7° after training, a 36% increase. Type 1 and 2 fiber area, representative of physiological cross-sectional area, increased more than anatomical cross-sectional area (ACSA). Based on the data of Aagaard et *al.* (3).

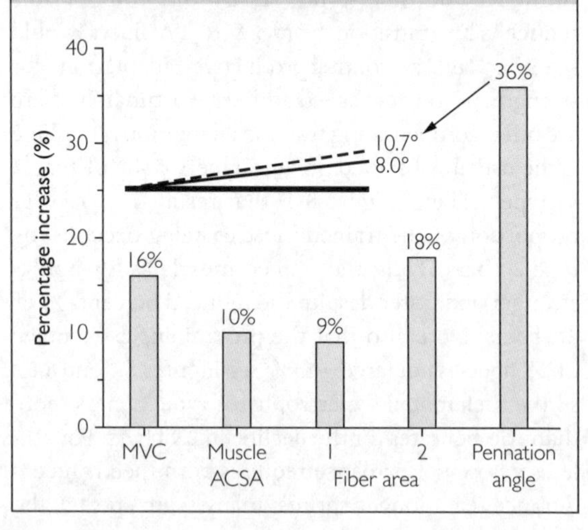

Fiber length Increasing fiber length increases fiber shortening velocity. As demonstrated in animals but not directly in humans, fiber length can be increased by increasing the number of sarcomeres in series along the length of myofibrils in the fiber (41,75). Figure 8.11 illustrates how sarcomere addition increases shortening velocity. Importantly, this can increase fiber velocity without increasing sarcomere velocity (41). Therefore, sarcomeres can shorten at lower velocities that enhance their force output according to the force–velocity relationship. Figures 8.12 and 8.13 show how two possible adaptations to training, increased fiber area and increased fiber length, can improve muscle contractile performance.

Sprint training (5,41,195) and explosive training (9,41) may increase muscle-fiber length and therefore increase contraction velocity and power. These increases in contraction speed are independent of fiber-type transitions, indicating that a slow-to-fast fiber-type transition is not obligatory for increased contractile speed and power (308). Figure 8.14 shows that sprinters have greater fiber length than distance runners and untrained subjects. Among sprinters, those with longer fiber length may (195) or may not (336) sprint faster, indicating that several factors, including muscle-fiber length, fiber-type distribution,

Figure 8.11 Schematic illustration of how adding sarcomeres in series increases fiber shortening velocity without increasing individual sarcomere shortening velocity. (A) A single sarcomere shortens through a range of 2 μm in a time *t*; therefore the shortening velocity is 2 μm/*t*. (B) Two sarcomeres in series shorten 4 μm (2 μm for each sarcomere) in the same time *t* as the single sarcomere in A, resulting in a velocity of 4 μm/*t*. Therefore, adding the second sarcomere doubled the overall velocity.

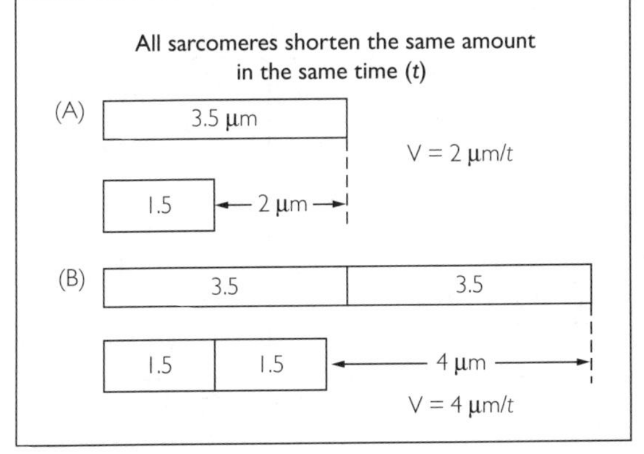

Figure 8.12 Schematic illustration of how fiber area and length can affect contractile performance. Inset depicts muscle fibers A, B, and C resolved into the number of sarcomeres in parallel, denoting fiber area and thus force-generating capacity, and the number of sarcomeres in series, denoting fiber length and thus shortening velocity. Each fiber's concentric force–velocity relationship is shown in the graph. Each sarcomere is given a maximum isometric (velocity = 0 on graph) force (ISO_{max})-generating capacity of 10 arbitrary units and a maximum unloaded shortening velocity (V_{max}) of 0.5 arbitrary units. Fiber A has one sarcomere in parallel and therefore produces an ISO_{max} of 10, and a V_{max} of 1 because there are two sarcomeres in series (0.5 + 0.5). Note that ISO_{max} is determined only by the number of sarcomeres in parallel; therefore, fiber A produces an ISO_{max} of only 10, not 20. Fiber B has two sarcomeres in parallel and thus produces an ISO_{max} of 20; on the other hand, this fiber has the same number of sarcomeres in series as fiber A and therefore the same V_{max}. Fiber C has the same number of sarcomeres in parallel as A and therefore the same ISO_{max} but twice the number of sarcomeres in series and therefore twice the V_{max}. Training that increases only fiber area but not fiber length will increase force over a range of velocities but will not increase V_{max} (compare fibers A and B). Training that increases only fiber length but not fiber area will increase V_{max} and force over a greater range of velocities but will not increase ISO_{max} (compare fibers A and C). See Figure 8.13 for the corresponding power–velocity relationships of fibers A, B, and C.

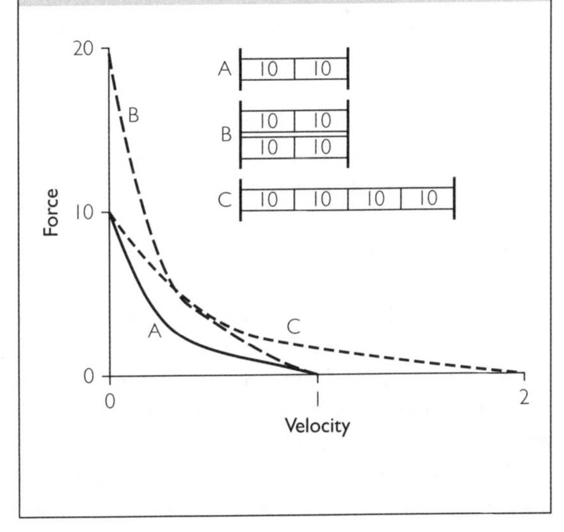

Figure 8.13 Depiction of power–velocity relationships of fibers A, B, and C based on their force–velocity relationships shown in Figure 8.12. Recall from Chapter 4 that power = force × velocity. Fiber B has greater peak power than fiber A not because of a greater V_{max} (same number of sarcomeres in series) but because of greater force-generating capacity (twice the number of sarcomeres in parallel). Fiber C has greater peak power than fiber A because of a greater V_{max} (twice the number of sarcomeres in series) rather than greater force-generating capacity (same number of sarcomeres in parallel).

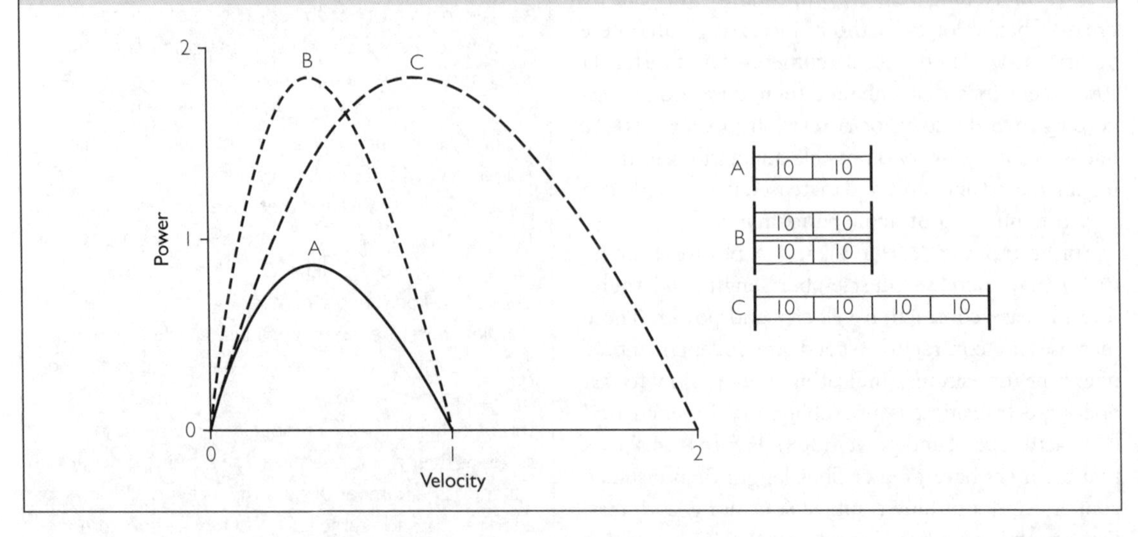

muscle tendon properties, and neural control all interact to determine sprint performance.

Interaction between fiber hypertrophy, fiber pennation angle, and fiber length Conventional resistance training (weight training) generally increases fiber area and thus pennation angle without affecting fiber length, whereas explosive and sprint training tend to increase fiber length without increasing or even decreasing pennation angle (41). However, resistance training with specialized equipment (isokinetic dynamometer or a flywheel apparatus) (42) or with accentuated eccentric contractions (95,324) has produced increases not only in hypertrophy and pennation angle but also in fiber length, and both have been observed after years of resistance training; for example, in powerlifters (who train and compete in the squat, deadlift, and bench-press exercises) (45) and Sumo wrestlers (172).

As already discussed (Figures 8.12 and 8.13), a training-induced increase in fiber length can enhance speed and power performance. An additional benefit of increased fiber length is that it can diminish or even abolish the increase in pennation angle that tends to occur with increased fiber area (9) (compare Figure 8.9B and C). Compared to slower sprinters, faster sprinters have a smaller pennation angle due to their greater fiber length despite equal or greater muscle size (195). The benefit of the reduced pennation angle is that a greater proportion of the increased

Figure 8.14 Muscle-fiber length, measured as fascicle length, in the vastus lateralis of untrained subjects, sprinters, and distance runners. Sprinters had longer and distance runners shorter fiber length than untrained control subjects, respectively. Longer fibers contribute to faster contraction (shortening) velocity. Based on the data of Abe et *al.* (5).

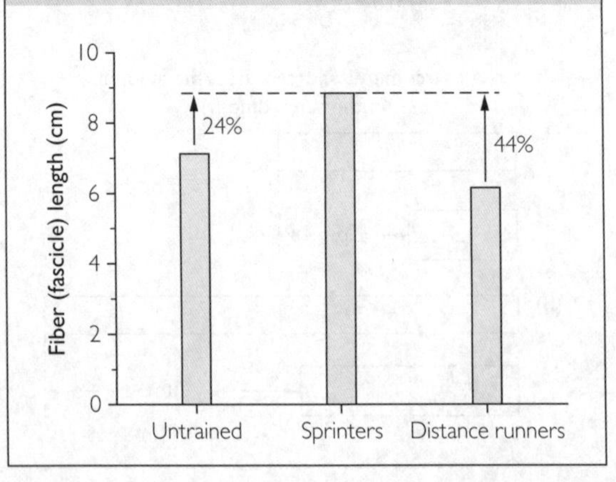

force-generating capacity of the enlarged fibers acts in the tendon's line of pull. Reduced increase in pennation angle also minimizes the decrease in specific force associated with increased pennation angle (172). It has been suggested that increased fiber length is a training adaptation that prevents or attenuates the loss of force that would occur if pennation angle became too large with fiber hypertrophy (45). This might explain the increased fiber length observed in powerlifters (45) and Sumo wrestlers (172). More elite powerlifters have both greater muscle size *and* fiber length (45). The greater fiber length would offset the tendency for greater muscle size to cause a greater pennation angle.

Force–length relationship Fiber area and length affect not only the force– and power–velocity relationships (Figures 8.12 and 8.13), but also the force–length relationship (Figure 8.15). Increased fiber length, resulting from an increased number of sarcomeres in series, increases the length range over which force can be developed (41). Therefore, there are three possible ways by which increased fiber length could enhance strength, power, and speed performance: (1) by directly increasing contractile speed and power (Figures 8.12 and 8.13); (2) by reducing or preventing the increase in pennation angle caused by muscle-fiber hypertrophy (Figure 8.9); and (3) by increasing the muscle-length range over which force can be produced (Figure 8.15).

Increasing fiber length by sarcomere addition may not always be beneficial to performance. As shown in Figure 8.14 and discussed in Box 8.2, distance runners have relatively short fiber length in some of the involved muscles, which would enhance endurance as a result of increased efficiency

Figure 8.15 Schematic illustration of how fiber area and length can affect the force–length relationship. Inset depicts muscle fibers A, B, and C resolved into the number of sarcomeres in parallel, denoting fiber area and thus force-generating capacity, and the number of sarcomeres in series, denoting fiber length and therefore the length range over which force can be produced. Each fiber's force–length relationship is shown in the graph. Each sarcomere is given a maximum isometric (velocity = 0 on graph) force (ISO_{max})-generating capacity of 10 arbitrary units (at optimal length) and a maximum operating length (L) of 0.5 arbitrary units. Fiber A has one sarcomere in parallel and therefore produces an ISO_{max} of 10, and a L of 1 because there are two sarcomeres in series (0.5 + 0.5). Note that ISO_{max} is determined only by the number of sarcomeres in parallel; therefore, fiber A produces an ISO_{max} of only 10, not 20. Fiber B has two sarcomeres in parallel and thus produces an ISO_{max} of 20; on the other hand, this fiber has the same number of sarcomeres in series as fiber A and therefore the same L. Fiber C has the same number of sarcomeres in parallel as A and therefore the same ISO_{max} but twice the number of sarcomeres in series and therefore twice the L. Training that increases only fiber area but not fiber length will increase force over a range of lengths but will not increase L (compare fibers A and B). Training that increases only fiber length but not fiber area will increase L and force over a greater range of lengths but will not increase ISO_{max} (compare fibers A and C).

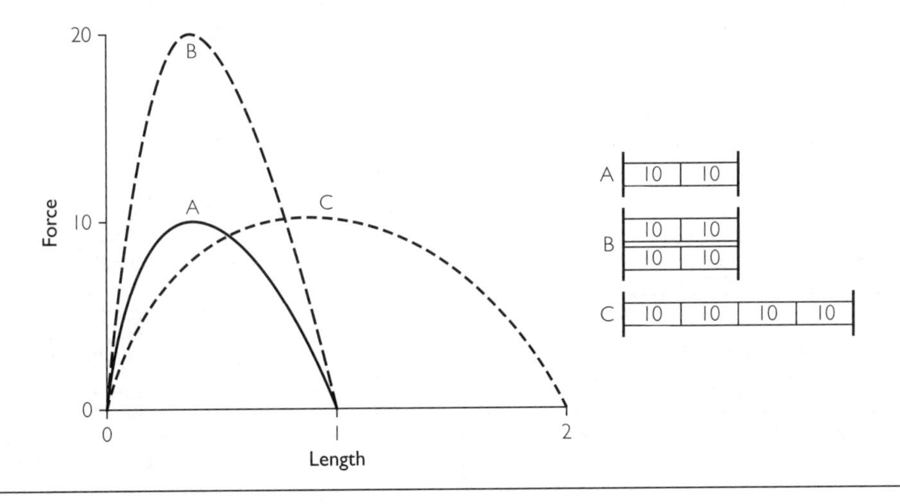

from having fewer sarcomeres in series. Whether or not a change in fiber length might improve performance depends on the specific movement pattern of the sport. Ideally, during the activity the sarcomeres of the active muscle fibers would be operating at or near their optimal length (L_0). As shown in Figures 8.16 and 8.17, in some cases sarcomere addition (i.e., increased fiber length) may be needed to allow individual sarcomeres to operate near L_0, whereas in other cases sarcomere deletion (i.e., decreased fiber length) may be needed. Figure 8.18 provides an example of how sarcomere addition or deletion could enhance performance.

BOX 8.2 EFFECT OF FIBER LENGTH ON ENERGY COST

In Figure 8.14 distance runners were shown to have shorter fiber length than both sprinters and untrained subjects. Given that short relative to long fiber length would decrease speed and power (Figure 8.12), what would be the advantage of short fibers in endurance performance? An increase in the number of sarcomeres in series does not increase force (compare fibers A and C in Figure 8.12) but does increase the energy cost of contraction because each contracting sarcomere consumes energy. In other words, shorter fibers are more efficient in maintaining a given force output, thus enhancing endurance performance (41).

Figure 8.16 Schematic illustration of how training at a specific muscle length could adjust the number of sarcomeres in series so that each sarcomere would be operating at its optimal length (L_0). (A) Before training, the set whole-muscle (fiber) length requires each sarcomere to operate at greater than L_0. If training at the set fiber length were to cause addition of the right number of sarcomeres, each sarcomere would then be operating at L_0. (B) In another muscle (fiber) the set length requires each sarcomere to operate at less than its L_0. In this case the appropriate adaptation would be a decrease in the number of sarcomeres in series (sarcomere deletion) to allow each sarcomere to operate at L_0. Note that in muscle fibers there would actually be thousands of sarcomeres in series rather than the few shown schematically in the figure. If necessary, review the sarcomere force–length relationship in Chapter 4 (see Figures 4.62 and 4.63).

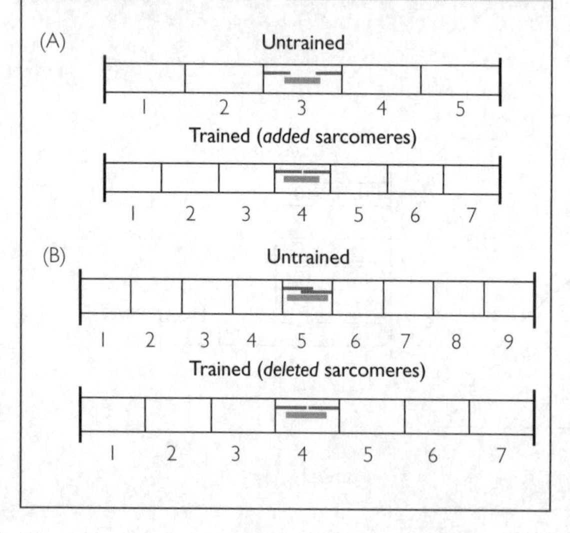

Figure 8.17 Schematic illustration of how a training-induced change in the number of sarcomeres in series could alter the force–length relationship of sarcomeres at a given muscle (fiber length). Pre-training each sarcomere is required to operate at greater than optimal length (L_0), reducing its force production at the muscle length set for training. If training at the set muscle length were to cause the addition of the right number of sarcomeres, each sarcomere would then be operating at L_0 post-training. Training at other muscle lengths might require sarcomere deletion to allow each sarcomere to operate at less than L_0, which would be corrected by sarcomere deletion as shown in Figure 8.16. Note that in muscle fibers there would actually be thousands of sarcomeres in series rather than the few shown schematically in the figure. If necessary, review the sarcomere force–length relationship in Chapter 4 (see Figures 4.62 and 4.63).

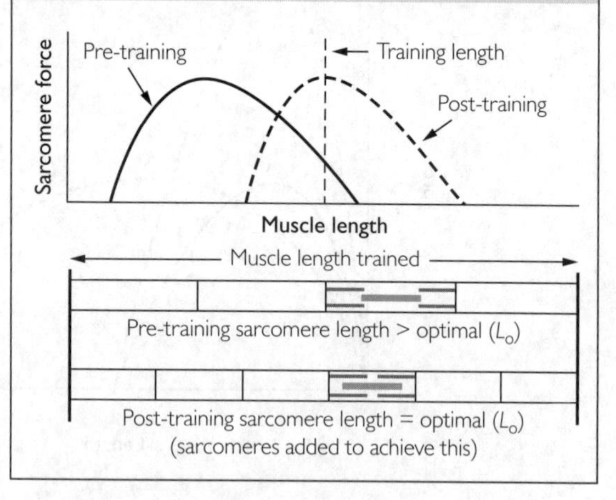

Figure 8.18 Schematic illustration of the knee-extension force–length relationship (FLR) of rectus femoris in cyclists and runners. Rectus femoris, as the only head of quadriceps that crosses both the hip and knee joints, has a force–length relationship that is affected by changes in hip angle, which cause changes in rectus femoris length. In cyclists rectus femoris must generate force at relatively short muscle lengths (small hip-joint angles). Apparently, a muscle adaptation has occurred to allow cyclists to produce their greatest rectus femoris force when the hip is flexed. Rectus femoris force decreases as hip extension proceeds, indicating that in cyclists it is the descending portion of the force–length relationship that is used while cycling. In terms of sarcomeres being able to operate at or near optimal length (L_o) when rectus femoris is relatively short, the training adaptation would be a reduction in the number of sarcomeres in series, as shown in Figure 8.16B and 8.17. In contrast to cyclists, runners need to generate greatest rectus femoris force at a relatively long muscle length (extended hip angle); therefore, the appropriate training adaptation would be sarcomere addition (Figure 8.16A) to ensure that sarcomeres would be operating near L_o when rectus femoris is relatively long (hip more extended). Consequently, in runners rectus femoris would be operating mainly on its ascending force–length relationship. Untrained subjects would have a sarcomere number and L_o intermediate between those of cyclists and runners. Based on the data of Rassier *et al.* (291). If necessary, review the sarcomere force–length relationship in Chapter 4 (see Figures 4.60, 4.62, and 4.63).

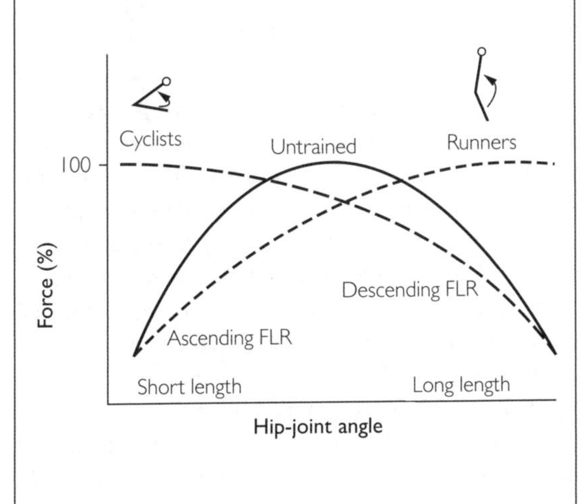

Muscle Contractile Properties

Conventional strength training, explosive training and sprint training all increase strength and power when assessed by measurements of *voluntary* muscle performance; however, it is unclear with these measures whether improved performance is the result of adaptations within the muscles or improved activation of muscles by the nervous system. The measurement of force, power, and velocity of *electrically evoked* contractions allows isolation of changes in muscle contractile performance. The effects of training on both whole-muscle and single-muscle-fiber evoked contractile properties are discussed below.

Whole muscle The force and rate of force development of isometric twitch and tetanic contractions either increases or remains unchanged after training (100,274). [Twitch contractions are considered to be a measure of excitation–contraction coupling rather than maximal contraction force (92); therefore, the focus here will be on maximal tetanic contractions.] If there is no change in evoked tetanic force despite increases in voluntary strength, increased strength would be attributed to neural adaptations. If training does increase force and rate of force development of evoked isometric tetanic contractions, the type of training influences the changes as shown in Figure 8.19, which compares the effects of isometric and ballistic training. Sprint cycle training has produced increases in maximal isometric tetanic force with no reported changes in rate of force development (138).

Although more relevant to athletic performance, few studies have examined the effect of training on whole-muscle maximum unloaded shortening velocity (V_{max}) and the force– and power–velocity relationships. Figure 8.20 compares the effects of isometric and ballistic training on the force– and power–velocity relationships. Note that isometric training (contraction velocity = 0) has little effect on V_{max} but increases maximum isometric tetanic force (ISO_{max}). In contrast, ballistic training increases V_{max} but has little effect on ISO_{max}. Also note that whereas isometric training primarily increases the velocity with which heavy loads can be moved, ballistic training increases the velocity with which light loads can be moved.

Figure 8.20 shows that both isometric and ballistic training increased power but, perhaps counterintuitively, the relative (percentage) increase in peak power was greater after isometric training. It should be

Figure 8.19 Effects of isometric and ballistic training on evoked isometric tetanic maximum force (ISO_{max}), rate of force development (RFD) and rate of force relaxation (RFR). Isometric training consisted of maximal isometric contractions whereas ballistic training consisted of maximal concentric contractions against loads corresponding to 30–40% ISO_{max}. Note that isometric training produced a greater increase in ISO_{max}; in contrast, ballistic training produced greater increases in rates of force development and force relaxation. Based on the data of Duchateau and Hainaut (92).

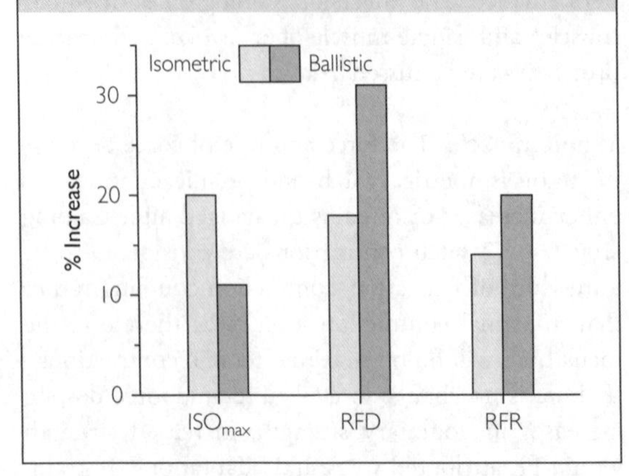

Figure 8.20 Effects of isometric (top) and ballistic (bottom) training on the concentric force– and power–velocity relationships. Isometric training consisted of maximal isometric contractions whereas ballistic training consisted of maximal concentric contractions against loads corresponding to 30–40% of maximum isometric force (ISO_{max}). V_{max} is maximum unloaded shortening velocity. Pre- (solid lines) and post-training (dashed lines) velocity and power values are shown. The power axis is omitted. See text for discussion of results. Based on the data of Duchateau and Hainaut (92).

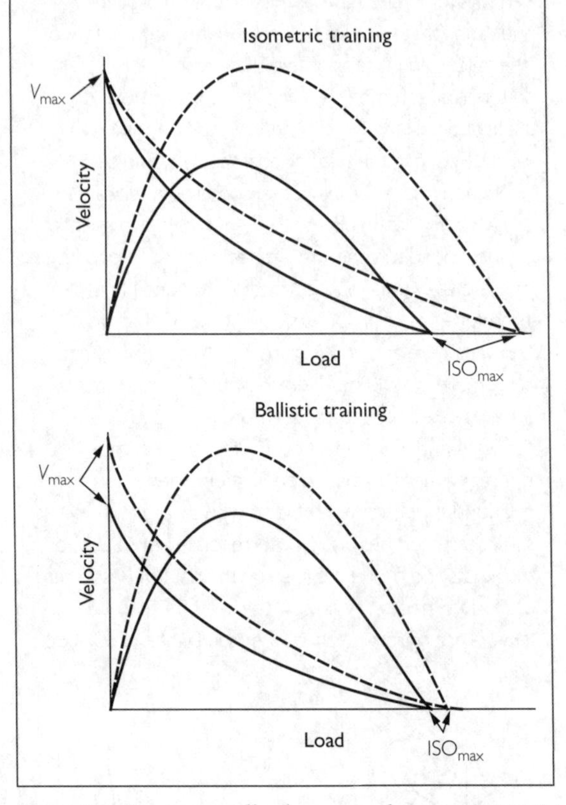

remembered, however, that power is the product of force and velocity (power = force × velocity). Therefore, the increased force-generating capacity (i.e., increased strength) produced by isometric training more than compensated for its failure to increase V_{max}. It is not obvious in Figure 8.20, but isometric training reduced the curvature of the force–velocity relationship, which further increased the velocity that could be attained with a given load. Finally, Figure 8.20 provides an example of velocity specificity in training. If a large load (e.g., an athlete's own body mass, or an opponent) must be moved as quickly as possible, then isometric or heavy resistance training can be effective in increasing the velocity attained with such heavy loads. On the other hand, if light loads (e.g., a baseball or javelin) must be moved quickly, then ballistic training would enhance performance. There are sports where both light and heavy loads must be moved quickly, in which case both types of training would be beneficial.

Single fibers Longitudinal strength training studies generally indicate that absolute isometric force of muscle fibers, mainly type 2 fibers, increases in proportion to the increase in fiber (cross-sectional) area; consequently, there is no change in specific force (227,375). V_{max} generally does not change (227,375), but, because of increased force, absolute peak power increases (375) in a way similar to that shown for whole muscles in Figure 8.20 (top). There is one report of a year of resistance training in young women, in which both specific force and V_{max} increased in type 1 and 2A fibers; surprisingly, the training employed did not increase fiber size (272). Cross-sectional studies have produced mixed results. Compared to untrained control subjects, male bodybuilders exhibited, in addition to increased absolute fiber force, increased and decreased specific force in type 2 and type 1 fibers, respectively; V_{max} was decreased (type 1) and not different (type 2) (80). In contrast, young men with several years of weight-training experience, but who were not bodybuilders or competitive weightlifters, had elevated absolute fiber force and power but not specific force or V_{max} (329). These findings (329) are in agreement with the results of longitudinal training studies. It is difficult to know whether the differences seen in bodybuilders (80) in specific force and V_{max} reflect specialized training or genetic factors.

Plyometric training, consisting of jumping exercises, is similar to conventional strength training in increasing fiber force with no change in specific force, but differs by increasing V_{max} in type 1, 2A, and 2A/2X fibers (225, 226, 227). Thus, in contrast to conventional strength training that increases power primarily by increasing fiber force, plyometric training increases power by increasing *both* force and V_{max}. Plyometric training's effect on the force– and power–velocity relationships is intermediate between the relative extremes for isometric and ballistic training shown in Figure 8.20. A study of sprint cycling found no change in specific force or V_{max} but also no effect on fiber size (134). As discussed earlier, sprint training can induce muscle-fiber hypertrophy. It is not known whether hypertrophy produced by sprint training is accompanied by changes in single-fiber contractile properties.

***In vitro* motility assay** In addition to measuring V_{max} of single muscle fibers, a method called the *in vitro* motility assay has been developed to measure the sliding velocity of actin filaments (V_f) on type 1 and 2 myosin. In a longitudinal study, 12 weeks of resistance training increased V_f in type 2 but not type 1 myosin (61). In a cross-sectional study comparing untrained subjects to male bodybuilders, bodybuilders had lower V_f in type 1 myosin but similar V_f in type 2 (80). The bodybuilders' V_f results corresponded to their single-fiber V_{max} results discussed above (80).

Tendon

Tendon size Although it had long been thought that tendon was metabolically inert and unresponsive to exercise or training (176), there is now evidence that collagen synthesis within muscle and tendon increases in response to single bouts of exercise (176,222) and after training (197); moreover, tendon collagen synthesis and muscle myofibrillar protein synthesis have been shown to be coordinated in their response to resistance exercise (246), although gender has an influence (Box 8.3). There is a correlation between muscle size and tendon size (325). The force acting through a tendon, divided by its CSA, is called **strain** (178). If muscle size and force were to increase with training without a corresponding increase in tendon CSA, tendon strain would increase and could lead to tendon injury. However, if tendon CSA were to increase along with muscle CSA and force, an increase in tendon strain would be minimized or prevented (178). Alternatively, training could alter tendon composition and mechanical properties to increase strain tolerance without an increase in CSA (102,325).

Although strength training usually causes an increase in muscle CSA, whether or not it also causes an increase in tendon CSA is somewhat controversial (176). This is probably the result of differences in the type of training used (24,26,178), the training history of the individuals, and the method of measuring tendon CSA (185,325). The last factor is particularly important because it is now known that increased tendon CSA may be regionally specific along the length of a tendon (77). Studies that failed to find increased tendon CSA may have made measurements at the nonspecific sites (176). On the other hand, when tendon CSA is measured at several sites along the length of a tendon, increases in tendon CSA are observed more often (26,178,325), if not every time (43). Figure 8.21 gives an example of the effect of heavy-resistance weight training on site-specific tendon CSA. In addition to longitudinal training studies, cross-sectional studies of athletes indicate that high loading of tendons leads to increased tendon CSA (77,326). It should be noted that research has focused on the Achilles tendon of triceps surae (gastrocnemius and soleus) and the patellar tendon of quadriceps. The effects of training on tendons of other muscles such as pectoralis major and latissimus dorsi have not been examined.

BOX 8.3 INFLUENCE OF GENDER ON THE RESPONSE OF TENDON TO TRAINING

In response to the same relative loading, the acute increase in collagen synthesis is less in women than in men. Correspondingly, the same training program will increase tendon CSA in men but not in women. The gender difference in tendon adaptability may contribute to women's greater susceptibility to tendon injury during training and performance (224).

Figure 8.21 Effects of heavy resistance training on knee-extensor isometric strength (MVC), quadriceps muscle CSA, proximal and distal patellar tendon CSA, and tendon stiffness. Training consisted of 10 sets of eight repetitions in a knee-extension exercise with 70% of 1 RM, done 3 days per week for 12 weeks. Tendon CSA increased at proximal and distal sites along the tendon's length but not in the mid-region (not shown). Based on the data of Kongsgaard *et al.* (178).

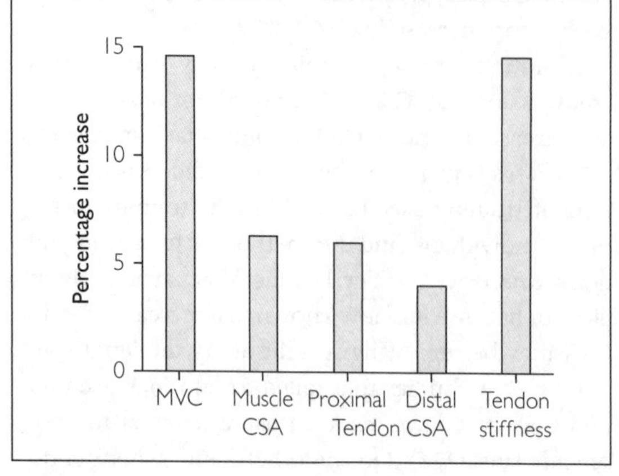

Figure 8.22 Schematic illustration of the effect of strength training on tendon stiffness. Patellar tendon stiffness was determined from the extent of tendon elongation during ramp isometric contractions of knee extensors to maximum force. Tendon force was determined from measured muscle force. The amount of tendon elongation increased as force increased both before (Pre) and after (Post) 12 weeks of isometric strength training. After training a given amount of tendon elongation (e.g., vertical dashed line) required a greater muscle/tendon force to be developed (arrow), indicating an increase in tendon stiffness. Based on the data of Kubo et *al.* (184).

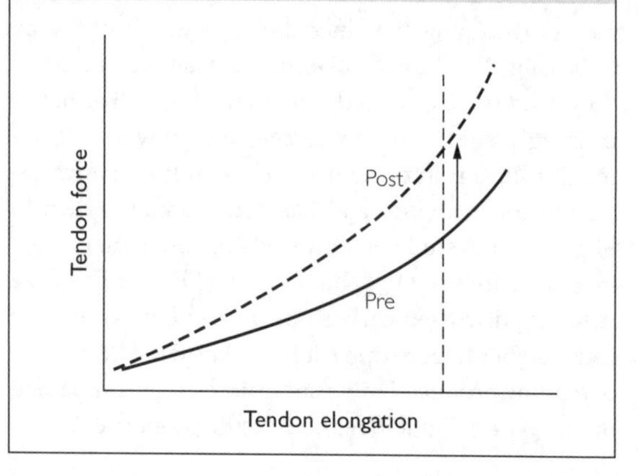

Plyometric training (jumping exercise) has not shown increased tendon CSA, either with or without corresponding increased muscle CSA (102,103,191). In contrast to these longitudinal training studies, cross-sectional studies demonstrate greater tendon CSA in volleyball athletes than in untrained individuals (177,194). This cross-sectional study suggests that long periods (years) of training may be necessary to increase tendon CSA. Plyometric training that does not result in muscle hypertrophy also appears less likely to increase tendon CSA (102,103). Because contraction times for jump exercises are so brief, they are probably less likely to increase tendon CSA than conventional strength training. Supporting evidence comes from a study in which 3 second isometric training contractions increased tendon CSA whereas 1 second contractions at the same intensity and total volume (total contraction time) failed to increase tendon CSA (24,26). To date the effects of sprint training on tendon CSA have not been investigated.

Tendon mechanical properties Training affects several mechanical properties of tendons (75,176,224,293). Such properties could be altered by increased tendon CSA or by altered material composition of tendons. Two examples of tendon mechanical properties are tendon stiffness and the tendon's ability to store and reutilize elastic energy. Stiffness refers to the resistive force of a tendon to a given amount of tendon elongation. Various modes of strength training typically but not always (103,186,190,192) increase tendon stiffness (24,95,178,184,185,187,188,190,191,192,325); an example is shown in Figure 8.22. Factors that influence strength training's effect on tendon stiffness include intensity and mode of training. Isometric training at muscle lengths allowing greater contraction force are more likely to increase tendon stiffness (192), as are contractions at a higher percentage of maximum strength at a given muscle length (24) and longer duration contractions (24,26). Isometric training is more conducive to increasing tendon stiffness than weight training at similar relative intensity (184).

Plyometric training has produced mixed effects on tendon or tendon–aponeurosis complex (Box 8.4) stiffness, with various jumping exercises either increasing (56,335,388), decreasing (129) or having no effect (191) on tendon stiffness. Resolving the inconsistent findings is difficult, partly because different methods of measuring stiffness have been used (Box 8.5). Some research indicates that plyometric training may increase

> **BOX 8.4 TENDON VS TENDON–APONEUROSIS STIFFNESS**
>
> Tendon often merges into one or more aponeuroses. Each aponeurosis is a band of connective tissue that extends from tendon into the muscle. Training studies have measured changes in the stiffness of the tendon–aponeurosis complex (95), indicative of whole-muscle stiffness, in addition to or instead of tendon stiffness. Training can increase both tendon–aponeurosis complex and tendon stiffness in the same muscle (184) or increase tendon–aponeurosis complex stiffness in a muscle without changing its tendon stiffness (184,190,193).

> **BOX 8.5 MUSCLE AND TENDON STIFFNESS MEASURED IN RELAXED VS CONTRACTING MUSCLE**
>
> In the text the focus has been on muscle and tendon stiffness measured during ramp isometric contractions to maximum force, as illustrated in Figure 8.22. Stiffness of muscle can also be measured in a *relaxed* muscle as it is forcibly lengthened. In this case, stiffness is the *passive* force with which a relaxed muscle-tendon unit resists a given amount of lengthening. Passive stiffness will be discussed in detail in Chapter 10, Stretching and Flexibility, with particular emphasis on the interaction of strength and flexibility training on passive stiffness.

the efficiency of storing and reutilizing elastic energy (102), which would improve performance in activities involving the stretch-shortening cycle such as jumping and sprinting.

There are apparently no longitudinal studies of the effect of sprint training on tendon or whole-muscle stiffness, and cross-sectional studies of sprinters indicate increased (25) or decreased (189) stiffness. Compared to untrained control subjects, sprinters had decreased whole-muscle stiffness of vastus lateralis but no difference in medial gastrocnemius (189); thus, adaptations to sprinting are likely muscle-/tendon-specific.

Tendon adaptations and performance Because increased tendon CSA could protect tendons from injury as muscle size and strength increase, such adaptations are advantageous. In addition, adaptations in mechanical properties could enhance the storage and recoil of elastic energy during jumping and sprinting. Increased tendon stiffness would allow the series elastic component (see Chapter 4) to be taken up more quickly in isometric and concentric contractions, promoting a faster rate of force development (188). Increased stiffness is especially important when isometric and concentric contractions are done in isolation; that is, without the benefit of an immediately preceding eccentric contraction (see also Chapter 4). For example, the increased tendon stiffness from strength training increased the height attained in squat jumps but not countermovement jumps (193). Squat jumps are not preceded by eccentric contractions and so would greatly benefit from the increased rate of force development afforded by increased stiffness. On the other hand, in countermovement jumps an eccentric contraction precedes the concentric contraction, so there would be less benefit from increased tendon stiffness.

Capillarization

Capillarization of muscle can be measured as the number of capillaries around each muscle fiber, the ratio of the total number of capillaries to the total number of muscle fibers, and capillary density, which is the number of capillaries per unit muscle CSA (Figure 8.23A). The first two measures determine the third measure, capillary density, which is the index of capillary service to the muscle. Endurance training is well known to increase all measures of capillarization (350). The effect of strength and power training on capillarization depends on the intensity and especially the volume of exercise (Figure 8.23B). High-load/low-volume training, as practiced for example by competitive weightlifters, will likely fail to induce the formation of new capillaries, called capillary **neoformation**. Consequently, as muscle-fiber hypertrophy develops during training, capillary density will decrease below the level seen in untrained individuals (350), as shown in Figures 8.23B and 8.24. Moderate load-higher volume training may result in capillary neoformation that will compensate for fiber hypertrophy and maintain capillary density at the level found in untrained individuals (Figures 8.23B and 8.24). Figure 8.24 shows cross-sectional capillary density data on groups of athletes. Figure 8.25 shows the results of a longitudinal resistance training study in which the volume of training was sufficient to cause capillary neoformation sufficient to prevent a decrease in capillary density that otherwise might have occurred because of hypertrophy.

There is little evidence on the effect of sprint or various forms of power training such as plyometrics on capillarization, although animal studies suggest that capillary neoformation could occur with high-intensity intermittent sprint training (199). As in resistance exercise, the volume (sets, repetitions, frequency) of sprint and power training is the determining factor as

Figure 8.23 (A) Capillarization can be quantified as the number of capillaries around or in contact with each muscle fiber (cap/around), as the ratio of the total number of capillaries to the total number of muscle fibers (cap/fiber), and as capillary density (cap/CSA), which is the number of capillaries per unit muscle CSA. (B) Effects of different types of training on capillarization. Endurance training induces the formation of new capillaries (filled capillaries) with little or no muscle hypertrophy; the result is an increase in capillary density. High-load/low-volume resistance training produces muscle hypertrophy with little or no new capillary formation; the result is a decrease in capillary density. Moderate-load/high-volume resistance exercise may induce new capillary formation that compensates for muscle hypertrophy; the result is no or little change in capillary density.

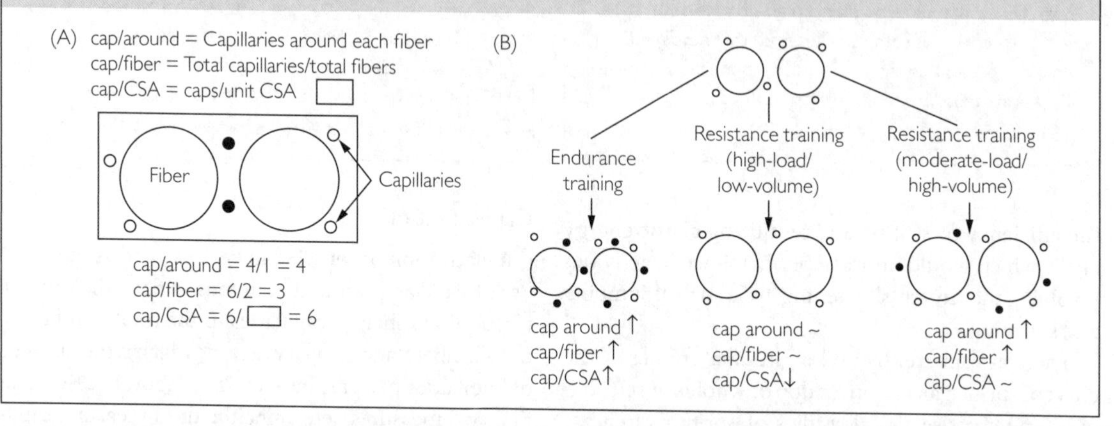

Figure 8.24 Effect of training on capillary density in vastus lateralis. The low-volume training of weightlifters (Olympic weightlifters and powerlifters) causes muscle hypertrophy without new capillary formation, resulting in reduced capillary density compared to untrained individuals. The higher-volume training of bodybuilders induces capillary formation along with muscle hypertrophy, with the result that capillary density is about the same as in untrained individuals. Rowers do aerobic, anaerobic, and strength training. In these athletes new capillary formation more than compensates for any muscle hypertrophy, with the result that capillary density is much greater than seen in untrained individuals. Based on the data of Schantz (316) and Tesch *et al.* (351).

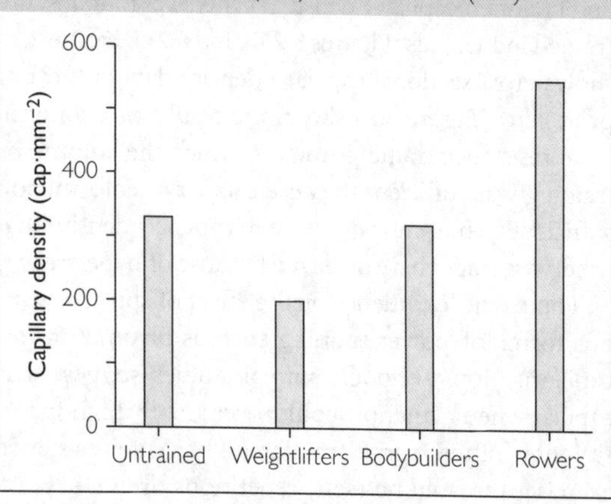

Figure 8.25 Effect of resistance training on capillarization in vastus lateralis. Training consisted of three sets of 6–8 RM in the squat, leg-press, and knee-extension exercises, done 3 days per week for 12 weeks. Training increased (*) muscle-fiber area and the number of capillaries per muscle fiber (Cap/fiber). The change in capillary density (Cap/mm^2) was not statistically significant; nevertheless, the formation of new capillaries prevented capillary density from falling below the pre-training level. The volume of training (nine sets done 3 days per week), similar to that done by bodybuilders (see Figures 8.23 and 8.24), was likely responsible for the formation of new capillaries. Based on the data of Green *et al.* (126).

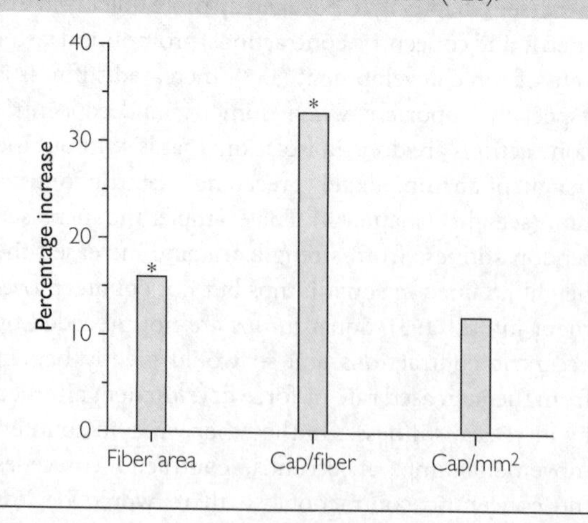

to whether capillary neoformation occurs, and the extent of hypertrophy determines whether there is an increase in capillary density. Increasing capillary density might not seem relevant to brief sprints or single jumps and throws; however, the resulting enhancement of muscle blood flow could delay fatigue in longer sprints or during a large number of repeated sprints or other power actions, and promote faster recovery from fatigue between bouts of exercise (352). The same applies to recovery from sets of resistance exercise.

Biochemical Adaptations

Biochemical adaptations to high-intensity intermittent training, including resistance training and sprint training, are discussed in detail in Chapters 2 and 7. The reader is also encouraged to consult reviews on biochemical adaptations to sprint (307) and heavy-resistance training (350).

Neural Adaptations

Most longitudinal strength-training studies show that strength increases faster and to a greater extent than muscle size (84,107). This is especially true when the measure of strength uses the same mode used in training; for example, using a 1 RM weight-lifting strength test before and after a weight-training program. If a less *specific* isometric strength test is also administered before and after a weight-training program, the increase in isometric strength is substantially lower. In fact, if the weight-training program consists of low-load/high-repetition sets without necessarily taking sets to failure to ensure full activation of muscles, specifically measured strength (1 RM) will nevertheless increase even though non-specific strength and muscle size fail to increase. All these observations are illustrated in Figure 8.26, and suggest indirectly (66,91) that adaptations in addition to those within muscles contribute to increased strength with training. Almost by default, **neural adaptations** are considered to be responsible for increases in strength that cannot be fully accounted for by muscular adaptations (91).

Neural adaptations to training have been the subject of several reviews (e.g., 1,34,66,91,93,94,100,107,111,112, 127,175,312) that consider factors such as altered central command to motoneurons, inter-muscle coordination, antagonist co-activation, altered reflex inputs to motoneurons, and altered intrinsic excitability of motoneurons (Figure 8.27; see also Box 8.6). The final common pathway of any neural adaptations is the MUs activated in a particular action. It is the muscle fibers of MUs that translate neural activation into mechanical output from the muscles, as illustrated in Figure 8.28. In maximal strength performance, the goal is to generate the greatest possible force in the intended direction of movement. This is achieved by increasing agonist activation as much as possible, reducing antagonist co-activation as much as possible without compromising joint stability, and activating synergists appropriately to provide an efficient, coordinated movement (66).

Figure 8.26 Effects of resistance training on specific (knee-extension 1 RM) and non-specific (isometric knee-extension strength, MVC) measures of strength, and muscle CSA of knee extensor muscles. Subjects trained one leg with a heavy resistance-training program consisting of 10 sets of eight repetitions in a knee-extension weight-training exercise with 70% of 1 RM. The other leg was trained with a light program consisting of 10 sets of 36 repetitions with a weight that equated the work (volume) done with the heavy leg. Training was done 3 days per week for 12 weeks. Note that specific strength increased more than non-specific strength and that strength increases exceeded increases in muscle CSA. For light training, only the 1 RM increase was statistically significant. Based on the data of Kongsgaard *et al.* (178).

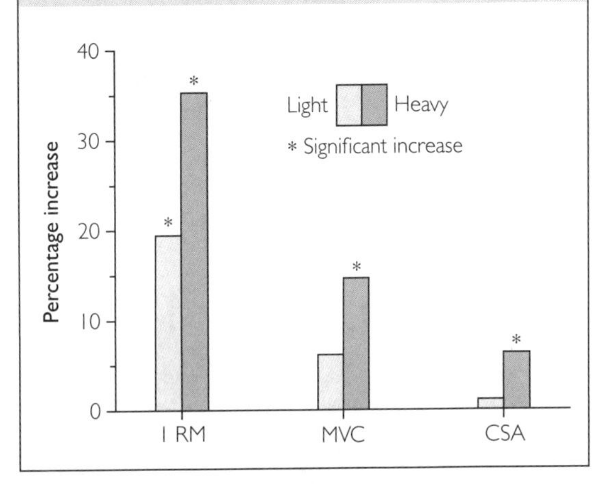

Of the three principal adaptations shown in Figure 8.28, the one that has received the most attention is increased agonist activation. As discussed in Chapter 5, most untrained individuals have difficulty fully activating muscles consistently (107). Activation tests have usually been done on unilateral, single-joint, isometric actions. It is more difficult to fully activate muscles in maximal eccentric contractions (91). Furthermore,

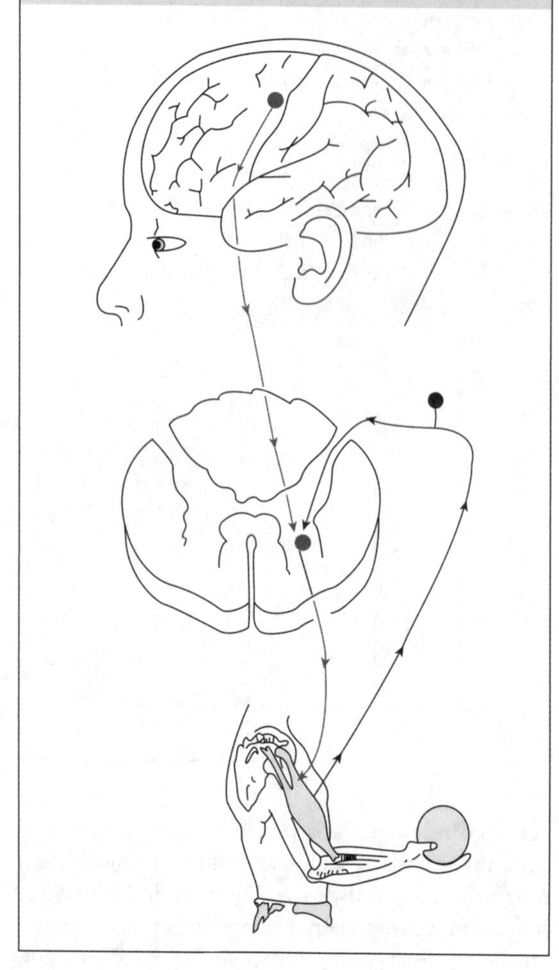

Figure 8.27 Potential neural adaptations to strength and power training could occur in the brain related to motor commands to motoneurons of agonists, antagonists, and synergists, and at the spinal level related to sensory and reflex inputs to motoneurons and altered properties of motoneurons that affect their excitability.

the challenge of fully activating muscles is greater with bilateral, multi-joint, dynamic movements (158). For example, it is likely more difficult to fully activate agonists in an overhead barbell shoulder press than in a seated unilateral isometric knee extension. In addition, more complicated movements may also provide a greater challenge in terms of antagonist and synergist coactivation. Unfortunately, it is difficult to apply the best methods for assessing the extent of activation to the more complicated movements used by athletes in training and performance.

Assuming that an athlete who is exposed to a training exercise for the first time has a deficit in agonist

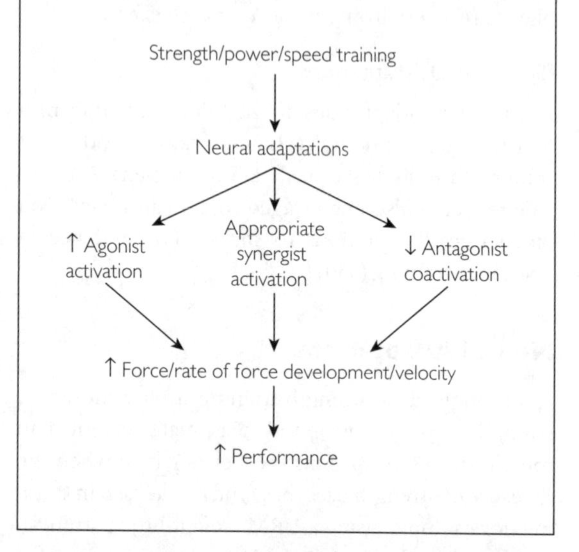

Figure 8.28 Outline of neural adaptations to training and their resultant effect on performance. The principal adaptations are (1) increased agonist activation, (2) decreased antagonist coactivation, and (3) appropriate activation of synergists. Collectively, these adaptations maximize the force in the intended direction of movement. From Sale (312). © John Wiley & Sons.

BOX 8.6 CROSS-TRAINING

As discussed in the text, neural adaptations to strength training have been implicated on the basis of increases in strength that cannot be attributed solely to muscular adaptations and by the marked specificity in the training response. An additional phenomenon that implicates neural adaptations is **cross-training**. Cross-training, also called cross-education, cross-transfer, and, in strength training, the contralateral strength training effect, refers to increased strength in an untrained limb when the contralateral limb is trained (65,203,391). Training one limb can increase strength in the contralateral untrained limb by approximately 7–8%, or up to approximately 25% of the increase in the trained limb (65,259). The basis of the contralateral strength training effect is neural adaptations induced from training one limb that spill over to the untrained limb (65), increasing the ability to activate the muscles of the untrained limb (204). For the athlete, the significance of cross-training is that training one limb can attenuate the loss of strength and perhaps muscle size in the contralateral limb rendered inactive or even immobilized by injury (141,223).

activation, three possible aspects of training-induced increased agonist activation and their effect on performance are illustrated in Figure 8.29. First, training could increase the consistency with which the highest-threshold MUs can be recruited (Figure 8.29, top). In large muscles almost all MUs are recruited when force reaches about 85% of maximum in isometric contractions (94); therefore, it would seem that even untrained subjects would be able to recruit all MUs. However, recruiting the highest-threshold MUs may be more difficult in more complex actions, so it remains a possible adaptation. Second, before training MU firing rates may not reach the plateau of the force–frequency relationship (94); increased agonist activation with training would allow the plateau to be more consistently attained (Figure 8.29, middle). Rapid (ballistic or explosive) contractions are associated with initial, brief bursts of high-frequency MU firing that serve to increase rate of force development. A training-induced increase in initial firing rate would increase rate of force development (Figure 8.29, bottom). There is evidence for and against increased agonist activation (66,107), probably because of the limitations in the measurement techniques used.

Because single-MU recordings can determine changes in recruitment thresholds, order, and firing rates, this is probably the most effective method for examining the effects of training on muscle activation. Nonetheless, only a few training studies have used this technique (107). With regard to MU firing rates, the results have not been consistent (94); however, a study of ballistic isometric contractions showed an increase in initial firing rate that corresponded to increased rate of force development (Figure 8.30), in agreement with the bottom part of Figure 8.29.

Depending upon the type of contraction and its speed, during contraction of an agonist, there may also be coactivation of an antagonist (174). In a single-joint action like knee extension, reducing activation of the knee flexors would contribute to the net extension force. However, this benefit must be balanced against the joint stability afforded by antagonist coactivation (107). In a multi-joint, triple-extension movement such as a squat or jump, the role of antagonist coactivation is more complex (75) because a muscle that crosses two joints can act simultaneously as an agonist at one joint and an antagonist at another. For example, in a squat, gastrocnemius is an agonist for ankle extension but an antagonist to quadriceps as it acts to extend the knee joint. The hamstrings act as agonists to extend the hip but simultaneously oppose quadriceps' action of knee extension. In the overhead barbell press, biceps brachii is an agonist for forward flexion at the shoulder joint but is simultaneously an antagonist to

Figure 8.29 Three aspects of increased agonist activation. (Top) Ramp isometric contractions to maximum possible force done before and after training. After training, high-threshold MUs can be more consistently recruited, contributing to greater force. (Middle) Force–frequency relationship is shown, indicating how an increase in maximum MU firing rate could increase force. (Bottom) Isometric contractions done with the intent to increase force as rapidly as possible. Training produces a higher initial firing rate, which increases rate of force development. From Sale (312). © John Wiley & Sons.

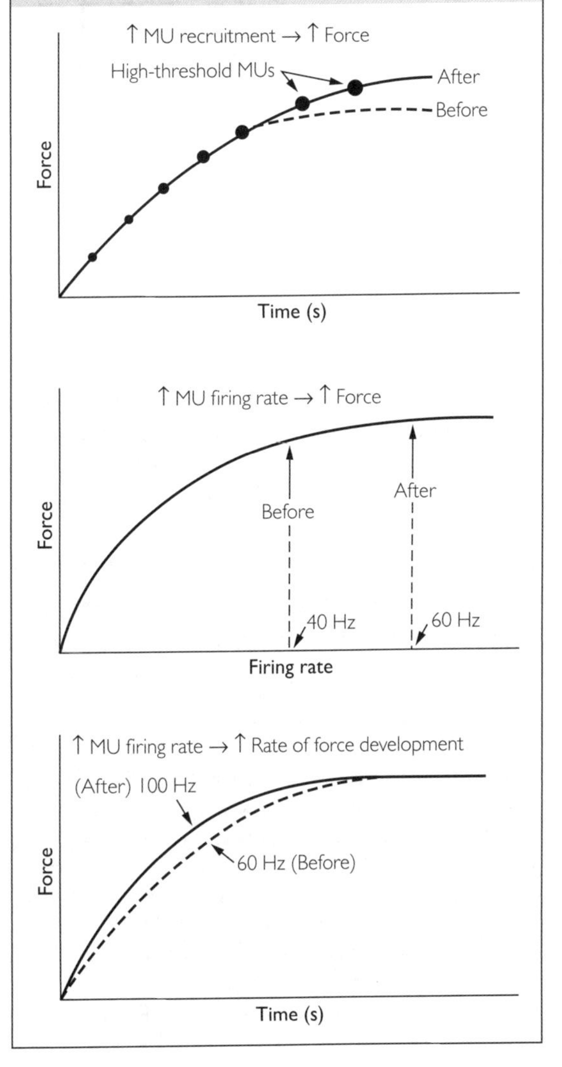

Figure 8.30 Effect of ballistic training on MU firing rate and rate of force development (RFD). Training of ankle dorsiflexors consisted of 10 sets of 10 ballistic *concentric* contractions with a load equal to 30–40% 1 RM, 5 days per week for 12 weeks. A schematic illustration of the results is shown. Top right shows ballistic *isometric* contractions done before and after training at the same percentage of maximum force, indicating a large increase in rate of force development after training. The rest of the figure represents the change in average MU firing rate calculated from the first three inter-spike intervals at the beginning of ballistic isometric contractions done before and after training. A spike is the action potential of a discharging MU. Firing rate is calculated from each inter-spike interval. Note the increase in firing rate after training, which should have contributed to the observed increase in rate of force development (see also bottom of Figure 8.29). From Sale (312), based on the data of Van Cutsem *et al.* (364). © John Wiley & Sons.

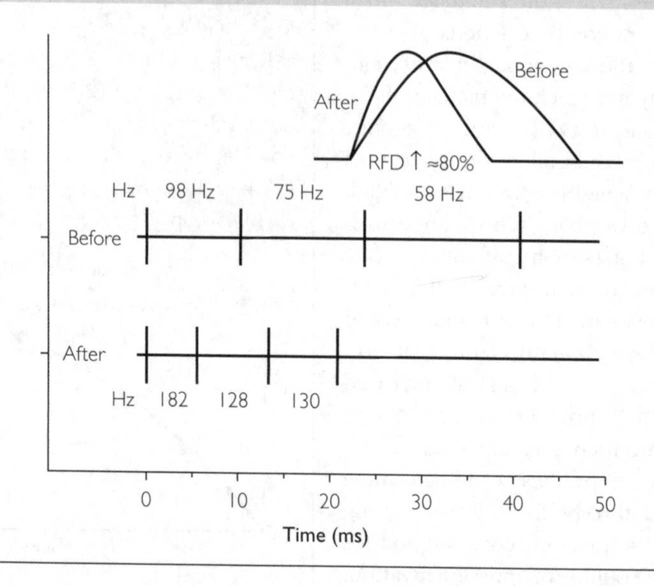

triceps in elbow extension. The nervous system must resolve these dilemmas to produce the most effective, efficient movement (Box 8.7). The findings on changes in antagonist coactivation after training are, not surprisingly, mixed (84,107). Similarly, synergists involved in actions could act simultaneously as agonists and antagonists at various joints to provide a solid base or posture for a particular movement. All of this inter-muscle coordination probably adapts with training but has received relatively little study (94), partly because of the difficulty in examining it. Nevertheless, adaptations in inter-muscle coordination likely play an important role when more complex exercises are used in training.

BOX 8.7 LOMBARD'S PARADOX

Lombard's paradox (290) refers to the apparent conflict between rectus femoris (one head of quadriceps), which acts to flex the hip and extend the knee, and hamstrings (knee flexors), which act to extend the hip and flex the knee. Thus, these muscles oppose each other and each muscle acts as both an agonist and antagonist in movements like a squat, jump, or sprinting. The paradox is resolved by rectus femoris having the greater moment arm at the knee joint whereas hamstrings have the greater moment arm at the hip; therefore, in simultaneous hip and knee extension, the hamstrings' extensor action dominates at the hip while the rectus femoris' extension action dominates at the knee.

Interaction of Muscular and Neural Adaptations

Because strength increases more rapidly over the course of a training program than can be accounted for by muscular adaptations, it is generally accepted that neural adaptations occur earlier and are followed by progressive muscular adaptations as training continues over a period of months or years. Figure 8.31 illustrates the relative roles of neural and muscular adaptations, and Figure 8.32 provides an example of "most training studies" as depicted in Figure 8.31. Since a single bout of resistance training can stimulate protein synthesis (280), it is not surprising that

Figure 8.31 Schematic illustration of the interaction between neural and muscular adaptations in increasing strength. The majority of training studies cover a period in which neural adaptations play the major but not exclusive role in strength increases. Athletes who train for many months or years with the same exercises will likely reach a plateau in neural adaptations before muscular adaptations reach their plateau. Based on Sale (312). © John Wiley & Sons.

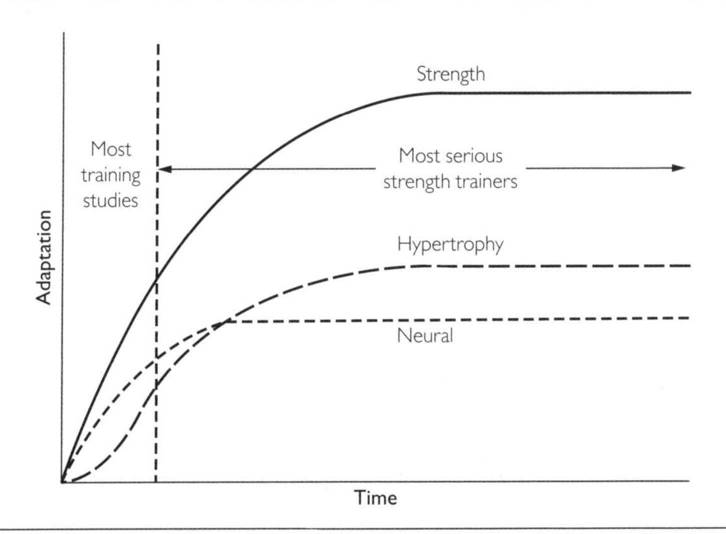

muscle hypertrophy begins early in a training program (83,370), as illustrated in Figures 8.31 and 8.32. Nevertheless, strength and some indicators of neural adaptation increase more rapidly than muscle size in the initial phase of training (Figures 8.31 and 8.32). In fact, in Figure 8.32 a non-specific strength test (isometric) was used to monitor strength gains from a dynamic (coupled concentric–eccentric contractions) training program. Had a specific strength test been used, strength increases and the corresponding neural adaptations would have been greater than the increase in isometric strength shown in the figure, and the results for specific and non-specific strength tests would have been similar to those shown in Figure 8.26.

The scheme shown in Figure 8.31 assumes that the athlete uses the same exercises throughout the training period. If the athlete adopts new, unfamiliar exercises, there will be a burst of neural adaptation as the new exercises are mastered. The resulting surge of strength in the new exercises could be viewed as renewed training progress, but the key concern is whether the new exercises cause a corresponding increase in muscle mass and an improvement in performance for which the training is being done. In the section on training principles it will be shown that different exercises can activate a particular muscle group differently. Different exercises activating and therefore inducing hypertrophy in different parts of a muscle are useful if the goal is maximum hypertrophy (e.g., bodybuilding). For many athletes, however, the goal is to increase strength and size in parts of muscles that enhance performance. In fact, hypertrophy of muscles or parts of muscles not needed for performance may be counterproductive, especially in sports where the athlete must support or move his/her own body mass.

Principles of Training for Strength, Power, and Speed

Overload

The **overload** principle is considered the cardinal principle of strength and power training. Simply expressed, it requires the trainee to make a maximal or near-maximal effort to complete a training exercise. One of the earliest demonstrations of the importance of the overload principle is shown in Figure 8.33, which indicates that an exercise task that is easy to complete is much less effective than the same amount of work done in a task that is difficult to complete. The physiological basis of the overload principle is that, to induce adaptation in a muscle's MUs, the MUs should be maximally or close to maximally activated; that is, all MUs should be recruited and firing at rates

Figure 8.32 Effect of strength training on neural and muscular adaptations. In a bilateral knee-extension exercise, training consisted of three sets of seven maximal, coupled concentric and eccentric contractions on a gravity-independent ergometer. Training was done 3 days per week over a period of 35 days. Maximal isometric strength (MVC) was measured and concurrent surface electromyography (EMG) was used as an indicator of neural adaptation. Quadriceps CSA was the measure of muscular adaptation. The results show that neural adaptation made the dominant contribution to increased strength during the training period, a pattern that conforms to the "most training studies" phase depicted in Figure 8.31. However, the increase (≈5%) in CSA was statistically significant after 20 days of training, demonstrating that notable hypertrophy can be induced by about approximately 3 weeks into a training program. Based on the data of Seynnes *et al.* (324).

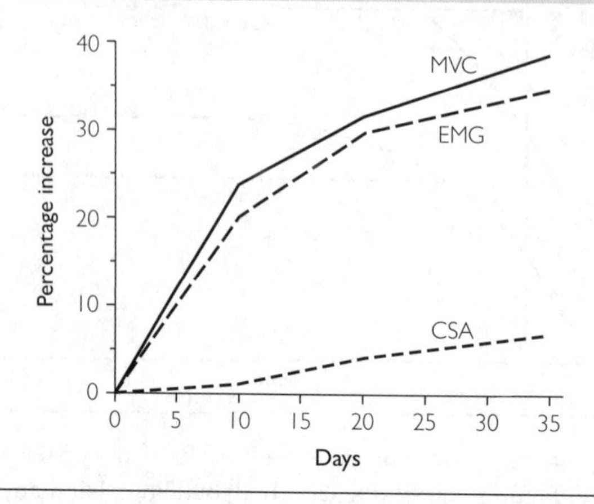

Figure 8.33 Experimental demonstration of the overload principle. Overload training was 10 sets of 25 RM in a wrist-flexion exercise; that is, maximal effort was required to complete each set of 25 repetitions. One form of underload was to do 20 sets with 50% of the 25 RM. Twenty sets were done to equal the relative work (resistance × repetitions) of the overload exercise. The second form of underload consisted of 40 sets with 25% of the 25 RM, again equaling the work done in the overload training. Compared to the overload training, the underload training took much less effort to complete. The results show that the greater the effort in completing the exercise, the greater the improvement in performance. Based on the data of Hellebrandt and Houtz (140).

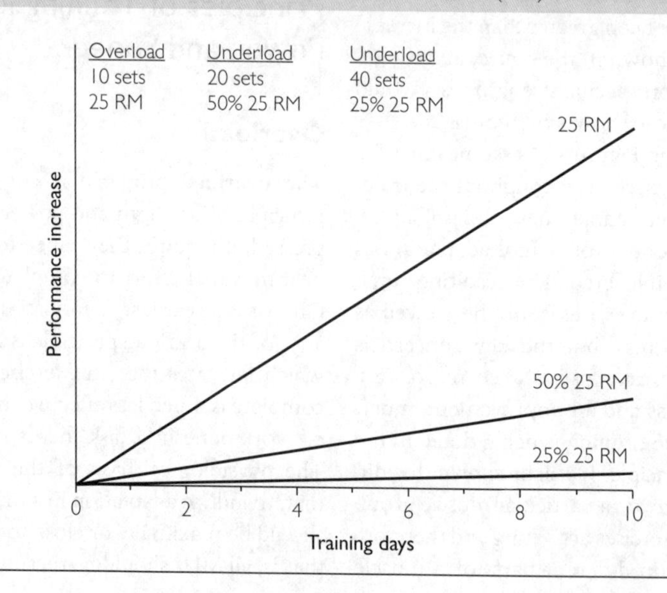

Figure 8.34 Achieving maximum possible MU activation in weight training either by doing single maximal lifts (1 RM) or by completing as many repetitions as possible with a weight that is a certain percentage of the 1 RM. In this example the chosen submaximal weight allowed 15 repetitions to be completed (15 RM). In weight training, the inability to complete an additional repetition, the sixteenth one in this case, is referred to as continuing a set to fatigue failure or simply failure. Repetitions in weight training consist of coupled concentric and eccentric contractions. Since eccentric contractions are stronger than concentric ones (Chapter 4) and fatigue at a slower rate (Chapter 5), it is concentric failure (fatigue) that terminates a set of repetitions; hence, the term "concentric failure" is commonly used in weight training.

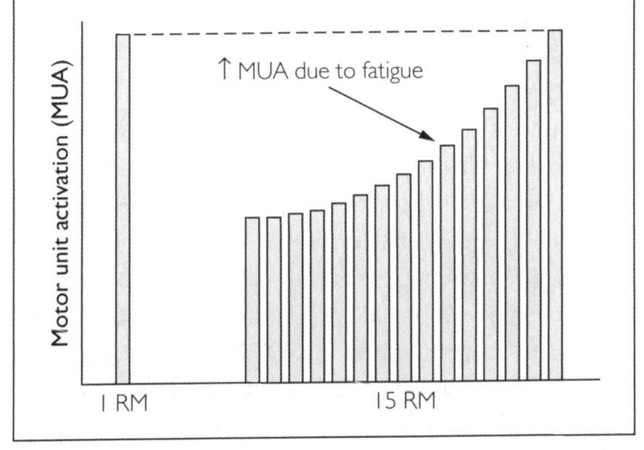

Figure 8.35 Achieving maximum possible MU activation, and therefore maximum overload, in isometric (ISO) training either by doing a series of brief maximal contractions (MVCs) or by sustaining a submaximal contraction to fatigue failure, in this case a contraction equal to 50% of maximum isometric strength (ISO_{max}). Note that fatigue also occurs in repeated maximal contractions.

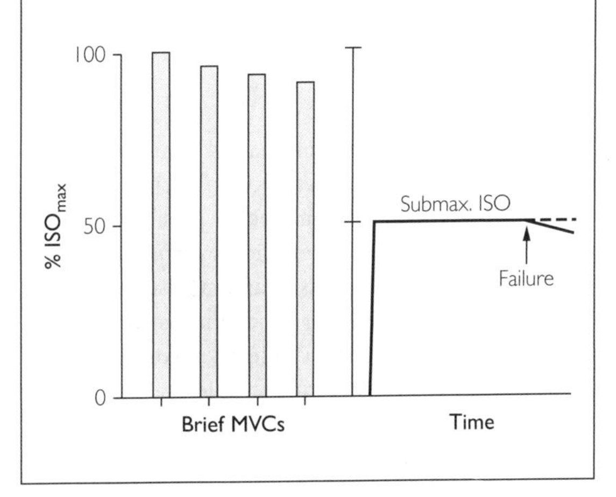

sufficient to attain maximum force and/or maximum rate of force development in the muscle fibers of the MUs. As discussed in Chapter 5, maximum MU activation can be achieved in one of two ways: (1) a series of brief, single maximal contractions or (2) submaximal contractions continued to fatigue failure (311,359). These two ways of achieving maximal MU activation are illustrated for conventional weight training in Figure 8.34 (see also Box 8.8). Similarly, isometric training could be a series of brief maximal contractions or submaximal contractions, at various percentages of maximum isometric contraction, sustained to fatigue failure (e.g., Figure 8.35). In isokinetic training, the trainee does sets of maximal-effort contractions at the selected velocity and contraction type (Figure 8.36).

Figure 8.36 Application of the overload principle to isokinetic training, with reference to the force–velocity relationship. The trainee or coach selects the contraction type (concentric or eccentric) and velocity and then the trainee does sets of maximum effort contractions to ensure full MU activation. Across contraction types and velocities there is a wide range of *absolute* force; however, *relative* force (percentage of maximum available force) is the same, provided that maximal effort is applied. ISO means isometric and corresponds to velocity 0.

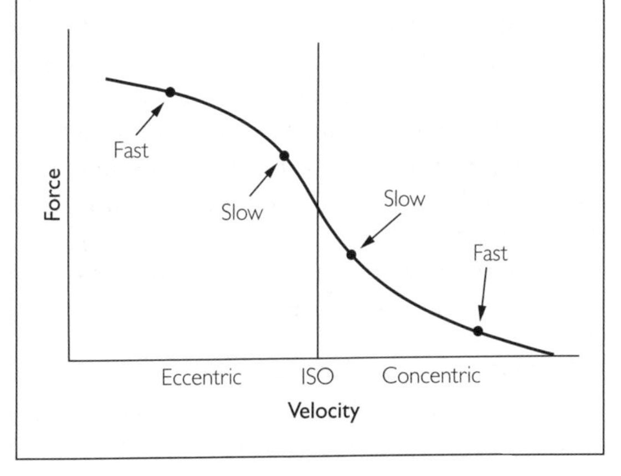

Quantifying Training Intensity

In exercise physiology, exercise or training intensity refers to the magnitude of force, power, and speed used in the exercise, often expressed as a percentage of a

BOX 8.8 IS HEAVY LIFTING NECESSARY TO RECRUIT HIGH-THRESHOLD MOTOR UNITS?

As discussed in Chapter 5, increasing contraction force, as in lifting heavier weights (see Figure 5.27), effectively recruits higher-threshold MUs in accordance with the *size principle*. This has led to the belief that heavy lifting is needed to recruit the highest-threshold MUs. However, as illustrated in Figures 5.31, 5.32, and 8.34, lifting lighter weights can also recruit high-threshold MUs, provided they are lifted repeatedly for as many repetitions as possible. In this case it is increasing fatigue that, by increasing the voluntary effort required to continue doing repetitions, increases the recruitment of MUs. Although it has been recognized for some time that lifting lighter weights to fatigue failure can effectively recruit high-threshold MUs (311), only recently has the idea that heavy lifting is needed to recruit high-threshold MUs been challenged (50,62). The distinction between lifting a heavy weight once and lifting a lighter weight repeatedly to failure is that significant fatigue occurs only in the latter. What is common to both types of lifting is the need for maximal or near-maximal voluntary effort.

maximum value obtained in testing. Thus, an athlete may cycle on an ergometer at a power level corresponding to 80% of the power attained in a test of maximal oxygen uptake (maximal aerobic power). In strength training, intensity is expressed as magnitude of force (e.g., isometric training) or load; that is, the weight lifted (weight training). If an athlete's maximum single-lift or 1 RM were 100 kg, for example, then a training prescription might include doing sets of repetitions with a load (intensity) equivalent to 80% of the 1 RM.

Intensity can also be defined as the degree of effort. The extent of effort applied in doing a bout of exercise at a given intensity (speed, power, force, load) can significantly influence its effectiveness as a training stimulus. In Figure 8.34, for example, the 15 RM set required maximal effort to complete the fifteenth repetition; that is, maximum intensity of effort leading to fatigue failure. In contrast, intensity expressed as load was not maximum but some fraction of the 1 RM. If the athlete had done only seven of a possible 15 repetitions as shown in Figure 8.34, intensity of load would have been the same in both cases but intensity of effort would have been greatly reduced. Maximum intensity of effort is certainly important because it ensures maximal or near-maximal MU activation, which is the essence of the overload principle. In fact, some argue that the term "intensity" should be applied only to effort and not to the magnitude of force, load, and power (99). The middle ground is to consider both intensity of load (or force, power, and speed) and effort in the design of training programs and to be aware of the distinction between the two expressions of intensity (Figure 8.37).

Figure 8.37 Quantifying training intensity. (A) Intensity can be expressed as percentage of maximum voluntary effort, percentage of maximum isometric force (ISO_{max}) or single maximum lift (1 RM), or percentage of maximum power (attained with a load or force that is less than half of maximum; e.g., Figure 8.20). (B) Three examples of activities are rated according to the three ways of expressing intensity. Doing a set of three repetitions (reps) with the 6 RM load would not require maximal effort, nor would power be near-maximal; however, the load is relatively high. In the three maximal squat jumps with a load of 30% 1 RM, effort and power would be high but the load low. Finally, a 30 rep squat with the 30 RM weight would require maximum effort on the last rep but load and power would be low. The three examples demonstrate the importance of understanding the different ways intensity can be expressed.

(A) *Quantifying Training Intensity*

Effort: % of maximum voluntary effort
Load/Force % of maximum force or load (1 RM)
Power % of maximum power
(≈30% of maximum force or load)

(B) *Intensity: Effort vs Load vs Power*

3 reps with the 6 RM
- effort submaximal
- load high (≈85% 1 RM)
- power low

3 squat jumps with 30% 1 RM
- effort maximal with each jump
- load low
- power high

30 reps in squat with 30 RM
- effort maximal on last rep
- load low (small % of 1 RM)
- power low (slow movement)

Is Training to Failure Necessary?

The ultimate application of the overload principle is to do a set of repetitions to failure; that is, until another repetition is not possible. However, Figure 8.33 shows that overload exhibits a dose–response pattern: there is some training stimulus even with *underload* training, which could also be termed *moderate overload* training. Figure 8.38 illustrates how stopping a set of repetitions before failure can nevertheless provide a training stimulus, if perhaps not a maximal stimulus. The importance of training to failure has been tested by comparing the response to the same number of repetitions with the same load but varying the development of fatigue to either prevent or cause failure. Failure is induced by performing sets of repetitions in the conventional way until no more repetitions could be completed. In "no failure" training, rest periods are allowed between repetitions or groups of repetitions. For example, training with a 6 RM (≈85% 1 RM) set to failure was compared to a set with the same load and repetitions but with 30 second rest periods between repetitions to prevent or reduce fatigue (no failure) (306). As a second example, doing four sets of six repetitions (each to failure) were compared to eight sets of three repetitions with the same load (no failure) (90). When volume and intensity (expressed as load) of training are equated and the intensity is a fairly high percentage of maximum force, there is evidence for the necessity of training to failure for maximum training results (90,123,306,319); however, no-failure training produces significant strength increases (90,101,155,319) that are sometimes equal to (101) or superior to (155) failure training. When relatively light loads are used, training to failure is more important to ensure that all MUs have been recruited (e.g., Figure 8.34).

In summary, training to failure may be needed to ensure full activation of the highest-threshold MUs and therefore to provide the greatest possible training stimulus; however, training close to, but not to, actual failure (Figure 8.38) can be effective. Thus, the athlete is not required to train to failure on every set or in every training session. Possible risks of training to failure are discussed in Box 8.9.

Figure 8.38 Implications of not continuing a set of repetitions (reps) to failure. A trainee does a set of reps with the pre-determined 10 RM load but stops after eight reps; thus the set was not continued to failure. MU activation (MUA) was clearly not maximal after eight reps; as discussed in Chapter 5 most MUs would have been recruited by eight reps but the highest-threshold MUs would not be firing at maximal rates. If mechanical load (force) is a stimulus to hypertrophy, then it was applied for 80% of the possible reps. Not achieving maximum possible force in the high-threshold MUs is significant because these MUs have the largest innervation ratio (more fibers per MU) and specific force. It would appear that the overload principle can be applied in a graded manner; it is not an all-or-nothing phenomenon.

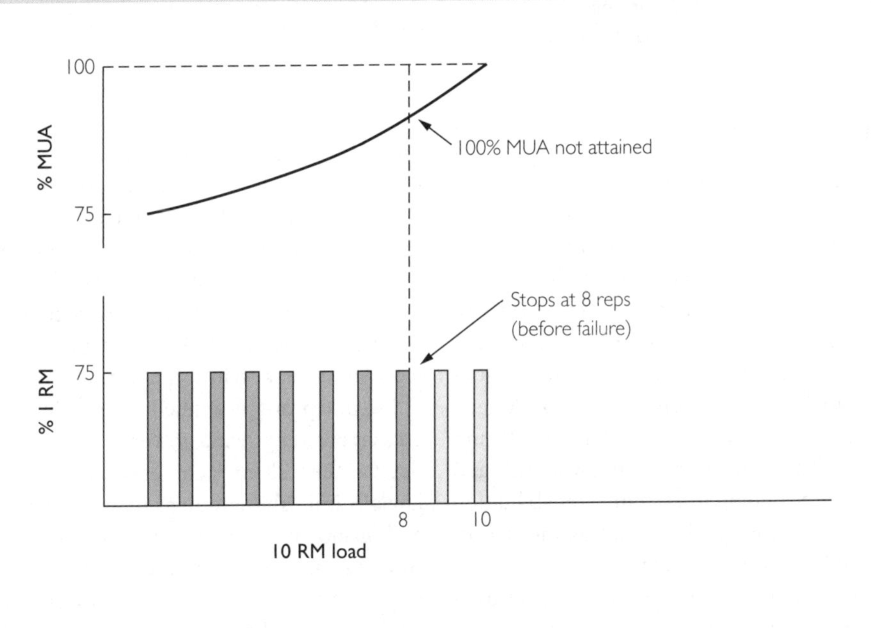

BOX 8.9 ARE THERE RISKS TO TRAINING TO FAILURE?

Blood pressure increases to high levels when a set of repetitions is taken to failure with loads equal to approximately 75%+ of the 1 RM (Figure 8.39). In the bilateral leg-press exercise, for example, average systolic and diastolic pressures at failure can increase to 320 and 250 mmHg, respectively, at failure (221), compared to normal resting blood pressure of approximately 120/80 mmHg. The Valsalva maneuver, an attempted forced expiration against a closed glottis, becomes more frequent and intense as failure is approached, and contributes to the extreme elevation in blood pressure during lifting (217,221). If the Valsalva maneuver can be replaced with a slow expiration, peak blood pressure is reduced (262); consequently, it has been suggested that the Valsalva maneuver be avoided (262). However, it may be difficult if not impossible for advanced strength trainers to avoid enlisting the Valsalva maneuver when doing sets to failure (217). In addition, the Valsalva maneuver may help stabilize the trunk during lifting and may actually protect cerebral blood vessels (221). It should also be noted that blood pressure reaches high levels even if a set is stopped before failure (217) (Figure 8.39). Although cerebral blood vessel injury associated with weight lifting has been reported in young, apparently healthy individuals (262), the incidence of such injury is fortunately rare.

Does training to failure increase the risk of muscle, tendon, or joint injury? If injury risk is related solely to the magnitude of muscle contraction force, continuing a set to failure may not pose a risk because muscle force actually decreases as failure approaches. For example, in a set of repetitions in a weight-lifting exercise, the speed of the concentric phase of each repetition typically decreases as fatigue develops and failure is approached; in the final repetition, lift speed may decrease to approximately 50% of the first repetition speed (156). The decrease in lifting speed indicates that the magnitude of accelerating force is decreasing as the set progresses (Figure 8.40). Failure occurs when fatigue causes the lifting force to decrease below that needed to raise the weight. The actual decline in contraction force contrasts to the perception of the trainee, who feels that the weight is becoming heavier because increasing voluntary effort is need to lift the weight as fatigue develops. Paradoxically, then, as actual contraction force decreases during the set, decreasing the risk of injury, the trainee perceives a need for increased force, and may therefore conclude that injury risk is increasing. This conclusion is based on two assumptions. One is that injury risk is determined only by magnitude of force and not by the product of force and its duration. If duration were equally or more important, however, doing several sets not to failure with a given load would pose a greater injury risk, due to the larger number of completed repetitions, than one set to failure. The second assumption is that the mechanics of the exercise remain the same throughout the set to failure. A change in mechanics, such as enlisting additional muscles to assist in the lift, might expose these newly or more extensively activated muscles to injury. Finally, the focus has been on the concentric (lifting) rather than eccentric (lowering) phase of each repetition. Eccentric contractions are stronger than concentric (Figure 8.36) and fatigue more slowly (Chapter 5). Thus, in contrast to the concentric phase of repetitions, the speed of the eccentric phase may remain unchanged or even speed up a bit. If the trainee were to let the weight drop and then stop it abruptly, a large eccentric force would be needed to decelerate the weight. This brief but large increase in eccentric force could pose an injury risk.

Progression

A training program consisting of a chosen intensity, volume, and frequency will in time fail to provide an adequate training stimulus because training adaptations will have increased strength to the point that the program no longer represents a maximal overload condition. Therefore there must be **progression** in training. The concept of progression is illustrated in Figure 8.41, and its application to isometric and isokinetic training in Figure 8.42 and to weight training in Figure 8.43. Athletes will most likely experience isometric and isokinetic strength training in an injury rehabilitation setting, whereas some form of weight training is most commonly used to improve performance.

Like the overload principle, the physiological basis of the progression is related to MU activation; in fact, progression ensures that the overload principle continues to be applied during a training program. Increases in strength from neural and muscular training adaptations make a given training task easier to complete, resulting in reduced activation of high-threshold MUs. Applying progression ensures continued activation of these MUs (Figure 8.44). Figure 8.45 illustrates how progression affects MU activation in a weight-training program.

Despite properly applied progression, continued increases in strength and power become increasingly difficult as the trainee becomes more advanced (76). This law of diminishing returns is illustrated in Figure 8.41 and, as shown in Figure 8.31, in very advanced stages of training improvement may be all but impossible. At this stage some frustrated athletes may be tempted to use performance-enhancing drugs to make further progress.

Figure 8.39 Schematic illustration of the blood pressure response to a set of repetitions to failure in a large-muscle-group exercise such as the bilateral leg press or squat. From a typical resting value of 120/80 mmHg, blood pressure increases sharply with the first repetition and continues to increase during the set until failure occurs. During each repetition blood pressure increases sharply at the beginning of each concentric phase (most effort required) and decreases somewhat during the eccentric phase (less effort required). The illustration shows the blood pressure response for the beginning of each concentric phase. For comparison, the typical blood pressure response at the end of a maximal aerobic power test is shown at the right. Weight-lifting blood pressure illustration is based on the data of MacDougall *et al.* (217,221).

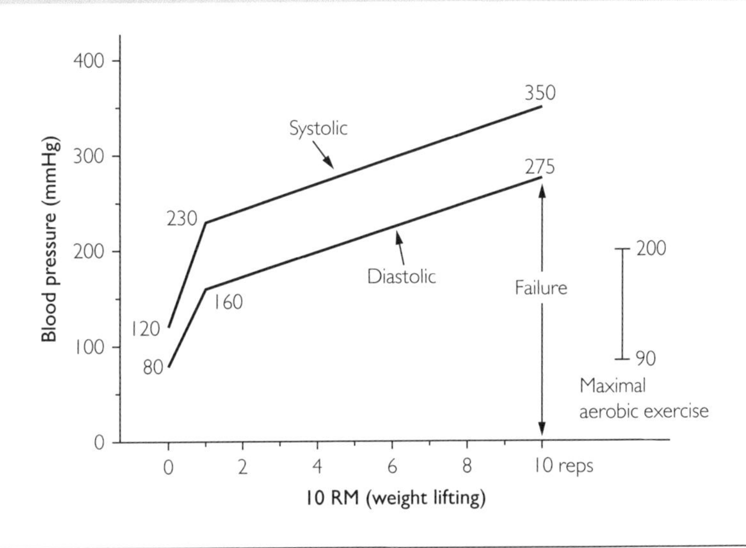

Figure 8.40 Training to failure and risk of injury. Depicted are the decrease in force applied to the weight, and consequently the velocity of lifting the weight, during a set of 10 repetitions to failure (10 RM). As fatigue develops during the set, the available force to accelerate the weight upward decreases, resulting in slower repetitions. If risk of injury is proportional to contraction force, it could be argued that the first repetition posed a greater injury risk than the last repetition. Paradoxically, the trainee perceives the weight as becoming heavier due to fatigue, and associates the increased effort required to lift the weight with an increase in contraction force. See Box 8.9 text for further discussion.

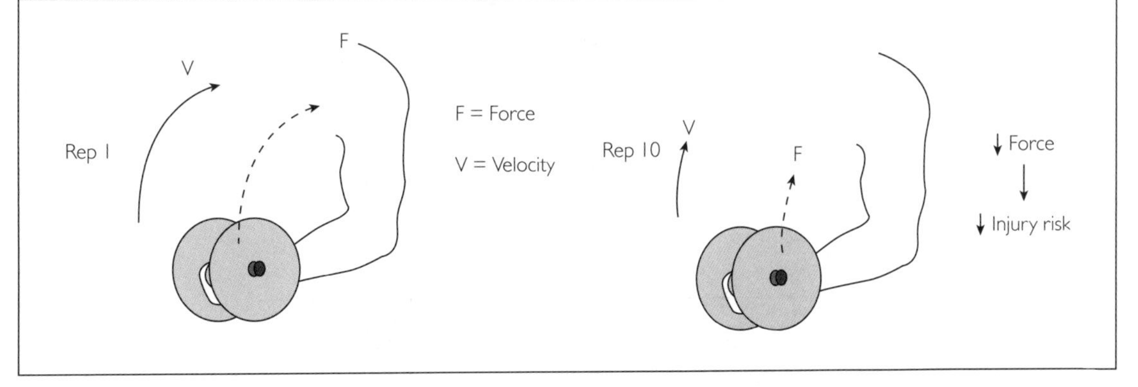

Figure 8.41 The concept of progression in strength training. The normal demand on strength in daily living is inadequate as a training stimulus to increase maximum strength. In the example shown, the trainee begins to train by doing contractions with a force equal to the pre-training maximum force. If this force continues to be used even as maximum strength increases, the chosen contraction force will soon fail to provide a maximal overload condition. To maintain maximal overload, progression must occur, in this case doing contractions at the new maximum force level. As maximum strength increases with the new training force, progression will have to be applied again and again as shown. Note that as training continues over a period of months or years, further increases in strength become more difficult to attain ("law of diminishing returns").

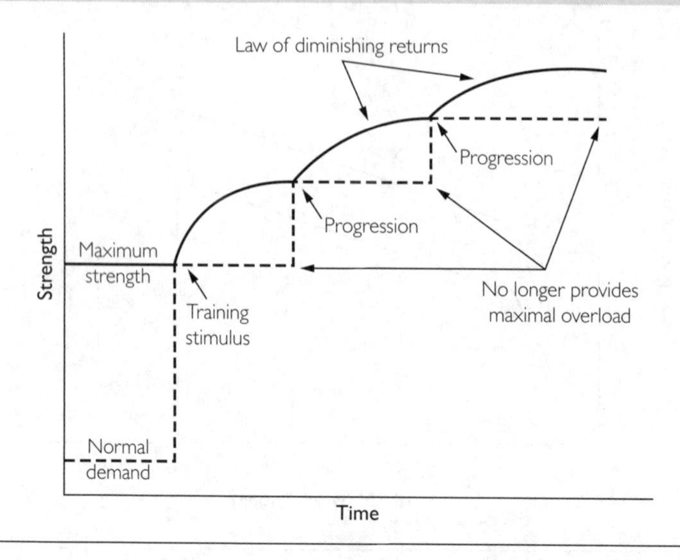

Figure 8.42 In isometric and isokinetic concentric and eccentric training, programs usually include doing sets of maximum voluntary contractions (MVCs). The training apparatus provides progression automatically by matching the increased contraction force as the trainee becomes stronger. In isometric training an alternative would be to train with contractions at a set percentage (e.g., 70%) of initial maximum strength; after a period of time there would be a test of maximum strength and a new 70% training level would be set in accordance with the new level of maximum strength.

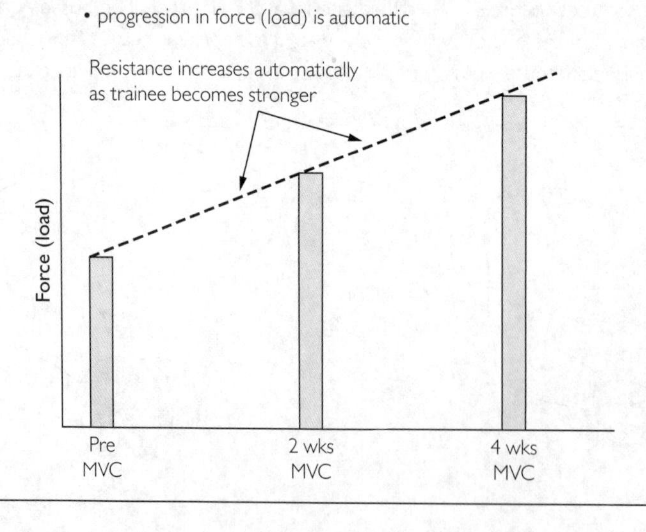

Figure 8.43 Progression in weight training. In weight training a "double progression" scheme is often used. The first part of progression is to increase the number of repetitions (reps) that can be done with a given load (weight). Once the upper limit of the designated range of reps can be achieved, the second part of progression is to increase the weight so that the number of completed reps falls to the lower limit of the designated range. The double progression scheme is continued for as long as possible over weeks and months of training. As depicted in Figure 8.41, progression becomes more difficult in the advanced stages of training. The illustration shows a rep range of three to six; in practice, a variety of rep ranges are used (6–8, 8–10, 10–12, etc.).

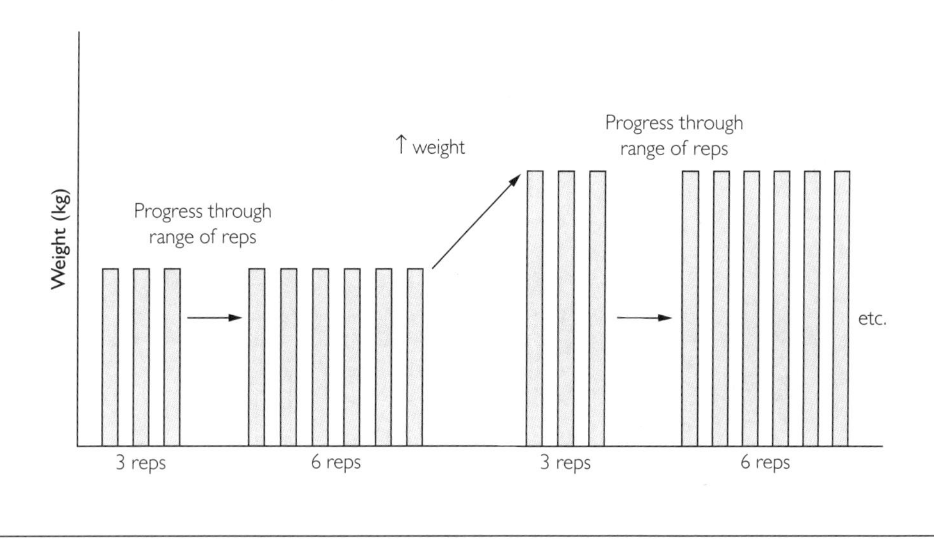

Figure 8.44 Physiological basis for progression in strength training. As the result of training adaptations, a given training load (intensity, volume, etc.) or task is easier to complete, resulting in decreased MU activation (MUA) that risks not fully activating the high-threshold MUs. By increasing the training load the trainee is forced to increase MU activation to maximum again. This progression is applied repeatedly as strength continues to increase.

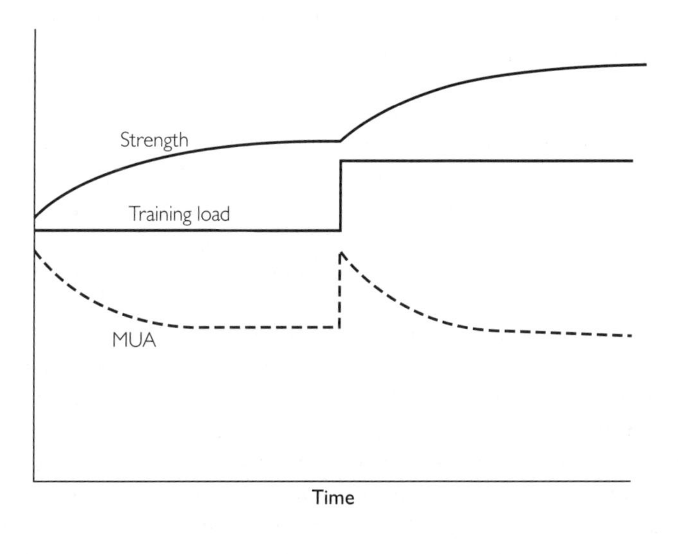

Figure 8.45 Illustration of how progression in weight training ensures full MU activation (MUA). In training session 1 full MU activation is needed to complete a set of five repetitions with the chosen weight (i.e., 5 RM). By training session 6, lifting the same weight for the same number of repetitions is easier, resulting in decreased MU activation needed to complete the set. To restore full MU activation, the weight must be increased so that completing a set of five repetitions again requires maximal effort. This progression would be repeated during the training program. In this illustration a single progression was used (increasing weight but keeping the number of repetitions the same). The same effect would be achieved with the double progression shown in Figure 8.43. In double progression, maintaining maximal MU activation would be achieved both by increasing repetitions through a designated range and by increasing weight.

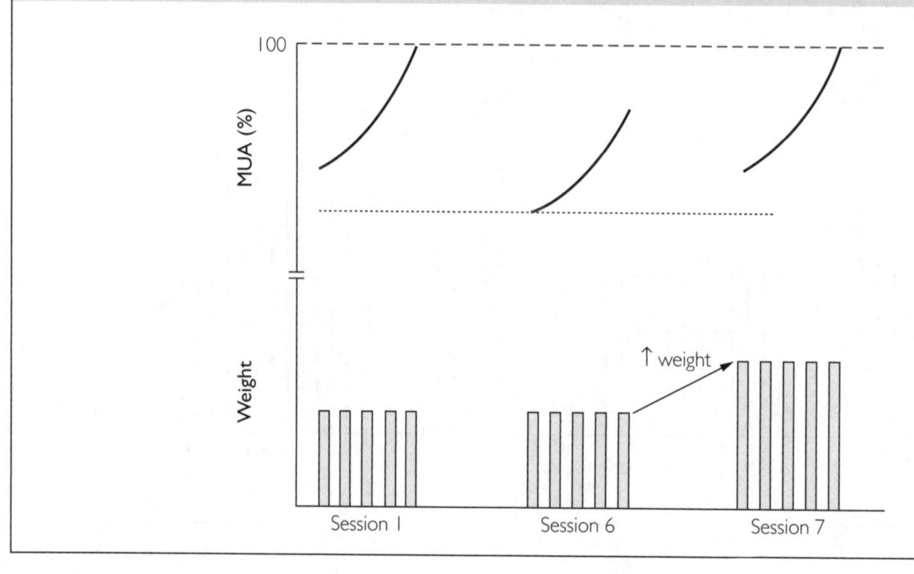

Specificity

Specificity refers to designing an athlete's training program to allow maximum transfer of training adaptations to the target athletic performance (76,100,252,311). For an athlete beginning a strength and power training program, non-specific training may provide a more generous transfer to improved performance; however, as training becomes advanced and further improvement is more difficult to achieve, specificity in training becomes more important (76). As discussed below there are several aspects to specificity in strength and power training.

Movement Pattern

A particular strength or power performance is partly a motor skill that requires a precise activation of agonists, antagonists, and synergists (Chapter 5). Within the relevant muscles, some movements may activate some groups of MUs while others may activate other groups of MUs (147). For example, professional male tennis players exhibit asymmetrical hypertrophy of rectus abdominis on the non-dominant side that is most evident in the distal (lower) part of the muscle (314). Selecting a training exercise that closely simulates a sport movement pattern may not be sufficient. In a sport action the major resistance to movement may occur at a particular point in the range of movement. Ideally, the selected training exercise would offer its greatest resistance at the same point in the range of movement (100). An example of the importance of specificity of movement pattern is shown in Figure 8.46, which indicates that strength tests similar to the training movement revealed greater increases in strength. Suppose in Figure 8.46 that the target sport action was the isometric knee-extension test. This would have been a poor return on the training investment. While some generality in the training response is possible (e.g., 43,231,232), particularly in the early stages of training, better returns are achieved when the training exercises simulate the sport movement pattern as closely as possible (76,100,252). Achieving effective specificity in some sports can be challenging. For example, it has been difficult to develop dry-land strength training exercises that transfer well to

Figure 8.46 Specificity of movement pattern and contraction type in strength training. Training consisted of three sets of 6 RM barbell squats done 3 days per week for 8 weeks. The strength test (1 RM squat) most similar to the training mode exhibited the greatest relative increase in strength. Smaller increases were found with isometric (ISO) leg-press tests at two knee joint angles, but involving both hip extensors and knee extensors. An even smaller increase was observed in an isolated knee-extension isometric test. Based on the data of Thorstensson *et al.* (357).

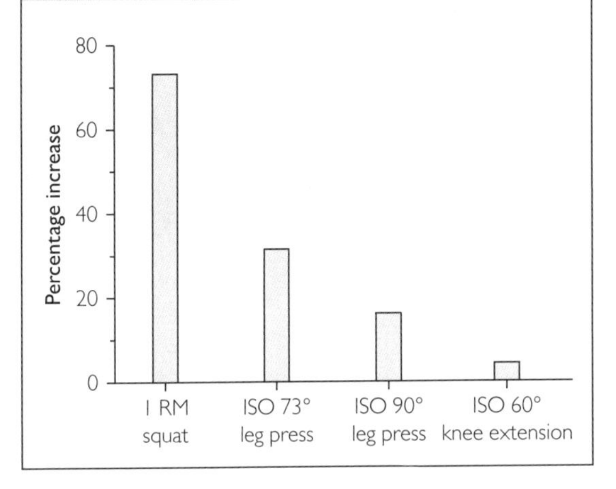

swimming performance; more specific in-water resistance training methods may be needed (344). Fortunately, effective transfer from conventional resistance training methods can be achieved in many other sports (344).

Velocity and Power

Many sport actions involve moving a load such as the athlete's own body mass (or that of an opponent) as fast as possible in what are called explosive or ballistic movements. As discussed in Chapter 4 (Figure 4.22), the velocity actually attained in explosive movements is proportional to the force acting on the load and inversely proportional to the mass of the load. The power developed is the product of the applied force and the resultant velocity (power = force × velocity). To achieve specificity of velocity and power in training, it is important to select explosive training loads that result in velocities and power outputs similar to those of the relevant sport actions. As illustrated in Figure 8.47, training with light and heavy loads would result in fast and slow movements, respectively, with the associated power output. A certain intermediate load, called the **optimal load**, would produce maximum power at an intermediate velocity.

While there is evidence of velocity and power specificity when training with isoinertial loads (i.e., lifting weights including body weight) (35,76,252), there is also evidence of transfer of high-load training to low-load performance and vice versa (11,76,135), although the mechanisms by which training with various loads increases explosive performance may vary (161). Training with the optimal load for power output (76,79,171) and with a combination of low and high loads (76,79,132) is also advocated. The common feature of all these training approaches is the *intent* to perform explosive actions. The effectiveness of explosive training with a variety of loads raises the question of whether it is the intended velocity or actual velocity of movement that is the key training stimulus for the high-velocity specific training response. The intent to perform a ballistic contraction results in a characteristic pattern of MU activation (Figure 8.30). If it is the repeated ballistic pattern of MU activation that improves ballistic performance, then the actual imposed load and resulting contraction velocity may not be crucial. One attempt to answer this question is shown in Figure 8.48, which indicates that the intent to contract explosively was more important than actual velocity. These results agree with the previous point that training with particular loads can improve performance with other loads provided that attempted ballistic actions are done with the selected load.

The extent of specificity or generality in the velocity and power training response may depend on the duration of the action. Up to a certain velocity and duration of movement, the nervous system can monitor and make adjustments as the movement progresses, much like guiding a missile after it has been fired. In the fastest ballistic movements, however, the motor command is sent to the MUs and the movement is executed without time for the feedback (e.g., from sensory receptors) required to alter the movement (Chapter 5). Thus, this type of pre-programmed movement is analogous to shooting a pistol: once the trigger is pulled, there is no way to alter the course of the bullet. For fast movements that allow time for adjustments while the movement is in progress, movements at specific velocities may be more important to allow adaptive adjustments to specific patterns of sensory feedback.

Because the magnitude of applied force determines the acceleration and velocity attained with a given

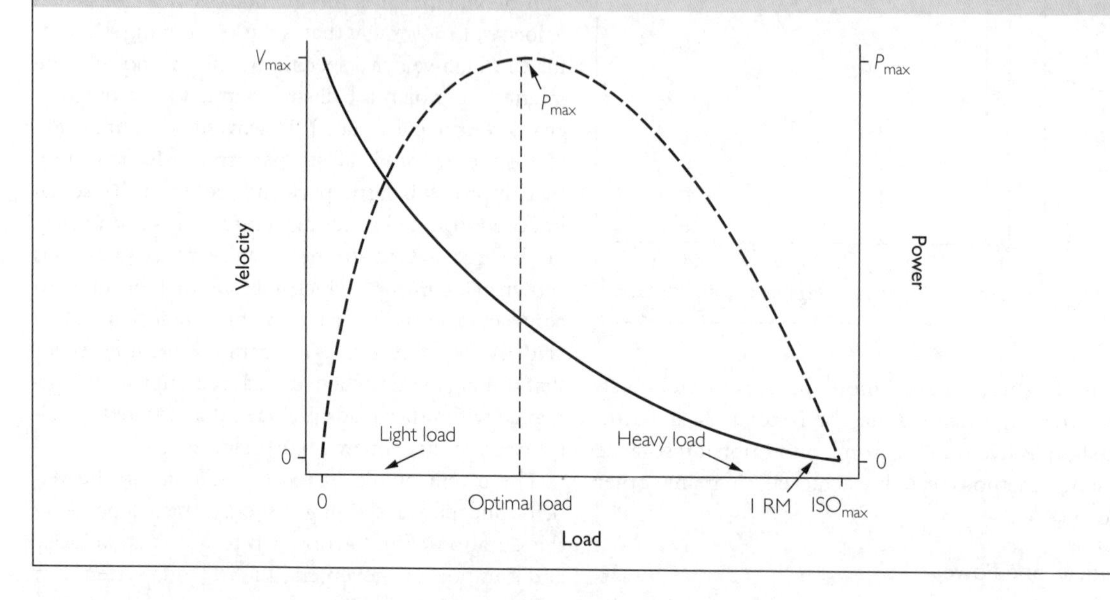

Figure 8.47 Specificity of velocity and power in strength and power training. In this schematic illustration the concentric contraction force– and power–velocity relationships are shown. Recall from Chapter 4 that power is the product of force and velocity. An athlete attempts to lift a range of loads (weights) as fast as possible. For each load, the force applied to the load, multiplied by the attained velocity, yields the power produced at each load. Maximum velocity (V_{max}) would be attained with zero load but power would also be zero. The load that would be just too heavy to move at all would correspond to maximum isometric force (ISO_{max}) and also result in zero power. A slightly lighter load, one that could be lifted slowly for one repetition only, would be the 1 RM. For practical reasons, training loads are usually expressed as a percentage of the 1 RM rather than ISO_{max}. An issue in designing training programs is deciding on the most effective training load for a particular sport. If there is specificity of velocity and power in training, then an athlete whose sport involves moving light loads as fast as possible should train with loads light enough so that the velocity and power of training exercises are similar to those of performance. Alternatively, if heavy loads must be moved in performance, then heavy loads should be used in training. Some advocate selecting the load (optimal load) that produces maximum power (P_{max}) regardless of the movement velocity of the performance. Finally, conventional resistance training with heavy loads may, by increasing strength, increase the speed that can be attained with lighter loads. See text for further discussion.

load, a conventional strength training program, with no attempt to lift explosively, can still improve speed and power performance (76,202,348), particularly with heavy loads. In fact, isometric training contractions (velocity = 0) with no intent for an explosive action, can greatly increase the velocity and power attained with a broad range of loads (Figure 8.20). Conventional heavy resistance training is probably most effective for athletes new to strength training because they will make large increases in strength in the early phase of training (76) (Figures 5.31 and 5.41). At more advanced stages of training when strength increases are smaller, it may be important to train more specifically, using training exercises with the intent to move explosively (76). Specificity is likely also more important when the sport performance involves moving light loads, such as throwing a ball or javelin or kicking a ball (365,390) (Figure 8.48). A series of slow, single lifts with a light load, not leading to fatigue failure, would not provide a training stimulus for strength or increased movement velocity (136). Instead, each lift should be a maximal effort to move the load explosively. Also, the force applied to a small load can be quite large if the load is lifted explosively (249). In other words, it is more important to emphasize the intent to move explosively when training with light loads. To cover all the bases, explosive training is often done with a range of loads (76,132,136,360,365). Another issue in explosive training with weights is whether the weight should be released at the end of a movement (Box 8.10).

BOX 8.10 THROWING WEIGHTS IN BALLISTIC TRAINING

Explosive weight lifting can be done with or without releasing the weight at the end of the range of movement. Releasing the weight is more specific to sport actions like throwing an object (e.g., shot put) or jumping. If the trainee is allowed to release (e.g., throw) the weight at the end of the range of motion, greater force, acceleration, and therefore velocity is attained than when the weight is held at the end of the movement (264). Throwing a weight may not be feasible or safe for some exercises except with special equipment (e.g., 264), but other exercises, such as weighted vertical jumps, can safely use the release method. In this case the release is jumping off the floor rather than releasing the added weight.

Figure 8.48 Intended vs actual movement velocity in strength training. (Top) The trainees attempted to do ballistic movements with one leg but movement was prevented, rendering the contractions isometric (velocity = 0). (Bottom) Ballistic contractions with the other leg were intended and an isokinetic dynamometer allowed fast concentric contractions to be done. Concentric force–velocity relationships tested before (Pre) and after (Post) training indicated that both types of training produced a high-velocity specific training effect; that is, high-velocity force increased more than low-velocity force. The results suggest that the intent to do ballistic movements was the key training stimulus, rather than the actual movement velocity. Based on the data of Behm and Sale (36).

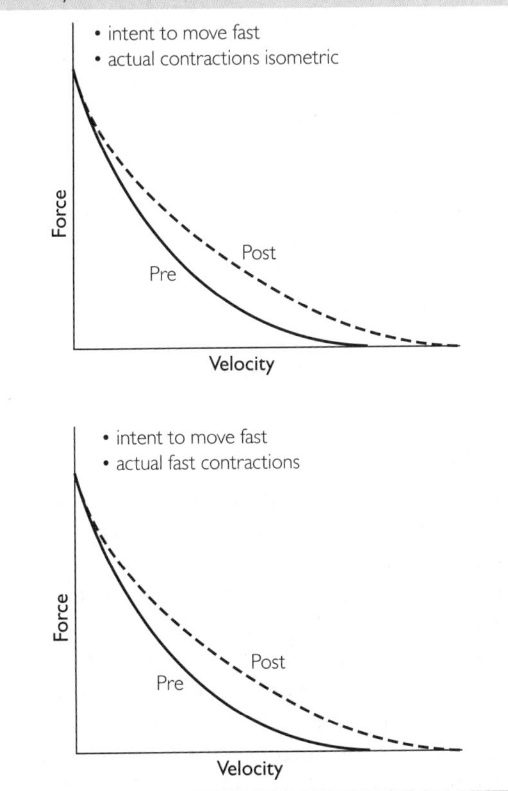

Load

In the application of load specificity to increase strength, muscle size, and endurance, sets of repetitions are usually done in the conventional way, to concentric contraction failure but with no attempt to move loads explosively. Lighter loads permit a greater number of repetitions to be completed before failure occurs. An example of the relationship between load and the number of repetitions to failure is shown in Figure 8.49.

In Figure 8.49 the values shown for the so-called strength training zone has been arbitrarily designated as using loads between 60 and 100% of the 1 RM, because a large number of strength training studies have used this range of loads (297). In Figure 8.49 the average number of repetitions completed with each percentage of the 1 RM are estimates. Several factors can affect the number of completed repetitions with a given percentage of the 1 RM. Exercises involving muscle groups normally involved in posture and locomotion have greater resistance to fatigue, allowing more repetitions to be completed; for example, at a given intensity more repetitions can be done in the squat than in the bench press (156). If repetitions are done more quickly (but not with ballistic intent), for example with one second concentric and eccentric phases vs three second phases, more repetitions can be done with faster repetitions, particularly with faster concentric phases (124). For a given exercise and load, women can complete more repetitions than men (234).

In load specificity applied to increasing endurance, training exclusively with heavy loads and low repetitions will after a time decrease *relative* endurance, that is, the number of repetitions that can be done with a given percentage of 1 RM, whereas training with lighter loads and more repetitions increases *relative* endurance (22,60). However, training with heavy loads can increase *absolute* endurance, that is, the number of repetitions that can be done with a given

Figure 8.49 (Left) The relationship between training load, expressed as a percentage of the 1 RM, and the number of completed repetitions. The strength training zone is arbitrarily designated as training with loads between 60 and 100% of the 1 RM. (Right) The strength training zone has been magnified. See text for further discussion.

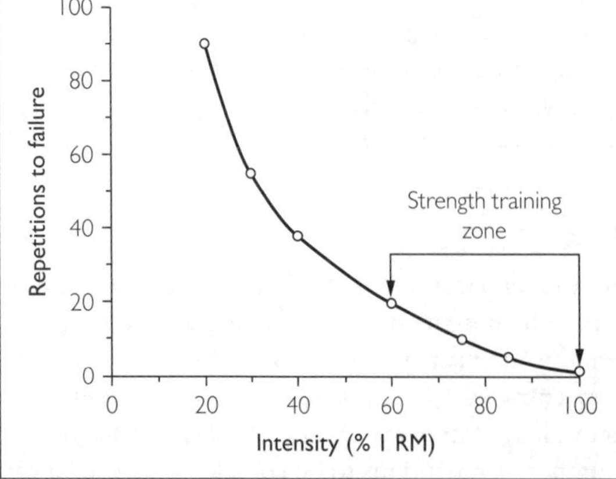

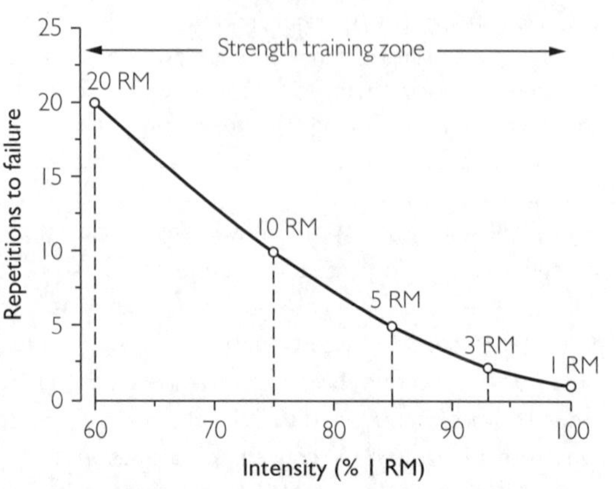

absolute load (e.g., 25 kg) (22). Since absolute endurance is most important to sport performance (Box 5.9), training with heavy loads can contribute to improved endurance performance.

Compared to heavier loads, training with lighter loads tends to cause smaller increases in strength but larger increases in relative and absolute endurance. Figure 8.50 provides an example of load specificity for increasing strength and endurance. The range of loads and repetitions depicted in Figure 8.50 is rather large (6–150 RM). Is load specificity still in effect when the range of loads compared is smaller, for example, comparing loads of 1–3 RM to 8–12 RM? With regard to increasing strength, the evidence is equivocal (370). For example, training with 3–5 RM was found to increase strength (1 RM) more than 9–11 RM (60), but training with 4 RM vs 10 RM produced similar increases in strength (71), as did training with 6–8 RM vs 15–20 RM (342). It is possible that load specificity for increasing strength becomes more important as trainees become more advanced in training and if the sport performance involves single maximal contractions against heavy loads. Lifting or moving heavy loads requires a specific pattern of MU activation in agonists (Figure 8.51), antagonists, and synergists that likely does not occur when lifting lighter loads repeatedly to failure. In addition to load specificity in relation to MU activation, there may also be specificity related to tendon adaptation; training with heavy loads may allow tendons to adapt to reduce injury risk when exposed to high loads in performance. Reviews of a large number of training studies have suggested that the optimal load for increasing strength, expressed as a percentage of the 1 RM, increases as trainees become more advanced (275,293).

If the primary goal of training is muscle hypertrophy, the tradition has been to use loads ≥60% 1 RM (Figure 8.49), although a wide range of loads have proved effective for stimulating muscle protein synthesis (27,53) and increasing muscle mass (370). Earlier in this chapter it was suggested that the stimulus for hypertrophy may be the force developed at the level of the MU rather than at the level of whole muscle. This would explain why training with a large range of loads, from 30 to 100% 1 RM, would effectively induce hypertrophy, and in some muscle groups to a similar extent (78,370). In fact, there is some evidence that using higher and lower loads in the same training session is more effective than training exclusively with high loads (121,122).

Contraction Type

If exercises are selected to closely simulate the sport movement patterns, then the appropriate contraction type (concentric, isometric, or eccentric) will also have been selected. For example, alpine skiing involves the development of large eccentric forces; therefore a strength training program that imposes large eccentric loads would be specific and effective (128). But in most conventional resistance exercises the same load is used for the lifting (concentric) and lowering (eccentric) phase of each repetition (see Figure 8.52 for

Figure 8.50 Specificity of training load in relation to increases in strength and endurance. Three groups of 15 men did the bench-press exercise 3 days per week for 9 weeks. One group trained with three sets of 6–8 RM, the second group with two sets of 30–40 RM, and the third group with one set of 150 RM. Strength (1 RM) and of endurance tests were administered. In a *relative* endurance test trainees did as many repetitions (reps) as possible with 40% of the pre-training 1 RM and then, after training, with 40% of the post-training 1 RM, which was a greater absolute weight. In an *absolute* endurance test, trainees did as many reps as possible with a weight of 27 kg, both pre- and post-training. Specificity of load was in effect, in that training with heavier loads produced greater increases in strength, whereas training with lighter loads produced greater increases in endurance. Note that training with the heaviest load (6–8 RM) actually decreased relative endurance but increased absolute endurance. Based on the data of Anderson and Kearney (22).

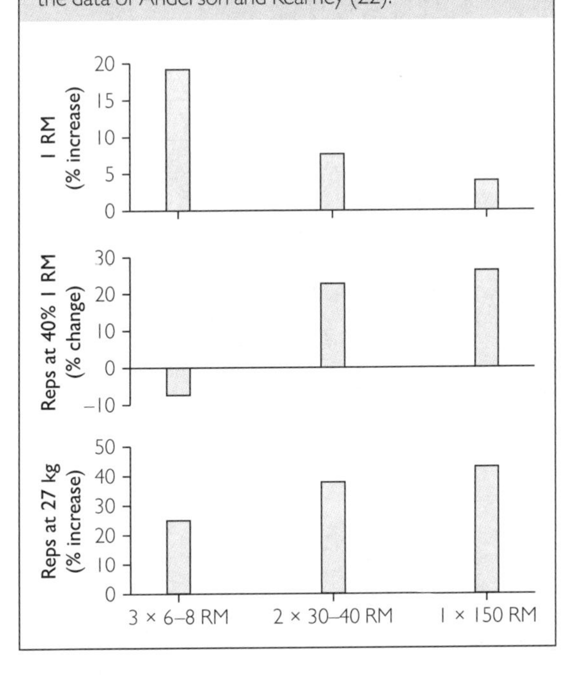

an example). Since eccentric contractions can produce more force, a relatively small percentage of maximum eccentric force will be taxed in a set of repetitions. Various methods can be used to train with large eccentric loads that exceed the concentric 1 RM. In one method assistants or spotters help the trainee lift the weight through the concentric phase and then watch as the trainee lowers the weight through the eccentric phase. Care must be taken not to use a weight that is too heavy, otherwise the weight would lower rapidly out of control with a possible risk of injury (see increasing velocity with heavier loads in Figure 8.52). In another method, experienced trainees can use a cheating style to lift the weight up through the difficult concentric phase but then lower the weight in the eccentric phase with a strict style. Some weight-training machines allow the trainee to lift the weight with both limbs (e.g., leg press) but lower it with one limb to impose a large eccentric load. High eccentric forces can be produced when training with isokinetic (130) and other special devices (128,266), but these devices are not readily available to most athletes. In addition, it is difficult to simulate many sport movement patterns with these devices. Thus, most athletes would have to use weight-training exercises modified as described above to impose large eccentric loads.

The interest in high-load eccentric training is based on the possibility that large eccentric forces act as potent training stimuli to increase strength and power through both neural and muscular (hypertrophy) adaptations. Advocates of this type of training can point to evidence of its effectiveness (130,298) but skeptics can cite evidence as well (370). The strongest positive evidence is that eccentric training increases eccentric strength more than isometric or concentric training (130), an example of contraction type specificity due largely to specific neural adaptations. The evidence is more divided on the issue of inducing muscle hypertrophy (370). When superiority of eccentric training for increasing muscle mass is observed, the greater increases in muscle mass are small compared to the much larger forces developed in eccentric training, suggesting that factors other than whole-muscle force contribute to the hypertrophic stimulus. One additional factor may be volume or total work (load × number of repetitions) completed in a training session. For example, when sets of isokinetic maximal eccentric contractions were compared to sets of a larger number of maximal concentric contractions to equate the work done per session, increases in whole-muscle and muscle-fiber size were similar despite eccentric force being approximately 30% greater than concentric force (251). In the case of isokinetic maximal eccentric contractions, speed seems to be a factor; faster contractions resulting in much less volume (force × time) per training session produce a moderate increase in force according to the force–velocity relationship, thus inducing greater hypertrophy (97,328).

Figure 8.51 Possible basis for load specificity in weight training. In a 1 RM all MUs must be activated in one maximal contraction, whereas in a 10 RM there is a progressive increase in MU activation (MUA) over the 10 repetitions. Enhancing MU activation with particular loads may require training with these loads. In addition, lifting different loads may also require different postural adjustments and a different pattern of activation of antagonists and synergists, patterns that would be enhanced by training with specific loads.

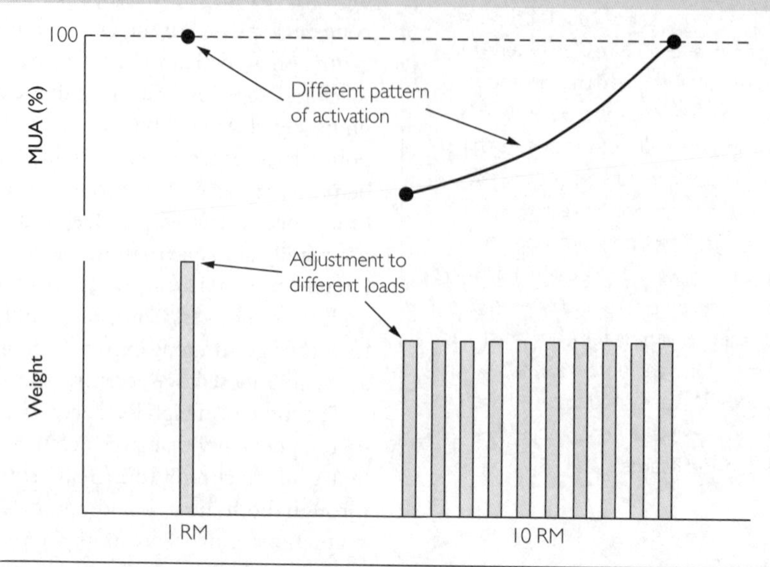

Figure 8.52 Schematic illustration of the load–velocity relationship, which would apply to weight-training exercises. In a conventional weight-training exercise such as doing a set of repetitions with a load equal to 80% of the 1 RM, the weight is lifted and lowered at a relatively slow velocity. Note that the concentric velocity attained is less than the maximum possible velocity with the 80% 1 RM load because no attempt was made to lift the weight explosively. Also, the 80% 1 RM load taxes only a small percentage of maximum eccentric strength, which could be approximately 140% of the concentric 1 RM. Challenging maximal eccentric strength requires loads that exceed the 1 RM; one example (≈130% 1 RM) is shown. Note that the load representing maximum isometric force (ISO_{max}) is slightly greater than the 1 RM load. Maximum shortening velocity (V_{max}) is the maximum velocity attained with zero load. See text for further discussion.

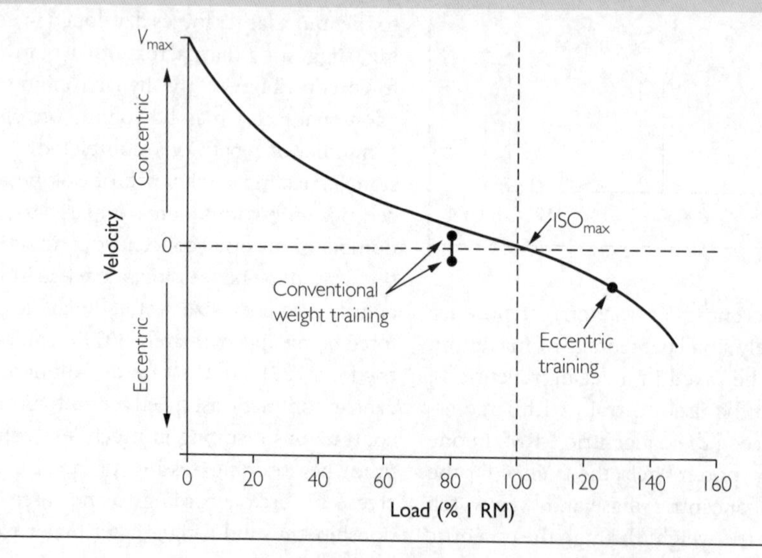

It appears that an array of interacting factors promote hypertrophy.

In summary, for athletes whose performance requires withstanding large eccentric loads, there is sufficient evidence to warrant a trial of specialized high-load eccentric training, which may induce specific neural and possible muscular adaptations to improve performance. For increasing muscle mass, the evidence is more equivocal, but if the trainee is advanced and has reached a plateau with conventional training, a period of high-load eccentric training could be introduced to see if it promotes further increases in muscle mass and strength.

Plyometrics The term **plyometrics** was introduced in Chapter 4 in connection with the stretch-shortening cycle (SSC) and SSC potentiation, which refers to the enhancement of concentric contraction force, speed, and power by immediately preceding eccentric contractions. Coupled eccentric–concentric contractions that produce SCC potentiation are common in sports that include jumping, throwing, and kicking. SSC potentiation also occurs when doing repetitions of many common weight-training exercises: the force of the concentric phase of a repetition is enhanced by the preceding eccentric phase. The term "plyometrics" is reserved for coupled eccentric–concentric contractions that are done quickly, as in sprinting, jumping, hopping, and throwing. Consequently, plyometric exercises include jumping, hopping, and throwing drills.

There is good evidence that plyometric training consisting of various jumping exercises increases vertical jump height, sprint speed, agility, and endurance (228,229) (see Box 8.11). The most common plyometric training exercises are: (1) countermovement jumps in which the athlete quickly crouches down (eccentric phase) and then jumps as high as possible (concentric phase) and (2) drop jumps in which the athlete drops down from various heights, decelerates his/her body mass (eccentric phase), and then jumps as high as possible (concentric phase). Because of the drop down from a height, large eccentric loads are applied in drop jumps. In another type of jump, the squat jump, the trainee lowers into the crouch and holds it before jumping upward. The holding phase eliminates or minimizes SSC potentiation; therefore, squat jumps could be considered non-plyometric exercises. Nevertheless, increasing squat-jump height is often a goal of plyometric training. Other plyometric exercises include hopping and bounding. The most effective plyometric training programs are those in which two or three sessions per week are continued for at least 6–15 weeks (229). Each session should include up to 50 high-intensity repetitions (jumps, hops, bounds) (89). A plyometric training program may be more effective if various types of jumps (countermovement jumps, drop jumps) are included rather than just one type (89).

Plyometric training, conventional weight training, and explosive training can all improve explosive performance (227,310), to some extent by different adaptations. Thus, there is probably merit in using all three types of training at various times in an overall training program.

BOX 8.11 PLYOMETRIC TRAINING AND ENDURANCE PERFORMANCE

Plyometric training is usually associated with increasing speed and power, so it might seem surprising that plyometric training could improve endurance performance. In distance runners, for example, plyometric training does not increase maximal aerobic power or raise the lactate threshold but improves running performance by increasing running economy (229); that is, decreasing the energy cost of running at a given speed. Decreased energy cost means that the runner can maintain a pace longer before becoming fatigued. Plyometric training improves running economy by adaptations that enhance SSC potentiation in running. SSC potentiation is discussed in detail in Chapter 4.

Volume

Training volume refers to the number of exercises used in a training session and the number of sets done with each exercise or muscle group. Most research has focused on the issue of the optimal number of sets of an exercise/muscle group for increasing strength and muscle mass. The issue has been raised repeatedly over the years and continues to be debated and reviewed, particularly in regard to conventional strength training (weight training) (63,64,99,108,182,183,275,276, 296,297,318,370). Even the methods used to review research on the question of optimal set number have been debated (276,383).

The debate has focused primarily on the question of whether one set per exercise/muscle group is sufficient to induce a maximal training response or whether multiple sets are more effective. Note that the number of sets at issue are training or work sets; that is, sets done to or near concentric contraction failure (see Box 8.12). The balance of research and related reviews suggest that although one training set is effective (64,99,108), multiple sets induce greater gains in muscle size and strength (108,182,183,275,276,296,297,370,384). The added benefit of multiple sets in increasing muscle size might be predicted from the larger and more prolonged acute increase in muscle protein synthesis after multiple sets vs one set (52). However, as Figure 8.53 illustrates, the additional benefit is not linear with additional sets; that is, four sets do not give four times the results of one set. Thus, the law of diminishing returns comes into play as set number increases (182,183,276,333,370). Supporting the law-of-diminishing-returns concept is the absence of a difference between three and six sets in the acute muscle protein synthesis response (196). Identifying the exact optimal number of sets remains difficult. Advanced trainees may do better with more sets (275,276,296) but it is not clear whether they need more sets or can better tolerate them. Notably, as discussed in the next section on training frequency, advanced trainees exercise particular muscle groups less frequently than novices, and may therefore tolerate more sets in each session.

BOX 8.12 TRAINING VS WARM-UP SETS

Training sets are often preceded by warm-up sets to prepare for the training sets. Several warm-up sets may be used by an advanced trainee before doing training sets in an exercise such as the back squat in which the training load could be high (e.g., 150 kg). On the other hand, a novice training with relatively light loads might find one or two warm-up sets adequate. Warm-up sets, which are not done to or near failure, have the benefits of increasing muscle temperature and inducing postactivation potentiation (Chapter 5), both of which can improve performance on training sets. Increased muscle temperature may also reduce the risk of injury. Warm-up sets also acclimate the trainee to the pattern of muscle activation used in the exercise. A disadvantage is that an excessive number of warm-up sets may induce fatigue that could impair performance in the training sets. With experience, most trainees learn to adjust the number of warm-up sets and effort expended to maximize the advantages and minimize the disadvantages.

Some of the research and related reviews have focused on the value of additional sets on increasing strength. It is *specific* strength that is usually assessed; for example, comparing the effectiveness of different numbers of sets of the bench press exercise on the

Figure 8.53 Schematic illustration of the relationship between the number of sets completed in a training session for a given exercise/muscle group and the magnitude of training effect. Some believe that adding sets increases the training effect in a linear fashion and even in proportion, so that four sets may be four times more effective than one set. Research indicates that there is a law of diminishing returns in which adding sets produces an increasingly smaller added training effect. See text for further discussion.

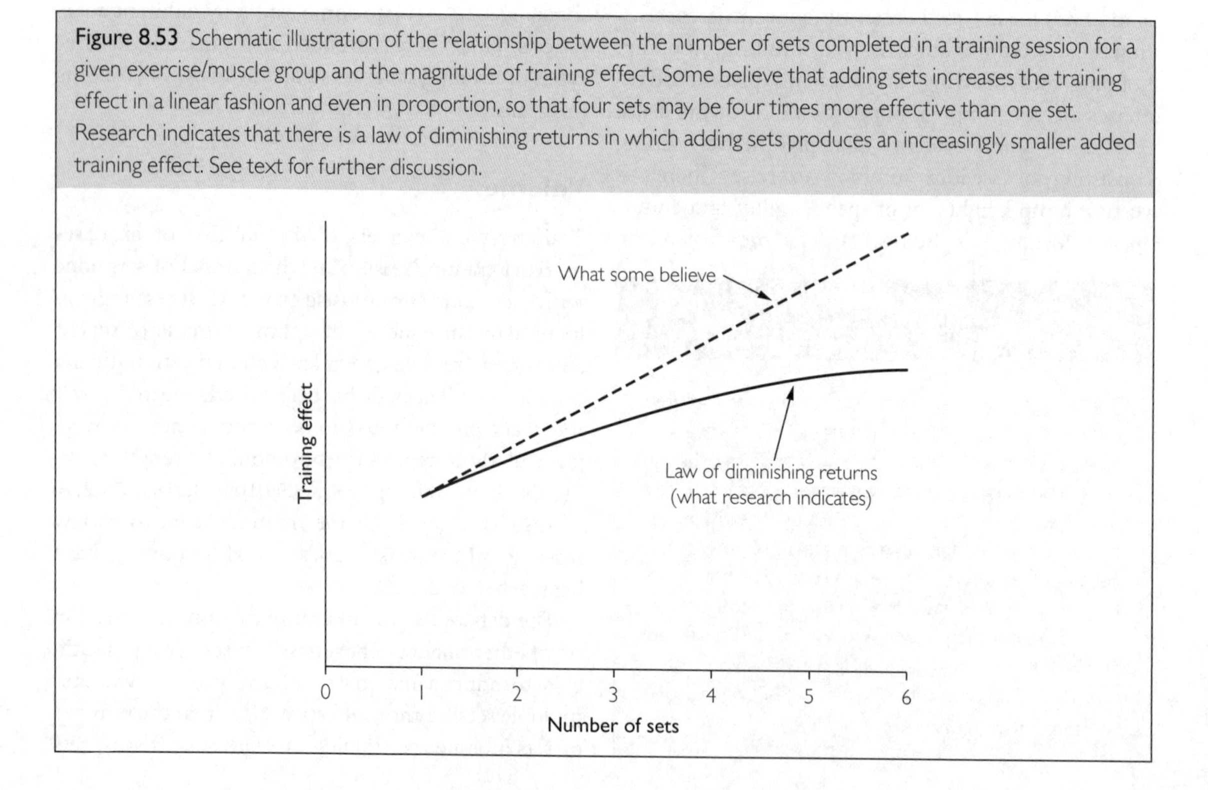

bench press 1 RM. In this case, the additional sets may enhance strength largely by providing more practice of the movement pattern involved in the strength test (1 RM). Evidence that this might be occurring comes from studies showing that, although additional sets cause greater increases in 1 RM tests, they do not produce greater increases in non-specific strength tests of the same muscle groups, such as isometric tests (230,248). In fact, non-specific strength may fail to increase even if hypertrophy has occurred (313). Similarly, it cannot be assumed that greater increases in 1 RM tests in the gym translate into greater increases in force generated in specific sport actions (Figure 8.46). These are the training effects of most interest to the athlete and coach. Little is known about the optimal number of sets for enhanced sport performance.

A final consideration is that, in contrast to many research studies in which the optimal number of sets is assessed for a single exercise, many athletes seek and train for strength and muscle size in several muscle groups. In addition to strength and power training, many athletes also train to increase aerobic and anaerobic power. It is quite possible that the optimal number of sets in a given exercise is affected (reduced?) by the overall volume of different types of training.

In summary, athletes will benefit from doing one to three training sets for each exercise/muscle group. The optimal set number may increase as the trainee advances, but the decision to add sets must take into account the athlete's overall training load. It may be unwise to add sets if the overall training volume is high and aimed at increasing aerobic (endurance) and anaerobic power as well as strength. Athletes should be cautious about assuming that training (more sets) that produces greater increases in strength, and in some cases even more hypertrophy, necessarily generates greater improvement in sport performance.

Frequency

Training frequency is usually indicated as the number of days per week that an exercise/muscle group is trained. For decades a frequency of 2–3 days per week has been the norm, in accordance with research findings and related reviews and recommendations (64,276,297,370). Note that a training frequency of 3 days per week implies that rest days will occur between training sessions; for example, with training on Monday, Wednesday, and Friday each week. A frequency of 2 days per week requires more rest days between sessions; for example, with training on Monday and Thursday each week. More frequent training may be successful over short periods (287,288) but is probably not well tolerated over many weeks or months. In fact, there is some evidence that advanced trainees who do more sets in each workout may do better with two rather than three training days per week (276). Also supporting the concept that "less is more" is a classic weight-training study in rats, in which every-other-day training inhibited muscle hypertrophy whereas training every third day produced a level of hypertrophy exceeding that usually seen in human training studies (386).

For many athletes the number of exercises/muscle groups targeted for training may be small, so strength training need occur only 2 or 3 days per week. In other sports athletes may need to train several muscle groups with more than one exercise, resulting in long training sessions. A potential solution is a **split routine**, in which about half the exercises are done on one day and the other half the next day. One common split routine is to train upper-body muscle groups on Mondays and Thursdays, and lower-body muscle groups on Tuesdays and Fridays, or vice versa. Note that this split routine involves four training sessions per week but the training frequency *per muscle group* is 2 days per week. There are many types of split routines. In some, all of the muscle groups or body parts to be trained are divided into three or even five groups. The trainee would thus train up to 5 days per week but each muscle group/body part would be trained once or twice each week. These multi-day split routines, popular in bodybuilding (131), may not be suitable for athletes who also have to train for aerobic and anaerobic power as well as attend sport practices. Examples of training frequency schemes are shown in Figure 8.54.

Split routines reduce the length of individual training sessions and therefore possibly reduce fatigue toward the end of sessions. But are split routines more effective? Possibly, but there have been few experimental tests of their effectiveness compared to whole routines (64). One test of split routines is shown in Figure 8.55. Over 20 weeks, whole-body and split routines produced similar increases in strength and lean tissue mass. A limitation of this study is that the subjects were initially untrained, whereas split routines are typically introduced at the intermediate stage of training. Nevertheless, the study showed that split training did not produce strikingly superior results over 5 months of training. In the study shown in Figure 8.55

Figure 8.54 Sample common frequency schemes for strength and power training. (1a, 1b) Training all exercises/muscle groups 2 or 3 days per week; (2a, 2b) dividing exercises/muscle groups into two groups and training each 2 or 3 days per week; (3) dividing all exercises/muscle groups into three groups and training each 2 days per week; (4) "every other day split" in which exercises/muscle groups are divided into two groups and alternated but a rest day occurs after each training day. Etc.: there are many other possible frequency schemes.

1. Train all exercises (whole body) 2–3 days/week
2. Train half of exercises 2–3 days/week; other half 2–3 other days/week
3. Train 1st, 2nd, and 3rd third of exercises 2 days/week
4. "Every other day" split

	Mon	Tues	Wed	Thrus	Fri	Sat	Sun
1a.	1			1			
1b.	1		1		1		
2a.	1	2		1	2		
2b.	1	2	1	2	1	2	
3.	1	2	3	1	2	3	
4.	1		2		1		2
		1		2		1	
	etc.						

each whole-body session consisted of 35 sets across seven exercises, and each session lasted 60–90 minutes. In contrast, for each muscle group a bodybuilder might do three to five sets of three to five exercises (131). A whole-body routine could consist of about 80 sets over 3–4 hours. It is not surprising, therefore, that split routines evolved in bodybuilding to reduce the duration of training sessions. Few if any athletes would need 20 or more *sport-specific* strength training exercises; therefore, most athletes would not have to resort to split routines to achieve training sessions of manageable duration (≈60 minutes). It would be a matter of choice based on convenience (length of sessions vs trips to the gym) and how the strength training would be integrated with other forms of training and sport practice.

In comparison to conventional strength training, there is little research on optimal training frequency for other forms of power and speed training. A plyometric training program consisting of drop jumps produced better results done 2 and 4 days per week than 1 day per week but 2 and 4 days produced similar results (88). A sprint training program done every second day improved performance whereas daily training did not (273). This limited information suggests that the optimal frequency for speed and power training, like conventional strength training, may be about 2–3 days per week.

Recovery Dictates Training Frequency

If the optimal training frequency is 2–3 days per week, then 4–5 days each week are needed for adaptive processes to run their course after each session and also for recovery from fatigue or muscle damage. One way to monitor recovery from a training session is to see if at least as many repetitions can be completed in the next session with the same number of sets and load. Figure 8.56 is an example of this monitoring. None of the subjects had recovered in 24 hours and the majority required 72–96 hours. However, being able to complete the same number of repetitions in a subsequent session may not represent full recovery because, if the previous session provided a training stimulus, an increased number of repetitions might be expected. In fact, several of the subjects in the study shown in Figure 8.56 were able to complete more repetitions in second sessions done either 72 or 96 hours after the first one. Thus, a distinction should be made between a training stimulus and training adaptations induced by a training session, and performance in a subsequent session. In Figure 8.56 the training stimulus was the same before each rest period; what varied was

Figure 8.55 Comparison between whole and split weight-training programs. One group of untrained young women did a whole routine 2 days per week, consisting of four upper-body (five sets of 6–10 RM) and three lower-body (five sets of 10–12 RM) exercises. Half of the whole group did the upper-body exercises first in the first weekly session and the lower-body exercises first in the second weekly session. The other half of the whole group did the reverse. A second group did a split routine in which the upper-body exercises were done on 2 days per week and the lower-body exercises on two other days per week. Half of the split group trained the upper body on Mondays and Thursdays and the lower body on Tuesdays and Fridays. The other half of the split group did the reverse. There were two 10 week training periods separated by a 2 week Christmas recess. Each whole training session lasted 90–120 minutes and each split session lasted 45–60 minutes. Whole and split routines produced similar increases in 1 RM (average for bench press, arm curl, and leg press), lean body mass (LBM; average of arms, legs and whole body), measured by dual-energy X-ray absorptiometry, and training weights (wts), averaged across all seven training exercises. Based on the data of Calder *et al.* (59).

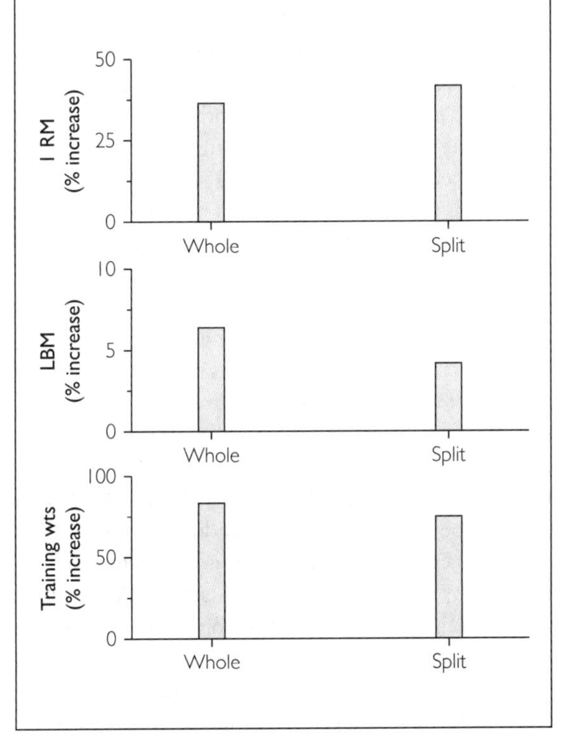

Figure 8.56 Recovery as an indicator of optimal training frequency. Ten young men did a series of experiments on separate days. At the beginning of each experiment the subjects did three sets of 10 RM. Then in different experiments subjects repeated the same number of sets with the same 10 RM load after resting 24, 48, 72, and 96 hours. Recovery was deemed complete if the subject could complete at least the same number of repetitions as done at the beginning of each experiment. As shown in the figure, none of the subjects had recovered in 24 hours. Most subjects needed 72 or 96 hours to recover, corresponding to a training frequency of 2 days per week. Based on the data of McLester *et al.* (243).

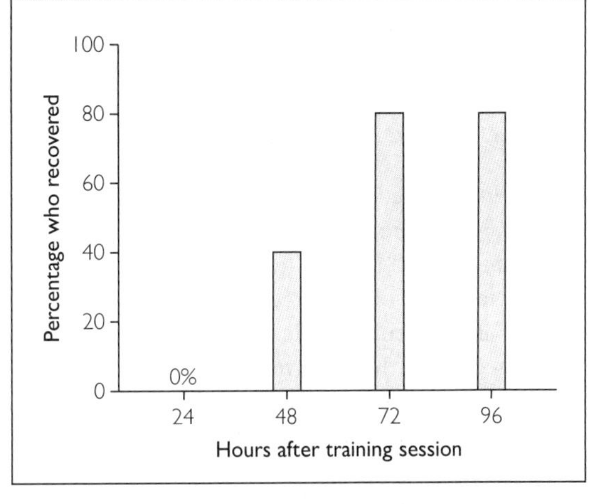

the time allowed for recovery. As a second example, in the sprint training study referred to earlier (273), daily training actually produced greater biochemical adaptations than every-other-day training, but performance improved only in the latter, presumably because the days of rest permitted more recovery. Finally, Figure 8.56 indicates the large inter-subject variation in the time needed for recovery. Some subjects did well with 48 hours of recovery but the majority needed 72 or 96 hours. Therefore, a coach should not assume that all athletes doing a particular training program will recover at the same rate.

There are several recovery and adaptive processes after a training session that could contribute to the 2–4 rest days needed between sessions. Each training session causes an acute muscle protein synthesis (MPS) response that may last up to 72 hours, but in most cases MPS decreases to near-resting values after approximately 24 hours (e.g., Figure 8.57). The time needed for the complete MPS response seems shorter

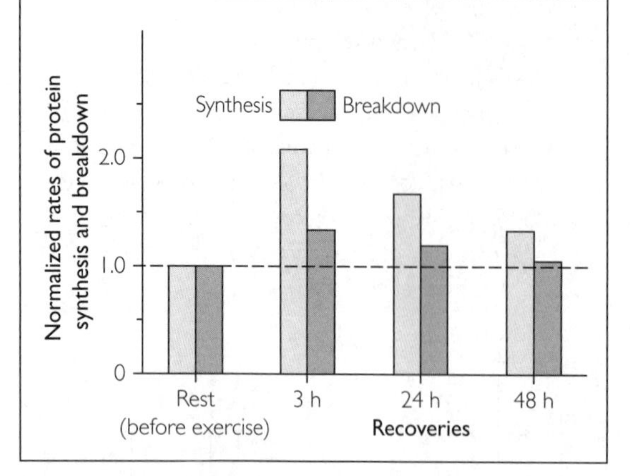

Figure 8.57 Time course of rates of muscle protein synthesis and breakdown after a resistance training session. In the knee-extension exercise, subjects did eight sets of eight repetitions with 2 minute rest periods between sets. Based on data of Phillips *et al.* (281).

than the time needed for full recovery. In fact, training apparently shortens the period of post-exercise elevated MPS (345). If the time course of MPS were the sole factor involved in recovery from training sessions, trainees should be able to train more rather than less frequently as they progress. But, as discussed, advanced trainees tend to train less frequently, suggesting that other factors besides MPS are involved in recovery.

Exercise-induced muscle damage is another factor that may contribute to prolonged recovery periods. Untrained subjects exposed to their first training session may experience muscle damage leading to delayed-onset muscle soreness (DOMS) and loss of strength (57,116). Full recovery of strength may take several days (57,116) and muscle damage is evident for at least 48 hours (116) and probably longer. However, just one such training session produces a repeated-bout effect that protects against damage in subsequent sessions (73,152,242). Regular training provides even more protection (115,267). However, even in experienced trainees an intense session results in muscle damage to about 50% of sampled muscle fibers (115). The damage seems to be repaired within 5 days (115) and perhaps sooner. Therefore, recovery from muscle damage in the absence of DOMS may play a role in recovery.

A third factor is fatigue. As discussed in Chapter 5, there are many possible mechanisms of fatigue. Failure to maintain activation of MUs is a potential cause of fatigue, but the ability to fully activate muscles recovers quite rapidly. Sets of several repetitions cause some depletion of muscle glycogen and a decrease in muscle pH due to lactic acid accumulation; however, pH and glycogen return to pre-exercise levels within about an hour and 24 hours, respectively. Perhaps the most relevant fatigue mechanism contributing to prolonged recovery is impaired excitation-contraction coupling leading to low-frequency fatigue (LFF). LFF can persist for at least 24 hours, but the one study that monitored LFF after a resistance training session found that recovery from LFF was complete by 33 hours (289).

Of the three factors discussed, the time course of muscle-damage repair seems to correspond most closely to the time needed to recover from a training session, but fatigue and the time course of muscle protein synthesis may also be involved.

Additional Training Considerations

Set Systems

Traditionally, trainees do all of the sets of an exercise before proceeding to the next exercise. This has been referred to as the **standard set** system or the **straight sets** system. As discussed in this section, rest periods between sets average 1.5–2 minutes. Rest periods often take longer than the sets, and the trainee spends more time resting than exercising. Increased access over the years to gyms with fixed-weight barbells and dumbbells as well as weight-stack machines has made it feasible and convenient to use a **circuit set** system. In this system, one set of the first exercise is performed, followed by one of the second exercise and so on until a circuit involving one set of each exercise has been completed. Three or more circuits may be performed. The use of circuits allows shorter rest periods between sets because each exercise uses a different muscle group. The result is a much shorter training session. Circuit training is usually done with light loads and many repetitions to increase endurance more than strength. However, it could be adapted to strength training by using the same loads, sets and repetitions as the straight sets system but doing circuits instead. Surprisingly, there has been little research comparing the two systems, particularly in relation to increasing non-specific strength and muscle mass and more importantly sport performance. Anecdotally, some advanced trainees like to concentrate on one exercise at a time and would find it difficult, for example, to do a set of squats with 150 kg and then walk to the arm-curl station for a set of curls. Other athletes would find circuits acceptable.

A compromise between straight sets and circuits is the **alternate set**, or super set, system. The alternate set system uses two exercises. The athlete performs one set of the first exercise and then the second exercise. This is repeated the desired number of times. Alternate sets are usually done with opposing muscle groups (e.g., bench press and row) to promote recovery. As with circuits, shorter rest periods can be used between sets. Here again there have been few comparisons between straight and alternate sets in training effectiveness.

Rest Periods Between Sets

The straight sets system is the most commonly used, particularly by advanced trainees, raising the question of the optimal rest period (or rest interval) between sets. In weight training to increase strength and muscle mass, rest periods can range from 30 seconds to 5 minutes. For increasing strength, a common recommendation is 2–3 minutes rest between sets (17,18), and perhaps longer after sets of exercises such as squats that involve large muscle groups. The majority (≈70%) of bodybuilders, whose primary goal is increasing muscle mass, prefer 1–2 minute rest periods, although approximately 30% use 2–3 minutes (131). Some bodybuilders use very short rest periods (30–60 seconds), but a recent survey shows that, when the focus of training is increasing muscle mass, only approximately 2% of bodybuilders use these short rest periods (131). It has been suggested that shorter rest periods are more conducive to promoting hypertrophy, primarily because of more elevated acute hormonal responses with short (e.g., 1 minute) vs long (e.g., 3 minute) rest periods (17,85). However, as discussed in this chapter, the role of acute hormonal responses in promoting hypertrophy has been challenged.

Shorter rest periods allow less recovery between sets so that, for a given load, fewer repetitions can be completed with successive sets. Figure 8.58 shows that most of the decrease in repetitions occurs over the first few sets, after which there is little further decline, indicating an equilibrium between the mechanisms of fatigue and recovery. Figure 8.58 also shows that for loads of 50 and 80% 1 RM the relative decline in repetitions over five sets with various rest periods was similar despite more repetitions per set with the 50% load. In contrast, recovery from single maximal repetitions is more rapid (377) than for the loads shown in the figure, indicating that strength recovers more rapidly than endurance (Figure 5.97). Longer rest periods lead to longer training sessions. For example, a session involving three sets of 10 exercises would have 20 inter-set rest periods. With 1, 2, and 3 minute rest periods, the total rest time would be 20, 40, and 60 minutes, respectively. The hope would be that longer rest periods would produce better training results to justify the longer training sessions.

Are the common recommendations for inter-set rest periods supported by evidence? The evidence is

Figure 8.58 Effect of rest periods between sets on the number of repetitions to failure over successive sets. In the bench-press exercise, subjects did five successive sets with inter-set rest periods of 1, 2, and 3 minutes on separate occasions with loads corresponding to 50% (A) and 80% (B) of the 1 RM. Based on the data of Willardson and Burkett (378). See text for further discussion.

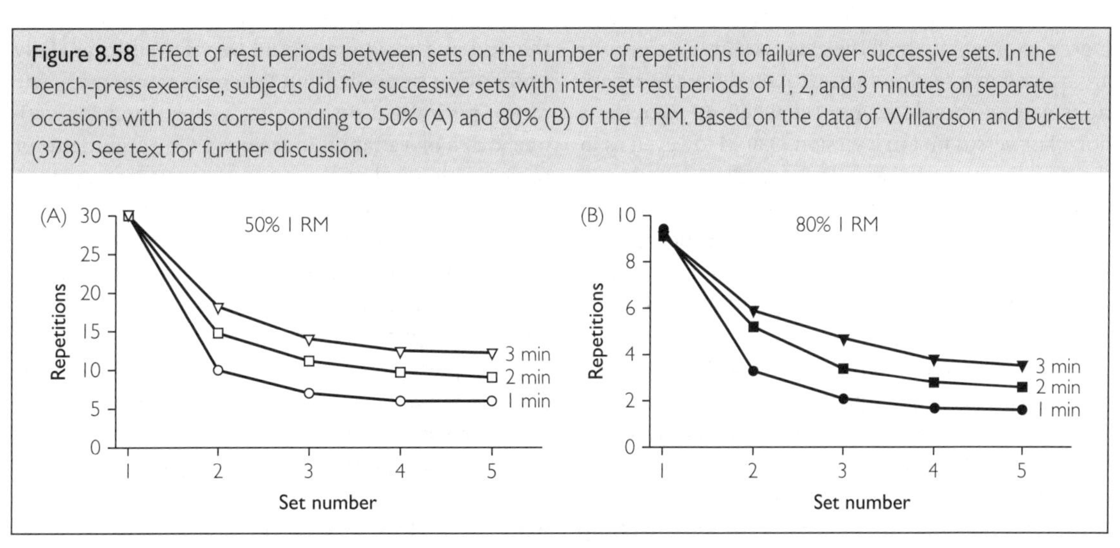

mixed at best (64,370). Longer rest periods, such as 5 vs 2 minutes (8), 2.5 vs 1 minutes (55), approximately 3.5 vs approximately 1.75 minutes (113), and 4 vs 2 minutes (379), produce similar increases in strength and muscle size. Two studies comparing constant rest periods of 2 minutes over 8 weeks of training to rest periods that decreased progressively from 2 minutes to 30 seconds over the 8 weeks found similar increases in strength and muscle size (86,334). Thus, a wide range of rest period durations can be effective. It had been hypothesized that longer rest periods would be more effective because of the greater total volume (load × repetitions) achieved for a given number of sets (85,377), as illustrated in Figure 8.58. The evidence suggests that, rather than volume, it is the number of sets taken to failure, regardless of the number of repetitions completed, that is the key training stimulus. This suggestion agrees with our earlier discussion about the stimulus for hypertrophy; inducing fatigue by taking sets to failure, shortening rest periods between sets, and vascular occlusion all encourage full recruitment of MUs.

An advanced trainee accustomed to long rest periods may initially find it difficult to adjust to shorter rest periods but training adaptations will quickly facilitate the adjustment. For example, bodybuilders who are accustomed to relatively short rest periods exhibit less fatigue and faster recovery with short rest periods than powerlifters who are accustomed to longer rest periods (181).

Order of Exercises

Some athletes need to train several muscle groups. Does the order in which the exercises/muscle groups are trained in a session matter? Acutely, fewer repetitions can be completed per set when an exercise is done last versus first in a session (330). However, as in the discussion of rest periods above, there is little evidence that ability to complete more repetitions per set translates into greater training-induced increases in strength and muscle mass (330). Nevertheless, it seems reasonable to do the most sport-specific exercises first in a training session.

It has been recommended (17) that large-muscle-group exercises (e.g., squat, leg press) be done before small-muscle-group exercises (e.g., biceps curls) on the basis that the large acute hormonal response produced by large muscle exercises promotes hypertrophy in subsequently trained small-muscle-group exercise (304). However, a recent test of this hypothesis showed that the prior large-muscle-group exercise did not increase hormone delivery to a subsequently exercised muscle group (372). Thus, the mechanism(s) by which a prior large-muscle-group exercise might enhance adaptations in a subsequently trained small muscle group are not known.

In summary, for those athletes whose training programs involve only a few sport-specific exercises, exercise order is likely not an issue. For athletes who use many exercises there may be merit in doing what are deemed the most important exercises first in a training session. Advocates of split routines, discussed in the section Principles of Training for Strength, Power, and Speed, might argue that a solution to the exercise order problem is to do some of the exercises one day and the others another day.

Periodization

Some athletes and coaches will decide on a particular program, in terms of exercises, intensity (e.g., 8–20 RM), frequency, and volume (number of sets), and stay with it indefinitely. A program may be selected primarily on the basis of the various aspects of specificity discussed earlier. However, there has also been interest in various forms of **variation** in training: variation in selected exercises, intensity, and volume over a period of weeks or months, or even within a week. Training variation has a long history. In the 1950s and 1960s, for example, a "heavy, light, medium" program was popular (270). Over three sessions during a week, the first, second, and third sessions might use loads of 1–3 RM, 10–12 RM, and 6–8 RM, respectively. Today, this training variation would be referred to as **daily undulating periodization** (see next paragraph). The term "**periodization**" is now widely used for various forms of variation in training. One argument for periodization is that it prevents the staleness and plateaus in training adaptations that might occur without variation (180). It has also been suggested that different phases or mesocycles of periodization achieve different adaptations; for example, a phase of lower-intensity, higher-volume training would promote hypertrophy whereas a higher-intensity, lower-volume phase would promote increases in strength and power (180). However, specificity of intensity in relation to strength and hypertrophy is currently under debate.

A number of forms of periodization are practiced. Perhaps the two most popular forms are *linear* periodization, in which intensity is increased and volume decreased over a period of weeks or months, and *undulating* periodization, in which intensity and volume are varied from week to week or even within a week (180). The effectiveness of different forms of periodization has been investigated in a number of studies (e.g., 49,145,247,250,284,285,331), with no clear indication of one form being superior to all others.

The key issue, however, is whether periodized training, in any form, is clearly superior to non-periodized training. Two reviews of periodization research, in which many of the same studies were reviewed using different approaches, came to opposite conclusions (64,295). Research conducted since these reviews (2004) continues to produce mixed results (145,250). One problem in comparing periodized and non-periodized training is controlling for factors other than variation itself. For example, if non-periodized training consisting of five sets of 8–10 RM over 6 months is compared to periodized training involving phases of five sets of 12–15 RM, 8–10 RM, and 3–5 RM over several weeks or months, a between-group study would require four training groups: 3–5 RM only (non-periodized), 8–10 RM only (non-periodized), 12–15 RM only (non-periodized), and periodized training (phases of 12–15, 8–10, and 3–5 RM). Such a design would distinguish between the value of variation itself or a particular phase of the training. Remarkably, few if any studies of this type have been conducted.

Given that both periodized and non-periodized training produce good results with no clear superiority of one over the other, or for one form of periodized training over the others, the key criteria for selecting non-periodized training or some form of periodized training should be specificity and choice based on experience. For example, rowers, paddlers, and kayakers would probably do best with loads of approximately 6–12 RM used indefinitely, applied to sport-specific exercises in a non-periodized training program. These athletes would have little need of 1–3 RM phases as part of a periodized training program. On the other hand, athletes whose performance consists of single maximal efforts against high resistance (e.g., American football linemen) would do well with a non-periodized training program with high intensity (e.g., 3–5 RM) or a periodized training program with a high-intensity phase, again emphasizing sport-specific exercises. Regarding the second criterion, personal choice, some athletes may indeed find the same program monotonous and would welcome the different phases inherent in periodized training. On the other hand, other athletes may find switching from one phase to another in a complicated periodized training program distracting and even annoying. The evidence suggests that athletes can do well with either choice.

Detraining, Maintenance Training, and Tapering

Detraining

Some athletes may elect to discontinue strength and power training for periods of time to focus on other forms of training (e.g., endurance) during the preseason or competition periods. These periods of detraining may last several days, weeks, or even months. Although some muscle-fiber atrophy can occur fairly quickly (255,307) and as soon as 10 days into detraining (160), strength training can be suspended for a few weeks with relatively little loss of strength and muscle mass; furthermore, any loss can be rapidly regained. An example of loss and rapid regain is shown in Figure 8.59. This figure also shows that the rapid regain can restore strength and muscle size to the level attained without any layoff from training. One reason behind the rapid regain of muscle mass is that the increased number of myonuclei that occur with hypertrophy are retained during a period of subsequent atrophy, allowing hypertrophy to develop more rapidly the second time (47). This rapid regain of muscle mass has been referred to by some as "muscle memory."

Longer periods of detraining will inevitably cause more significant losses of strength and especially muscle mass (254) (Figures 8.60 and 8.61). Figure 8.60 illustrates the effects of detraining over a period of 3 months. As shown, strength is typically relatively well preserved for some time. The preserved strength is largely attributable to sustained skill and muscle activation in performing the strength tests (e.g., activation in Figure 8.60). In contrast, loss of muscle mass (e.g., CSA in Figures 8.59 and 8.60) is relatively more rapid.

Well-preserved strength during detraining (254) may seem to justify discontinuing strength training for periods of time; however, sport-specific strength may decline even if a non-specific strength test shows no loss. For example, after 4 weeks of detraining, forces produced during swimming decreased despite maintained strength as measured on a swim bench (263). Similarly, eccentric strength may decline more rapidly than concentric strength during detraining (151).

Figure 8.59 Comparison of training continuously (group Ctr) for 15 weeks versus stopping training between weeks 6 and 9 and then retraining from weeks 9 to 15 (group Rtr). Figure starts after 6 weeks of training. Note that despite a 3 week lay-off, after 6 weeks of retraining, group Rtr achieved results similar to Ctr in triceps and pectoralis major (averaged) muscle CSA (top), bench-press 1 RM (middle), and isometric elbow extension strength (MVC). Training was three sets of approximately 10 RM of the bench-press exercise done 3 days per week. Based on the data of Ogasawara et *al.* (269).

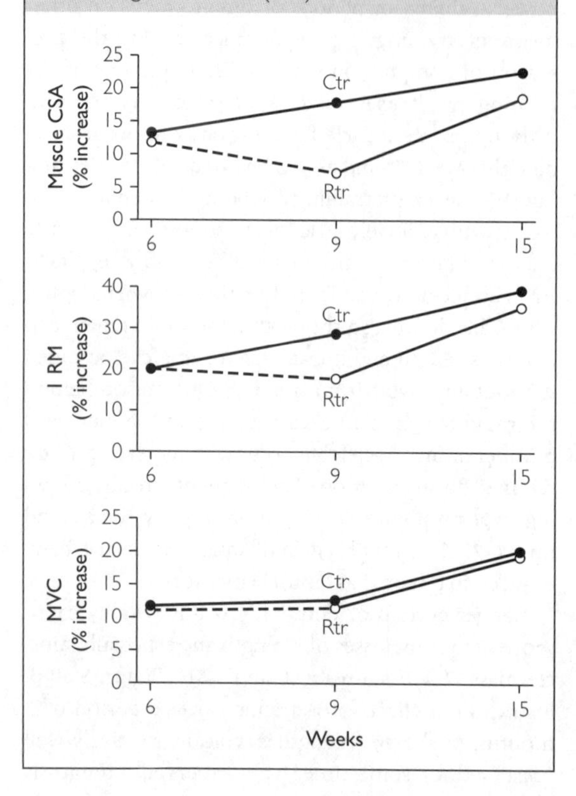

Figure 8.60 Effects of 3 months of isometric strength training (Post) and 3 months of detraining (D1, D2, D3) on maximal isometric strength, neural activation, and muscle CSA. Note during detraining the rapid loss of strength and CSA compared to activation. Based on the data of Kubo et *al.* (185).

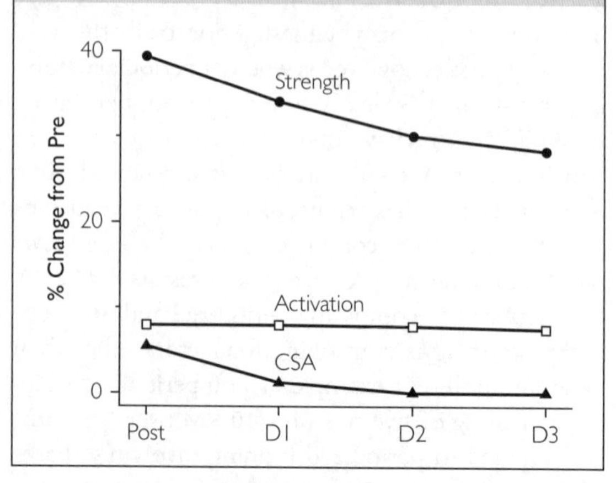

There is also some evidence that training-induced increases in power are lost more rapidly than strength (110,157). Finally, the notable losses in muscle mass seen with detraining (e.g., Figures 8.60 and 8.61) may impair sport performance, even if some measures of strength are preserved. If the primary reason for discontinuing strength training is to save time for greater emphasis on other types of training, a better approach would be to reduce rather than completely stop the strength training, as discussed in the next section.

It was noted earlier in this chapter (Figures 8.7 and 8.8) that strength and power training decreases the percentage of type 2X muscle fibers, the fibers with the fastest maximum *unloaded* shortening velocity (V_{max}). In contrast, with detraining the percentage of 2X fibers recovers to and may exceed the pre-training level (2X overshoot; Figure 8.8). Can it therefore be concluded that detraining is a viable strategy to enhance speed performance by promoting a 2X overshoot? A 2X overshoot with detraining has been linked to an increase in V_{max}. This might be relevant to sport actions with minimal loads such as punching and kicking. However, most sport actions are done against large loads (e.g., body mass in jumping and sprinting). With these loads any advantage of a 2X overshoot would be nullified by type 2 fiber atrophy that would impair the ability to accelerate loads to achieve high movement velocities. Detraining might also adversely affect neural adaptations associated with explosive actions. Unfortunately, even a greatly reduced training volume while maintaining intensity prevents the restoration of the percentage of 2X to pre-training levels (Figure 8.7). But, as discussed under Maintenance Training, reduced training volume has other merits.

Maintenance Training

Is it possible to greatly reduce the time spent strength training and still maintain the gains made during a period of more time-consuming training? A large reduction in training can prevent the losses seen with complete detraining (e.g., 110). The most effective

Figure 8.61 Comparison of complete detraining (DT) vs reduced training over a 32 week period. Before detraining or reduced training, three groups of young men trained the leg muscles 3 days per week for 16 weeks, doing three sets of approximately 8–12 RM in knee-extension, squat, and leg-press exercises in each session. Figure begins with increases after 16 weeks of training (Post). One group did no training (DT) for the following 32 weeks. A second group (1/3) trained as described above but trained only 1 day per week. A third group (1/9) also trained 1 day per week but did only one set of each exercise. (Left) Thigh lean mass (TLM) decreased to slightly below pre-training values (0) with detraining but was maintained with both forms of reduced training. (Right) Knee-extension 1 RM not only was well preserved but tended to increase with reduced training. Based on the data of Bickel *et al.* (39).

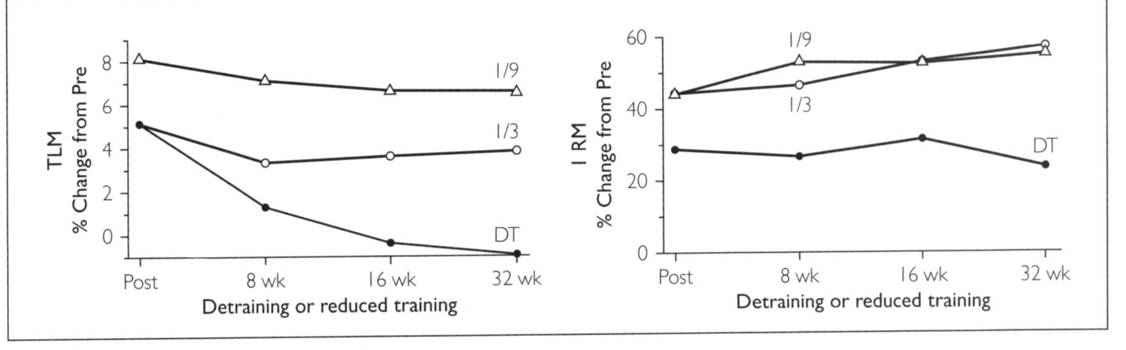

approach is to maintain intensity (e.g., 8–10 RM) but decrease the number of exercises or sets per training session and the number of training sessions per week (e.g., 299). In other words, intensity is maintained but volume is decreased (253,256). Figure 8.61 shows how a greatly reduced volume, down to approximately 10–30% of the volume before maintenance training began, can sustain the gains made from a prior period of higher volume training. In fact, depending on the training history of the individual, there may even be continued progress with reduced training volume (e.g., 39).

Tapering

As will be discussed in more detail in Chapter 9, tapering is a specific form of reduced training done for a period before competition and can be applied to all types of training: strength and power/speed, and aerobic and anaerobic power. Thus, tapering is used by most athletes, including the extremes of shot putters and marathon runners. Like the maintenance training discussed previously, tapering consists of a marked reduction in training volume while maintaining or even increasing intensity (253). The taper may continue for up to 3 weeks or more; for example, highly trained swimmers may taper for 21 days (362). Tapering is designed to allow recovery from accumulated fatigue from a previous period of high-volume, intense training and to promote positive training adaptations leading up to competition (253). That a taper reduces accumulated fatigue is not in dispute. On the other hand, there is some debate over the extent to which a taper promotes further adaptations as opposed to simply maintaining adaptations (256,257,354). An example of tapering producing further training adaptations is shown in Figure 8.62. There is some evidence that a moderate increase in training volume at the end of the taper may promote additional adaptations without risking fatigue (356). A taper may be more effective when it is preceded by a period of intense, high-volume training that borders on over-reaching or overtraining (354,355). Presumably, the period of high-volume training maximizes training adaptations, whereas the taper dissipates the accumulated fatigue. A taper that lasts too long may pass the borderline between fatigue reduction and positive adaptations into detraining (253,256); thus, the art of the taper is finding the right combination of taper intensity, volume, and duration to maximize performance.

Interaction between Strength and Endurance Training

Although performance in many sports requires high levels of both strength and endurance, some, such as weight lifting, jumping, or throwing, require minimal levels of endurance, while others, such as road

Figure 8.62 Effects of a 21 day swim taper in highly trained swimmers. During the taper, volume (total swim distance per day) decreased approximately 70% but intensity was maintained or increased. Percentage increases pre- to post-taper include swim power and deltoid type 2A fiber CSA, maximum isometric force (ISO_{max}), maximum shortening velocity (V_{max}), and power. These data indicate that positive training adaptations occurred during the taper. Based on the data of Trappe *et al.* (362).

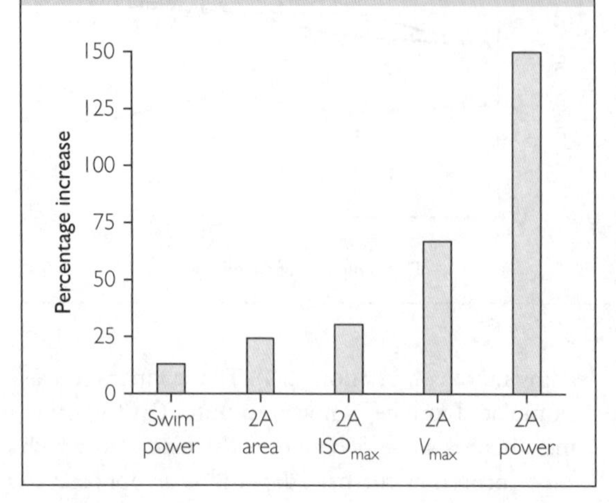

cycling or distance running, require minimal levels of strength. Three questions arise about the training programs for such athletes. (1) Can endurance training improve the performance of strength/power and speed athletes? (2) Can strength training benefit endurance athletes? (3) Does concurrent strength and endurance training interact in a positive or negative way because they induce very different training adaptations? These questions are addressed below.

Endurance Training for Strength/Power/Speed Athletes

The addition of endurance training to the training programs of athletes like 100 m sprinters and high jumpers appears to offer little or no benefit and may even have negative consequences. Although strength training by itself causes a transition from the fastest type 2X muscle fibers to slower 2A fibers, endurance training tends to accelerate this transition (19). High-volume endurance training also delays recovery from strength/power training sessions, and has been shown to diminish the hypertrophic response to strength training (19). Although some types of endurance training may accentuate (212) or at least not limit (86) the increase in muscle mass produced by strength training, the general recommendation would be for strength/power/speed athletes to avoid high-volume, high-frequency endurance training.

Strength Training for Endurance Athletes

Endurance athletes competing in events lasting from less than 15 to more than 30 minutes can benefit from the addition of strength training (2). That is, it can improve endurance performance beyond that achieved by endurance training alone (2). For example, appropriately designed strength training programs improve performance in middle- and long-distance runners, cross-country skiers, cyclists, rowers, kayakers, and canoeists (2,305,344). Figure 8.63 shows how the addition of strength training increased endurance performance in trained cyclists. On the other hand, swimmers do not benefit as much from dry-land strength training (309,344), perhaps because of the difficulty in specifically matching exercise movements to in-water swimming actions. Perhaps surprisingly, training with high loads (e.g., Figure 8.63) and lifting high loads with explosive actions typical for strength and power athletes have been found more beneficial to endurance athletes than conventional (e.g., 8–12 RM done at slow velocity) training (2,305).

Strength training can enhance endurance performance in several ways (2,109,305). Although unlikely to increase an endurance athlete's maximal aerobic power ($\dot{V}O_{2max}$), strength training can increase the speed or power attained at $\dot{V}O_{2max}$. It also induces a transition from more fatigable 2X fibers into more fatigue-resistant 2A fibers. One effect of increased strength is that a given *absolute* exercise intensity (e.g., running or skiing speed) taxes a smaller percentage of maximum strength; as a result, lower threshold, more fatigue-resistant MUs (1 and 2A) can meet the exercise demand longer and the activation of higher-threshold, more fatigable MUs can be delayed. Put another way, increased strength results in a given absolute intensity becoming a smaller *relative* intensity that is therefore easier to sustain for long periods. Since blood-flow restriction (BFR) in exercise is related to relative rather than absolute intensity, BFR related to intramuscular tension would be less at a given absolute intensity as strength increases. Strength training increases the economy or efficiency of exercising at a given intensity, thus delaying fatigue. Increased economy may result from greater use of more efficient lower threshold MUs, altered muscle activation patterns

Figure 8.63 Effect of adding 16 weeks of strength training to the endurance training program of top-level cyclists. Two groups of cyclists did their usual 14–18 hours per week of endurance training. One group (SE) added strength training while the other group (E) did not. Strength training was done 2–3 days per week. Exercises included knee extension, leg press, knee flexion, and calf raises, each done for four sets. Loads increased from 10–12 RM to 5–6 RM over the 16 week training period. In doing repetitions of exercises trainees were instructed to lift the weight in a controlled, non-explosive manner. Isometric knee-extension strength increased only in the SE group. The maximum average power during a 5 minute cycling trial (endurance test) increased in both SE and E with no significant difference between groups. In contrast, average power in a 45 minute trial increased only in the SE group. Not included in the figure were the results of no increase in muscle-fiber area in either group, although only the SE group increased lean body mass; no change in capillary density in either group; and in the SE group only, a significant increase and decrease in the percentage of muscle occupied by type 2A and 2X muscle fibers, respectively. * Statistically significant increase pre- to post-training. Based on the data of Aagaard *et al.* (4).

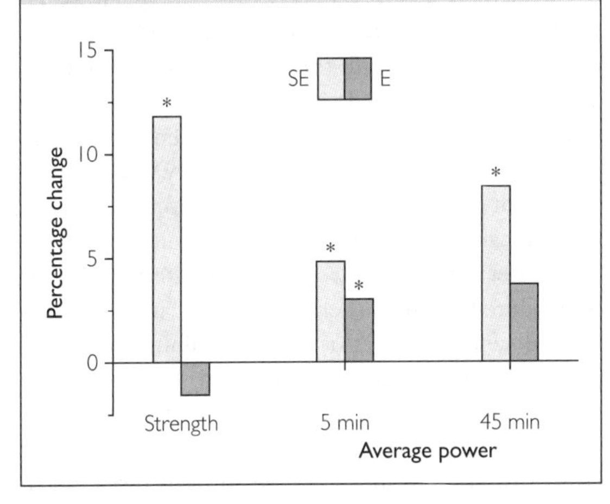

that improve coordination, and tendon adaptations (e.g., 301) such as stiffness changes that would for example enhance stretch-shortening cycle potentiation. One sign of increased efficiency is a reduction in the amount of muscle blood flow needed when exercising at a given intensity (33).

A common adaptation to strength training is muscle hypertrophy leading to an increase in fat-free body mass. Hypertrophy may contribute to increased endurance by increasing strength as discussed above. On the other hand, an increase in body weight, if the increase in muscle mass is not compensated for by a decrease in fat mass (e.g., 300), would be detrimental to endurance events in which body weight must be supported (e.g., running and cross-country skiing) (2,305). Another disadvantage to increasing muscle and muscle-fiber CSA is that the diffusion distance from capillaries to the core of muscle fibers increases. This could compromise, for example, the exchange of oxygen and carbon dioxide between blood and muscle tissue. Therefore it is not surprising that many endurance athletes are reluctant to add strength training to their endurance training (2). However, it has been repeatedly shown that when endurance athletes add strength training to their high-volume endurance training programs, strength usually increases mainly as a result of neural adaptations, with relatively little (Figure 8.64) or no muscle hypertrophy (2,305). There are two possible explanations for the small hypertrophic response. One is that endurance athletes, particularly those who compete in the longest duration events, tend to be poor gainers in response to strength training. A second explanation is that when endurance training and strength training are done concurrently, the endurance training may interfere with the hypertrophy response to the strength training. This latter possibility is discussed in the next section.

Interference Between Concurrent Strength and Endurance Training

Concurrent strength and endurance training does not appear to interfere with endurance-related adaptations and probably improves endurance performance beyond that seen with endurance training alone (e.g., Figure 8.63). In contrast, concurrent strength and endurance training can under certain conditions impair the development of strength and power (109,382). The probability of concurrent strength and endurance training interfering with strength and power adaptations is influenced by the type, volume, and frequency of endurance training. Moderate-volume interval or continuous endurance training done 2 days per week may not cause interference (87,212,245). In fact, there have been reports of added low-volume continuous and low-frequency interval endurance training augmenting the hypertrophy response to strength training (212,245). But many athletes who must develop high levels of both strength and endurance (e.g., rowers, kayakers) train with higher volumes and frequencies of strength training

Figure 8.64 Effect of concurrent strength and endurance training on strength, power, and muscle mass. A group of top-level cyclists (SE) added, for the first time, a 12 week resistance training program to their regular endurance training (9.9 hours per week). Resistance training was done 2 days per week and consisted of half squat, one-leg leg press, one-leg hip flexion, and calf raises. Four sets of each exercise were done in each session. The relative training load increased from 6–10 RM to 4–6 RM over the 12 weeks. In doing repetitions of exercises trainees were instructed to lift the weight (concentric phase) as quickly as possible (≈1 second) and lower the weight (eccentric phase) more slowly (≈2–3 seconds). A second group (S) of previously untrained subjects did the same resistance training but no other training. Measures shown include the absolute training load (weights) used over the 12 weeks, averaged 1 RM for squat and leg press, isometric squat rate of force development (RFD), height attained in a squat jump (SJ), and thigh muscle CSA. * Statistically significant increase pre- to post-training. # Greater increase in S than SE. Based on the data of Ronnestad *et al.* (302).

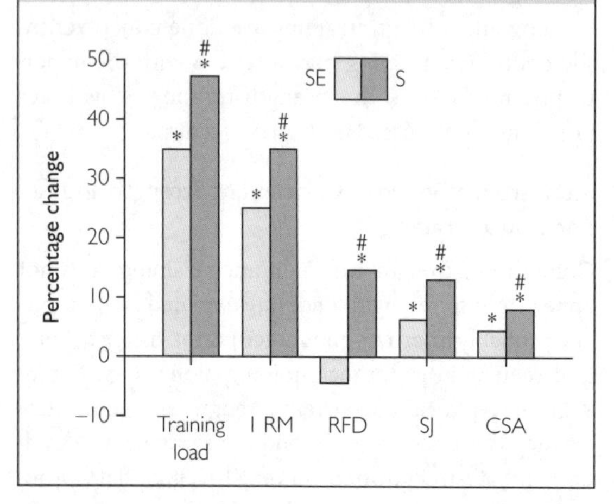

and especially endurance training than just described. For these athletes the potential for interference increases (109). An example of the deleterious effect of high-volume endurance training on strength and related adaptations is shown in Figure 8.64. Top-level cyclists cannot develop the same level of strength, power, and muscle size with strength and endurance training as a group of non-athletes who do strength training only.

There are several strategies to minimize the interference caused by concurrent high-volume strength and endurance training. One was suggested in a previous discussion on maintenance training (see Figure 8.61). At the start of a high-volume phase of endurance training, strength training volume and frequency can be reduced while maintaining intensity, allowing for better recovery from and adaptations to the increased endurance training. Importantly, this maintenance strength training holds the adaptations attained with the previous phase of high-volume strength training and has been successful in the training of cyclists (299). Related to the maintenance strategy is phase or block periodization in which over time different adaptations (aerobic power, anaerobic power, strength) are the prime focus in relatively short phases (e.g., 5 weeks) (109). The off-phase for a particular attribute would be a maintenance phase; as shown previously, taking a few weeks off from one type of training may cause little loss, and any such loss can be made up quickly (Figure 8.60). Another potential strategy involves sequencing the strength and endurance training on different days or, if done on the same day, separating the strength and endurance sessions by several hours to promote recovery (109). For many athletes who need only strength training to train a limited number of muscle groups, 2–3 days per week should be sufficient, leaving 5 days per week for rest and other types of training. However, other athletes such as hockey and basketball players may need four strength-training days per week (e.g., 2 days upper body, 2 days lower body). This poses a greater challenge and care must be taken to avoid cumulative fatigue and insufficient recovery. One strategy to prevent plateaus in strength adaptations is to maintain intensity but avoid taking sets to failure, allowing for greater recovery between sessions (109). Instead of non-failure training, others may prefer failure training but reduce the number of sets and training frequency. Regardless of the strategies adopted, they will probably have to be adjusted for individual athletes (109).

Besides insufficient recovery from concurrent strength and endurance training, are there other mechanisms that may be responsible for the interference effect? Strength and endurance training induce different adaptations at the muscle level. Strength training stimulates myofibrillar protein synthesis, leading to increased strength and power, whereas endurance training stimulates mitochondrial protein synthesis, which increases aerobic capacity (74). As discussed in Chapter 9, it may be difficult for a muscle to undergo both types of protein synthesis simultaneously. In fact, an acute bout of resistance exercise stimulates both myofibrillar and mitochondrial protein synthesis in an untrained subject (368,376) but after a period of resistance training, the acute response to a bout of resistance exercise is more specific, with only myofibrillar protein synthesis being stimulated (376). In contrast, in both the untrained and trained states, a bout of endurance exercise stimulates mitochondrial protein synthesis exclusively (376). Thus, the protein synthetic response to strength and endurance training becomes more specific and distinct as training progresses, perhaps increasing the risk of interference.

Myofibrillar and mitochondrial protein synthesis are driven by different molecular signaling pathways (137), and activating one pathway may inhibit the other (28,29,260). For example, aerobic exercise may inhibit the pathways leading to satellite cell activation (30) and myofibrillar protein synthesis (28,29,260). It is not clear why the interference effect is greater in impairing strength than endurance development. Perhaps the myofibrillar protein synthesis pathway is more sensitive to competition from the mitochondrial synthesis pathway than vice versa. Also, there have been reports that the expected specific protein synthesis responses to strength and endurance training can occur with only partial activation of the expected associated specific molecular signaling pathways (87,376). Future research will likely clarify these discrepancies.

Genetic Influence on Training Response

Many coaches and their athletes have noticed that the response to a particular training program is unpredictable; that is, there is a large inter-individual variation in progress despite equivalent efforts to improve. The large variation has been well documented (32,81,153,374) and one example is given in Figure 8.65. At one extreme, poor or non-responders, who may represent up to a quarter of trainees (277), may gain little or no strength and muscle mass, whereas extreme positive responders may make gains that are three or four times greater than average (153). Although there may be genetic influences on neural adaptations to strength training, research has focused on factors affecting the hypertrophy response (32). For example, easy gainers have muscles that increase myonuclei and satellite cell numbers more

Figure 8.65 Illustration of the large inter-individual variation in the response to the same strength training program. Subjects were 585 young adults (342 women, 243 men) who trained the elbow flexors of the non-dominant arm for 12 weeks. Exercises included preacher curl, concentration curl, and standing biceps curl. In each training session each exercise was done for three sets of 12 RM, 8 RM, and 6 RM in weeks 1–4, 5–9, and 10–12, respectively. Training frequency was not described but was presumably 2–3 days per week. The Figure shows the average change with range after training in preacher curl 1 RM, elbow flexion maximum isometric strength (ISO_{max}), and biceps CSA. Note that 1 RM and ISO_{max} are referenced to the left vertical axis whereas CSA is referenced to the right one. Based on the data of Hubal *et al.* (153).

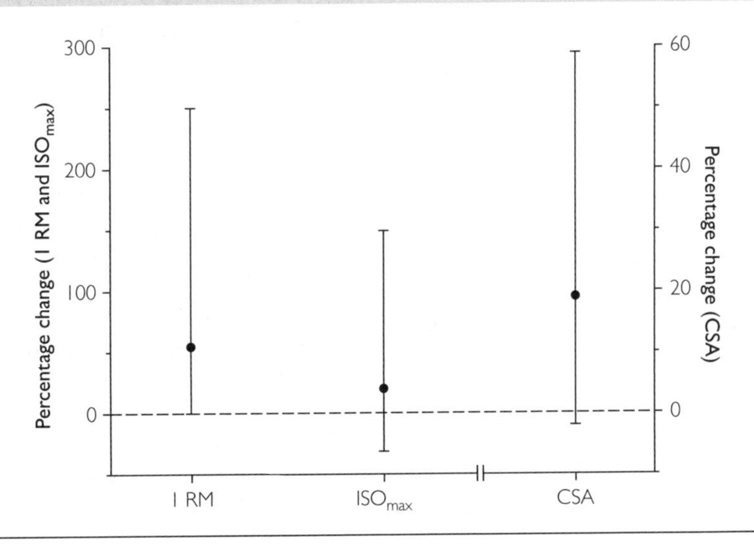

than poor gainers (277). In addition, some molecular signaling factors and processes promoting muscle protein synthesis are more pronounced in extreme positive than in poor responders or non-responders (32,81,237,286,292,348). More research will be needed to determine all of the factors influencing trainability but there is no doubt that the genetic influence is strong. It may seem unfair, but in terms of trainability it may be a case of the rich getting richer and the poor staying poor. For example, untrained people already possessing above-average strength and muscle mass tend also to gain, both absolutely and relatively, more strength and muscle mass after training (366).

An interesting observation is that consuming the same energy and protein (≈1 g/kg body mass per day) and doing the same training program produces the same large variation in hypertrophy response (353). This underscores the importance of genetics compared to training and nutrition on the training response. It also suggests that force-feeding protein may not solve the problem of inherently poor responders.

In the future it may be possible to tailor training programs, in terms of intensity, volume, and frequency, to an athlete's genetic profile. At present, a trial-and-error approach to program alterations is recommended to determine the most productive program for an individual athlete.

Throughout this chapter, reference has been made to contradictory findings on particular training issues (e.g., failure vs non-failure training). Many studies use a *between-group* design with one group training one way and a second group another. Usually, subjects are assigned to groups randomly, sometimes after subjects are initially matched on some measures (e.g., pre-training strength level). Despite these procedures, a particular group may have more than its share of either poor or extreme positive responders. These outliers may swing the group average training response one way or the other, especially with small groups. The outliers may have a greater influence than the particular training program. In the future it may be possible to match subjects genetically before group assignment to minimize the outlier effect. In the meantime, an alternative is to use larger groups to minimize the outlier effect, but this is costly. A second alternative is to use a *within-subject* design (e.g., 248). One version of this design assigns one arm or leg to one type of training and the other limb to another type. This design, called the unilateral training model, eliminates the genetic influence. A poor responder will respond poorly in both limbs, and a positive responder will do well in both. A potential problem with this design is the cross-training effect discussed earlier in this chapter. However, the cross-training effect for strength gains is small and minimal for muscle adaptations. Therefore, a clearly superior training method will be revealed with the unilateral training model.

Sample Strength Training Programs

Soccer

Soccer demands strength and power for jumping, kicking, short-distance sprints, and repeated changes of direction; anaerobic power to support longer and repeated sprints; and aerobic power for sustained running (142). On-field soccer practice and various sprint and change-of-direction or agility drills can improve soccer performance (48,163,327,361), but off-field strength training, including plyometrics, can produce further improvements in jumping and sprinting ability and also increase endurance (142).

Pre-Season Training

Although strength training with many exercises has been used for soccer (241,258,385), one exercise, the half squat, has proven very successful for increasing sprint speed and jump height in soccer players (44,69,142,139,173,303,315). Furthermore, a one-exercise strength training program can be more easily integrated with other types of soccer training. The squat exercise should be done 2 days per week, for example on Mondays and Thursdays. Sometimes the load (RM) is increased throughout the weeks of training; for example, increasing from approximately 10 RM to approximately 3 RM over a period of 8–12 weeks (69,173,304,315). Equally good results can be achieved with high loads (≈3–4 RM) from the beginning of training (142,139,303,385). An example of a pre-season program would be four or five sets of 3–5 RM in the back squat with approximately 3 minute rest periods between sets, done 2 days per week. Good results with this exercise have been obtained when the athlete is instructed to lift the weight as quickly as possible (≈1 second) in the concentric phase and then lower it slowly (≈2–3 seconds) in the eccentric phase of each repetition (142). High loads with few repetitions per set are often chosen in the belief that this type of training increases strength exclusively through neural adaptations, minimizing the negative effects of increasing body mass caused by bodybuilding training (e.g., sets of 8–12 RM) (142). High-load, low-volume training (e.g., 3–4 RM) has increased strength over a period of weeks without hypertrophy (142) when combined with concurrent endurance training and regular soccer training but, as discussed earlier in this chapter, it is not clear whether the lack of hypertrophy is due to the chosen loads, the responder status of soccer players, or interference from concurrent endurance and other types of training.

Plyometrics can also be used to increase jump height and sprint speed (70,303,363) (Figure 8.66). Sample plyometric exercises would be drop jumps from various heights (e.g., 40 cm) and hurdle jumps with hurdles set at various heights (40–90 cm) (70,363). A sample program would have the athlete do five sets of six drop jumps and five sets of six hurdle jumps in each session, with two sessions done each week. The athlete should be instructed to attempt maximum height with each jump (70).

Figure 8.66 Effects of plyometric training on jump and sprint ability in soccer players. Before regular soccer training sessions, a plyometric training group (Plyo) did hurdle jumps and drop jumps 2 days per week for 8 weeks. A control group did only the soccer training. During the first 4 weeks athletes did five to 10 sets of hurdle jumps with the hurdles set to 40 or 60 cm. In the final 4 weeks each session consisted of four sets of 10 drop jumps from a height of 40 cm. After 8 weeks of Plyo, squat jump (SJ) and countermovement jump (CMJ) height, and 40 m sprint velocity, increased. * Statistically significant increase. Based on the data of Chelly *et al.* (70).

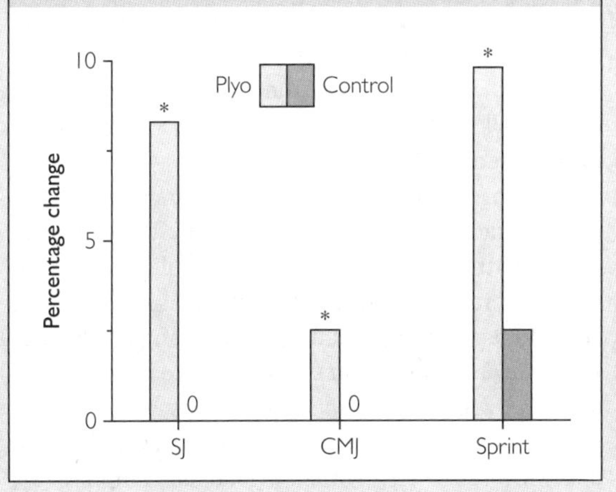

In the pre-season it may be difficult for the athlete to tolerate all-out training for aerobic and anaerobic endurance as well as strength and power. A solution is to emphasize just two types or even one type of training at a time (109). For example, when plyometrics are to be emphasized for a phase and strength training de-emphasized, intensity and volume of each strength session could be maintained but frequency reduced from 2 to 1 day per week (304). The same approach would be used when plyometrics is to be de-emphasized.

In-Season Training

The demands of in-season games and related stress such as travel may require many forms of training to be cut back to a maintenance level. In strength training and plyometric exercises, decreasing frequency (from 2 to 1 day per week) while maintaining the intensity and volume of each session will maintain adaptations achieved during the pre-season (304).

Cross-Country Skiing

Cross-country skiers race in sprint (1.5 km) and distance (30–50 km) events. Sprint rather than distance racers might seem to gain more from strength and power training; however, as discussed, strength training can increase short- and long-term endurance in various sports, including cross-country skiing (142). Unlike runners and cyclists, skiers must develop strength and endurance in both upper- and lower-body muscle groups; in fact, upper-body strength training has received the most attention in research. In addition to the strength training described here, skiers must also train to increase aerobic and anaerobic power (Chapters 6 and 7).

Off-Season/Pre-Season

A pulley weight apparatus that allows the skier to simulate single or double poling is the most specific and time-efficient method for training the upper body. This one exercise, done 2–3 days per week for three to five sets of approximately 6–8 RM, has proven quite effective (143,144,211,271), particularly when the concentric phase of each repetition is done as quickly as possible despite the heavy load (144,271). The resultant increase in both rate of force development and strength

Figure 8.67 Effects of strength training in highly trained cross-country skiers. The strength training program consisted of one exercise, a pulley weight exercise that simulated the double-poling action in skiing. The exercise was done for three sets of 6 RM three days per week for 9 weeks. Each strength training session lasted approximately 15 minutes. The strength training program was combined with regular ski training. A second control group of skiers did only the ski training. The strength training program increased strength (1 RM in the pulley weight exercise). On a ski ergometer, economy, endurance time at maximal aerobic power, and the peak power determined from a load-velocity test all increased significantly (*). Note that maximal aerobic power ($\dot{V}O_{2max}$) and anaerobic threshold did not increase significantly. None of the changes in the control group was significant (not shown). Based on the data of Osteras *et al.* (271).

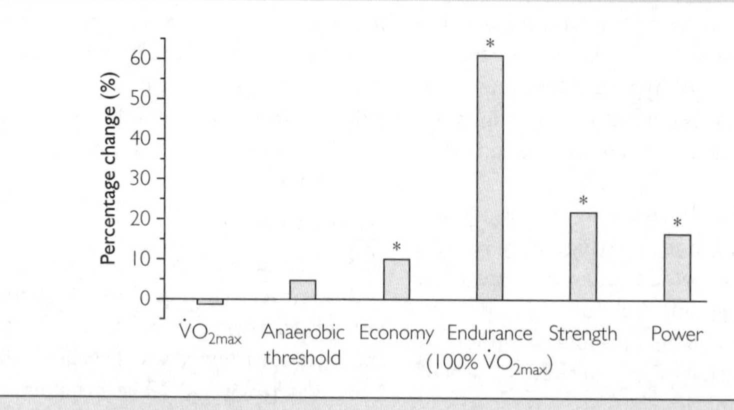

contributes to increased skiing efficiency and endurance (143,144,271). Figure 8.67 illustrates the positive effects of the specific pulley weight exercise. Prescribing skiing-specific lower-body exercises is more challenging, so standard exercises such as squats are done as described above for upper-body training (211,244). Like soccer players, skiers must train concurrently for strength and endurance, and the same techniques for minimizing potential interference with concurrent training can be used if necessary; for example, reducing strength training frequency when emphasis is to be put on training for aerobic and anaerobic power.

In Season

During the competitive season the risk of cumulative fatigue increases. More than one type of training may have to be reduced simultaneously. For strength training, five sets of each exercise done twice weekly could be reduced to three sets done once weekly. Intensity would be maintained.

Bodybuilding Training for Athletes

In the sample programs for soccer and cross-country skiing and in other sports discussed (e.g., cycling, running), it is considered beneficial to increase strength with minimal increases in muscle and body mass that might adversely affect efficiency. These athletes would be reluctant to adopt a bodybuilding style of strength training designed to maximize increases in muscle mass. To increase muscle mass, most bodybuilders will do four or five exercises per muscle group, with three to six sets of each exercise, loads of 7–12 RM, and 1–2 minutes rest between sets (131). The total volume of such a program requires the use of split routines in which only one or two muscle groups are trained each training day. Typically, each muscle group will be trained once or twice each week but 4–6 training days are needed to cover all the muscle groups (131). Individual training sessions last about 40–90 minutes (131). The training practices described would be at the upper end, in terms of volume, of what some (18,370) but not all (64,99) recommend for increasing muscle mass, based on the available research.

For athletes who want to minimize gains in body mass while increasing strength mainly through neural adaptations, an apparently successful approach has been to use a limited number of sport-specific exercises done with high loads (3–6 RM) and a relatively small number of sets (142). In addition to possibly minimizing hypertrophy, especially in muscle groups not crucial to performance, there is some evidence that doing few repetitions with high loads and emphasizing maximum possible speed in the concentric phase of each repetition improves power and endurance more than the moderate loads (8–12 RM) typical of bodybuilding routines (44,138). For athletes who must do concurrent strength and endurance training, the interference effect (Figure 8.64) will help minimize unwanted gains in muscle and body mass.

There are many athletes, however, who seek increases in muscle and body mass and the resulting increase in absolute strength and power. Hockey, basketball, and American football players and weight throwers in track and field are examples. The increase in body inertia and momentum can aid the performance of these athletes. But there is a compromise in that increased body mass reduces the training-induced increase in the strength/body mass ratio that allows the athlete to jump high or accelerate quickly in running and skating. To what extent should these athletes include bodybuilding-type training in their overall program? For athletes needing a general increase in strength and muscle mass, loads (6–10 RM) and sets per exercise (three to five) might be similar to those used by bodybuilders. The difference would be the number and type of exercises. While bodybuilders focus on developing various muscle groups with many exercises, some of them isolated, single-joint movements, athletes should select bilateral, multi-joint exercises (e.g., squat, leg press, bench press, etc.) and, if possible, exercises that simulate the target sport movement patterns. Earlier in this chapter it was noted that relatively light loads (20–30 RM) can produce considerable hypertrophy; however, neural adaptations related to handling heavy loads may not be optimal with light-load training. Therefore, athletes seeking maximum increases in both size *and* strength might do best with loads in the upper range for so-called hypertrophy training (e.g., 6–8 RM) with the occasional use of even higher loads (3–4 RM).

Summary of Key Points

- Muscular adaptations to strength and power training include the following.
  - Muscle hypertrophy occurs due to increased muscle-fiber CSA, which results from an increased number and size of myofibrils, in turn resulting from increased number of actin and myosin filaments.
  - Fiber hypertrophy, more pronounced in type 2 than in type 1 fibers, is generally associated with decreased mitochondrial density and increased glycogen concentration. There is an absolute but not relative increase in sarcoplasmic reticulum and the transverse tubular network, leaving unchanged the proportion of a muscle fiber occupied by myofibrils.
  - There is an increased absolute but not relative amount of inter-fiber connective tissue.
  - Stimuli for hypertrophy may include a minimum level of contractile force or load, muscle damage, acute hormonal response to exercise, and acute muscle-fiber swelling.
  - The large range of training loads that effectively induce hypertrophy suggests that muscle fiber rather than whole-muscle force is the crucial hypertrophy stimulus.
  - The principal fiber-type transition is an increased percentage of type 2A muscle fibers at the expense of 2X fibers. Training decreases the percentage of hybrid fibers.
  - Conventional strength training tends to increase muscle-fiber pennation angle without affecting fiber length, whereas explosive and sprint training tend to increase fiber length without increasing or even decreasing pennation angle. Increased pennation angle reduces the resolved force in the line of pull of the tendon but this is offset by the increase in absolute fiber force. Increased fiber length increases absolute fiber shortening velocity. Training with specific ranges of joint motion may increase or decrease fiber length so that sarcomeres are operating at close to optimal length.
  - Measurements of whole-muscle and single-fiber evoked contractile properties indicate that isometric or conventional weight training increases maximum absolute isometric force and rate of force development with little change in specific force or maximum shortening velocity (V_{max}). Explosive training may increase both fiber force and V_{max}.
  - Tendon adaptations include increased CSA, which prevents an excessive increase in tendon

strain as muscles become stronger, and changes in tendon stiffness that either allow force to be developed more rapidly or allow more efficient storage and reutilization of elastic energy in activities such as jumping and sprinting.

- Marked muscle-fiber hypertrophy is associated with decreased capillary density or no change if high-volume strength training is done or strength training is combined with endurance training.
- Neural adaptations include increased agonist activation and altered antagonist and synergist activation to promote the greatest possible force in the intended direction of movement. These adaptations are produced by a combination of changes in central commands to motoneurons, altered reflex inputs to motoneurons and altered intrinsic excitability of motoneurons.
- The greater increase in strength than in muscle size suggests that neural adaptations predominate in the early phase of training, particularly when trainees are exposed to new exercises. At later stages when neural adaptations plateau, the final level of attained strength and power is dictated by the extent of muscular adaptations.

- The primary principles of training include the following.
  - Overload, in which the exercise task requires maximal or near-maximal effort to complete. Overload ensures that the high-threshold MUs comprising type 2 muscle fibers are fully activated in training. Overload can be achieved by doing a series of single maximal contractions or repeating submaximal force contractions to fatigue failure.
  - Progression, in which the training load is progressively increased to ensure that the highest-threshold MUs will always be activated in the training exercises.
  - Specificity, including specificity of movement pattern, actual or intended velocity, and load. Specificity promotes the maximum transfer of gains made in training to improved sport performance. In specificity of velocity, the intent to contract at maximum velocity may be as important as the actual training velocity, the latter dictated by the magnitude of the imposed load. In specificity of load for increasing strength there may be a wide range of loads effective for producing muscle hypertrophy. There may be a narrower range of optimal loads for increasing strength; high loads induce neural and tendon adaptations beneficial to increasing strength. Specificity of contraction type (concentric, isometric, eccentric) dictates that the contraction types involved in particular sport actions be emphasized in training. However, supplementary high-force eccentric training may induce greater hypertrophy that transfers to other contraction types.
  - Volume includes the number of exercises used in a training session and the number of sets done with each exercise. Although one set is effective, multiple sets are more effective for increasing strength and muscle mass. However, the added benefit of additional sets diminishes progressively.
  - Frequency refers to the number of days a week a muscle group/exercise is trained. The optimal frequency is 2–3 days per week per muscle group or exercise. In many sports the relatively small number of specific exercises will allow strength training to be restricted to two or three sessions per week. Athletes who need to train several muscles groups can use split routines in which different exercises are done on different days. Training frequency is dictated by the time course of adaptation processes such as muscle protein synthesis and repair of muscle damage, and recovery from fatigue.
- Additional training considerations include the following.
  - The most common set system is straight sets, in which all sets of an exercise are completed before doing the next exercise. Circuit sets are an alternative in which one set of each exercise is done in succession. The circuit is repeated one or more times. If the circuit is arranged so that different muscle groups are used in successive exercises, circuits save time because, compared to straight sets, shorter inter-set rest periods can be used. A compromise between straight and circuit sets are alternate sets in

which exercises devoted to different muscle groups are grouped in pairs. Sets are alternated between exercises of each pair. Like circuits, alternate sets can save time by reducing inter-set rest periods.

- The duration of rest periods between sets affects the extent of recovery from completed sets. Rest periods average 1.5 to 2 minutes when straight sets are done. Shorter rest periods reduce the number of repetitions completed in sets but have nevertheless proven effective. Longer rest periods may be necessary in some exercises because cumulative fatigue could risk injury by impairing proper exercise execution. Longer rest periods (3–5 minutes) significantly increase the duration of training sessions.
- There may be merit in doing the most important, sport-specific exercises first in a training session while the trainee is less fatigued. It is often recommended that large-muscle-group, multi-joint exercises be done before small-muscle-group, single-joint exercises. Many athletes may not have to include the latter exercises in their program.
- Some athletes may favor maintaining the same program indefinitely whereas others prefer variation or periodization in training. The most common form of periodization is to systematically change training intensity (e.g., from 10 RM to 6 RM to 3 RM) over the training period. The superiority of periodized over non-periodized training is under debate but some athletes may welcome changes in training. Other athletes may favor a fixed but effective program and actually be annoyed by program changes.

- With cessation of training, strength and muscle mass will be relatively well preserved for a few weeks, after which detraining occurs. Neural adaptations such as learning the skill of doing exercises are preserved for long periods but muscle atrophy occurs more rapidly. The training-induced transition from 2X to 2A muscle fibers is reversed in detraining and there may be a 2X overshoot in which the percentage of 2X fibers increases above the pre-training level.
- Training cessation often occurs to free time for other forms of training or in-season competition. To avoid detraining, an alternative is a maintenance program in which training volume and frequency are reduced but intensity is maintained.
- A specialized form of reduced training is tapering, in which training volume is reduced but intensity is maintained or even increased. The taper is scheduled over the last few weeks before an important competition to promote both recovery from cumulative fatigue and additional training adaptations.
- Athletes requiring a high level of both strength and endurance must do both types of training concurrently. There could be interference between concurrent strength and endurance training because they induce different muscular adaptations. Concurrent training is more likely to adversely affect strength than endurance adaptations. The potential for interference is influenced by the total volume of concurrent training. Strategies to minimize interference include combining reduced, maintenance-type strength training with full-volume endurance training or vice versa.
- A homogeneous group of athletes doing the same training program with the same effort and dedication exhibit a large range in the response to training. Some genetic factors linked to responder status have been identified and in the future it may be possible to tailor training programs in accordance with responder status to maximize adaptations.

PART III

Additional Factors Affecting Performance

9

Peaking, Tapering, and Overtraining

Introduction

The most effective training program is one that allows the athlete to achieve the maximum possible adaptation prior to what are considered the most important competitions for that year. However, as outlined in Chapter 1, when athletes have been training for several years or more, the magnitude of the training stimulus must be greater each year in order for further adaptation to occur. Performance gains at this level are therefore often achieved only at the cost of increased risk of overuse injury or illness. The process is further complicated by the fact that many sports require adaptations in a number of different physiological systems and that these adaptations occur at different rates in different systems. Fortunately, once a training adaptation has been achieved, less training stimulus is required to maintain it than was initially required to achieve it. It is thus less stressful to focus training on a given set of physiological adaptations (e.g., endurance training for increased oxygen delivery and uptake) and then, once maximal adaptation has occurred, to reduce the training stimulus to a level that maintains the effect, while a different form of training (e.g., sprint interval training to increase maximal anaerobic power) is introduced. **Peaking** occurs when an athlete completes a training program that incorporates the correct sequencing of different forms of training, with the optimal balance between the duration, intensity, and volume of the stimulus, so that maximum adaptation coincides with the competition date.

During the 1970s and early 1980s, anecdotal reports began to surface from athletes who were forced to suspend training for a week or more as a result of illness, injury, or travel just prior to an important competition. In many cases, when they chose to compete despite the interruption in training, they were surprised by their gains in performance and often achieved personal bests. By the mid-1980s, it had become common practice for many athletes to stop or to reduce their training significantly over the 4 or 5 days preceding competition, in what was often described as a catch-up or super-compensation period. The process (now known as **tapering**) has more recently been shown to result in a number of physiological adjustments that can enhance physical performance.

Overtraining is the condition that arises when the magnitude of the training stimulus chronically exceeds the athlete's capacity to adapt to it. When this occurs, there is a decline in performance, usually accompanied by a feeling of general fatigue. Overtraining is also known to have a suppressive effect on certain components of the immune system, and can make the athlete less resistant to infection and common illnesses, such as upper-respiratory-tract infections (URTI). Recognition of the early indications that an athlete may be approaching the overtrained state is a major challenge for the coach, since the symptoms often vary among individuals. In addition, non-training-related stressors may well combine with those directly related to training to accelerate the process.

Figure 9.1 The effect of peaking several times during the training year (Athlete B) vs a single peak (Athlete A). MAC, maximal aerobic capacity. See text for further details.

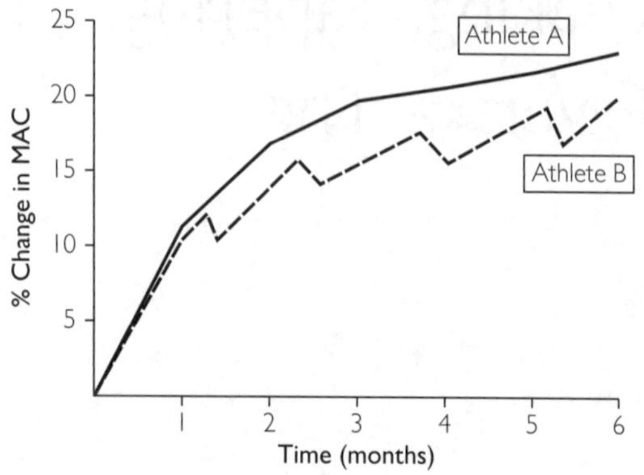

Peaking

Deciding When to Peak

Although the time frames for packaging training programs may vary in length, it is customary to think in terms of a yearly program. One of the most important decisions that an athlete and his / her coach can make is the key calendar date(s) on which the athlete should be at peak performance in the upcoming year. *Once they have determined this date (or dates), they can then work backwards to design the year's training program*. The simplest situation is obviously one where the athlete can aim to peak on a single date. In reality, however, such situations are quite rare, since, in most instances, athletes must succeed in various qualifying competitions before they can move on to further competition. The frequency of the multiple peaks required varies with different sports and different levels of competition, but the fact remains that, from the physiological perspective, the fewer peaks the better. This point is illustrated in Figure 9.1, where athletes A and B begin their training program at a similar fitness level, but A is required to peak only once during the year, whereas B must peak (and taper) for four separate competitions. Since training must be suspended or modified for each of these events, the magnitude of the training effect (in this example, the gain in maximal aerobic capacity) will obviously be greater for athlete A than for athlete B by the end of the season.

Training the Right Systems at the Right Time

While some sports rely on energy delivered almost exclusively via a single pathway (e.g., jumping or marathon running), most involve two or more energy-delivery systems. Similarly, as outlined in Chapter 1, the importance of muscular strength and power to overall performance varies widely in different sports.

Table 9.1 Summary of the recommended training times needed for various adaptations.

Adaptation	Training stimulus	Time course
Increased ventricular volume and wall thickness	Long-duration continuous training	2.5–3 months
Increased total blood volume	Long-duration continuous training Endurance interval training	2.5–3 months
Increased mitochondrial density, oxidative enzyme activity, and muscle capillarization	Endurance interval training	3–5 weeks
Increased glycolytic enzyme activity, muscle buffering capacity, and ion-pump capacity	Sprint interval training	2–3 weeks
Muscle size and strength	Resistance training (<80% MVC)	3–4 months
Muscle power	High-velocity resistance training	3–5 weeks

In any given year, the emphasis to be placed on training for improvements in a particular energy-delivery system or for increased muscular strength and power will depend not only on the demands of the sport, but also on the individual characteristics of the athlete concerned. For example, consider a 400 m runner with a personal-best time of 48 seconds. As outlined in Chapter 2, approximately 80% of the athlete's energy in this event is derived from the two anaerobic pathways and approximately 20% from oxidative metabolism. It would be an oversimplification, however, to relegate the oxidative training component automatically to 20% of the total training time. This is because the time course required for adaptations in this system is considerably longer than for adaptations in the anaerobic pathways. In addition, testing (see Chapter 11) may indicate that, although our runner scores very highly on measures of maximal anaerobic capacity and power, his maximal aerobic power and capacity are only average. In this situation, the most effective program would devote more time to achieving adaptations in oxygen delivery and utilization.

Table 9.1 presents a summary of the recommended training times needed for the various adaptations to occur. In general, those that involve synthesis of structural proteins (cardiac and skeletal muscle) require several months (see Chapters 6 and 8) before approaching their maximum. In contrast, increases in enzyme activity approach their peak within 2 to 4 weeks (see Chapters 6 and 7). It has also been shown that, the higher the training intensity, the greater the stress on the athlete's immune system. As a result, it is possible for athletes to combine some forms of training during the same period, but not others. For example, in the off-season, an athlete can normally perform continuous endurance training in combination with resistance training for increased muscle size and strength during the same time period, but not in combination with high-intensity interval training.

Consider, for example, the program for a soccer player. This sport requires high levels of anaerobic power and capacity for the maximal bursts that occur repeatedly throughout the game, plus high aerobic power and capacity for recovery from these bursts and for moderate-intensity running (a forward may run more than 10 km in a game). It also requires high levels of strength and power in the muscles of the lower body. Consequently, the off-season program should ideally begin a minimum of 18 weeks before the first game. The first 10 weeks would involve continuous running (Chapter 6) and resistance training to increase muscle size and strength (Chapter 8). The running component would be performed 4 or 5 times per week and the resistance training 3 times per week. Week 11 phases in endurance interval training 3 times per week, with a transition to resistance training for power and speed 3 times per week. In week 15, athletes would begin sprint interval training, starting at 3 times per week and increasing to 4. Two days per week would be devoted to resistance training to maintain the previous gains in strength and power. The sprint intervals would take the form of all-out 30-second drills involving game-specific skills such as changing direction, dribbling, passing and kicking. Once the season begins in week 19, the training emphasis will be on maintaining the adaptations that have been

> **BOX 9.1**
>
> The authors have observed professional hockey coaches conducting sprint-skating conditioning drills at the end of a team practice session. These drills typically involve all-out skating, with frequent stops and starts, for intervals of approximately 30 seconds, followed by 1–2 minutes of low-intensity recovery. Although the groups were often split into defensemen and forwards, the conditioning coach would normally require each player in the group to complete the same number of drills. Invariably, towards the end of each session, it would become obvious that some individuals were fatiguing much more rapidly than others. It was then common for the coach to single out these players and have them do additional drills. In fact, for this form of training, the opposite approach should be taken: these players should do fewer drills, since they are often particularly vulnerable to travel fatigue and minor illnesses, such as URTI.

made with no additional overload on the system (see Chapter 7).

Individual Tolerance

The training frequency, intensity, or volume that is ideal for one athlete may exceed the tolerance level for some of their fellow athletes (see Box 9.1). Although it is easier to recognize the symptoms of overtraining (see Recognizing Overtraining) and to re-adjust the programs for athletes in individual sports, it is often more difficult to do so for athletes in team sports. During in-season training, these athletes normally train as a group, performing the same number of the same drills with the same recovery time, and individual differences often go undetected.

Tapering

It is now well known that reducing the training stimulus for a week or more before competition will lead to improved performance in endurance events (3,28,33), performance in repeated bouts of sprinting (2), muscular strength (8,13), and power (35). Theoretically, the magnitude of the stimulus could be reduced by decreasing training volume, training intensity, or any combination of the two. Although slight improvements in performance can be achieved by reducing both volume and intensity, studies show that the most effective approach is to reduce training volume significantly while maintaining training intensity.

Physiological Adaptations to Tapering

Endurance-trained athletes In an investigation where highly trained runners significantly reduced their training volume but retained training intensity over a period of 7 days, there was a significant increase in muscle glycogen concentration, mitochondrial enzyme (citrate synthase) activity, total blood volume, and red cell volume compared to pre-taper values (33). Although the increase in glycogen would probably have little effect on performance in middle-distance endurance events, it would be a major benefit in longer events such as the triathlon or marathon. On the other hand, the increase in oxidative enzyme activity could be expected to increase maximal aerobic capacity and performance in any endurance event. Similarly, as discussed in Chapter 3, the expanded blood volume would help to maintain cardiac output and sweat rates as body temperature increases. An increase in red cell volume would also have a beneficial effect on blood buffering capacity in middle-distance events.

In a similar study with trained cyclists following a 7-day taper, significant increases were found in the activity of a number of mitochondrial enzymes, as well as in myosin ATPase (29). Other changes that have been reported at the muscle level following similar tapers include an upregulation in Na^+/K^+ pump activity (1) and an increase in strength in the endurance-trained muscles (33).

After a considerably longer taper period (3 weeks), a group of competitive distance runners showed significant improvements in performance but no change in citrate synthase activity (19). The authors did, however, find increases in type 2A fiber diameter, peak force, and absolute power when individual gastrocnemius fibers were examined.

Strength-trained athletes There have been relatively few studies of tapering by strength-trained athletes. A study in which bodybuilders greatly reduced their volume of training (a 62% decrease in the number of sets to failure) over a period of 10 days found significantly greater voluntary low-velocity and isometric strength over this period, but little or no change in evoked contractile properties or high-velocity strength. When the same subjects performed a rest-only taper, voluntary strength increased significantly after 2 days but began to fall below pre-taper values after 4 days (8). It is known that significant myofibrillar disruption and damage occurs, even in trained athletes, with each training session (7) and that protein synthesis is elevated in these muscles for 24–36 hours following the session (20). The increase in strength is thus probably the result of a significant repair of contractile protein damage occurring over the two days of rest.

A study in which subjects tapered their resistance training for a period of 4 weeks found that voluntary strength (but not power) increased, as did resting blood levels of insulin-like growth factor 1 (IGF-1) (13). Some investigators have suggested that IGF-1 may be a major stimulator of muscle hypertrophy following resistance training, while others question its relative importance in this role.

Team-sport athletes In a study where 11 trained team-sport athletes underwent a 10-day taper, a significant improvement was found in performance with repeated short sprints, each followed by 30 seconds of recovery (2). Although the study does not report adaptations at the muscle level, it can be speculated that any improvement in muscle oxidative capacity (see above) would have had a beneficial effect on between-bout recovery time.

Designing the Taper Program

Fundamental to any taper is the assumption that the athlete has attained near-maximum adaptation within the physiological systems vital to performance within their sport. The taper then provides a catch-up period for systems that are slower in adapting, while providing enough training stimulus in others to prevent detraining. Some of the adaptations that take longer to occur following stimulation include an increase in red blood cell volume [a minimum of 7 days (33)], muscle glycogen concentration [3–4 days to reach peak (34)], and mitochondrial enzyme activity [more than 6 days (10) but fewer than 14 (4)]. It is thus not surprising that, once peaking has been attained, 2–3 days of complete rest prior to a competition can improve performance. Similarly, large reductions in training volume and intensity over longer periods of time can be even more effective. The most effective taper, however, appears to be one in which total volume and training time are greatly reduced but intensity is maintained (24,25,33).

Duration of the taper If the taper period is too brief, maximum adaptation in the systems affecting performance will not occur. If the taper period is too long, the adaptations will begin to diminish. In addition, in some situations, the exact duration of the taper will be influenced by such factors as travel, adjustment to new accommodations, and access to training facilities. Although improvements in performance have been found following as much as 3 weeks of tapering, the majority of studies (3) and studies conducted in the authors' laboratory (8,33) support a tapering period of 1–2 weeks. If circumstances preclude a true taper, 3 days of complete recovery from the last training session would be the next best alternative.

Sample Taper Program

In this example, a highly trained 1500 m runner will undergo 8 days of tapering prior to the event that he and his coach consider his most important competition of the year. (For previous regional competitions the same year, they have decided not to perform a full taper, but simply to reduce training significantly or to rest for 3 days before the race). Prior to this point, he has completed 6 months of training, consisting of approximately 4 months of progressive high-volume submaximal continuous training (5 or 6 days per week) followed by a gradual phasing-in of progressive high-intensity endurance interval training on three of the 5 days for approximately 6 weeks. (For a more detailed description, see Chapter 6.) In the week prior to the taper, his program was as follows.

Monday	Ten 45-second intervals at race pace or above, separated by 3–4 minutes of walking recovery
Tuesday	Sixty minutes of continuous running at 70–75% $\dot{V}O_{2max}$, and two 10 RM sets of leg-press exercise on a universal gym
Wednesday	As on Monday
Thursday	As on Tuesday
Friday	As on Monday and Wednesday

For his taper, he will eliminate any resistance training and his program will be as follows.

Day 1 Six 45 second intervals as above
Day 2 Thirty minutes of continuous running as above
Day 3 Four 45 second intervals
Day 4 Twenty minutes of continuous running
Day 5 Three 45 second intervals
Day 6 Twenty minutes of continuous running
Day 7 Two 45 second intervals
Day 8 No training or 20 minutes of continuous low-intensity running (optional)
Day 9 Competition day

This taper represents an approximate 60% reduction in continuous running volume and a 50% reduction in the volume of interval training. The authors have found that many athletes find it difficult to forgo training completely on the day before competition because it has become such a major part of their biorhythms that its elimination contributes to sleeping problems that night. It is therefore recommended that these athletes perform 30–45 minutes of light continuous exercise at their normal training time.

Overtraining

When an individual experiences a physical or psychological event, real or imagined, that the brain perceives as stressful (see Box 9.2), the hypothalamus responds by stimulating the pituitary gland to release corticotropin. This in turn causes the adrenal glands to release the stress-response hormones epinephrine, norepinephrine, cortisol, and glucagon, in what is sometimes referred to as the fight-or-flight response. It has long been known that even moderate-intensity exercise (60% $\dot{V}O_{2max}$) causes an increase in these circulating hormones. As reviewed in Chapter 2, this response increases linearly with exercise duration and exponentially with exercise intensity. While these hormones perform an essential role in the cardiovascular and metabolic response to exercise, the fact remains that the brain has perceived exercise as a stressor. It is also known, however, that repeated exposure to exercise (training) blunts the response so that exercise at the same absolute intensity is no longer considered to be as stressful. The problem, of course, is that an athlete in training must keep increasing the magnitude of the stimulus to improve performance. As a consequence, he/she may reach a point at which the stimulus begins to exceed his/her capacity to adapt to it. If this point is recognized early enough, the training stimulus can be modified and the athlete can recover with only a minimal effect on performance. Such situations are referred to as overreaching. If not detected, however, they can lead to major declines in performance, suppression of immune function and vulnerability to common illnesses. This condition is known as overtraining.

BOX 9.2

The concept of stress and a predictable series of reactions to it was first advanced by Canadian scientist Hans Selye in the 1930s and 1940s. He observed that the body will adapt to a stressor following repeated exposure. However, if the stressor becomes too intense, if it is combined with other stressors, or if exposure continues for too long, the body's resistive mechanisms will eventually collapse, leading to exhaustion and illness (Selye's general adaptation syndrome).

Exercise and Immune Function

As reviewed in Chapter 3, the body's major source of protection against invading organisms and infection lies in the five different types of white blood cells or leukocytes. Four of these (neutrophils, monocytes, eosinophils, and basophils) represent the body's first line of defense and have phagocytic capabilities that allow them to digest and destroy damaged tissue or bacteria. The cell membrane and diameter of white cells allow them to pass through the capillary membrane; as a result, until they are activated, they are stored largely in the spleen and other vascular compartments, with low concentrations in the general circulation.

The fifth type of white blood cell (lymphocytes) consists of three subgroups: T-cells, B-cells, and null cells. B-lymphocytes can program themselves to recognize foreign proteins (antigens) and to form antibodies, so that when these proteins enter the body again they can be destroyed. T-lymphocytes do not form antibodies, but some types, known as cytotoxic T-cells, can attach themselves directly to an antigen and destroy it. Other types of T-cells include helper T-cells (T4 cells) and suppressor T-cells (T8 cells). The helper cells play an important role in recognizing antigens and in programming other T-cells, while the suppressor cells have a downregulating effect that prevents possible over-damage to other body systems (16). Null cells are lymphocytes other than T- and B-cells. One type of null cell, the natural killer cell, plays an important role in destroying abnormal cells such as tumor cells.

When an individual goes from the resting state to the exercise state, all of these leukocytes are rapidly mobilized from their storage sites, and within minutes their numbers show a significant increase in circulating blood. The magnitude of this increase is related to exercise intensity and duration but is independent of gender or fitness (17,23). Following exercise, most leukocytes return to normal baseline levels within approximately 45 minutes.

Many immunologists consider a change in the proportions of the different types of circulating T-cells (especially T4 and T8 cells) to be an indicator of suppressed immune function. When athletes in training suddenly increase their daily volume of training, blood samples taken immediately after exercise show reductions in the T4/T8 ratio, primarily as a result of decreases in T4 (helper) cell numbers. When athletes suddenly increase the intensity of their training, this reduction is even more marked. In both situations, however, after several days of training at the new load, this reduction is attenuated (14). Although normal T4/T8 ratios are usually re-established about 2 hours after exercise, natural killer-cell numbers may be depressed for 8 hours or more (15). This brief period of suppressed immune function after a session of heavy exercise is sometimes referred to as the open window for susceptibility to infection.

Perhaps the most effective method of determining the practical significance of the possible link between exercise and reduced immune function is to examine the incidence of minor illnesses such as URTI in athletes and non-athletes. Studies of this kind (9,22,30,31) indicate that individuals who perform regular moderate exercise and athletes at the lower-intensity stages of their training have fewer URTIs than sedentary individuals. When they increase training intensity in preparation for a competition (or immediately following a competition), however, they have a higher rate of URTI than less active individuals (Figure 9.2).

Figure 9.2 The J-shaped curve illustrating the incidence of minor illnesses such as URTI in sedentary individuals and athletes who are adapting to training vs athletes who are approaching the overtrained state.

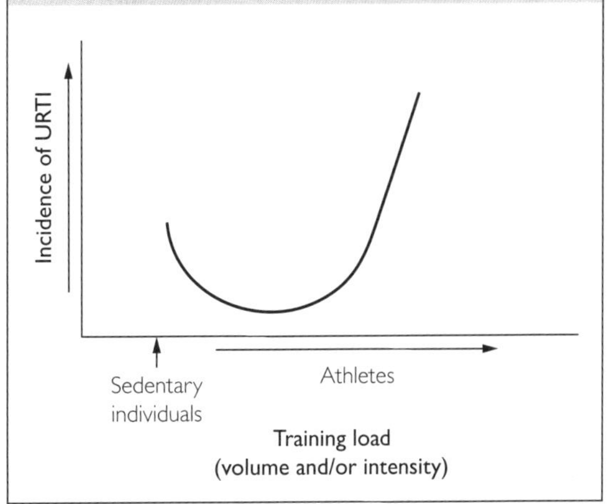

Recognizing Overtraining

In theory, constant monitoring of leukocyte numbers and ratios, stress hormone levels, and testosterone in the athlete's blood and urine should provide some indication that he or she is approaching the overtrained state. In most real-life situations, however, this process would be expensive and impractical. As well, interpretation of these variables is complicated by individual differences and diurnal variations in these markers (21). From the practical standpoint, the most effective indicators of overtraining are probably changes in performance, body mass (weight), sleep patterns, incidence and duration of minor illnesses and overuse injuries, and menstrual irregularities in female athletes.

Changes in performance Inability to adapt fully to an increasing training stimulus will eventually manifest itself as a decline in performance. In individual sports such as sprinting or middle-distance events, performance can be assessed through weekly time trials. In jumping, throwing or lifting events, this can be accomplished by two or three maximal efforts at the end of each week's training. It is important, however, that these tests of performance be given only after a day of complete recovery and a standardized warm-up.

Coaches are sometimes reluctant to incorporate time trials of this kind into the program for middle-distance athletes because they see it as time taken away from training. An alternative is to monitor heart rate (as in Chapter 6) at the same point in the daily training schedule: for example, during the fifth 3 minute interval, during recovery from a given interval or during minute 35 of continuous training, etc. Although slight variations in daily heart rate are common, persistently higher rates at the same point in time are a good indication of overtraining.

Changes in performance in team-sport athletes are usually more difficult to detect because of the number of variables at play. In most instances, however, the most effective indicator is probably fatigue noted in an individual during practice sessions involving repeated sprint-type drills. This can be detected by watching the athlete's performance in comparison to that of their teammates and to their performance earlier in the season.

Changes in body weight Following the pre-season training phase, an athlete's body weight is normally very consistent; any rapid or progressive decline in weight is therefore a good indication of stress-mediated changes in appetite or digestive processes. Of course, it is important that weight be measured at the same time each day, preferably first thing in the morning, without clothing, and before any food or fluid intake.

Changes in sleeping patterns Elevations in circulating stress hormones, especially epinephrine and norepinephrine, are known to interfere with normal sleeping patterns. Although it is normal for concentrations of these hormones to increase during exercise, they usually return to baseline levels within an hour or so following each training session. Elevations persisting longer than this can affect the quality of sleep, causing fatigue and mood changes. A brief self-analysis questionnaire to be completed by the athlete two or three times per week should help to identify these symptoms (21).

Incidence and duration of illnesses Group travel and sharing of common dining, changing, and shower facilities can increase the exposure of athletes (especially those in team sports) to common minor illnesses such as URTI, sore throats, and gastrointestinal disorders. Individuals who have a higher incidence of infection than their teammates or longer recovery times following these illnesses may be approaching the overtrained state.

The incidence of overuse injuries such as tendonitis or inflammation in hip, knee, or ankle joints is also a good marker for possible overtraining. Although such injuries do not necessarily reflect reduced immune function, they are relatively common in endurance athletes and can only be resolved by reductions in training.

Menstrual irregularities in female athletes Amenorrhea in previously normally menstruating athletes is often an indication of overtraining. A combination of amenorrhea with eating disorders such as anorexia or bulimia and decreases in bone mineral density is known as the **female athlete triad syndrome**. While this syndrome is most common in athletes who compete in weight-classified events and gymnastics, it also occurs in runners and even in team-sport athletes. The primary underlying cause of the syndrome is generally believed to be an imbalance between energy expenditure and energy intake (27). In most cases, the solution thus lies in resolving the eating disorder problems through psychological counseling and/or encouraging increased energy intake.

Interference of Training Modes

Several studies suggest that, when athletes perform heavy-resistance training and endurance training at the same point in their training cycle, the magnitude of the training effect may be less than if they performed the same training at different points in the season (see Chapter 8). The question of whether the adaptive response to one type of training may interfere with the response to another type of training remains controversial (18,26). Based on the literature, however, it appears that, when athletes perform strength and endurance training *on the same day*, their gains in strength (but not their gains in endurance performance) are less than those achieved with strength or endurance training alone (6,12). When the two forms of training are performed on alternate days, this interference effect is largely eliminated (32).

As reviewed in Chapters 6 and 8, one of the major functional adaptations to high-intensity endurance training is an increase in mitochondrial protein, while heavy-resistance training leads to an increase in contractile protein. The genetic and molecular mechanisms that result in increased synthesis of these two types of protein are quite different and it has recently been suggested that the mechanisms stimulating mitochondrial protein synthesis may exert an inhibitory effect on the signaling pathways that stimulate myofibrillar protein synthesis (5,11,36). In contrast, another recent study suggests that concurrent strength and endurance training may actually *enhance* the molecular signaling pathways for mitochondrial synthesis (37).

The issue of possible interference between the two training modes is of no major practical significance to athletes relying entirely on strength and power or entirely on endurance. For sports that require high levels of both strength and endurance (e.g., combative sports and many team sports), it is still easy to avoid. One obvious solution is to perform the different types of training on alternate days. However, although resistance training three times per week would be adequate for most sports, 3 days of endurance training would not be as effective as 5 days per week.

Fortunately, as noted repeatedly in this text, it takes considerably less training stimulus to maintain a training adaptation than to produce it in the first place. The solution for these sports, then, is to periodize the overall training program so that the heavy-resistance training phase coincides with the continuous, lower-intensity endurance training phase (see Chapters 6 and 7). Since the major adaptation to continuous endurance training involves central factors (increases in blood and heart volume) rather than peripheral factors (mitochondrial protein synthesis), interference between the two training modes will be minimal. The subsequent high-intensity endurance interval training phase will focus simply on *maintaining* the gains in strength and power that have been achieved, a goal that can be attained with as few as two training sessions per week and reduced training volume. In addition, minimal interference will occur when athletes are doing high-velocity power training, since the adaptations here are in the central nervous system and not at the muscle level (see Chapter 8).

Summary of Key Points

- When an athlete completes a training program that causes maximum adaptation to coincide with the most important competition(s) of the year, the process is known as peaking. Designing such a program is more complicated in sports where performance is affected by more than one physiological system, since adaptations occur at different rates in different systems. The process begins with the athlete and her/his coach identifying what they consider to be the key calendar date(s) on which the athlete should be at peak performance. Once these dates have been determined, they can then work backwards to design the year's training program.
- Peaking thus requires training the appropriate system(s) at the appropriate time and adjusting the training stimulus so that the adaptations that have been gained are maintained. From the physiological perspective, the fewer times that the athlete must peak the better, since training must be suspended or modified (tapered) for each peak performance.
- Once an athlete has peaked, performance can be further enhanced by reducing the training stimulus over the week or two immediately prior to competition. The most effective form of taper appears to be one where training volume is sharply reduced but intensity is maintained. This process has been shown to be effective for improving both endurance performance and muscle strength and power.
- Overtraining is the condition that occurs when the training stimulus increases to a point that exceeds the athlete's capacity to adapt to it. This will lead to a decline in performance and, unless training is modified, an increase in certain stress-response hormones and impaired immune function.
- Although individual responses among athletes can vary, the most effective indicators of overtraining are probably changes in performance, body weight, sleeping patterns, and the incidence and duration of minor illnesses and overuse injuries.
- In female athletes, amenorrhea is also a probable indicator of overtraining. The combination of amenorrhea with an eating disorder such as anorexia and a decrease in bone mineral density is known as the female athlete triad syndrome.
- Some studies indicate that, when athletes perform strength and endurance training on the same days, their gains in strength are less than if they performed strength training alone. It has been suggested that this interference

between training modes may be caused by the signaling pathways for mitochondrial protein synthesis inhibiting those for myofibrillar protein synthesis.

- For athletes in sports requiring high levels of both endurance and strength, the possible interference issue can be avoided largely by performing the different forms of training on alternate days or by periodizing the training program so that the heavy-resistance phase coincides with the lower-intensity continuous training phase.

10

Stretching and Flexibility

Most athletes perform some type of muscle-stretching exercises as part of their warm-up before training sessions and competition, in the belief that stretching will prevent injury and possibly enhance performance (50,51). Some athletes also do regular stretch training to increase their flexibility, which can be defined as the maximum range of movement achievable without injury at a joint or series of joints (108). A high level of flexibility is important in many types of performance (e.g., diving, gymnastics, hurdling). The issues of whether pre-activity stretching prevents injury or enhances (or hinders) performance will be discussed in subsequent sections. Also covered will be methods of flexibility training. However, first to be considered are the different types of stretching and their effect on the muscle–tendon unit.

Stretching

Types of Stretching

Static stretching The most commonly performed stretch is called a **static stretch**. A static stretch begins with a slow stretch of the **muscle–tendon unit** (**MTU**) to, or close to, the maximum tolerable length (point of discomfort) and then held at this length for several seconds (hence the term "static"). Accordingly, a static stretch encompasses the slow stretch followed by a held stretch. A static stretch to maximum tolerable length would correspond to the tolerable **maximum range of movement** (**ROM$_{max}$**) at a particular joint, or flexibility. Figure 10.1 is an illustration of a static stretch with the assumption that the muscle is relaxed. As the

MTU is stretched, passive force increases to the end point of the stretch at maximum tolerable length. If the final stretch length is maintained, passive force begins to decline; this decline is called **stress relaxation** (15,69). The rate of stress relaxation is initially fast but then progressively slows (Figure 10.1). Stress relaxation may continue for as long as the stretch is maintained, for example, as long as 2 minutes (26). Most of the stress relaxation occurs during the first 20–30 seconds of maintained stretch, a duration which is more typical of a stretch routine used by athletes. For a stretch duration of 20–30 s, the magnitude of stress relaxation is about 10–15% (26,67). As an alternative to stretching the MTU to a fixed length and observing stress relaxation, a fixed stretch load can be imposed on the MTU. The result is a relatively rapid lengthening of the MTU followed by a slower lengthening called **creep**.

Repeated static stretches Rather than a single stretch, most athletes do a series of stretches before training or competition or as part of a stretch training program. If a series of stretches to the maximum tolerable length is performed, there is a progressive increase in both the maximum tolerable length and the peak passive force attained (Figure 10.2). Maximum tolerable length may continue to increase over a large number of stretches, whereas peak passive force plateaus after three or four stretches (26,64,67). The continued increase in tolerable length without a corresponding increase in peak passive force indicates that passive force for a given length has decreased (see Box 10.1). This is best seen when a series of stretches

Figure 10.1 Schematic illustration of a slow stretch to the maximum tolerable length of the muscle–tendon unit (MTU). The stretch is then held (static) for several seconds. Passive force rises to a peak at the end of the stretch, but then decreases as the stretch is held. The decrease in passive force during a sustained stretch is called **stress relaxation**.

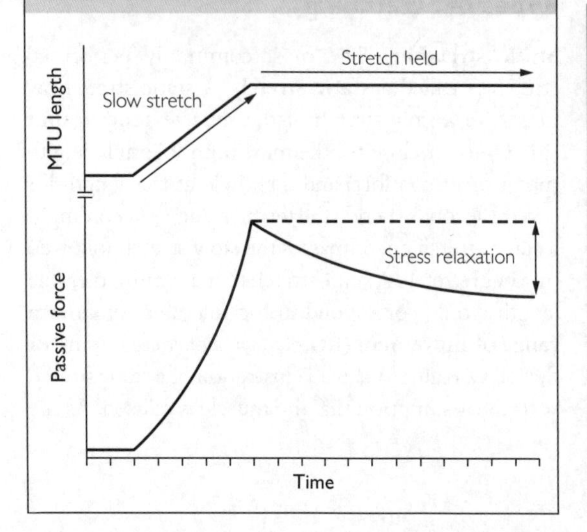

Figure 10.2 Schematic illustration of three successive static stretches to the maximum tolerable length of the MTU, each held for about 2 minutes. Note the increase in maximum tolerable length and peak passive force over the three stretches. Stress relaxation became smaller with each successive stretch. Based on data from Fowles *et al.* (26).

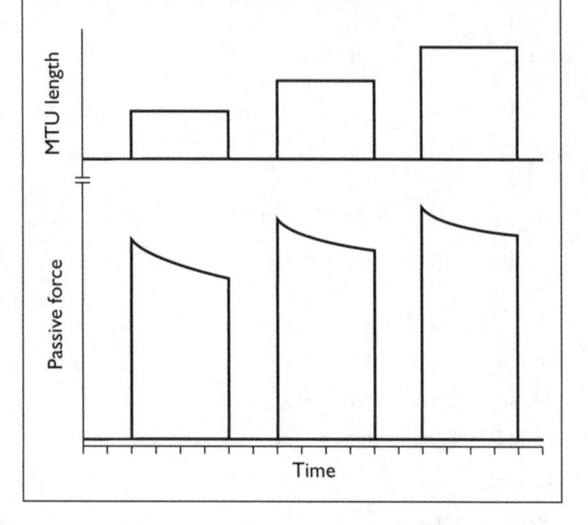

BOX 10.1 PASSIVE FORCE, STIFFNESS, AND COMPLIANCE OF THE MUSCLE–TENDON UNIT

A stretching session may transiently decrease passive force of an MTU at a given muscle length (Figure 10.4A). The decrease in passive force is referred to as a decrease in **stiffness** or an increase in **compliance**. Stiffness is the ratio of the change in resistive force to the change in imposed length (stretch) ($\Delta F/\Delta L$). A greater increase in force (F) in response to a given amount of stretch (L) indicates greater stiffness. Compliance is the reciprocal of stiffness (33); that is, the ratio of the change in length in response to a change in imposed (stretching) force ($\Delta L/\Delta F$). Thus, if stiffness has decreased after a stretching session (Figure 10.7A), compliance has increased (Figure 10.7B).

Figure 10.3 In contrast to Figure 10.2, which showed the effect of three successive static stretches all extended to the maximum tolerable length, this figure schematically illustrates the effect of three successive static stretches to the *same* final length of the MTU, each held for about 30 seconds. Note the decrease in peak passive force and magnitude of stress relaxation over the three stretches. Based on data from Magnusson *et al.* (67).

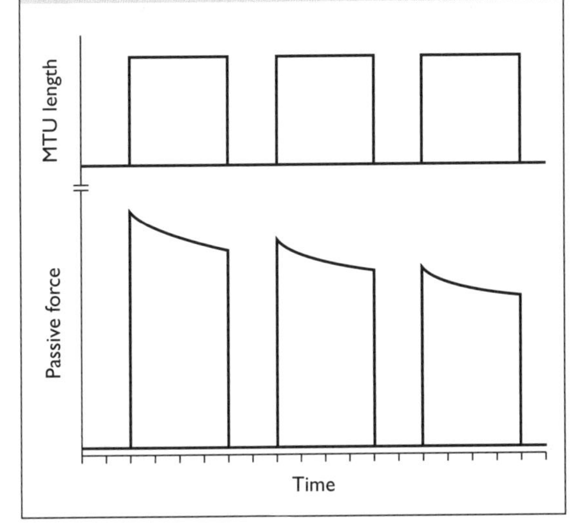

Figure 10.4 Schematic illustration of two possible effects of a series of static stretches on the passive force–length relationship. (A) After the stretches, maximum tolerable length and maximum passive force increase whereas the passive force at a given length decreases. (B) After the stretches, maximum tolerable length and maximum passive force increase whereas the passive force at a given length is unchanged.

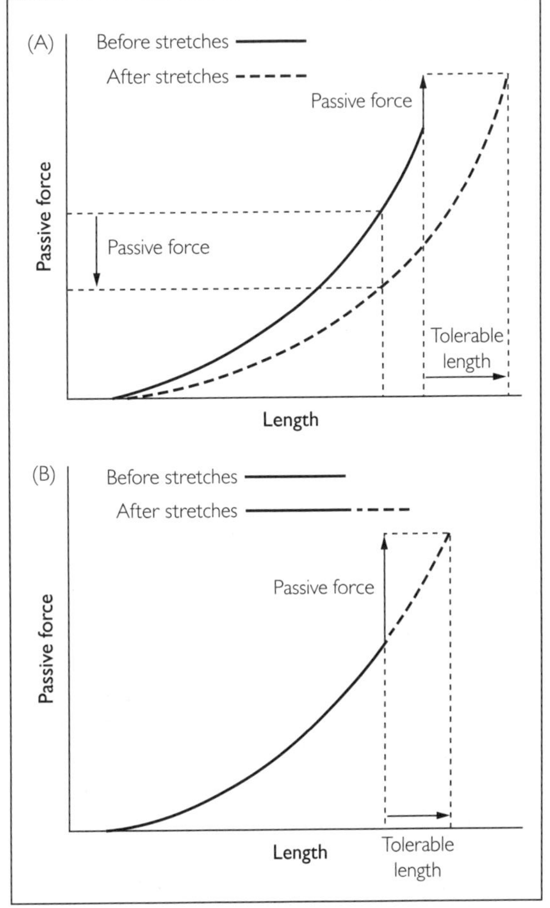

is done to the *same* length (Figure 10.3). Peak passive force and the magnitude of stress relaxation decrease with each successive stretch. The acute effect of a series of stretches can also be shown by plotting the passive force–length relationship before and immediately after the series of stretches (Figure 10.4). Two different effects have been observed. One effect is that a series of maximum stretches of about 0.5–2 minutes duration increases maximal tolerable length and passive force, and also decreases the passive force at a given length (26,67,81,88) (Figure 10.4A). An alternative effect is that a series of maximum stretches of about 30 seconds in duration increases maximal tolerable length and passive force, but has no effect on passive force at a given length (40) (Figure 10.4B).

Ballistic stretching As the name implies, ballistic stretches are rapid stretches usually initiated by a strong, fast contraction of muscles (antagonists) that oppose the muscle to be stretched. Two examples are kicking the leg up with the hip flexors to stretch the hamstrings and throwing the arms back with the shoulder horizontal extensors (e.g., posterior deltoid) to stretch the horizontal flexors (anterior deltoid, pectoralis major). Unlike static stretches, in ballistic stretches no attempt is made to maintain the stretched position; thus, there is little time for the stress relaxation typical of static stretches to develop (review Figure 10.1). Ballistic stretching is less popular than other stretching methods, perhaps in part because of the belief that it may cause injury (91,107,108). There is some justification for this belief because, for a given magnitude of stretch, passive force increases to a higher level in a fast than in a slow stretch (Figure 10.5).

Figure 10.5 Comparison of the passive force response to a fast (ballistic) and slow stretch through the same range of muscle length in rabbit tibialis anterior muscle. The fast stretch produces a faster and larger increase in passive force. The response was similar for innervated and denervated (no possible stretch reflex response) muscles. Schematic illustration based on data from Taylor *et al.* (107).

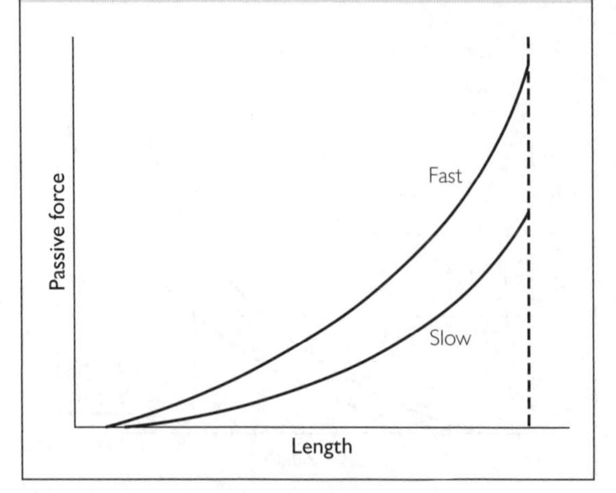

Figure 10.6 Schematic illustration of one variation of PNF stretching. With the help of an assistant, the trainee begins by slowly stretching the relaxed target muscle (TM) to the maximum tolerable length and passive force. The maximum length is held (static stretch) for several seconds but during this period the trainee does a voluntary isometric contraction (Vol_{ISO}) of the TM against resistance provided by the assistant. The trainee relaxes the TM just before the assistant applies an additional stretching force to a *new* maximal tolerable length and passive force. Based on data from Magnusson *et al.* (66).

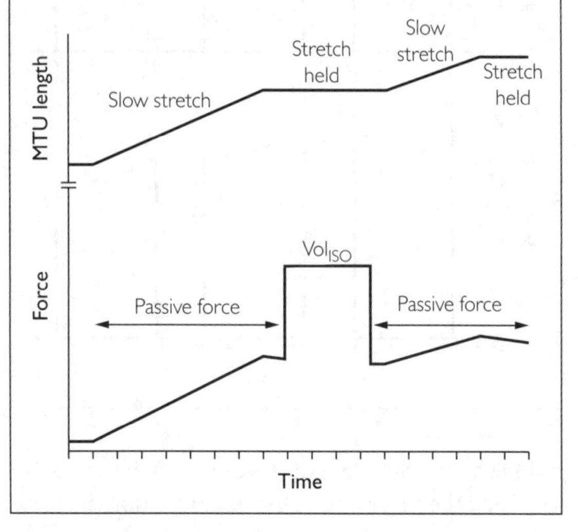

Perhaps because of its lesser popularity, the acute effects of a series of ballistic stretches have received less study than other forms of stretching (98). One investigation showed that during 10 relatively fast stretches there was a decrease in passive force at a given length; however, when a test was done 10 minutes after the stretches, the effect on passive force was gone. What remained was increased stretch tolerance; that is, an increase in the maximal tolerable stretch length and stretch force (63). Thus, the results were like those shown in Figure 10.4B.

Peripheral neuromuscular facilitation stretching Peripheral neuromuscular facilitation (PNF) stretching has many variations (3,98). In addition, the names of some variations, such as **contract–relax** and **contract–relax–agonist contract** (48), have been applied to different stretch protocols (98). Here, two common variations will be described as examples, with the muscle to be stretched referred to as the **target muscle** (TM), and the muscle that is the antagonist to the TM referred to as the **opposing muscle** (OM) (15,98). For example, if the hamstrings are stretched, they are the TM; the hip flexors are the OM.

One version of the contract–relax is as follows (see Figure 10.6). First, often with the help of an assistant, a slow stretch of the relaxed TM is done to maximal or near-maximal tolerable length and held for several seconds (static stretch). Second, with the assistant providing resistance, the trainee isometrically contracts the TM for a few seconds (e.g., 3–6 s) at the held length. Third, another slow stretch is done to a yet longer length (if possible) and again held for several seconds. If possible, an additional contract–stretch cycle is then performed.

A second common PNF variation is called **contract–relax–agonist contract**, one version of which is as follows. First, the trainee contracts the OM to stretch the TM (the OM may initially contract concentrically and then transition into an isometric contraction); an assistant usually helps stretch the TM to the maximal tolerable length, where it is held for several seconds. Second, with the assistant providing resistance, the trainee isometrically contracts the TM for a few seconds (e.g., 3–6 s) at the held length. Third, the trainee contracts the OM again to stretch the TM, with the help of the assistant. If possible, the three-phase process will be repeated. Note that in the traditional description of the contract–relax–agonist contract, the agonist is actually the OM, whereas its antagonist is the TM.

Several PNF variations are effective in acutely increasing maximum tolerable MTU length and maximum tolerable force (69,98). Like static stretching, the stretches done as part of a PNF maneuver exhibit stress relaxation (Figure 10.1) (66). Several cycles of one PNF variation reduced passive force at a given MTU length for 90 minutes (109), producing the pattern seen in Figure 10.4A.

In the various PNF variations, the isometric contractions of the TM done as part of the maneuver may be maximal or submaximal and may last for 3–15 seconds (98). Since submaximal contractions are as effective as maximal contractions (98), they may be preferred. Similarly, the benefit of PNF stretching is largely independent of contraction duration, so a brief contraction (e.g., 3 s) can be used (14,98). In the example cited above in which passive force at a given MTU length was reduced for 90 minutes after the PNF stretching, the isometric contractions were maximal and sustained for 8 seconds (109). For PNF variations that include contractions of the OM, there are no data on the optimal intensity or duration of contraction (98).

Dynamic stretching Dynamic stretching, which is perhaps less well known than static, ballistic, and PNF stretching, is a controlled movement through the active range of motion (24). Dynamic stretching consists of exercises designed to stretch TMs, for example, upward leg kicks to stretch the hamstrings (90) and flick-backs to stretch the quadriceps (24). To the unpracticed eye, it may be difficult to distinguish dynamic stretching from typical warm-up exercises such as drills and calisthenics (77). In many dynamic stretching exercises, there is no particular attempt to stretch the muscle to ROM_{max} (77). Although dynamic stretching is distinguished from ballistic stretching by the inclusion of controlled actions (24), some dynamic stretching movements (e.g., high kicks to stretch hamstrings) seem similar to ballistic stretching, and the two terms have been used synonymously (10,124). Dynamic stretching can increase ROM_{max} acutely but not to the same extent as static stretching (10). It is not known whether dynamic stretching decreases passive force at a given MTU length.

Figure 10.7 Passive force, stiffness, and compliance of the MTU. (A) After a stretching session, the passive force (PF) at a given muscle length (L_G) may decrease. In this case, the MTU stiffness, the increase in resistive force to an imposed increase in length ($\Delta F/\Delta L$), is reduced. (B) Alternatively, the effect of the stretching session can be assessed as an increase in MTU compliance, which is the ratio of the change (increase) in MTU length to a change (increase) in imposed force ($\Delta L/\Delta F$). Compliance is the reciprocal of stiffness (33). Compliance may also be expressed as the increase in length (L) in response to a given imposed force (F_G). From a practical standpoint, a stretching session that decreases passive force at a given MTU length can also be said to have reduced stiffness and increased compliance.

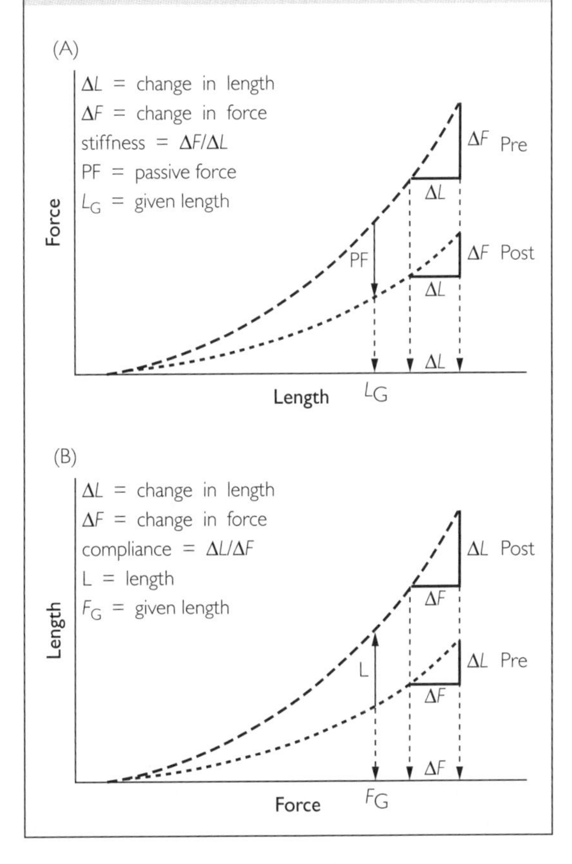

Neural Response to Stretching

A potential impediment to a successful muscle stretch, whether it is part of a static, ballistic, or PNF protocol, is a stretch reflex that opposes the stretch (Figure 10.8).

Figure 10.8 Schematic illustration of the neurophysiological basis of the stretch reflex. When a muscle is stretched, stretch receptors within the muscle, called **muscle spindles**, are also stretched. The spindles respond to the stretch by sending a train of nerve impulses along primary (IA) afferent axons to the motoneurons, which are thereby excited. Sufficient excitation causes reflex contraction of the stretched muscle.

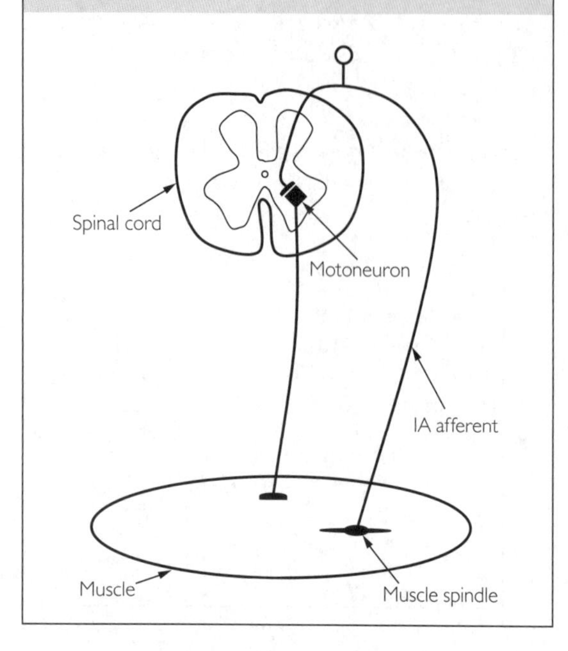

Figure 10.9 Schematic comparison of the neural response to a slow (joint angular velocity of ≈25°·s^{-1}, dashed line) versus a faster (≈70°·s^{-1}, solid line) stretch of human calf muscles. In the slow stretch there was no reflex response; that is, the electromyogram (EMG) was silent. In contrast, the faster stretch evoked a stretch reflex response, indicated by the EMG, that added active force to the passive force produced by the slow stretch. The additional active force opposed the stretch. Based on the data of Nicol and Komi (85).

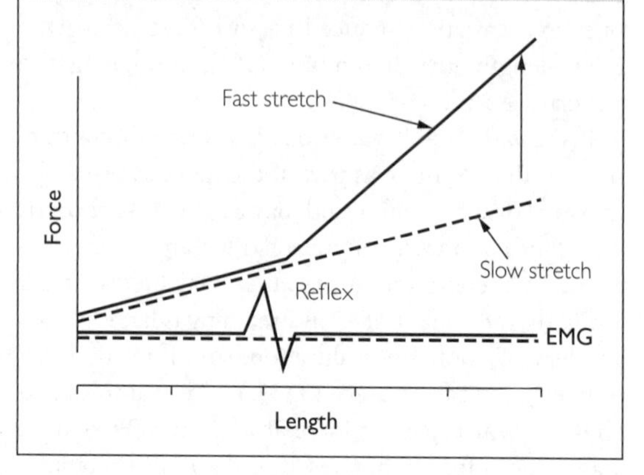

The extent to which the stretch reflex opposes stretches, and whether the success of the various stretching methods depends on successful suppression of the stretch reflex, have been matters of debate (15,38,48,69). Slow stretches that end in a held (static) stretch are usually associated with no or very little motor unit activity in the stretched muscle, suggesting that the stretch reflex is very weak or overridden by other neural mechanisms (15,38,48,81,98,115) (Figure 10.9). When detected in a slow stretch, the force produced by the observed motor unit activity is very small, equal to or less than 3% of the force of a maximal voluntary isometric contraction (15,69,93). Whether a stretch reflex is evoked in a nominally fast stretch depends on the definition of "fast." In human calf muscles, a stretch corresponding to an ankle joint angular velocity of approximately 25°·s^{-1} did not evoke a stretch reflex but the reflex was evoked at a velocity of approximately 70°·s^{-1} (Figure 10.9). In stretching of human hamstring muscles, knee extension velocities of 5 and 20°·s^{-1} did not evoke stretch reflexes (63). It should be pointed out that for different joints the same angular velocity does not necessarily correspond to the same velocity of muscle stretch (lengthening). The velocity threshold for evoking a stretch reflex is likely directly linked to the velocity of muscle stretch, and the threshold velocity may vary in different muscles.

Earlier it was shown that a faster stretch over a given muscle length causes a greater rise in passive force, even in denervated muscles in which a stretch reflex response is not possible (Figure 10.5). Now it has been shown that a ballistic stretch is more likely to cause an opposing stretch reflex force (Figure 10.9). The combined effects of a greater increase in passive force during stretch and the possibility of a stretch reflex provide a rationale for the lack of enthusiasm for ballistic stretching (82).

The very name for the PNF stretching method, peripheral neuromuscular facilitation, implies that in a PNF maneuver the neural response to stretching is modified to ensure that the TM remains relaxed during stretch, specifically that the possibility of a stretch reflex is minimized (15,98). When a PNF maneuver includes an isometric contraction of the TM

(Figure 10.6), the force of contraction stimulates muscle tendon receptors called the **Golgi tendon organs**. When stimulated, Golgi tendon organs can exert an inhibitory effect on the motoneurons innervating the TM. If the inhibitory effect were to persist after the TM relaxed and into the subsequent stretch, the possibility of a stretch reflex would be reduced. If the OM contracts during a stretch of the TM, the activation of the OM exerts an inhibitory effect on the TM, by a neural mechanism called **reciprocal inhibition**. The inhibitory effect reduces the probability of a stretch reflex. However, any Golgi tendon organ-mediated inhibitory effects on the motoneurons of the TM after it contracts are small and of very short duration (15,98) and therefore might not have any influence on a possible stretch reflex response in TM after an isometric contraction. On the contrary, an increase rather than a decrease in TM motor unit activation may occur during stretch after the TM has contracted (48). Although a concentric followed by isometric contraction of the OM may cause reciprocal inhibition of the motoneurons innervating the TM, there are other neural pathways acting on these motoneurons. The result is that the OM contraction does not reduce motor unit activation in the TM below levels observed in a static stretch (48), which are non-existent or minimal (15).

In summary, a slow muscle stretch held at final length (static stretch) causes no or minimal stretch reflex opposition to the stretch. If muscles are already fully relaxed in static stretches there is little to be gained, in terms of inducing muscle relaxation, from including the TM and OM contractions typical of PNF maneuvers. In contrast, fast (ballistic) stretches, if done at high enough speeds, increase the probability of stretch reflexes that would oppose the stretch.

Acute Effects of Stretching

Maximum Range of Motion and Passive Force

Ballistic, dynamic, static, and PNF stretching are all effective in acutely increasing maximum range of motion (ROM_{max}) at a joint; however, PNF stretching has, in most cases, been shown to be the superior method (15,69,98). Why is PNF superior? According to the previous discussion on the neural response to stretching, PNF's superiority, compared to static stretching, does not appear to be the result of TM and OM contractions causing greater TM muscle relaxation during stretch. One possible mechanism is that PNF stretching reduces muscle stiffness (reduced passive muscle force at a given length; see Figures 10.4B and 10.7) to a greater extent than static stretching. However, in the case of a single PNF maneuver (Figure 10.6) compared to a single static stretch, PNF produces a greater transient increase in ROM_{max} but has a similar effect on passive muscle force at a given length (66) (Figure 10.10). A series of static stretches can decrease passive force at a given length for a period of several minutes after completion of the stretches (26,67,93). Similarly, several cycles of PNF stretching may decrease MTU stiffness for up to 90 minutes (109). In the absence of a greater effect on TM relaxation during stretch or a greater reduction in passive force at a given muscle length, the greater effectiveness of PNF stretching has been attributed to an altered stretch sensation that produces a greater tolerance to stretch, and thus permits a greater amount of stretch and a greater ROM_{max} (69). The mechanism by which PNF increases stretch tolerance to a greater extent than static stretching has not been identified.

Figure 10.10 Schematic illustration of the effects of a single static stretch (SS) compared with a single PNF maneuver (see Figure 10.6). PNF produced a greater increase in maximum tolerable passive force and length, but did not differ from SS in its effect on passive force at a given muscle length. Based on the data of Magnusson *et al.* (66).

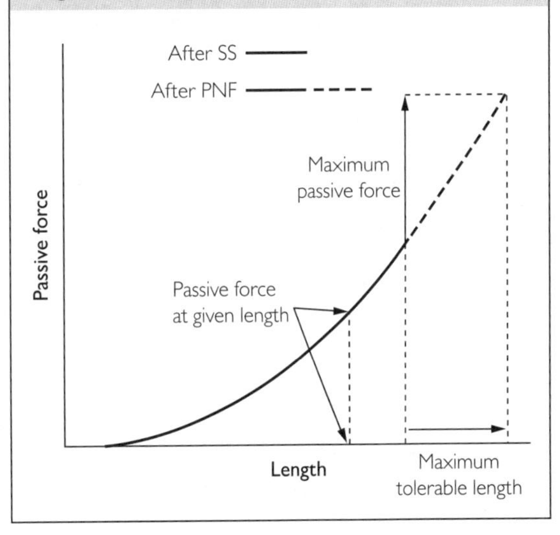

How Long do the Effects of a Stretching Session Last?

Whether an acute bout of stretching prevents injury or affects performance (positively or negatively) depends on how long the effects of pre-activity stretching last into a training session or performance. Increased ROM_{max} after a static stretching session, of the type that might be done by athletes (e.g., three 30 second stretches per muscle group) before a training session or performance, may persist from 20 minutes up to 24 hours (22,25,116). Of more importance to the possible effects of stretching on injury prevention and performance may be the magnitude and persistence of any decrease in passive force (stiffness) after a stretching session. Two examples of the magnitude and persistence of passive force and stiffness decrease after static stretching are shown in Figure 10.11. The two examples involve the same muscle group but different subjects. Nevertheless, the two examples taken together suggest a dose–response effect: a larger number of longer stretches produces a larger and more persistent decrease in passive force and stiffness.

How much time would have to be spent stretching with the minimum routine (8 × 30 seconds with 20 second rest periods between stretches; Figure 10.11B) needed to reduce MTU stiffness for at least 10 minutes after the routine? It would take 7 minutes to do the routine on one muscle group. If six muscle groups need to be stretched using this routine, 42 minutes of stretching would be required. For some sports, even more muscle groups and thus more time would be needed to complete the stretching routine (74).

A PNF stretching session that included five repetitions of stretch–contract–relax–stretch variation reduced passive force at a given angle by approximately 5% 90 minutes after the stretching session (109). Based on the pattern shown in Figure 10.11, the decrease in passive force was likely greater over the first 30 minutes after the stretching session. However, assuming a 20 second rest period between repetitions of the PNF routine described above, it would take approximately 4 minutes to stretch each muscle group. If six muscle groups need to be stretched, the whole stretching routine would take 24 minutes.

While a persistent decrease in passive force after a stretching session has been documented (see above), a decrease in passive force is by no means guaranteed (39,40). There may be individual differences in the response to the same stretching session, and different muscle groups may respond differently (Box 10.2). If a particular stretching session achieves a persistent decrease in passive force (stiffness), this might conceivably reduce the risk of injury, but the decrease in passive force might also contribute to impaired performance,

Figure 10.11 (A) Time course of decreased MTU passive force at three angles (combined) after 13 static stretches of approximately 2 minutes duration in human plantarflexor muscles. Compared to a control condition (not shown), the decreases were statistically significant up to 30 minutes. Based on the data of Fowles *et al.* (26). (B) Time course of decreased MTU stiffness at four one-degree angular displacements (combined) after 4, 8, and 16 30 second stretches of the human plantarflexors. There were 20 second rests between stretches. For the 8 × 30 seconds and 16 × 30 seconds stretch routines, MTU stiffness was significantly decreased for 10 min. Based on the data of Ryan *et al.* (93).

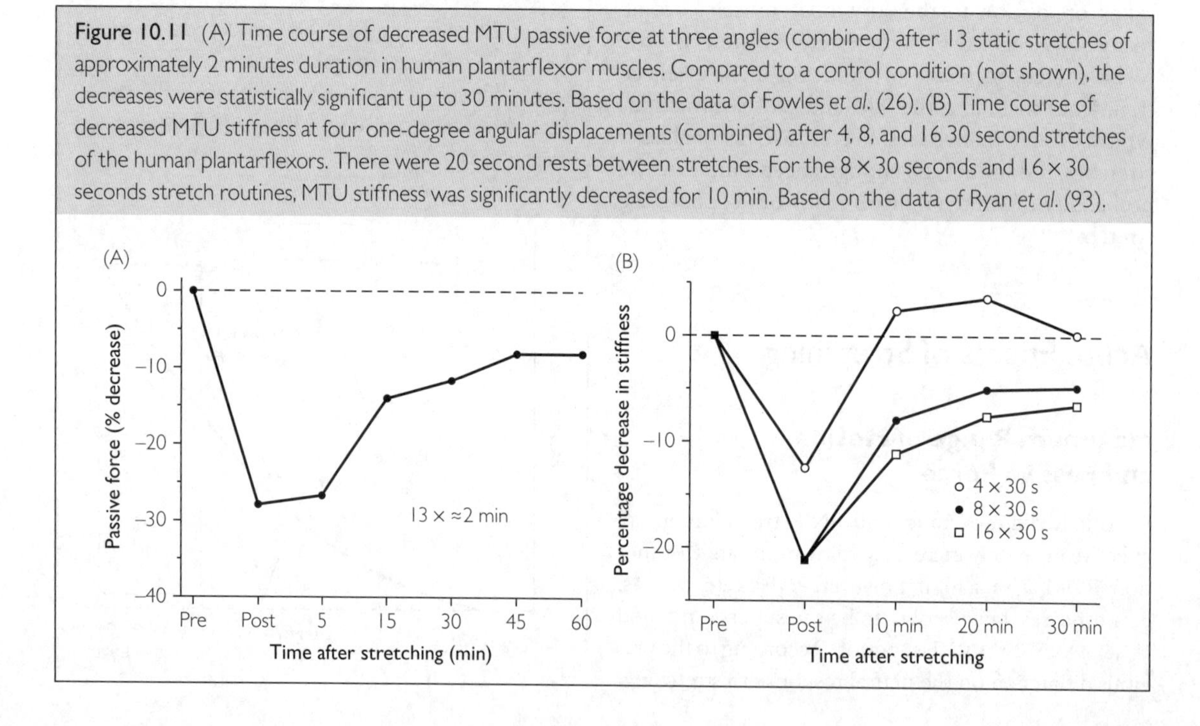

BOX 10.2 WHAT PART OF THE MUSCLE–TENDON UNIT IS LENGTHENED DURING STRETCH?

When a relaxed muscle is stretched, both the muscle and its tendons are stretched. A muscle plus its tendons are called an MTU. Thus, the term **muscle stretch** often refers to a stretch of the MTU. The relative contributions of muscle and tendon to MTU elongation during stretch have been measured in some human muscles, with varying results. In gastrocnemius, for example, tendon elongation has accounted for the majority (45) or less than the majority (1) of the elongation of the MTU to maximal or near-maximal length. There may also be individual differences in the relative contribution of tendon and muscle elongation during stretch (1). It has also been shown that not all **fascicles** (bundles of muscle fibers) within a muscle begin to elongate at the same MTU length. They are successively recruited as the MTU unit elongates (45). Tendons are actually less compliant (offer greater resistance to a given increase in length) than muscle fascicles (43), but tendons' much greater length than muscle fascicles' allows tendons to make a significant contribution to MTU elongation during stretch.

After several static stretches, does the relative contribution of tendon and muscle to the acute increase in MTU length change? This question was addressed in one study, and it was found that the muscle contributed more than the tendon to the acute increase in elongation of the MTU. Furthermore, within the muscle it was the connective tissue rather than muscle fascicles (muscle fibers) that elongated after the stretches (81).

for example decreased strength (26). Thus, the athlete may face a dilemma when stretching before training or competition: a stretching session sufficient to prevent injury by decreasing passive force may also impair subsequent performance, especially strength (10). Alternatively, if the athlete moderates the stretching session to avoid performance impairment, any decrease in passive force may not be large enough or sufficiently persistent to protect against injury, assuming that reduced muscle stiffness prevents injury. Then, as previously noted, there is the time factor. For example, a stretching routine sufficient to reduce MTU stiffness for at least 10 minutes would require at least 30 minutes of stretching if several muscle groups are stretched. This issue will be discussed in more detail later in this chapter.

Strength, Power, and Speed

Static stretching is the most common form of stretching; therefore, its effect on performance has received the most investigation (52). Static stretching may impair subsequent strength, power, and speed performance (10,52). Any negative effect appears to be greater on strength than on power or speed (10,52,74). As shown in Figure 10.12, there is a dose–response effect in relation to the duration of pre-activity static stretching and its effect on performance (10,52). A total time under stretch up to 45 seconds has a minimal effect on performance, whereas a total time under stretch of 60 seconds or longer is more likely to have a negative effect and a greater negative effect (10,52). Strength deficits produced by static stretching may persist for 1–2 hours after the end of the stretching session (10). Most pre-activity stretching routines used by athletes have a total time under stretch of 45 seconds or less per muscle group (10,52), so according to Figure 10.11 such routines would pose little risk of impairing performance but are capable of producing transient increases in ROM_{max} (10).

The intensity of stretch is also a factor. Stretches to ROM_{max} are more likely than submaximal $(<\mathrm{ROM}_{max})$ stretches to cause performance impairment (10). Training status generally and flexibility specifically do not appear to affect stretching-induced performance impairment (10).

Figure 10.12 Effect of duration of static stretching (i.e., total time under stretch) on the extent (bars) and probability (arrows and percentage numbers) of performance decrease. Performance includes strength, power, and speed. Time after stretching to performance was ≤20 min. Based on the review by Kay and Blazevich (52).

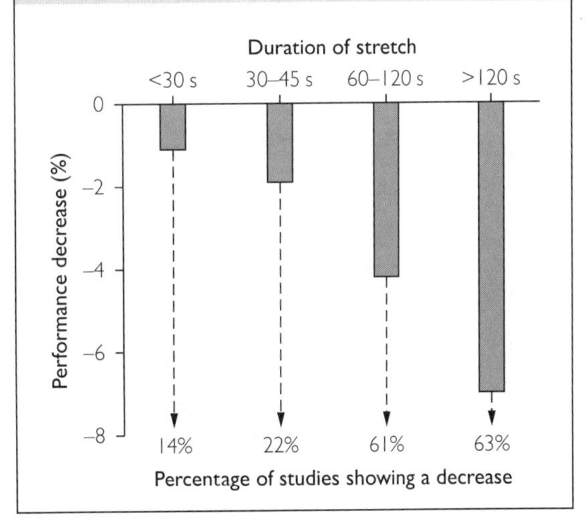

How might static stretching cause a transient decrease in strength, power, and speed? Static stretching can reduce the ability to fully activate muscles (10,74), and can thus obviously decrease performance. If static stretching affects the passive force–length relationship by reducing passive force at a given muscle length (i.e., less stiff or more compliant muscle), the rate of force development, and force, speed, and power of isolated concentric contractions may be reduced (18).

An increased MTU compliance, by shifting the passive force–length relationship to the right—that is, to longer MTU lengths (Figure 10.13)—may also shift the active force–length relationship to the right (113). The result would be a relatively greater force loss at shorter muscle lengths (46,83) (Figure 10.13). A more compliant series-elastic component (SEC; see Chapter 4) would cause muscle fibers to shorten to a greater extent, for a given MTU length, to lengths farther from their optimal length (L_o).

Decreased muscle stiffness and altered sensation of stretch might affect sensory feedback from muscles to the central nervous system and thereby negatively impact motor control during performance (11). On the other hand, decreased muscle stiffness could have a positive effect on performance by permitting a greater range of motion without the impediment of MTU passive resistive force (10,74).

Figure 10.13 Schematic illustration of the change in passive and active force–length relationships after a series of static stretches. After the stretches, passive force begins to rise above zero at a longer MTU length (rightward shift) and is smaller at any given length. Active force is disproportionately decreased at shorter lengths and the optimal length (L_o) is shifted rightward (see arrow). Active force–length relationships are based on the data of Herda *et al.* (46), Nelson *et al.* (83), and Weir *et al.* (113).

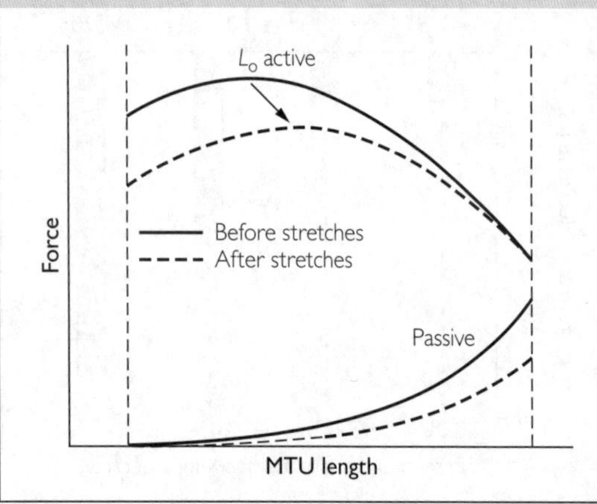

In contrast to static stretching, dynamic stretching is more likely to improve strength, power, and speed performance with little risk of impairing performance (10,74). However, the term dynamic stretching has been applied to ballistic stretching and various exercises (10), some of which mimic the actions of the performance to be tested (42). Thus, in addition to any stretch (e.g., passive force, stretch tolerance) effects, elements of so-called dynamic stretching (independent of any stretch) would also be expected to increase muscle temperature and induce postactivation potentiation, both of which can improve performance (discussed in Chapter 5). Furthermore, incorporating exercises that mimic the target performance into dynamic stretching would constitute a practice effect. It is also not clear what effect dynamic stretching might have on ROM_{max}, passive force of the MTU, and stretch tolerance, although an increase in muscle temperature might be expected to reduce muscle stiffness (87).

In contrast to the neutral or positive effects of dynamic stretching on strength, ballistic stretching may impair strength to the same extent as static stretching (82).

Stretching is rarely used alone before training sessions or competition. More commonly, a pre-activity warm-up consists of low-intensity aerobic exercise, followed by stretching, and finishing with sport-specific drills (10,28). Either alone or in combination with other components of a warm-up routine, static stretching may, if of sufficient duration and intensity, cause a transient strength deficit (10).

Motor Control

Compared to strength, power, and speed, relatively little research has investigated the effect of stretching on aspects of motor control. In young men, three 45 second static stretches impaired balance and increased reaction and movement time without decreasing isometric strength (11). In contrast, a similar static stretching routine did not impair balance in young women, whereas a shorter stretching routine (three 15 second static stretches) improved balance (17). A dynamic stretching routine, consisting of various exercises and drills, improved agility tested with a sprinting task that included changes in direction and acceleration (111). None of the studies tested for any possible change in MTU compliance.

Endurance

One factor affecting endurance performance is exercise efficiency or economy. For example, distance runners who have greater **running economy** (reduced energy cost running at a given velocity) will have a competitive advantage (96). One of many factors that can affect running economy is flexibility (96). Less flexible distance runners have greater running economy (20,47,49,110). If, as discussed in a previous section, an acute bout of static stretching can transiently increase flexibility (ROM_{max}) for several minutes after completion of the stretching session, is static stretching advisable before a training run or competition? In male distance runners, static stretching (4 × 30 seconds) of the knee extensors, ankle plantarflexors, and four other lower-limb muscle groups caused a decrease in running economy and reduced the distance covered in a 30 minute run (121). In contrast, static stretching of shorter duration (2 × 30 seconds) had no effect on running economy (41). Taken together, these two observations suggest a dose–response effect of static stretching. Greater MTU stiffness in some muscle groups involved in running is associated with increased running economy (6); therefore, if static stretching were to decrease running economy by transiently reducing MTU stiffness, then longer durations of stretching would be more likely to reduce running economy (see Figure 10.11B).

Female distance runners are more flexible (110) but have lower running economy (21) compared to male distance runners. Lesser MTU stiffness (36) may account for females' greater flexibility. Within a group of female distance runners, the less flexible runners may (110) or may not (8) have lower running economy. Does static stretching acutely affect running economy similarly in female and male distance runners? Non-elite female runners did one stretch of unspecified duration in five muscle groups; there was no effect on running economy (79). However, it is difficult to compare these results to those reported in males (see above) because single stretches of unspecified duration may not have transiently decreased MTU stiffness (see Figure 10.11B). The issue of whether female and male distance runners respond differently to static stretching in its effect on running economy may be resolved when both groups of runners perform the same number and duration of static stretches, and when the time course of any changes in MTU stiffness is monitored, as shown in Figure 10.11B.

Less is known about the effect of ballistic, PNF, or dynamic stretching on running economy. One type of dynamic stretching that causes a transient increase in ROM_{max} does not reduce running economy in trained male runners (125); however, it is not known whether MTU stiffness was altered by dynamic stretching.

Short-term endurance Some static and PNF stretching routines may reduce the number of repetitions completed in subsequently performed sets of weight-training exercises (29,84). A likely mechanism for a decrease in this kind of exercise, which may last for approximately 30–60 seconds, would be a decrease in strength caused by the stretching. The consequence of this would be that a given weight would represent a higher percentage of maximum strength after stretching, reducing the number of completed repetitions.

Prevention of Injury

Among athletes and coaches, a low level of flexibility and a high level of MTU stiffness are thought to be risk factors for MTU injury (103). Consequently, pre-activity stretching that transiently increases flexibility (ROM_{max}) and reduces MTU stiffness is believed to prevent injuries (99,103). A distinction should be made between the effects of an acute bout of stretching on injury risk in the subsequent training session or competition, versus the effects of stretch (flexibility) training over a period of weeks or months, the timing of which could occur before or after training sessions or even in separate sessions on the same or different days (48,101). This section will consider the effects of an acute bout of stretching on injury risk in the subsequent training session or competition. The effects of stretch (flexibility) training on injury prevention will be considered in a subsequent section on flexibility training.

Stretching is believed to prevent acute **muscle strain injuries (MSIs)** (53). Such injuries can be prevalent in some sports (e.g., sprinting, jumping, gymnastics) (53,74,114), whereas overuse injuries, considered to be less affected by stretching (74), are more prevalent in other sports (e.g., distance running, swimming) (74). An MSI consists of a partial or complete tear of the MTU (114). The typical location of injury is just proximal to the distal muscle–tendon junction (34,53,94), a location that has been confirmed in animal experiments (53). Muscles particularly at risk for MSI include biceps and triceps brachii, pectoralis major, gastrocnemius, hamstrings, rectus femoris, adductor longus, iliopsoas, and flexor pollicus longus (94).

Most MSIs occur during eccentric contractions in the normal range of motion (108) but muscles may be more vulnerable during eccentric contractions at long muscle lengths (74). Fast, strong eccentric contractions that are part of the stretch-shortening cycle (SSC, see Chapter 4) may impose particular stress on the MTU, particularly tendons (123). Recall from Chapter 4 that eccentric contractions can produce the greatest force compared to isometric and concentric contractions, and eccentric force increases and then plateaus as velocity of lengthening increases. Thus, actions with fast, strong eccentric contractions (e.g., jumping, sprinting, changing direction while sprinting) are most likely to produce MSI. For example, hamstring strains may result when these muscles contract eccentrically at their longest length during the gait cycle in sprinting (89,97). MSI tends to occur late in training sessions or competition (53,89), suggesting that fatigue may be a predisposing factor (53).

Accepted risk factors for MSI include age, previous injury, and muscle weakness (101). There is contradictory evidence on whether low flexibility is a risk factor (74,89,103). Part of the contradiction may be because both the most and least flexible athletes are prone to injury (108), the former being at greater risk for some forms of acute injury and the latter at greater risk for overuse injury (103).

For many athletes and coaches, the rationale for pre-activity stretching's role in preventing MSI is straightforward. If it can be assumed that stiffer MTUs are more susceptible to MSI, then stretching, by transiently decreasing MTU stiffness, decreases the risk of injury (see Box 10.3). A less stiff MTU allows a greater passive or active (eccentric contraction) force to be tolerated before strain injury occurs (however, see Box 10.3). There is no doubt that stretching of sufficient intensity and duration can decrease MTU stiffness for at least several minutes (Figure 10.11). Even if stretching has the potential to prevent MSI by decreasing MTU stiffness, the dilemma facing the coach and athlete is that stretching sufficiently intense and prolonged to reduce MTU stiffness carries the risk of impairing performance (Figure 10.12). It should also be cautioned that a stretching routine that transiently increases flexibility (ROM_{max}) does not ensure there has been a decrease in MTU stiffness; instead, there may be a temporary increase in stretch tolerance with no decrease in MTU stiffness (no reduced passive force at a given MTU length, see Figure 10.4B). It has been suggested that altered sensation of stretch (analgesic effect of stretching) may predispose toward rather than protect against injury (100,101).

BOX 10.3 IS THE ASSUMPTION THAT REDUCED MUSCLE STIFFNESS (INCREASED MUSCLE COMPLIANCE) PROTECTS AGAINST MUSCLE STRAIN INJURY VALID?

That decreased MTU stiffness protects against injury is intuitively appealing; however, a decrease in resting muscle stiffness is associated with greater injury risk in animal muscles (100). It has also been questioned whether decreased stiffness of a relaxed MTU (or increased ROM_{max}) is even relevant to injury risk, since most MSIs occur in active muscles undergoing forced stretch (eccentric contractions) (108). More relevant would be muscle compliance/stiffness during contraction, which is determined by the number of attached myosin cross-bridges. In contrast, resting muscle compliance is determined by the compliance of the cytoskeleton, intramuscular connective tissue, and tendon (100,101).

Much research has focused on the question of whether pre-activity stretching prevents MSI and injuries in general, and in turn this research has been evaluated in many reviews (28,44,74,99,100,101,103, 105,108,114,122,123,124). Unfortunately, these reviews offer no consensus on the effectiveness of pre-activity stretching in preventing injuries. Several factors likely account for the lack of consensus, including variation in the assessed quality of studies, the various sports and activities examined, the different classifications of injuries, and variation in the intensity, duration, and type of stretching used. An additional problem in assessing the effectiveness of stretching in preventing injuries is that stretching is seldom done alone as the pre-activity warm-up. Instead, stretching is typically part of a warm-up procedure that includes, in order, low-intensity aerobic exercise, some form of stretching, and sport-specific drills. The warm-up as a whole is thought to prevent injuries in a subsequent training session or competition. The relative importance of each component of the warm-up in preventing injury is difficult to assess (108) and has not been systematically investigated (28).

Here are some examples of how the other components (aerobic exercise, sport-specific drills) of a typical warm-up make it difficult to evaluate the effectiveness of the stretching component in preventing injury. First, the aerobic exercise and drills increase core and muscle temperature; the latter could decrease muscle stiffness independently of and perhaps more than the stretching component (53,94). One review noted that when warm-up emphasized exercises that increased core (body) temperature, there was greater injury prevention compared to warm-ups that emphasized stretching (28). Secondly, the aerobic component and particularly the more intense sport drills could cause a magnitude of muscle and tendon stretch that might rival that of the formal stretching component. Strong isometric, eccentric, and even concentric contractions will stretch tendons and intramuscular connective tissue (58,94) (see Figure 4.36, Chapter 4). Thirdly, the practice effect of pre-activity drills might reduce the incidence of injury during training sessions and competition by reducing the probability of awkward movements that might cause injury.

To avoid the previously discussed complications in assessing the role of stretching as a pre-activity injury preventive measure, the following suggestions have been proposed for future research (74). The study design should include these interventions: (1) stretching alone, (2) warm-up alone, (3) stretching plus warm-up, and (4) neither. Apply an intensity and duration of stretches (e.g., 4–5 × 60 seconds) or 8 × 30 seconds (see Figure 10.11B) with a high probability of reducing MTU stiffness. Study a variety of sports with a high incidence of MSIs. Also, place more emphasis on studies with trained athletes (108).

In the face of a lack of consensus on the value of pre-activity stretching for the prevention of injuries, what should the athlete do? As part of the conventional three-component warm-up described above, there is insufficient evidence to endorse or advise against stretching for the prevention of injuries. Consequently, most athletes can probably continue with the warm-up routine they have been using, bearing in mind that intense and prolonged stretching routines may impair performance.

Flexibility

Definition

Flexibility has been defined as the maximum range of movement (ROM_{max}) achievable without injury at a joint or series of joints (108). ROM_{max} is usually considered to be limited by extensibility of the MTU rather than by joint structure (33). Various forms of flexibility have been described, including active, passive, dynamic, static, and functional flexibility (3). For athletes, the important distinction is between a range of movement (ROM) achieved with little effort (i.e., little resistance to stretch and minimal discomfort) in comparison to a ROM achieved only with assistance and some discomfort. Examples of the former would be a gymnast easily sliding into splits, a diver easily assuming a tight pike position, and a hurdler easily abducting the hip joint sufficiently to clear a hurdle. Obviously, these actions cannot be done with assistance, and performance would be impaired if great effort were required to achieve the needed ROM. For the athlete, the key is *ease* of moving through the required ROM. This is in contrast to some tests of flexibility, in which the MTU is stretched beyond the range through which an athlete can easily move. Similarly, in typical stretching sessions the goal is to *extend* the ROM through which the athlete can easily move. To achieve this goal, a stretch overload must be applied, which consists of stretching to the current limit of stretch tolerance.

Determinants of Flexibility

In previous sections, attention was given to the effects of MTU stiffness (compliance) and stretch tolerance on ROM_{max}. Since flexibility is defined as ROM_{max}, it follows that MTU stiffness and stretch tolerance must be important determinants of flexibility, as shown in Figure 10.14. In contrast, reflex contractions of stretched muscles do not appear to affect flexibility in untrained and trained subjects free of neurological limitations (65,75). Contrary to common belief, more flexible individuals achieve their greater ROM_{max} not only as a result of their greater MTU compliance, but also because of a greater maximum tolerable stretch force; in fact, at the limit of stretch, flexible subjects have greater MTU stiffness than inflexible subjects (61) (Figure 10.14).

Gender Difference in Flexibility

Women are generally more flexible than men (3). The gender difference is already apparent in prepubescent children (73). Greater ROM_{max} in females is related at least in part to a smaller passive force at a given joint

Figure 10.14 Schematic comparison of range of motion (ROM) and passive force during stretch (increase in range of motion and thus muscle length) in a flexible and an inflexible subject. As expected, the flexible subject attains a greater ROM (flexibility = ROM_{max}) and, as might also be expected, has less MTU stiffness (and less passive force) at a given MTU length in the ROM (L_G). Perhaps unexpected is that the flexible subject can tolerate a greater peak stretch force, resulting in a greater MTU stiffness at the limit of stretch. Based on data from Magnusson *et al.* (61,65) and McHugh *et al.* (75).

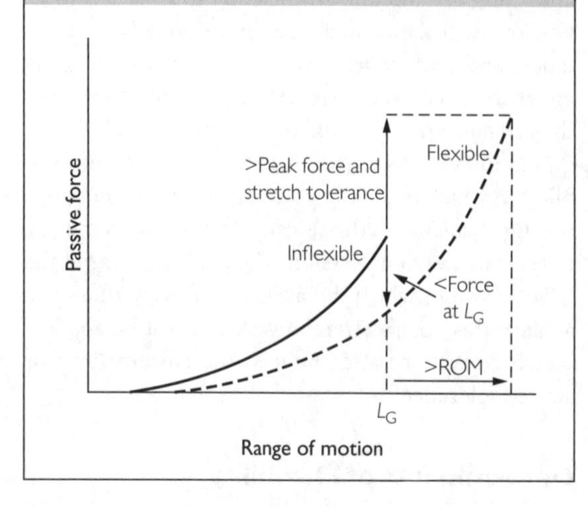

angle (i.e., lower MTU stiffness) (32,76). The greater MTU stiffness in males may be partly the result of their greater muscle mass (117), since larger muscle mass is associated with greater stiffness in the mid-range of motion and a smaller ROM_{max} (65).

Specificity of Flexibility

Contrary to the common belief that flexibility is a general trait—that is, that a person is flexible or inflexible at all joints—research indicates a high level of specificity in flexibility (3,78). Thus, high flexibility at the hip joint is a poor predictor of shoulder flexibility. A genetic component related to general flexibility cannot be completely discounted (3); however, in large samples of tested subjects, flexibility is specifically related to subjects' sport and activity involvement. For example, compared to non-tennis athletes in sports involving mainly the lower body, tennis athletes have greater shoulder flexibility in dominant-arm external rotation, but lower flexibility in dominant-arm internal rotation (16). Therefore, flexibility may be specific not only to a particular joint, but also to a particular movement of the joint and to whether the joint is in the dominant or non-dominant limb.

Flexibility and Performance

The influence of flexibility on performance can be assessed in two ways. In one approach, called a cross-sectional study, a group of subjects is tested for both flexibility and some type of performance, and then correlations are calculated to determine if flexibility is positively or negatively correlated with performance. In the second approach, a longitudinal or intervention study, a group of subjects is prescribed a flexibility training program for a period of time, after which tests of flexibility and performance are conducted to determine if increases in flexibility are correlated, positively or negatively, with changes in performance. This section will cover the cross-sectional type of study, whereas intervention studies will be covered in a subsequent section on flexibility training.

Some aesthetic sports, such as gymnastics, diving, and figure skating, contain elements that are demonstrations of flexibility (e.g., splits in gymnastics). The importance of (relatively extreme) flexibility in performing these elements is obvious. Less obvious is the role that flexibility might play in sports like distance running, sprint running, jumping, swimming, racquet sports, and many team sports. Once an athlete can move easily through a sport-specific range of motion, does an additional level of flexibility (ROM_{max}) enhance or impair performance? Both enhancement and impairment are possibilities. In tennis players, for example, *decreased* ROM of shoulder internal rotation was suggested to be an adaptation to repeated forehand strokes (enhancement?), whereas *increased* ROM of shoulder external rotation was an adaptation to repeated serves (enhancement?) (16). On the other hand, does tight (less flexible) internal rotation predispose toward injury? Would a flexibility training program that decreases injury risk by increasing internal rotation ROM_{max} have the unintended side effect of impairing forehand stroke performance? These questions have not been definitively answered.

Activities like cycling, boxing, skating, or swimming that emphasize more or less isolated concentric contractions might not benefit from flexibility beyond that needed to move easily through the required ROM, particularly if the increased ROM is accompanied by reduced MTU stiffness. Similarly, a relatively compliant MTU would be a disadvantage in lifting a weight with an isolated concentric contraction (Figure 10.15). Instead, a relatively stiff MTU would allow a rapid transfer of muscle force to bone (123), thereby increasing rate of force development in the early phase of an isometric or isolated concentric contraction (57,120) (see Figure 10.15).

In contrast, activities such as jumping and running that utilize the SSC might benefit from increased flexibility that is accompanied by reduced MTU stiffness (greater MTU compliance), because more elastic energy could be stored and released from more compliant tendons (123). The same benefit would apply to weight-lifting exercise if lifts (concentric contractions) are immediately preceded by lowering the weight (eccentric contractions). A more compliant MTU allows for a greater SSC potentiation (Figure 10.15). However, optimal MTU stiffness for SSC potentiation may vary depending on the speed or frequency of movement (106). Whereas a relatively compliant MTU may be optimal for a relatively slow movement such as a barbell bench press (120) (Figure 10.15), a relatively stiff MTU may be optimal for faster SSC movements like running. Thus, economy in distance running is associated with a stiff MTU (6). Therefore, each sport action must be analyzed to determine whether a relatively stiff or compliant MTU is optimal for performance.

Flexibility may also influence a muscle's *active* **force–length relationship**. Figure 10.16 shows how the optimal length is shifted to a shorter muscle length in the inflexible subject. This leftward shift in optimal length might affect performance if the force is not maximal at the appropriate muscle length (joint angle). The biomechanics of a sport action could therefore be adversely affected. The converse is also possible, however, in that extreme flexibility might

Figure 10.15 Schematic illustration of force generated in an isolated concentric (CON) bench-press lift compared to a lift in which the concentric contraction was preceded by an eccentric contraction (ECC), resulting in SSC potentiation (i.e., greater force during concentric contraction). Subjects with a more compliant MTU (top) exhibited more SSC potentiation than subjects with a stiffer MTU (bottom). However, inspection of the top and bottom panels indicates that the stiff subjects performed better in the concentric condition. Based on data from Wilson *et al.* (120). See text for further discussion.

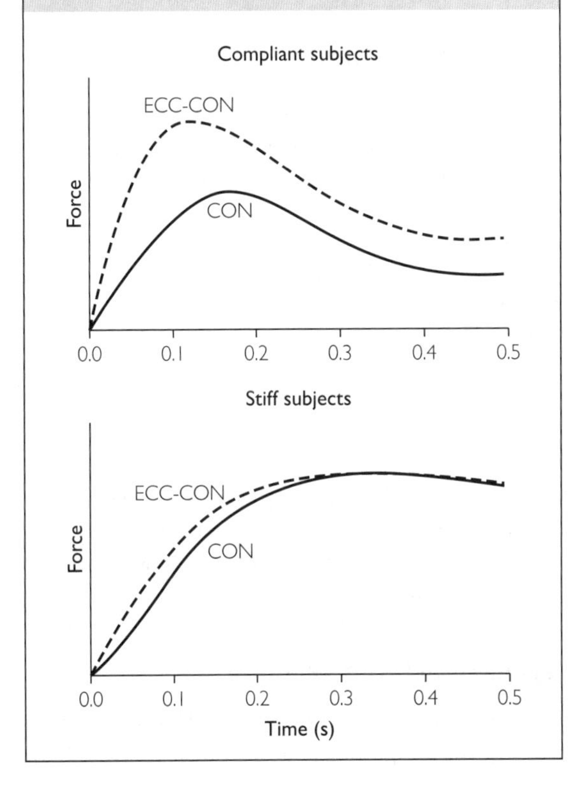

Figure 10.16 Schematic illustration of how flexibility could affect the *active* force–length relationship. The inflexible subject with an associated shorter muscle length would have the force–length relationship shifted leftward so that the optimal muscle length would occur at a joint angle corresponding to a shorter muscle length. Based on knee flexor (hamstring) data from Alonzo *et al.* (2).

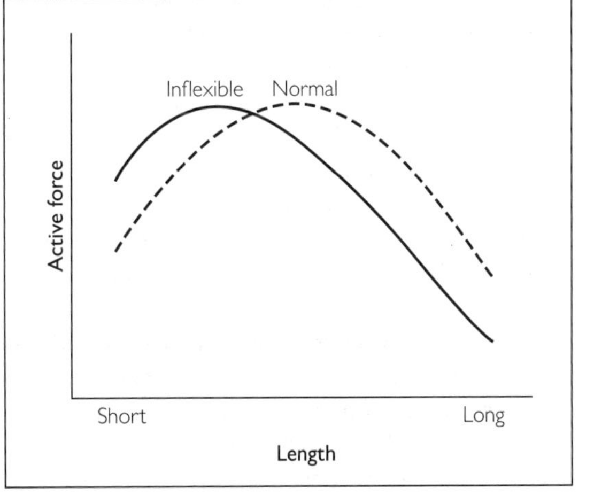

cause an excessive rightward shift of the force–length relationship, which might impair performance.

Flexibility and Injury Prevention

As noted earlier in the section on the Acute Effects of Stretching, a low level of flexibility and a high level of MTU stiffness are thought by many to be risk factors for MTU injury (103). However, there is contradictory evidence on whether low flexibility is a risk factor for injury (74,89,103), partly because both the most and least flexible athletes are prone to injury (108), the former being at greater risk for some forms of acute injury and the latter at greater risk for overuse injury (103). The role that flexibility plays in preventing injuries likely varies across sports. For example, increased flexibility appears to reduce the incidence of MSI in soccer players (105). In contrast, shoulder flexibility was unrelated to injury history in baseball pitchers (62).

Flexibility Testing

In a training or practice session, an athlete or coach may assess flexibility simply by visual inspection and sensation of stretch at a particular position. A more precise measurement of flexibility (ROM_{max}) in a laboratory is typically done with some form of **goniometer** that provides a value for joint angular displacement, expressed in degrees or radians (30,108), or with a **fleximeter** to measure the degree of bending. In testing athletes, the first consideration is to select tests that are specific and important to performance. Then attention should be given to reliability (also called reproducibility or repeatability) of measurement. In the previous section on the Acute Effects of Stretching, it has been shown that a bout of stretching can cause a transient increase in ROM_{max}. An increase in muscle temperature can also increase ROM_{max}. Therefore, the reliability of flexibility measurements is improved if repeated testing is conducted under the same controlled conditions. Reliability of measurement is particularly important when monitoring changes in flexibility during training programs.

Stretch tolerance vs muscle–tendon unit stiffness Changes in ROM_{max} could be due to changes in stretch tolerance, MTU stiffness or both (Figure 10.4). Therefore, there would be value, when testing flexibility in the laboratory, to include measures of both ROM_{max} and MTU stiffness (or compliance) (Figure 10.7). Actual changes in MTU stiffness would likely have a greater effect on performance. Test protocols combining ROM_{max} and MTU stiffness measurements have been developed for knee flexors (hamstrings) (69,116) and ankle plantarflexors (calf muscles) (26,93). It may be challenging to develop combined test protocols for other joint motions (e.g., shoulder inward and outward rotation) (62).

Active vs passive range-of-movement measurement One way of measuring ROM_{max} is to have a tester or apparatus apply the stretching force while the subject relaxes (2,65). This is called a **passive** test of ROM_{max}. Alternatively, the subject is instructed to stretch the TM(s) by contracting antagonists (2). For example, the subject contracts the knee extensors and/or hip flexors to stretch the knee flexors (2). This is called an **active** test of ROM_{max}. One consideration in choosing between a passive or active ROM_{max} test is that the result of an active ROM_{max} test could be determined by the strength of antagonist muscles rather than MTU stiffness and stretch tolerance in the TM. An active test may still be preferred if the emphasis is on *ease* of moving through a ROM.

Flexibility Training

Flexibility training, also called stretch training, consists of regular bouts of stretching over a period of weeks or months. To increase flexibility (ROM_{max}), stretch **overload** should be applied; that is, a stretch should be taken to the point of mild discomfort or a sensation of tightness. In its simplest form, stretch training is the daily repetition of the stretching routine done as part of the warm-up before training sessions. For some athletes, the stretch routines used as part of warm-up may be sufficient for increasing ROM_{max} to the desired level. For other athletes, more elaborate stretching sessions done before, during or after training sessions may be necessary to achieve the flexibility goals. This section will cover stretch routines specifically designed to increase flexibility (ROM_{max}) over a training period.

Types of Stretch Training

The same types of stretching are used for stretch training and warm-up: static, ballistic, dynamic, and PNF.

The different types of stretching are described in detail in the section on the Acute Effects of Stretching. Static stretching is most commonly used because of its simplicity and the fact that the athlete can usually do it without assistance. Ballistic stretching can also be done without assistance, but there is often concern about the safety of this form of stretching. As will be shown, however, ballistic stretching can be effective at increasing ROM_{max}. PNF stretching may be the most effective type of stretching but is more difficult to do and usually requires assistance. Dynamic stretching is increasingly popular, but this form of stretching often includes sport-specific drills that have effects in addition to possibly providing a stretch stimulus.

Stretch Training Routines

The components of a stretch training routine include type of stretching, number of stretches per muscle group, duration of each stretch, total time under stretch, number of stretch training sessions per day, and number of stretch training sessions per week.

Static stretch training Static stretch training sessions consisting of a total of 60 seconds of stretch per muscle group, done once per day, 2–4 days per week over a period of 3–12 weeks, will increase ROM_{max}. The 60 seconds of stretch can be divided into individual stretches of 10–30 seconds duration (4). There is some evidence of a dose–response effect, however. Increased total time under stretch (e.g., 180 seconds), longer individual stretch duration (e.g., 60 seconds) and greater training frequency (e.g., 5–7 days per week) will produce better results (4,112). On the other hand, if training frequency is high, the added effect of an increased daily total stretch duration may be minimal (7).

Ballistic stretch training Ballistic stretch training sessions consist of a number of stretches of each muscle group, for a total duration of 1–2 minutes. Strong contractions of OMs may be used to create momentum in the limb or trunk to stretch the TM group. Alternatively, body weight can be used to produce the stretching force, for example in stretching the calf muscles. As in the case of static stretching, a greater frequency of ballistic stretching produces better results (112).

Peripheral neuromuscular facilitation stretch training PNF stretch training has several variations. The most effective form of PNF involves contraction of both the TM and OM (98). In this form, the trainee first contracts the OM to stretch the TM (the OM may initially contract concentrically and then transition into an isometric contraction); an assistant usually helps stretch the TM to the maximal tolerable length, where it is held for several seconds. Second, with the assistant providing resistance, the trainee isometrically contracts the TM for a few seconds at the held length. Third, the trainee contracts the OM again to stretch the TM, with the help of the assistant. This three-phase cycle, comprising one repetition, will be repeated one or more times. To increase ROM_{max}, one repetition of PNF done a minimum of twice per week is sufficient; furthermore, the isometric contraction of the TM does not have to be held longer than approximately 3 seconds and the intensity of contraction of the TM does not have to exceed 20% of maximum isometric force (ISO_{max}) (98). However, to achieve greater increases in ROM_{max}, TM contractions at a greater percentage of ISO_{max} (up to 75%) and longer duration (up to 6 seconds) have been recommended, and the duration of the assisted stretch should be 10–30 seconds (4). A greater frequency of PNF training sessions (112) and more PNF repetitions per session may also be beneficial.

Dynamic stretch training Dynamic stretch training, which consists of controlled movements through the active range of motion (24), has several variations. A simple variation has the trainee contract the OM to stretch the TM, more slowly than in ballistic stretching. The end-stretch position is held for several seconds (e.g., 5 seconds). The trainee then relaxes the OM to allow the TM to shorten slowly (e.g., 5 s) to the starting position. This process is repeated several (e.g., six) times (7). Another variation of dynamic stretching involves rapid movements similar to ballistic stretches; for example, upward leg kicks to stretch the hamstrings (90) and flick-backs to stretch the quadriceps (24). In fact, the terms ballistic and dynamic have been used synonymously (10,124). Yet another form of dynamic stretching consists of sport-specific drills in which there is no particular attempt to stretch the muscle to ROM_{max} (77).

Effectiveness of different types of stretch training PNF stretch training, particularly when it includes contraction of the TM and OM, is generally but not always most effective for increasing ROM_{max} (4,23,98,108). The slightly greater flexibility gains of

PNF stretching come at the cost of a more time-consuming and complicated method that requires assistance. Ballistic stretching can be as effective as static stretching, and PNF and static stretching are generally more effective than dynamic stretching (4,7).

When to stretch? Stretching before or after training sessions appears to produce similar acute increases in ROM_{max} (9) and may also provide a similar stretch training stimulus, but there are other factors to consider. The section Acute Effects of Stretching discussed the advantages and disadvantages of stretching before a training session (aerobic, anaerobic, strength, and power training). If the stretch training is done before other types of training, the stretching may prevent some kinds of injuries. On the other hand, pre-activity stretching may impair performance in other types of training, perhaps reducing the training stimulus. A stretch training session with a total time under stretch of less than 45 seconds per muscle group is unlikely to impair performance (Figure 10.12); however, longer stretch durations will have the opposing effects of increasing the flexibility training stimulus but decreasing, for example, the strength training stimulus of a subsequent strength training session. Therefore, if an optimal training stimulus for both flexibility and strength is desired, it may be preferable to do the flexibility training after the strength training. Some athletes do not have this option; they must alternate between stretch and other types of training in the same training session. For example, gymnasts doing a training routine often alternate between applying stretch and strength training stimuli for the same muscle group. Gymnasts and other athletes (e.g., wrestlers) must develop extreme levels of both flexibility and strength, as well as power and speed, in the same muscle groups.

Effect of stretch training and detraining on ROM_{max} As with other types of training (aerobic power, anaerobic power, strength, etc.), increases in flexibility (ROM_{max}) are initially rapid but are then subject to the law of diminishing returns; in other words, the rate of increase declines over time. Figure 10.17 illustrates this pattern of increased ROM_{max}. Whereas the pattern shown in the figure would apply to most stretch training programs (95), the amount of increase in ROM_{max} would vary based on the type of stretching used, the total time under stretch in each training session, the frequency of training, the muscle group stretched, and the athlete's initial level of flexibility. When a period of stretch training ends, ROM_{max} begins to decline over a period of weeks (Figure 10.17). If stretch training resumes after a period of detraining, ROM_{max} can be regained at about the rate of the first period of training (Figure 10.18). It is possible that a maintenance program with a reduced volume and frequency of stretch training will reduce or eliminate the loss of ROM_{max} after a higher-volume and frequency training period.

Mechanisms of increased ROM_{max} with stretch training Intuitively, the most likely mechanism for increased ROM_{max} (flexibility) after stretch training would be a decrease in MTU stiffness, which would cause a decrease in passive force at a given muscle length (joint angle) and make it easier to lengthen the MTU. This adaptation has often been observed

Figure 10.17 Sample time course of increase in flexibility (ROM_{max}) during a stretch training program lasting 6 weeks, followed by a 4 week detraining period. Static stretching of ankle plantarflexors (calf muscles) was done 5 days per week. In each training session the total stretch duration was 10 minutes: for four exercises, 5 × 30 seconds stretches were done. ROM_{max} was measured after 2, 4, and 6 weeks of training and after 4 weeks of detraining. Note the decline in the rate of ROM_{max} increase over the 6 weeks of stretch training. After 4 weeks of detraining ROM_{max} was still approximately 20% above the pre-training level. Based on the data of Guissard and Duchateau (37).

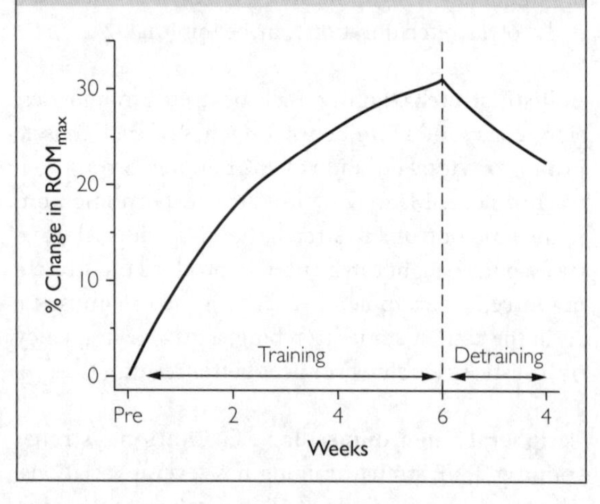

Figure 10.18 Effect of a period of 6 weeks of stretch training, 4 weeks of detraining, and 6 weeks of retraining on flexibility (ROM_{max}). Training sessions of two 30 seconds static hamstring stretches were done 5 days per week. ROM_{max} was measured after 6 weeks of training, 4 weeks of detraining, and 6 weeks of retraining. All of the increase in ROM_{max} after 6 weeks of training was lost after 4 weeks of detraining. A resumption of training for 6 weeks regained the ROM_{max} lost after detraining. Based on the data of Willy *et al.* (118).

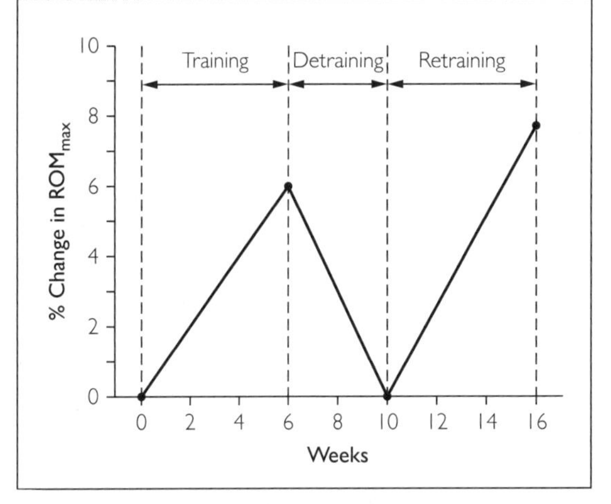

Figure 10.19 Schematic representation of the three possible effects of stretch training on the passive force–length relationship. The solid line represents the pre-training force–length relationship. Stretch training almost always increases ROM_{max} (flexibility), as shown. Intuitively, the likely mechanism for the increased ROM_{max} would be a decrease in MTU stiffness, which would decrease the passive force at a given muscle length (joint angle). However, sometimes the increase in ROM_{max} occurs without a change in the passive force at a given muscle length (joint angle); instead, there is simply an increase in the maximum passive force and muscle length (joint angle) attained during stretch. Counter-intuitively but rarely, the passive force at a given muscle length (joint angle) may increase.

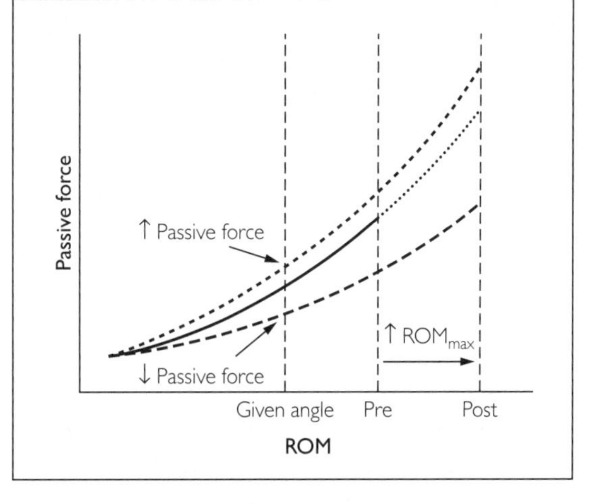

(37,56,71,72,109); however, sometimes the increase in ROM_{max} occurs without a change in MTU stiffness (no decrease in passive force at a given joint angle). Instead, there is simply an increase in the maximum passive force and muscle length (joint angle) attained during stretch (12,13,60,68,70,92). Rarely, and counter-intuitively, a trend toward an increase in passive force at a given joint angle has been observed (31). The possible changes in passive force at a given joint angle (MTU stiffness) after stretch training are illustrated in Figure 10.19.

How could ROM_{max} increase without a decrease in muscle–tendon unit stiffness? In the absence of decreased MTU stiffness, increased ROM_{max} has been attributed to altered sensation of stretch that allows a trainee to tolerate a greater magnitude of stretch force and thus a greater amount of stretch (12,68,98). The altered stretch tolerance is a neural adaptation to stretch training, a feature of which is that it is specific to the muscle group stretched. For example, if the hamstrings of only one leg are stretch-trained, only the trained leg exhibits the altered stretch sensation along with increased ROM_{max}. The neural mechanism for the increased stretch tolerance is unknown, but probably involves changes in sensory neural pathways. Another neural adaptation that has been traditionally attributed to stretch training is a reduction in stretch reflex contraction opposing stretch (37,38); however, in the majority of training studies, reflex activity (indicated by very low EMG activity) is already minimal in testing before training begins, particularly when ROM_{max} is measured with static stretches (69,98). Nevertheless, stretch training is similar to strength, power, and speed training in inducing both neural and muscular (MTU) adaptations. The relative roles of neural and MTU adaptations and their time course have not been determined in long-term training studies. MTU adaptations may require a large volume of stretching in each training session (72). In addition, most stretch training studies last a few to several weeks, whereas many athletes train for many months and years. Possibly, and as in the case of strength training (see Chapter 8), neural adaptations may tend to dominate early in a stretch training program, whereas MTU adaptations dominate later in the training period (37).

Adaptations in the muscle–tendon unit The decrease in MTU stiffness sometimes observed after stretch training (Figure 10.19) could be the result of adaptations in the muscle, intramuscular connective tissue, or tendon (i.e., any or all parts of the MTU). Specific adaptations have not been identified in human subjects, but possible adaptations have been hypothesized based on animal research (33). Intramuscular connective tissue (perimysium), which is considered most influential in determining passive force in response to stretch (75), is a probable site for adaptation. Alterations in collagen fibers might explain changes in tendon stiffness (71). Extreme stretch training over a long period may increase muscle length by the addition of sarcomeres in series at the ends of the myofibrils of muscle fibers (19,33).

Do different types of stretching cause different training adaptations? Different tests have been used to measure training-induced changes in the MTU after stretch training. One test, which measures the passive force response of the whole MTU to a slow stretch, is illustrated in Figure 10.19. Another test that specifically measures changes in tendon stiffness requires the subject to perform a ramp isometric contraction, increasing force from submaximal to maximal over a few seconds (e.g., 5 s) (58). Employing these two tests, static stretch training has been found to have no effect on tendon stiffness but to decrease passive force of the whole MTU at a given MTU length (joint angle) (58,71). In contrast, ballistic stretch training reduces tendon stiffness but has no effect on the passive force–length relationship (71). Based on the different effects of static and ballistic stretch training on the MTU, it has been recommended that both types of stretching be used in a training program (71). Six weeks of PNF stretch training increased ROM_{max} but had no effect on either tendon stiffness or the passive force–length relationship; therefore, the increased ROM_{max} was attributed to increased stretch tolerance (70).

Interaction between stretch (flexibility) and strength training Does combining strength and stretch training affect gains in strength? A static stretch training program concurrent with an 8 week weight-training program has been reported to augment strength increases (55). No measurements (e.g., muscle size) were made to account for the amplifying effect of the stretch training. In contrast, a combination of weight-training and static stretch training resulted in the same increases in strength and muscle volume as strength training alone (59). Based on animal research, increases in strength after stretch training might result from muscle stretch acting as a stimulus to protein synthesis (35) and consequent muscle growth (5). However, a test (27) in a human calf muscle (soleus) showed that a session of approximately 30 minutes of maximal tolerated passive stretch, a duration of stretch far exceeding that occurring in a typical stretch training session, failed to increase muscle protein synthesis. The failure of passive stretch to increase protein synthesis in human muscle may be partly because the maximum tolerated passive force during stretch is only a fraction of the force produced by voluntary contraction. For example, in human soleus muscle the maximum tolerated stretch was equivalent to only approximately 30% of the voluntary contraction force at the same joint angle (27). More importantly, when a maximum tolerated stretch force is compared to an equivalent voluntary isometric contraction force, only the latter stimulates protein synthesis (27). Therefore, conventional strength training involving muscle contractions is far more potent than passive stretch as a stimulus to protein synthesis and muscle growth; this fact would explain why combining stretch training with strength training does not increase muscle mass more than strength training alone (59).

Does combining stretch and strength training affect increases in flexibility and changes in MTU stiffness? As noted above, stretch training alone may decrease MTU stiffness. Strength training may cause MTU stiffness to increase (59), and this may be attributable in part to an increase in muscle size (cross-sectional area) (65). Therefore, the addition of strength training may prevent or diminish the decrease in MTU stiffness produced by stretch training (see Box 10.4).

BOX 10.4 DOES STRENGTH TRAINING MAKE AN ATHLETE MUSCLE-BOUND?

Half a century ago, many if not most athletes and coaches shunned or were at least wary of strength training for fear that increased muscle strength and muscle size would make an athlete inflexible or muscle-bound. The fear persisted despite readily observed examples of athletes who exhibited both extreme flexibility and great strength combined with marked hypertrophy. Male gymnasts were prime examples. In training and in competition, these athletes combine extreme strength and flexibility in the same muscle groups. How do gymnasts and other athletes escape becoming muscle-bound despite developing great strength and muscle mass? An obvious approach is to combine strength training with stretch training. The addition of stretch training can reduce or eliminate the tendency of strength training to increase MTU stiffness (59), and thus allow a flexibility increase similar to stretch training done without concurrent strength training (86). Another approach is to select strength training exercises that emphasize a full range of motion. These exercises may increase flexibility even in the absence of concurrent stretch training (80,102). Today it is generally recognized that great strength and flexibility can co-exist with proper training; in fact, younger athletes and coaches may not even be familiar with the term "muscle-bound".

Cross-sectional vs longitudinal muscle growth Stretch training over a period of many months or years may induce elongation of muscle fibers by serial sarcomere addition. This is a form of muscle growth that will increase muscle volume and mass. However, as discussed in detail in Chapter 8, increasing the number of sarcomeres *in series* does not increase strength. To increase strength, training must increase the number of sarcomeres *in parallel*; that is, there must be an increase in muscle fiber and thus whole-muscle cross-sectional area. Therefore, even if stretch training were to stimulate muscle growth, it cannot be assumed that the growth results in increased strength. On the other hand, if stretch training were to increase muscle fiber length by serial sarcomere addition, there would be performance implications. A longer fiber expands the active force–length relationship, allowing force to be produced over a greater joint range of motion. Also, a longer fiber translates into a faster maximum shortening velocity that not only can contribute to speed of movement but can also increase power even in the absence of an increase in strength, because power is the product of force *and* velocity. See Chapter 8 for further discussion on the effect of training on sarcomeres in series and in parallel.

Peripheral neuromuscular facilitation may combine stretch and strength training Versions of PNF stretch training that include maximal or near-maximal isometric contractions of the TM are actually combining isometric strength training and static stretch training. Strength training by itself tends to increase MTU stiffness, expressed as an increase in passive force at a given muscle length (joint angle), whereas if resistance training is combined with static stretch training, the increase in MTU stiffness is prevented (59). This interaction between stretch and strength training may explain why PNF stretch training, which is quite successful in increasing ROM_{max}, fails to decrease MTU stiffness (70).

Effect of flexibility training on performance Whether stretch training by itself can enhance performance depends on the nature of the performance. Reduced MTU stiffness, by increasing the elastic energy stored and released (123), may enhance some activities involving the SSC, such as jumping, sprinting and weight lifting. Thus, stretch training could benefit these activities. For example, 10 weeks of static stretch training increased weight-lifting strength and endurance, vertical jump height, and 20 m sprint speed, in addition to ROM_{max}; possible changes in MTU stiffness were not measured (54). Similarly, 8 weeks of static stretch training increased SSC potentiation in the bench-press exercise; an oscillation measurement indicated reduced MTU stiffness (119).

In contrast, increased rather than reduced MTU stiffness may enhance other types of performance. A relatively stiff MTU would allow a rapid transfer of muscle force to bone (123), thereby increasing rate of force development in the early phase of an isometric or isolated concentric contraction (57,59,120). In weight lifting, for example, a stiffer MTU is better for isolated concentric contractions, whereas a more compliant MTU is better for a lift that begins with an eccentric phase followed by a concentric phase (120) (see Figure 10.15). In running, a more compliant MTU may be better for sprinting (54) but a stiffer MTU enhances distance running performance by increasing running economy (6,104). Thus, the contribution of stretch training to performance will vary widely depending on the role that greater or lesser MTU stiffness has in the performance.

Box 10.5 provides a case study of stretching and flexibility training in an athlete.

BOX 10.5 THE CASE OF A NOVICE ICE HOCKEY GOALTENDER

The coach of a novice goaltender contacts a sport scientist for advice about pre-practice and pre-game stretching and about stretch training. The coach tells the scientist that the goaltender has general talent but has difficulty doing front splits when making saves. Front splits must be done with either the right or left leg forward. Splits require a high level of flexibility in the hip extensors, knee flexors, and hip adductors of the forward leg, and in the hip flexors, hip adductors, and knee extensors of the rearward leg. For pre-activity stretching, the sport scientist recommends a three-component warm-up consisting of skating (aerobic activity), static stretching, and goaltending drills that would include full-range movements described by some as being dynamic stretching. Static stretching would be three 10 second stretches to the point of tightness for splits with the right and left leg forward. The total time under stretch of 30 seconds should minimize the probability of performance impairment (Figure 10.12) and, combined with the increased muscle temperature produced by the skating and drills, be sufficient to decrease MTU stiffness and therefore possibly reduce the risk of acute MSI. For stretch training, the dose–response effect dictates that more frequent, higher-volume stretch training will be more effective. For example, in static stretching aim for 2–4 minutes under stretch in each training session and five or six stretch training sessions per week. The higher-volume stretching carries a greater risk of performance impairment (Figure 10.12); therefore, this volume of stretch training would be best done at the end of a practice or in separate training sessions. It should not be done just before a game.

In the game situation, splits are a fast, almost explosive action. Therefore, once full-range splits are possible with controlled slower and sustained stretches, faster ballistic-type stretches should be progressively added to the static stretching in training. This could be done as a splits drill at the end of a practice. Two sets of 10 splits could be done, alternating the forward leg. After a few weeks increase to three sets of 10 splits. On the basis of some evidence that static stretching mainly decreases the stiffness of the whole MTU whereas ballistic stretching mainly decreases tendon stiffness (70,71), the scientist recommends that both static and ballistic stretches be used regularly in stretch training. Once the athlete can perform the full splits at maximum speed, doing the splits frequently in practice would become the maintenance stretch training. Maintenance static stretching would continue to be done after practices or in an off-ice training session. Maintenance stretch training is important to prevent a gradual loss of flexibility (Figure 10.17).

Summary of Key Points

- Pre-activity static stretching has the potential to transiently impair strength, power, and speed performance, especially if total stretch duration per muscle group exceeds about 60 seconds. Power and speed appear to be less affected than strength. Balance, reaction time, and efficiency could also be adversely affected.
- Lower intensity and duration of static stretching may prevent or attenuate adverse effects, but may also reduce the transient increase in flexibility. Alternatively, stretching could be done after training or in separate sessions, although this strategy eliminates any benefit to performance and injury prevention from a transient increase in flexibility.
- Less is known about the acute effects of ballistic, PNF, and dynamic stretching on performance. So-called dynamic stretching contains elements that are essentially sport-specific drills, making it difficult to determine whether effects of dynamic stretching are due to the actual stretching or to other factors such as increased muscle temperature or skill practice.
- There is no consensus on whether pre-activity stretching prevents injuries or increases the risk of injury. Unfortunately, the probability of both injury prevention and performance impairment increases with increased intensity and duration of stretching.
- Static, ballistic, PNF, and dynamic stretching methods are effective in transiently increasing flexibility.
- Static stretching with a total time under stretch of 30–40 seconds per muscle group will transiently increase flexibility and possibly reduce muscle-tendon stiffness with a low risk of impairing performance. The total time under stretch can be achieved in different ways; for example, 1 × 30 seconds or 3 × 10 seconds. It is uncertain whether this amount of stretching will prevent injury.
- The optimal number of ballistic stretches per muscle group is not known, but a conservative recommendation would be performing 10–20

ballistic stretches per muscle group; for example, doing two sets of 10 stretches per muscle group. The intensity of the stretches should be gradually increased over a set of stretches to minimize the risk of injury.

- One or two repetitions of PNF stretching per muscle group may be optimal. The version of PNF involving contraction of both the muscle to be stretched (TM) and its antagonist (OM) is recommended. In PNF, isometric contractions of the TM at 20–100% of maximum sustained for 3–15 seconds would be effective. The static stretch phase should be held for approximately 8–10 seconds. There may be a dose–response effect with more intense, prolonged contractions and longer-held stretches being more effective, but this must be weighed against a greater possibility of impaired performance. PNF stretching may be the most effective for transiently increasing flexibility but requires assistance and is time-consuming.
- Dynamic stretching, consisting of ballistic stretching, controlled movements, and sport-specific drills, carries little risk of subsequent performance impairment; however, little is known about the effect of dynamic stretching on MTU stiffness.
- Over time athletes should experiment with all forms of stretching, and targeted muscle groups and movements should be sport-specific.
- Stretch training programs over several weeks or months can increase flexibility by increasing stretch tolerance and reducing MTU stiffness. Short-term (a few weeks) stretching programs mainly increase stretch tolerance.
- Long-term static stretching decreases MTU stiffness as a whole but has little effect on tendon stiffness; in contrast, ballistic stretching mainly decreases tendon stiffness. Therefore, combining static and ballistic stretching is recommended.
- PNF stretch training increases flexibility although PNF's effect on MTU stiffness is not well documented. The version of PNF combining contraction of both the TM and the OM is recommended. The isometric contraction of the TM can be effective at a range of intensities but contractions at or near maximal strength done for 6–8 seconds may be most effective. The static stretch phase of PNF should be held for 8–10 seconds. Three to four repetitions of the PNF procedure should be done for each muscle group.
- With any type of stretch training, increasing the time under stretch in each session and increasing the number of days per week of stretching will accelerate the training response and lead to greater final increases in flexibility. However, the intensity and volume of stretching optimal for increasing flexibility may acutely impair training or competition performance; therefore, it may be beneficial to place flexibility training sessions after other types of training or on separate days. Some athletes (e.g., gymnasts) may have to train for flexibility and strength concurrently in the same sessions.
- After a period of several months of high-volume stretch training, further increases in flexibility and MTU compliance will be slower and eventually reach a plateau. It is not clear whether continued high-volume stretch training is needed to maintain the plateau or whether a reduced maintenance program is sufficient. Therefore, the athlete should experiment by gradually reducing the volume of stretching to determine the minimum needed to maintain the plateau. Complete cessation of stretch training will result in a gradual loss of flexibility.

11

Other Considerations

Nutritional Factors for Performance and Training

Carbohydrate

Although pathways exist in muscle for deriving energy from fats and proteins, as documented in Chapter 2, carbohydrate (muscle glycogen and blood glucose) is the prime energy source for ATP production during exercise. Blood glucose is also the primary energy source for cells in the brain and central nervous system and the only energy source for red blood cells. Since blood glucose is constantly being taken up and metabolized over the course of a day, it must be constantly replaced by a dietary intake of carbohydrate or by release from its liver storage site. An hour of heavy continuous exercise training, competing in a team sport such as soccer, rugby, or ice hockey, or as few as four 30 second sprint-training intervals can reduce muscle glycogen concentration by more than 60%. Since very little glycogen can be resynthesized until new carbohydrate is ingested, daily carbohydrate intake by athletes must be considerably greater than that by sedentary individuals.

Whether ingested in the form of simple sugars such as fructose, sucrose, or lactose, or as complex carbohydrates (starches), all carbohydrates are eventually converted by digestive enzymes to the six-carbon sugar, glucose, before they can enter the glycolytic pathway or be stored as glycogen. Simple sugars are digested more quickly than starches and appear sooner in the blood. They also cause greater elevations in blood sugar (mostly glucose) than the same amount of carbohydrate ingested in the form of starch.

Nutritionists have derived a numerical scale known as the **glycemic index (GI)** to rate various foods by how quickly and to what extent they increase blood glucose. Foods such as most fruits and vegetables (with the exception of potatoes) have a relatively low GI, whereas most breads and desserts containing table sugar (sucrose) have a high GI. The higher the GI, the greater the subsequent increase in the release of insulin from the pancreas. Insulin is a potent stimulator for liver, muscle, and fat cells to take up and thus reduce blood glucose. Although blood glucose concentration and insulin levels are tightly coupled, in some individuals there may be a brief window in which blood glucose levels drop to values below normal while insulin levels are still higher than normal. Beginning exercise at this point in time could exaggerate this hypoglycemia and have a negative effect on endurance performance (20). It has therefore been recommended that athletes refrain from ingesting foods with a high GI in the 45 to 15 minutes prior to exercise (20). Once exercise begins, however, carbohydrate intake does not pose this problem because of the glucagon-enhancing and insulin-dampening effect of the norepinephrine that accompanies exercise (Chapter 2).

Major dietary sources for carbohydrates include pasta, breads, cereals, vegetables, rice, fruits, desserts, fruit juices, soft drinks, sweeteners such as honey and syrups, and sport drinks and energy bars. Although some people view carbohydrates as fattening, they have a relatively low calorie content [approximately 4.2 kilocalories (kcal) or Calories (see Box 11.1) per gram, compared to approximately 9.5 kcal per gram of fat]. To put the carbohydrate needs of an athlete in training into perspective, consider the energy expenditure of an endurance athlete who is training for 90 minutes each day. Let's assume that our athlete has a $\dot{V}O_{2max}$ of 5.0 L·min^{-1} and that his oxygen consumption over this time averages 75% of $\dot{V}O_{2max}$, or 3.75 L·min^{-1}. Over the 90 minutes, his total $\dot{V}O_2$ would be 337.5 L. Assuming a caloric consumption of approximately 5 kcal per liter of oxygen consumed (Chapter 3), he would have expended approximately 1688 kcal over the course of the training session. Since, at this training intensity, most of the energy would have come from muscle glycogen and liver glycogen (via blood glucose), his body carbohydrate store would be depleted by approximately 402 g at the end of the training session. Thus, to maintain this store, our athlete must consume almost 1700 kcal more of carbohydrates each day than a sedentary individual of the same body weight.

BOX 11.1

A calorie is the heat energy required to increase the temperature of 1 g of water by 1 °C. (In SI units this equals ≈4.2 joules, J). The energy content of foods is conventionally expressed in kilocalorie units (kcal) or Calories (with an upper-case C). Thus 1 kcal (or 1 Calorie) is the heat energy required to increase the temperature of 1 kg of water by 1 °C. To scientists and Europeans, calorie (with a lower-case c) means one-thousandth of a kilocalorie or one-thousandth of a Calorie. In the USA and Canada, on the other hand, calorie (with a lower-case c) is commonly used as a synonym for kilocalorie. Thus, a candy bar labeled in North America as containing 200 calories will be described in Europe as containing 200 kcal.

Fat

As documented in Chapter 2, almost half of the energy consumed in the resting state or during mild exercise such as walking is derived from the oxidation of fatty acids. Whether mobilized from their storage sites in muscle or transported to the muscle as free fatty acids, fats are a rich source of energy, with 1 g of fat producing almost 9.5 kcal. They do, however, require about 12% more oxygen per unit of ATP produced and cannot be used for high-intensity anaerobic exercise. As a result, their contribution to energy production during high-intensity endurance events that are completed in 60 minutes or less will be quite small. However, in longer-duration events, where liver glycogen and blood glucose begin to become depleted, their contribution will increase to the point that approximately 50% of the energy consumed over a 2.5 hour marathon would be derived from lipid oxidation. In an ultra-distance event lasting 5 hours, more than 70% of the total energy would be derived from fats. Since body stores of fat are many times larger than those of carbohydrate, there is never a problem with their depletion even during ultra-long-distance events.

Dietary sources for fat include meats, milk and dairy products, eggs, and cooking oils. Although excessive intake is clearly linked with obesity and cardiovascular disease (especially if it is in the form of saturated fatty acids), fat is an indispensable part of our diet. In addition to its energy-supplying role, it is a vital structural component for all nervous tissue, cell membranes, and hormones. While standards vary somewhat between countries, most purchased foods are nutritionally labeled to indicate their total energy content and breakdown as to carbohydrate, fat, and protein. In addition, labels usually include the proportion of saturated fats, sugar, sodium, common vitamins, and fiber, expressed as a percentage of the daily recommended value. For fat intake, this recommended value is usually equivalent to 30% of the total caloric intake, based on the assumption of a 2000 kcal daily intake. Because of the higher daily carbohydrate intake required by athletes, as described above, adjustment of this recommended daily value for fat to a target of 25% of total would be a healthier yet still adequate goal.

Protein

Again as documented in Chapter 2, during exercise certain amino acids can be funneled into the Krebs cycle to generate ATP or be converted to alanine, which the liver can then convert to glucose to help maintain blood glucose during prolonged exercise. These pathways are vital during starvation but are used only minimally by athletes in the fed state. For example, even during 3 h of heavy exhaustive exercise, the contribution from protein would still account for less than 10% of the total energy generated (33). Although it is a small proportion of the total, this amount can still be significant for endurance athletes who are training 5 or 6 days per week. As a result, in these individuals daily dietary protein requirements may be in the 1.2–1.4 $g{\cdot}kg^{-1}$ range, compared to the 0.8–1.0 $g{\cdot}kg^{-1}$ normally recommended for the general population. Despite this difference, these values are usually reached or exceeded in the typical athlete's diet (74).

Amino acids are the main structural component for all cells in the body. In muscle fibers in particular, actin and myosin protein filaments and their cytoskeleton attachments make up almost 85% of the total cell (Chapter 4). Different types of cells have widely varying lifespans, but are constantly turning over and being replaced. In addition, amino acids are continuously involved in the formation of enzymes, mitochondria, hemoglobin, antibodies, and hormones. As a result, and because proteins cannot be stored in the body to the same extent as fats, they must be constantly taken in on a daily basis. Proteins consist of various combinations of amino acids. Of the 20 amino acids, nine are considered *essential* since they cannot be synthesized from other amino acids within the body and must come from a dietary source. Foods such as eggs, milk products, fish, and meats contain all these essential amino acids, whereas protein from grains and vegetables does not.

Although the caloric cost for a session of resistance training is quite low compared to other forms of training, the increased turnover rate required for repair of contractile protein and synthesis of larger fibers that occurs following such training (Chapter 8) results in increases in amino acid uptake beginning immediately following the session. Muscle protein synthetic rate is significantly elevated within 4 hours, reaching a peak of more than double the resting rate and remaining elevated for more than 36 hours (14,57). As a result, protein requirements for athletes who are performing heavy resistance training are higher than for the general population, but by no means as high as most athletes believe them to be.

A study conducted in the authors' laboratory found that healthy young male bodybuilders who performed 90 minutes of heavy-resistance training 6 days per week could maintain nitrogen balance with a protein intake averaging 1.35 $g{\cdot}kg^{-1}{\cdot}day^{-1}$. Increasing this value to 2.62 $g{\cdot}day^{-1}$ with protein supplements was no more effective in increasing muscle mass and strength than the lower protein intake (54). The training program for the subjects in this study involved all major muscle groups in a 3-day split routine (day 1, chest and back; day 2, legs; day 3, shoulders and arms) with one day of rest per week. For each exercise, subjects performed four sets to failure at approximately 80% of 1 RM. Since the number of muscle groups and volume of training involved in this study exceed those normally occurring for athletes (other than weightlifters or bodybuilders), we can conclude that protein requirements for the resistance-training phase of most athletes' programs can be met with daily intakes in the 1.4 $g{\cdot}kg^{-1}$ range. For bodybuilders and athletes such as heavyweight weightlifters, who are not concerned about increases in body mass and whose primary goal is to maximize the increases in muscle size in a large number of muscle groups, intakes up to 1.8 $g{\cdot}kg^{-1}$ may be advantageous (66). Although suppliers of protein supplements for athletes would like them to believe otherwise, this amount is almost always exceeded in their normal diet.

The Athlete's Diet

Total caloric intake for athletes will depend upon body size, type and frequency of training, and whether they are competing in weight-class events. However, whatever their sport, reasonable target values for dietary components would be approximately 55% of total energy intake in the form of carbohydrate, 25% in the form of fat, and 20% in the form of protein. In addition, where possible, athletes should select carbohydrates with a lower GI and fats with lower saturated fatty acid content.

Electrolytes and Fluid Intake

At low sweating rates, the concentrations of sodium (Na^+), chloride (Cl^-), and potassium (K^+) in the sweat released to the skin are quite low, since these electrolytes are reabsorbed as the sweat moves through the ducts that conduct it from the secretory coil to the skin, and their concentrations become similar to those in the interstitial space (Chapter 3). At high sweat rates, the increased blood flow to the sweat glands causes the electrolyte concentration in the secretory coil to increase towards the levels found in plasma. In addition, at high sweating rates, the time available for re-absorption into the interstitial space decreases. As a result, sweat electrolyte concentrations do not decrease to the same extent before the sweat reaches the skin. Since sweat rates for most athletes performing prolonged heavy exercise typically exceed 1 $L \cdot h^{-1}$ (or double in a hot humid environment), significant amounts of Na^+ and Cl^- and (to a lesser extent) K^+ can be lost during competition or training. In spite of such losses, electrolyte concentrations in interstitial fluid and plasma following exercise will actually be higher than before exercise because of the large loss in body water, and will remain high until rehydration occurs. With rehydration, however, electrolyte concentrations will become hypotonic until replacement electrolytes are ingested during recovery.

A loss of 1 L of sweat results in an approximate 220 mL reduction in plasma volume (Chapter 3), and thus a decrease in ventricular filling rate and stroke volume. As stroke volume falls, heart rate must increase to maintain cardiac output so that exercise can continue at the same intensity. However, as heart rate approaches maximal, diastolic filling time decreases and stroke volume falls even more, causing cardiac output to decrease to the extent that exercise intensity cannot be maintained. It is thus important that water be taken in at the same rate at which it is being lost if cardiac output and body temperatures are to be maintained. Because of the logistical problems associated with supplying and drinking fluids during exercise, this is more difficult than it sounds. In addition, even under ideal conditions when athletes are encouraged to ingest fluids during endurance events, team sports, or practice sessions, body weight following such exercise is invariably less than before, indicating that thirst drive is not adequate for fluid replacement.

Nonetheless, athletes should make an effort to ingest as much fluid as they comfortably can during the event or practice session. For events involving less than 1 hour of total exercise time, plain cool water is probably the best fluid replacement. For events lasting longer than 1 hour, especially ultra-endurance events, it is recommended that the fluid contain carbohydrate (4–8% by volume) and Na^+ (≈ 0.5 $g \cdot L^{-1}$), and be ingested at a rate of 600–1200 $mL \cdot h^{-1}$ (18). Because the concentration of K^+ in plasma and sweat is about one-tenth that of Na^+, a much lower proportion of total body K^+ is lost during prolonged heavy exercise. Consequently, it is normally not necessary to include it in fluid replacement during exercise. During recovery, most of the K^+ that has been lost can be replaced by relatively small quantities of K^+-rich foods such as potatoes, bananas, and raisins. The carbohydrate and Na^+ concentrations recommended above are contained in most commercial sports drinks but can also easily be achieved by athletes mixing their own fluid-replacement drinks with the indicated proportions of table salt and sugar. The caveat here, however, is that the drink must taste good, since taste is an important factor in determining how much the athlete will drink. Athletes should also avoid attempting to hyperload fluids in the hour before the event. When plasma volume is normal, any additional fluid will simply be filtered out by the kidneys and passed on to the bladder as an extra and unusable increase in body mass.

Timing of Food Intake

Carbohydrate Although fluid-intake stations are usually very efficiently designed and administered and athletes can become very skilled at consuming fluids during road running, cycling, and cross-country skiing, the physical act of receiving and consuming an exercise drink causes a brief disruption in motor patterns. In continuous endurance events, even these few seconds could affect the outcome of the race. For athletes on a normal high-carbohydrate diet who have eaten 2 or 3 hours before competition, the trade-off is probably not worth it for continuous endurance events lasting less than 1 hour. However, because of the stoppages in play in racquet sports, most team sports, and training sessions, this is not a problem and athletes should be encouraged to take in carbohydrate and fluid even when exercise time is less than 1 hour (65).

For longer-duration events, especially those exceeding 90 minutes, carbohydrate feeding will help maintain blood glucose and delay muscle glycogen depletion, thus permitting the athlete to maintain a higher exercise intensity than would have been possible using fat as the energy source (Chapter 2). In addition, as discussed, fluid intake will help to preserve plasma volume and body temperature. In these events, athletes should begin intake within the first 30 minutes. The volume of fluid that can be quickly consumed at a time without nausea or stomach cramps varies widely between athletes. As a result, a target of 900 $mL{\cdot}h^{-1}$ might require six 150 mL intakes per hour (i.e., every 10 minutes) for some athletes but only four 225 mL intakes (every 15 minutes) for others.

Protein Most of the interest in the timing of protein intake is related to resistance-training programs designed to increase muscle mass. There may be a window of anabolic opportunity (16) that refers to ingesting protein at specified times before, during, and after a bout of resistance exercise in order to maximize muscle protein synthesis (MPS). Timing of protein intake is considered important because the magnitude of acute increase in MPS after a bout of such exercise is predictive of both the MPS response over at least the following 24 hours (77) and increases in muscle mass and strength over the course of a training program (48). The evidence in favor of immediate pre- and peri-exercise (during-exercise) protein intake is currently equivocal in the view of some (52,66); nevertheless, consuming protein before (within 1–2 hours) and during a bout of resistance exercise has been recommended by others (5,13,39,48,51). In contrast, there is consistent evidence that protein feeding soon after a training session (within ≈1–3 hours) amplifies the post-exercise increase in MPS more than later ingestion (39,52,66,73). In addition, because MPS may remain elevated for up to 48 hours (16,49,52) or even 72 hours (39) after each training session, there may be value in extending and exploiting the window of anabolic opportunity as long as possible with repeated protein feedings.

It is recommended that athletes consume protein primarily from whole foods rather than supplements (13), with emphasis on high-quality protein from animal sources (meat, fish, eggs, and milk). Plant-based soy protein is also of high quality, although milk protein (5,16), and perhaps other animal source protein may be superior to soy in the form of whole foods and supplements (13). The superiority of milk over soy protein may relate to the amino acid profile or differences in digestion and absorption (16,77). The essential amino acids of ingested protein are mainly responsible for stimulating MPS (16). Of these, **leucine** in particular may act as a trigger in stimulating MPS (16,39,66); however, isolated leucine supplements would be of little additional benefit over intake of high-quality proteins (66).

For athletes with a similar body composition, the optimal amount of protein for the immediate post-exercise feeding and subsequent feedings is likely related to body mass and state of training. For example, young men with an average body mass of approximately 85 kg and with recreational resistance-training experience ranging from 4 months to 8 years exhibited the greatest increase in MPS with immediate (within 2 hours) post-exercise ingestion of 20 g of egg protein containing approximately 10 g of essential amino acid; doubling the dose of protein did not increase MPS further. It was speculated that five or six additional similar intakes over the ensuing hours may be required for maximal MPS rates (61). If these men were meeting the most liberal recommendation for daily protein intake (2.0 g of protein per kilogram of body mass = 170 g for an 85 kg athlete) (13,51), each dose (20 g) would be approximately 12% of the recommended daily protein intake, whereas six feedings would be approximately 70% of the recommended intake.

The optimal post-exercise protein dose for athletes outside the 80–90 kg range has not been determined (16), but athletes above and below this range may require more or less protein, respectively (66). Regarding state of training, athletes in a phase of relatively rapid and large increases in muscle mass would need more protein than those on a plateau with little increase in muscle mass (52). However, even in rapid gainers, increases in muscle mass require weeks or months of training and supporting nutrition, reflecting the slow turnover rate of muscle proteins ($\approx 1\% \cdot day^{-1}$ for contractile proteins) (49). For example, it has been calculated that for an 80 kg athlete to gain 5 kg of muscle mass over 1 year he would need an additional 12.8 g of protein per day; a 400 mL glass of milk contains 14.4 g of protein (77). It should also be emphasized that protein in excess of the optimal amount cannot force muscle growth; instead, it is oxidized (16,51) or could contribute to a positive energy balance possibly leading to an increase in body fat. On the other hand, some degree of positive energy balance (energy intake > energy expenditure) is required for muscle growth. No amount of protein can cause muscle growth if there is a negative energy balance (77).

Protein in combination with carbohydrate may be beneficial for stimulating MPS (48) but is of lesser importance if protein is sufficient (16,66). On the other hand, high-volume resistance exercise can cause considerable muscle glycogen depletion, so post-exercise meals containing both carbohydrate and protein would assist recovery (16,66,77). Milk has been suggested as an effective means of co-ingesting carbohydrate and high-quality protein (66).

What are the protein timing practices of athletes? Athletes keen on gaining muscle mass typically eat five or six times per day and each meal or snack most probably contains carbohydrate, fat, and protein. One of these feedings would occur within 1–2 hours before a training session, and one within 1–2 hours afterward. Some athletes may begin consuming a protein supplement as they leave the gym while others may choose to delay their first post-exercise feeding until after they have had a shower. Some athletes will also consume a liquid protein supplement during a long training session. In addition, a few athletes may even set the alarm for protein feedings at 2 and 4 a.m. This latter practice has received little study (5) but athletes would probably benefit more from a night of uninterrupted sleep. As for the amount of protein, most athletes easily meet if not exceed the most liberal recommendation for daily protein intake (77). In summary, experienced athletes have been ahead of the science; that is, for years they have been doing what is now receiving some scientific support. In fact, as noted above, many athletes have been consuming more than the needed amount of protein. However, novice athletes would benefit from tutoring on the optimal amount and timing of protein intake.

Ergogenic Aids

For many decades athletes have experimented with various substances and procedures in an attempt to enhance physical performance over and above the levels that can be achieved through normal training and diet. Some that have shown varying degrees of effectiveness include anabolic steroids, amphetamines, blood boosting, blood buffering, creatine, caffeine, growth hormone, beta-blockers, and altitude training. Because of their perceived effectiveness (which would thereby place other athletes at a disadvantage) and their potentially harmful side effects, sports governing bodies have prohibited many of these substances and practices: some at all times, some during competition, and some only in certain sports. Of the above list, caffeine, creatine, and altitude training are not prohibited. Caffeine and creatine will be discussed in this section and altitude training elsewhere in the chapter.

Caffeine

The mental stimulatory and wakefulness-promoting effects of caffeine in beverages such as coffee and tea have been known for several centuries. Caffeine (chemically known as trimethylxanthine) is thought to exert its effect by acting on the central nervous system, causing an increase in sympathetic nervous system activity and an elevation in circulating norepinephrine. Whether as a direct effect of the caffeine or a secondary effect of the increased norepinephrine, there is also an upregulation of the enzyme activity of hormone-sensitive lipase and lipoprotein lipase (see Chapter 2), causing an increase in mobilization of plasma free fatty acids.

The relationship between the ingestion of caffeine and increases in circulating free fatty acid concentrations has spawned a number of studies to determine whether it might be an effective ergogenic aid in long-duration endurance events. It has been hypothesized that increasing the proportion of energy derived from fat oxidation early in the event would result in a sparing effect on muscle and liver glycogen that could enhance performance late in the event. While a few earlier studies tended to support this hypothesis (21,30,45), most newer studies do not (35,37,42,43,46), suggesting that, although caffeine elevates free fatty acids in the resting (pre-exercise) state, this response is no more effective once exercise begins than that which would have occurred naturally (without caffeine) as a result of increased circulating norepinephrine. In addition, it is now known that, since the amount of free fatty acids delivered to the muscle during endurance exercise is many times greater than the amount taken up, any further increases in their concentration do not increase their rate of utilization (71).

Although glycogen-sparing is probably not the mechanism, there is little doubt that caffeine can aid performance in endurance exercise (for recent reviews, see 32,34,75). The mechanism(s) by which it provides this ergogenic effect probably relates to its inhibitory effect on adenosine receptors, thereby blocking the relaxation-promoting effect of dopamine and reducing the perception of fatigue by the central nervous system (17). An additional effect might be on neural transmission through enhanced motor unit recruitment (22,76).

Whether or not caffeine improves performance in other forms of exercise is less clear. A few studies have found that, in some individuals, it can have an ergogenic effect in activities that involve repeated brief bouts of sprint-type exercise (26,82). Its effect on performance by athletes in resistance exercise such as weight lifting is still controversial. Its most useful contribution in activities of these kinds, however, would probably be during training sessions, where its effect on fatigue perception helps to motivate the athlete to complete more high-intensity intervals or sets of weight lifting than would otherwise be done.

The caffeine dosages that improved performance in the studies quoted here ranged from approximately 2 to 6 $mg \cdot kg^{-1}$ of body weight, taken within 1 hour prior to exercise. To put this into perspective, and depending upon brewing time and the individual's body weight, this is equivalent to the caffeine contained in one to three cups of coffee, six or seven cups of tea, four or five 355 mL bottles of cola, or two 250 mL bottles of an energy drink (10). Since these concentrations represent normal daily intakes in habitual caffeine consumers and because urinary excretion rates vary widely between individuals, urine testing for caffeine was discontinued by the World Anti-Doping Agency in 2004.

There is no question that caffeine use becomes habit-forming, and for many years its chronic consumption was associated with negative health effects. More recently, however, there have been an increasing number of studies which strongly suggest that caffeine and its associated antioxidants from coffee sources may actually have some positive preventive benefits for diseases such as diabetes, Parkinsonism, Alzheimer disease, and certain types of cancer (11).

Creatine

The International Society of Sports Nutrition considers creatine (Cr) to be the most effective nutritional supplement currently available for increasing the capacity to perform repeated high-intensity exercise and for promoting adaptations to strength and power training (9,12,51). Popular and widely used since the early 1990s (9,25,67), Cr supplementation has been the subject of hundreds of scientific studies, the majority (≈70%) of which have shown positive results on performance and training adaptations (9,50). Of the remaining studies, none has shown a performance decrease (9), and in some of the studies showing no benefit, muscle Cr levels were either not measured or did not increase after supplementation (12). After Cr supplementation, the average gain in performance is 10–15% depending on the activity (9). Examples of acute performance enhancement are increased strength and power (5–15%), increased work output (resistance/intensity × repetitions/time) (5–15%), single sprint performance (1.5%), and multiple sprint performance (5–15%) (8,9,12,50). Enhanced training adaptations include increased fat-free mass, strength, power, sprint performance, and muscle size (12,50). As an example of improved training adaptations, during 4–8 weeks of resistance training subjects taking a Cr supplement gain about twice as much fat-free mass (e.g., ≈1–2 kg) as those taking a placebo (9,12). Importantly, Cr supplementation not only improves performance in laboratory tests, it also improves actual sport performance (e.g., football, ice hockey, squash) (8,12,50).

In contrast to the positive effects of Cr supplementation on performance of predominantly anaerobic activities, there is little evidence in support of positive effects on predominantly aerobic activities (8). This might not seem surprising since increasing muscle Cr and phosphocreatine (PCr) levels would be much less important for aerobic than anaerobic activities (see Chapter 2). Some mechanisms by which Cr supplementation might enhance aerobic performance have been advanced (19) but it appears that Cr supplementation offers a much diminished benefit to activities lasting longer than about 2.5 minutes (8).

The effectiveness of Cr supplementation has been attributed primarily to its effect of increasing muscle stores of Cr and PCr (9,19,50). Recall from Chapter 2 that PCr plays an important role in providing energy for brief, high-intensity exercise. Muscle PCr is rapidly depleted during single and multiple bouts of intense exercise. Recovery between bouts is partly dependent on how rapidly PCr can be resynthesized. In a study from our laboratory, we found that, although 3 days of Cr loading did not improve maximum power output over a single 30 second Wingate test, muscle biopsies indicated that it accelerated PCr resynthesis during recovery (64). By facilitating more rapid PCr resynthesis (7), Cr supplementation would increase the capacity to perform single and multiple bouts of high-intensity exercise. Increased muscle PCr stores may also reduce the contribution of anaerobic glycolysis during intense exercise, thereby delaying fatigue by reducing the rate of glycogen depletion and lactic acid accumulation (7). Increased muscle Cr levels may also increase oxygen uptake during high-intensity exercise by transferring high-energy phosphates to mitochondria (7). The acute effects of Cr supplementation on exercise capacity are also thought to promote adaptations to training. By supporting more intense, higher-volume training sessions, Cr supplementation could contribute to greater increases in strength and muscle mass during strength and power training programs (7,9,12,51).

However, there may be other mechanisms by which Cr supplementation could improve performance and training adaptations. Cr supplementation may increase muscle glycogen storage mediated by glucose transporter (GLUT4) expression (40). Cr supplementation may promote muscle hypertrophy by effects on satellite cell proliferation, myogenic transcription factors, and insulin-like growth factor 1 (IGF-1) signaling (40). Cr supplementation in conjunction with resistance exercise may decrease serum levels of myostatin, a muscle growth inhibitor (19).

Cr is a naturally occurring compound. Most (95%) of the body's Cr store is in skeletal muscle (7), where it is stored as free Cr (≈30–40%) and PCr (≈60–70%) (9,19). Fast (type 2) muscle fibers can store more Cr and PCr than slow (type 1) fibers (7). The total Cr muscle pool (Cr + PCr) would be about 120–140 g for a 70 kg individual (19) but can be increased to about 160 g (9). Individuals with more muscle mass and a higher percentage of type 2 muscle fibers would tend to have a larger Cr muscle pool.

Each day approximately 1 g of Cr is synthesized in the body and an equivalent amount is consumed in the diet, primarily from meat and fish (19). The total amount (≈2 g is ≈2% of the Cr pool) of Cr synthesized and consumed daily is matched by an equivalent breakdown of Cr into creatinine, which is excreted in the urine (9). As will be shown, the amount of Cr that must be ingested daily for rapid maximization of muscle levels of Cr and PCr (e.g., 20–25 g) far exceeds the amount that can be easily obtained from the diet, hence the development and widespread use of Cr supplements. Cr supplements are an efficient way of consuming large amounts of Cr without having to consume an excessive amount of food (mainly meat and fish) (9). It is not unethical or illegal to use Cr supplements. The International Olympic Committee considers Cr a food and does not test for it. Cr loading (see below) has been compared to another nutritional strategy, carbohydrate loading (9).

Supplemental doses of Cr are related to body mass and expressed in grams of Cr per kilogram of body mass per day ($g{\cdot}kg^{-1}{\cdot}day^{-1}$). The most common dosage regimen of Cr supplementation for rapidly increasing muscle levels of Cr and PCr consists of a loading phase in which 0.3 $g{\cdot}kg^{-1}{\cdot}day^{-1}$ is consumed for 3–7 days (7,9,12), followed by a maintenance phase of 0.03 $g{\cdot}kg^{-1}{\cdot}day^{-1}$ (12), which may be continued for weeks or months. The maintenance dose is recommended because the body breaks down Cr at a rate of approximately 1–2 $g{\cdot}day^{-1}$ (7). For an athlete with a body mass of 70 kg, the loading phase would be approximately 20 $g{\cdot}day^{-1}$ divided into four or five equal doses. The maintenance dose would be approximately 2 $g{\cdot}day^{-1}$. This dosage regimen could cause a 10–40% increase in muscle Cr and PCr stores (7,9,50). The magnitude of the increase in muscle Cr and PCr stores is significant because it correlates positively with acute performance increases and training adaptations (9,12). The response to Cr loading varies among individuals. Responders who experience the greatest increases in muscle Cr and PCr tend to have the lowest initial levels of Cr and PCr, the highest percentage of type 2 muscle fibers, and the largest muscle fiber cross-sectional area (72). Vegetarians, who may have lower muscle Cr and PCr stores because of limited or no meat and fish consumption, may show a greater increase in stores with supplementation (19).

Other dosage regimens that effectively increase muscle Cr and PCr levels include 3 $g{\cdot}day^{-1}$ for 28 days, 6 $g{\cdot}day^{-1}$ for 12 weeks, and an ongoing 0.25 $g{\cdot}kg^{-1}{\cdot}day^{-1}$

(≈18 g·day^{-1} for a 70 kg athlete) (9). Regimens that increase Cr and PCr levels more slowly take longer to produce an ergogenic effect (9).

Some Cr supplements are combined with carbohydrate or carbohydrate plus protein (7). The addition of carbohydrate and protein to Cr intake appears to increase retention of Cr, allowing the loading phase to be reduced to 2–3 days (9). Athletes who would be ingesting Cr supplements would likely also be consuming large amounts of carbohydrate and protein in food, so the addition of carbohydrate and protein in the actual supplement may not be necessary.

Many forms of Cr are currently available (12), including but not limited to Cr monophosphate, Cr phosphate, Cr + sodium bicarbonate, Cr magnesium-chelate, Cr + glycerol, Cr + glutamine, Cr + β-alanine, Cr ethyl ester, and Cr with cinulin extract, as well as effervescent and serum formulations. Of these, Cr monohydrate (CM) is the most extensively researched form of Cr (9). To the extent that alternatives to CM have been studied, most are no more effective than CM in enhancing performance (9), although there is some evidence that CM combined with β-alanine is more effective than CM alone for increasing strength and lean body mass (9).

As Cr supplementation became more widely used, concerns were raised about possible unhealthy side effects such as dehydration and stress exercising in a hot environment, cramping, kidney and liver damage, musculoskeletal injury, and gastrointestinal distress (9). These concerns were based on anecdotal reports and a few case studies of patients already suffering from renal disease (9,19,67). Based on data from short-term (e.g., 4 weeks) and long-term (up to 5 years) studies, Cr supplementation at conventional doses (see above) is considered safe (7,9,19,25,50,51,67). In fact, in the case of dehydration and heat stress, recent evidence indicates that Cr supplementation may actually improve performance in hot, humid environments (25). The judgment that Cr supplementation is safe extends to young athletes past puberty provided there is parental consent, supplementation is supervised, and recommended doses are not exceeded (9). The last point is important because excess Cr must be eliminated, mostly by the kidneys. If Cr is to be ingested over the long term, it would be prudent to have regular check-ups to detect potential problems (67).

A potential side effect in the first few weeks of supplementation is a rapid increase in body mass due primarily to an increase in intracellular water (7,9,51). While this poses no health risk, performances in which the athlete must lift his or her body mass (e.g., jumping) could be adversely affected. On the other hand, the weight gain may be an advantage in sports where increased body mass and momentum improve performance (e.g., in a football lineman). It should be noted that the increase in muscle mass following resistance training combined with Cr supplementation is a true hypertrophy rather than simply a relative increase in total body water (50).

Athlete Testing and Monitoring

As the level of competition increases, it is becoming more and more common for coaches and athletes to enlist the assistance of exercise scientists in testing and monitoring the physiological parameters that are key to performance in their particular sport. As documented in this section, a well-planned and properly administered program of laboratory or field testing can provide valuable information for use in designing the optimal training program for a given athlete. It should be recognized, however, that such a program is only a training aid and not a magical tool for identifying future gold medalists. This is because the science of determining genetic limits is still in its infancy, and it is difficult to predict the extent to which an athlete has the potential to improve.

The Purposes of Testing (56)

- An effective physiological testing program offers insight into the individual athlete's strengths or weaknesses and provides important baseline data for designing (or modifying) an overall training program. It can also supply important feedback on the adaptive response of that particular athlete to the program, since what is most effective for one athlete may be less so for another.
- Many sports and activities involve more than one physiological component. Although it may be relatively easy to evaluate the sum result in the field setting, it is often difficult to assess the athlete's status on each of these components. With specific laboratory tests, the sport scientist can isolate an individual component and objectively assess the athlete's performance on

that variable, so that training can be designed to concentrate on any areas of weakness.

- A testing program can also provide information on the athlete's health status. Comparison of an athlete's performance on a given test with that on previous tests may indicate the possible onset of overtraining, which might otherwise have gone undetected (56).

What Constitutes Effective Testing

- The key to an effective testing program is to test parameters that are relevant to performance to a particular sport and to use tests that are a valid measure of these parameters. For example, there would little or no benefit in testing jumpers or throwers for maximal aerobic power or capacity. Likewise, measurements of muscular strength and power in these athletes are of practical value only if the muscles tested are specific to the sport.
- The tests must be sufficiently reliable to detect actual physiological changes that have occurred since the last testing session. For example, a measurement with a test-retest method error of 5% would be of little use in detecting a 3% improvement or decline in an athlete's performance over that time. A reliable testing program requires that tests be administered consistently at all times. This means standardizing the instructions given to the athlete, practice and warm-up procedures, order of test items, recovery time between items, environmental temperature and humidity, and equipment calibration procedures. Since testing equipment (gas collection systems and analyzers, ergometers, dynamometers, and blood chemistry analyzers) is expensive and testing must be conducted by experienced technicians, accredited athlete-testing laboratories are usually associated with universities or government-supported sports institutes.
- The purpose of each test must be clearly explained to the athlete and coach, and test results must be reported promptly and interpreted in language that they relate to and understand. This final step is crucial, but is often the one that is most poorly handled by the scientist. As the authors have observed on a number of occasions, when athletes understand why they are being tested and how they can benefit from the testing, they are more inclined to make true maximal efforts.
- Testing must be repeated at regular intervals and before and after different phases of training in order to measure the effectiveness of the program for that individual athlete. One-shot (or even once-per-year) testing, although of potential interest to the scientist, is of little practical value to the athlete. In addition, whenever possible, repeat tests should be performed in the same laboratory. Although tests in a given laboratory may be extremely reliable, slight differences in procedures and equipment can significantly reduce reliability *between* laboratories.

Laboratory vs Field Tests

A laboratory test is a measurement conducted in a controlled environment using protocols and equipment that simulate the sport or activity. A field test is a measurement that is conducted while the athlete is performing in a simulated competitive situation. Measurement of a rower's maximal oxygen consumption on a rowing ergometer is a laboratory test, while the same measurement performed while rowing over water in the course of a performance trial is a field test. Similarly, an anaerobic test for an ice hockey player on a cycle ergometer to measure units of work or power is considered a laboratory test, while an anaerobic test of the player's skating speed over a prescribed course (measuring units of time and distance) is considered a field test.

In general, results obtained from field tests are less reliable than those gained from laboratory tests but are often more valid because of their greater specificity. Because the scientist cannot control variables such as wind velocity, temperature, and humidity, or playing-surface or track conditions, athlete performance varies more in the field setting. In addition, the portable measuring systems necessary for field testing are generally not as accurate as the fixed equipment used in the laboratory. However, for those sports that cannot be effectively simulated in a laboratory setting, field testing is the only solution (56).

Some Common Performance Tests

A survey of the scientific literature and exercise physiology textbooks reveals that dozens of performance tests and procedures have been developed over the years, many of which have been adapted for specific sports. Some of the most common are reviewed briefly in the following section.

Tests for Maximal Aerobic Power and Capacity

Maximal oxygen consumption Oxygen consumption ($\dot{V}O_2$) is traditionally defined as the difference between the amount of oxygen inspired and expired per minute. It is conventionally determined in humans by measuring the volume of gas that is expired (V_e), and its oxygen and carbon dioxide fractional content (F_eO_2 and F_eCO_2). After gas volumes are converted to STPD (as explained in Chapter 3), these figures are used to calculate the volume of gas inspired (V_i), and $\dot{V}O_2$ is then calculated using the formula:

$$\dot{V}O_2 = (V_i \times F_iO_2) - (V_e \times F_eO_2)$$

Maximal oxygen consumption ($\dot{V}O_{2max}$) can be determined by measuring $\dot{V}O_2$ during progressive exercise to the point of temporary exhaustion (Figure 11.1). The exercise mode should be as sport-specific as possible, with runners and most team sport athletes being tested on a treadmill, cyclists on a cycle ergometer, rowers on a rowing ergometer, and so on. The way in which exercise intensity is increased should also be consistent. In a treadmill running test, for example, some laboratories will use a constant velocity and increase the grade every minute or two, while others will keep the treadmill at a low incline (2–4%) and progressively increase the velocity. Whatever the protocol, it must be consistent and, after a brief warm-up, should ideally lead to exhaustion in approximately 10–12 minutes.

As illustrated in Figure 11.1 and documented in Chapter 3, $\dot{V}O_2$ will increase linearly with exercise intensity and begin to plateau as the subject approaches exhaustion. A true plateau occurs when there is no further increase in $\dot{V}O_2$ when exercise intensity is increased. The oxygen consumption at this point is considered to be the athlete's $\dot{V}O_{2max}$ and is expressed in absolute units (L·min^{-1}) or relative to the athlete's body weight (mL·kg^{-1}·min^{-1}). Occasionally even highly trained and motivated athletes will fail to demonstrate a plateau in $\dot{V}O_2$ despite reaching maximum heart rate and exhaustion. In such instances, the highest value that they reach is defined as their **peak oxygen consumption** ($\dot{V}O_{2peak}$). Although this value is probably the same as $\dot{V}O_{2max}$, further testing would be required for confirmation.

Figure 11.1 The relationship between oxygen uptake ($\dot{V}O_2$) and increasing exercise intensity. When $\dot{V}O_2$ ceases to increase in spite of an increase in exercise intensity, this value is considered to be $\dot{V}O_{2max}$.

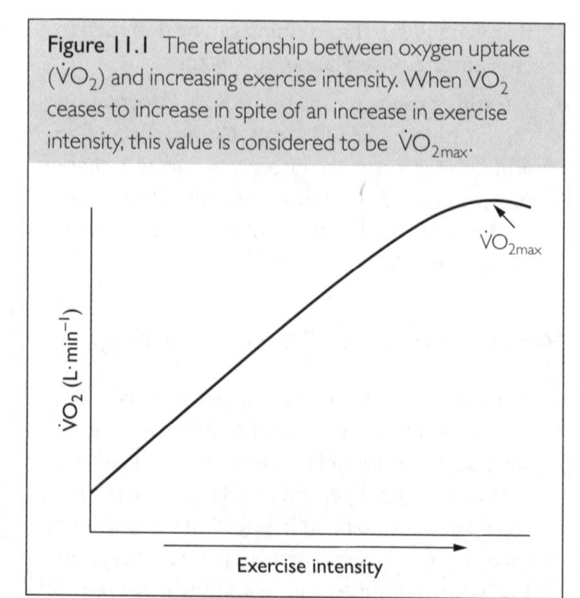

Because the measured value for $\dot{V}O_{2max}$ is dependent upon the muscle mass that is active during the test and its state of training, the same athlete will show different values in different exercise modes. For example, when runners or untrained subjects are tested on a treadmill, their $\dot{V}O_{2max}$ is usually 2 or 3 mL·kg^{-1} higher than when they are tested on a cycle ergometer. In contrast, when cyclists are tested, their $\dot{V}O_{2max}$ is the same or slightly higher on a cycle ergometer than on a treadmill.

The lactate (anaerobic) threshold As discussed in Chapter 2, the lactate threshold (also known as the anaerobic threshold) is normally defined as the exercise intensity above which there is a non-linear increase in blood lactic acid (Figure 11.2). Measurement of the lactate threshold thus involves frequent sampling of the athlete's blood during progressive exercise, as in a $\dot{V}O_{2max}$ test. Although some laboratories prefer to sample venous blood by means of repeated venipunctures or an indwelling catheter, most sample capillary blood from a fingertip or earlobe. Protocols for measuring the lactate threshold also vary widely between laboratories; here again consistency is the key factor. In addition, most laboratories include measurement of $\dot{V}O_2$ during the test as an indicator that steady state has been achieved at each workload and to provide a

Figure 11.2 The relationship between blood lactate concentration and increasing exercise intensity. The lactate threshold is the exercise intensity above which there is an exponential increase in blood lactate concentration.

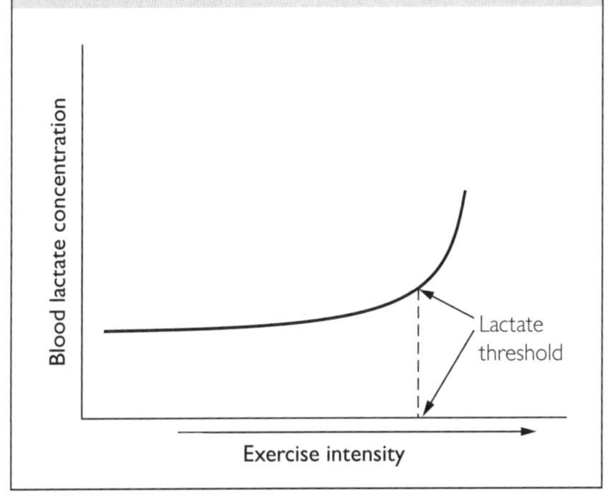

reference so that the lactate threshold can be expressed as a percentage of $\dot{V}O_{2max}$.

Although capillary blood can be sampled quite easily during exercise on a cycle ergometer, it is more difficult to do so on a treadmill and impossible on a rowing ergometer. As a result, most protocols involve brief (15–20 second) stoppages in exercise at the end of each workload for sampling. When retesting, it is important that blood be sampled at the same workloads as in previous tests.

Time to fatigue at $\dot{V}O_{2max}$ Another effective test for maximal aerobic capacity involves simply determining how long the athlete can exercise at the same power output at which $\dot{V}O_{2max}$ was previously achieved. Following a consistent progressive warm-up and short recovery, the athlete begins exercise on the appropriate ergometer, which is adjusted as quickly as possible to achieve the desired power output. Exercise continues until this power output can no longer be maintained (usually 8–12 minutes in endurance-trained athletes). Some laboratories prefer to modify this test by using time to fatigue at 95% of $\dot{V}O_{2max}$.

Economy of exercise A common method of testing an athlete's efficiency of movement is simply measuring $\dot{V}O_2$ during steady-state exercise at different submaximal velocities or power outputs. The lower the athlete's $\dot{V}O_2$, heart rate, and blood lactic acid at a given power output, the greater their economy. Slight differences in economy can generate major differences in performance in longer-duration events. As a result, this can be an important measurement since it can flag up the existence of certain flaws in mechanics. Through subsequent film and biomechanical analysis, these flaws can then be isolated so that mechanics can be altered to improve economy.

As previously discussed, measurements of $\dot{V}O_2$ are generally more accurate and reliable when performed in a laboratory setting because of possible limitations in the portable gas-collection systems required for field measurements. In addition, the validity of these measurements is questionable if the gas-collection procedure alters exercise mechanics in any way, or if the ergometer does not accurately simulate the movement mechanics of the sport.

Tests for Maximal Anaerobic Power and Capacity

Wingate test Probably the best-known test of anaerobic performance is the Wingate 30 second cycle ergometer test. For this measurement, athletes pedal as fast as they can against a constant resistance load on the ergometer. On mechanically braked ergometers such as the Monark, this resistance is normally 75 $g \cdot kg^{-1}$ of body weight, or slightly higher for well-trained athletes. Typical power output for this test is illustrated in Figure 11.3. Calculations based on the

Figure 11.3 Power output in a male hockey player over a 30 second Wingate test. From the authors' laboratory.

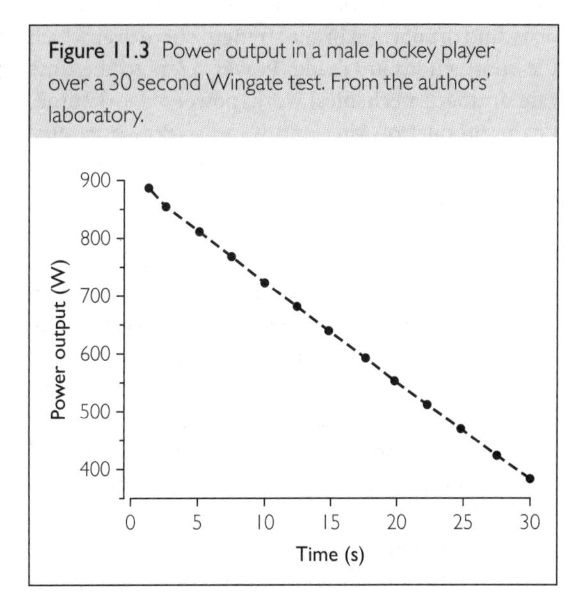

number of revolutions per second and the braking resistance applied are used to determine **peak power** (usually achieved in the first few seconds), **mean power** over the 30 seconds, and **percent fatigue** (the difference between peak and lowest power output divided by peak power). Peak power thus reflects the athlete's maximal anaerobic power and mean power his maximal anaerobic capacity.

Cunningham and Faulkner treadmill test Following a standardized warm-up, the subject begins running on a treadmill set to a speed of 8 mph and an incline grade of 20%. Time to exhaustion (the point at which the subject is forced to grab the handrails) is recorded to the nearest half second. Theoretically, for a test to isolate **anaerobic** capacity, exercise intensity should be high enough to cause fatigue in under 60 seconds. Tests lasting any longer cause a progressively greater involvement of **aerobic** metabolism. Since highly trained anaerobic athletes usually run considerably longer than 60 seconds at this grade and speed (36,68), some laboratories have modified this test to use a running speed of 9 mph at a 20% grade or 10 mph at 10%.

Tests for Muscle Strength and Power

Maximal strength is the peak force or torque generated by a maximal voluntary contraction under a given set of conditions that include contraction type (isometric, concentric, eccentric), contraction velocity, and the portion of the strength curve over which the contractions are made. **Power** refers to the time rate of doing mechanical work (power = work/time) or the product of force and velocity (power = force × velocity). For a more detailed discussion of these terms, see Chapter 4.

Testing for strength and power can be challenging because of difficulties in applying specificity in test protocols (24), establishing the correlation between laboratory test results and sport performance, developing tests that will monitor training adaptations, and limitations in testing equipment (1,23). As a result, this section will be restricted to presenting general testing principles (with some examples) rather than detailed testing protocols for a large number of sports.

Types of strength tests There are three types of strength tests: isometric, isokinetic (isovelocity), and isoinertial. The prefix iso- means "same" or "constant;" thus, isometric, isokinetic, and isoinertial mean the same or constant muscle length/joint position, velocity, and inertia (mass), respectively.

Figure 11.4 Schematic illustration of an isometric force (or torque) recording. Computer analysis of the force signal can give values of peak force and maximum rate of force development (RFD).

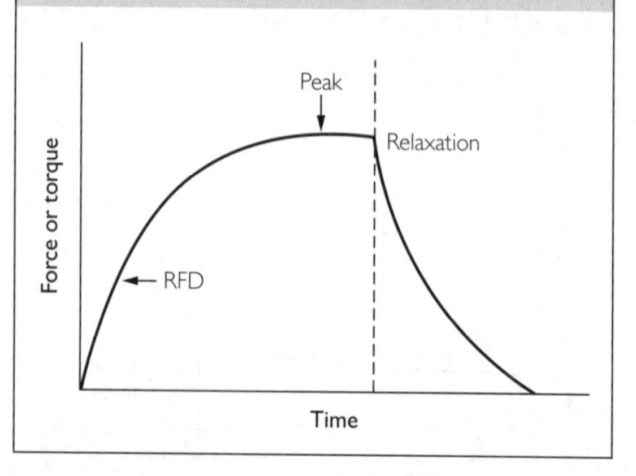

Isometric strength tests utilize custom-made dynamometers that have been developed to test unilateral or bilateral single- or multi-joint movements at various joint angles through a range of motion. These dynamometers can also be instrumented to measure rate of force development, in which case the athlete is instructed to build up force as rapidly as possible (Figure 11.4). Results are expressed in units of force (**newtons, N**) or torque (**newton meters, N·m**). Rate of force development is expressed in $N{\cdot}s^{-1}$ or $N{\cdot}m{\cdot}s^{-1}$.

Isokinetic tests are done with dynamometers that restrict movements to various pre-set velocities. In a concentric contraction test, a subject contracts maximally in an attempt to accelerate the dynamometer's resistance mechanism (usually connected to a lever arm), but the dynamometer prevents acceleration and keeps the velocity constant. In an eccentric contraction test, the dynamometer moves the lever arm at a set constant velocity while the subject attempts to stop the lever arm with what becomes a maximal eccentric contraction. Because set velocities are kept constant during contractions, such tests are also called **isovelocity** tests. Commercially available isokinetic dynamometers permit the testing of only unilateral single-joint movements, although they have

Figure 11.5 Schematic illustration of fast and slow isokinetic (isovelocity) concentric contraction recordings made on an isokinetic dynamometer. The most obvious measurement would be the peak torque produced by each contraction, but computer analysis of the torque signal can provide additional measurements. Multiplying peak torque by the set angular velocity (in radians per second) would give peak power. The area under the torque/time or torque/displacement recording is equal to impulse and work, respectively. Average power is either work/time or average force × velocity. The average torque over the whole contraction would be impulse/time. All of these measurements could also be made if the contractions were eccentric rather than concentric; however, in accordance with the force–velocity relationship (see Figure 4.22), peak torque would be greater at fast than at slow velocities, the opposite of what is shown here for concentric contractions.

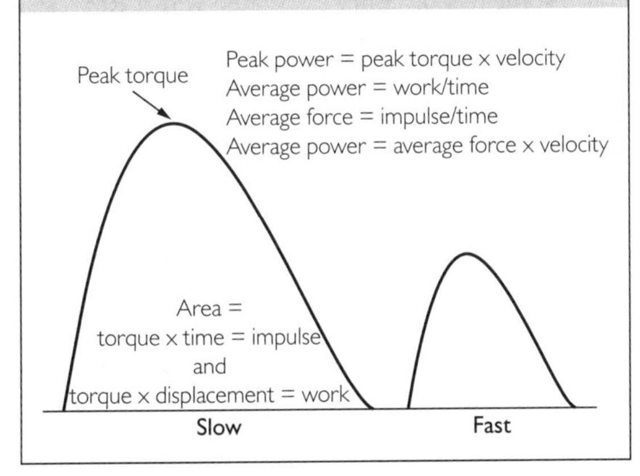

occasionally been modified to permit testing of bilateral multi-joint movements (e.g., 79). The results of isokinetic strength measurements are usually expressed in units of torque (N·m) because isokinetic dynamometers test joint angular motions (Figure 11.5).

Isoinertial tests are basically weight-lifting tests. **Inertia**, the property of an object to remain at rest or in motion at a constant velocity unless acted upon by an external force, is quantified as mass. The unit for mass is the **kilogram (kg)**. Because the mass of an object (e.g., a barbell) remains constant regardless of any forces exerted upon it, the term "isoinertial" is used when lifting a mass. Some may confuse **mass** with **weight** (see Appendix A). Weight, whose unit is the newton (N), quantifies the effect of gravity on a mass; specifically, weight is equal to mass multiplied by the acceleration due to gravity, which is 9.8 m·s^{-2}. Therefore an object with a mass of 1 kg would weigh 9.8 N. However, by convention, even among sport scientists, the kilogram is permitted for use as the unit for weight as well as mass. Thus, an athlete is said to weigh 75 kg even though his or her weight would really be approximately 750 N. Similarly, the result of a *weight*-lifting test is usually given in kilograms rather than newtons.

The standard isoinertial strength test is the **one repetition maximum [1 RM]**, the heaviest weight that can be lifted (concentric contraction) just once. A 1 RM test would seem to be straightforward; however, as illustrated in several figures in Chapter 4, the force of a maximal concentric contraction is greatly affected by contractile history. A preceding eccentric (see Figure 4.33) or isometric contraction (Figure 4.49) will increase the force of a subsequent concentric contraction. In a back-squat 1 RM test, for example, a common procedure is to instruct the athlete to lower (eccentric contraction) the weight to a set position, hold (isometric contraction) the position for 1–2 seconds, then rise up (concentric contraction). The pause of 1–2 seconds after the eccentric phase would allow time for stretch-shortening cycle potentiation to dissipate (Figure 4.33), but there would still be some isometric–concentric potentiation, although its effect would be smaller (Figure 4.49). In addition, an eccentric contraction increases the force of a subsequent isometric contraction (eccentric–isometric force enhancement, Figure 4.52), which in turn would increase the subsequent concentric force. Therefore, a properly administered 1 RM test should either eliminate or control for the effects of contractile history. The effects of contractile history can be eliminated if the weight is supported on a rack and the athlete lifts it with an isolated concentric contraction. If the test were to include eccentric, isometric, and concentric phases, it would be important to standardize the velocity of the eccentric phase and the duration of the isometric phase (as in the back-squat example described in this paragraph).

Force- and load-velocity tests In isokinetic tests, it is possible to measure peak torque at a series of velocities from which a **force–velocity relationship** can be constructed (Figure 4.23, left); however, the highest velocity that can be tested on most dynamometers corresponds to only approximately 50% of maximum contraction velocity (V_{max}). Despite this limitation, testing at a series of velocities may provide useful information, such as whether an athlete is stronger relative to other athletes at a particular velocity.

Instead of setting velocity and measuring force or torque, another approach is to set different loads and measure the maximum velocity attained when the athlete attempts to lift the load as fast as possible. Various loads are set either as a percentage of the 1 RM or isometric maximum force (ISO_{max}). If the velocity attained with a series of loads is measured, a **load–velocity relationship** can be constructed (Figure 4.23, right). A technical problem with load/velocity tests is that an athlete cannot be allowed to simply accelerate and release a load (weight) without accounting for how and where it will land. This problem has been solved by the development of special power racks (24) in which the path of the thrown weight (usually a barbell) can be guided and the weight brought to a safe stop. In load/velocity tests the key measurement is velocity. The force exerted to attain a given velocity could also be measured if the apparatus is instrumented for that purpose (24,81; see Figure 4.33). Alternatively, the force could be calculated from the mass of the object and its acceleration using the formula: force equals mass times acceleration ($F = ma$). Unlike isokinetic tests that can test only up to approximately 50% V_{max}, load/velocity tests can use loads light enough that V_{max} can be approached (see Figure 4.25).

Types of power test There are several options for power tests. **Isokinetic dynamometers** can provide values of power by multiplying torque values by the set angular velocity; for example, a torque of 200 N·m produced at a velocity of 1.5 rad·s^{-1} (rad = radians, unit of angular displacement) would correspond to a power of 300 W (W = watts, unit of power) (Figure 11.5). Isoinertial power tests have become increasingly popular because of specificity: isoinertial loads rather than isokinetic resistances are experienced in most sports (31). With appropriate instrumentation, values of power can also be derived from the load/velocity tests described previously.

Several types of **jump power** tests are used to measure power, including squat jumps, countermovement jumps, and drop jumps, the latter two involving stretch-shortening cycle potentiation (Figure 4.42). In a jump test, the athlete's body mass serves as the isoinertial load. The height of the jump can be used as the indirect measure of power, especially in a field test. In the laboratory, actual values of peak power (in watts) can be obtained if the jumps are done on a force platform or on an apparatus that measures displacement, although measurement of both force and position data gives the best results (29). Sometimes jump tests are done with a series of added loads so that load– and power–velocity relationships can be constructed. If added loads are used, peak power should be calculated based on the total load (body mass + added mass) and expressed in relation to body mass (29); for example, if the athlete's body mass is 80 kg and the added load is 20 kg, the total load would be 1.25 times body mass ($(80 + 20)/80 = 1.25$).

Upper body power tests include the seated shot put, medicine ball chest pass (31), and bench-press throw (24).

Predictive tests Isoinertial 1 RM tests are time-consuming because of the warm-up sets required. There is also some risk of injury associated with 1 RM tests. To avoid these problems, predictive tests have been developed. The number of repetitions completed with a given absolute weight is used to predict the 1 RM. The basis of the prediction is that a stronger athlete should be able to complete more repetitions with a given weight because it is a smaller percentage of the athlete's 1 RM (Figure 11.6). For example, in an exercise like the bench press or squat, an athlete able to do 10 repetitions with 135 kg would have a predicted 1 RM of 182 kg, whereas an athlete able to do only two repetitions with 135 kg would have a predicted 1 RM of 143 kg (62). Prediction error is smallest when the weights selected are a relatively high percentage (e.g., >70%) of the 1 RM of the athletes tested (62). The prediction error is also reduced if the athletes being tested are in the same sport and have a similar training history. The prediction error is never zero, however, so a decision has to be made whether the time and safety gained with a predictive test outweighs the loss of precision.

Common predictive tests of power include jump height, sprint speed, seated shot put, and medicine ball chest pass. An athlete or coach will likely be more interested in whether these tests predict sport performance as opposed to predicting power values expressed in watts.

Standardization of tests As previously noted, tests must be carefully standardized if they are to provide valid and reliable results. A number of factors must

Figure 11.6 Correlation between the number of completed repetitions with heavy (A) and light (B) loads. If a heavy (i.e., a high percentage of 1 RM for all athletes tested) load is used for the test, the number of completed repetitions is a good predictor (high correlation) of the 1 RM. In contrast, the prediction is poor if a light load is used. However, even when the correlation between repetitions and 1 RM is high, there is still prediction error. For example, note that the athletes labeled 1 and 2 have almost the same 1 RM, but athlete 2 completed notably more repetitions with the heavy load. Based on the prediction equation derived from the correlation (regression line), athlete 2's predicted 1 RM would have been higher than the real 1 RM. See text for further discussion.

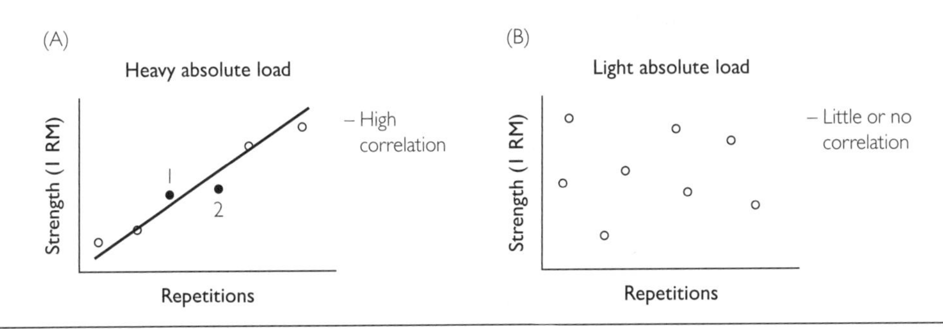

be controlled for when testing strength and power, as follows.

- Variation in warm-up procedures before the test could affect results by inducing differences in muscle temperature and postactivation potentiation (Chapter 5).
- In 1 RM tests, athletes with less resistance-training experience will need more practice with the mechanics of the test before an actual testing session (69).
- Since strength can vary considerably through a range of motion (see Figure 4.7), the same starting position and range of motion must be used in every test.
- In jumping tests, the starting position, the extent to which stretch-shortening cycle potentiation is permitted, and the use or non-use of arm swing must be consistent.
- Controlling for stretch-shortening cycle potentiation is also necessary in isokinetic testing: if the test consists of alternating concentric and eccentric contractions, concentric torque will be greater than in concentric contractions done in isolation (Figure 4.49) and this effect will be amplified at higher velocities (Figures 4.40 and 4.50). Instructions must therefore be consistent and precise (1,29,83).
- Finally, the number of test trials must be consistent to control for factors that could enhance (muscle temperature, postactivation potentiation, skill practice) or impair (fatigue) test performance.

Expressing results The results of strength and power tests can be expressed as absolute values (e.g., kilograms, newton meters, watts) or as relative values, usually relative to body mass. For example, a 90 kg athlete achieves a bench-press 1 RM of 140 kg. The athlete's absolute strength would be 140 kg, whereas the relative strength (strength to body mass ratio) would be $1.6\ \text{kg}\cdot\text{kg}^{-1}$ ($140\ \text{kg}\cdot 90\ \text{kg}^{-1} = 1.6\ \text{kg}\cdot\text{kg}^{-1}$). The decision to express test results in absolute or relative values depends on whether the sport performance requires the athlete primarily to support, lift, or accelerate his/her own body mass, primarily to support, lift, and accelerate an external mass (e.g., an object or opponent), or both. Gymnasts, rock climbers, and jumpers require a high relative strength and power because their own body mass is the primary load encountered; in contrast, weight throwers (e.g., hammer throwers) and American football players need high absolute strength and power to accelerate and move external loads. In many sports, it may be useful to provide both absolute and relative values of strength and power. Sports like rugby, ice hockey, and basketball require an athlete to have a high level of absolute strength and power to contend with external masses (opponents) but also the highest possible level of relative strength to effectively move and accelerate his/her own body mass in actions like jumping and sprinting.

Selection of tests based on the purpose of testing There are several possible purposes of strength and power testing: talent identification based on prediction of performance, strengths and weaknesses within athletes, testing for muscle imbalances that may lead to injuries, and monitoring training progress. Some purposes (talent identification, monitoring training progress) are best fulfilled by applying the specificity principle: tests should simulate as closely as possible the movement pattern, contraction type, and velocity of the sport action (31). For some sports this can be a challenge because of multiple movement patterns and the limitations of testing equipment. Tests designed to indicate specific weaknesses or reveal susceptibility to injury need not be as specific.

There are proponents and detractors of isometric, isokinetic, and isoinertial testing methods (1). In most sports, isoinertial tests are considered to be the most specific to performance (23). On the other hand, isometric and isokinetic tests can be useful for some applications. The following examples show how two sports have established strength and power tests.

During games, **soccer players** exhibit strength and power with sprints, quick changes in direction, kicking, and jumping. For elite soccer players, one strength test, the half-squat 1 RM, has proved sufficient. The half-squat 1 RM correlates highly with jump height, sprint times over 10–30 m, and time of a 10 m shuttle run, and successfully monitors the progress of a high-intensity squat training program, which in turn improves sprint performance and jump height. In addition to the half-squat 1 RM, predictive tests of power (sprint and shuttle run times, jump height) are regularly administered (41).

In contrast to soccer players, **rowers** perform one skill requiring a high level of strength and power in both the upper and lower body. Nevertheless, many tests for several purposes have been developed for rowers (53). The most specific test of rowing strength is peak force generated during a 2000 m on-water race (53). A rowing ergometer can be used for more feasible but still highly specific tests, such as peak force produced on a seven-stroke test and peak power generated on a five-stroke test; these tests correlate highly with ergometer rowing speed over 2000 m (44,63). Less specific but still useful tests, such as isoinertial bilateral leg press 1 RM and single-joint, unilateral isokinetic knee extension and flexion, can distinguish between high and low rowing performance (53). In contrast, isometric tests at specific joint angles do not correlate highly with rowing performance or force and power production during ergometer rowing (53). The less specific tests may have other uses. Unilateral strength tests have indicated that less experienced rowers have less bilateral symmetry; the oarside leg is stronger than the non-oarside leg (53). In addition, single-joint, unilateral isokinetic knee flexion tests have indicated that rowers with poor hamstring strength are susceptible to low back pain (53).

Testing Flexibility

For a more detailed discussion of flexibility testing, readers are referred to Chapter 10. The key points can be summarized as follows.

- Flexibility tests can range from visual inspection of maximum range of motion (ROM_{max}) during a training session to precise measurement of ROM_{max} with a goniometer in a laboratory.
- ROM_{max} is determined not only by the stiffness of the muscle–tendon unit (MTU), but also by stretch tolerance. Since performance and possibly injury prevention are affected mainly by changes in MTU stiffness, it is recommended that, when feasible, laboratory measurements of ROM_{max} be accompanied by measurements of MTU stiffness.
- Increased muscle temperature will increase ROM_{max} by decreasing MTU stiffness; therefore, it is important to standardize the type and amount of activity before ROM_{max} testing.
- ROM_{max} tests should be applied to ranges of motion that are sport-specific. Specific tests will correlate more highly with sport performance and will better monitor changes in ROM_{max} during specific stretch training programs.
- ROM_{max} tests may be passive, with a tester or apparatus applying the stretching force while the athlete relaxes, or active, with the athlete contracting opposing muscles (antagonists) to stretch the target muscles. Both types of tests are useful. Passive tests can provide a more accurate measurement of MTU stiffness, whereas active tests indicate the range of movement through which an athlete can easily move.

Measuring Body Composition

Although fat is an essential component of the human body, excess body fat will have a negative effect on performance in most sports. This is particularly true in body-weight-supported activities such as running, jumping, and team sports, and less so in swimming, throwing, and lifting events. Knowledge of an athlete's fat and lean body mass is crucial for determining his/her ideal competition weight and division for weight-class events.

Fat content is conventionally expressed as a proportion of the subject's body weight (mass), or **percentage body fat**. The three most common measurements used to determine this value in athletes are skinfold fat thickness, body density as determined by underwater weighing (densitometry), and electrical impedance in the body.

Skinfold techniques A standard caliper is used to measure the thickness of a double layer of skin and subcutaneous fat at a number of sites: sometimes as few as three, but more often six or more. Using formulae derived through correlation with measurements of body density, these skinfold measurements are then converted to an estimate of percentage body fat. The procedure is simple and, when performed by an experienced individual, quite reliable. Its validity, however, is based on the assumption that there is a direct relationship between subcutaneous and internal body fat and that the sites selected are representative of the subject's total fat deposition pattern. Because this is not always the case, the estimate of percentage fat cannot be expected to be as accurate as the estimates produced by the other two techniques. Nonetheless, any change in the sum value of these measurements is an effective indicator of a change in total body fat.

Densitometry For this procedure, the subject's volume is calculated by measuring his/her weight while completely submerged in water. According to Archimedes' principle, the difference between this weight and the subject's weight in air is equal to the weight of the volume of water that is displaced. After correcting for air in the lungs and airways, intestinal gas and water temperature, the subject's body density is determined by dividing mass by volume. Percentage fat is then calculated using formulae based on the density of fat and lean tissue in a typical human body. This step is probably the major limitation of the procedure, since these densities can vary slightly between individuals.

Electrical impedance With this method, a low-intensity electric current is passed through the body, using conducting and receiving surface electrodes that are usually attached to the ankle and wrist. Since this current flows most rapidly through extra- and intracellular body water and their electrolytes, and slowest through fat (which has a very low water content), its rate of flow can be used to calculate total body water. Fat-free body mass is then calculated from total body water and total fat from the difference between this value and the subject's body weight.

Altitude

Problems with Competition at Altitude

Since the partial pressure of inspired oxygen (P_iO_2) declines at higher altitudes (Figure 11.7), the O_2 content of arterial blood drops as altitude increases. As reviewed in Chapter 3, because of the shape of the oxyhemoglobin dissociation curve, this reduction in arterial O_2 content is not linearly related to the drop in P_iO_2 at low to moderate altitudes. Nonetheless, it is a given that $\dot{V}O_{2max}$, the lactate threshold, and performance in events that rely on oxidative energy supply will decline as athletes ascend to higher altitudes (Figure 11.8). Performance in the shorter sprinting events may actually be better at altitude, because of reduced air resistance. Similarly, performance in jumping events and shot put may be slightly enhanced because of reduced gravity. Any improvement would be less pronounced, however, in throwing events like discus and javelin, since the change in gravity is

Figure 11.7 The decline in the partial pressure of oxygen from sea level to moderately high altitude.

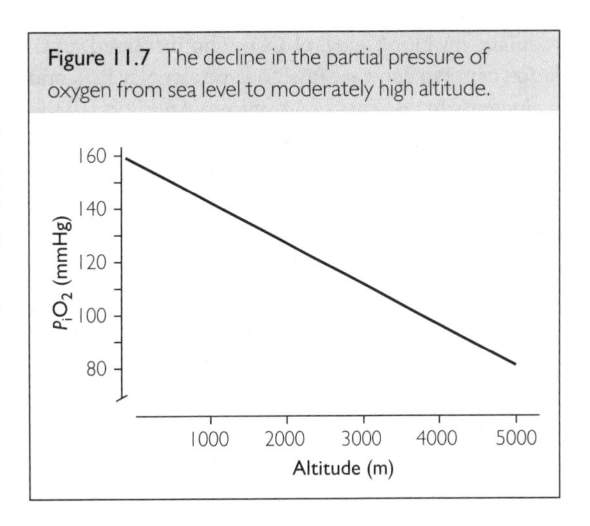

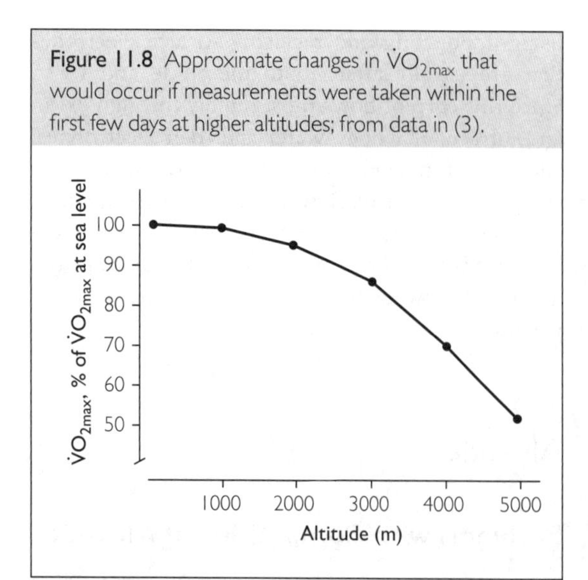

Figure 11.8 Approximate changes in $\dot{V}O_{2max}$ that would occur if measurements were taken within the first few days at higher altitudes; from data in (3).

partially counteracted by a reduced planing effect in the less dense air.

Acclimatization to Altitude

Although there are wide individual differences in respiratory center sensitivity, most individuals will experience a rapid increase in resting ventilation shortly after arriving at altitudes greater than about 2500 m. This hyperventilatory response is caused by a sudden increased drive from the peripheral and central chemoreceptors (discussed in Chapter 3) in response to the decrease in arterial P_{O2}. If this is the athlete's first exposure to altitude, it will probably be the first time that these receptors have responded to a drop in P_{O2}, since, up to this point, ventilation has always been regulated by blood levels of CO_2. This increased ventilation causes a decrease in blood and alveolar P_{CO2} and an increase in respiratory water loss. In addition to the water lost through ventilation, there is an increase in evaporative body water loss due to the lower humidity at altitude. As a result, even with increased fluid intake, most athletes become dehydrated.

Although reduced blood levels of CO_2 partially offset the reduction in P_iO_2 by increasing alveolar P_{O2}, they also cause the blood to become more alkaline. The kidneys react to this increased alkalinity by excreting more bicarbonate in an effort to restore normal pH. Since plasma bicarbonate is an important buffer of H^+, this adaptation has a negative impact on anaerobic capacity. Dehydration manifests itself by reductions in plasma volume, which will impair sweating during exercise but increase the concentration of red blood cells for a given volume of blood and therefore oxygen-carrying capacity. Heart rate is higher and stroke volume lower during exercise at altitude than at sea level, even at the same exercise intensity. Finally, as discussed in Chapter 3, in some individuals, acute altitude exposure may cause vasoconstriction of pulmonary vessels, leading to pulmonary hypertension and edema.

Fortunately, most of these problems begin to subside after a few days at altitude. The hyperventilatory response becomes blunted, and resting and exercise ventilatory volumes decline towards (but never achieve) those at sea level, helping to relieve the headaches and dizziness that most individuals experience during their first day at altitude. As blood P_{CO2} returns towards normal, the kidneys excrete less bicarbonate, returning blood bicarbonate to sea-level values as well. After about 10 days, an increase in the appearance rate of additional red blood cells can be detected. Although the decrease in liver and kidney P_{O2} elevates erythropoietin on day 1 of exposure, erythropoiesis requires a minimum of 6 days before final maturation of a red blood cell (as reviewed in Chapter 3). These changes in red blood cell and bicarbonate concentration lead to gradual improvements in $\dot{V}O_{2max}$ and exercise performance, but sea-level values are never regained.

Training for Competition at Altitude

After the problems experienced by athletes at the Mexico City Olympics (2300 m) in 1968, it is unlikely that national and international sports governing bodies will ever again sanction major competitions above 2000 m. Nonetheless, it is not uncommon for Winter Olympics athletes and those in team sports to compete in tournaments at locations that approach this altitude. In such cases, it is important that they arrive at the location (or preferably at a site that is slightly higher) a week to 10 days before competition. For these altitudes, this is adequate time for acclimatization to occur. Training should be resumed on the second day at altitude but with a lower volume than at sea level. Although longer periods of training at altitude may enhance some of the muscle adaptations achieved while training at sea level (60), they may actually result in the loss of some of these adaptations as well. Because stroke volume and maximal cardiac output are significantly lower while training at altitude

(58) and the stretch on cardiac muscle fibers caused by ventricular filling is an important stimulus for increasing chamber size (Chapter 6), it is probable that some detraining occurs in the heart.

Altitude Training for Sea-Level Competition

Although it has become quite common for endurance athletes to incorporate some form of altitude training or hypoxic exposure prior to competing at sea level, it is still unclear whether this approach improves performance more than training at sea level itself. Various strategies have been used to maximize the positive adaptations while minimizing negative adaptations in an effort to improve performance at sea level. Examples include living and training for several weeks at altitude (the live-high/train-high strategy), living at sea level but training at altitude or in a hypoxic environment (live-low/train-high strategy), and sleeping in a hypoxic environment but training at sea level (live-high/train-low strategy). Under the latter strategy, some athletes live at altitude and descend to a much lower altitude each day for training, but a more common approach is to have them stay at sea level and sleep in chambers that simulate altitude by adding nitrogen to inspired air or evacuating air to reduce barometric pressure.

The literature includes studies that support all of these strategies for improving $\dot{V}O_{2max}$ and/or performance at sea level (15,38,55,70). Many studies, however, have found little or no improvement (2,27,28,47,78). In addition, it appears that there can be significant individual differences in response to altitude training, with some athletes responding positively, others not at all and still others negatively. For a recent review of the efficacy of different forms of altitude/hypoxic training for sea-level competitions, together with a discussion of the ethical considerations that they pose for the World Anti-Doping Agency, readers are referred to the article by Wilber (80).

Travel Across Time Zones

For any international competition, it is inevitable that many of the athletes will have to travel by air across a number of time zones to participate. As a result, their normal circadian (24–27 hour) rhythms will be disrupted. This condition (often referred to as jet lag) continues until their biological clocks have adapted to the new location and may have negative effects on physical performance. Symptoms of jet lag include problems with sleeping and gastrointestinal function, loss of appetite, mood changes, reduced ability to concentrate, and decreases in fine motor coordination. The severity of these symptoms is directly related to the number of time zones crossed and is influenced as well by the direction of travel, with westward travel permitting easier adaptation because it prolongs daylight exposure, thus extending the normal cycle and accelerating the catch-up process.

A number of hormones show daily fluctuation patterns, but the two that appear most closely linked to control of the body's circadian rhythm are **melatonin** and **cortisol**. The release of melatonin from the pineal gland is inversely related to exposure to light: levels decline during daylight hours but increase at night. Since melatonin helps to initiate and control the quality of sleep, any changes in its release pattern will affect normal rhythms. Cortisol is secreted by the adrenal glands in response to stress and causes increases in blood pressure and blood glucose levels. Even in non-stressful situations, its relatively lower blood concentrations still exert an effect on body temperature, blood pressure, appetite, and wakefulness. Diurnal variations in cortisol are inversely related to those in melatonin, being lowest at night and highest in most individuals about mid-morning. With travel across time zones, the athlete's pre-flight rhythms become confused by differences in the new hours of daylight and darkness and there will be a delay (in most cases, approximately 1 day for each time zone crossed) before these hormones re-establish their patterns.

Procedures for Reducing the Effects of Jet Lag

Theoretically, the time required to re-adjust diurnal rhythms can be accelerated by gradually changing pre-departure bedtimes, meal times, and training times towards what they will be in the new location. However, since the influence of daylight exposure cannot be changed and because of the resulting disruptions in social patterns, most athletes have difficulty with this procedure: they find that they do not get adequate sleep and cannot train as effectively. As a result, this approach should be attempted only in the few days prior to take-off and only when traveling east (59).

While some athletes have attempted to alleviate jet lag by ingesting melatonin on arrival at their destination, this procedure does not seem to be very effective (4). It thus appears that the best solution still lies in careful planning of the travel schedule and timing of the athletes' arrival to allow adequate time for natural biorhythms to be restored. Every attempt should be made to reduce ground transport time, the number of connecting flights, and airport waiting time so that total travel time is as short as possible.

Once in flight, athletes should begin adapting their sleeping and eating times to those of the time zone where they will be competing. When awake, they should be encouraged to spend a few minutes walking about the cabin and stretching every few hours. To avoid dehydration, they should also increase their fluid intake above that at sea level (except for caffeinated beverages, which should be taken at the same time as they would be in the new destination). On arrival day, some light exercise should be scheduled at the same time at which they will be competing, with normal training resuming on day 2. Ideally, athletes will have peaked just prior to departure, so that the tapering procedures presented in Chapter 9 can be incorporated over the days leading up to their first competition.

Wheelchair and Other Paralympic Athletes

Depending upon the level of their lesions, athletes with spinal cord injuries will have some degree of impaired function in several areas that can affect sport performance. These include problems with temperature regulation, reduced or lack of sympathetic input to the heart, and pooling of blood in the lower body and limbs. Since blood flow to the skin and sweating is controlled by input from the anterior hypothalamus via the sympathetic system (Chapter 3), these functions will be impaired below the level of the lesion. Wheelchair athletes are thus more susceptible to exercise hyperthermia during longer-distance track events. Similarly, lack of sympathetic drive to the heart below the lesion blocks or reduces the increase in heart rate, stroke volume, and cardiac output that normally occurs with exercise. Finally, the lack of venous vasoconstriction and removal of the muscle pump provided by the legs results in reduced venous return to the heart, further compromising stroke volume and cardiac output (6).

Despite these limitations, wheelchair and other Paralympic athletes (including athletes with disabilities such as blindness, cerebral palsy, and amputations) compete regularly and perform impressively in many of the same sports as the able-bodied. These include such individual sports as track and field and road racing, swimming, archery, and tennis, as well as team sports such as basketball and rugby. Although in most cases different muscle groups must be targeted, the principles involved in designing their training programs to increase muscle strength and power, aerobic and anaerobic power, and capacity will be the same as those for other athletes.

Summary of Key Points

Nutritional Factors

- Carbohydrate (muscle glycogen and blood glucose) is the primary energy source for ATP production during exercise. Since muscle glycogen can be reduced over 60% by an hour of training or competition in a team sport and cannot be resynthesized until new carbohydrate is ingested, daily intake must be considerably greater in athletes than in sedentary individuals.
- Fats are a rich source of energy, with 1 g of fat producing almost 9.5 kcal.
- Fats are over twice as efficient as carbohydrate but require approximately 12% more oxygen per unit of ATP produced. In the resting state or mild exercise, approximately 50% of all energy is derived from fatty acid oxidation. However, as exercise intensity increases, this contribution declines so that in high-intensity competition or training sessions, almost all energy comes from carbohydrate.
- In longer-duration events where liver glycogen and blood glucose become depleted, lipid contribution increases, providing approximately 50% of total energy in a 2.5 hour marathon and over 70% in ultra-distance events exceeding 5 hours. Body stores of fat are considerably larger than those of carbohydrate, and are not depleted even during long-distance events.
- Protein is not normally used to produce energy except during starvation or when blood glucose declines during prolonged exercise. For example, during 3 hours of heavy exhaustive exercise, protein provides less than 10% of all energy generated.

- Endurance athletes training for several hours, 5–6 days / week, may require daily protein intakes of 1.2–1.4 $g \cdot kg^{-1}$.
- A typical heavy-resistance training session causes an immediate increase in muscle protein synthetic rate for repair of contractile protein and synthesis of larger fibers. Protein requirements for athletes performing such training are therefore higher than for the general population but can normally be met with daily intakes of approximately 1.4 $g \cdot kg^{-1}$.
- Daily caloric intake for athletes depends upon body size, type, and frequency of training and whether they are competing in weight-class events. Reasonable target values are 55% of total energy intake from carbohydrates, 25% from fat, and 20% from protein.
- As sweat rates increase, the electrolyte concentration in sweat also increases. Significant amounts of sodium, chloride, and, to a lesser extent, potassium are therefore lost during training and competition.
- Sweat rates for athletes performing prolonged heavy exercise typically exceed 1 $L \cdot h^{-1}$, or 2 $L \cdot h^{-1}$ in a hot humid environment.
- A loss of 1 L of sweat results in an approximately 220 mL reduction in plasma volume, reducing ventricular filling rate and stroke volume. Heart rate increases to maintain cardiac output but this reduces diastolic filling time and causes further reductions in stroke volume. As heart rate reaches maximal, cardiac output can no longer be maintained and exercise intensity must be reduced.
- To prevent this and to help maintain adequate sweating rates, water must be taken in at the same rate at which it is lost. For events involving less than 1 hour of total exercise time, plain cool water is the best fluid replacement. For longer events or practice sessions, the fluid should contain carbohydrate (4–8% by volume) and sodium ($\approx$0.5 $g \cdot L^{-1}$), and be ingested at a rate of 600–1200 $mL \cdot h^{-1}$.
- Athletes on a normal high carbohydrate diet who have eaten 2–3 hours before competition do not normally require carbohydrate intake for events lasting less than 1 hour. For longer duration events, athletes should begin carbohydrate intake within the first 30 minutes.
- For athletes doing resistance training to increase muscle mass, protein intake within 1–2 hours after the training session increases muscle protein synthetic rate more than the same amount ingested later.

Ergogenic Aids

- Ergogenic aids are substances or procedures that enhance physical performance beyond the levels achievable through normal training and diet. Some potentially effective aids include anabolic steroids, amphetamines, blood boosting, blood buffering, caffeine, Cr, growth hormone, beta-blockers, and altitude training. Of these, Cr, caffeine and altitude training are not prohibited by sports governing bodies.
- Studies show that 2–6 mg of caffeine per kg of body weight can improve performance in endurance exercise, and possibly in repeated bouts of sprint-type exercise, by inhibiting adenosine receptors, thus blocking the relaxation-promoting effect of dopamine and reducing the perception of fatigue. Consequently, caffeine can help promote longer training sessions or more high-intensity intervals than would normally be tolerated.
- Most studies show that Cr can effectively enhance performance in repeated high-intensity bouts of exercise and adaptations to strength and power training. Additional Cr intake increases muscle Cr and PCr concentrations and accelerates the rate of PCr resynthesis following a brief intense bout of exercise.
- Cr supplementation may benefit strength and power training by allowing athletes to perform more repetitions in a training session and possibly by counteracting the muscle growth inhibitor myostatin.

Athlete Testing and Monitoring

- A testing program can provide important physiological data for use in designing or modifying training programs for individual athletes. The parameters tested must be determinants of performance in that particular sport and the tests must be valid and reliable measures of these parameters. Testing must be repeated before and after different phases of training to assess the program's effectiveness for that individual athlete.

- Common performance tests include those for maximal oxygen consumption, lactate threshold, time to fatigue at $\dot{V}O_{2max}$, economy of exercise, maximal anaerobic power and capacity, muscle strength and power, flexibility, and body fat.

Altitude

- The partial pressure of inspired oxygen declines at higher altitudes, causing a decrease in arterial oxygen content. As a result, $\dot{V}O_{2max}$, lactate threshold, and performance in aerobic events will also decline. At altitudes over 2500 m this drop in arterial P_{O2} stimulates a rapid increase in resting ventilation, causing a decrease in blood and alveolar P_{CO2} and an increase in respiratory water loss.
- Because of the lower humidity at altitude, evaporative water loss increases and most athletes become dehydrated even with increased fluid intake. As CO_2 concentrations decline, the blood becomes more alkaline and the kidneys excrete bicarbonate in an attempt to restore normal pH. This reduction in plasma bicarbonate reduces the buffering effect on H^+ and thus the athlete's anaerobic capacity. The reduced P_{O2} levels in blood in the liver and kidneys increase erythropoietin, but new red blood cells require a week or more to mature.
- After a few days, the hyperventilatory response decreases, P_{CO2} begins to return toward normal and the kidneys excrete less bicarbonate, but sea level values are never attained.
- Athletes should arrive at high-altitude competition sites a week to 10 days before competition to permit acclimatization, and resume training on the second day, but with a lower volume than at sea level.
- Many endurance athletes add some form of altitude training or hypoxic exposure to their program before competition at sea level. Strategies to maximize sea-level performance include: living and training at altitude for several weeks before competing at sea level (live high/train high); living at sea level but training at altitude or in a hypoxic environment (live low/train high); and sleeping in a hypoxic environment but training at sea level (live high/train low).
- Although studies exist to support all of these strategies, many have found them to be no more effective in increasing $\dot{V}O_{2max}$ or endurance performance at sea level than training at sea level.

Travel Across Time Zones

- Air travel across time zones upsets the body's normal circadian rhythm and can negatively affect athletic performance. This jet lag is caused primarily by disruption in normal diurnal variations of the hormones melatonin and cortisol and results in sleeping and gastrointestinal problems, loss of appetite, reduced ability to concentrate and decreases in fine motor coordination.
- The severity of these symptoms is directly related to the number of time zones crossed; restoration of normal hormonal patterns requires approximately one day per time zone. Athletes should therefore arrive at the competition site far enough in advance for recovery.

Wheelchair and Other Paralympic Athletes

- Athletes with spinal lesions will have varying degrees of problems with temperature regulation, reduced or lack of sympathetic input to the heart and blood pooling in the lower limbs. Despite these challenges, these athletes compete regularly and impressively in many of the same sports as the able-bodied.
- Although different muscle groups must often be targeted, the principles involved in designing training programs to increase muscle strength and power, aerobic and anaerobic power, and capacity are the same as for other athletes.

Appendix A

Units for Expressing Measurements of Physiological Parameters and Factors Affecting Exercise Performance

Measurements of physiological parameters and human performance are usually expressed in *Système international d'unités* (SI) units. This appendix reviews the units that appear in this book, including SI and SI-compatible units, together with some that have been used because they are more familiar to the general public than their SI counterparts.

Expression and Use of SI and SI-Compatible Units

All unit symbols are expressed *without* a period (e.g., m not m.), except at the end of a sentence, and are used only with numerical values (e.g., there are 10^3 mm in 1 m, but a thousand millimeters in one meter). Symbols are not altered in the plural (e.g., 60 mL) and there is always a space between the numeral and the unit (60 mL, not 60mL).

When reporting results of measurements, numerals and symbols are clearer than words. For example, it is better to describe an athlete's maximal oxygen uptake as "5 L·min^{-1}" than "five liters per minute" because it is more easily understood by readers with a limited knowledge of English.

Units Formed by Multiplication and Division

Multiplication

The product of two or more units in symbolic form is indicated by a centered dot or a space. For example, an athlete's knee extension torque as measured on an isokinetic dynamometer can be expressed as 280 N·m or 280 N m. The dot is preferable to the space because it is less likely to cause confusion, and so is used in this book. If the dot cannot be raised, the expression 280 N.m is permissible, but not 280 Nm or 280 N × m. When names of units are used, multiplication is indicated by a space, not a hyphen (e.g., newton meter, not newton-meter).

Division

Division can be indicated by a solidus (oblique stroke, /), negative power (exponent), or horizontal line. For example, 100 m sprint velocity would normally be reported as 10 m/s or 10 m·s^{-1}. However, to prevent confusion, the solidus should not be repeated; negative powers are preferable (e.g., maximal oxygen uptake of 60 ml·kg^{-1}·min^{-1}, rather than 60 mL/kg/min).

Table A1 Units used in this book

Quantity	Name	Symbol
SI units		
length	meter	m
area	square meter	m^2
volume	cubic meter	m^3
mass	kilogram	kg
amount of substance	mole	mol
time	second	s
frequency	hertz	Hz (s^{-1})
speed, velocity	meter per second	$m \cdot s^{-1}$
acceleration	meter per second squared	$m \cdot s^{-2}$
angle	radian	rad
angular velocity	radian per second	$rad \cdot s^{-1}$
pressure	pascal	Pa
energy, work	joule	J
power	watt	W
force	newton	N
moment of force, torque	newton meter	N·m
temperature	kelvin	K
SI-compatible units		
time	minute	min
	hour	h
	day	d
angle	degree	° (1 rad = 57.3°)
volume	liter	l, L* (1 L = 1 dm^3)
pressure	millimeter of mercury	mmHg (1 mmHg = 133.3 Pa)
temperature	degree Celsius	°C (0 °C = 273.15 K)
Non-SI units in common use		
energy	calorie (physics)	cal (1 cal = 4.1868 J)
	Calorie (dietary)	Cal (1 Cal = 1 kcal = 4.186 kJ)
length	foot	ft (1 ft = 30.48 cm)
	mile	mi (1 mile = 1.609 km)
mass	pound†	lb (1 lb = 0.454 kg)
velocity	miles per hour	mph (1 mph = 1.609 $km \cdot h^{-1}$)
time	week	wk
	month	mo
	year	yr, y

*The official symbol for liter is l, but because it can be confused with the number 1, the upper case L is commonly used.

†Use of the term weight and the unit pound can cause confusion because both can be used as the equivalent of mass and also as the force of gravity acting on a mass. In commercial and everyday use, weight is often taken as the equivalent of mass and the pound is a unit of mass. For example, if an athlete weighed 220 lb and achieved a 300 lb maximum lift (1 RM), in SI units the athlete's mass and 1 RM would be 99.9 kg and 136.2 kg, respectively. In SI, the force of gravity acting on a mass (kg) is calculated as the mass multiplied by the acceleration due to gravity, which is 9.806 $m \cdot s^{-2}$. Thus, the force of gravity acting on a 100 kg mass would be 980.6 N (100 × 9.806). To calculate the force (weight) acting on a one pound mass, the acceleration due to gravity is 32 $ft \cdot s^{-2}$, and the resulting force is expressed in another unit of force, the poundal. For example, the weight of a 1 lb mass would be 32 poundals. Fortunately, all of this confusion can be avoided by using SI units for mass (kg) and force (N).

Table A2 SI prefixes used in this book

Factor	Prefix name	Symbol	Example
10^{3}	kilo	k	kg (kilogram)
10^{-2}	centi	c	cm (centimeter)
10^{-3}	milli	m	mL (milliliter)
10^{-6}	micro	μ	μm (micrometer)
10^{-9}	nano	n	nm (nanometer)
10^{-12}	pico	p	pn (piconewton)

References

Chapter 1

1 **Astrand I, Astrand PO, Christensen EH, and Hedman R.** Intermittent muscular work. *Acta Physiol. Scand.* 48: 448–453; 1960.

2 **Bouchard C, Taylor AW, Simoneau JA, and Dulac S.** Testing anaerobic power and capacity. In: *Physiological Testing of the High-Performance Athlete*, edited by MacDougall JD, Wenger HA, and Green HJ. Champaign, IL: Human Kinetics, pp. 175–221; 1991.

3 **Bouchard C, An P, Rice T, Skinner JS, Wilmore JH, Gagnon J, Perusse L, Leon AS, and Rao DC.** Familial aggregation of $\dot{V}O_{2max}$ response to exercise training: results from the HERITAGE Family Study. *J. Appl. Physiol.* 87(3): 1003–1008; 1999.

4 **Bouchard C, Sarzynski MA, Rice TK, Kraus WE, Church TS, Sung YJ, Rao DC, and Rankinen T.** Genomic predictors of the maximal O uptake response to standardized exercise training programs. *J. Appl. Physiol.* 110(5): 1160–1170; 2011.

5 **Bouchard C, Simoneau JA, Lortie G, Boulay MR, Marcotte M, and Thibault MC.** Genetic effects in human skeletal muscle fiber type distribution and enzyme activities. Can. *J. Physiol. Pharmacol.* 64(9): 1245–1251; 1986.

6 **Brynteson P and Sinning WE.** The effects of training frequencies on the retention of cardiovascular fitness. *Med. Sci. Sports* 5(1): 29–33; 1973.

7 **Coyle EF, Martin 3rd WH, Sinacore DR, Joyner MJ, Hagberg JM, and Holloszy JO.** Time course of loss of adaptations after stopping prolonged intense endurance training. *J. Appl. Physiol.* 57(6): 1857–1864; 1984.

8 **Edwards RH.** Human muscle function and fatigue. *Ciba Found. Symp.* 82: 1–18; 1981.

9 **Essen B.** Studies on the regulation of metabolism in human skeletal muscle using intermittent exercise as an experimental model. *Acta Physiol. Scand. Suppl.* 454: 1–32; 1978.

10 **Green HJ.** What do tests measure? In: *Physiological Testing of the High-Performance Athlete*, edited by MacDougall JD, Wenger HA, and Green HJ. Champaign, IL: Human Kinetics, pp. 7–19; 1991.

11 **Hamel P, Simoneau JA, Lortie G, Boulay MR, and Bouchard C.** Heredity and muscle adaptation to endurance training. *Med. Sci. Sports Exerc.* 18(6): 690–696; 1986.

12 **Hickson RC, Foster C, Pollock ML, Galassi TM, and Rich S.** Reduced training intensities and loss of aerobic power, endurance, and cardiac growth. *J. Appl. Physiol.* 58(2): 492–499; 1985.

13 **Linossier MT, Dormois D, Perier C, Frey J, Geyssant A, and Denis C.** Enzyme adaptations of human skeletal muscle during bicycle short-sprint training and detraining. *Acta Physiol. Scand.* 161(4): 439–445; 1997.

14 **Malina RM and Bouchard C, eds.** *Sport and Human Genetics*. Champaign, IL: Human Kinetics; 1986.

15 **Neufer PD, Costill DL, Fielding RA, Flynn MG, and Kirwan JP.** Effect of reduced training on muscular strength and endurance in competitive swimmers. *Med. Sci. Sports Exerc.* 19(5): 486–490; 1987.

Chapter 2

1 **Abernethy PJ, Thayer R, and Taylor AW.** Acute and chronic responses of skeletal muscle to endurance and sprint exercise. A review. *Sports Med.* 10(6): 365–389; 1990.

2 **Alway SE, MacDougall JD, Sale DG, Sutton JR, and McComas AJ.** Functional and structural adaptations in skeletal muscle of trained athletes. *J. Appl. Physiol.* 64(3): 1114–1120; 1988.

3 **Baldwin KM, Fitts RH, Booth FW, Winder WW, and Holloszy JO.** Depletion of muscle and liver glycogen during exercise. Protective effect of training. *Pflugers Arch.* 354(3): 203–212; 1975.

4 **Balsom PD, Soderlund K, Sjodin B, and Ekblom B.** Skeletal muscle metabolism during short duration high-intensity exercise: influence of creatine supplementation. *Acta Physiol. Scand.* 154(3): 303–310; 1995.

5 **Bell GJ and Wenger HA.** The effect of one-legged sprint training on intramuscular pH and nonbicarbonate buffering capacity. *Eur. J. Appl. Physiol. Occup. Physiol.* 58(1–2): 158–164; 1988.

6 **Berthon PM, Howlett RA, Heigenhauser GJ, and Spriet LL.** Human skeletal muscle carnitine palmitoyltransferase I activity determined in isolated intact mitochondria. *J. Appl. Physiol.* 85(1): 148–153; 1998.

7 **Boros-Hatfaludy S, Fekete G, and Apor P.** Metabolic enzyme activity patterns in muscle biopsy samples in different athletes. *Eur. J. Appl. Physiol. Occup. Physiol.* 55(3): 334–338; 1986.

8 **Burgomaster KA, Hughes SC, Heigenhauser GJ, Bradwell SN, and Gibala MJ.** Six sessions of sprint interval training increases muscle oxidative potential and cycle endurance capacity in humans. *J. Appl. Physiol.* 98(6): 1985–1990; 2005.

9 **Cadefau J, Casademont J, Grau JM, Fernandez J, Balaguer A, Vernet M, Cusso R, and Urbano-Marquez A.** Biochemical and histochemical adaptation to sprint training in young athletes. *Acta Physiol. Scand.* 140(3): 341–351; 1990.

10 **Chilibeck PD, Bell GJ, Farrar RP, and Martin TP.** Higher mitochondrial fatty acid oxidation following intermittent versus continuous endurance exercise training. *Can. J. Physiol. Pharmacol.* 76(9): 891–894; 1998.

11 **Costill DL.** *Inside Running: Basics of Sport Physiology*. Indianapolis: Benchmark; 1986.

12 **Costill DL, Coyle EF, Fink WF, Lesmes GR, and Witzmann FA.** Adaptations in skeletal muscle following strength training. *J. Appl. Physiol.* 46(1): 96–99; 1979.

13 **Costill DL, Fink WJ, and Pollock ML.** Muscle fiber composition and enzyme activities of elite distance runners. *Med. Sci. Sports* 8(2): 96–100; 1976.

14 **Dawson B, Fitzsimons M, Green S, Goodman C, Carey M, and Cole K.** Changes in performance, muscle metabolites, enzymes and fiber types after short sprint training. *Eur. J. Appl. Physiol. Occup. Physiol.* 78(2): 163–169; 1998.

15 **Fournier M, Ricci J, Taylor AW, Ferguson RJ, Montpetit RR, and Chaitman BR.** Skeletal muscle adaptation in adolescent boys: sprint and endurance training and detraining. *Med. Sci. Sports Exerc.* 14(6): 453–456; 1982.

16 **Gollnick PD, Armstrong RB, Sembrowich WL, Shepherd RE, and Saltin B.** Glycogen depletion pattern in human skeletal muscle fibers after heavy exercise. *J. Appl. Physiol.* 34(5): 615–618; 1973.

17 **Gollnick PD and Saltin B.** Significance of skeletal muscle oxidative enzyme enhancement with endurance training. *Clin. Physiol.* 2(1): 1–12; 1982.

18 **Graham TE, Rush JWE, and MacLean DA.** Skeletal muscle amino acid metabolism and ammonia production during exercise. In: *Exercise Metabolism*, edited by Hargreaves M. Champaign, IL: Human Kinetics, pp. 154–156; 1995.

19 **Greiwe JS, Hickner RC, Shah SD, Cryer PE, and Holloszy JO.** Norepinephrine response to exercise at the same relative intensity before and after endurance exercise training. *J. Appl. Physiol.* 86(2): 531–535; 1999.

20 **Hargreaves M, Finn JP, Withers RT, Halbert JA, Scroop GC, Mackay M, Snow RJ, and Carey MF.** Effect of muscle glycogen availability on maximal exercise performance. *Eur. J. Appl. Physiol. Occup. Physiol.* 75(2): 188–192; 1997.

21 **Harris RC, Edwards RH, Hultman E, Nordesjo LO, Nylind B, and Sahlin K.** The time course of phosphorylcreatine resynthesis during recovery of the quadriceps muscle in man. *Pflugers Arch.* 367(2): 137–142; 1976.

22 **Henriksson J.** Effects of physical training on the metabolism of skeletal muscle. *Diabetes Care* 15(11): 1701–1711; 1992.

23 **Henriksson J and Reitman JS.** Time course of changes in human skeletal muscle succinate dehydrogenase and cytochrome oxidase activities and maximal oxygen uptake with physical activity and inactivity. *Acta Physiol. Scand.* 99(1): 91–97; 1977.

24 **Holloszy JO.** Adaptation of skeletal muscle to endurance exercise. *Med. Sci. Sports* 7(3): 155–164; 1975.

25 **Hoppeler H, Howald H, Conley K, Lindstedt SL, Claassen H, Vock P, and Weibel ER.** Endurance training in humans: aerobic capacity and structure of skeletal muscle. *J. Appl. Physiol.* 59(2): 320–327; 1985.

26 **Hoppeler H, Luthi P, Claassen H, Weibel ER, and Howald H.** The ultrastructure of the normal

human skeletal muscle. A morphometric analysis on untrained men, women and well-trained orienteers. *Pflugers Arch.* 344(3): 217–232; 1973.

27 **Howald H, Hoppeler H, Claassen H, Mathieu O, and Straub R.** Influences of endurance training on the ultrastructural composition of the different muscle fiber types in humans. *Pflugers Arch.* 403(4): 369–376; 1985.

28 **Hultman E and Sahlin K.** Acid-base balance during exercise. *Exerc. Sport Sci. Rev.* 8: 41–128; 1980.

29 **Liljedahl ME, Holm I, Sylven C, and Jansson E.** Different responses of skeletal muscle following sprint training in men and women. *Eur. J. Appl. Physiol. Occup. Physiol.* 74(4): 375–383; 1996.

30 **Linossier MT, Denis C, Dormois D, Geyssant A, and Lacour JR.** Ergometric and metabolic adaptation to a 5-s sprint training program. *Eur. J. Appl. Physiol. Occup. Physiol.* 67(5): 408–414; 1993.

31 **Luthi JM, Howald H, Claassen H, Rosler K, Vock P, and Hoppeler H.** Structural changes in skeletal muscle tissue with heavy-resistance exercise. *Int. J. Sports Med.* 7(3): 123–127; 1986.

32 **MacDougall JD.** Hypertrophy and Hyperplasia. In: *Strength and Power in Sport*, 2nd edn, edited by Komi PV. Malden, MA: Blackwell, 2003; pp. 252–264.

33 **MacDougall JD, Hicks AL, MacDonald JR, McKelvie RS, Green HJ, and Smith KM.** Muscle performance and enzymatic adaptations to sprint interval training. *J. Appl. Physiol.* 84(6): 2138–2142; 1998.

34 **MacDougall JD, Ray S, Sale DG, McCartney N, Lee P, and Garner S.** Muscle substrate utilization and lactate production during weightlifting. *Can. J. Appl. Physiol.* 24(3): 209–215; 1999.

35 **MacDougall JD, Sale DG, Moroz JR, Elder GC, Sutton JR, and Howald H.** Mitochondrial volume density in human skeletal muscle following heavy resistance training. *Med. Sci. Sports* 11(2): 164–166; 1979.

36 **MacDougall JD, Ward GR, Sale DG, and Sutton JR.** Biochemical adaptation of human skeletal muscle to heavy resistance training and immobilization. *J. Appl. Physiol.* 43(4): 700–703; 1977.

37 **McGilvery RW and Goldstein GW.** *Biochemistry: a Functional Approach*. Philadelphia: W.B. Saunders; 1983.

38 **Melissa L, MacDougall JD, Tarnopolsky MA, Cipriano N, and Green H.** Skeletal muscle adaptations to training under normobaric hypoxic versus normoxic conditions. *Med. Sci. Sports Exerc.* 29(2): 238–243; 1997.

39 **Meyer RA and Foley JM.** Cellular processes integrating the metabolic response to exercise. In: *Handbook of Physiology*, edited by Rowell LB and Shepherd JT. New York: Oxford University Press/ American Physiological Society, 1996; pp. 841–870.

40 **Murakami T, Shimomura Y, Fujitsuka N, Sokabe M, Okamura K, and Sakamoto S.** Enlargement glycogen store in rat liver and muscle by fructose-diet intake and exercise training. *J. Appl. Physiol.* 82(3): 772–775; 1997.

41 **Nakatani A, Han DH, Hansen PA, Nolte LA, Host HH, Hickner RC, and Holloszy JO.** Effect of endurance exercise training on muscle glycogen supercompensation in rats. *J. Appl. Physiol.* 82(2): 711–715; 1997.

42 **Newsholm EA and Leech AR.** *Biochemistry for the Medical Sciences*. Toronto: Wiley; 1988.

43 **Nilsson LH and Hultman E.** Liver glycogen in man—the effect of total starvation or a carbohydrate-poor diet followed by carbohydrate refeeding. *Scand. J. Clin. Lab. Invest.* 32(4): 325–330; 1973.

44 **Odland LM, MacDougall JD, and Tarnopolsky M.** Energy delivery pathways during brief maximum intensity exercise in humans. *Med. Sci. Sports Exerc.* 27(5): S38; 1995.

45 **Parkhouse WS and McKenzie DC.** Possible contribution of skeletal muscle buffers to enhanced anaerobic performance: a brief review. *Med. Sci. Sports Exerc.* 16(4): 328–338; 1984.

46 **Pernow B and Saltin B.** Availability of substrates and capacity for prolonged heavy exercise in man. *J. Appl. Physiol.* 31(3): 416–422; 1971.

47 **Perry CG, Heigenhauser GJ, Bonen A, and Spriet LL.** High-intensity aerobic interval training increases fat and carbohydrate metabolic capacities in human skeletal muscle. *Appl. Physiol. Nutr. Metab.* 33(6): 1112–1123; 2008.

48 **Pette D and Vrbova G.** Adaptation of mammalian skeletal muscle fibers to chronic electrical stimulation. *Rev. Physiol. Biochem. Pharmacol.* 120: 115–202; 1992.

49 **Phillips SM, Green HJ, Tarnopolsky MA, Heigenhauser GJ, and Grant SM.** Progressive effect of endurance training on metabolic adaptations in working skeletal muscle. *Am. J. Physiol.* 270(2, Pt 1): E265–E272; 1996.

50 **Roberts AD, Billeter R, and Howald H.** Anaerobic muscle enzyme changes after interval training. *Int. J. Sports Med.* 3(1): 18–21; 1982.

51 **Sahlin K.** Intracellular pH and energy metabolism in skeletal muscle of man. With special reference to exercise. *Acta Physiol. Scand. Suppl.* 455: 1–56; 1978.

52 **Sahlin K, Katz A, and Broberg S.** Tricarboxylic acid cycle intermediates in human muscle during prolonged exercise. *Am. J. Physiol.* 259(5, Pt 1): C834–C841; 1990.

53 **Saltin B and Astrand PO.** Free fatty acids and exercise. *Am. J. Clin. Nutr.* 57(5 Suppl): 752S–757S; 1993.

54 **Sharp RL, Costill DL, Fink WJ, and King DS.** Effects of eight weeks of bicycle ergometer sprint training on human muscle buffer capacity. *Int. J. Sports Med.* 7(1): 13–17; 1986.

55 **Spriet LL.** Phosphofructokinase activity and acidosis during short-term tetanic contractions. *Can. J. Physiol. Pharmacol.* 69(2): 298–304; 1991.

56 **Staron RS, Karapondo DL, Kraemer WJ, Fry AC, Gordon SE, Falkel JE, Hagerman FC, and Hikida RS.** Skeletal muscle adaptations during early phase of heavy-resistance training in men and women. *J. Appl. Physiol.* 76(3): 1247–1255; 1994.

57 **Staron RS, Malicky ES, Leonardi MJ, Falkel JE, Hagerman FC, and Dudley GA.** Muscle hypertrophy and fast fiber type conversions in heavy resistance-trained women. *Eur. J. Appl. Physiol. Occup. Physiol.* 60(1): 71–79; 1990.

58 **Suter E, Hoppeler H, Claassen H, Billeter R, Aebi U, Horber F, Jaeger P, and Marti B.** Ultrastructural modification of human skeletal muscle tissue with 6-month moderate-intensity exercise training. *Int. J. Sports Med.* 16(3): 160–166; 1995.

59 **Tarnopolsky MA.** Protein metabolism in strength and endurance activities. In: *Perspectives in Exercise Science and Sports Medicine*, edited by Lamb DR and Murray R. Carmel, IN: Cooper Publishing Group, 1999; pp. 125–164.

60 **Tesch AP.** Training for bodybuilding. In: *Strength and Power in Sport*, edited by Komi PV. Oxford: Blackwell, 1992; pp. 370–380.

61 **Tesch PA.** Skeletal muscle adaptations consequent to long-term heavy resistance exercise. *Med. Sci. Sports Exerc.* 20(5 Suppl): S132–S134; 1988.

62 **Tesch PA, Komi PV, and Hakkinen K.** Enzymatic adaptations consequent to long-term strength training. *Int. J. Sports Med.* 8(Suppl 1): 66–69; 1987.

63 **Thorstensson A, Hulten B, von Dobeln W, and Karlsson J.** Effect of strength training on enzyme activities and fiber characteristics in human skeletal muscle. *Acta Physiol. Scand.* 96(3): 392–398; 1976.

64 **Vandenberghe K, Hespel P, Vanden Eynde B, Lysens R, and Richter EA.** No effect of glycogen level on glycogen metabolism during high intensity exercise. *Med. Sci. Sports Exerc.* 27(9): 1278–1283; 1995.

65 **Wibom R, Hultman E, Johansson M, Matherei K, Constantin-Teodosiu D, and Schantz PG.** Adaptation of mitochondrial ATP production in human skeletal muscle to endurance training and detraining. *J. Appl. Physiol.* 73(5): 2004–2010; 1992.

66 **Wilkinson SB, Phillips SM, Atherton PJ, Patel R, Yarasheski KE, Tarnopolsky MA, and Rennie MJ.** Differential effects of resistance and endurance exercise in the fed state on signaling molecule phosphorylation and protein synthesis in human muscle. *J. Physiol.* 586(15): 3701–3717; 2008.

67 **Zendzian-Piotrowska M and Gorski J.** Metabolic adaptation to daily exercise of moderate intensity to exhaustion in the rat. *Eur. J. Appl. Physiol. Occup. Physiol.* 67(1): 77–82; 1993.

Chapter 3

1 **Andersen P and Henriksson J.** Capillary supply of the quadriceps femoris muscle of man: adaptive response to exercise. *J. Physiol.* 270(3): 677–690; 1977.

2 **Antonio J and Gonyea WJ.** Progressive stretch overload of skeletal muscle results in hypertrophy before hyperplasia. *J. Appl. Physiol.* 75(3): 1263–1271; 1993.

3 **Armstrong LE, Hubbard RW, Jones BH, and Daniels JJ.** Preparing Alberto Salazar for the heat of the 1984 Olympic marathon. *Physician Sportsmed.* 14: 73–81; 1986.

4 **Astrand PO, Rodahl K, Dahl HA, and Strome SB.** *Textbook of Work Physiology: Physiological Bases of Exercise*. Champaign, IL: Human Kinetics; 2003.

5 **Astrand PO, Cuddy TE, Saltin B, and Stenberg J.** Cardiac output during submaximal and maximal work. *J. Appl. Physiol.* 19: 268–274; 1964.

6 **Berne RM and Levy MN.** *Physiology*. St. Louis: C.V. Mosby; 1988.

7 **Bishop JE and Lindahl G.** Regulation of cardiovascular collagen synthesis by mechanical load. *Cardiovasc. Res.* 42(1): 27–44; 1999.

8 **Bonde-Petersen F, Mork AL, and Nielsen E.** Local muscle blood flow and sustained contractions of human arm and back muscles. *Eur. J. Appl. Physiol. Occup. Physiol.* 34(1): 43–50; 1975.

9 **Brodal P, Ingjer F, and Hermansen L.** Capillary supply of skeletal muscle fibers in untrained and endurance-trained men. *Am. J. Physiol.* 232(6): H705–H712; 1977.

10 **Claremont AD, Nagle F, Reddan WD, and Brooks GA.** Comparison of metabolic, temperature, heart rate and ventilatory responses to exercise at extreme ambient temperatures (0 degrees and 35 degrees C). *Med. Sci. Sports* 7(2): 150–154; 1975.

11 **Dempsey JA.** JB Wolffe Memorial Lecture. Is the lung built for exercise? *Med. Sci. Sports Exerc.* 18(2): 143–155; 1986.

12 **Ehsani AA.** Loss of cardiovascular adaptations after cessation of training. *Cardiol. Clin.* 10(2): 257–266; 1992.

13 **Fagard RH.** Athlete's heart: a meta-analysis of the echocardiographic experience. *Int. J. Sports Med.* 17(Suppl 3): S140–S144; 1996.

14 **Giada F, Bertaglia E, De Piccoli B, Franceschi M, Sartori F, Raviele A, and Pascotto P.** Cardiovascular adaptations to endurance training and detraining in young and older athletes. *Int. J. Cardiol.* 65(2): 149–155; 1998.

15 **Gledhill N, Cox D, and Jamnik R.** Endurance athletes' stroke volume does not plateau: major advantage is diastolic function. *Med. Sci. Sports Exerc.* 26(9): 1116–1121; 1994.

16 **Haykowsky MJ, Eves ND, Warburton DER, and Findlay MJ.** Resistance exercise, the Valsalva maneuver, and cerebrovascular transmural pressure. *Med. Sci. Sports Exerc.* 35(1): 65–68; 2003.

17 **Heinicke K, Wolfarth B, Winchenbach P, Biermann B, Schmid A, Huber G, Friedmann B, and Schmidt W.** Blood volume and hemoglobin mass in elite athletes of different disciplines. *Int. J. Sports Med.* 22(7): 504–512; 2001.

18 **Henriksen E, Landelius J, Wesslen L, Arnell H, Nystrom-Rosander C, Kangro T, Jonason T, Rolf C, Lidell C, Hammarstrom E, Ringqvist I, and Friman G.** Echocardiographic right and left ventricular measurements in male elite endurance athletes. *Eur. Heart J.* 17(7): 1121–1128; 1996.

19 **Hoppeler H.** Vascular growth in hypoxic skeletal muscle. *Adv. Exp. Med. Biol.* 474: 277–286; 1999.

20 **Ingjer F.** Effects of endurance training on muscle fiber ATPase activity, capillary supply and mitochondrial content in man. *J. Physiol.* 294: 419–432; 1979.

21 **Ingjer F.** Maximal aerobic power related to the capillary supply of the quadriceps femoris muscle in man. *Acta Physiol. Scand.* 104(2): 238–240; 1978.

22 **Kano Y, Shimegi S, Masuda K, Sakato H, Ohmori H, and Katsuta S.** Effects of different intensity endurance training on the capillary network in rat skeletal muscle. *Int. J. Microcirc. Clin. Exp.* 17(2): 93–96; 1997.

23 **Klausen K, Secher NH, Clausen JP, Hartling O, and Trap-Jensen J.** Central and regional circulatory adaptations to one-leg training. *J. Appl. Physiol.* 52(4): 976–983; 1982.

24 **MacDougall JD.** Blood pressure responses to resistive, static and dynamic exercise. In: *Cardiovascular Response to Exercise*, edited by Fletcher GF. Mount Kisko, NY: Futura Publishing, 1994; pp. 155–173.

25 **MacDougall JD, Green HJ, Sutton JR, Coates G, Cymerman A, Young P, and Houston CS.** Operation Everest II: structural adaptations in skeletal muscle in response to extreme simulated altitude. *Acta Physiol. Scand.* 142(3): 421–427; 1991.

26 **MacDougall JD, Reddan WG, Layton CR, and Dempsey JA.** Effects of metabolic hyperthermia on performance during heavy prolonged exercise. *J. Appl. Physiol.* 36(5): 538–544; 1974.

27 **MacDougall JD, Tuxen D, Sale DG, Moroz JR, and Sutton JR.** Arterial blood pressure response to heavy resistance exercise. *J. Appl. Physiol.* 58(3): 785–790; 1985.

28 **Melissa L, MacDougall JD, Tarnopolsky MA, Cipriano N, and Green H.** Skeletal muscle adaptations to training under normobaric hypoxic versus normoxic conditions. *Med. Sci. Sports Exerc.* 29(2): 238–243; 1997.

29 **Mitchell JH, Payne FC, Saltin B, and Schibye B.** The role of muscle mass in the cardiovascular response to static contractions. *J. Physiol.* 309: 45–54; 1980.

30 **Mitchell JH, Reardon WC, and McCloskey DI.** Reflex effects on circulation and respiration from contracting skeletal muscle. *Am. J. Physiol.* 233(3): H374–H378; 1977.

31 **Montain SJ and Coyle EF.** Influence of graded dehydration on hyperthermia and cardiovascular drift

during exercise. *J. Appl. Physiol.* 73(4): 1340–1350; 1992.

32 **Nadel ER, Cafarelli E, Roberts MF, and Wenger CB.** Circulatory regulation during exercise in different ambient temperatures. *J. Appl. Physiol.* 46(3): 430–437; 1979.

33 **Pelliccia A, Maron BJ, De Luca R, Di Paolo FM, Spataro A, and Culasso F.** Remodeling of left ventricular hypertrophy in elite athletes after long-term deconditioning. *Circulation* 105(8): 944–949; 2002.

34 **Prior BM, Yang HT, and Terjung RL.** What makes vessels grow with exercise training? *J. Appl. Physiol.* 97(3): 1119–1128; 2004.

35 **Richardson RS.** Oxygen transport: air to muscle cell. *Med. Sci. Sports Exerc.* 30(1): 53–59; 1998.

36 **Rowell LB.** Muscle blood flow in humans: how high can it go? *Med. Sci. Sports Exerc.* 20(5 Suppl): S97–S103; 1988.

37 **Sawka MN, Convertino VA, Eichner ER, Schnieder SM, and Young AJ.** Blood volume: importance and adaptations to exercise training, environmental stresses, and trauma / sickness. *Med. Sci. Sports Exerc.* 32(2): 332–348; 2000.

38 **Sawka MN and Coyle EF.** Influence of body water and blood volume on thermoregulation and exercise performance in the heat. *Exerc. Sport Sci. Rev.* 27: 167–218; 1999.

39 **Segal SS and Kurjiaka DT.** Coordination of blood flow control in the resistance vasculature of skeletal muscle. *Med. Sci. Sports Exerc.* 27(8): 1158–1164; 1995.

40 **Shen W, Zhang X, Zhao G, Wolin MS, Sessa W, and Hintze TH.** Nitric oxide production and NO synthase gene expression contribute to vascular regulation during exercise. *Med. Sci. Sports Exerc.* 27(8): 1125–1134; 1995.

41 **Shono N, Urata H, Saltin B, Mizuno M, Harada T, Shindo M, and Tanaka H.** Effects of low intensity aerobic training on skeletal muscle capillary and blood lipoprotein profiles. *J. Atheroscler. Thromb.* 9(1): 78–85; 2002.

42 **Smith JJ and Kampine JP.** *Circulatory Physiology—The Essentials.* Baltimore, MD: Williams & Wilkins; 1980.

43 **Suter E, Hoppeler H, Claassen H, Billeter R, Aebi U, Horber F, Jaeger P, and Marti B.** Ultrastructural modification of human skeletal muscle tissue with 6-month moderate-intensity exercise training. *Int. J. Sports Med.* 16(3): 160–166; 1995.

44 **Ward MP.** *High Altitude Medicine and Physiology.* London: Chapman and Hall Medical; 1995.

45 **West JB.** *Respiratory Physiology: The Essentials.* Philadelphia, PA: Lippincott Williams & Wilkins; 1990.

46 **Zumstein A, Mathieu O, Howald H, and Hoppeler H.** Morphometric analysis of the capillary supply in skeletal muscles of trained and untrained subjects—its limitations in muscle biopsies. *Pflugers Arch.* 397(4): 277–283; 1983.

Chapter 4

1 **Aagaard P, Simonsen EB, Anderson JL, Magnussen J, Kalkjer-Kristensen J, and Dyhre-Poulsen P.** Neural inhibition during maximal eccentric and concentric quadriceps contraction: effects of resistance training. *J. Appl. Physiol.* 89: 2249–2257; 2000.

2 **Abbott BC, Bigland B, and Ritchie JM.** The physiological cost of negative work. *J. Physiol.* 117: 380–390; 1952.

3 **Asmussen E and Bonde-Petersen F.** Storage of elastic energy in skeletal muscles in man. *Acta Physiol. Scand.* 91(3): 385–392; 1974.

4 **Asmussen E and Bonde-Petersen F.** Apparent efficiency and storage of elastic energy in human muscles during exercise. *Acta Physiol. Scand.* 92(4): 537–545; 1974.

5 **Biewener AA and Roberts TJ.** Muscle and tendon contributions to force, work, and elastic energy savings: a comparative perspective. *Exerc. Sport Sci. Rev.* 28(3): 99–107; 2000.

6 **Billeter R and Hoppler H.** Muscular basis of strength. In: *Strength and Power in Sport*, edited by Komi PV. Oxford: Blackwell, 1992; pp. 39–63.

7 **Bobbert MF and Casius LJ.** Is the effect of a countermovement on jump height due to active state development? *Med. Sci. Sports Exerc.* 37(3): 440–446; 2005.

8 **Bobbert MF, Gerritsen KG, Litjens MC, and Van Soest AJ.** Why is countermovement jump height greater than squat jump height? *Med. Sci. Sports Exerc.* 28(11): 1402–1412; 1996.

9 **Bosco C, Tarkka I, and Komi PV.** Effect of elastic energy and myoelectrical potentiation of triceps surae during stretch shortening cycle exercise. *Int. J. Sports Med.* 3: 137–140; 1982.

10 **Brooks GA, Fahey TD, and Baldwin KM.** *Exercise Physiology. Human Bioenergetics and Its Applications*, 4th edn. New York: McGraw-Hill; 2005.

11 **Cabell L and Zebas CJ.** Resistive torque validation of the Nautilus multi-biceps machine. *J. Strength Cond. Res.* 13(1): 20–23; 1999.

12 **Cavagna GA, Dusman B, and Margaria R.** Positive work done by a previously stretched muscle. *J. Appl. Physiol.* 24(1): 21–32; 1968.

13 **Cavanagh PR and Komi PV.** Electromechanical delay in human skeletal muscle under concentric and eccentric contractions. *Eur. J. Appl. Physiol.* 42: 159–163; 1979.

14 **Clarke HH, Elkins EC, Martin GM, and Wakim KG.** Relationship between body position and the application of muscle power to movements of the joints. *Arch. Phys. Med. Rehabil.* 31(2): 81–89; 1950.

15 **Cooke R.** The sliding filament model: 1972–2004. *J. Gen. Physiol.* 123: 643–656; 2004.

16 **Dawson TJ and Taylor CR.** Energetic cost of locomotion in kangaroos. *Nature* 246(30): 313–314; 1973.

17 **de Ruiter CJ and de Haan A.** Shortening-induced depression of voluntary force in unfatigued and fatigued human adductor pollicis muscle. *J. Appl. Physiol.* 94: 69–74; 2003.

18 **de Ruiter CJ, de Haan A, Jones DA, and Sargeant AJ.** Shortening-induced depression in human adductor pollicis muscle. *J. Physiol.* 507(2): 583–591; 1998.

19 **Dietz V, Schmidtbleicher D, and Noth J.** Neuronal mechanisms of human locomotion. *J. Neurophysiol.* 42(5): 1212–1222; 1979.

20 **di Prampero PE.** Metabolic and circulatory limitations to VO_{2max} at the whole animal level. *J. Exp. Biol.* 115: 319–332; 1985.

21 **Edman KAP.** The force bearing capacity of frog muscle fibers during stretch: its relation to sarcomere length and fiber width. *J. Physiol.* 519(2): 515–526; 1999.

22 **Edman KAP.** Contractile performance of skeletal muscle fibers. In: *Strength and Power in Sport*, 2nd edn, edited by Komi PV. Malden, MA: Blackwell, 2003; pp. 114–133.

23 **Edman KAP, Elzinga G, and Noble MIM.** Residual force enhancement after stretch of contracting frog single muscle fibers. *J. Gen. Physiol.* 80(5): 769–784; 1982.

24 **Ettema GJC.** Effects of contraction history on control and stability in explosive actions. *J. Electromyogr. Kinesiol.* 12(6): 455–461; 2002.

25 **Faulkner JA.** Terminology for contractions of muscles during shortening, while isometric, and during lengthening. *J. Appl. Physiol.* 95(2): 455–459; 2003.

26 **Folland J and Morris B.** Variable-cam resistance training machines: do they match the angle-torque relationship in humans? *J. Sports Sci.* 26(2): 163–169; 2008.

27 **Fukunaga T, Kawakami Y, Kubo K, and Kahehisa H.** Muscle and tendon interaction during human movements. *Exerc. Sport Sci. Rev.* 30(3): 106–110; 2002.

28 **Fusi L, Reconditi M, Linari M, Brunello E, Elangovan R, Lombardi V, and Piazzesi G.** The mechanism of the resistance to stretch of isometrically contracting single muscle fibers. *J. Physiol.* 588(3): 495–510; 2010.

29 **Gordon AM, Homsher E, and Regnier M.** Regulation of contraction in striated muscle. *Physiol. Rev.* 80(2): 853–924; 2000.

30 **Hahn D, Seiberl W, Schmidt S, Schweizer K, and Schwirtz A.** Evidence of residual force enhancement for multi-joint leg extension. *J. Biomech.* 43(8): 1503–1508; 2010.

31 **Hamada T, Sale DG, MacDougall JD, and Tarnopolsky MA.** Postactivation potentiation, fiber type, and twitch contraction time in human knee extensor muscles. *J. Appl. Physiol.* 88: 2131–2137; 2000.

32 **Harman E.** Resistive torque of 5 Nautilus exercise machines. *Med. Sci. Sports Exerc.* 15(2): 113; 1983.

33 **Herzog W.** The relation between the resultant moments at a joint and the moments measured by an isokinetic dynamometer. *J. Biomech.* 21(1): 5–12; 1988.

34 **Herzog W.** History dependence of skeletal muscle force production: implications for movement control. *Hum. Mov. Sci.* 23: 591–604; 2004.

35 **Herzog W, Hasler E, and Abrahamse SK.** A comparison of knee extensor strength curves obtained theoretically and experimentally. *Med. Sci. Sports Exerc.* 23(1): 108–114; 1991.

36 **Herzog W, Lee EJ, and Rassier DE.** Residual force enhancement in skeletal muscle. *J. Physiol.* 574(3): 635–642; 2006.

37 **Herzog W and Leonard TR.** The history dependence of force production in mammalian skeletal

muscle following stretch-shortening and shortening-stretch cycles. *J. Biomech.* 33: 531–542; 2000.

38 **Herzog W, Schachar R, and Leonard TR.** Characterization of the passive component of force enhancement following active stretching of skeletal muscle. *J. Exp. Biol.* 206: 3635–3643; 2003.

39 **Highsmith S.** Lever arm model of force generation by actin-myosin-ATP. *Biochemistry* 38(31): 9791–9797; 1999.

40 **Hill AV.** The series elastic component of muscle. *Proc. R. Soc. Lond. B* 137(887): 273–280; 1950.

41 **Hof AL, Geelen BA, and van den Berg JW.** Calf muscle moment, work and efficiency in level walking: role of series elascticity. *J. Biomech.* 16(7): 523–537; 1983.

42 **Holmes K and Geeves MA.** The structural basis of muscle contraction. *Phil. Trans. R. Soc. Lond. B* 355: 419–431; 2000.

43 **Houston ME, Norman RW, and Froese EA.** Mechanical measures during maximal velocity knee extension exercise and their relation to fiber composition of the human vastus lateralis muscle. *Eur. J. Appl. Physiol.* 58: 1–7; 1988.

44 **Houtz SJ, LeBow MJ, and Beyer FR.** Effect of posture on strength of the knee flexor and extensor muscles. *J. Appl. Physiol.* 11(3): 475–480; 1957.

45 **Huxley AF.** Muscle structure and theories of contraction. *Prog. Biophys. Chem.* 7: 255–318; 1957.

46 **Huxley AF.** Biological motors: energy storage in myosin molecules. *Current Biol.* 8: R485–R488; 1998.

47 **Ishikawa M and Komi PV.** Effects of different dropping intensities on fascicle and tendinous tissue behavior during stretch-shortening cycle exercise. *J. Appl. Physiol.* 96(3): 848–852; 2004.

48 **Ishikawa M, Niemela E, and Komi PV.** The interaction between fascicle and tendinous tissues in short contact stretch-shortening cycle exercise with varying eccentric intensities. *J. Appl. Physiol.* 99(2): 603–608; 2005.

49 **Jensen RC, Warren B, Laursen C, and Morrissey MC.** Static preload effect on knee extensor isokinetic concentric and eccentric performance. *Med. Sci. Sports Exerc.* 23(1): 10–14; 1991.

50 **Johnson JH, Colodny S, and Jackson D.** Human torque capability versus resistive torque for four Eagle resistance machines. *J. Appl. Sport Sci. Res.* 4: 83–87; 1990.

51 **Joumaa V and Herzog W.** Force depression in single myofibrils. *J. Appl. Physiol.* 108: 356–362; 2010.

52 **Kawakami Y, Kubo K, Kanehisa H, and Fukunaga T.** Effect of series elasticity on isokinetic torque-angle relationship in humans. *Eur. J. Appl. Physiol.* 87: 381–387; 2002.

53 **Kawakami Y, Muroka T, Kanehisa H, and Fukunaga T.** *In vivo* muscle fiber behavior during counter-movement exercise in humans reveals a significant role for tendon elasticity. *J. Physiol.* 540(2): 635–646; 2002.

54 **Kitai TA and Sale DG.** Specificity of joint angle in isometric training. *Eur. J. Appl. Physiol.* 58(7): 744–748; 1989.

55 **Komi PV.** Physiological and biomechanical correlates of muscle function: effects of muscle structure and stretch-shortening cycle on force and speed. *Exerc. Sport Sci. Rev.* 12: 81–121; 1984.

56 **Komi PV.** Training of muscle strength and power: interaction of neuromotoric, hypertrophic, and mechanical factors. *Int. J. Sports Med.* 7: S10–S15; 1986.

57 **Komi PV.** Stretch-shortening cycle. In: *Strength and Power in Sport*, 2nd edn, edited by Komi PV. Malden, MA: Blackwell, 2003; pp. 184–202.

58 **Krstić RV.** *Die Gewebe des Menschen und der Säugetiere.* Berlin: Springer Verlag; 1978.

59 **Kubo K, Kanahisa H, and Fukunaga T.** Effects of different duration isometric contractions on tendon elasticity in human quadriceps muscles. *J. Physiol.* 536(2): 649–655; 2001.

60 **Kubo K, Ohgo K, Yoshinaga K, Tsunoda N, Kanehisa H, and Fukunaga T.** Effects of series elasticity on the human knee extension torque-angle relationship in vivo. *Res. Q. Exerc. Sport* 77(4): 408–416; 2006.

61 **Kubo K, Morimoto M, Komuro T, Tsunoda N, and Fukunaga T.** Influences of tendon stiffness, joint stiffness, and electromyographic activity on jump performances using single joint. *Eur. J. Appl. Physiol.* 99: 235–243; 2007.

62 **Kulig K, Andrews JG, and Hay JG.** Human strength curves. *Exerc. Sport Sci. Rev.* 12: 417–466; 1984.

63 **Lee HD and Herzog W.** Force enhancement following muscle stretch of electrically and voluntarily activated human adductor pollicis. *J. Physiol.* 545(1): 321–330; 2002.

64 **Lee HD and Herzog W.** Force depression following muscle shortening of voluntarily activated and

electrically stimulated human adductor pollicis. *J. Physiol.* 551: 993–1003; 2003.

65 **Lee HD, Suter E, and Herzog W.** Force depression in human quadriceps femoris following voluntary shortening contractions. *J. Appl. Physiol.* 87(5): 1651–1655; 1999.

66 **Lee HD, Herzog W, and Leonard T.** Effects of cyclic changes in muscle length on force production in in-situ cat soleus. *J. Biomech.* 34: 979–987; 2001.

67 **Leonard TR and Herzog W.** Regulation of muscle force in the absence of actin-myosin-based cross-bridge interaction. *Am. J. Physiol.* 299: C14–C20; 2010.

68 **Lieber RL and Ward SR.** Skeletal muscle design to meet functional demands. *Phil. Trans. R. Soc. B* 366: 1466–1476; 2011.

69 **Linari M, Bottinelli R, Pellegrino MA, Reconditi M, Reggiani C, and Lombardi V.** The mechanism of the force response to stretch in human skinned muscle fibers with different myosin isoforms. *J. Physiol.* 554(2): 335–352; 2004.

70 **Linari M, Lucii L, Reconditi M, Casoni ME, Amenitsch H, Bernstorff S, Piazzesi G, and Lombardi V.** A combined mechanical and X-ray diffraction study of stretch potentiation in single frog muscle fibers. *J. Physiol.* 526(3): 589–596; 2000.

71 **Lindsted SL, LaStayo PC, and Reich TE.** When active muscles lengthen: properties and consequences of eccentric contractions. *News Physiol. Sci.* 16: 256–261; 2001.

72 **Lombardi V and Piazzesi G.** The contractile response during steady lengthening of stimulated frog muscle fibers. *J. Physiol.* 431: 141–171; 1990.

73 **Lombardi V, Piazzesi G, Reconditi M, Linari M, Lucii L, Stewart A, Sun Y, Boesecke P, Narayanan T, Irving T, and Irving M.** X-ray diffraction studies of the contractile mechanism in single muscle fibers. *Phil. Trans. R. Soc. B* 359: 1883–1893; 2004.

74 **Magnaris CN.** Force-length characteristics of *in vivo* human skeletal muscle. *Acta Physiol. Scand.* 172: 279–285; 2001.

75 **Magnaris CN, Baltzopoulos V, and Sargeant AJ.** Changes in Achilles tendon moment arm from rest to maximum isometric plantarflexion: *in vivo* observations in man. *J. Physiol.* 510(3): 977–985; 1998.

76 **Manning RJ, Graves JE, Carpenter DM, Leggett SH, and Pollock ML.** Constant vs variable resistance knee extension training. *Med. Sci. Sports Exerc.* 22(3): 397–401; 1990.

77 **McCartney N, Heigenhauser GJ, and Jones NL.** Power output and fatigue of human muscle in maximal cycling exercise. *J. Appl. Physiol.* 55(1): 218–224; 1983.

78 **McDaniel J, Elmer SJ, and Martin JC.** The effect of shortening history on isometric and dynamic muscle function. *J. Biomech.* 43(4): 606–611; 2010.

79 **McGowan CP, Neptune RR, and Herzog W.** A phenomenological model and validation of shortening induced force depression during muscle contractions. *J. Biomech.* 43(3): 449; 2010.

80 **Mesentean S, Koppole S, Smith JC, and Fischer S.** The principal motions involved in the coupling mechanism of the recovery stroke of the myosin motor. *J. Mol. Biol.* 367: 591–602; 2007.

81 **Morgan DL.** An explanation for residual increased tension in striated muscle after stretch during contraction. *Exp. Physiol.* 79: 831–838; 1994.

82 **Morgan DL, Whitehead NP, Wise AK, Gregory JE, and Proske U.** Tension changes in the cat soleus muscle following slow stretch or shortening of the contracting muscle. *J. Physiol.* 522(3): 503–513; 2000.

83 **Murray WM, Buchanan TS, and Delp SL.** The isometric capacity of muscles that cross the elbow. *J. Biomech.* 33(8): 943–952; 2000.

84 **Murray WM, Buchanan TS, and Delp SL.** Scaling of peak moment arms of elbow muscles with upper extremity bone dimensions. *J. Biomech.* 35(1): 19–26; 2002.

85 **Murray WM, Delp SL, and Buchanan TS.** Variation of muscle moment arms with elbow and forearm position. *J. Biomech.* 28(5): 513–525; 1995.

86 **Nicol C and Komi PV.** Significance of passively induced stretch reflexes on Achilles tendon force enhancement. *Muscle Nerve* 21(11): 1546–1548; 1998.

87 **Oskouei AE and Herzog W.** Observations on force enhancement in submaximal voluntary contractions of human adductor pollicis muscle. *J. Appl. Physiol.* 98: 2087–2095; 2005.

88 **Oskouei AE and Herzog W.** Force enhancement at different levels of voluntary contraction in human adductor pollicis. *Eur. J. Appl. Physiol.* 97: 280–287; 2006.

89 **Peeters M, Svantesson U, and Grimby G.** Effect of prior isometric muscle action on concentric

torque output during plantar flexion. *Eur. J. Appl. Physiol.* 71(2–3): 272–275; 1995.

90 **Perrine JJ.** Isokinetic exercise and the mechanical energy potential of muscle. *J. Health Phys. Educ. Rec.* 39: 40–48; 1968.

91 **Piazzesi G, Lucii L, and Lombardi V.** The size and speed of the working stroke of muscle myosin and its dependence on the force. *J. Physiol.* 545(1): 145–151; 2002.

92 **Piazzesi G, Reconditi M, Linari M, Lucii L, Bianco P, Brunello E, Decostre V, Stewart A, Gore DB, Irving TC, Irving M, and Lombardi V.** Skeletal muscle performance determined by modulation of number of myosin motors rather than motor force or stroke size. *Cell* 131: 784–795; 2007.

93 **Pizzimenti MA.** Mechanical analysis of the Nautilus leg curl machine. *Can. J. Sport Sci.* 17(1): 41–48; 1992.

94 **Ranatunga KW, Roots H, Pinniger GJ, and Offer GW.** Crossbridge and non-crossbridge contributions to force in shortening and lengthening muscle. *Adv. Exp. Med. Biol.* 682: 207–221; 2010.

95 **Rassier DE.** The effects of length on fatigue and twitch potentiation in human skeletal muscle. *Clin. Physiol.* 20(6): 474–482; 2000.

96 **Rassier DE and Herzog W.** Considerations on the history dependence of muscle contraction. *J. Appl. Physiol.* 96(2): 419–427; 2004.

97 **Rassier DE and Herzog W.** Force enhancement and relaxation rates after stretch of activated muscle fibers. *Proc. R. Soc. Lond. B Biol. Sci.* 272: 475–480; 2005.

98 **Rassier DE, MacIntosh BR, and Herzog W.** Length dependence of active force production in skeletal muscle. *J. Appl. Physiol.* 86(5): 1445–1457; 1999.

99 **Rousanoglou EN, Oskouei AE, and Herzog W.** Force depression following muscle shortening in submaximal voluntary contractions of human adductor pollicis. *J. Biomech.* 40(1): 1–8; 2007.

100 **Svantesson U and Grimby G.** Stretch-shortening cycle during plantarflexion in young and elderly women and men. *Eur. J. Appl. Physiol.* 71(5): 381–385; 1995.

101 **Svantesson U, Grimby G, and Thomee R.** Potentiation of concentric plantar flexion torque following eccentric and isometric muscle actions. *Acta Physiol. Scand.* 152(3): 287–293; 1994.

102 **Takarada Y, Hirano Y, Ishige Y, and Ishii N.** Stretch-induced enhancement of mechanical power output in human multijoint exercise with countermovement. *J. Appl. Physiol.* 83(5): 1749–1755; 1997.

103 **Takarada Y, Iwamoto H, Sugi H, Hirano Y, and Ishii N.** Stretch-induced enhancement of mechanical work production in frog single fibers and human muscle. *J. Appl. Physiol.* 83(5): 1741–1748; 1997.

104 **Tilp M, Steib S, Schoppacher-Tilp G, and Herzog W.** Changes in fascicle lengths and pennation angles do not contribute to residual force enhancement/depression in voluntary contractions. *J. Appl. Biomech.* 27: 64–73; 2011.

105 **Tsapoulos DE, Baltzopoulos V, Richards PJ, and Magnaris CN.** *In vivo* changes in the human patellar tendon moment arm length with different modes and intensities of contraction. *J. Biomech.* 40: 3325–3332; 2007.

106 **Tskhovrebova L and Trinick J.** Roles of titin in the structure and elasticity of the sarcomere. *J. Biomed. Biotechnol.* article 612482; 2010.

107 **Tsunoda N, O'Hagan F, Sale DG, and MacDougall JD.** Elbow flexion strength curves in untrained men and women and male bodybuilders. *Eur. J. Appl. Physiol.* 66: 235–239; 1993.

108 **Tyska MJ and Warshaw DM.** The myosin motor. *Cell Mot. Cytoskeleton* 51(1): 1–15; 2002.

109 **Vander AJ, Sherman JH, and Luciano DS.** *Human Physiology: The Mechanisms of Body Function*, 8th edn. New York: McGraw-Hill; 2001.

110 **van Ingen Schenau GJ, Bobbert MF, and de Haan A.** Does elastic energy enhance work and efficiency in the stretch-shortening cycle? *J. Appl. Biomech.* 13(4): 389–415; 1997.

111 **Walshe AD, Wilson GJ, and Ettema GJ.** Stretch-shorten cycle compared with isometric preload: contributions to enhanced muscular performance. *J. Appl. Physiol.* 84(1):97–106; 1998.

112 **Wang K and Ramirez-Mitchell R.** A network of transverse and longitudinal intermediate filaments is associated with sarcomeres of adult vertebrate skeletal muscle. *J. Cell Biol.* 96(2): 562–570; 1983.

113 **Wilson GJ, Wood GA, and Elliott BC.** Optimal stiffness of series elastic component in a stretch shorten cycle activity. *J. Appl. Physiol.* 70(2): 825–832; 1991.

114 **Winter DA, Wells RP, and Orr GW.** Errors in the use of isokinetic dynamometers. *Eur. J. Appl. Physiol.* 46: 397–408; 1981.

115 **Winter SL and Challis JH.** The force-length curves of the human rectus femoris and gastrocnemius muscles. *J. Appl. Biomech.* 26: 45–51; 2010.

Chapter 5

1 **Aagaard P, Simonsen EB, Andersen JL, Magnusson SP, Halkjaer-Kristensen J, and Dyhre-Poulsen P.** Neural inhibition during maximal eccentric and concentric quadriceps contraction: effects of resistance training. *J. Appl. Physiol.* 89(6): 2249–2257; 2000.

2 **Abbate F, Sargeant AJ, Verdijk PWL, and de Haan A.** Effects of high-frequency initial pulses and post-tetanic potentiation on power output of skeletal muscle. *J. Appl. Physiol.* 88(1): 35–40; 2000.

3 **Abbis CR and Laursen PB.** Models to explain fatigue during prolonged endurance cycling. *Sports Med.* 35(10): 865–898; 2005.

4 **Abbis CR and Laursen PB.** Is part of the mystery surrounding fatigue complicated by context? *J. Sci. Med. Sport* 10(5): 277–279; 2007.

5 **Adams GR, Duvoisin MR, and Dudley GA.** Magnetic resonance imaging and electromyography as indexes of muscle function. *J. Appl. Physiol.* 73(4): 1578–1583; 1992.

6 **Adams GR, Hather BM, Baldwin KM, and Dudley GA.** Skeletal muscle myosin heavy chain composition and resistance training. *J. Appl. Physiol.* 74(2): 911–915; 1993.

7 **Allen DG, Lamb GD, and Westerblad H.** Impaired calcium release during fatigue. *J. Appl. Physiol.* 104(1): 296–305; 2008.

8 **Allen DG, Lamb GD, and Westerblad H.** Skeletal muscle fatigue: cellular mechanisms. *Physiol. Rev.* 88(1): 287–332; 2008.

9 **Alway SE, Stray-Gundersen J, Grumbt WH, and Gonyea WJ.** Muscle cross-sectional area and torque in resistance-trained subjects. *Eur. J. Appl. Physiol.* 60(2): 86–90; 1990.

10 **Amann M and Secher NH.** Point: afferent feedback from fatigued locomotor muscles is an important determinant of endurance exercise performance. *J. Appl. Physiol.* 108(2): 452–454; 2010.

11 **Arampatzis A, Karamanidis K, Stafilidis S, Morey-Klapsing G, DeMonte G, and Bruggemann GP.** Effect of different ankle- and knee-joint positions on gastrocnemius medialis fascicle length and EMG activity during isometric plantar flexion. *J. Biomech.* 39(10): 1891–1902; 2006.

12 **Asmussen E and Boje O.** Body temperature and capacity for work. *Acta Physiol. Scand.* 10: 1–22; 1945.

13 **Asmussen E, Bonde-Petersen F, and Jorgensen K.** Mechano-elastic properties of human muscles at different temperatures. *Acta Physiol. Scand.* 96(1): 83–93; 1976.

14 **Asp S, Daugaard JR, Rohde T, Adamo K, and Graham T.** Muscle glycogen accumulation after a marathon: roles of fiber type and pro- and macroglycogen. *J. Appl. Physiol.* 86(2): 474–478; 1999.

15 **Babault N, Pousson M, Ballay Y and van Hoecke JV.** Activation of human quadriceps femoris during isometric, concentric, and eccentric contractions. *J. Appl. Physiol.* 91(6): 2628–2634; 2001.

16 **Babault N, Pousson M, Michaut A, Ballay Y and van Hoecke JV.** EMG activity and voluntary activation during knee-extensor concentric torque generation. *Eur. J. Appl. Physiol.* 86(6): 541–547; 2002.

17 **Babault N, Pousson M, Michaut A, Ballay Y and van Hoecke JV.** Effect of quadriceps femoris muscle length on neural activation during isometric and concentric contractions. *J. Appl. Physiol.* 94(3): 983–990; 2003.

18 **Bailey SJ, Vanhatlo A, Wilkerson DP, DiMenna FJ, and Jones AM.** Optimizing the "priming" effect: influence of prior exercise intensity and recovery duration on O_2 uptake kinetics and severe-intensity exercise tolerance. *J. Appl. Physiol.* 107(7): 1743–1756; 2009.

19 **Baker AJ, Kostov KG, Miller RG, and Weiner MW.** Slow force recovery after long-duration exercise: metabolic and activation factors in muscle fatigue. *J. Appl. Physiol.* 74(5): 2294–2300; 1993.

20 **Baldwin J, Snow RJ, Gibala MJ, Garnham A, Howarth K, and Febbraio MA.** Glycogen availability does not affect the TCA cycle or TAN pools during prolonged, fatiguing exercise. *J. Appl. Physiol.* 94(6): 2181–2187; 2003.

21 **Ball D, Burrows C and Sargeant AJ.** Human power output during repeated sprint cycle exercise: the influence of thermal stress. *Eur. J. Appl. Physiol.* 79(4): 360–366; 1999.

22 **Balog EM.** Excitation-contraction coupling and minor triadic proteins in low-frequency fatigue. *Exerc. Sport Sci. Rev.* 38(3): 135–142; 2010.

23 **Balsom PD, Gaitanos GC, Soderlund K, and Ekblom B.** High-intensity exercise and muscle glycogen availability in humans. *Acta Physiol. Scand.* 165(4): 337–345; 1999.

24 **Bamman MM, Newcomer BR, Larson-Meyer DE, Weinsier RL, and Hunter GR.** Evaluation of the strength-size relationship *in vivo* using various muscle size indices. *Med. Sci. Sports Exerc.* 32(7): 1307–1313; 2000.

25 **Barnett C, Kippers V, and Turner P.** Effects of variations in the bench press exercise on the EMG activity of five shoulder muscles. *J. Strength Cond. Res.* 9(4): 222–227; 1995.

26 **Bassett Jr DR and Howley ET.** Limiting factors for maximum oxygen uptake and determinants of endurance performance. *Med. Sci. Sports Exerc.* 32(1): 70–84; 2000.

27 **Baudry S and Duchateau J.** Postactivation in human muscle is not related to the type of maximal conditioning contraction. *Muscle Nerve* 30(3): 328–336; 2004.

28 **Baudry S and Duchateau J.** Postactivation potentiation in a human muscle: effect on the rate of torque development of tetanic and voluntary isometric contractions. *J. Appl. Physiol.* 102(4): 1394–1401; 2007.

29 **Baudry S and Duchateau J.** Postactivation potentiation in a human muscle: effect on the load-velocity relation of tetanic and voluntary shortening contractions. *J. Appl. Physiol.* 103(4): 1318–1325; 2007.

30 **Becker R and Awiszus F.** Physiological alterations of maximal voluntary quadriceps activation by changes of knee joint angle. *Muscle Nerve* 24(5): 667–672; 2001.

31 **Behm DG, Button DC, Barbour G, Butt JC, and Young WB.** Conflicting effects of fatigue and potentiation on voluntary force. *J. Strength Cond. Res.* 18(2): 365–372; 2004.

32 **Behm DG, St-Pierre MM, and Perez D.** Muscle inactivation: assessment of interpolated twitch technique. *J. Appl. Physiol.* 81(5): 2267–2273; 1996.

33 **Behm DG, Whittle J, Button D, and Power K.** Intermuscle differences in activation. *Muscle Nerve* 25(2): 236–243; 2002.

34 **Beltman JG, de Haan A, Haan H, Gerrits HL, van Mechelen W, and Sargeant AJ.** Metabolically assessed muscle fiber recruitment in brief isometric contractions at different intensities. *Eur. J. Appl. Physiol.* 92(4–5): 485–492; 2004.

35 **Beltman JGM, Sargeant AJ, van Mechelen W and de Haan A.** Voluntary activation level and muscle fiber recruitment of human quadriceps during lengthening contractions. *J. Appl. Physiol.* 97(2): 619–626; 2004.

36 **Bergh U and Ekblom B.** Influence of muscle temperature on maximal muscle strength and power output in human skeletal muscles. *Acta Physiol. Scand.* 107(1): 33–37; 1979.

37 **Bigland B and Lippold OCJ.** The relation between force, velocity and integrated electrical activity in human muscles. *J. Physiol.* 123(1): 214–224; 1954.

38 **Biral D, Betto R, Danielo-Betto D, and Salviati G.** Myosin heavy chain composition of single fibers from normal human muscle. *Biochem. J.* 250(1): 307–308; 1988.

39 **Bishop D.** Warm-up I. Potential mechanisms and the effects of passive warm-up. *Sports Med.* 33(6): 439–454; 2003.

40 **Bishop D, Bonetti D, and Dawson B.** The effect of three different warm-up intensities on kayak ergometer performance. *Med. Sci. Sports Exerc.* 33(6): 1026–1032; 2001.

41 **Bogdanis GC, Nevill ME, Boobis LH, Lakomy HKA, and Nevill AM.** Recovery of power output and muscle metabolites following 30 s of maximal sprint cycling in man. *J. Physiol.* 482(2): 467–480; 1995.

42 **Bogdanis GC, Nevill ME, Lakomy HK, and Boobis LH.** Power output and muscle metabolism during and following recovery from 10 and 20 s of maximal sprint exercise in humans. *Acta Physiol. Scand.* 163(3): 261–272; 1998.

43 **Borg J, Grimby L, and Haanerz J.** The fatigue of voluntary contraction and the peripheral electrical propagation of single motor units in man. *J. Physiol.* 340: 435–444; 1983.

44 **Bottinelli R, Canapari M, Pellegrino MA, and Reggiani C.** Force-velocity properties of human skeletal muscle fibers: myosin heavy chain isoform and temperature dependence. *J. Physiol.* 495(2): 573–586; 1996.

45 **Bottinelli R, Pellegrino MA, Canapari M, Rossi R, and Reggiani C.** Specific contributions of various muscle fiber types to human muscle performance: an *in vitro* study. *J. Electromyogr. Kinesiol.* 9(2): 87–95; 1999.

46 **Brechue WF and Abe T.** The role of FFM accumulation and skeletal muscle architecture in powerlifting performance. *Eur. J. Appl. Physiol.* 86(4): 327–336; 2002.

47 **Bruce SA, Phillips SK, and Woledge RC.** Interpreting the relation between force and cross-sectional area in human muscle. *Med. Sci. Sports Exerc.* 29(5): 677–683; 1997.

48 **Buchthal F, Erminio F, and Rosenfalck P.** Motor unit territory in different human muscles. *Acta Physiol. Scand.* 45(1): 72–87; 1959.

49 **Burke RE, Levine DN, Tsairis P, and Zajac FE 3rd.** Physiological types and histochemically profiles in motor units of cat gastrocnemius. *J. Physiol.* 234(3): 723–748; 1973.

50 **Butterfield TA.** Eccentric exercise *in vivo*: strain-induced muscle damage and adaptation in a stable system. *Exerc. Sport Sci. Rev.* 38(2): 51–60; 2010.

51 **Cairns SP.** Lactic acid and exercise performance. Culprit or friend? *Sports Med.* 36(4): 279–291; 2006.

52 **Casey A, Constantin-Teodosiu D, Howell S, Hultman E, and Greenhaff PL.** Metabolic response of type I and II muscle fibers during repeated bouts of maximal exercise in humans. *Am. J. Physiol.* 271(1, Pt 1): E38–E43; 1996.

53 **Castle PC, Macdonald AL, Philp A, Webborn A, Watt PW, and Maxwell NS.** Precooling leg muscle improves intermittent sprint exercise in hot, humid conditions. *J. Appl. Physiol.* 100(4): 1377–1384; 2006.

54 **Caterini D, Gittings W, Huang J, and Vandenboom R.** The effect of work cycle frequency on the potentiation of dynamic force in mouse fast-twitch skeletal muscle. *J. Exp. Biol.* 214(23): 3915–3923; 2011.

55 **Chalmers GR.** Can fast-twitch muscle fibers be selectively recruited during lengthening contractions? Review and applications to sport movements. *Sports Biomech.* 7(1): 137–157; 2008.

56 **Cheung SS and Sleivert GG.** Multiple triggers for hyperthermic fatigue and exhaustion. *Exerc. Sport Sci. Rev.* 32(3): 100–106; 2004.

57 **Clarke RSJ, Hellon RF, and Lind AR.** The duration of sustained contractions of the human forearm at different muscle temperatures. *J. Physiol.* 143(3): 454–473; 1958.

58 **Close RI.** Dynamic properties of mammalian skeletal muscles. *Physiol. Rev.* 52(1): 129–197; 1972.

59 **Colombini B, Nocella M, Benelli G, Cecchi G, and Bagni MA.** Effect of temperature on cross-bridge properties in intact frog muscle fibers. *Am. J. Physiol.* 294(4): C1113–C1117; 2008.

60 **Conley MS, Meyer RA, Bloomberg JJ, Feeback DL, and Dudley GA.** Noninvasive analysis of human neck muscle function. *Spine* 20(23): 2505–2512; 1995.

61 **Davies CTM, Mecrow IK, and White MJ.** Contractile properties of the human triceps surae with some observations on the effects of temperature and exercise. *Eur. J. Appl. Physiol.* 49(2): 255–269; 1982.

62 **Davies CTM and Young K.** Effect of temperature on the contractile properties and muscle power of triceps surae in humans. *J. Appl. Physiol.* 55(1): 191–195; 1983.

63 **Decostre V, Bianco P, Lombardi V, and Piazzesi G.** Effect of temperature on the working stroke of muscle myosin. *Proc. Natl. Acad. Sci. USA* 102(39): 13927–13932; 2005.

64 **de Haan A.** The influence of stimulation frequency on the force-velocity characteristics of *in situ* rat medial gastrocnemius muscle. *Exp. Physiol.* 83(1): 77–84; 1998.

65 **de Haan A, Gerrits KHL, and de Ruiter CJ.** Counterpoint: the interpolated twitch does not provide a valid measure of the voluntary activation of muscle. *J. Appl. Physiol.* 107(1): 355–357; 2009.

66 **de Haan A, Gerrits KHL, and de Ruiter CJ.** Last Word on Point:Counterpoint: the interpolated twitch does / does not provide a valid measure of the voluntary activation of muscle. *J. Appl. Physiol.* 107(1): 368; 2009.

67 **De Luca CJ, LeFever RS, McCue MP, and Xenakis AP.** Behavior of human motor units in different muscles during linearly varying contractions. *J. Physiol.* 329: 113–128; 1982.

68 **Del Valle DO and Thomas CK.** Firing rates of motor units during strong dynamic contractions. *Muscle Nerve* 32(3): 316–325; 2005.

69 **de Ruiter CJ and de Haan A.** Temperature effect on the force / velocity relationship of the fresh and fatigued human adductor pollicis muscle. *Pflugers Arch.* 440(1): 163–170; 2000.

70 **de Ruiter CJ and de Haan A.** Similar effects of cooling and fatigue on eccentric and concentric force-velocity relationships in human muscle. *J. Appl. Physiol.* 90(6): 2109–2116; 2001.

71 **de Ruiter CJ, Didden WJM, Jones DA, and de Haan A.** The force-velocity relationship of human adductor pollicis muscle during stretch and the effects of fatigue. *J. Physiol.* 526(3): 671–681; 2000.

72 **de Ruiter CJ, Jones DA, Sargeant AJ, and de Haan A.** Temperature effect on rates of isometric force development and relaxation in the fresh and fatigued human adductor pollicis muscle. *Exp. Physiol.* 84(6): 1137–1150; 1999.

73 **Desbrosses K, Babault N, Scaglioni G, Meyer J-P, and Pousson M.** Neural activation after maximal isometric contractions at different muscle lengths. *Med. Sci. Sports Exerc.* 38(5): 937–944; 2006.

74 **Desmedt JE and Godaux E.** Ballistic contractions in man: characteristic recruitment pattern of single motor units of the tibialis anterior muscle. *J. Physiol.* 264(3): 673–693; 1977.

75 **Desmedt JE and Godaux E.** Voluntary motor commands in human ballistic movements. *Ann. Neurol.* 5(5): 415–421; 1979.

76 **Docherty D, Robbins D, and Hodgson M.** Complex training revisited: a review of its current status as a viable training approach. *Strength Cond. J.* 26(6): 52–57; 2004.

77 **Doyle JA, Sherman WM, and Strauss RL.** Effects of eccentric exercise and concentric exercise on muscle glycogen replenishment. *J. Appl. Physiol.* 74(4): 1848–1855; 1993.

78 **Drinkwater E.** Effects of peripheral cooling on characteristics of local muscle. *Med. Sport Sci.* 53: 74–88; 2008.

79 **Drust B, Rasmussen P, Mohr M, Nielsen B, and Nybo L.** Elevations in core and muscle temperature impairs repeated sprint performance. *Acta Physiol. Scand.* 183(2): 181–190; 2005.

80 **Duchateau J and Enoka RM.** Neural control of shortening and lengthening contractions: influence of task constraints. *J. Physiol.* 586(24): 5853–5864; 2008.

81 **Duchateau J and Enoka RM.** Human motor unit recordings: origins and insights into the integrated motor system. *Brain Res.* 1409: 42–61; 2011.

82 **Duchateau J, Le Bozec S, and Hainaut K.** Contributions of slow and fast muscles of triceps surae to a cyclic movement. *Eur. J. Appl. Physiol.* 55(5): 476–481; 1986.

83 **Dudley GA, Harris RT, Duvoisin MR, Hather BM, and Buchanan P.** Effect of voluntary vs artificial activation on the relationship of muscle torque to speed. *J. Appl. Physiol.* 69(6): 2215–2221; 1990.

84 **Ebben WP.** Complex training: a brief review. *J. Sports Sci. Med.* 1: 42–46; 2002.

85 **Edwards RHT, Harris RC, Hultman E, Kaijser L, Koh D, and Nordesjo L-O.** Effect of temperature on muscle energy metabolism and endurance during successive isometric contractions, sustained to fatigue, of the quadriceps muscle in man. *J. Physiol.* 220(2): 335–352; 1972.

86 **Edwards RHT, Hill DK, Jones DA, and Merton PA.** Fatigue of long duration in human skeletal muscle after exercise. *J. Physiol.* 272(3): 769–778; 1977.

87 **Enoka RM.** Morphological features and activation patterns of motor units. *J. Clin. Neurophysiol.* 12(6): 538–559; 1995.

88 **Enoka RM.** Eccentric contractions require unique activation strategies by the nervous system. *J. Appl. Physiol.* 81(6): 2339–2346; 1996.

89 **Enoka RM and Duchateau J.** Muscle fatigue: what, why and how it influences muscle function. *J. Physiol.* 586(1): 11–23; 2008.

90 **Enoka RM and Fuglevand AJ.** Motor unit physiology: some unresolved issues. *Muscle Nerve* 24(1): 4–17; 2001.

91 **Erskin RM, Jones DA, Magnaris CN, and Degens H.** *In vivo* specific tension of the human quadriceps femoris muscle. *Eur. J. Appl. Physiol.* 106(6): 827–838; 2009.

92 **Farina D, Arendt-Nielsen L, and Graven-Nielsen T.** Effect of temperature on spike-triggered average torque and electrophysiological properties of low-threshold motor units. *J. Appl. Physiol.* 99(1): 197–203; 2005.

93 **Faulkner SH, Ferguson RA, Gerrett N, Hupperets M, Hodder SG, and Havenith G.** Reducing muscle temperature drop post warm-up improves cycling performance. *Med. Sci. Sports Exerc.* 45(2): 359–365; 2013.

94 **Ferguson RA, Ball D, and Sargeant AJ.** Effect of muscle temperature on rate of oxygen uptake during exercise in humans at different contraction frequencies. *J. Exp. Biol.* 205(7): 981–987; 2002.

95 **Fitts RH.** The cross-bridge cycle and skeletal muscle fatigue. *J. Appl. Physiol.* 104(2): 551–558; 2008.

96 **Folland JP and Williams AG.** Methodological issues with the interpolated twitch technique. *J. Electromyogr. Kinesiol.* 17(3): 317–327; 2007.

97 **Fradkin AJ, Zazryn TR, and Smoliga JM.** Effects of warming-up on physical performance: a

systematic review with meta-analysis. *J. Strength Cond. Res.* 24(1): 140–148; 2010.

98 **Frey Law LA and Avin KG.** Endurance time is joint-specific: a modelling and meta-analysis investigation. *Ergonomics* 53(1): 109–129; 2010.

99 **Ftaiti F, Grelot L, Coudreuse JM, and Nicol C.** Combined effect of heat stress, dehydration and exercise on neuromuscular function in humans. *Eur. J. Appl. Physiol.* 84(1–2): 87–94; 2001.

100 **Fukunaga T, Roy RR, Shellock FG, Hodgson JA, and Edgerton VR.** Specific tension of human plantar flexors and dorsiflexors. *J. Appl. Physiol.* 80(1): 158–165; 1996.

101 **Gaitanos GC, Williams C, Boobis LH, and Brooks S.** Human muscle metabolism during intermittent maximal exercise. *J. Appl. Physiol.* 75(2): 712–719; 1993.

102 **Gandevia SC.** Spinal and supraspinal factors in fatigue. *Physiol. Rev.* 81(4): 1725–1789; 2001.

103 **Gandevia SC, Herbert RD, and Leeper JB.** Voluntary activation of human elbow flexor muscles during maximal concentric contractions. *J. Physiol.* 512(2): 595–602; 1998.

104 **Gandevia SC and McKenzie DK.** Activation of human muscles at short muscle lengths during maximal static efforts. *J. Physiol.* 407: 599–613; 1988.

105 **Garnett RA, O'Donovan MJ, Stephens JA, and Taylor A.** Motor unit organization of human medial gastrocnemius. *J. Physiol.* 287: 33–43; 1979.

106 **Gibala MJ, Interisano SA, Tarnopolsky MA, Roy BD, MacDonald JR, Yarasheski KE, and MacDougall JD.** Myofibrillar disruption following acute concentric and eccentric exercise in strength-trained men. *Can. J. Physiol. Pharmacol.* 78(8): 656–661; 2000.

107 **Glaister M.** Multiple sprint work. Physiological responses, mechanisms of fatigue and the influence of aerobic fitness. *Sports Med.* 35(9): 757–777; 2005.

108 **Gokhin DS, Kim NE, Lewis SA, Hoenecke HR, D'Lima DD, and Fowler VM.** Thin-filament length correlates with fiber type in human skeletal muscle. *Am. J. Physiol.* 302(3): C555–C565; 2012.

109 **Gollnick PD, Armstrong RB, Sembrowich WL, Shepherd RE, and Saltin B.** Glycogen depletion pattern in human skeletal muscle fibers after heavy exercise. *J. Appl. Physiol.* 34(5): 615–618; 1973.

110 **Gossen ER and Sale DG.** Effect of postactivation potentiation on dynamic knee extension performance. *Eur. J. Appl. Physiol.* 83(6): 524–530; 2000.

111 **Grange RW and Houston ME.** Simultaneous potentiation and fatigue in quadriceps after a 60-second maximal voluntary isometric contraction. *J. Appl. Physiol.* 70(2): 726–731; 1991.

112 **Gray SR, De Vito G, Nimmo MA, Farina D, and Ferguson RA.** Skeletal muscle ATP turnover and muscle fiber conduction velocity are elevated at higher muscle temperatures during maximal power output development in humans. *Am. J. Physiol.* 290(2): R376–R382; 2006.

113 **Greenhaff PL, Nevill ME, Soderlund K, Bodin K, Boobis LH, Williams C, and Hultman E.** The metabolic responses of human type I and II muscle fibers during maximal treadmill sprinting. *J. Physiol.* 478(1): 149–155; 1994.

114 **Grimby L and Hannerz J.** Firing rate and recruitment order of toe extensor motor units in different modes of voluntary contraction. *J. Physiol.* 264(3): 865–879; 1977.

115 **Grimby L, Hannerz J, and Hedman B.** Contraction time and voluntary discharge properties of individual short toe extensor motor units in man. *J. Physiol.* 289: 191–201; 1979.

116 **Grimby L, Hannerz J, and Hedman B.** The fatigue and voluntary discharge properties of single motor units in man. *J. Physiol.* 316: 545–554; 1981.

117 **Gullich A and Schmidtbleicher D.** Short-term potentiation of power performance induced by maximal voluntary contractions. *XVth Congress of the International Society of Biomechanics* pp. 348–349; 1995.

118 **Hahn D.** Lower extremity extension force and electromyography properties as a function of knee angle and their relation to joint torques: implications for strength diagnostics. *J. Strength Cond. Res.* 25(6): 1622–1631; 2011.

119 **Hamada T, Sale DG, and MacDougall JD.** Postactivation potentiation in endurance-trained athletes. *Med. Sci. Sports Exerc.* 32(3): 403–411; 2000.

120 **Hamada T, Sale DG, MacDougall JD, and Tarnopolsky MA.** Postactivation potentiation, fiber type and twitch contraction time in the knee extensor muscles. *J. Appl. Physiol.* 88(6): 2131–2137; 2000.

121 **Harridge SDR, Bottinelli R, Canapari M, Pellegrino MA, Reggiani C, Esbjornsson M, and Saltin B.** Whole muscle and single-fiber contractile properties and myosin heavy chain isoforms in humans. *Pflugers Arch.* 432(5): 913–920; 1996.

122 **Hawley JA, Schabort EJ, Noakes TD, and Dennis SC.** Carbohydrate-loading and exercise performance. An update. *Sports Med.* 24(2): 73–81; 1997.

123 **Heckman CJ and Binder MD.** Computer simulations of the effects of different synaptic input systems on motor unit recruitment. *J. Neurophysiol.* 70(5): 1827–1840; 1993.

124 **Henneman E, Somjen G, and Carpenter DO.** Functional significance of cell size in spinal motoneurons. *J. Neurophysiol.* 28: 560–580; 1965.

125 **Hermansen L and Osnes J-B.** Blood and muscle pH after maximal exercise in man. *J. Appl. Physiol.* 32(3): 304–308; 1972.

126 **Hickner RC, Fisher JS, Hansen PA, Racette SB, Mier CM, Turner MJ, and Holloszy JO.** Muscle glycogen accumulation after endurance exercise in trained and untrained individuals. *J. Appl. Physiol.* 83(3): 897–903; 1997.

127 **Hodgson M, Docherty D, and Robbins D.** Post-activation potentiation. Underlying physiology and implications for motor performance. *Sports Med.* 35(7): 585–595; 2005.

128 **Horstman AM and others.** Comments on Point: Counterpoint: the interpolated twitch does / does not provide a valid measure of the voluntary activation of muscle. *J. Appl. Physiol.* 107(1): 359–366; 2009.

129 **Houston ME and Grange RW.** Myosin phosphorylation, twitch potentiation, and fatigue in human skeletal muscle. *Can. J. Physiol. Pharmacol.* 68(7): 908–913; 1990.

130 **Huber A, Suter E, and Herzog W.** Inhibition of quadriceps muscles in elite male volleyball players. *J. Sports Sci.* 16(3): 281–289; 1998.

131 **Hulten B, Thorstensson A, Sjodin B, and Karlsson J.** Relationship between isometric endurance and fiber types in human leg muscles. *Acta Physiol. Scand.* 93(1): 135–138; 1975.

132 **Jones AM, Koppo K, and Burnley M.** Effects of prior exercise on metabolic and gas exchange responses to exercise. *Sports Med.* 33(13): 949–971; 2003.

133 **Jones AM, Wilkerson DP, Burnley M, and Koppo K.** Prior heavy exercise enhances performance during subsequent perimaximal exercise. *Med. Sci. Sports Exerc.* 35(12): 2085–2092; 2003.

134 **Jones DA.** Changes in the force-velocity relationship of fatigued muscle: implications for power production and possible causes. *J. Physiol.* 588(16): 2977–2986; 2010.

135 **Jones DA, de Ruiter CJ, and de Haan A.** Change in contractile properties of human muscle in relationship to the loss of power and slowing or relaxation seen with fatigue. *J. Physiol.* 576(3): 913–922; 2006.

136 **Jones DA, Newham DJ, and Torgan C.** Mechanical influences on long-lasting human muscle fatigue. *J. Physiol.* 412: 415–427; 1989.

137 **Josephson RK and Edman KA.** The consequences of fiber heterogeneity on the force-velocity relation of skeletal muscle. *Acta Physiol. Scand.* 132(3): 341–352; 1988.

138 **Karatzaferi C, de Haan A, Ferguson RA, van Mechelen W, and Sargeant AJ.** Phosphocreatine and ATP content in human single muscle fibers before and after dynamic exercise. *Pflugers Arch.* 442(3): 467–474; 2001.

139 **Kennedy PM and Cresswell AG.** The effect of muscle length on motor-unit recruitment during isometric plantar flexion in humans. *Exp. Brain Res.* 137(1): 58–64; 2001.

140 **Kenny GP, Reardon FD, Zaleski W, Reardon ML, Haman F, and Ducharme MB.** Muscle temperature transients before, during, and after exercise measured using an intramuscular multisensor probe. *J. Appl. Physiol.* 94(6): 2350–2357; 2003.

141 **Kernell D.** Principles of force gradation in skeletal muscles. *Neural. Plast.* 10(1–2): 69–76; 2003.

142 **Kernell D and Monster AW.** Motoneurone properties and motor fatigue. *Exp. Brain Res.* 46(2): 197–204; 1982.

143 **Kohn TA, Essen-Gustavsson B, and Myburgh KH.** Exercise pattern influences skeletal muscle hybrid fibers of runners and nonrunners. *Med. Sci. Sports Exerc.* 39(11): 1977–1984; 2007.

144 **Kooistra RD, de Ruiter CJ, and de Haan A.** Conventionally assessed voluntary activation does not represent relative voluntary torque production. *Eur. J. Appl. Physiol.* 100(3): 309–320; 2007.

145 **Kraemer WJ.** A series of studies—the physiological basis for strength training in American football: fact over philosophy. *J. Strength Cond. Res.* 11(3): 131–142; 1997.

146 **Kukulka CG and Clamann HP.** Comparison of the recruitment and discharge properties of motor units in human brachial biceps and adductor pollicis

during isometric contractions. *Brain Res.* 219(1): 45–55; 1981.

147 **Linari M, Bottinelli R, Pellegrino MA, Reconditi M, Reggiani C, and Lombardi V.** The mechanism of the force response to stretch in human skinned muscle fibers with different myosin isoforms. *J. Physiol.* 554(2): 335–352; 2004.

148 **Linari M, Brunello E, Sun Y-B.**, Panine P, Narayanan T, Piazzesi G, Lombardi V, and Irving M. The structural basis of the increase in isometric force with temperature in frog skeletal muscle. *J. Physiol.* 567(2): 459–469; 2005.

149 **Lind AR and Petrofsky JS.** Isometric tension from rotary stimulation of fast and slow muscles. *Muscle Nerve* 1(3): 213–218; 1978.

150 **Loeb GE.** Motoneurone task groups: coping with kinematic heterogeneity. *J. Exp. Biol.* 115: 137–146; 1985.

151 **Lorenz D.** Postactivation potentiation: an introduction. *Int. J. Sports Phys. Ther.* 6(3): 234–240; 2011.

152 **MacDougall JD, Reddan WG, Layton CR, and Dempsey JA.** Effects of metabolic hyperthermia on performance during heavy prolonged exercise. *J. Appl. Physiol.* 36(5): 538–544; 1974.

153 **MacDougall JD, Ward GR, Sale DG, and Sutton JR.** Muscle glycogen depletion after high-intensity intermittent exercise. *J. Appl. Physiol.* 42(2): 129–132; 1977.

154 **MacIntosh BR.** Role of calcium sensitivity modulation in skeletal muscle performance. *News Physiol. Sci.* 18: 222–225; 2003.

155 **MacIntosh BR.** Cellular and whole muscle studies of activity dependent potentiation. *Adv. Exp. Med. Biol.* 682: 315–342; 2010.

156 **MacIntosh BR, Robillard M-E, and Tomaras EK.** Should postactivation potentiation be the goal of your warm-up? *Appl. Physiol. Nutr. Metab.* 37(3): 546–550; 2012.

157 **MacIntosh BR, Taub EC, Dormer GN, and Tomaras EK.** Potentiation of isometric and isotonic contractions during high-frequency stimulation. *Pflugers Arch.* 456(2): 449–458; 2008.

158 **Magnaris CN, Baltzopoulis V, Ball D, and Sargeant AJ.** *In vivo* specific tension of human skeletal muscle. *J. Appl. Physiol.* 90(3): 865–872; 2001.

159 **Malisoux L, Francaux M, Nielens H, and Theisen D.** Stretch-shortening cycle exercises: an effective training paradigm to enhance power output of human single muscle fibers. *J. Appl. Physiol.* 100(3): 771–779; 2006.

160 **Malisoux L, Francaux M, and Thiesen D.** What do single-fiber studies tell us about exercise training? *Med. Sci. Sports Exerc.* 39(7): 1051–1060; 2007.

161 **Marcora S.** Counterpoint: afferent feedback from fatigued locomotor muscles is not an important determinant of endurance exercise performance. *J. Appl. Physiol.* 108(2): 456–457; 2010.

162 **Maughan RJ, Harmon M, Leiper JB, Sale DG, and Delman A.** Endurance capacity of untrained males and females in isometric and dynamic muscular contractions. *Eur. J. Appl.* 55(4): 395–400; 1986.

163 **Maughan RJ, Nimmo MA, and Harmon M.** The relationship between muscle myosin ATPase activity and isometric endurance in untrained subjects. *Eur. J. Appl. Physiol.* 54(3): 291–296; 1985.

164 **McComas AJ.** *Skeletal Muscle. Form and Function.* Champaign, IL: Human Kinetics; 1996.

165 **McLeod JG and Wray SH.** Conduction velocity and fiber diameter of the median and ulnar nerves of the baboon. *J. Neurol. Neurosurg. Psychiat.* 30(3): 240–247; 1967.

166 **McMahon S and Jenkins D.** Factors affecting the rate of phosphocreatine resynthesis following intense exercise. *Sports Med.* 32(12): 761–784; 2002.

167 **Mettler JA and Griffin L.** Postactivation potentiation and muscular endurance training. *Muscle Nerve* 45(3): 416–425; 2012.

168 **Miller AE, MacDougall JD, Tarnopolsky MA, and Sale DG.** Gender differences in strength and muscle fiber characteristics. *Eur. J. Appl. Physiol.* 66(3): 254–262; 1993.

169 **Miller RG, Mirka A, and Maxfield M.** Rate of tension development in isometric contractions of a human hand muscle. *Exp. Neurol.* 73(1): 267–285; 1981.

170 **Milner-Brown HS, Stein RB, and Yemm R.** Changes in firing rate of human motor units during linearly changing voluntary contractions. *J. Physiol.* 230(2): 371–390; 1973.

171 **Minett GM, Duffield R, Marino FE, and Portus M.** Volume-dependent response of precooling for intermittent-sprint exercise in the heat. *Med. Sci. Sports Exerc.* 43(9): 1760–1769; 2011.

172 **Mitchell CJ and Sale DG.** Enhancement of jump performance after a 5-RM squat is associated with

postactivation potentiation. *Eur. J. Appl. Physiol.* 111(8): 1957–1963; 2011.

173 **Miyamoto N, Kanahisa H, Fukunaga T, and Kawakami Y.** Effect of postactivation potentiation on the maximal voluntary isokinetic concentric torque in humans. *J. Strength Cond. Res.* 25(1): 186–192; 2011.

174 **Mohr M, Krustrup P, Nybo L, Nielsen JJ, and Bangsbo J.** Muscle temperature and sprint performance during soccer matches—beneficial effect of re-warm-up at half time. *Scand. J. Med. Sci. Sports* 14(3): 156–162; 2004.

175 **Montgomery JC and Macdonald JA.** Effects of temperature on nervous system: implications for behavioral performance. *Am. J. Physiol.* 259(2, Pt 2): R191–R196; 1990.

176 **Morana C and Perrey S.** Time course of postactivation potentiation during intermittent submaximal fatiguing contractions in endurance- and power-trained athletes. *J. Strength Cond. Res.* 23(5): 1456–1464; 2009.

177 **Moritz CT, Barry BK, Pascoe MA, and Enoka RM.** Discharge rate variability influences the variation in force fluctuations across the working range of a hand muscle. *J. Neurophysiol.* 93(5): 2449–2459; 2005.

178 **Morrison S, Sleivert GG, and Cheung SS.** Passive hyperthermia reduces voluntary activation and isometric force production. *Eur. J. Appl. Physiol.* 91(5–6): 729–736; 2004.

179 **Morse CI, Thom JM, Reeves ND, Birch KM, and Narici MV.** *In vivo* physiological cross-sectional area and specific force are reduced in the gastrocnemius of elderly men. *J. Appl. Physiol.* 99(3): 1050–1055; 2005.

180 **Nardone A and Schieppati M.** Selective recruitment of high-threshold motor units during voluntary isotonic lengthening of active muscles. *J. Physiol.* 409: 451–471; 1989.

181 **Newham DJ, McCarthy T, and Turner J.** Voluntary activation of human quadriceps during and after isokinetic exercise. *J. Appl. Physiol.* 71(6): 2122–2126; 1991.

182 **Nybo L.** Hyperthermia and fatigue. *J. Appl. Physiol.* 104(3): 871–878; 2008.

183 **Nybo L and Secher NH.** Cerebral perturbations provoked by prolonged exercise. *Prog. Neurobiol.* 72(4): 223–261; 2004.

184 **Nyitrai M, Rossi R, Adamek N, Pellegrino MA, Bottinelli R, and Geeves MA.** What limits the velocity of fast-skeletal muscle contraction in mammals? *J. Mol. Biol.* 355(3): 432–442; 2006.

185 **Olson MC, Kruger M, Meyer LH, Ahnlund L, Linke WA, and Larsson L.** Fiber type-specific increase in passive muscle tension in spinal cord-injured subjects with spasticity. *J. Physiol.* 577(1): 339–352; 2006.

186 **Oya T, Riek S, and Cresswell AG.** Recruitment and rate coding organization for soleus motor units across entire range of voluntary plantar flexions. *J. Physiol.* 587(19): 4737–4748; 2009.

187 **Paasuke M, Saapar L, Ereline J, Gapeyeva H, Requena B, and Ööpik V.** Postactivation potentiation of knee extensor muscles in power- and endurance-trained, and untrained women. *Eur. J. Appl. Physiol.* 101(5): 577–585; 2007.

188 **Parcell AC, Sawyer RD, and Craig Poole R.** Single muscle fiber myosin heavy chain distribution in elite female track athletes. *Med. Sci. Sports Exerc.* 35(3): 434–438; 2003.

189 **Park JH, Brown RL, Park CR, Cohn M, and Chance B.** Energy metabolism of the untrained muscle of elite runners as observed by ^{31}P magnetic resonance spectroscopy: evidence suggesting a genetic endowment for endurance exercise. *Proc. Natl. Acad. Sci. USA* 85(23): 8780–8784; 1988.

190 **Parkin JM, Carey MF, Zhao S, and Febbraio MA.** Effect of ambient temperature on human skeletal muscle metabolism during fatiguing submaximal exercise. *J. Appl. Physiol.* 86(3): 902–908; 1999.

191 **Pascoe DD, Costill DL, Fink WJ, Robergs RA, and Zachwieja JJ.** Glycogen resynthesis in skeletal muscle following resistance exercise. *Med. Sci. Sports Exerc.* 25(3): 349–354; 1993.

192 **Pascoe DD and Gladden LB.** Muscle glycogen resynthesis after short term, high intensity exercise and resistance exercise. *Sports Med.* 21(2): 98–118; 1996.

193 **Pasquet B, Carpentier A, Duchateau J, and Hainaut K.** Muscle fatigue during concentric and eccentric contractions. *Muscle Nerve* 23(11): 1727–1735; 2000.

194 **Peterson MD, Alvar BA, and Rhea MR.** The contribution of maximal force production to explosive movement among young collegiate athletes. *J. Strength Cond. Res.* 20(4): 867–873; 2006.

195 **Phillips SM, Sproule J, and Turner AP.** Carbohydrate ingestion during team games exercise. Current knowledge and areas for future research. *Sports Med.* 41(7): 559–585; 2011.

196 **Piehl K.** Time course for refilling of glycogen stores in human muscle fibers following exercise-induced glycogen depletion. *Acta Physiol. Scand.* 90(2): 297–302; 1974.

197 **Place N, Yamada T, Bruton JD, and Westerblad H.** Muscle fatigue: from observations in humans to underlying mechanisms studied in intact single muscle fibers. *Eur. J. Appl. Physiol.* 110(1): 1–15; 2010.

198 **Place N, Lepers R, Deley G, and Millet GY.** Time course of neuromuscular alterations during prolonged running exercise. *Med. Sci. Sports Exerc.* 36(8): 1347–1356; 2004.

199 **Powers RK and Binder MD.** Input-output functions of mammalian motoneurons. *Rev. Physiol. Biochem. Pharmacol.* 143: 137–263; 2001.

200 **Powers SC, Lawler J, Dempsey JA, Dodd S, and Landry G.** Effects of incomplete pulmonary gas exchange on VO_{2max}. *J. Appl. Physiol.* 66(6): 2491–2495; 1989.

201 **Prado LG, Makarenko I, Andresen C, Kruger M, Opitz CA, and Linke WA.** Isoform diversity of giant proteins in relation to passive and active contractile properties of rabbit skeletal muscles. *J. Gen. Physiol.* 126(5): 461–480; 2005.

202 **Quod MJ, Martin DT, and Laursen PB.** Cooling athletes before competition in the heat. Comparison of techniques and practical considerations. *Sports Med.* 36(8): 671–682; 2006.

203 **Rall JA and Woledge RC.** Influence of muscle temperature on mechanics of muscle contraction. *Am. J. Physiol.* 259(2, Pt 2): R197–R203; 1990.

204 **Ranatunga KW.** Force and power generating mechanism(s) in active muscle as revealed from temperature perturbation studies. *J. Physiol.* 588(19): 3657–3670; 2010.

205 **Rassier DE and MacIntosh BR.** Coexistence of potentiation and fatigue in skeletal muscle. *Braz. J. Med Biol. Res.* 33(5): 499–508; 2000.

206 **Raue U, Terpstra B, Williamson DL, Gallagher PM, and Trappe SW.** Effects of short-term concentric vs eccentric resistance training on single muscle fiber MHC distribution in humans. *Int. J. Sports Med.* 26(5): 339–343; 2005.

207 **Reid B, Slater CR, and Bewick GS.** Synaptic vesicle dynamics in rat fast and slow motor nerve terminals. *J. Neurosci.* 19(7): 2511–2521; 1999.

208 **Rijkelijkhuizen JM, de Ruiter CJ, Huijing PA, and de Haan A.** Low-frequency fatigue is fiber type related and most pronounced after eccentric activity in rat medial gastrocnemius muscle. *Pflugers Arch.* 447(2): 239–246; 2003.

209 **Robergs RA, Pearson DR, Costill DL, Fink WJ, Pascoe DD, Benedict MA, Lambert CP, and Zachweija JJ.** Muscle glycogenolysis during differing intensities of weight-resistance exercise. *J. Appl. Physiol.* 70(4): 1700–1706; 1991.

210 **Ross MLR, Garvican LA, Jeacocke NA, Laursen PB, Abbiss CR, Martin DT, and Burke LM.** Novel precooling strategy enhances time trial cycling in the heat. *Med. Sci. Sports Exerc.* 43(1): 123–133; 2011.

211 **Ruff RL.** Sodium channel slow inactivation and the distribution of sodium channels on skeletal muscle fibers enable the performance properties of different skeletal muscle fiber types. *Acta Physiol. Scand.* 156(3): 159–168; 1996.

212 **Sahlin K and Ren JM.** Relationship of contraction capacity to metabolic changes during recovery from a fatiguing contraction. *J. Appl. Physiol.* 67(2): 648–654; 1989.

213 **Sale DG.** Postactivation potentiation: role in human performance. *Exerc. Sport Sci. Rev.* 30(3): 138–143; 2002.

214 **Sale DG, MacDougall JD, Alway SE, and Sutton JR.** Voluntary strength and muscle characteristics in untrained men and women and male bodybuilders. *J. Appl. Physiol.* 62(5): 1787–1793; 1987.

215 **Saltin B.** Metabolic fundamentals in exercise. *Med. Sci. Sports* 5(3): 137–146; 1973.

216 **Saltin B, Gagge AP, and Stolwijk JAJ.** Muscle temperature during submaximal exercise in man. *J. Appl. Physiol.* 25(6): 679–688; 1968.

217 **Sargeant AJ.** Effect of muscle temperature on leg extension force and short-term power output in humans. *Eur. J. Appl. Physiol.* 56(6): 693–698; 1987.

218 **Sargeant AJ.** Structural and functional determinants of power. *Exp. Physiol.* 92(2): 323–331; 2007.

219 **Schiaffino S and Reggiani C.** Fiber types in mammalian skeletal muscles. *Physiol. Rev.* 91(4): 1447–1531; 2011.

220 **Schluter JM and Fitts RH.** Shortening velocity and ATPase activity of rat skeletal muscle fibers: effects of endurance exercise training. *Am. J. Physiol.* 266(6, Pt 1): C1699–C1713; 1994.

221 **Sherman WM.** Recovery from endurance exercise. *Med. Sci. Sports Exerc.* 24(Suppl 9): S336–S339; 1992.

222 **Sherman WM, Armstrong LE, Murray TM, Hagerman FC, Costill DL, Staron RC, and Ivy JL.** Effect of a 42.2 km footrace and subsequent rest or exercise on muscular strength and work capacity. *J. Appl. Physiol.* 57(6): 1668–1673; 1984.

223 **Sherman WM, Costill DL, Fink WJ, Hagerman FC, Armstrong LE, and Murray TF.** Effect of a 42.2 km footrace and subsequent rest or exercise on muscle glycogen and enzymes. *J. Appl. Physiol.* 55(4): 1219–1224; 1983.

224 **Sherman WM, Doyle JA, Lamb DR, and Strauss RH.** Dietary carbohydrate, muscle glycogen, and exercise performance during 7 d of training. *Am. J. Clin. Nutr.* 57(1): 27–31; 1993.

225 **Shield A and Zhou S.** Assessing voluntary muscle activation with the twitch interpolation technique. *Sports Med.* 34(4): 253–267; 2004.

226 **Siegel R and Laursen PB.** Keeping your cool. Possible mechanisms for enhanced exercise performance with internal cooling methods. *Sports Med.* 42(2): 89–98; 2012.

227 **Simoneau JA and Bouchard C.** Human variation in skeletal muscle fiber-type proportion and enzyme activities. *Am. J. Physiol.* 257(4, Pt 1): E567–E572; 1989.

228 **Skein M, Duffield R, Cannon J, and Marino FE.** Self-paced intermittent-sprint performance and pacing strategies following respective pre-cooling and heating. *Eur. J. Appl. Physiol.* 112(1): 253–266; 2012.

229 **Sokoloff AJ, Siegel SG, and Cope TC.** Recruitment order among motoneurons from different motor nuclei. *J. Neurophysiol.* 81(5): 2485–2492; 1999.

230 **Spielmann JM, Laouris Y, Nordstrom MA, Robinson GA, Reinking RM, and Stuart DG.** Adaptation of cat motoneurons to sustained and intermittent extracellular activation. *J. Physiol.* 464: 75–120; 1993.

231 **Srinivasan RC, Lungren MP, Langenderfer JE, and Hughes RE.** Fiber type composition and maximum shortening velocity of muscles crossing the shoulder joint. *Clin. Anat.* 20(2): 144–149; 2007.

232 **Staron RS, Hagerman FC, Hikida RS, Murray TF, Hostler DP, Crill MT, Ragg KE, and Toma K.** Fiber type composition of the vastus lateralis muscle of young men and women. *J. Histochem. Cytochem.* 48(5): 623–629; 2000.

233 **Stuart DS, Lingley MD, Grange RW, and Houston ME.** Myosin light chain phosphorylation and contractile performance of human skeletal muscle. *Can. J. Physiol. Pharmacol.* 66(1): 49–54; 1988.

234 **Stull JT, Kamm KE, and Vandenboom R.** Myosin light chain kinase and the role of myosin light chain phosphorylation in skeletal muscle. *Arch. Biochem. Biophys.* 510(2): 120–128; 2011.

235 **Sweeney HL, Bowman BF, and Stull JT.** Myosin light chain phosphorylation in vertebrate striated muscle: regulation and function. *Am. J. Physiol.* 264(5, Pt 1): C1085–C1095; 1993.

236 **Taylor JL.** Point:Counterpoint: The interpolated twitch does/does not provide a valid measure of the voluntary activation of muscle. Point: The interpolated twitch does provide a valid measure of the voluntary activation of muscle. *J. Appl. Physiol.* 107(1): 354–358; 2009.

237 **Taylor JL.** Last Word on Point:Counterpoint: The interpolated twitch does/does not provide a valid measure of the voluntary activation of muscle. *J. Appl. Physiol.* 107(1): 367; 2009.

238 **ter Haar Romeny BM, Denier van der Gon JJ, and Gielen CCAM.** Relation of location of a motor unit in the human biceps brachii and its critical firing levels for different tasks. *Exp. Neurol.* 85(3): 631–650; 1984.

239 **Tesch PA.** *Target Bodybuilding*. Windsor, Ont: Human Kinetics; 1999.

240 **Tesch PA and Karlsson J.** Muscle fiber types and size in trained and untrained muscles of elite athletes. *J. Appl. Physiol.* 59(6): 1716–1720; 1985.

241 **Thomas MM, Cheung SS, Elder GC, and Sleivert GG.** Voluntary muscle activation is impaired by core temperature rather than local muscle temperature. *J. Appl. Physiol.* 100(4): 1361–1369; 2006.

242 **Thornley LJ, Maxwell NS, and Cheung SS.** Local tissue temperature effects on peak torque and muscular endurance during isometric knee extension. *Eur. J. Appl. Physiol.* 90(5–6): 588–594; 2003.

243 **Thorstensson A and Karlsson J.** Fatigability and fiber composition of human skeletal muscle. *Acta Physiol. Scand.* 98(3): 318–322; 1976.

244 **Tillin NA and Bishop D.** Factors modulating post-activation potentiation and its effect on performance of subsequent explosive activities. *Sports Med.* 39(2): 147–166; 2009.

245 **Todd G, Butler JE, and Gandevia SC.** Hypothermia: a failure of the motor cortex and the muscle. *J. Physiol.* 563(2): 621–631; 2005.

246 **Todd G, Taylor JL, and Gandevia SC.** Measurement of voluntary activation of fresh and fatigued muscles using transcranial magnetic stimulation. *J. Physiol.* 551(2): 661–671; 2003.

247 **Tomaras EK and MacIntosh BR.** Less is more: standard warm-up causes fatigue and less warm-up permits greater cycling power output. *J. Appl. Physiol.* 111(1): 228–235; 2011.

248 **Van Cutsem M, Feiereisen P, Duchateau J, and Hainaut K.** Mechanical properties and behavior of motor units in the tibialis anterior during voluntary contractions. *Can. J. Appl. Physiol.* 22(6): 585–597; 1997.

249 **Vandervoort AA, Quinlan J, and McComas AJ.** Twitch potentiation after voluntary contraction. *Exp. Neurol.* 81(1): 141–152; 1983.

250 **Vollestad NK, Vaage O, and Hermansen L.** Muscle glycogen depletion patterns in type I and subgroups of type II fibers during prolonged severe exercise. *Acta Physiol. Scand.* 122(4): 433–441; 1984.

251 **Warhol MJ, Siegel AJ, Evans WJ, and Silverman LM.** Skeletal muscle injury and repair in marathon runners after competition. *Am. J. Pathol.* 118(2): 331–339; 1985.

252 **Wegmann M, Faude O, Poppendieck W, Hecksteden A, Frolich M, and Meyer T.** Pre-cooling and sports performance. A meta-analytic review. *Sports Med.* 42(7): 545–564; 2012.

253 **Westerblad H, Allen DG, and Lannergren J.** Muscle fatigue: lactic acid or inorganic phosphate the major cause? *News Physiol. Sci.* 17: 17–21; 2002.

254 **Westing SH, Seger JY, and Thorstensson A.** Effects of electrical stimulation on eccentric and concentric torque-velocity relationships during knee extension in man. *Acta Physiol. Scand.* 140(1): 17–22; 1990.

255 **Weston AR, Myburgh KH, Lindsay FH, Dennis SC, Noakes TD, and Hawley JA.** Skeletal muscle buffering capacity and endurance performance after high-intensity interval training by well-trained cyclists. *Eur. J. Appl. Physiol.* 75(1): 7–13; 1997.

256 **Widrick JJ, Trappe SW, Blaser CA, Costill DL, and Fitts RH.** Isometric force and maximum shortening velocity of single muscle fibers from elite master runners and sedentary men. *Am. J. Physiol.* 271(2, Pt 1): C666–C675; 1996.

257 **Williamson DL, Gallagher PM, Carroll CC, Raue U, and Trappe SW.** Reduction in hybrid single muscle fiber proportions with resistance training in humans. *J. Appl. Physiol.* 91(5): 1955–1961; 2001.

258 **Wilson JM, Duncan NM, Marin PJ, Brown LE, Loenneke JP, Wilson SM, Jo E, Lowery RP, and Ugrinowitsch C.** Meta-analysis of postactivation potentiation and power: effects of conditioning activity, volume, gender, rest periods, and training status. *J. Strength Cond. Res.* 27(3): 854–859; 2013.

259 **Wyman RJ, Waldron I, and Wachtel GM.** Lack of fixed order of recruitment in cat motoneuron pools. *Exp. Brain Res.* 20(2): 101–114; 1974.

260 **Yang Z, Stull JT, Levine RJC, and Sweeney HL.** Changes in interfilament spacing mimic the effects of myosin regulatory light chain phosphorylation in rabbit psoas fibers. *J. Struct. Biol.* 122(1–2): 139–148; 1998.

261 **Young K and Davies CTM.** Effect of diet on human muscle weakness following prolonged exercise. *Eur. J. Appl. Physiol.* 53(1): 81–85; 1984.

Chapter 6

1 **Antonio J and Gonyea WJ.** Progressive stretch overload of skeletal muscle results in hypertrophy before hyperplasia. *J. Appl. Physiol.* 75(3): 1263–1271; 1993.

2 **Bell GJ, Syrotuik D, Martin TP, Burnham R, and Quinney HA.** Effect of concurrent strength and endurance training on skeletal muscle properties and hormone concentrations in humans. *Eur. J. Appl. Physiol.* 81(5): 418–427; 2000.

3 **Bishop JE and Lindahl G.** Regulation of cardiovascular collagen synthesis by mechanical load. *Cardiovasc. Res.* 42(1): 27–44; 1999.

4 **Brooks S, Nevill ME, Meleagros L, Lakomy HK, Hall GM, Bloom SR, and Williams C.** The hormonal responses to repetitive brief maximal exercise in humans. *Eur. J. Appl. Physiol. Occup. Physiol.* 60(2): 144–148; 1990.

5 **Burgomaster KA, Heigenhauser GJ, and Gibala MJ.** Effect of short-term sprint interval training on

human skeletal muscle carbohydrate metabolism during exercise and time-trial performance. *J. Appl. Physiol.* 100(6): 2041–2047; 2006.

6 **Burgomaster KA, Hughes SC, Heigenhauser GJ, Bradwell SN, and Gibala MJ.** Six sessions of sprint interval training increases muscle oxidative potential and cycle endurance capacity in humans. *J. Appl. Physiol.* 98(6): 1985–1990; 2005.

7 **Convertino VA, Greenleaf JE, and Bernauer EM.** Role of thermal and exercise factors in the mechanism of hypervolemia. *J. Appl. Physiol.* 48(4): 657–664; 1980.

8 **Costill DL.** *Inside Running: Basics of Sport Physiology.* Indianapolis: Benchmark; 1986.

9 **Costill DL, Daniels J, Evans W, Fink W, Krahenbuhl G, and Saltin B.** Skeletal muscle enzymes and fiber composition in male and female track athletes. *J. Appl. Physiol.* 40(2): 149–154; 1976.

10 **Costill DL, Thomas R, Robergs RA, Pascoe D, Lambert C, Barr S, and Fink WJ.** Adaptations to swimming training: influence of training volume. *Med. Sci. Sports Exerc.* 23(3): 371–377; 1991.

11 **Davies KJ, Packer L, and Brooks GA.** Biochemical adaptation of mitochondria, muscle, and whole-animal respiration to endurance training. *Arch. Biochem. Biophys.* 209(2): 539–554; 1981.

12 **Edge J, Bishop D, Goodman C, and Dawson B.** Effects of high- and moderate-intensity training on metabolism and repeated sprints. *Med. Sci. Sports Exerc.* 37(11): 1975–1982; 2005.

13 **Fagard RH.** Athlete's heart: a meta-analysis of the echocardiographic experience. *Int. J. Sports Med.* 17(Suppl 3): S140–S144; 1996.

14 **Gibala MJ, Little JP, van Essen M, Wilkin GP, Burgomaster KA, Safdar A, Raha S, and Tarnopolsky MA.** Short-term sprint interval versus traditional endurance training: similar initial adaptations in human skeletal muscle and exercise performance. *J. Physiol.* 575(3): 901–911; 2006.

15 **Gledhill N, Cox D, and Jamnik R.** Endurance athletes' stroke volume does not plateau: major advantage is diastolic function. *Med. Sci. Sports Exerc.* 26(9): 1116–1121; 1994.

16 **Gollnick P, Piel K, Karlsson J, and Saltin B.** Glycogen depletion patterns in human skeletal muscle fibers after varying types and intensities of exercise. In: *Metabolic Adaptations to Prolonged Physical Exercise*, edited by Howald H and Poortmans J. Basel: Birkhauser Verlag, 1975; pp. 416–421.

17 **Green HJ, Ball-Burnett M, Symon S, Grant S, and Jamieson G.** Short-term training, muscle glycogen, and cycle endurance. *Can. J. Appl. Physiol.* 20(3): 315–324; 1995.

18 **Green HJ, Thomson JA, Ball ME, Hughson RL, Houston ME, and Sharratt MT.** Alterations in blood volume following short-term supramaximal exercise. *J. Appl. Physiol.* 56(1): 145–149; 1984.

19 **Henriksson J and Reitman JS.** Time course of changes in human skeletal muscle succinate dehydrogenase and cytochrome oxidase activities and maximal oxygen uptake with physical activity and inactivity. *Acta Physiol. Scand.* 99(1): 91–97; 1977.

20 **Hickson RC.** Interference of strength development by simultaneously training for strength and endurance. *Eur. J. Appl. Physiol. Occup. Physiol.* 45(2–3): 255–263; 1980.

21 **Hood DA.** Invited Review: contractile activity-induced mitochondrial biogenesis in skeletal muscle. *J. Appl. Physiol.* 90(3): 1137–1157; 2001.

22 **Hoppeler H and Fluck M.** Plasticity of skeletal muscle mitochondria: structure and function. *Med. Sci. Sports Exerc.* 35(1): 95–104; 2003.

23 **Hoppeler H, Vogt M, Weibel ER, and Fluck M.** Response of skeletal muscle mitochondria to hypoxia. *Exp. Physiol.* 88(1): 109–119; 2003.

24 **Jensen L, Bangsbo J, and Hellsten Y.** Effect of high intensity training on capillarization and presence of angiogenic factors in human skeletal muscle. *J. Physiol.* 557(2): 571–582; 2004.

25 **Johansen L and Quistorff B.** 31P-MRS characterization of sprint and endurance trained athletes. *Int. J. Sports Med.* 24(3): 183–189; 2003.

26 **Kraemer WJ, Patton JF, Gordon SE, Harman EA, Deschenes MR, Reynolds K, Newton RU, Triplett NT, and Dziados JE.** Compatibility of high-intensity strength and endurance training on hormonal and skeletal muscle adaptations. *J. Appl. Physiol.* 78(3): 976–989; 1995.

27 **MacDougall JD.** Blood pressure responses to resistive, static and dynamic exercise. In: *Cardiovascular Response to Exercise*, edited by Fletcher GF. Mount Kisko, NY: Futura Publishing, 1994; pp. 155–173.

28 **MacDougall JD, Hicks AL, MacDonald JR, McKelvie RS, Green HJ, and Smith KM.** Muscle

performance and enzymatic adaptations to sprint interval training. *J. Appl. Physiol.* 84(6): 2138–2142; 1998.

29 **MacDougall JD, Ward GR, and Sutton JR.** Muscle glycogen repletion after high-intensity intermittent exercise. *J. Appl. Physiol.* 42(2): 129–132; 1977.

30 **Melissa L, MacDougall JD, Tarnopolsky MA, Cipriano N, and Green HJ.** Skeletal muscle adaptations to training under normobaric hypoxic versus normoxic conditions. *Med. Sci. Sports Exerc.* 29(2): 238–243; 1997.

31 **Mujika I and Padilla S.** Detraining: loss of training-induced physiological and performance adaptations. Part II: Long term insufficient training stimulus. *Sports Med.* 30(3): 145–154; 2000.

32 **Niess AM, Fehrenbach E, Strobel G, Roecker K, Schneider EM, Buergler J, Fuss S, Lehmann R, Northoff H, and Dickhuth HH.** Evaluation of stress responses to interval training at low and moderate altitudes. *Med. Sci. Sports Exerc.* 35(2): 263–269; 2003.

33 **Parra J, Cadefau JA, Rodas G, Amigo N, and Cusso R.** The distribution of rest periods affects performance and adaptations of energy metabolism induced by high-intensity training in human muscle. *Acta Physiol. Scand.* 169(2): 157–165; 2000.

34 **Pelliccia A, Culasso F, Di Paolo FM, and Maron BJ.** Physiologic left ventricular cavity dilatation in elite athletes. *Ann. Intern. Med.* 130(1): 23–31; 1999.

35 **Pelliccia A, Maron BJ, Culasso F, Spataro A, and Caselli G.** Athlete's heart in women. Echocardiographic characterization of highly trained elite female athletes. *JAMA* 276(3): 211–215; 1996.

36 **Pluim BM, Zwinderman AH, van der Laarse A, and van der Wall EE.** The athlete's heart. A meta-analysis of cardiac structure and function. *Circulation* 101(3): 336–344; 2000.

37 **Prior BM, Yang HT, and Terjung RL.** What makes vessels grow with exercise training? *J. Appl. Physiol.* 97(3): 1119–1128; 2004.

38 **Rockl KS, Witczak CA, and Goodyear LJ.** Signaling mechanisms in skeletal muscle: acute responses and chronic adaptations to exercise. *IUBMB Life* 60(3): 145–153; 2008.

39 **Rodas G, Ventura JL, Cadefau JA, Cusso R, and Parra J.** A short training program for the rapid improvement of both aerobic and anaerobic metabolism. *Eur. J. Appl. Physiol.* 82(5–6): 480–486; 2000.

40 **Sahlin K and Henriksson J.** Buffer capacity and lactate accumulation in skeletal muscle of trained and untrained men. *Acta Physiol. Scand.* 122(3): 331–339; 1984.

41 **Sawka MN and Coyle EF.** Influence of body water and blood volume on thermoregulation and exercise performance in the heat. *Exerc. Sport Sci. Rev.* 27: 167–218; 1999.

42 **Sherman WM.** Recovery from endurance exercise. *Med. Sci. Sports Exerc.* 24(9 Suppl): S336–S339; 1992.

43 **Sundstedt M, Hedberg P, Jonason T, Ringqvist I, Brodin LA, and Henriksen E.** Left ventricular volumes during exercise in endurance athletes assessed by contrast echocardiography. *Acta Physiol. Scand.* 182(1): 45–51; 2004.

44 **van Wessel T, de Haan A, van der Laarse WJ, and Jaspers RT.** The muscle fiber type-fiber size paradox: hypertrophy or oxidative metabolism? *Eur. J. Appl. Physiol.* 110(4): 665–694; 2010.

45 **Wackerhage H and Woods NM.** Exercise-induced signal transduction and gene regulation in skeletal muscle. *J. Sports Sci. Med.* 1: 103–114; 2002.

46 **Zhou B, Conlee RK, Jensen R, Fellingham GW, George JD, and Fisher AG.** Stroke volume does not plateau during graded exercise in elite male distance runners. *Med. Sci. Sports Exerc.* 33(11): 1849–1854; 2001.

Chapter 7

1 **Ballantyne CS, Bahn M, Sale DG, Tarnopolsky MA, and MacDougall JD.** Endurance training does not impair muscle strength or power. *Can. J. Appl. Physiol.* 2b(5): 460; 2001.

2 **Bell GJ and Wenger HA.** The effect of one-legged sprint training on intramuscular pH and nonbicarbonate buffering capacity. *Eur. J. Appl. Physiol. Occup. Physiol.* 58(1–2): 158–164; 1988.

3 **Bogdanis GC, Nevill ME, Boobis LH, Lakomy HK, and Nevill AM.** Recovery of power output and muscle metabolites following 30 s of maximal sprint cycling in man. *J. Physiol.* 482(2): 467–480; 1995.

4 **Burgomaster KA, Heigenhauser GJ, and Gibala MJ.** Effect of short-term sprint interval training on human skeletal muscle carbohydrate metabolism during exercise and time-trial performance. *J. Appl. Physiol.* 100(6): 2041–2047; 2006.

5 **Burgomaster KA, Howarth KR, Phillips SM, Rakobowchuk M, Macdonald MJ, McGee SL, and Gibala MJ.** Similar metabolic adaptations during exercise after low volume sprint interval and traditional endurance training in humans. *J. Physiol.* 586(1): 151–160; 2008.

6 **Burgomaster KA, Hughes SC, Heigenhauser GJ, Bradwell SN, and Gibala MJ.** Six sessions of sprint interval training increases muscle oxidative potential and cycle endurance capacity in humans. *J. Appl. Physiol.* 98(6): 1985–1990; 2005.

7 **Cadefau J, Casademont J, Grau JM, Fernandez J, Balaguer A, Vernet M, Cusso R, and Urbano-Marquez A.** Biochemical and histochemical adaptation to sprint training in young athletes. *Acta Physiol. Scand.* 140(3): 341–351; 1990.

8 **Costill DL.** *Inside Running: Basics of Sport Physiology.* Indianapolis: Benchmark; 1986.

9 **Davies KJ, Packer L, and Brooks GA.** Exercise bioenergetics following sprint training. *Arch. Biochem. Biophys.* 215(1): 260–265; 1982.

10 **Dawson B, Fitzsimons M, Green S, Goodman C, Carey M, and Cole K.** Changes in performance, muscle metabolites, enzymes and fiber types after short sprint training. *Eur. J. Appl. Physiol. Occup. Physiol.* 78(2): 163–169; 1998.

11 **Edge J, Bishop D, Goodman C, and Dawson B.** Effects of high- and moderate-intensity training on metabolism and repeated sprints. *Med. Sci. Sports Exerc.* 37(11): 1975–1982; 2005.

12 **Gibala MJ, Little JP, van Essen M, Wilkin GP, Burgomaster KA, Safdar A, Raha S, and Tarnopolsky MA.** Short-term sprint interval versus traditional endurance training: similar initial adaptations in human skeletal muscle and exercise performance. *J. Physiol.* 575(3): 901–911; 2006.

13 **Green H, Bishop P, Houston M, McKillop R, Norman R, and Stothart P.** Time-motion and physiological assessments of ice hockey performance. *J. Appl. Physiol.* 40(2): 159–163; 1976.

14 **Green HJ, Daub BD, Painter DC, and Thomson JA.** Glycogen depletion patterns during ice hockey performance. *Med. Sci. Sports* 10(4): 289–293; 1978.

15 **Harmer AR, McKenna MJ, Sutton JR, Snow RJ, Ruell PA, Booth J, Thompson MW, Mackay NA, Stathis CG, Crameri RM *et al.*** Skeletal muscle metabolic and ionic adaptations during intense exercise following sprint training in humans. *J. Appl. Physiol.* 89(5): 1793–1803; 2000.

16 **Holloszy JO.** Adaptation of skeletal muscle to endurance exercise. *Med. Sci. Sports* 7(3): 155–164; 1975.

17 **Johansen L and Quistorff B.** 31P-MRS characterization of sprint and endurance trained athletes. *Int. J. Sports Med.* 24(3): 183–189; 2003.

18 **Linossier MT, Denis C, Dormois D, Geyssant A, and Lacour JR.** Ergometric and metabolic adaptation to a 5-s sprint training program. *Eur. J. Appl. Physiol. Occup. Physiol.* 67(5): 408–414; 1993.

19 **Linossier MT, Dormois D, Perier C, Frey J, Geyssant A, and Denis C.** Enzyme adaptations of human skeletal muscle during bicycle short-sprint training and detraining. *Acta Physiol. Scand.* 161(4): 439–445; 1997.

20 **MacDougall JD.** Thermoregulatory problems encountered in ice hockey. *Can. J. Appl. Sport Sci.* 4(1): 35–38; 1979.

21 **MacDougall JD, Hicks AL, MacDonald JR, McKelvie RS, Green HJ, and Smith KM.** Muscle performance and enzymatic adaptations to sprint interval training. *J. Appl. Physiol.* 84(6): 2138–2142; 1998.

22 **MacDougall JD, Ward GR, and Sutton JR.** Muscle glycogen repletion after high-intensity intermittent exercise. *J. Appl. Physiol.* 42(2): 129–132; 1977.

23 **McKenna MJ, Harmer AR, Fraser SF, and Li JL.** Effects of training on potassium, calcium and hydrogen ion regulation in skeletal muscle and blood during exercise. *Acta Physiol. Scand.* 156(3): 335–346; 1996.

24 **McKenna MJ, Heigenhauser GJ, McKelvie RS, MacDougall JD, and Jones NL.** Sprint training enhances ionic regulation during intense exercise in men. *J. Physiol.* 501(3): 687–702; 1997.

25 **Montgomery DL.** Physiology of ice hockey. *Sports Med.* 5(2): 99–126; 1988.

26 **Nevill ME, Boobis LH, Brooks S, and Williams C.** Effect of training on muscle metabolism during treadmill sprinting. *J. Appl. Physiol.* 67(6): 2376–2382; 1989.

27 **Parkhouse WS and McKenzie DC.** Possible contribution of skeletal muscle buffers to enhanced anaerobic performance: a brief review. *Med. Sci. Sports Exerc.* 16(4): 328–338; 1984.

28 **Parolin ML, Chesley A, Matsos MP, Spriet LL, Jones NL, and Heigenhauser GJ.** Regulation of

skeletal muscle glycogen phosphorylase and PDH during maximal intermittent exercise. *Am. J. Physiol.* 277(5, Pt 1): E890–E900; 1999.

29 **Parra J, Cadefau JA, Rodas G, Amigo N, and Cusso R.** The distribution of rest periods affects performance and adaptations of energy metabolism induced by high-intensity training in human muscle. *Acta Physiol. Scand.* 169(2): 157–165; 2000.

30 **Pascoe DD and Gladden LB.** Muscle glycogen resynthesis after short term, high intensity exercise and resistance exercise. *Sports Med.* 21(2): 98–118; 1996.

31 **Ross A and Leveritt M.** Long-term metabolic and skeletal muscle adaptations to short-sprint training: implications for sprint training and tapering. *Sports Med.* 31(15): 1063–1082; 2001.

32 **Trappe S, Harber M, Creer A, Gallagher P, Slivka D, Minchev K, and Whitsett D.** Single muscle fiber adaptations with marathon training. *J. Appl. Physiol.* 101(3): 721–727; 2006.

33 **Widrick JJ, Trappe SW, Blaser CA, Costill DL, and Fitts RH.** Isometric force and maximal shortening velocity of single muscle fibers from elite master runners. *Am. J. Physiol.* 271(2, Pt 1): C666–C675; 1996.

Chapter 8

1 **Aagaard P.** Making muscles "stronger": exercise, nutrition, drugs. *J. Musculoskel. Neuron Interact.* 4(2): 165–174; 2004.

2 **Aagaard P and Andersen JL.** Effects of strength training on endurance capacity in top-level endurance athletes. *Scand. J. Med. Sci. Sports* 20(Suppl 2): 39–47; 2010.

3 **Aagaard P, Andersen JL, Dyre-Poulsen P, Leffers A, Wagner A, Magnusson SP, Halkjaer-Kristensen J, and Simonsen EB.** Mechanism of increased contractile strength of human pennate muscle in response to strength training: changes in muscle architecture. *J. Physiol.* 534(2): 613–623; 2001.

4 **Aagaard P, Andersen JL, Bennekon M, Larson B, Olesen JL, Magnusson SP, and Kjaer M.** Effects of resistance training on endurance capacity and muscle fiber composition in young top-level cyclists. *Scand. J. Med. Sci. Sports* 21(6): e298–e307; 2011.

5 **Abe T, Kumagai K, and Brechue WF.** Fascicle length of leg muscles is greater in sprinters than distance runners. *Med. Sci. Sports Exerc.* 32(6): 1125–1129; 2000.

6 **Abe T, Loenneke JP, Fahs CA, Rossow LM, Thiebaud RS, and Bemben MG.** Exercise intensity and muscle hypertrophy in blood flow-restricted limbs and non-restricted muscles: a brief review. *Clin. Physiol. Funct. Imaging* 32(4): 247–252; 2012.

7 **Adams GR.** Satellite cell proliferation and skeletal muscle hypertrophy. *Appl. Physiol. Nutr. Metab.* 31(6): 782–790; 2006.

8 **Ahtiainen JP, Pakarinen A, Alen M, Kraemer WJ, and Hakkinen K.** Short vs long rest periods between the sets in hypertrophic resistance training: influence on muscle strength, size, and hormonal adaptations in trained men. *J. Strength Cond. Res.* 19(3): 572–582; 2005.

9 **Alegre LM, Jimenez F, Gonzalo-Orden JM, Martin-Acero R, and Aguado X.** Effects of dynamic resistance training on fascicle length and isometric strength. *J. Sports Sci.* 24(5): 501–508; 2006.

10 **Allemeier CA, Fry AC, Johnson P, Hikida RS, Hagerman FC, and Staron RS.** Effects of sprint cycle training on human skeletal muscle. *J. Appl. Physiol.* 77(5): 2385–2390; 1994.

11 **Almasbakk B and Hoff J.** Coordination, the determinant of velocity specificity? *J. Appl. Physiol.* 80(5): 2046–2052; 1996.

12 **Alway SE.** Is fiber mitochondrial volume density a good indicator of muscle fatigability to isometric exercise? *J. Appl. Physiol.* 70(5): 2111–2119; 1991.

13 **Alway SE, Gonyea WJ, and Davis ME.** Muscle fiber formation and fiber hypertrophy during the onset of stretch-overload. *Am. J. Physiol.* 259(1, Pt 1): C92–C102; 1990.

14 **Alway SE, Grumbt WH, Stray-Gundersen J, and Gonyea WJ.** Effects of resistance training on elbow flexors of highly competitive bodybuilders. *J. Appl. Physiol.* 72(4): 1512–1521; 1992.

15 **Alway SE, MacDougall JD, Sale DG, Sutton JR, and McComas AJ.** Functional and structural adaptations in skeletal muscle of trained athletes. *J. Appl. Physiol.* 64(3): 1114–1120; 1988.

16 **Alway SE, Winchester PK, Davis ME, and Gonyea WJ.** Regionalized adaptations and muscle fiber proliferation in stretch-induced enlargement. *J. Appl. Physiol.* 66(2): 771–781; 1989.

17 **American College of Sports Medicine.** American College of Sports Medicine position stand. Progression

models in resistance training for healthy adults. *Med. Sci. Sports Exerc.* 34(2): 364–380; 2002.

18 **American College of Sports Medicine.** American College of Sports Medicine position stand. Progression models in resistance training for healthy adults. *Med. Sci. Sports Exerc.* 41(3): 687–708; 2009.

19 **Andersen JL and Aagaard P.** Effects of strength training on muscle fiber types and size; consequences for athletes training for high-intensity sport. *Scand. J. Med. Sci. Sports* 20(Suppl 2): 32–38; 2010.

20 **Andersen JL and Aagaard P.** Myosin heavy chain IIX overshoot in human skeletal muscle. *Muscle Nerve* 23(7): 1095–1104; 2000.

21 **Andersen LL, Andersen JL, Magnusson SP, Suetta C, Madsen JL, Christensen LR, and Aagaard P.** Changes in the human force-velocity relationship in response to resistance training and subsequent detraining. *J. Appl. Physiol.* 99(1): 87–94; 2005.

22 **Anderson T and Kearney JT.** Effects of three resistance training programs on muscular strength and absolute and relative endurance. *Res. Q. Exerc. Sport* 53(1): 1–7; 1982.

23 **Antonio J and Gonyea WJ.** Progressive stretch overload of skeletal muscle results in hypertrophy before hyperplasia. *J. Appl. Physiol.* 75(3): 1263–1271; 1993.

24 **Arampatzis A, Karamanidis K, and Albrecht K.** Adaptational responses of the human Achilles tendon by modulation of the applied cyclic strain magnitude. *J. Exp. Biol.* 210: 2743–2753; 2007.

25 **Arampatzis A, Karamanidis K, Morey-Klapsing G, De Monte G, and Stafilidis S.** Mechanical properties of the triceps surae tendon and aponeurosis in relation to intensity of sport activity. *J. Biomech.* 40(9): 1946–1952; 2007.

26 **Arampatzis A, Peper A, Bierbaum S, and Albrecht K.** Plasticity of human Achilles tendon mechanical and morphological properties in response to cyclic strain. *J. Biomech.* 43(16): 3073–3079; 2010.

27 **Atherton PJ and Smith K.** Muscle protein synthesis in response to nutrition and exercise. *J. Physiol.* 590(5): 1049–1057; 2012.

28 **Baar K.** Training for endurance and strength: lessons from cell signaling. *Med. Sci. Sports Exerc.* 38(11): 1939–1944; 2006.

29 **Baar K.** The signaling underlying FITness. *Appl. Physiol. Nutr. Metab.* 34(3): 411–419; 2009.

30 **Babcock L, Escano M, D'Lugos A, Todd K, Murach K, and Luden N.** Concurrent aerobic exercise interferes with the satellite cell response to acute resistance exercise. *Am. J. Physiol.* 302(12): R1458–R1465; 2012.

31 **Ballantyne CS, Phillips SM, MacDonald JR, Tarnopolsky MA, and MacDougall JD.** The acute effects of androstenedione supplementation in healthy young males. *Can. J. Appl. Physiol.* 25(1): 68–78; 2000.

32 **Bamman MM, Petrella JK, Kim J, Mayhew DL, and Cross JM.** Cluster analysis tests the importance of myogenic gene expression during myofiber hypertrophy in humans. *J. Appl. Physiol.* 102(6): 2232–2239; 2007.

33 **Barrett-O'Keefe Z, Helgerud J, Wagner PD, Richardson RS.** Maximal strength training and increased work efficiency: contribution from the trained muscle bed. *J. Appl. Physiol.* 113(12): 1846–1851; 2012.

34 **Bawa P.** Neural control of motor output: Can training change it? *Exerc. Sport Sci. Rev.* 30(2): 59–63; 2002.

35 **Behm DG and Sale DG.** Velocity specificity of resistance training. *Sports Med.* 15(6): 374–388; 1993.

36 **Behm DG and Sale DG.** Intended rather than actual movement velocity determines velocity-specific training response. *J. Appl. Physiol.* 74(1): 359–368; 1993.

37 **Bhasin S, Storer TW, Berman N, Callegari C, Clevenger B, Phillips J, Bunnell TJ, Tricker R, Shirazi A, and Casaburi R.** The effects of supraphysiologic doses of testosterone on muscle size and strength in normal men. *N. Engl. J. Med.* 335(1): 1–7; 1996.

38 **Bhasin S, Woodhouse L, Casaburi R, Singh AB, Mac RP, Lee M, Yarasheski KE, Sinha-Hikim I, Dzekov C, Dzekov J, Magliano L, and Storer TW.** Older men are as responsive as young men to the anabolic effects of graded doses of testosterone on the skeletal muscle. *J. Clin. Endocrinol. Metab.* 90(2): 678–688; 2005.

39 **Bickel CS, Cross JM, and Bamman MM.** Exercise dosing to retain resistance training adaptations in young and older adults. *Med. Sci. Sports Exerc.* 43(7): 1177–1187; 2011.

40 **Bischoff R.** Analysis of muscle regeneration using single myofibers in culture. *Med. Sci. Sports Exerc.* 21(5 Suppl): S164–S172; 1989.

41 **Blazevich AJ.** Effects of physical training and detraining, immobilization, growth and aging on human fascicle geometry. *Sports Med.* 36(12): 1003–1017; 2006.

42 **Blazevich AJ, Cannavan D, Coleman DR, and Horne S.** Influence of concentric and eccentric resistance training on architectural adaptation in human quadriceps muscles. *J. Appl. Physiol.* 103(5): 1565–1575; 2007.

43 **Bloomquist K, Langberg H, Karlsen S, Madsgaard S, Boesen M, and Raastad T.** Effect of range of motion in heavy load squatting on muscle and tendon adaptations. *Eur. J. Appl. Physiol.* 113(8): 2133–2142; 2013.

44 **Bogdanis GC, Papaspyrou A, Souglis AG, Theos A, Sotiropoulos A, and Maridaki M.** Effects of two different half-squat training programs on fatigue during repeated cycling sprints in soccer players. *J. Strength Cond. Res.* 25(7): 1849–1856; 2011.

45 **Brechue WF and Abe T.** The role of FFM accumulation and skeletal muscle architecture in powerlifting performance. *Eur. J. Appl. Physiol.* 86: 327–336; 2002.

46 **Brodsky IG, Balagopal P, and Nair KS.** Effects of testosterone replacement on muscle mass and muscle protein synthesis in hypogonadal men—a clinical research center study. *J. Clin. Endocrinol. Metab.* 81(10): 3469–3475; 1996.

47 **Bruusgaard JC, Johansen IB, Egner IM, Rana ZA, and Gundersen K.** Myonuclei acquired by overload exercise precede hypertrophy and are not lost on detraining. *Proc. Natl. Acad. Sci. USA* 107(34): 15111–15116; 2010.

48 **Buchheit M, Mendez-Villanueva A, Delhomel G, Brughelli M, and Ahmaidi S.** Improving repeated sprint ability in young elite soccer players: repeated shuttle sprints vs explosive strength training. *J. Strength Cond. Res.* 24(10): 2715–2722; 2010.

49 **Buford TW, Rossi SJ, Smith DB, and Warren AJ.** A comparison of periodization models during nine weeks with equated volume and intensity for strength. *J. Strength Cond. Res.* 21(4): 1245–1250; 2007.

50 **Burd NA, Moore DR, Mitchell CJ, and Phillips SM.** Big claims for big weights but with little evidence. *Eur. J. Appl. Physiol.* 113(1): 267–268; 2013.

51 **Burd NA, Tang JE, Moore DR, and Phillips SM.** Exercise training and protein metabolism: influences of contraction, protein intake, and sex-based differences. *J. Appl. Physiol.* 106(5): 1692–1701; 2009.

52 **Burd NA, Holwerda AM, Selby KC, West DWD, Staples AW, Cain NE, Cashaback JGA, Potvin JR, Baker SK, and Phillips SM.** Resistance exercise volume affects myofibrillar protein synthesis and anabolic signaling molecule phosphorylation in young men. *J. Physiol.* 588(16): 3119–3130; 2010.

53 **Burd NA, Mitchell CJ, Churchward-Venne TA, and Phillips SM.** Bigger weights may not beget bigger muscles: evidence from acute muscle protein synthetic responses after resistance exercise. *Appl. Physiol. Nutr. Metab.* 37(3): 551–554; 2012.

54 **Burd NA, West DWD, Staples AW, Atherton PJ, Baker JM, Moore DR, Holwerda AM, Parise G, Rennie MJ, Baker SK, and Phillips SM.** Low-load high volume resistance exercise stimulates muscle protein synthesis more than high-load low volume resistance exercise in young men. *PLoS ONE* 5(8): e2033; 2010.

55 **Buresh R, Berg K, and French J.** The effect of resistive exercise rest interval on hormonal response, strength, and hypertrophy with training. *J. Strength Cond. Res.* 23(1): 62–71; 2009.

56 **Burgess KE, Connick MJ, Graham-Smith P, and Pearson SJ.** Plyometric vs isometric training influences on tendon properties and muscle output. *J. Strength Cond. Res.* 21(3): 986–989; 2007.

57 **Byrne C, Twist C, and Eston R.** Neuromuscular function after exercise-induced muscle damage: theoretical and applied implications. *Sports Med.* 34(1): 49–69; 2004.

58 **Cadefau J, Casademont J, Grau JM, Fernadez J, Balaguer A, Vernet M, and Urbano-Marquez A.** Biochemical and histochemical adaptation to sprint training in young athletes. *Acta Physiol. Scand.* 140(3): 341–351; 1990.

59 **Calder AW, Chilibeck PD, Webber CE, and Sale DG.** Comparison of whole and split weight training routines in young women. *Can. J. Appl. Physiol.* 19(2): 185–199, 1994.

60 **Campos GER, Luecke TJ, Wendeln HK, Toma K, Hagerman FC, Murray TF, Ragg KE, Ratamess NA, Kraemer WJ, and Staron RS.** Muscular adaptations in response to three different resistance-training regimens: specificity of repetition maximum training zones. *Eur. J. Appl. Physiol.* 88(1–2): 50–60; 2002.

61 **Canepari M, Rossi R, Pellegrino MA, Orrell RW, Cobbold M, Harridge S, and Bottinelli R.** Effects of resistance training on myosin function studied by the *in vitro* motility assay in young and older men. *J. Appl. Physiol.* 98(6): 2390–2395; 2005.

62 **Carpinelli RN.** Challenging the American College of Sports Medicine 2009 Position Stand on resistance training. *Med. Sport.* 13(2): 131–137; 2009.

63 **Carpinelli RN and Otto RM.** Strength training. Single versus multiple sets. *Sports Med.* 26(2): 73–84; 1998.

64 **Carpinelli RN, Otto RM, and Winett RA.** A critical analysis of the ACSM Position Stand on resistance training: insufficient evidence to support recommended training protocols. *JEP Online* 7(3): 1–60; 2004.

65 **Carroll TJ, Herbert RD, Munn J, Lee M, and Gandevia SC.** Contralateral effects of unilateral strength training: evidence and possible mechanisms. *J. Appl. Physiol.* 101(5): 1514–1522; 2006.

66 **Carroll TJ, Selvanayagam VS, Riek S, and Semmler JG.** Neural adaptations to strength training: moving beyond transcranial magnetic stimulation and reflex studies. *Acta Physiol.* 202(2): 119–140; 2011.

67 **Cermak NM, Snijders T, McKay BR, Parise G, Verdijk LB, Tarnopolsky MA, Gibala MJ, and van Loon LJ.** Eccentric exercise increases satellite cell content in type II muscle fibers. *Med. Sci. Sports Exerc.* 45(2): 230–237; 2013.

68 **Chambers RL and McDermott JC.** Molecular basis of skeletal muscle regeneration. *Can. J. Appl. Physiol.* 21(3): 155–184; 1996.

69 **Chelly MS, Fathloun M, Cherif N, Amar MB, Tabka Z, and Van Praagh E.** Effects of a back squat training program on leg power, jump, and sprint performances in junior soccer players. *J. Strength Cond. Res.* 23(8): 2241–2249; 2009.

70 **Chelly MS, Ghenem MA, Abid K, Hermassi S, Tabka Z, and Shephard RJ.** Effects of in-season short-term plyometric training program on leg power, jump- and sprint performance of soccer players. *J. Strength Cond. Res.* 24(10): 2670–2676; 2010.

71 **Chestnut JL and Docherty D.** The effects of 4 and 10 repetition maximum weight-training protocols on neuromuscular adaptations in untrained men. *J. Strength Cond. Res.* 13(4): 353–359; 1999.

72 **Claassen H, Gerber C, Hoppeler H, Luthi JM, and Vock P.** Muscle filament spacing and short-term heavy-resistance exercise in humans. *J. Physiol.* 409: 491–495; 1989.

73 **Clarkson PM and Hubal MJ.** Exercise-induced muscle damage in humans. *Am. J. Phys. Med. Rehabil.* 81(11 Suppl): S52–S69; 2002.

74 **Coffey VG and Hawley JA.** The molecular bases of training adaptation. *Sports Med.* 37(9): 737–763; 2007.

75 **Cormie P, McGuigan MR, and Newton RU.** Developing maximal neuromuscular power Part 1—Biological basis of maximal power production. *Sports Med.* 41(1): 17–38; 2011.

76 **Cormie P, McGuigan MR, and Newton RU.** Developing maximal neuromuscular power Part 2—Training considerations for improving maximal power production. *Sports Med.* 41(1): 17–38; 2011.

77 **Couppe C, Kongsgaard M, Aagaard P, Hansen P, Bojsen-Moller J, Kjaer M, and Magnusson SP.** Habitual loading results in tendon hypertrophy and increased stiffness of the human patellar tendon. *J. Appl. Physiol.* 105(3): 805–810; 2008.

78 **Crewther B, Cronin J, and Keogh J.** Possible stimuli for strength and power adaptation. Acute mechanical responses. *Sports Med.* 35(11): 967–989; 2005.

79 **Cronin J and Sleivert G.** Challenges in understanding the influence of maximal power training on improving athletic performance. *Sports Med.* 35(3): 213–234; 2005.

80 **D'Antona G, Lanfranconi F, Pellegrino MA, Brocca L, Adami R, Rossi R, Moro G, Miotti D, Canepari M, and Roberto Bottinelli R.** Skeletal muscle hypertrophy and structure and function of skeletal muscle fibers in male bodybuilders. *J. Physiol.* 570(3): 611–627; 2006.

81 **Davidsen PK, Gallagher IJ, Hartman JW, Tarnopolsky MA, Dela F, Helge JW, Timmons JA, and Phillips SM.** High responders to resistance exercise training demonstrate differential regulation of skeletal muscle microRNA expression. *J. Appl. Physiol.* 110(2): 309–317; 2011.

82 **Dawson B, Fitzsimons M, Green S, Goodman C, Carey M, and Cole K.** Changes in performance, muscle metabolites, enzymes and fiber types after short sprint training. *Eur. J. Appl. Physiol.* 78(2): 163–169; 1998.

83 **DeFreitas JM, Beck TW, Stock MS, Dillon MA, and Kasishke II PR.** An examination of the time course of training-induced skeletal muscle hypertrophy. *Eur. J. Appl. Physiol.* 111(11): 2785–2790; 2011.

84 **Degens H, Erskine RM, and Morse CI.** Disproportionate changes in skeletal muscle strength and size with resistance training and ageing. *J. Musculoskelet. Neuronal Interact.* 9(3): 123–129; 2009.

85 **de Salles BF, Simao R, Miranda F, Novaes JDS, Lemos A, and Willardson JM.** Rest interval between sets in strength training. *Sports Med.* 39(9): 765–777; 2009.

86 **De Sousa EO, Tricoli V, Roschel H, Brum PC, Bacurau AVN, Ferreira JCB, Aoki MS, Neves-Jr M, Aihara AY, Fernandas ADRC, and Ugrinowitsch C.** Molecular adaptations to concurrent training. *Int. J. Sports Med.* 34(3): 207–213; 2013.

87 **De Souza Jr TP, Fleck SJ, Simao R, Dubas JP, Pereira B, De Brito Pacheco EM, Da Silva AC, and De Oliveira PR.** Comparison between constant and decreasing rest intervals: influence on maximal strength and hypertrophy. *J. Strength Cond. Res.* 24(7): 1843–1850; 2010.

88 **De Villarreal ES, González-Badillo JJ, and Izquierdo M.** Low and moderate plyometric training frequency produces greater jumping and sprinting gains compared with high frequency. *J. Strength Cond. Res.* 22(3): 715–725; 2008.

89 **De Villarreal ES, Kellis E, Kraemer WJ, and Izquierdo M.** Determining variables of plyometric training for improving vertical jump height performance: a meta-analysis. *J. Strength Cond. Res.* 23(2): 495–506; 2009.

90 **Drinkwater EJ, Lawton TW, Lindsell RP, Pyne DB, Hunt PH, and McKenna MJ.** Training leading to repetition failure contributes to strength gains in elite junior athletes. *J. Strength Cond. Res.* 19(2): 382–388; 2005.

91 **Duchateau J and Enoka RM.** Neural adaptations with chronic activity patterns in able-bodied humans. *Am. J. Phys. Med. Rehabil.* 81(Suppl): S17–S27; 2002.

92 **Duchateau J and Hainaut K.** Isometric or dynamic training: differential effects on mechanical properties of a human muscle. *J. Appl. Physiol.* 56 (2): 296–301; 1984.

93 **Duchateau J and Hainaut K.** Mechanisms of muscle and motor unit adaptation to explosive power training. In *Strength and Power in Sport*, 2nd edn, edited by Komi PV. Malden, MA: Blackwell, 2003; pp. 315–330.

94 **Duchateau J, Semmler JG, and Enoka RM.** Training adaptations in the behavior of human motor units. *J. Appl. Physiol.* 101(6): 1766–1775; 2006.

95 **Duclay J, Martin A, Dulay A, Cometti G, and Pousson M.** Behavior of fascicles and the myotendinous junction of human medial gastrocnemius following eccentric strength training. *Muscle Nerve* 39(6): 819–827; 2009.

96 **Egan B and Zierath JR.** Exercise metabolism and the molecular regulation of skeletal muscle adaptation. *Cell Metab.* 17(2): 162–184; 2013.

97 **Farthing JP and Chilibeck PD.** The effects of eccentric and concentric training at different velocities on muscle hypertrophy. *Eur. J. Appl. Physiol.* 89(6): 578–586; 2003.

98 **Farup J, Kjolhede T, Sorensen H, Dalgas U, Moller AB, Vestergaard PF, Ringgaard S, Bojsen-Moller J, and Vissing K.** Muscle morphological and strength adaptations to endurance vs resistance training. *J. Strength Cond. Res.* 26(2): 398–407; 2012.

99 **Fisher J, Steele J, Bruce-Low S, and Smith D.** Evidence-based resistance training recommendations. *Med. Sport.* 15(3): 147–162; 2011.

100 **Folland JP and Williams AG.** The adaptations to strength training morphological and neurological contributions to increased strength. *Sports Med.* 2007; 37(2): 145–168; 2007.

101 **Folland JP, Irish CS, Roberts JC, Tarr JE, and Jones DA.** Fatigue is not a necessary stimulus for strength gains during resistance training. *Br. J. Sports Med.* 36(5): 370–374; 2002.

102 **Foure A, Nordez A, McNair P, and Cornu C.** Effects of plyometric training on both active and passive parts of the plantarflexors series elastic component stiffness of muscle-tendon complex. *Eur. J. Appl. Physiol.* 111(3): 539–548; 2011.

103 **Foure A, Nordez A, and Cornu C.** Effects of plyometric training on passive stiffness of gastrocnemii muscles and Achilles tendon. *Eur. J. Appl. Physiol.* 112(8): 2849–2857; 2012.

104 **Fowles JR, MacDougall JD, Tarnopolsky MA, Sale DG, Roy BD, and Yarasheski KE.** The acute effects of passive stretch on muscle protein synthesis in humans. *Can. J. Appl. Physiol.* 25(3): 165–180; 2000.

105 **Fry AC.** The role of resistance exercise intensity on muscle fiber adaptations. *Sports Med.* 34(10): 663–679; 2004.

106 **Fry AC, Allemeier CA, and Staron RS.** Correlation between percentage fiber type area and

myosin heavy chain content in human skeletal muscle. *Eur. J. Appl. Physiol.* 68(3): 246–251; 1994.

107 **Gabriel DA, Kamen G, and Frost G.** Neural adaptations to resistive exercise. Mechanisms and recommendations for training practices. *Sports Med.* 36(2): 133–149; 2006.

108 **Galvao DA and Taaffe DR.** Single- vs multiple-set resistance training: recent developments in the controversy. *J. Strength Cond. Res.* 18(3): 660–667; 2004.

109 **Garcia-Pallares J and Izquierdo M.** Strategies to optimize concurrent training of strength and aerobic fitness for rowing and canoeing. *Sports Med.* 41(4): 329–343; 2011.

110 **Garcia-Pallares J, Sanchez-Medina L, Perez CE, Izquierdo-Gabarren M, and Izquierdo M.** Physiological effects of tapering and detraining in world-class kayakers. *Med. Sci. Sports Exerc.* 42(6): 1209–1214; 2010.

111 **Gardiner PF.** Changes in α-motoneuron properties with altered physical activity levels. *Exerc. Sport Sci. Rev.* 34(2): 54–58; 2006.

112 **Gardiner P, Beaumont E, and Cormery B.** Motoneurones "learn" and "forget" physical activity. *Can. J. Appl. Physiol.* 30(3): 352–370; 2005.

113 **Gentil P, Bottaro M, Oliveira E, Veloso J, Amorim N, Saiuri A, and Wagner DR.** Chronic effects of different between-set rest durations on muscle strength in nonresistance trained young men. *J. Strength Cond. Res.* 24(1): 37–42; 2010.

114 **Gibala MJ.** Molecular responses to high-intensity interval exercise. *Appl. Physiol. Nutr. Metab.* 34(3): 428–432; 2009.

115 **Gibala MJ, Interisano SA, Tarnopolsky MA, Roy BD, MacDonald JR, Yarasheski KE, and MacDougall JD.** Myofibrillar disruption following acute concentric and eccentric resistance exercise in strength-trained men. *Can. J. Physiol. Pharmacol.* 78(8): 656–661; 2000.

116 **Gibala MJ, MacDougall JD, Tarnopolsky MA, Stauber WT, and Elorriaga A.** Changes in human skeletal muscle ultrastructure and force production after acute resistance exercise. *J. Appl. Physiol.* 78(2): 702–708; 1995.

117 **Giddings CJ and Gonyea WJ.** Morphological observations supporting muscle fiber hyperplasia following weight-lifting exercise in cats. *Anat. Rec.* 233(2): 178–195; 1992.

118 **Gillies AR and Lieber RL.** Structure and function of the skeletal muscle extracellular matrix. *Muscle Nerve* 44(3): 318–331; 2011.

119 **Goldspink G.** Changes in muscle mass and phenotype and the expression of autocrine and systemic growth factors by muscle in response to stretch and overload. *J. Anat.* 194(3):P 323–334; 1999.

120 **Gonyea WJ, Sale DG, Gonyea FB, and Mikesky A.** Exercise induced increases in muscle fiber number. *Eur. J. Appl. Physiol. Occup. Physiol.* 55(2): 137–141; 1986.

121 **Goto K, Sato K, and Takamatsu K.** A single set of low intensity resistance exercise immediately following high intensity resistance exercise stimulates growth hormone secretion in men. *J. Sports Med. Phys. Fitness* 43(2): 243–249; 2003.

122 **Goto K, Nagasawa M, Yanagisawa O, Kizuka T, Ishii N, and Takamatsu K.** Muscular adaptations to combinations of high- and low-intensity resistance exercises. *J. Strength Cond. Res.* 18(4): 730–737; 2004.

123 **Goto K, Ishii N, Kizuka T, and Takamatsu K.** The impact of metabolic stress on hormonal responses and muscular adaptations. *Med. Sci. Sports Exerc.* 37(6): 955–963; 2005.

124 **Goto K, Ishii N, Kizuka T, Kraemer RR, Honda Y, and Takamatsu K.** Hormonal and metabolic responses to slow movement resistance exercise with different durations of concentric and eccentric actions. *Eur. J. Appl. Physiol.* 106(5): 731–739; 2009.

125 **Green HJ, Thomson JA, Daub WD, Houston ME, and Ranney DA.** Fiber composition, fiber size and enzyme activities in vastus lateralis of elite athletes involved in high intensity exercise. *Eur. J. Appl. Physiol.* 41(2): 109–117; 1979.

126 **Green H, Goreham C, Ouyang J, Ball-Burnett M, and Ranney D.** Regulation of fiber size, oxidative potential, and capillarization in human muscle by resistance exercise. *Am. J. Physiol.* 276(2, Pt 2): R591–R596; 1999.

127 **Griffin L and Cafarelli E.** Resistance training: cortical, spinal, and motor unit adaptations. *Can. J. Appl. Physiol.* 30(3): 328–340; 2005.

128 **Gross M, Luthy F, Muller E, Hoppeler H, and Vogt M.** Effects of eccentric cycle ergometry in alpine skiers. *Int. J. Sports Med.* 31(8): 572–576; 2010.

129 **Grosset J-F.**, Piscione J, Lambertz D, and Perot C. Paired changes in electromechanical delay and mus-

culo-tendinous stiffness after endurance or plyometric training. *Eur. J. Appl. Physiol.* 105(1): 131–139; 2009.

130 **Guilhem G, Cornu C, and Guevel A.** Neuromuscular and muscle-tendon system adaptations to isotonic and isokinetic eccentric exercise. *Ann. Phys. Rehabil. Med.* 53(5): 319–341; 2010.

131 **Hackett DA, Johnson NA, and Chow C-M.** Training practices and ergogenic aids used by male bodybuilders. *J. Strength Cond. Res.* 27(6): 1609–1617; 2013.

132 **Haff GG and Nimphius S.** Training principles for power. *Strength Cond. J.* 34(6): 2–12; 2013.

133 **Harridge SDR.** Plasticity of human skeletal muscle: gene expression to *in vivo* function. *Exp. Physiol.* 92(5): 783–797; 2007.

134 **Harridge SDR, Bottinelli R, Canepari M, Pellegrino M, Reggiani C, Esbjornsson M, Balsom PD, and Saltin B.** Sprint training, *in vitro* and *in vivo* muscle function, and myosin heavy chain expression. *J. Appl. Physiol.* 84(2): 442–449; 1998.

135 **Harris NK, Cronin JB, Hopkins WG, and Hansen KT.** Squat jump training at maximal power loads vs heavy loads: effect on sprint ability. *J. Strength Cond. Res.* 22(6): 1742–1749; 2008.

136 **Harris N, Cronin J, and Keogh J.** Contraction force specificity and its relationship to functional performance. *J. Sports Sci.* 25(2): 201–212; 2007.

137 **Hawley JA.** Molecular responses to strength and endurance training: are they incompatible? *Appl. Physiol. Nutr. Metab.* 34(3): 355–361; 2009.

138 **Heggelund J, Fimland MS, Helgerud J, and Hoff J.** Maximal strength training improves work economy, rate of force development and maximal strength more than conventional strength training. *Eur. J. Appl. Physiol.* 113(6): 1565–1573; 2013.

139 **Helgerud J, Rodas G, Kemi OJ, and Hoff J.** Strength and endurance in elite football players. *Int. J. Sports Med.* 32(9): 677–682; 2011.

140 **Hellebrandt FA and Houtz SJ.** Mechanisms of muscle training in man: experimental demonstration of the overload principle. *Phys. Ther. Rev.* 36(6): 371–383; 1956.

141 **Hendy AM, Spittle M, and Kidgell DJ.** Cross education and immobilization: mechanisms and implications for injury rehabilitation. *J. Sci. Med. Sport* 15(2): 94–101; 2012.

142 **Hoff J.** Training and testing physical capacities for elite soccer players. *J. Sports Sci.* 23(6): 573–582; 2005.

143 **Hoff J, Helgerud J, and Wisloff U.** Maximal strength training improves work economy in trained female cross-country skiers. *Med. Sci. Sports Exerc.* 31(6): 870–877; 1999.

144 **Hoff J, Gran A, and Helgerud J.** Maximal strength training improves aerobic endurance performance. *Scand. J. Med. Sci.* Sports 12(5): 288–295; 2002.

145 **Hoffman JR, Ratamess NA, Klatt M, Faigenbaum AD, Ross RE, Tranchina NM, McCurley RC, Kang J, and Kraemer WJ.** Comparison between different off-season resistance training programs in Division III American college football players. *J. Strength Cond. Res.* 23(1): 11–19; 2009.

146 **Holm L, Reitelseder S, Pedersen TG, Doessing S, Petersen SG, Flyvbjerg A, Andersen JL, Aagaard P, and Kjaer M.** Changes in muscle size and MHC composition in response to resistance exercise with heavy and light loading intensity. *J. Appl. Physiol.* 105(5): 1454–1461; 2008.

147 **Holtermann A, Roeleveld K, Mork PJ, Gronlund C, Karlsson JS, Andersen LL, Olsen HB, Zebis MK, Sjogaard G, and Sogaard K.** Selective activation of neuromuscular compartments within the human trapezius muscle. *J. Electromyogr. Kinesiol.* 19(5): 896–902; 2009.

148 **Hoppeler H, Luthi P, Claassen H, Weibel ER, and Howald H.** The ultrastructure of the normal human skeletal muscle. A morphometric analysis on untrained men, women and well-trained orienteers. *Pflugers Arch.* 344(3): 217–232; 1973.

149 **Hornberger T.** Mechanotransduction and the regulation of mTORC1 signaling in skeletal muscle. *Int. J. Biochem. Cell Biol.* 43(9): 1267–1276; 2011.

150 **Hornberger TA, Sukhija KB, and Chien S.** Regulation of mTOR by mechanically induced signaling events in skeletal muscle. *Cell Cycle* 5(13): 1391–1396; 2006.

151 **Hortobagyi T, Houmard JA, Stevenson JR, Fraser DD, Johns RA, and Israel RG.** The effects of detraining on power athletes. *Med. Sci. Sports Exerc.* 25(8): 929–935; 1993.

152 **Howatson G and Van Someren KA.** The prevention and treatment of exercise-induced muscle damage. *Sports Med.* 38(6): 483–503; 2008.

153 **Hubal MJ, Gordish-Dressman H, Thompson PD, Price TB, Hoffman EP, Angelopoulos TJ,**

Gordon PM, Moyna NM, Pescatello LS *et al.* Variability in muscle size and strength gain after unilateral resistance training. *Med. Sci. Sports Exerc.* 37(6): 964–972; 2005.

154 **Ikegawa S, Funato K, Tsunoda N, Kanehisa H, Fukunaga T, and Kawakami Y.** Muscle force per cross-sectional area is inversely related with pennation angle in strength trained athletes. *J. Strength Cond. Res.* 22(1): 128–131; 2008.

155 **Izquierdo M, Ibañez J, Gonzalez-Badillo JJ, Hakkinen K, Ratamess NA, Kraemer WJ, French DN, Eslava J, Altadill A, Asiain X, and Gorostiaga EM.** Differential effects of strength training leading to failure versus not to failure on hormonal responses, strength, and muscle power gains. *J. Appl. Physiol.* 100(5): 1647–1656; 2006.

156 **Izquierdo M, González-Badillo JJ, Hakkinen K, Ibañez J, Kraemer WJ, Altadill A, Eslava J, and Gorostiaga EM.** Effect of loading on unintentional lifting velocity declines during single sets of repetitions to failure during upper and lower extremity muscle actions. *Int. J. Sports Med.* 27(9): 718–724; 2006.

157 **Izquierdo M, Ibañez J, Gonzalez-Badillo JJ, Ratamess NA, Kraemer WJ, Hakkinen K, Bonnabau H, Granados C, French DN, and Gorostiaga EM.** Detraining and tapering effects on hormonal responses and strength performance. *J. Strength Cond. Res.* 21(3): 768–775; 2007.

158 **Jacobi JM and Chilibeck PD.** Bilateral and unilateral contractions: possible differences in maximal voluntary force. *Can. J. Appl. Physiol.* 26(1): 12–33; 2001.

159 **Jacobs I, Esbjornsson M, Sylven C, Holm I, and Jansson E.** Sprint training effects on muscle myoglobin, enzymes, fiber types, and blood lactate. *Med. Sci. Sports Exerc.* 19(4): 368–374; 1987.

160 **Jespersen JG, Nedergaard A, Andersen LL, Schjerling P, and Andersen JL.** Myostatin expression during human muscle hypertrophy and subsequent atrophy: increased myostatin with detraining. *Scand. J. Med. Sci. Sports.* 21(2): 215–223; 2011.

161 **Jidovtseff B, Crosier JL, Scimar N, Demoulin C, Maquet D, and Crielaard JM.** The ability of isoinertial assessment to monitor specific training effects. *J. Sports Med. Phys. Fitness* 48(1): 55–64; 2008.

162 **Jones DA, Newham DJ, Round JM, and Tolfree SE.** Experimental human muscle damage: morphological changes in relation to other indices of damage. *J. Physiol.* 375: 435–448; 1986.

163 **Jovanovic M, Sporis G, Omrcen D, and Fiorentini F.** Effects of speed, agility, quickness training method on power performance in elite soccer players. *J. Strength Cond. Res.* 25(5): 1285–1292; 2011.

164 **Jurimae J, Abernethy PJ, Blake K, and McEniery MT.** Changes in the myosin heavy chain isoform profile of the triceps brachii muscle following 12 weeks of resistance training. *Eur. J. Appl. Physiol.* 74(3): 287–292; 1996.

165 **Kadi F and Thornell LE.** Training affects myosin heavy chain phenotype in the trapezius muscle of women. *Histochem. Cell Biol.* 112(1): 73–78; 1999.

166 **Kadi F, Charifi N, Denis C, Lexell J, Andersen JL, Schjerling P, Olsen S, and Kjaer M.** The behavior of satellite cells in response to exercise: what have we learned from human studies? *Pflugers Arch.* 451(2): 319–327; 2005.

167 **Kadi F, Schjerling P, Andersen LL, Charifi N, Madsen JL, Christensen LR, and Andersen JL.** The effects of heavy resistance training and detraining on satellite cells in human skeletal muscles. *J. Physiol.* 558(3): 1005–1012; 2004.

168 **Kawakami Y, Abe T, and Fukunaga T.** Muscle-fiber pennation angles are greater in hypertrophied than in normal muscles. *J. Appl. Physiol.* 74(2): 590–595; 1993.

169 **Kawakami Y, Abe T, Kuno SY, and Fukunaga T.** Training-induced changes in muscle architecture and specific tension. *Eur. J. Appl. Physiol.* 72 (1–2): 37–43; 1995.

170 **Kawakami Y, Ichinose Y, and Fukunaga T.** Architectural and functional features of human triceps surae muscles during contraction. *J. Appl. Physiol.* 85(2): 398–404; 1998.

171 **Kawamori N and Haff GG.** The optimal training load for the development of muscular power. *J. Strength Cond. Res.* 18(3): 675–684; 2004.

172 **Kearns CF, Abe T, and Brechue WF.** Muscle enlargement in Sumo wrestlers includes increased muscle fascicle length. *Eur. J. Appl. Physiol.* 83(4–5): 289–296; 2000.

173 **Keiner M, Sander A, Wirth K, and Schmidtbleicher D.** Long term strength training effects on change-of-direction sprint performance. *J. Strength Cond. Res.* 28(1): 223–231; 2014.

174 **Kellis E.** Quantification of quadriceps and hamstring antagonist activity. *Sports Med.* 25(1): 37–62; 1998.

175 **Kidgell DJ and Pearce AJ.** What has transcranial magnetic stimulation taught us about neural adaptations to strength training? A brief review. *J. Strength Cond. Res.* 25(11): 3208–3217; 2011.

176 **Kjaer M, Langberg H, Heinemeier K, Bayer ML, Hansen M, Holm L, Doessing S, Kongsgaard M, Krogsgaard MR, and Magnusson SP.** From mechanical loading to collagen synthesis, structural changes and function in human tendon. *Scand. J. Med. Sci. Sports.* 19(4): 500–510; 2009.

177 **Kongsgaard M, Aagaard P, Kjaer M, and Magnusson SP.** Structural Achilles tendon properties in athletes subjected to different exercise modes and in Achilles tendon rupture patients. *J. Appl. Physiol.* 99(5): 1965–1971; 2005.

178 **Kongsgaard M, Reitelseder S, Pedersen TG, Holm L, Aagaard P, Kjaer M, and Magnusson SP.** Region specific patellar tendon hypertrophy in humans following resistance training. *Acta Physiol. (Oxf.)* 191(2): 111–121; 2007.

179 **Kosek DJ and Bamman MM.** Modulation of the dystrophin-associated protein complex in response to resistance training in young and older men. *J. Appl. Physiol.* 104(5): 1476–1484; 2008.

180 **Kraemer WJ and Ratamess NA.** Fundamentals of resistance training: progression and exercise prescription. *Med. Sci. Sports Exerc.* 36(4): 674–688; 2004.

181 **Kraemer WJ, Noble BJ, Clark MJ, and Culver BW.** Physiologic responses to heavy-resistance exercise with very short rest periods. *Int. J. Sports Med.* 8(4): 247–252; 1987.

182 **Kreiger JW.** Single versus multiple sets of resistance exercise: a meta-regression. *J. Strength Cond. Res.* 23(6): 1890–1901; 2009.

183 **Kreiger JW.** Single vs multiple sets of resistance exercise for muscle hypertrophy: a meta-analysis. *J. Strength Cond. Res.* 24(4): 1150–1159; 2010.

184 **Kubo K, Ikebukuro T, Yaeshima K, Yata H, Tsunoda N, and Kanehisa H.** Effects of static and dynamic training on the stiffness and blood volume of tendon *in vivo*. *J. Appl. Physiol.* 106(2): 412–417; 2009.

185 **Kubo K, Ikebukuro T, Yata H, Tsunoda N, and Kanehisa H.** Time course of changes in muscle and tendon properties during strength training and detraining. *J. Strength Cond. Res.* 24(2): 322–331; 2010.

186 **Kubo K, Kanehisa H, and Fukunaga T.** Effects of different duration isometric contractions on tendon elasticity in human quadriceps muscles. *J. Physiol.* 536(2): 649–655; 2001.

187 **Kubo K, Kanehisa H, Ito M, and Fukunaga T.** Effects of resistance and stretching training programs on the viscoelastic properties of human tendon structures *in vivo*. *J. Physiol.* 538(1): 219–226; 2002.

188 **Kubo K, Kanehisa H, Ito M, and Fukunaga T.** Effects of isometric training on the elasticity of human tendon structures *in vivo*. *J. Appl. Physiol.* 91(1): 26–32; 2001.

189 **Kubo K, Kanehisa H, Kawakami Y, and Fukunaga T.** Elasticity of tendon structures of the lower limbs in sprinters. *Acta Physiol. Scand.* 168(2): 327–235; 2000.

190 **Kubo K, Komuro T, Ishiguro N, Tsunoda N, Sato Y, Ishii N, Kanehisa H, and Fukunaga T.** Effects of low-load resistance training with vascular occlusion on the mechanical properties of muscle and tendon. *J. Appl. Biomech.* 22(2): 112–119; 2006.

191 **Kubo K, Morimoto M, Komuro T, Yata H, Tsunoda N, Kanehisa H, and Fukunaga T.** Effects of plyometric and weight training on muscle-tendon complex and jump performance. *Med. Sci. Sports Exerc.* 39(10): 1801–1810; 2007.

192 **Kubo K, Ohgo K, Takeishi R, Yoshinaga K, Tsunoda N, Kanehisa H, and Fukunaga T.** Effects of isometric training at different knee angles on the muscle-tendon complex *in vivo*. *Scand. J. Med. Sci. Sports.* 16(3): 159–167; 2006.

193 **Kubo K, Yata H, Kanehisa H, and Fukunaga T.** Effects of isometric squat training on the tendon stiffness and jump performance. *Eur. J. Appl. Physiol.* 96(3): 305–314; 2006.

194 **Kulig K, Landel R, Chang YJ, Hannanvash N, Reischl SF, Song P, and Bashford GR.** Patellar tendon morphology in volleyball athletes with and without patellar tendinopathy. *Scand. J. Med. Sci. Sports.* 23(2): e81–e88; 2013.

195 **Kumagai K, Abe T, Brechue WF, Ryushi T, Takano S, and Mizuno M.** Sprint performance is related to muscle fascicle length in male 100 m sprinters. *J. Appl. Physiol.* 88(3): 811–816; 2000.

196 **Kumar V, Atherton PJ, Selby A, Rankin D, Williams J, Smith K, Hiscock N, and Rennie MJ.** Muscle protein synthetic responses to exercise: effects of age, volume, and intensity. *J. Gerontol. A Biol. Sci. Med. Sci.* 67(11): 1170–1177; 2012.

197 **Langberg H, Rosendal L, and Kjaer M.** Training-induced changes in peritendinous type I collagen turnover determined by microdialysis in humans. *J. Physiol.* 534(1): 297–302; 2001.

198 **Larsson L and Tesch PA.** Motor unit fiber density in extremely hypertrophied skeletal muscles in man. Electrophysiological signs of muscle fiber hyperplasia. *Eur. J. Appl. Physiol.* 55(2): 130–136; 1986.

199 **Laughlin MH and Roseguini B.** Mechanisms for exercise training-induced increases in skeletal muscle blood flow capacity: differences with interval sprint training versus aerobic endurance training. *J. Physiol. Pharmacol.* 59(Suppl 7): 71–88; 2008.

200 **Laurentino G, Ugrinowitsch C, Aihara AY, Fernandes AR, Parcell AC, Ricard M, and Tricoli V.** Effects of strength training and vascular occlusion. *Int. J. Sports Med.* 29(8): 664–667; 2008.

201 **Lauritzen F, Paulsen G, Raastad T, Bergersen LH, and Owe SG.** Gross ultrastructural changes and necrotic fiber segments in elbow flexor muscles after maximal voluntary eccentric action in humans. *J. Appl. Physiol.* 107(6): 1923–1934; 2009.

202 **Leal ML, Lamas L, Aoki MS, Ugrinowitsch C, Ramos MS, Tricoli V, and Moriscot AS.** Effect of different resistance-training regimens on the WNT-signaling pathway. *Eur. J. Appl. Physiol.* 111(10): 2535–2545; 2011.

203 **Lee M and Carroll TJ.** Cross education: possible mechanisms for the contralateral effects of unilateral resistance training. *Sports Med.* 37(1): 1–14; 2007.

204 **Lee M, Gandevia SC, and Carroll TJ.** Unilateral strength training increases voluntary activation of the opposite untrained limb. *Clin. Neurophysiol.* 120(4): 802–808; 2009.

205 **Linossier MT, Dormois D, Geyssant A, and Denis C.** Performance and fiber characteristics of human skeletal muscle during short sprint training and detraining on a cycle ergometer. *Eur. J. Appl. Physiol.* 75(6): 491–498; 1997.

206 **Loenneke JP, Fahs CA, Rossow LM, Abe T, and Bemben MG.** The anabolic benefits of venous blood flow restriction training may be induced by muscle cell swelling. *Med. Hypotheses* 78(1): 151–154; 2012.

207 **Loenneke JP, Fahs CA, Wilson JM, and Bemben MG.** Blood flow restriction: the metabolite/volume threshold theory. *Med. Hypotheses* 77(5): 748–752; 2011.

208 **Loenneke JP, Wilson JM, Marín PJ, Zourdos MC, and Bemben MG.** Low intensity blood flow restriction training: a meta-analysis. *Eur. J. Appl. Physiol.* 112(5): 1849–1859; 2012.

209 **Loenneke JP, Wilson JM, Pujol TJ, and Bemben MG.** Acute and chronic testosterone response to blood flow restricted exercise. *Horm. Metab. Res.* 43(10): 669–673; 2011.

210 **Loenneke JP, Wilson GJ, and Wilson JM.** A mechanistic approach to blood flow occlusion. *Int. J. Sports Med.* 31(1): 1–4; 2010.

211 **Losnegard T, Mikkelsen K, Rønnestad BR, Hallén J, Rud B, and Raastad T.** The effect of heavy strength training on muscle mass and physical performance in elite cross country skiers. *Scand. J. Med. Sci. Sports.* 21(3): 389–401; 2011.

212 **Lundberg TR, Fernandez-Gonzalo R, Gustafsson T, and Tesch PA.** Aerobic exercise does not compromise muscle hypertrophy response to short-term resistance training. *J. Appl. Physiol.* 114(1): 81–89; 2013.

213 **Luthi JM, Howald H, Claassen H, Rösler K, Vock P, and Hoppeler H.** Structural changes in skeletal muscle tissue with heavy-resistance exercise. *Int. J. Sports Med.* 7(3): 123–127; 1986.

214 **Macaluso F, Isaacs AW, and Myburgh KH.** Preferential type II muscle fiber damage from plyometric exercise. *J. Athl. Train.* 47(4): 414–420; 2012.

215 **MacDougall JD.** Hypertrophy and hyperplasia. In: *Strength and Power in Sport*, 2nd edn, edited by Komi PV. Malden, MA: Blackwell, 2003; pp. 252–264.

216 **MacDougall JD, Gibala MJ, Tarnopolsky MA, MacDonald JR, Interisano SA, and Yarasheski KE.** The time course for elevated muscle protein synthesis following heavy resistance exercise. *Can. J. Appl. Physiol.* 20(4): 480–486; 1995.

217 **MacDougall JD, McKelvie RS, Moroz DE, Sale DG, McCartney N, and Buick F.** Factors affecting blood pressure during heavy weight lifting and static contractions. *J. Appl. Physiol.* 73(4): 1590–1597; 1992.

218 **MacDougall JD, Sale DG, Alway SE, and Sutton JR.** Muscle fiber number in biceps brachii in bodybuilders and control subjects. *J. Appl. Physiol.* 57(5): 1399–1403; 1984.

219 **MacDougall JD, Sale DG, Elder GC, and Sutton JR.** Muscle ultrastructural characteristics of elite

powerlifters and bodybuilders. *Eur. J. Appl. Physiol.* 48(1): 117–126; 1982.

220 **MacDougall JD, Sale DG, Moroz JR, Elder GC, Sutton JR, and Howald H.** Mitochondrial volume density in human skeletal muscle following heavy resistance training. *Med. Sci. Sports.* 11(2): 164–166; 1979.

221 **MacDougall JD, Tuxen D, Sale DG, Moroz JR, and Sutton JR.** Arterial blood pressure response to heavy resistance exercise. *J. Appl. Physiol.* 58(3): 785–790; 1985.

222 **Mackey AL, Heinemeier KM, Koskinen SO, and Kjaer M.** Dynamic adaptation of tendon and muscle connective tissue to mechanical loading. *Connect Tissue Res.* 49(3): 165–168; 2008.

223 **Magnus CR, Barss TS, Lanovaz JL, and Farthing JP.** Effects of cross-education on the muscle after a period of unilateral limb immobilization using a shoulder sling and swathe. *J. Appl. Physiol.* 109(6): 1887–1894; 2010.

224 **Magnusson SP, Hansen M, Langberg H, Miller B, Haraldsson B, Westh EK, Koskinen S, Aagaard P, and Kjaer M.** The adaptability of tendon to loading differs in men and women. *Int. J. Exp. Pathol.* 88(4): 237–240; 2007.

225 **Malisoux L, Francaux M, Nielens H, Renard P, Lebacq J, and Theisen D.** Calcium sensitivity of human single muscle fibers following plyometric training. *Med. Sci. Sports Exerc.* 38(11): 1901–1908; 2006.

226 **Malisoux L, Francaux M, Nielens H, and Theisen D.** Stretch-shortening cycle exercises: an effective training paradigm to enhance power output of human single muscle fibers. *J. Appl. Physiol.* 100(3): 771–779; 2006.

227 **Malisoux L, Francaux M, and Theisen D.** What do single-fiber studies tell us about exercise training? *Med. Sci. Sports Exerc.* 39(7): 1051–1060; 2007.

228 **Markovic G.** Does plyometric training improve vertical jump height? A meta-analytical review. *Br. J. Sports Med.* 41(6): 349–355; 2007.

229 **Markovic G and Mikulic P.** Neuro-musculoskeletal and performance adaptations to lower-extremity plyometric training. *Sports Med.* 40(10): 859–895; 2010.

230 **Marshall PW, McEwen M, and Robbins DW.** Strength and neuromuscular adaptation following one, four, and eight sets of high intensity resistance exercise in trained males. *Eur. J. Appl. Physiol.* 111(12): 3007–3016; 2011.

231 **Massey CD, Vincent J, Maneval M, and Johnson JT.** Influence of range of motion in resistance training in women: early phase adaptations. *J. Strength Cond. Res.* 19(2): 409–411; 2005.

232 **Massey CD, Vincent J, Maneval M, Moore M, and Johnson JT.** An analysis of full range of motion vs partial range of motion training in the development of strength in untrained men. *J. Strength Cond. Res.* 18(3): 518–521; 2004.

233 **Matta T, Simao R, de Salles BF, Spineti J, and Oliveira LF.** Strength training's chronic effects on muscle architecture parameters of different arm sites. *J. Strength Cond. Res.* 25(6): 1711–1717; 2011.

234 **Maughan RJ, Harmon M, Leiper JB, Sale DG, and Delman A.** Endurance capacity of untrained males and females in isometric and dynamic muscular contractions. *Eur. J. Appl. Physiol.* 55(4): 395–400; 1986.

235 **Mauras N.** Growth hormone and sex steroids. Interactions in puberty. *Endocrinol. Metab. Clin. North Am.* 30(3): 529–544; 2001.

236 **Mauro A.** Satellite cell of skeletal muscle fibers. *J. Biophys. Biochem. Cytol.* 9: 493–495; 1961.

237 **Mayhew DL, Hornberger TA, Lincoln HC, and Bamman MM.** Eukaryotic initiation factor 2B epsilon induces cap-dependent translation and skeletal muscle hypertrophy. *J. Physiol.* 589(12): 3023–3037; 2011.

238 **McCall GE, Byrnes WC, Dickinson A, Pattany PM, and Fleck SJ.** Muscle fiber hypertrophy, hyperplasia, and capillary density in college men after resistance training. *J. Appl. Physiol.* 81(5): 2004–2012; 1996.

239 **McCall GE, Byrnes WC, Fleck SJ, Dickinson A, and Kraemer WJ.** Acute and chronic hormonal responses to resistance training designed to promote muscle hypertrophy. *Can. J. Appl. Physiol.* 24(1): 96–107; 1999.

240 **McCarthy JJ, Mula J, Miyazaki M, Erfani R, Garrison K, Farooqui AB, Srikuea R, Lawson BA, Grimes B, Keller C *et al.*** Effective fiber hypertrophy in satellite cell-depleted skeletal muscle. *Development* 138(17): 3657–3666; 2011.

241 **McGawley K and Andersson PI.** The order of concurrent training does not affect soccer-related performance adaptations. *Int. J. Sports Med.* 34(11): 983–990; 2013.

242 **McHugh MP.** Recent advances in the understanding of the repeated bout effect: the protective effect against muscle damage from a single bout of eccentric exercise. *Scand. J. Med. Sci. Sports*. 13(2): 88–97; 2003.

243 **McLester JR, Bishop PA, Smith J, Wyers L, Dale B, Kozusko J, Richardson M, Nevett ME, and Lomax R.** A series of studies—a practical protocol for testing muscular endurance recovery. *J. Strength Cond. Res*. 17(2): 259–273; 2003.

244 **Mikkola JS, Rusko HK, Nummela AT, Paavolainen LM, and Hakkinen K.** Concurrent endurance and explosive type strength training increases activation and fast force production of leg extensor muscles in endurance athletes. *J Strength Cond. Res*. 21(2): 613–620; 2007.

245 **Mikkola J, Rusko H, Izquierdo M, Gorostiaga EM, and Hakkinen K.** Neuromuscular and cardiovascular adaptations during concurrent strength and endurance training in untrained men. *Int. J. Sports Med*. 33(9): 702–710; 2012.

246 **Miller BF, Olesen JL, Hansen M, Døssing S, Crameri RM, Welling RJ, Langberg H, Flyvbjerg A, Kjaer M, Babraj JA, Smith K, and Rennie MJ.** Coordinated collagen and muscle protein synthesis in human patella tendon and quadriceps muscle after exercise. *J. Physiol*. 567(3): 1021–1033; 2005.

247 **Miranda F, Simao R, Rhea M, Bunker D, Prestes J, Leite RD, Miranda H, de Salles BF, and Novaes J.** Effects of linear vs daily undulatory periodized resistance training on maximal and submaximal strength gains. *J. Strength Cond. Res*. 25(7): 1824–1830; 2011.

248 **Mitchell CJ, Churchward-Venne TA, West DW, Burd NA, Breen L, Baker SK, and Phillips SM.** Resistance exercise load does not determine training-mediated hypertrophic gains in young men. *J. Appl. Physiol*. 113(1): 71–77; 2012.

249 **Mohamad NI, Cronin JB, and Nosaka KK.** Difference in kinematics and kinetics between high- and low-velocity resistance loading equated by volume: implications for hypertrophy training. *J. Strength Cond. Res*. 26(1): 269–275; 2012.

250 **Monteiro AG, Aoki MS, Evangelista AL, Alveno DA, Monteiro GA, Piçarro Ida C, and Ugrinowitsch C.** Nonlinear periodization maximizes strength gains in split resistance training routines. *J. Strength Cond. Res*. 23(4): 1321–1326; 2009.

251 **Moore DR, Young M, and Phillips SM.** Similar increases in muscle size and strength in young men after training with maximal shortening or lengthening contractions when matched for total work. *Eur. J. Appl. Physiol*. 112(4): 1587–1592; 2012.

252 **Morrissey MC, Harman EA, and Johnson MJ.** Resistance training modes: specificity and effectiveness. *Med. Sci. Sports Exerc*. 27(5): 648–660; 1995.

253 **Mujika I.** Intense training: the key to optimal performance before and during the taper. *Scand. J. Med. Sci. Sports* 20(Suppl 2): 24–31; 2010.

254 **Mujika I and Padilla S.** Detraining: loss of training-induced physiological and performance adaptations. Part II: Long term insufficient training stimulus. *Sports Med*. 30(3): 145–154; 2000.

255 **Mujika I and Padilla S.** Muscular characteristics of detraining in humans. *Med. Sci. Sports Exerc*. 33(8): 1297–1303; 2001.

256 **Mujika I and Padilla S.** Scientific bases for precompetition tapering strategies. *Med. Sci. Sports Exerc*. 35(7): 1182–1187; 2003.

257 **Mujika I, Padilla S, Pyne D, and Busso T.** Physiological changes associated with the pre-event taper in athletes. *Sports Med*. 34(13): 891–927; 2004.

258 **Mujika I, Santisteban J, and Castagna C.** In-season effect of short-term sprint and power training programs on elite junior soccer players. *J. Strength Cond. Res*. 23(9): 2581–2587; 2009.

259 **Munn J, Herbert RD, and Gandevia SC.** Contralateral effects of unilateral resistance training: a meta-analysis. *J. Appl. Physiol*. 96(5): 1861–1866; 2004.

260 **Nader GA.** Concurrent strength and endurance training: from molecules to man. *Med. Sci. Sports Exerc*. 38(11): 1965–1970; 2006.

261 **Narici MV, Hoppeler H, Kayser B, Landoni L, Claassen H, Gavardi C, Conti M, and Cerretelli P.** Human quadriceps cross-sectional area, torque and neural activation during 6 months strength training. *Acta Physiol. Scand*. 157(2): 175–186; 1996.

262 **Narloch JA and Brandstater ME.** Influence of breathing technique on arterial blood pressure during heavy weight lifting. *Arch. Phys. Med. Rehabil*. 76(5): 457–462; 1995.

263 **Neufer PD, Costill DL, Fielding RA, Flynn MG, and Kirwan JP.** Effect of reduced training on muscular strength and endurance in competitive swimmers. *Med. Sci. Sports Exerc*. 19(5): 486–490; 1987.

264 **Newton RU, Kraemer WJ, Hakkinen K, Humphries B, and Murphy AJ.** Kinematics, kinetics, and muscle activation during explosive upper body movements. *J. Appl. Biomech.* 12(1): 31–43; 1996.

265 **Nielsen JL, Aagaard P, Bech RD, Nygaard T, Hvid LG, Wernbom M, Suetta C, and Frandsen U.** Proliferation of myogenic stem cells in human skeletal muscle in response to low-load resistance training with blood flow restriction. *J. Physiol.* 590(17): 4351–4361; 2012.

266 **Norrbrand L, Fluckey JD, Pozzo M, and Tesch PA.** Resistance training using eccentric overload induces early adaptations in skeletal muscle size. *Eur. J. Appl. Physiol.* 102(3): 271–281; 2008.

267 **Nosaka K and Newton M.** Concentric or eccentric training effect on eccentric exercise-induced muscle damage. *Med. Sci. Sports Exerc.* 34(1): 63–69; 2002.

268 **Nygren AT, Karlsson M, Norman B, and Kaijser L.** Effect of glycogen loading on skeletal muscle cross-sectional area and T2 relaxation time. *Acta Physiol. Scand.* 173(4): 385–390; 2001.

269 **Ogasawara R, Yasuda T, Sakamaki M, Ozaki H, and Abe T.** Effects of periodic and continued resistance training on muscle CSA and strength in previously untrained men. *Clin. Physiol. Funct. Imaging.* 31(5): 399–404; 2011.

270 **O'Shea JP.** *Scientific Principles and Methods of Strength Fitness.* Boston: Addison Wesley; 1969.

271 **Osteras H, Helgerud J, and Hoff J.** Maximal strength-training effects on force-velocity and force-power relationships explain increases in aerobic performance in humans. *Eur. J. Appl. Physiol.* 88(3): 255–263; 2002.

272 **Pansarasa O, Rinaldi C, Parente V, Miotti D, Capodaglio P, and Bottinelli R.** Resistance training of long duration modulates force and unloaded shortening velocity of single muscle fibers of young women. *J. Electromyogr. Kinesiol.* 19(5): e290–e300; 2009.

273 **Parra J, Cadefau JA, Rodas G, Amigo N, and Cusso R.** The distribution of rest periods affects performance and adaptations of energy metabolism induced by high-intensity training in human muscle. *Acta Physiol. Scand.* 169(2): 157–165; 2000.

274 **Pensini M, Martin A, and Maffiuletti NA.** Central versus peripheral adaptations following eccentric resistance training. *Int. J. Sports Med.* 23(8): 567–574; 2002.

275 **Peterson MD, Rhea MR, and Alvar BA.** Maximizing strength development in athletes: a meta-analysis to determine the dose-response relationship. *J. Strength Cond. Res.* 18(2): 377–382; 2004.

276 **Peterson MD, Rhea MR, and Alvar BA.** Applications of the dose-response for muscular strength development: a review of meta-analytic efficacy and reliability for designing training prescription. *J. Strength Cond. Res.* 19(4): 950–958; 2005.

277 **Petrella JK, Kim JS, Mayhew DL, Cross JM, and Bamman MM.** Potent myofiber hypertrophy during resistance training in humans is associated with satellite cell-mediated myonuclear addition: a cluster analysis. *J. Appl. Physiol.* 104(6): 1736–1742; 2008.

278 **Pette D and Staron RS.** Myosin isoforms, muscle fiber types, and transitions. *Microsc. Res. Tech.* 50(6): 500–509; 2000.

279 **Pette D and Vrbova G.** Neural control of phenotypic expression in mammalian muscle fibers. *Muscle Nerve.* 8(8): 676–689; 1985.

280 **Phillips SM.** Short-term training: when do repeated bouts of resistance exercise become training? *Can. J. Appl. Physiol.* 25(3): 185–193; 2000.

281 **Phillips SM, Tipton KD, Aarsland A, Wolf SE, and Wolfe RR.** Mixed muscle protein synthesis and breakdown after resistance exercise in humans. *Am. J. Physiol.* 273(1, Pt 1): E99–E107; 1997.

282 **Philp A, Hamilton DL, and Baar K.** Signals mediating skeletal muscle remodeling by resistance exercise: PI3-kinase independent activation of mTORC1. *J. Appl. Physiol.* 110(2): 561–568; 2011.

283 **Pope ZK, Willardson JM, and Schoenfeld BJ.** Exercise and blood flow restriction. *J. Strength Cond. Res.* 27(10): 2914–2926; 2013.

284 **Prestes J, Frollini AB, de Lima C, Donatto FF, Foschini D, de Cássia Marqueti R, Figueira A Jr and Fleck SJ.** Comparison between linear and daily undulating periodized resistance training to increase strength. *J. Strength Cond. Res.* 23(9): 2437–2442; 2009.

285 **Prestes J, de Lima C, Frollini AB, Donatto FF, and Conte M.** Comparison of linear and reverse linear periodization effects on maximal strength and body composition. *J. Strength Cond. Res.* 23(1): 266–274; 2009.

286 **Puthucheary Z, Skipworth JR, Rawal J, Loosemore M, Van Someren K, and Montgomery HE.** The ACE gene and human performance: 12 years on. *Sports Med.* 41(6): 433–448; 2011.

287 **Raastad T, Glomsheller T, Bjøro T, and Hallen J.** Changes in human skeletal muscle contractility and hormone status during 2 weeks of heavy strength training. *Eur. J. Appl. Physiol.* 84(1–2): 54–63; 2001.

288 **Raastad T, Glomsheller T, Bjøro T, and Hallen J.** Recovery of skeletal muscle contractility and hormonal responses to strength exercise after two weeks of high-volume strength training. *Scand. J. Med. Sci. Sports.* 13(3): 159–168; 2003.

289 **Raastad T and Hallen J.** Recovery of skeletal muscle contractility after high- and moderate-intensity strength exercise. *Eur. J. Appl. Physiol.* 82(3): 206–214; 2000.

290 **Rasch PJ and Burke RK.** *Kinesiology and Applied Anatomy*, 3rd edn. Philadelphia: Lea & Febiger; 1967.

291 **Rassier DE, MacIntosh BR, and Herzog W.** Length dependence of active force production in skeletal muscle. *J. Appl. Physiol.* 86(5): 1445–1457; 1999.

292 **Raue U, Trappe TA, Estrem ST, Qian HR, Helvering LM, Smith RC, and Trappe S.** Transcriptome signature of resistance exercise adaptations: mixed muscle and fiber type specific profiles in young and old adults. *J. Appl. Physiol.* 112(10): 1625–1636; 2012.

293 **Reeves ND.** Adaptation of the tendon to mechanical usage. *J. Musculoskelet. Neuronal Interact.* 6(2): 174–180; 2006.

294 **Rennie MJ, Wackerhage H, Spangenburg EE, and Booth FW.** Control of the size of the human muscle mass. *Annu. Rev. Physiol.* 66: 799–828; 2004.

295 **Rhea MR and Alderman BL.** A meta-analysis of periodized versus nonperiodized strength and power training programs. *Res. Q. Exerc. Sport.* 75(4): 413–422; 2004.

296 **Rhea MR, Alvar BA, and Burkett LN.** Single versus multiple sets for strength: a meta-analysis to address the controversy. *Res. Q. Exerc. Sport.* 73(4): 485–488; 2002.

297 **Rhea MR, Alvar BA, Burkett LN, and Ball SD.** A meta-analysis to determine the dose response for strength development. *Med. Sci. Sports Exerc.* 35(3): 456–464; 2003.

298 **Roig M, O'Brien K, Kirk G, Murray R, McKinnon P, Shadgan B, and Reid WD.** The effects of eccentric versus concentric resistance training on muscle strength and mass in healthy adults: a systematic review with meta-analysis. *Br. J. Sports Med.* 43(8): 556–568; 2009.

299 **Ronnestad BR, Hansen EA, and Raastad T.** In-season strength maintenance training increases well-trained cyclists' performance. *Eur. J. Appl. Physiol.* 110(6): 1269–1282; 2010.

300 **Ronnestad BR, Hansen EA, and Raastad T.** Effect of heavy strength training on thigh muscle cross-sectional area, performance determinants, and performance in well-trained cyclists. *Eur. J. Appl. Physiol.* 108(5): 965–975; 2010.

301 **Ronnestad BR, Hansen EA, and Raastad T.** Strength training affects tendon cross-sectional area and freely chosen cadence differently in noncyclists and well-trained cyclists. *J. Strength Cond. Res.* 26(1): 158–166; 2012.

302 **Ronnestad BR, Hansen EA, and Raastad T.** High volume of endurance training impairs adaptations to 12 weeks of strength training in well-trained endurance athletes. *Eur. J. Appl. Physiol.* 112(4): 1457–1466; 2012.

303 **Ronnestad BR, Kvamme NH, Sunde A, and Raastad T.** Short-term effects of strength and plyometric training on sprint and jump performance in professional soccer players. *J. Strength Cond. Res.* 22(3): 773–780; 2008.

304 **Ronnestad BR, Nygaard H, and Raastad T.** Physiological elevation of endogenous hormones results in superior strength training adaptation. *Eur. J. Appl. Physiol.* 111(9): 2249–2259; 2011.

305 **Ronnestad BR and Mujika I Optimizing strength training for running and cycling endurance performance: a review.** *Scand. J. Med. Sci. Sports.* 2013 Aug 5 [Epub ahead of print].

306 **Rooney KJ, Herbert RD, and Balnave RJ.** Fatigue contributes to the strength training stimulus. *Med. Sci. Sports Exerc.* 26(9): 1160–1164; 1994.

307 **Ross A and Leveritt M.** Long-term metabolic and skeletal muscle adaptations to short-sprint training: implications for sprint training and tapering. *Sports Med.* 31(15): 1063–1082; 2001.

308 **Sacks RD and Roy RR.** Architecture of the hind limb muscles of cats: functional significance. *J. Morphol.* 173(2): 185–195; 1982.

309 **Sadowski J, Mastalerz A, Gromisz W, and Niyinikowski T.** Effectiveness of the power dry-land training programs in youth swimmers. *J. Hum. Kinet.* 32: 77–86; 2012.

310 **Saez-Saez deVillarreal E, Requena B and Newton RU.** Does plyometric training improve strength performance? A meta-analysis. *J. Sci. Med. Sport.* 13(5): 513–522; 2010.

311 **Sale DG.** Influence of exercise and training on motor unit activation. *Exerc. Sport Sci. Rev.* 15: 95–151; 1987.

312 **Sale DG.** Neural adaptation to strength training. In *Strength and Power in Sport*, 2nd edn, edited by Komi PV. Malden, MA: Blackwell, 2003; pp. 281–314.

313 **Sale DG, Martin JE, and Moroz DE.** Hypertrophy without increased isometric strength after weight training. *Eur. J. Appl. Physiol.* 64(1): 51–55; 1992.

314 **Sanchis-Moysi J, Idoate F, Dorado C, Alayón S, and Calbet JA.** Large asymmetric hypertrophy of rectus abdominis muscle in professional tennis players. *PLoS One* 5(12): e15858; 2010.

315 **Sander A, Keiner M, Wirth K, and Schmidtbleicher D.** Influence of a 2-year strength training program on power performance in elite youth soccer players. *Eur. J. Sport Sci.* 13(5): 445–451; 2013.

316 **Schantz P.** Capillary supply in hypertrophied human skeletal muscle. *Acta Physiol. Scand.* 114(4): 635–637; 1982.

317 **Schiaffino S and Reggiani C.** Fiber types in mammalian skeletal muscles. *Physiol Rev.* 91(4): 1447–1531; 2011.

318 **Schoenfeld BJ.** The mechanisms of muscle hypertrophy and their application to resistance training. *J. Strength Cond. Res.* 24(10): 2857–2872; 2010.

319 **Schott J, McCully K, and Rutherford OM.** The role of metabolites in strength training. II. Short versus long isometric contractions. *Eur. J. Appl. Physiol.* 71(4): 337–341; 1995.

320 **Schuenke MD, Herman JR, Gliders RM, Hagerman FC, Hikida RS, Rana SR, Ragg KE, and Staron RS.** Early-phase muscular adaptations in response to slow-speed versus traditional resistance-training regimens. *Eur. J. Appl. Physiol.* 112(10): 3585–3595; 2012.

321 **Schuenke MD, Herman J, and Staron RS.** Preponderance of evidence proves "big" weights optimize hypertrophic and strength adaptations. *Eur. J. Appl. Physiol.* 113(1): 269–271; 2013.

322 **Schultz E.** Satellite cell behavior during skeletal muscle growth and regeneration. *Med. Sci. Sports Exerc.* 21(5 Suppl): S181–S186; 1989.

323 **Schultz E, Jaryszak DL, Gibson MC, and Albright DJ.** Absence of exogenous satellite cell contribution to regeneration of frozen skeletal muscle. *J. Muscle Res. Cell. Motil.* 7(4): 361–367; 1986.

324 **Seynnes OR, de Boer M, and Narici MV.** Early skeletal muscle hypertrophy and architectural changes in response to high-intensity resistance training. *J. Appl. Physiol.* 102(1): 368–373; 2007.

325 **Seynnes OR, Erskine RM, Maganaris CN, Longo S, Simoneau EM, Grosset JF, and Narici MV.** Training-induced changes in structural and mechanical properties of the patellar tendon are related to muscle hypertrophy but not to strength gains. *J. Appl. Physiol.* 107(2): 523–530; 2009.

326 **Seynnes OR, Kamandulis S, Kairaitis R, Helland C, Campbell EL, Brazaitis M, Skurvydas A, and Narici MV.** Effect of androgenic-anabolic steroids and heavy strength training on patellar tendon morphological and mechanical properties. *J. Appl. Physiol.* 115(1): 84–89; 2013.

327 **Shalfawi SA, Haugen T, Jakobsen TA, Enoksen E, and Tonnessen E.** The effect of combined resisted agility and repeated sprint training vs strength training on female elite soccer players. *J. Strength Cond. Res.* 27(11): 2966–2972; 2013.

328 **Shepstone TN, Tang JE, Dallaire S, Schuenke MD, Staron RS, and Phillips SM.** Short-term high- vs low-velocity isokinetic lengthening training results in greater hypertrophy of the elbow flexors in young men. *J. Appl. Physiol.* 98(5): 1768–1776; 2005.

329 **Shoepe TC, Stelzer JE, Garner DP, and Widrick JJ.** Functional adaptability of muscle fibers to long-term resistance exercise. *Med. Sci. Sports Exerc.* 35(6): 944–951; 2003.

330 **Simao R, de Salles BF, Figueiredo T, Dias I, and Willardson JM.** Exercise order in resistance training. *Sports Med.* 42(3): 251–265; 2012.

331 **Simao R, Spineti J, de Salles BF, Matta T, Fernandes L, Fleck SJ, Rhea MR, and Strom-Olsen HE.** Comparison between nonlinear and linear periodized resistance training: hypertrophic and strength effects. *J. Strength Cond. Res.* 26(5): 1389–1395; 2012.

332 **Sleivert GG, Backus RD, and Wenger HA.** The influence of a strength-sprint training sequence on multi-joint power output. *Med. Sci. Sports Exerc.* 27(12): 1655–1665; 1995.

333 **Sooneste H, Tanimoto M, Kakigi R, Saga N, and Katamoto S.** Effects of training volume on

strength and hypertrophy in young men. *J. Strength Cond. Res.* 27(1): 8–13; 2013.

334 **Souza-Junior TP, Willardson JM, Bloomer R, Leite RD, Fleck SJ, Oliveira PR, and Simao R.** Strength and hypertrophy responses to constant and decreasing rest intervals in trained men using creatine supplementation. *J. Int. Soc. Sports Nutr.* 8(1): 17; 2011.

335 **Spurrs RW, Murphy AJ, and Watsford ML.** The effect of plyometric training on distance running performance. *Eur. J. Appl. Physiol.* 89(1): 1–7; 2003.

336 **Stafilidis S and Arampatzis A.** Muscle-tendon unit mechanical and morphological properties and sprint performance. *J. Sports Sci.* 25(9): 1035–1046; 2007.

337 **Staron RS, Herman JR, and Schuenke MD.** Misclassification of hybrid fast fibers in resistance-trained human skeletal muscle using histochemical and immunohistochemical methods. *J. Strength Cond. Res.* 26(10): 2616–2622; 2012.

338 **Staron RS, Hikida RS, Murray TF, Nelson MM, Johnson P, and Hagerman F.** Assessment of skeletal muscle damage in successive biopsies from strength-trained and untrained men and women. *Eur. J. Appl. Physiol. Occup. Physiol.* 65(3): 258–264; 1992.

339 **Staron RS, Karapondo DL, Kraemer WJ, Fry AC, Gordon SE, Falkel JE, Hagerman FC, and Hikida RS.** Skeletal muscle adaptations during early phase of heavy-resistance training in men and women. *J. Appl. Physiol.* 76(3): 1247–1255; 1994.

340 **Staron RS, Leonardi MJ, Karapondo DL, Malicky ES, Falkel JE, Hagerman FC, and Hikida RS.** Strength and skeletal muscle adaptations in heavy-resistance-trained women after detraining and retraining. *J. Appl. Physiol.* 70(2): 631–640; 1991.

341 **Staron RS, Malicky ES, Leonardi MJ, Falkel JE, Hagerman FC, and Dudley GA.** Muscle hypertrophy and fast fiber type conversions in heavy resistance-trained women. *Eur. J. Appl. Physiol.* 60(1): 71–79; 1990.

342 **Stone WJ and Coulter SP.** Strength/endurance effects from three resistance training protocols with women. *J. Strength Cond. Res.* 8(4): 231–234; 1994.

343 **Takarada Y and Ishii N.** Effects of low-intensity resistance exercise with short interset rest period on muscular function in middle-aged women. *J. Strength Cond. Res.* 16(1): 123–128; 2002.

344 **Tanaka H and Swensen T.** Impact of resistance training on endurance performance. A new form of cross-training? *Sports Med.* 25(3): 191–200; 1998.

345 **Tang JE, Perco JG, Moore DR, Wilkinson SB, and Phillips SM.** Resistance training alters the response of fed state mixed muscle protein synthesis in young men. *Am. J. Physiol.* 294(1): R172–R178; 2008.

346 **Tanimoto M and Ishii N.** Effects of low-intensity resistance exercise with slow movement and tonic force generation on muscular function in young men. *J. Appl. Physiol.* 100(4): 1150–1157; 2006.

347 **Tanimoto M, Sanada K, Yamamoto K, Kawano H, Gando Y, Tabata I, Ishii N, and Miyachi M.** Effects of whole-body low-intensity resistance training with slow movement and tonic force generation on muscular size and strength in young men. *J. Strength Cond. Res.* 22(6): 1926–1938; 2008.

348 **Terzis G, Stratakos G, Manta P, and Georgiadis G.** Throwing performance after resistance training and detraining. *J. Strength Cond. Res.* 22(4): 1198–1204; 2008.

349 **Tesch PA.** Skeletal muscle adaptations consequent to long-term heavy resistance exercise. *Med. Sci. Sports Exerc.* 20(5 Suppl): S132–S134; 1988.

350 **Tesch PA and Alkner BA.** Acute and chronic muscle metabolic adaptations to strength training. In *Strength and Power in Sport*, 2nd edn, edited by Komi PV. Malden, MA: Blackwell, 2003; pp. 265–280.

351 **Tesch PA, Thorsson A, and Kaiser P.** Muscle capillary supply and fiber type characteristics in weight and powerlifters. *J. Appl. Physiol. Respir. Environ. Exerc. Physiol.* 56(1): 35–38; 1984.

352 **Tesch PA and Wright JE.** Recovery from short term intense exercise: its relation to capillary supply and blood lactate concentration. *Eur. J. Appl. Physiol. Occup. Physiol.* 52(1): 98–103; 1983.

353 **Thalacker-Mercer AE, Petrella JK and Bamman MM.** Does habitual dietary intake influence myofiber hypertrophy in response to resistance training? A cluster analysis. *Appl. Physiol. Nutr. Metab.* 34(4): 632–639; 2009.

354 **Thomas L and Busso T.** A theoretical study of taper characteristics to optimize performance. *Med. Sci. Sports Exerc.* 37(9): 1615–1621; 2005.

355 **Thomas L, Mujika I and Busso T.** A model study of optimal training reduction during pre-event taper in elite swimmers. *J. Sports Sci.* 26(6): 643–652; 2008.

356 **Thomas L, Mujika I and Busso T.** Computer simulations assessing the potential performance benefit of a final increase in training during pre-event taper. *J. Strength Cond. Res.* 23(6): 1729–1736; 2009.

357 **Thorstensson A, Karlsson J, Viitasalo JH, Luhtanen P, and Komi PV.** Effect of strength training on EMG of human skeletal muscle. *Acta Physiol. Scand.* 98(2): 232–236; 1976.

358 **Thorstensson A, Sjodin B, and Karlsson J.** Enzyme activities and muscle strength after "sprint training" in man. *Acta Physiol. Scand.* 94(3): 313–318; 1975.

359 **Toigo M and Boutellier U.** New fundamental resistance exercise determinants of molecular and cellular muscle adaptations. *Eur. J. Appl. Physiol.* 97(6): 643–663; 2006.

360 **Toji H and Kaneko M.** Effect of multiple-load training on the force-velocity relationship. *J. Strength Cond. Res.* 18(4): 792–795; 2004.

361 **Tonnessen E, Shalfawi SA, Haugen T, and Enoksen E.** The effect of 40 m repeated sprint training on maximum sprinting speed, repeated sprint speed endurance, vertical jump, and aerobic capacity in young elite male soccer players. *J. Strength Cond. Res.* 25(9): 2364–2370; 2011.

362 **Trappe S, Costill D, and Thomas R.** Effect of swim taper on whole muscle and single muscle fiber contractile properties. *Med. Sci. Sports Exerc.* 32(12): 48–56; 2000.

363 **Vaczi M, Tollar J, Meszler B, Juhasz I, and Karsai I.** Short-term high intensity plyometric training program improves strength, power and agility in male soccer players. *J. Hum. Kinet.* 36: 17–26; 2013.

364 **Van Cutsem M, Duchateau J, and Hainaut K.** Changes in single motor unit behavior contribute to the increase in contraction speed after dynamic training in humans. *J. Physiol.* 513(1): 295–305; 1998.

365 **Van Den Tillaar R.** Effect of different training programs on the velocity of overarm throwing: a brief review. *J. Strength Cond. Res.* 18(2): 388–396; 2004.

366 **Van Etten LM, Verstappen FT, and Westerterp KR.** Effect of body build on weight-training-induced adaptations in body composition and muscular strength. *Med. Sci. Sports Exerc.* 26(4): 515–521; 1994.

367 **Verdijk LB, Gleeson BG, Jonkers RA, Meijer K, Savelberg HH, Dendale P and van Loon LJ.** Skeletal muscle hypertrophy following resistance training is accompanied by a fiber type-specific increase in satellite cell content in elderly men. *J. Gerontol. A Biol. Sci. Med. Sci.* 64(3): 332–339; 2009.

368 **Wang L, Mascher H, Psilander N, Blomstrand E, and Sahlin K.** Resistance exercise enhances the molecular signaling of mitochondrial biogenesis induced by endurance exercise in human skeletal muscle. *J. Appl. Physiol.* 111(5): 1335–1344; 2011.

369 **Wang N, Hikida RS, Staron RS, and Simoneau JA.** Muscle fiber types of women after resistance training—quantitative ultrastructure and enzyme activity. *Pflugers Arch.* 424(5–6): 494–502; 1993.

370 **Wernbom M, Augustsson J, and Thomeé R.** The influence of frequency, intensity, volume and mode of strength training on whole muscle cross-sectional area in humans. *Sports Med.* 37(3): 225–264; 2007.

371 **West DW, Burd NA, Tang JE, Moore DR, Staples AW, Holwerda AM, Baker SK, and Phillips SM.** Elevations in ostensibly anabolic hormones with resistance exercise enhance neither training-induced muscle hypertrophy nor strength of the elbow flexors. *J. Appl. Physiol.* 108(1): 60–67; 2010.

372 **West DW, Cotie LM, Mitchell CJ, Churchward-Venne TA, MacDonald MJ, and Phillips SM.** Resistance exercise order does not determine postexercise delivery of testosterone, growth hormone, and IGF-1 to skeletal muscle. *Appl. Physiol. Nutr. Metab.* 38(2): 220–226; 2013.

373 **West DW, Kujbida GW, Moore DR, Atherton P, Burd NA, Padzik JP, De Lisio M, Tang JE, Parise G, Rennie MJ, Baker SK, and Phillips SM.** Resistance exercise-induced increases in putative anabolic hormones do not enhance muscle protein synthesis or intracellular signaling in young men. *J. Physiol.* 587(21): 5239–5247; 2009.

374 **West DW and Phillips SM.** Associations of exercise-induced hormone profiles and gains in strength and hypertrophy in a large cohort after weight training. *Eur. J. Appl. Physiol.* 112(7): 2693–2702; 2012.

375 **Widrick JJ, Stelzer JE, Shoepe TC, and Garner DP.** Functional properties of human muscle fibers after short-term resistance exercise training. *Am. J. Physiol.* 283(2): R408–R416; 2002.

376 **Wilkinson SB, Phillips SM, Atherton PJ, Patel R, Yarasheski KE, Tarnopolsky MA, and Rennie MJ.** Differential effects of resistance and endurance exercise in the fed state on signaling molecule phosphorylation and protein synthesis in human muscle. *J. Physiol.* 586(15): 3701–3717; 2008.

377 **Willardson JM.** A brief review: factors affecting the length of the rest interval between resistance exercise sets. *J. Strength Cond. Res.* 20(4): 978–984; 2006.

378 **Willardson JM and Burkett LN.** The effect of rest interval length on bench press performance with heavy vs light loads. *J. Strength Cond. Res.* 20(2): 396–399; 2006.

379 **Willardson JM and Burkett LN.** The effect of different rest intervals between sets on volume components and strength gains. *J. Strength Cond. Res.* 22(1): 146–152; 2008.

380 **Williamson DL, Gallagher PM, Carroll CC, Raue U, and Trappe SW.** Reduction in hybrid single muscle fiber proportions with resistance training in humans. *J. Appl. Physiol.* 91(5): 1955–1961; 2001.

381 **Wilson JM, Loenneke JP, Jo E, Wilson GJ, Zourdos MC, and Kim JS.** The effects of endurance, strength, and power training on muscle fiber type shifting. *J. Strength Cond. Res.* 26(6): 1724–1729; 2012.

382 **Wilson JM, Marin PJ, Rhea MR, Wilson SM, Loenneke JP, and Anderson JC.** Concurrent training: a meta-analysis examining interference of aerobic and resistance exercises. *J. Strength Cond. Res.* 26(8): 2293–2307; 2012.

383 **Winett RA.** Meta-analyses do not support performance of multiple sets or high volume resistance training. *JEP Online* 7(5): 10–20; 2004.

384 **Wolfe BL, LeMura LM, and Cole PJ.** Quantitative analysis of single- vs multiple-set programs in resistance training. *J. Strength Cond. Res.* 18(1): 35–47; 2004.

385 **Wong PL, Chaouachi A, Chamari K, Dellal A, and Wisloff U.** Effect of preseason concurrent muscular strength and high-intensity interval training in professional soccer players. *J. Strength Cond. Res.* 24(3): 653–660; 2010.

386 **Wong TS and Booth FW.** Skeletal muscle enlargement with weight-lifting exercise by rats. *J. Appl. Physiol.* 65(2): 950–954; 1988.

387 **Woolstenhulme MT, Conlee RK, Drummond MJ, Stites AW and Parcell AC.** Temporal response of desmin and dystrophin proteins to progressive resistance exercise in human skeletal muscle. *J. Appl. Physiol.* 100(6): 1876–1882; 2006.

388 **Wu YK, Lien YH, Lin KH, Shih TT, Wang TG, and Wang HK.** Relationships between three potentiation effects of plyometric training and performance. *Scand. J. Med. Sci. Sports* 20(1): e80–e86; 2010.

389 **Yasuda T, Loenneke JP, Thiebaud RS, and Abe T.** Effects of blood flow restricted low-intensity concentric or eccentric training on muscle size and strength. *PLoS One* 7(12): e52843; 2012.

390 **Young WB and Rath DA.** Enhancing foot velocity in football kicking: the role of strength training. *J. Strength Cond. Res.* 25(2): 561–566; 2011.

391 **Zhou S.** Chronic neural adaptations to unilateral exercise: mechanisms of cross education. *Exerc. Sport Sci. Rev.* 28(4): 177–184; 2000.

Chapter 9

1 **Bangsbo J, Gunnarsson TP, Wendell J, Nybo L, and Thomassen M.** Reduced volume and increased training intensity elevate muscle Na^{+}-K^{+} pump alpha2-subunit expression as well as short- and long-term work capacity in humans. *J. Appl. Physiol.* 107(6): 1771–1780; 2009.

2 **Bishop D and Edge J.** The effects of a 10-day taper on repeated-sprint performance in females. *J. Sci. Med. Sport* 8(2): 200–209; 2005.

3 **Bosquet L, Montpetit J, Arvisais D, and Mujika I.** Effects of tapering on performance: a meta-analysis. *Med. Sci. Sports Exerc.* 39(8): 1358–1365; 2007.

4 **Burgomaster KA, Hughes SC, Heigenhauser GJ, Bradwell SN, and Gibala MJ.** Six sessions of sprint interval training increases muscle oxidative potential and cycle endurance capacity in humans. *J. Appl. Physiol.* 98(6): 1985–1990; 2005.

5 **Coffey VG and Hawley JA.** The molecular bases of training adaptation. *Sports Med.* 37(9): 737–763; 2007.

6 **Dudley GA and Djamil R.** Incompatibility of endurance- and strength-training modes of exercise. *J. Appl. Physiol.* 59(5): 1446–1451; 1985.

7 **Gibala MJ, Interisano SA, Tarnopolsky MA, Roy BD, MacDonald JR, Yarasheski KE, and MacDougall JD.** Myofibrillar disruption following acute concentric and eccentric resistance exercise in strength-trained men. *Can. J. Physiol. Pharmacol.* 78(8): 656–661; 2000.

8 **Gibala MJ, MacDougall JD, and Sale DG.** The effects of tapering on strength performance in trained athletes. *Int. J. Sports Med.* 15(8): 492–497; 1994.

9 **Gleeson M.** Immune function in sport and exercise. *J. Appl. Physiol.* 103(2): 693–699; 2007.

10 **Green HJ, Helyar R, Ball-Burnett M, Kowalchuk N, Symon S, and Farrance B.** Metabolic adaptations to training precede changes in muscle mitochondrial capacity. *J. Appl. Physiol.* 72(2): 484–491; 1992.

11 **Hawley JA.** Molecular responses to strength and endurance training: are they incompatible? *Appl. Physiol. Nutr. Metab.* 34(3): 355–361; 2009.

12 **Hickson RC.** Interference of strength development by simultaneously training for strength and endurance. *Eur. J. Appl. Physiol. Occup. Physiol.* 45(2–3): 255–263; 1980.

13 **Izquierdo M, Ibañez J, Gonzalez-Badillo JJ, Ratamess NA, Kraemer WJ, Hakkinen K, Bonnabau H, Granados C, French DN, and Gorostiaga EM.** Detraining and tapering effects on hormonal responses and strength performance. *J. Strength Cond Res.* 21(3): 768–775; 2007.

14 **Kajiura JS, MacDougall JD, Ernst PB, and Younglai EV.** Immune response to changes in training intensity and volume in runners. *Med. Sci. Sports Exerc.* 27(8): 1111–1117; 1995.

15 **Kakanis MW, Peake J, Brenu EW, Simmonds M, Gray B, Hooper SL, and Marshall-Gradisnik SM.** The open window of susceptibility to infection after acute exercise in healthy young male elite athletes. *Exerc. Immunol. Rev.* 16: 119–137; 2010.

16 **Keast D, Cameron K, and Morton AR.** Exercise and the immune response. *Sports Med.* 5(4): 248–267; 1988.

17 **Kendall A, Hoffman-Goetz L, Houston M, MacNeil B, and Arumugam Y.** Exercise and blood lymphocyte subset responses: intensity, duration, and subject fitness effects. *J. Appl. Physiol.* 69(1): 251–260; 1990.

18 **Leveritt M, Abernethy PJ, Barry BK, and Logan PA.** Concurrent strength and endurance training. A review. *Sports Med.* 28(6): 413–427; 1999.

19 **Luden N, Hayes E, Galpin A, Minchev K, Jemiolo B, Raue U, Trappe TA, Harber MP, Bowers T, and Trappe S.** Myocellular basis for tapering in competitive distance runners. *J. Appl. Physiol.* 108(6): 1501–1509; 2010.

20 **MacDougall JD, Gibala MJ, Tarnopolsky MA, MacDonald JR, Interisano SA, and Yarasheski KE.** The time course for elevated muscle protein synthesis following heavy resistance exercise. *Can. J. Appl. Physiol.* 20(4): 480–486; 1995.

21 **MacKinnon LT.** Special feature for the Olympics: effects of exercise on the immune system: over-training effects on immunity and performance in athletes. *Immunol. Cell Biol.* 78(5): 502–509; 2000.

22 **Moreira A, Delgado L, Moreira P, and Haahtela T.** Does exercise increase the risk of upper respiratory tract infections? *Br. Med. Bull.* 90: 111–131; 2009.

23 **Moyna NM, Acker GR, Weber KM, Fulton JR, Goss FL, Robertson RJ, and Rabin BS.** The effects of incremental submaximal exercise on circulating leukocytes in physically active and sedentary males and females. *Eur. J. Appl. Physiol. Occup. Physiol.* 74(3): 211–218; 1996.

24 **Mujika I.** Intense training: the key to optimal performance before and during the taper. *Scand. J. Med. Sci. Sports* 20(Suppl 2): 24–31; 2010.

25 **Mujika I and Padilla S.** Scientific bases for precompetition tapering strategies. *Med. Sci. Sports Exerc.* 35(7): 1182–1187; 2003.

26 **Nader GA.** Concurrent strength and endurance training: from molecules to man. *Med. Sci. Sports Exerc.* 38(11): 1965–1970; 2006.

27 **Nattiv A, Loucks AB, Manore MM, Sanborn CF, Sundgot-Borgen J, Warren MP, and American College of Sports Medicine.** American College of Sports Medicine position stand. The female athlete triad. *Med. Sci. Sports Exerc.* 39(10): 1867–1882; 2007.

28 **Neary JP, Bhambhani YN, and McKenzie DC.** Effects of different stepwise reduction taper protocols on cycling performance. *Can. J. Appl. Physiol.* 28(4): 576–587; 2003.

29 **Neary JP, Martin TP, and Quinney HA.** Effects of taper on endurance cycling capacity and single muscle fiber properties. *Med. Sci. Sports Exerc.* 35(11): 1875–1881; 2003.

30 **Nieman DC.** Current perspective on exercise immunology. *Curr. Sports Med. Rep.* 2(5): 239–242; 2003.

31 **Nieman DC.** Exercise, upper respiratory tract infection, and the immune system. *Med. Sci. Sports Exerc.* 26(2): 128–139; 1994.

32 **Sale DG, MacDougall JD, Jacobs I, and Garner S.** Interaction between concurrent strength and endurance training. *J. Appl. Physiol.* 68(1): 260–270; 1990.

33 **Shepley B, MacDougall JD, Cipriano N, Sutton JR, Tarnopolsky MA, and Coates G.** Physiological effects of tapering in highly trained athletes. *J. Appl. Physiol.* 72(2): 706–711; 1992.

34 **Sherman WM, Costill DL, Fink WJ, and Miller JM.** Effect of exercise-diet manipulation on muscle glycogen and its subsequent utilization during performance. *Int. J. Sports Med.* 2(2): 114–118; 1981.

35 **Trappe S, Costill D, and Thomas R.** Effect of swim taper on whole muscle and single muscle fiber contractile properties. *Med. Sci. Sports Exerc.* 33(1): 48–56; 2001.

36 **van Wessel T, de Haan A, van der Laarse WJ, and Jaspers RT.** The muscle fiber type-fiber size paradox: hypertrophy or oxidative metabolism? *Eur. J. Appl. Physiol.* 110(4): 665–694; 2010.

37 **Wang L, Mascher H, Psilander N, Blomstrand E, and Sahlin K.** Resistance exercise enhances the molecular signaling of mitochondrial biogenesis induced by endurance exercise in human skeletal muscle. *J. Appl. Physiol.* 111(5): 1335–1344; 2011.

Chapter 10

1 **Abellaneda S, Guissard N and Duchateau J.** The relative lengthening of the myotendinous structures in the medial gastrocnemius during passive stretching differs among individuals. *J. Appl. Physiol.* 106: 169–177; 2009.

2 **Alonso J, McHugh MP, Mullaney MJ, and Tyler TF.** Effect of hamstring flexibility on isometric knee flexion angle-torque relationship. *Scand. J. Med. Sci. Sports* 19: 252–256; 2009.

3 **Alter MJ.** *Science of Flexibility*, 3rd edn. Champaign, IL: Human Kinetics; 2004.

4 **American College of Sports Medicine (ACSM) Position Stand.** Quantity and quality of exercise for developing and maintaining cardiorespiratory, musculoskeletal, and neuromotor fitness in apparently healthy adults: guidance for prescribing exercise. *Med. Sci. Sports Exerc.* 43(7): 1334–1359; 2011.

5 **Antonio J and Gonyea WJ.** Role of muscle fiber hypertrophy and hyperplasia in intermittently stretched avian muscle. *J. Appl. Physiol.* 74(4): 1893–1898; 1993.

6 **Arampatiz A, De Monte G, Karamanidis K, Morey-Klapsing G, Stafilidis S, and Bruggeman G-P.** Influence of the muscle-tendon unit's mechanical and morphological properties on running economy. *J. Exp. Biol.* 209: 3345–3357; 2006.

7 **Bandy WD, Irion JM, and Briggler M.** The effect of static and dynamic range of motion training on the flexibility of the hamstring muscles. *J. Orthop. Sports Phys. Ther.* 27(4): 295–300; 1998.

8 **Beaudoin CM and Blum JW.** Flexibility and running economy in female collegiate track athletes. *J. Sports Med. Phys. Fitness* 45(3): 295–300; 2005.

9 **Beedle BB, Leydig SN, and Carnucci JM.** No difference in pre- and postexercise stretching on flexibility. *J. Strength Cond. Res.* 21(3): 780–783; 2007.

10 **Behm DG and Chaouachi A.** A review of the acute effects of static and dynamic stretching on performance. *Eur. J. Appl. Physiol.* 111(11): 2633–2651; 2011.

11 **Behm DG, Bambury A, Cahill F, and Power K.** Effect of acute static stretching on force, balance, reaction time, and movement time. *Med. Sci. Sports Exerc.* 36(8): 1397–1402; 2004.

12 **Ben M and Harvey LA.** Regular stretch does not increase muscle extensibility: a randomized control trial. *Scand. J. Med. Sci. Sports* 20: 136–144; 2010.

13 **Bjorklund M, Hamberg J, and Crenshaw AG.** Sensory adaptation after a 2-week stretching regimen of the rectus femoris muscle. *Arch. Phys. Med. Rehabil.* 82: 1245–1250; 2001.

14 **Bonnar BP, Deivert RG, and Gould TE.** The relationship between isometric contraction durations during hold-relax stretching and improvement of hamstring flexibility. *J. Sports Med. Phys. Fitness* 44: 258–261; 2004.

15 **Chalmers G.** Re-examination of the possible role of Golgi tendon organ and muscle spindle reflexes in proprioceptive neuromuscular facilitation muscle stretching. *Sports Biomech.* 3(1): 159–183; 2004.

16 **Chandler TJ, Kibler WB, Uhl TL, Wooten B, Kiser A, and Stone E.** Flexibility comparisons of junior elite tennis players to other athletes. *Am. J. Sports Med.* 18(2): 134–136; 1990.

17 **Costa PB, Graves BS, Whitehurst M, and Jacobs PL.** The acute effects of different durations of static stretching on dynamic balance performance. *J. Strength Cond. Res.* 23(1): 141–147; 2009.

18 **Costa PB, Ryan ED, Herda TJ, Walter AA, Hoge KM, and Cramer JT.** Acute effects of passive stretching on the electromechanical delay and evoked twitch properties. *Eur. J. Appl. Physiol.* 108: 301–310; 2010.

19 **Coutinho EL, Gomes ARS, Franca CN, Oishi J, and Salvini TF.** Effect of passive stretching on the immobilized soleus muscle fiber morphology. *Braz. J. Med. Biol. Res.* 37: 1853–1861; 2004.

20 **Craib MW, Mitchell VA, Fields KB, Cooper TR, Hopewell R, and Morgan DW.** The association between flexibility and running economy in sub-elite male distance runners. *Med. Sci. Sports Exerc.* 28(6): 737–743; 1996.

21 **Daniels J and Daniels N.** Running economy of elite male and elite female runners. *Med. Sci. Sports Exerc.* 24(4): 483–489; 1992.

22 **de Weijer VC, Gorniak GC, and Shamus E.** The effect of static stretching and warm-up exercise on hamstring length over the course of 24 hours. *J. Orthop. Sports Phys. Ther.* 33(12): 727–733; 2003.

23 **Decoster LC, Cleland J, Altieri C, and Russell P.** The effects of hamstring stretching on range of motion: A systematic literature review. *J. Orthop. Sports Phys. Ther.* 35(6): 377–387; 2006.

24 **Fletcher IM and Jones B.** The effect of different warm-up stretch protocols on 20 meter sprint performance in trained rugby union players. *J. Strength Cond. Res.* 18(4): 885–888; 2004.

25 **Ford P and McChesney J.** Duration of maintained hamstring ROM following termination of three stretching protocols. *J. Sports Rehabil.* 16(1): 18–27; 2007.

26 **Fowles JR, Sale DG, and MacDougall JD.** Reduced strength after passive stretch of the human plantarflexors. *J. Appl. Physiol.* 89: 1179–1188; 2000.

27 **Fowles JR, MacDougall JD, Tarnopolsky MA, Sale DG, Roy BD, and Yarasheski KE.** The effects of acute passive stretch on muscle protein synthesis in humans. *Can. J. Appl. Physiol.* 25(3): 165–180; 2000.

28 **Fradkin AJ, Gabbe BJ, and Cameron PA.** Does warming up prevent injury in sport? The evidence from randomized controlled trials. *J. Sci. Med. Sport* 9: 214–220; 2006.

29 **Franco BL, Signorelli GR, Trajano GS, and De Oliveira CG.** Acute effects of different stretching exercises on muscular endurance. *J. Strength Cond. Res.* 22(6): 1832–1837; 2008.

30 **Gajdosik RL and Bohannon RW.** Clinical measurement of range of motion. Review of goniometry emphasizing reliability and validity. *Phys. Ther.* 67(12): 1867–1872; 1987.

31 **Gajdosik RL, Allred JD, Gabbert HL, and Sonsteng BA.** A stretching program increases the dynamic passive length and passive resistive properties of the calf muscle-tendon unit of unconditioned younger women. *Eur. J. Appl. Physiol.* 99: 449–454; 2007.

32 **Gajdosik RL, Giuliani CA, and Bohannon RW.** Passive compliance and length of the hamstring muscles in healthy men and women. *Clin. Biomech.* 5: 23–29; 1990.

33 **Gajdosik RL.** Passive extensibility of skeletal muscle: review of the literature with clinical implications. *Clin. Biomech.* 16: 87–101; 2001.

34 **Garrett WE.** Muscle strain injuries. *Am. J. Sports Med.* 24(6 Suppl): S2–S8; 1996.

35 **Goldspink DF, Cox VM, Smith SK, Eaves LA, Osbaldeston NJ, Lee DM, and Mantle D.** Muscle growth in response to mechanical stimuli. *Am. J. Physiol.* 268(31): E288–E297; 1995.

36 **Granata KP, Wilson SE, and Padua DA.** Gender differences in active musculoskeletal stiffness. Part I. Quantification in controlled measurements of knee joint dynamics. *J. Electromyographr. Kinesiol.* 12: 119–126; 2002.

37 **Guissard N and Duchateau J.** Effect of static stretch training on neural and mechanical properties of the human plantar-flexor muscles. *Muscle Nerve* 29: 248–255; 2004.

38 **Guissard N and Duchateau J.** Neural aspects of stretching. *Exerc. Sport Sci. Rev.* 34(4): 154–158; 2006.

39 **Halbertsma JPK, Mulder I, Goeken LNH, and Eisma WH.** Repeated passive stretching: acute effect on the passive muscle moment and extensibility of short hamstrings. *Arch. Phys. Med. Rehabil.* 80: 407–414; 1999.

40 **Halbertsma JPK, van Bolhuis AI, and Goeken LNH.** Sport stretching: effect on passive muscle stiffness of short hamstrings. *Arch. Phys. Med. Rehabil.* 77: 688–692; 1996.

41 **Hayes PR and Walker A.** Pre-exercise stretching does not impact upon running economy. *J. Strength Cond. Res.* 21(4): 1227–1232; 2007.

42 **Hedrick A.** Dynamic flexibility training. *Strength Cond. J.* 22(5): 33–38; 2000.

43 **Herbert RD and Crosbie J.** Rest length and compliance of non-immobilized and immobilized soleus muscle and tendon. *Eur. J. Appl. Physiol.* 76: 472–479; 1997.

44 **Herbert RD and Gabriel M.** Effects of stretching before and after exercising on muscle soreness and risk of injury: systematic review. *Br. Med. J.* 325: 468–472; 2002.

45 **Herbert RD, Clarke J, Kwah LK, Diong J, Martin J, Clarke EC, Bilston LE, and Gandevia SC.** *In vivo* passive mechanical behavior of muscle fascicles and tendons in human gastrocnemius muscle-tendon units. *J. Physiol.* 589(21): 5257–5267; 2011.

46 **Herda TJ, Cramer JT, Ryan ED, McHugh MP, and Stout JR.** Acute effects of static versus dynamic stretching on isometric peak torque, electromyography, and mechanomyography of the biceps femoris muscle. *J. Strength Cond. Res.* 22(3): 809–817; 2008.

47 **Hunter GR, Katsoulis K, McCarthy JP, Ogard WK, Bamman MM, Wood DS, Den Hollander JA, Blaudeau TE, and Newcomer BR.** Tendon length and joint flexibility are related to running economy. *Med. Sci. Sports Exerc.* 43(8): 1492–1499; 2011.

48 **Hutton RS.** Neuromuscular basis of stretching exercise. In: *Strength and Power in Sport*, edited by Komi PV. Malden, MA: Blackwell, 1992; pp. 29–38.

49 **Jones AM.** Running economy is negatively related to sit-and-reach test performance in international standard distance runners. *Int. J. Sports Med.* 23(1): 40–43; 2002.

50 **Judge LW, Bellar D, Craig B, Petersen J, Camerota J, Wanless E, and Bodey K.** An examination of preactivity and postactivity flexibility practices of national collegiate athletic association division I tennis coaches. *J. Strength Cond. Res.* 26(1): 184–191; 2012.

51 **Judge LW, Craig B, Baudendistal S, and Bodey KJ.** An examination of the stretching practices of division I and division III college football programs in the midwestern United States. *J. Strength Cond. Res.* 23(4): 1091–1096; 2009.

52 **Kay AD and Blazevich AJ.** Effect of acute static stretch on maximal muscle performance: a systematic review. *Med. Sci. Sports Exerc.* 44(1): 154–164; 2012.

53 **Kirkendall DT and Garrett WE.** Clinical perspectives regarding eccentric muscle injury. *Clin. Orthopaed. Rel. Res.* 403S: S81–S89; 2002.

54 **Kokkonen J, Nelson AG, Eldredge C, and Winchester JB.** Chronic static stretching improves exercise performance. *Med. Sci. Sports Exerc.* 39(10): 1825–1831; 2007.

55 **Kokkonen J, Nelson AG, Tarawhiti T, Buckingham P, and Winchester JB.** Early-phase resistance training strength gains in novice lifters are enhanced by doing static stretching. *J. Strength Cond. Res.* 24(2): 502–506; 2010.

56 **Kubo K, Kanehisa H, and Fukunaga T.** Effect of stretching training on the viscoelastic properties of human tendon structures *in vivo*. *J. Appl. Physiol.* 92: 595–601; 2002.

57 **Kubo K, Kanehisa H, and Fukunaga T.** Effects of different duration isometric contractions on tendon elasticity in human quadriceps muscles. *J. Physiol.* 536(2): 649–655; 2001.

58 **Kubo K, Kanehisa H, and Fukunaga T.** Effects of transient muscle contractions and stretching on the tendon structures *in vivo*. *Acta Physiol. Scand.* 175(2): 157–64; 2002.

59 **Kubo K, Kanehisa H, and Fukunaga T.** Effects of resistance and stretching training programs on the viscoelastic properties of human tendon structures *in vivo*. *J. Physiol.* 538(1): 219–226; 2002.

60 **LaRoche DP and Connolly DAJ.** Effects of stretching on passive muscle tension and response to eccentric exercise. *Am. J. Sports Med.* 34(6): 1000–1007; 2006.

61 **Magnusson SP, Aagaard P, Simonsen EB, and Bojsen-Moller F.** Passive tensile stress and energy of the human hamstring muscles *in vivo*. *Scand. J. Med. Sci. Sports* 10: 351–359; 2000.

62 **Magnusson SP, Gleim GW, and Nicholas JA.** Shoulder weakness in professional baseball pitchers. *Med. Sci. Sports Exerc.* 26(1): 5–9; 1994.

63 **Magnusson SP, Simonsen E, and Bojsen-Moller F.** A biomechanical evaluation of cyclic and static stretch in human skeletal muscle. *Int. J. Sports Med.* 19(5): 310–316; 1998.

64 **Magnusson SP, Simonsen EB, Aagaard P, and Kjaer M.** Biomechanical responses to repeated stretches in human hamstring muscle *in vivo*. *Am. J. Sports Med.* 24(5): 622–628; 1996.

65 **Magnusson SP, Simonsen EB, Aagaard P, Boesen J, Johannsen F, and Kjaer M.** Determinants of musculoskeletal flexibility: viscoelastic properties, cross-sectional area, EMG and stretch tolerance. *Scand. J. Med. Sci. Sports* 7: 195–202; 1997.

66 **Magnusson SP, Simonsen EB, Aagaard P, Dyhre-Poulsen P, McHugh MP, and Kjaer M.** Mechanical and physiological responses to stretching with and without preisometric contraction in human skeletal muscle. *Arch. Phys. Med. Rehabil.* 77: 373–378; 1996.

67 **Magnusson SP, Simonsen EB, Aagaard P, Gleim GW, McHugh MP, and Kjaer M.** Viscoelastic response to repeated static stretching in the human hamstring muscle. *Scand. J. Med. Sci. Sports* 5: 342–347; 1995.

68 **Magnusson SP, Simonsen EB, Aagaard P, Sorensen H, and Kjaer M.** A mechanism for altered

flexibility in human skeletal muscle. *J. Physiol.* 497(1): 291–298; 1996.

69 **Magnusson SP.** Passive properties of human skeletal muscle during stretch maneuvers. *Scand. J. Med. Sci. Sports* 8: 65–77; 1998.

70 **Mahieu NN, Cools A, De Wilde B, Boon M, and Witvrouw E.** Effect of proprioceptive neuromuscular facilitation stretching on the plantar flexor muscle-tendon tissue properties. *Scand. J. Med. Sci. Sports* 19: 553–560; 2009.

71 **Mahieu NN, McNair P, De Muynck M, Stevens V, Blanckaert I, Smits N, and Witvrouw E.** Effect of static and ballistic stretching on the muscle-tendon tissue properties. *Med. Sci. Sports Exerc.* 39(3): 494–501; 2007.

72 **Marshall PMW, Cashman A, and Cheema BS.** A randomized controlled trial for the effect of passive stretching on measures of hamstring extensibility, passive stiffness, strength, and stretch tolerance. *J. Sci. Med. Sport* 14: 535–540; 2011.

73 **Marta CC, Marinho DA, Barbosa TM, Izquierdo M, and Marques MC.** Physical fitness differences between prepubescent boys and girls. *J. Strength Cond. Res.* 26(7): 1756–1766; 2012.

74 **McHugh MP and Cosgrove CH.** To stretch or not to stretch: the role of stretching in injury prevention and performance. *Scand. J. Med. Sci. Sports* 20: 169–181; 2010.

75 **McHugh MP, Kremenic IJ, Fox MB, and Gleim GW.** The role of mechanical and neural restraints to joint range of motion during passive stretch. *Med. Sci. Sports Exerc.* 30(6): 928–932; 1998.

76 **McHugh MP, Magnusson SP, Gleim GW, and Nicholas JA.** Viscoelastic stress relaxation in human skeletal muscle. *Med. Sci. Sports Exerc.* 24(12): 1375–1382; 1992.

77 **McMillan DJ, Moore JH, Hatler BS, and Taylor DC.** Dynamic vs static-stretching warm-up: the effect on power and agility performance. *J. Strength Cond. Res.* 20(3): 492–499; 2006.

78 **McNeil JR and Sands WA.** Stretching for performance enhancement. *Curr. Sports Med. Rep.* 5: 141–146; 2006.

79 **Mojock CD, Kim J-S, Eccles DW, and Panton LB.** The effects of static stretching on running economy and endurance performance in female distance runners during treadmill running. *J. Strength Cond. Res.* 25(8): 2170–2176; 2011.

80 **Monteiro WD, Simao R, Polito MD, Santana CA, Chaves RB, Bezerra E, and Fleck SJ.** Influence of strength training on adult women's flexibility. *J. Strength Cond. Res.* 22(3): 672–677; 2008.

81 **Morse CI, Degens H, Seynnes OR, Magnaris CN, and Jones DA.** The acute effect of stretching on the passive stiffness of the human gastrocnemius muscle tendon unit. *J. Physiol.* 586(1): 97–106; 2008.

82 **Nelson AG and Kokkonen J.** Acute ballistic muscle stretching inhibits maximal strength performance. *Res. Q. Exerc. Sport* 72(4): 415–419; 2001.

83 **Nelson AG, Allen JD, Cornwell A, and Kokkonen J.** Inhibition of maximal voluntary torque production by acute stretching is joint-angle specific. *Res. Q. Exerc. Sport* 72(1): 68–70; 2001.

84 **Nelson AG, Kokkonen J, and Arnall DA.** Acute muscle stretching inhibits muscle strength endurance performance. *J. Strength Cond. Res.* 19(2): 338–343; 2005.

85 **Nicol C and Komi PV.** Significance of passively induced stretch reflexes on achilles tendon force enhancement. *Muscle Nerve* 21: 1546–1548; 1998.

86 **Nobrega ACL, Paula KC, and Carvalho ACG.** Interaction between resistance training and flexibility training in healthy young adults. *J. Strength Cond. Res.* 19(4): 842–846; 2005.

87 **Noonan TJ, Best TM, Seaber AV, and Garrett WE.** Thermal effects on skeletal muscle tensile behavior. *Am. J. Sports Med.* 21(4): 517–522; 1993.

88 **Nordez A, McNair PJ, Casari P, and Cornu C.** Static and cyclic stretching: their different effects on the passive torque-angle curve. *J. Sci. Med. Sport* 13: 156–160; 2010.

89 **Opar DA, Williams MD, and Shield AJ.** Hamstring strain injuries. Factors that lead to injury and re-injury. *Sports Med.* 42(3): 209–226; 2012.

90 **O'Sullivan K, Murray E, and Sainsbury D.** The effect of warm-up, static stretching and dynamic stretching on hamstring flexibility in previously injured subjects. *BMC Musculo. Dis.* 10: 37–46; 2009.

91 **Page P.** Current concepts in muscle stretching for exercise and rehabilitation. *Int. J. Sports Phys. Ther.* 7(1): 109–119; 2012.

92 **Reid DA and McNair PJ.** Passive force, angle, and stiffness changes after stretching of hamstring muscles. *Med. Sci. Sports Exerc.* 36(11): 1944–1948; 2004.

93 **Ryan ED, Beck TW, Herda TJ, Hull HR, Hartman MJ, Costa PB, Defreitas JM, Stout JR, and Cramer JT.** Time course of musculotendinous stiffness responses following different durations of passive stretching. *J. Orthopaed. Sports Phys. Ther.* 38(10): 632–639; 2008.

94 **Safran MR, Garrett WE, Seaber AV, Glisson RR, and Ribbeck BM.** The role of warm-up in muscular injury prevention. *Am. J. Sports Med.* 16(2): 123–129; 1988.

95 **Sainz de Baranda P, and Ayala F.** Chronic flexibility improvement after 12 week of stretching program utilizing the ACSM recommendations: hamstring flexibility. *Int. J. Sports Med.* 31(6): 389–396; 2010.

96 **Saunders PU, Pyne DB, Telford RD, and Hawley JA.** Factors affecting running economy in trained distance runners. *Sports Med.* 34(7): 465–485; 2004.

97 **Schache AG, Kim H-J, Morgan DL, and Pandy MG.** Hamstring muscle forces prior to and immediately following an acute sprinting-related muscle strain injury. *Gait Posture* 32: 136–140; 2010.

98 **Sharman MJ, Cresswell AG, and Riek S.** Proprioceptive neuromuscular facilitation stretching. Mechanisms and clinical implications. *Sports Med.* 36(11): 929–939; 2006.

99 **Shehab R, Mirabelli M, Gorenflo D, and Fetters MD.** Pre-exercise stretching and sports related injuries: Knowledge, attitudes and practices. *Clin. J. Sport Med.* 16(3): 228–231; 2006.

100 **Shrier I.** Stretching before exercise does not reduce the risk of local muscle injury: A critical review of the clinical and basic science literature. *Clin. J. Sport Med.* 9: 221–227; 1999.

101 **Shrier I.** Stretching perspectives. *Curr. Sports Med. Rep.* 4: 237–238; 2005.

102 **Simao R, Lemos A, Salles B, Leite T, Oliveira E, Rhea M, and Reis VM.** The influence of strength, flexibility, and simultaneous training on flexibility and strength gains. *J. Strength Cond. Res.* 25(5): 1333–1338; 2011.

103 **Small K, McNaughton L, and Matthews M.** A systematic review into the efficacy of static stretching as part of a warm-up for the prevention of exercise-related injury. *Res. Sports Med.* 16: 213–231; 2008.

104 **Spurrs RW, Murphy AJ, and Watsford ML.** The effect of plyometric training on distance running performance. *Eur. J. Appl. Physiol.* 89: 1–7; 2003.

105 **Stojanovic MD and Ostojic SM.** Stretching and injury prevention in football: Current perspectives. *Res. Sports Med.* 19: 73–91; 2011.

106 **Taylor CR.** Force development during sustained locomotion: a determinant of gait, speed and metabolic power. *J. Exp. Biol.* 115: 253–262; 1985.

107 **Taylor DC, Dalton JD, Seaber AV, and Garrett WE.** Viscoelastic properties of muscle-tendon units. *Am. J. Sports Med.* 18(3): 300–309; 1990.

108 **Thacker SB, Gilchrist J, Stroup DF, and Kimsey CD.** The impact of stretching on sports injury risk: a systematic review of the literature. *Med. Sci. Sports Exerc.* 36(3): 371–378; 2004.

109 **Toft E, Sinkjaer T, Kalund S, and Espersen GT.** Biomechanical properties of the human ankle in relation to passive stretch. *J. Biomech.* 11(12): 1129–1132; 1989.

110 **Trehearn TL and Buresh RJ.** Sit-and-reach flexibility and running economy of men and women collegiate distance runners. *J. Strength Cond. Res.* 23(1): 158–162; 2009.

111 **Van Gelder LH and Bartz SD.** The effect of acute stretching on agility performance. *J. Strength Cond. Res.* 25(11): 3014–3021; 2011.

112 **Wallin D, Ekblom B, Grahn R, and Nordenborg T.** Improvement of flexibility. A comparison between two techniques. *Am. J. Sports Med.* 13(4): 263–268; 1985.

113 **Weir DE, Tingley J, and Elder GCB.** Acute passive stretching alters the mechanical properties of human plantar flexors and the optimal angle for maximal voluntary contraction. *Eur. J. Appl. Physiol.* 93: 614–623; 2005.

114 **Weldon SM and Hill RH.** The efficacy of stretching for prevention of exercise-related injury: a systematic review of the literature. *Man. Ther.* 8(3): 141–150; 2003.

115 **Weppler CH and Magnusson SP.** Increasing muscle extensibility: a matter of increasing length or modifying sensation? *Phys. Ther.* 90(3): 438–449; 2010.

116 **Whatman C, Knappstein A, and Hume P.** Acute changes in passive stiffness and range of motion post-stretching. *Phys. Ther. Sport* 7: 195–200; 2006.

117 **Wiegner AW and Watts RL.** Elastic properties of muscles measured at the elbow in man: I. normal controls. *J. Neurol. Neurosurg. Psychiat.* 49: 1171–1176; 1986.

118 **Willy RW, Kyle BA, Moore SA, and Chleboun GS.** Effect of cessation and resumption of static hamstring muscle stretching on joint range of motion. *J. Orthop. Sports Phys. Ther.* 31(3): 138–144; 2001.

119 **Wilson GJ, Elliott BC, and Wood GA.** Stretch shorten cycle performance enhancement through flexibility training. *Med. Sci. Sports Exerc.* 24(1): 116–123; 1992.

120 **Wilson GJ, Wood GA, and Elliott BC.** Optimal stiffness of series elastic component in a stretch-shorten cycle activity. *J. Appl. Physiol.* 70(2): 825–833; 1991.

121 **Wilson JM, Hornbuckle LM, Kim J-S, Ugrinowitsch C, Lee S-R, Zoundos MC, Sommer B, and Panton LB.** Effects of static stretching on energy cost and running endurance performance. *J. Strength Cond. Res.* 24(9): 2274–2279; 2010.

122 **Witvrouw E, Mahieu N, Danneels L, and McNair P.** Stretching and injury prevention. An obscure relationship. *Sports Med.* 34(7): 443–449; 2004.

123 **Witvrouw E, Mahieu N, Roosen P, and McNair P.** The role of stretching in tendon injuries. *Br. J. Sports Med.* 41: 224–226; 2007.

124 **Woods K, Bishop P, and Jones E.** Warm-up and stretching in the prevention of muscular injury. *Sports Med.* 37(12): 1089–1099; 2007.

125 **Zourdos MC, Wilson JM, Lee S-R, Park Y-M, Henning PC, Panton LB, and Kim J-S.** Effects of dynamic stretching on energy cost and running endurance performance in trained male runners. *J. Strength Cond. Res.* 26(2): 335–341; 2012.

Chapter 11

1 **Abernethy P, Wilson G, and Logan P.** Strength and power assessment. Issues, controversies and challenges. *Sports Med.* 19(6): 401–417; 1995.

2 **Adams WC, Bernauer EM, Dill DB, and Bomar Jr JB.** Effects of equivalent sea-level and altitude training on $\dot{V}O_{2max}$ and running performance. *J. Appl. Physiol.* 39(2): 262–266; 1975.

3 **Astrand PO, Rodahl K, Dahl H, and Stromme SB.** *Textbook of Work Physiology: Physiological Bases of Exercise*, 4th edn. Champaign, IL: Human Kinetics; 2003.

4 **Atkinson G, Buckley P, Edwards B, Reilly T, and Waterhouse J.** Are there hangover-effects on physical performance when melatonin is ingested by athletes before nocturnal sleep? *Int. J. Sports Med.* 22(3): 232–234; 2001.

5 **Beelen M, Burke LM, Gibala MJ, and van Loon LJC.** Nutritional strategies to promote postexercise recovery. *Int. J. Sport Nutr. Exerc. Metab.* 20(6): 515–532; 2010.

6 **Bhambhani Y.** Physiology of wheelchair racing in athletes with spinal cord injury. *Sports Med.* 32(1): 23–51; 2002.

7 **Bishop D.** Dietary supplements and team-sport performance. *Sports Med.* 40(12): 995–1017; 2010.

8 **Branch JD.** Effect of creatine supplementation on body composition and performance: a meta-analysis. *Int. J. Sport Nutr. Exerc. Metab.* 13(2): 198–226; 2003.

9 **Buford TW, Kreider RB, Stout JR, Greenwood M, Campbell B, Spano M, Ziegenfuss T, Lopez H, Landis J, and Antonio J.** International Society of Sports Nutrition position stand: creatine supplementation and exercise. *J. Int. Soc. Sports Nutr.* 4: 6; 2007.

10 **Burke LM.** Caffeine and sports performance. *Appl. Physiol. Nutr. Metab.* 33(6): 1319–1334; 2008.

11 **Butt MS and Sultan MT.** Coffee and its consumption: benefits and risks. *Crit. Rev. Food Sci. Nutr.* 51(4): 363–373; 2011.

12 **Campbell BI, Wilborn CD, and La Bounty PM.** Supplements for strength-power athletes. *Str. Cond. J.* 32(1): 93–100; 2010.

13 **Campbell B, Kreider RB, Ziegenfuss T, La Bounty P, Roberts M, Burke FD, Landis J, Lopez H, and Antonio J.** International Society of Sports Nutrition position stand: protein and exercise. *J. Int. Soc. Sports Nutr.* 4: 8; 2007.

14 **Chesley A, MacDougall JD, Tarnopolsky MA, Atkinson SA, and Smith K.** Changes in human muscle protein synthesis after resistance exercise. *J. Appl. Physiol.* 73(4): 1383–1388; 1992.

15 **Christoulas K, Karamouzis M, and Mandroukas K.** "Living high - training low" vs "living high - training high": erythropoietic responses and performance of adolescent cross-country skiers. *J. Sports Med. Phys. Fitness* 51(1): 74–81; 2011.

16 **Churchward-Venne TA, Burd NA, and Phillips SM.** Nutritional regulation of muscle protein synthesis with resistance exercise: strategies to enhance anabolism. *Nutr. Metab.(Lond.)* 9(1): 40; 2012.

17 **Cole KJ, Costill DL, Starling RD, Goodpaster BH, Trappe SW, and Fink WJ.** Effect of caffeine ingestion on perception of effort and subsequent work production. *Int. J. Sport Nutr.* 6(1): 14–23; 1996.

18 **Convertino VA, Armstrong LE, Coyle EF, Mack GW, Sawka MN, Senay Jr LC, and Sherman WM.** American College of Sports Medicine position stand. Exercise and fluid replacement. *Med. Sci. Sports Exerc.* 28(1): i–vii; 1996.

19 **Cooper R, Naclerio F, Allgrove J, and Jimenez A.** Creatine supplementation with specific view to exercise / sports performance: an update. *J. Int. Soc. Sports Nutr.* 9(1): 33; 2012.

20 **Costill DL.** Carbohydrate nutrition before, during, and after exercise. *Fed. Proc.* 44(2): 364–368; 1985.

21 **Costill DL, Dalsky GP, and Fink WJ.** Effects of caffeine ingestion on metabolism and exercise performance. *Med. Sci. Sports* 10(3): 155–158; 1978.

22 **Cox GR, Desbrow B, Montgomery PG, Anderson ME, Bruce CR, Macrides TA, Martin DT, Moquin A, Roberts A, Hawley JA, and Burke LM.** Effect of different protocols of caffeine intake on metabolism and endurance performance. *J. Appl. Physiol.* 93(3): 990–999; 2002.

23 **Cronin J and Sleivert G.** Challenges in understanding the influence of maximal power training on improving athletic performance. *Sports Med.* 35(3): 213–234; 2005.

24 **Cronin JB and Owen GJ.** Upper-body strength and power assessment in women using a chest pass. *J. Strength Cond Res.* 18(3): 401–404; 2004.

25 **Dalbo VJ, Roberts MD, Stout JR, and Kerksick CM.** Putting to rest the myth of creatine supplementation leading to muscle cramps and dehydration. *Br. J. Sports Med.* 42(7): 567–573; 2008.

26 **Davis JK and Green JM.** Caffeine and anaerobic performance: ergogenic value and mechanisms of action. *Sports Med.* 39(10): 813–832; 2009.

27 **de Paula P and Niebauer J.** Effects of high altitude training on exercise capacity: fact or myth. *Sleep Breath* 16(1): 233–239; 2012.

28 **Debevec T, Amon M, Keramidas ME, Kounalakis SN, Pisot R, and Mekjavic IB.** Normoxic and hypoxic performance following 4 weeks of normobaric hypoxic training. *Aviat. Space Environ. Med.* 81(4): 387–393; 2010.

29 **Dugan EL, Doyle TL, Humphries B, Hasson CJ, and Newton RU.** Determining the optimal load for jump squats: a review of methods and calculations. *J. Strength Cond Res.* 18(3): 668–674; 2004.

30 **Essig D, Costill DL, and Van Handel PJ.** Effects of caffeine ingestion on utilization of muscle glycogen and lipid during leg ergometer cycling. *Int. J. Sports Med.* 1: 86–90; 1980.

31 **Falvo MJ, Schilling BK, and Weiss LW.** Techniques and considerations for determining isoinertial upper-body power. *Sports Biomech.* 5(2): 293–311; 2006.

32 **Ganio MS, Klau JF, Casa DJ, Armstrong LE, and Maresh CM.** Effect of caffeine on sport-specific endurance performance: a systematic review. *J. Strength Cond Res.* 23(1): 315–324; 2009.

33 **Graham TE, Rush JWE, and MacLean DA.** Skeletal muscle amino acid metabolism and ammonia production during exercise. In *Exercise Metabolism*, edited by Hargreaves M. Champaign, IL: Human Kinetics, 1995; pp. 154–156.

34 **Graham TE, Battram DS, Dela F, El-Sohemy A, and Thong FS.** Does caffeine alter muscle carbohydrate and fat metabolism during exercise? *Appl. Physiol. Nutr. Metab.* 33(6): 1311–1318; 2008.

35 **Graham TE, Helge JW, MacLean DA, Kiens B, and Richter EA.** Caffeine ingestion does not alter carbohydrate or fat metabolism in human skeletal muscle during exercise. *J. Physiol.* 529(3): 837–847; 2000.

36 **Green HJ and Houston ME.** Effect of a season of ice hockey on energy capacities and associated functions. *Med. Sci. Sports* 7(4): 299–303; 1975.

37 **Greer F, Friars D, and Graham TE.** Comparison of caffeine and theophylline ingestion: exercise metabolism and endurance. *J. Appl. Physiol.* 89(5): 1837–1844; 2000.

38 **Hahn AG, Gore CJ, Martin DT, Ashenden MJ, Roberts AD, and Logan PA.** An evaluation of the concept of living at moderate altitude and training at sea level. *Comp. Biochem. Physiol. A Mol. Integr. Physiol.* 128(4): 777–789; 2001.

39 **Hawley JA, Burke LM, Phillips SM, and Spriet LL.** Nutritional modulation of training-induced skeletal muscle adaptations. *J. Appl. Physiol.* 110(3): 834–845; 2011.

40 **Hespel P and Derave W.** Ergogenic effects of creatine in sports and rehabilitation. *Subcell. Biochem.* 46: 245–259; 2007.

41 **Hoff J.** Training and testing physical capacities for elite soccer players. *J. Sports Sci.* 23(6): 573–582; 2005.

42 **Hulston CJ and Jeukendrup AE.** Substrate metabolism and exercise performance with caffeine and carbohydrate intake. *Med. Sci. Sports Exerc.* 40(12): 2096–2104; 2008.

43 **Hunter AM, St Clair Gibson A, Collins M, Lambert M, and Noakes TD.** Caffeine ingestion does not alter performance during a 100 km cycling time-trial performance. *Int. J. Sport Nutr. Exerc. Metab.* 12(4): 438–452; 2002.

44 **Ingham SA, Whyte GP, Jones K, and Nevill AM.** Determinants of 2000 m rowing ergometer performance in elite rowers. *Eur. J. Appl. Physiol.* 88(3): 243–246; 2002.

45 **Ivy JL, Costill DL, Fink WJ, and Lower RW.** Influence of caffeine and carbohydrate feedings on endurance performance. *Med. Sci. Sports* 11(1): 6–11; 1979.

46 **Jacobson TL, Febbraio MA, Arkinstall MJ, and Hawley JA.** Effect of caffeine co-ingested with carbohydrate or fat on metabolism and performance in endurance-trained men. *Exp. Physiol.* 86(1): 137–144; 2001.

47 **Jensen K, Nielsen TS, Fiskastrand A *et al.*** High-altitude training does not increase maximal oxygen uptake or work capacity at sea level in rowers. *Scand. J. Med. Sci. Sports* 3: 256–262; 1993.

48 **Kerksick C, Harvey T, Stout J, Campbell B, Wilborn C, Kreider R, Kalman D, Ziegenfuss T, Lopez H, Landis J, Ivy JL, and Antonio J.** International Society of Sports Nutrition position stand: nutrient timing. *J. Int. Soc. Sports Nutr.* 5: 17; 2008.

49 **Koopman R, Saris WH, Wagenmakers AJ, and van Loon LJ.** Nutritional interventions to promote post-exercise muscle protein synthesis. *Sports Med.* 37(10): 895–906; 2007.

50 **Kreider RB.** Effects of creatine supplementation on performance and training adaptations. *Mol. Cell. Biochem.* 244 (1–2): 89–94; 2003.

51 **Kreider RB, Wilborn CD, Taylor L, Campbell B, Almada AL, Collins R, Cooke M, Earnest CP, Greenwood M, Kalman DS *et al.*** ISSN exercise & sport nutrition review: research & recommendations. *J. Int. Soc. Sports Nutr.* 7: 7; 2010.

52 **Kumar V, Atherton P, Smith K, and Rennie MJ.** Human muscle protein synthesis and breakdown during and after exercise. *J. Appl. Physiol.* 106(6): 2026–2039; 2009.

53 **Lawton TW, Cronin JB, and McGuigan MR.** Strength testing and training of rowers: a review. *Sports Med.* 41(5): 413–432; 2011.

54 **Lemon PW, Tarnopolsky MA, MacDougall JD, and Atkinson SA.** Protein requirements and muscle mass/strength changes during intensive training in novice bodybuilders. *J. Appl. Physiol.* 73(2): 767–775; 1992.

55 **Levine BD and Stray-Gundersen J.** "Living high-training low": effect of moderate-altitude acclimatization with low-altitude training on performance. *J. Appl. Physiol.* 83(1): 102–112; 1997.

56 **MacDougall JD and Wenger HA.** The purpose of physiological testing. In: *Physiological Testing of the High-Performance Athlete*, edited by MacDougall JD, Wenger HA, and Green HJ. Champaign, IL: Human Kinetics, 1991; pp. 1–5.

57 **MacDougall JD, Gibala MJ, Tarnopolsky MA, MacDonald JR, Interisano SA, and Yarasheski KE.** The time course for elevated muscle protein synthesis following heavy resistance exercise. *Can. J. Appl. Physiol.* 20(4): 480–486; 1995.

58 **MacDougall JD, Reddan WG, Dempsey JA, and Forster H.** Acute alterations in stroke volume during exercise at 3100 m altitude. *J. Hum. Ergol. (Tokyo)* 5(2): 103–111; 1976.

59 **Manfredini R, Manfredini F, Fersini C, and Conconi F.** Circadian rhythms, athletic performance, and jet lag. *Br. J. Sports Med.* 32(2): 101–106; 1998.

60 **Melissa L, MacDougall JD, Tarnopolsky MA, Cipriano N, and Green HJ.** Skeletal muscle adaptations to training under normobaric hypoxic versus normoxic conditions. *Med. Sci. Sports Exerc.* 29(2): 238–243; 1997.

61 **Moore DR, Robinson MJ, Fry JL, Tang JE, Glover EI, Wilkinson SB, Prior T, Tarnopolsky MA, and Phillips SM.** Ingested protein dose response of muscle and albumin protein synthesis after resistance exercise in young men. *Am. J. Clin. Nutr.* 89(1): 161–168; 2009.

62 **Morales J and Sobonya S.** Use of submaximal repetition tests for predicting 1 RM strength in class athletes. *J. Strength Cond. Res.* 10(3): 186–189; 1996.

63 **Nevill AM, Allen SV, and Ingham SA.** Modeling the determinants of 2000 m rowing ergometer performance: a proportional, curvilinear allometric approach. *Scand. J. Med. Sci. Sports* 21(1): 73–78; 2011.

64 **Odland LM, MacDougall JD, Tarnopolsky MA, Elorriaga A, and Borgmann A.** Effect of oral creatine supplementation on muscle [PCr] and short-term maximum power output. *Med. Sci. Sports Exerc.* 29(2): 216–219; 1997.

65 **Phillips SM, Sproule J, and Turner AP.** Carbohydrate ingestion during team games exercise: current knowledge and areas for future investigation. *Sports Med.* 41(7): 559–585; 2011.

66 **Phillips SM and van Loon LJ.** Dietary protein for athletes: from requirements to optimum adaptation. *J. Sports Sci.* 29(Suppl 1): S29–S38; 2011.

67 **Poortmans JR and Francaux M.** Adverse effects of creatine supplementation: fact or fiction? *Sports Med.* 30(3): 155–170; 2000.

68 **Rhodes EC, Mosher RE, McKenzie DC, Franks IM, Potts JE, and Wenger HA.** Physiological profiles of the Canadian Olympic Soccer Team. *Can. J. Appl. Sport Sci.* 11(1): 31–36; 1986.

69 **Ritti-Dias RM, Avelar A, Salvador EP, and Cyrino ES.** Influence of previous experience on resistance training on reliability of one-repetition maximum test. *J. Strength Cond Res.* 25(5): 1418–1422; 2011.

70 **Robertson EY, Saunders PU, Pyne DB, Gore CJ, and Anson JM.** Effectiveness of intermittent training in hypoxia combined with live high/train low. *Eur. J. Appl. Physiol.* 110(2): 379–387; 2010.

71 **Saltin B and Astrand PO.** Free fatty acids and exercise. *Am. J. Clin. Nutr.* 57(5 Suppl): 752S–757S; 1993.

72 **Syrotuik DG and Bell GJ.** Acute creatine monohydrate supplementation: a descriptive physiological profile of responders vs nonresponders. *J. Strength Cond Res.* 18(3): 610–617; 2004.

73 **Tang JE and Phillips SM.** Maximizing muscle protein anabolism: the role of protein quality. *Curr. Opin. Clin. Nutr. Metab. Care* 12(1): 66–71; 2009.

74 **Tarnopolsky MA.** Protein requirements for endurance athletes. *Nutrition* 20: 7–8: 662–668; 2004.

75 **Tarnopolsky MA.** Caffeine and creatine use in sport. *Ann. Nutr. Metab.* 57(Suppl 2): 1–8; 2010.

76 **Tarnopolsky MA.** Effect of caffeine on the neuromuscular system—potential as an ergogenic aid. *Appl. Physiol. Nutr. Metab.* 33(6): 1284–1289; 2008.

77 **Tipton KD and Witard OC.** Protein requirements and recommendations for athletes: relevance of ivory tower arguments for practical recommendations. *Clin. Sports Med.* 26(1): 17–36; 2007.

78 **Truijens MJ, Toussaint HM, Dow J, and Levine BD.** Effect of high-intensity hypoxic training on sea-level swimming performances. *J. Appl. Physiol.* 94(2): 733–743; 2003.

79 **Vandervoort AA, Sale DG, and Moroz J.** Comparison of motor unit activation during unilateral and bilateral leg extension. *J. Appl. Physiol.* 56(1): 46–51; 1984.

80 **Wilber RL.** Application of altitude/hypoxic training by elite athletes. *J. Hum. Sport Exerc.* 6(2): 271–286; 2011.

81 **Wilson GJ, Wood GA, and Elliott BC.** Optimal stiffness of series elastic component in a stretch-shorten cycle activity. *J. Appl. Physiol.* 70(2): 825–833; 1991.

82 **Woolf K, Bidwell WK, and Carlson AG.** The effect of caffeine as an ergogenic aid in anaerobic exercise. *Int. J. Sport Nutr. Exerc. Metab.* 18(4): 412–429; 2008.

83 **Young WB, Pryor JF, and Wilson GJ.** Effect of instructions on characteristics of countermovement and drop jump performance. *J. Strength Cond. Res.* 9(4): 232–236; 1995.

Index

C

D

E

N

O

P

U

V

W